Advances in Veterinary Medicine

Volume 41
Veterinary Vaccines and Diagnostics

Advances in Veterinary Medicine

Edited by

W. Jean Dodds
HEMOPET
Santa Monica, California

Advisory Board

Kenneth C. Boveé
William S. Dernell
Carolton Gyles
Robert O. Jacoby
Ann B. Kier
Raymond F. Nachreiner
Carl A. Osborne
Fred W. Quimby
Alan H. Rebar
Ronald D. Schultz

Advances in Veterinary Medicine

Volume 41

Veterinary Vaccines and Diagnostics

Edited by

Ronald D. Schultz

Department of Pathobiological Sciences
School of Veterinary Medicine
University of Wisconsin–Madison
Madison, Wisconsin

Academic Press
*San Diego London Boston
New York Sydney Tokyo Toronto*

This book is printed on acid-free paper.

Copyright © 1999 by ACADEMIC PRESS

All Rights Reserved.
No part of this publication may be reproduced or transmitted in any form or by any means, electronic or mechanical, including photocopy, recording, or any information storage and retrieval system, without permission in writing from the Publisher.
The appearance of the code at the bottom of the first page of a chapter in this book indicates the Publisher's consent that copies of the chapter may be made for personal or internal use of specific clients. This consent is given on the condition, however, that the copier pay the stated per copy fee through the Copyright Clearance Center, Inc. (222 Rosewood Drive, Danvers, Massachusetts 01923), for copying beyond that permitted by Sections 107 or 108 of the U.S. Copyright Law. This consent does not extend to other kinds of copying, such as copying for general distribution, for advertising or promotional purposes, for creating new collective works, or for resale. Copy fees for pre-1999 chapters are as shown on the title pages. If no fee code appears on the title page, the copy fee is the same as for current chapters.
1093-975X/99 $30.00

Academic Press
a division of Harcourt Brace & Company
525 B Street, Suite 1900, San Diego, California 92101-4495, USA
http://www.apnet.com

Academic Press Limited
24-28 Oval Road, London NW1 7DX, UK
http://www.hbuk.co.uk/ap/

International Standard Book Number: 0-12-039242-9

PRINTED IN THE UNITED STATES OF AMERICA
98 99 00 01 02 03 EB 9 8 7 6 5 4 3 2 1

CONTENTS

Contributors xxv
Preface xxxv

I
VACCINES AND DIAGNOSTICS
Historic and Contemporary Perspectives

Vaccination: A Philosophical View
Marian C. Horzinek

I. Introduction	1
II. Will There Be Vaccination in the Next Millenium?	2
III. What Is the Future of Veterinary Vaccinology?	4
IV. Vaccination in the Twenty-First Century	5
V. Outlook	6

Grease, Anthraxgate, and Kennel Cough: A Revisionist History of Early Veterinary Vaccines
Ian Tizard

I. Introduction	7
II. The Origin of Vaccinia	9
III. Anthraxgate: A Minor Nineteenth-Century Scandal	11
IV. Early Canine Distemper Vaccines	17
V. Summary	22
References	23

Diagnostic Medicine: The Challenge of Differentiating Infection from Disease and Making Sense for the Veterinary Clinician

JAMES F. EVERMANN AND INGE S. ERIKS

I.	Introduction	25
II.	Differentiating Infection Detection from Disease Diagnosis	26
III.	How Early Do We Want to Detect Infection?	31
IV.	What Are the Consequences of the Results?	32
V.	Where Are We Heading with Veterinary Diagnostics?	34
	References	36

II
CONCEPTS IN IMMUNOLOGY AND VACCINOLOGY

Genetic Effects on Vaccination

BRUCE N. WILKIE AND BONNIE A. MALLARD

I.	Introduction	39
II.	Genetic Effects on Health and Vaccination	40
III.	Strategies for Genetic Enhancement of Health	41
IV.	High Immune Response Phenotype	44
V.	Discussion and Summary	48
	References	50

Nutritional Effects on Vaccination

MARK E. COOK

I.	Vaccination Cost to Performance	53
II.	Biochemical Mechanisms in Immune-Induced Wasting	54
III.	Conjugated Linoleic Acid	55
IV.	Cholecystokinin and Immune-Induced Anorexia	56
V.	Summary	57
	References	57

Effects of Stress on Leukocyte Trafficking and Immune Responses: Implications for Vaccination

M. E. KEHRLI, J. L. BURTON, B. J. NONNECKE, AND E. K. LEE

I.	Introduction	61
II.	Leukocyte Trafficking	65
III.	Effects of Stress on Immunity	70
IV.	Summary	74
	References	74

Role of Macrophage Cytokines in Mucosal Adjuvanticity

DENNIS L. FOSS AND MICHAEL P. MURTAUGH

I.	Introduction	83
II.	Mucosal Adjuvanticity of Cholera Toxin	84
III.	Mechanisms of Mucosal Adjuvanticity	92
IV.	Summary	98
	References	99

Cholera Toxin B Subunit as an Immunomodulator for Mucosal Vaccine Delivery

MICHAEL W. RUSSELL, HONG-YIN WU, GEORGE HAJISHENGALLIS, SUSAN K. HOLLINGSHEAD, AND SUZANNE M. MICHALEK

I.	Introduction	105
II.	Responses to Mucosal Immunization with SBR-CTA2/B	106
III.	Responses to Mucosal Immunization with *Salmonella* Expressing SBR-CTA2/B	108
IV.	Discussion and Summary	110
	References	112

Deceptive Imprinting: Insights into Mechanisms of Immune Evasion and Vaccine Development

PETER L. NARA

I.	Introduction and Background	115
II.	Deceptive Imprinting	117
III.	Refocusing the Immune Response by Masking Epitopes Responsible for Deceptive Imprinting: Novel Approach to Vaccination	129
	References	130

Vaccination against Tuberculosis: Recent Progress

IAN M. ORME

I.	Introduction	135
II.	Acquired Immunity and Memory Immunity to Tuberculosis Infection	136
III.	Types of Vaccines	138
IV.	Can We Increase Herd Resistance to Bovine Tuberculosis?	141
	References	141

Viral Vectors for Veterinary Vaccines

MICHAEL SHEPPARD

I.	Introduction and Background	145
II.	Viral Vector Construction	147
III.	Advantages and Disadvantages of Viral Vectors for Vaccine Delivery	148
IV.	Construction of Safer Viral Vectors for Vaccine Delivery	149
V.	Examples of Reported Viral Veterinary Vaccine Vectors	151
VI.	Commercially Available Viral Vaccine Vectors for Veterinary Use	151
VII.	Summary	153
	References	155

DNA Immunization: Present and Future

L. A. BABIUK, J. LEWIS, S. VAN DEN HURK, AND R. BRAUN

I.	Introduction	163
II.	Universality of DNA Immunization	164

III.	Induction of Immunity	166
IV.	Role of Different Antibodies in Clearing Viruses	168
V.	Vaccine Delivery	169
VI.	DNA Immunization in the Face of Passive Antibody	171
VII.	Regulatory	172
VIII.	Epilogue	174
	References	176

Contribution of Advances in Immunology to Vaccine Development

W. I. Morrison, G. Taylor, R. M. Gaddum, and S. A. Ellis

I.	Introduction	181
II.	Advances in Immunology Relevant to Vaccine Development	182
III.	Mechanisms of Immune Protection against Bovine Respiratory Syncytial Virus	187
IV.	Summary	192
	References	192

III
BOVINE VACCINES AND DIAGNOSTICS

Bovine Viral Vaccines, Diagnostics, and Eradication: Past, Present, and Future

Jan T. van Oirschot

I.	Introduction	197
II.	Foot-and-Mouth Disease	198
III.	Infectious Bovine Rhinotracheitis	201
IV.	Bovine Virus Diarrhea	207
V.	Posteradication Period	210
	References	211

Immunization and Diagnosis in Bovine Reproductive Tract Infections

Lynette B. Corbeil

I.	Introduction and Background	217
II.	*Brucella abortus* Infection	218
III.	*Haemophilus somnus* Infection	219
IV.	*Campylobacter fetus* subsp. *venerealis* Infection	221
V.	*Tritrichomonas foetus* Infection	224
VI.	Summary and Future Directions	232
	References	233

Progress and Expectations for Helminth Vaccines

Els N. T. Meeusen and Jillian F. Maddox

I.	Introduction	241
II.	Vaccination Using Defined Parasite Antigens	242
III.	Vaccine-Induced Immune Responses	245
IV.	Simulation Models for Host–Parasite Population Dynamics	247
V.	Host-Immunity and Population Dynamics of Gastrointestinal Nematode Infections	250
VI.	Development of a Simple Model for Vaccination against *Haemonchus*	251
VII.	Summary	253
	References	254

Vaccines and Diagnostic Methods for Bovine Mastitis: Fact and Fiction

Robert J. Yancey, Jr.

I.	Introduction	257
II.	Vaccines for Contagious Pathogens	258
III.	Vaccines for Environmental Mastitis Pathogens	263
IV.	Diagnostic Methods	267
V.	Summary	268
	References	269

T-Cell Responses and the Influence of Dendritic Cells in Cattle

C. J. HOWARD, R. A. COLLINS, P. SOPP, G. P. BROOKE, L. S. KWONG, K. R. PARSONS, V. WEYNANTS, J.-J. LETESSON, AND G. P. BEMBRIDGE

I. Introduction	275
II. Identification of the Major T-Cell Subpopulations in Cattle	276
III. Role of Different T-Cell Populations *in Vivo*	276
IV. Identification of Subpopulations of CD4 and CD8 T Cells That Differ in Function	278
V. Activation Requirements and Function of $\gamma\delta$ T Cells	282
VI. T-Cell Responses Induced by Dendritic Cells	284
References	285

IV
CANINE AND FELINE VACCINES

Canine Viral Vaccines at a Turning Point—A Personal Perspective

L. E. CARMICHAEL

I. General Remarks	289
II. Veterinary Vaccines	292
III. Comments on Selected Vaccines	293
IV. Summary	302
References	305

Forty Years of Canine Vaccination

M. J. G. APPEL

I. Introduction	309
II. Rabies Virus	311
III. Canine Distemper Virus	311
IV. Canine Parvovirus	312
V. Canine Coronavirus	313
VI. Canine Adenovirus Type 1 (Infectious Canine Hepatitis Virus)	314
VII. Canine Adenovirus Type 2	314
VIII. Canine Parainfluenza Virus	315

IX. *Bordetella bronchiseptica*	315
X. *Borrelia burgdorferi*	316
XI. Leptospirosis	317
XII. Summary	318
References	319

Analysis of the Protective Immunity Induced by Feline Immunodeficiency Virus Vaccination

MARGARET J. HOSIE AND OSWALD JARRETT

I. Introduction and Background	325
II. Whole Inactivated Virus Vaccines	326
III. Subunit Vaccines	328
IV. DNA Vaccination	329
References	331

Vaccination of Cats against Emerging and Reemerging Zoonotic Pathogens

CHRISTOPHER W. OLSEN

I. Introduction	333
II. *Toxoplasma gondii*	336
III. *Bartonella henselae*	338
IV. *Helicobacter pylori*	341
V. Other Agents	341
VI. Summary	342
References	343

Evaluation of Risks and Benefits Associated with Vaccination against Coronavirus Infections in Cats

FRED W. SCOTT

I. Historical Perspectives of FIP	347
II. Current Status of FIP	348
III. Causative Agent of FIP	348

IV.	Pathogenesis of Feline Coronavirus Infections	349
V.	Immunology of Feline Coronavirus Infections	350
VI.	Antibody-Dependent Enhancement	351
VII.	FIP Vaccine	351
VIII.	Risks of FIP Vaccination	352
IX.	Benefits of FIP Vaccination	353
	References	356

V
EQUINE VACCINES AND DIAGNOSTICS

Diagnosis and Prevention of Equine Infectious Diseases: Present Status, Potential, and Challenges for the Future
PHILIPPE DESMETTRE

I.	Introduction	359
II.	Viral Diseases	360
III.	Bacterial Diseases	368
IV.	Other Viral and Bacterial Diseases	371
V.	Summary	372
	References	373

The Equine Influenza Surveillance Program
J. A. MUMFORD

I.	Introduction	379
II.	1983: WHO Informal Workshop on Vaccination against Equine Influenza	380
III.	1992: WHO/OIE Consultation on Newly Emerging Strains of Equine Influenza	381
IV.	1995: Consultation of OIE and WHO Experts on Progress in Surveillance of Equine Influenza and Application to Strain Selection	382
V.	1996: Actions Taken by the OIE	383
VI.	Findings of the Expert Surveillance Panel	384
VII.	Action Taken by the European Pharmacopoeia	385
VIII.	Actions Taken by the Committee for Veterinary Medicinal Products	385
IX.	Actions Taken by the National Institute of Biological Standardization and Control	386
X.	Action Taken by the USDA	386
	References	387

Vaccination against *Strongylus vulgaris* in Ponies: Comparison of the Humoral and Cytokine Responses of Vaccinates and Nonvaccinates

C. E. SWIDERSKI, T. R. KLEI, R. W. FOLSOM, S. S. POURCIAU, A. CHAPMAN, M. R. CHAPMAN, R. M. MOORE, J. R. MCCLURE, H. W. TAYLOR, AND D. W. HOROHOV

I.	Introduction	390
II.	Methods	391
III.	Results	395
IV.	Discussion	397
	References	402

ISCOM: A Delivery System for Neonates and for Mucosal Administration

B. MOREIN, M. VILLACRÉS-ERIKSSON, J. EKSTRÖM, K. HU, S. BEHBOUDI, AND K. LÖVGREN-BENGTSSON

I.	Introduction: Immune Stimulating Complex	405
II.	The ISCOM Concept	406
III.	Formation of ISCOM	406
IV.	Antigen Presentation and Targeting by ISCOMs	407
V.	Adjuvant Influences on the Transport of Antigen	408
VI.	Adjuvants and Delivery Systems for Induction of Mucosal Immunity	409
VII.	ISCOMs Induce a Cytokine Th1 Type Response But Also Th2	410
VIII.	Antigens Loaded in ISCOMs Induce Immune Response in Neonates	410
IX.	Protective Immunity	411
	References	412

An Epidemiologic Approach to Evaluating the Importance of Immunoprophylaxis 415

PAUL S. MORLEY

VI
SWINE VACCINES

Present Uses of and Experiences with Swine Vaccines
LENNART BÄCKSTRÖM

I. Introduction 419
II. Material 421
III. Results 422
IV. Discussion 425
References 427

Enteric Viral Infections of Pigs and Strategies for Induction of Mucosal Immunity
LINDA J. SAIF

I. Introduction and Background 429
II. Characteristics of Enteropathogenic Viruses 430
III. Mucosal Immunity to Enteropathogenic Viruses 434
References 442

Use of Interleukin 12 to Enhance the Cellular Immune Response of Swine to an Inactivated Herpesvirus Vaccine
FEDERICO A. ZUCKERMANN, STEPHEN MARTIN, ROBERT J. HUSMANN, AND JULIE BRANDT

I. Introduction 447
II. Cytokines as Vaccine Adjuvants 448
III. Interleukin 12 and Its Ability to Modulate Acquired Immunity 449
IV. Porcine Model to Examine the Adjuvant Effect of Interleukin 12 451
V. Immune Mechanism(s) of Protective Immunity against Herpesvirus 452
VI. Interleukin 12-Mediated Enhancement of the Cell-Mediated Immune Response to an Inactivated PrV Vaccine 455
VII. Summary and Conclusion 456
References 458

Swinepox Virus as a Vaccine Vector for Swine Pathogens
DEOKI N. TRIPATHY

```
   I. Introduction ........................................................... 463
  II. Conventional Vaccines ................................................. 465
 III. Recombinant Virus Vectored Vaccines ................................... 467
  IV. Summary ............................................................... 476
      References ............................................................ 477
```

VII
POULTRY VACCINES

Introduction to Poultry Vaccines and Immunity
J. M. SHARMA

```
   I. Introduction ........................................................... 481
  II. Disease Prevention by Vaccination ..................................... 482
 III. Vaccination Programs .................................................. 484
  IV. Vaccine Delivery ...................................................... 485
   V. New Developments in Poultry Vaccines .................................. 486
  VI. General Features of the Avian Immune System .......................... 487
 VII. Humoral Immunity ...................................................... 488
VIII. Cell-Mediated Immunity ................................................ 489
  IX. Summary ............................................................... 490
      References ............................................................ 490
```

In Ovo Vaccination Technology
C. A. RICKS, A. AVAKIAN, T. BRYAN, R. GILDERSLEEVE, E. HADDAD, R. ILICH, S. KING, L. MURRAY, P. PHELPS, R. POSTON, C. WHITFILL, AND C. WILLIAMS

```
   I. Introduction ........................................................... 495
  II. Technology Discovery .................................................. 496
 III. Commercialization of In Ovo Marek's (HVT/SB1) Vaccination ............ 497
  IV. Commercialization of Other In Ovo Live Viral Vaccines ................ 503
   V. Maternal Antibody Effects on Viral Vaccine Efficacy ................... 505
  VI. Safe and Effective Vaccination in Presence of Maternal Antibodies ..... 506
 VII. Bacterial Vaccines .................................................... 511
VIII. Coccidial Vaccines .................................................... 512
```

IX. Summary 512
References 513

Current and Future Recombinant Viral Vaccines for Poultry
MARK W. JACKWOOD

I. Introduction and Background 517
II. Virus Vectors 517
III. Future Recombinant Vaccines for Poultry 519
IV. Summary 521
References 522

VIII
FISH, EXOTIC, AND WILDLIFE VACCINES

Development and Use of Modified Live *Edwardsiella ictaluri* Vaccine against Enteric Septicemia of Catfish
PHILLIP H. KLESIUS AND CRAIG A. SHOEMAKER

I. Introduction 523
II. Materials and Methods 525
III. Results 527
IV. Discussion 532
V. Summary 534
References 535

Fish Vaccines
SOMSAK VINITNANTHARAT, KJERSTI GRAVNINGEN, AND EVAN GREGER

I. Introduction 539
II. Immunization Methods in Fish 541
III. Benefit of Using Vaccine 542

IV.	Vaccines against Some Specific Diseases	543
V.	Summary	549
	References	550

Cross-Species Vaccination in Wild and Exotic Animals
JOSEPH F. CURLEE, JR.

I.	Introduction	551
II.	Canine Distemper Virus	551
III.	*Clostridium botulinum* Type C	553
IV.	Summary	555
	References	555

Some Experiments and Field Observations of Distemper in Mink and Ferrets
JOHN R. GORHAM

I.	Introduction	557
II.	Vaccines	558
III.	Routes of Vaccination	562
IV.	Maternal Antibody and Vaccination	562
V.	Vaccination of Pregnant Female Mink	563
VI.	Transplacental and Neonatal Attempts to Immunize Ferrets against CDV	564
VII.	Time Interval Required to Infect Ferrets by Direct Contact	564
VIII.	Experimental Epidemiology	565
IX.	Future Research	567
	References	568

Vaccination of Wildlife against Rabies: Successful Use of a Vectored Vaccine Obtained by Recombinant Technology
M. MACKOWIAK, J. MAKI, L. MOTES-KREIMEYER, T. HARBIN, AND K. VAN KAMPEN

I.	Introduction and Background	571
II.	Raboral V-RG: A Rabies Vaccine Created by Recombinant Technology	572

III.	USDA Testing of Recombinant Vaccines	573
IV.	Controlling Raccoon Rabies	576
V.	Rabies in Texas: Coyotes and Gray Foxes	580
VI.	Summary	581
	References	582

IX
REGULATION, LICENSING, AND STANDARDIZATION OF VACCINES AND DIAGNOSTICS

Authorities and Procedures for Licensing Veterinary Biological Products in the United States
DAVID A. ESPESETH AND THOMAS J. MYERS

I.	Introduction	585
II.	Organization	586
III.	Licensing Procedures for Conventional Vaccines and Bacterins	587
IV.	Licensing Procedures for Nonconventional Products	590
V.	Conditional Licenses	591
VI.	Licenses for Further Manufacture	591
VII.	Sublicensing	592
VIII.	Exemptions to Licensure	592
IX.	Autogenous Products	592
X.	Summary	593
	Reference	593

Licensing Procedures for Immunological Veterinary Medicinal Products in the European Union
P. P. PASTORET AND F. FALIZE

I.	Introduction	595
II.	Role of the European Agency for the Evaluation of Medicinal Products	596
III.	Available European Procedures	597
IV.	New Definitions of Veterinary Biologicals	598
V.	Revision-Validation of Vaccines Already on the Market	600
VI.	Manufacturing Authorization	601
VII.	Batch Control/Release	602
VIII.	Special Case of Equine Influenza Vaccines	602
IX.	Role of the European Pharmacopoeia	604

X. Summary	607
References	607

International Association of Biological Standardization and International Harmonization
Daniel Gaudry

I. Introduction	609
II. Introducing the IABS	610
III. A Case Study: Report of the Avian Products Standardization Committee (March 1979)	612

Technical Requirements for the Licensing of Pseudorabies (Aujeszky's Disease) Vaccines in the European Union
P. Vannier

I. Introduction	615
II. Safety Testing	616
III. Efficacy Testing	620
IV. Batch Release Controls	624
V. Summary	624
References	625

Specific Licensing Considerations for Modified Live Pseudorabies Vaccines in the United States
Diane L. Sutton

I. Introduction	627
II. Licensing Considerations	628
Reference	632

Why Do Vaccine Labels Say the Funny Things They Do?

DAVID R. HUSTEAD

I. Introduction 633
II. Vaccine Label Expectations 634
III. Vaccine Label Reality 634
IV. Recommendations for Standardization Improvements 640

Standardization of Diagnostic Assays for Animal Acute Phase Proteins

P. DAVID ECKERSALL, SUSAN DUTHIE,
MATHILDA J. M. TOUSSAINT, ERIK GRUYS, PETER HEEGAARD,
MARIA ALAVA, CORNELIA LIPPERHEIDE,
AND FRANCOIS MADEC

I. Introduction and Background 643
II. Acute Phase Protein in Animals 645
III. Methods of Acute Phase Protein Assay 647
IV. Standardization of Acute Phase Protein Assays 649
   References 653

Vaccination Practices in Veterinary Medicine: Standardization versus Tailored to Needs?

SERGE MARTINOD

I. Introduction 657
II. Can We Standardize Vaccination Practices? 658
III. Are We Vaccinating Too Much? 662
IV. Consequences for the Animal Health Industry 665
V. Summary 668

International Harmonization of Standards for Diagnostic Tests and Vaccines: Role of the Office International des Epizooties

Peter F. Wright

I.	Introduction	669
II.	Organization and Structure	670
III.	Specialist Commissions	672
IV.	Standards Commission	672
V.	Summary	677
	References	679

X
ADVERSE VACCINE REACTIONS, FAILURES, AND POSTMARKETING SURVEILLANCE

Mechanistic Bases for Adverse Vaccine Reactions and Vaccine Failures

James A. Roth

I.	Introduction	682
II.	Adverse Vaccine Reactions	684
III.	Vaccine Failure	691
IV.	Summary	696
	References	696

Weighing the Risks and Benefits of Vaccination

Larry T. Glickman

I.	Defining the Problem	701
II.	Vaccine Risk Assessment	702
III.	Current Sources of Data on Adverse Reactions	703
IV.	Suggested Improvements in Postmarketing Surveillance	706
V.	Risk Management and Risk Communication	709
VI.	Summary	710
	References	712

More Bumps on the Vaccine Road
W. JEAN DODDS

I. Introduction and Background 715
II. Overview of Adverse Effects of Vaccines 716
III. Breed Study Examples 719
IV. Periodicity of Booster Vaccination 724
V. Alternative Strategies to Conventional Vaccination 727
VI. Summary and Future Directions 729
References 729

Vaccine-Induced Autoimmunity in the Dog
HARM HOGENESCH, JUAN AZCONA-OLIVERA, CATHARINE SCOTT-MONCRIEFF, PAUL W. SNYDER, AND LARRY T. GLICKMAN

I. Introduction 733
II. Materials and Methods 735
III. Results 737
IV. Discussion 740
References 745

An Introduction to Analytical Methods for the Postmarketing Surveillance of Veterinary Vaccines
DAVID SIEV

I. Introduction 749
II. Postmarketing Surveillance and Public Policy 750
III. Postmarketing Surveillance Datum 751
IV. Fathoming Spontaneous Adverse Event Report Data 754
V. Quantitative Analysis of Adverse Event Report Data 758
VI. Summary 769
References 773

INDEX 775

CONTRIBUTORS

Numbers in parentheses indicate the pages on which the authors' contributions begin.

MARIA ALAVA, Universidad de Zaragoza, 12 Pedro Cerbuno, 50009 Zaragoza, Spain (643)

M. J. G. APPEL, James A. Baker Institute for Animal Health, College of Veterinary Medicine, Cornell University, Ithaca, New York 14853 (309)

A. AVAKIAN, Embrex, Inc., P.O. Box 13989, Research Triangle Park, North Carolina 27709 (495)

JUAN AZCONA-OLIVERA, Departments of Veterinary Pathobiology and Veterinary Clinical Sciences, Purdue University, West Lafayette, Indiana 47907 (733)

L. A. BABIUK, Veterinary Infectious Disease Organization, Saskatoon, Saskatchewan, Canada S7N 5E3 (163)

LENNART BÄCKSTRÖM, School of Veterinary Medicine, University of Wisconsin, Madison, Wisconsin 53706 (419)

S. BEHBOUDI, Department of Veterinary Microbiology, Section of Virology, Swedish University of Agricultural Sciences, Biomedical Centre Box 585, S-751 23 Uppsala, Sweden (405)

G. P. BEMBRIDGE, Institute for Animal Health, Compton, Newbury RG20 7NN, United Kingdom (275)

JULIE BRANDT, Department of Veterinary Pathobiology, University of Illinois, Urbana, Illinois 61801 (447)

R. BRAUN, Veterinary Infectious Disease Organization, Saskatoon, Saskatchewan, Canada S7N 5E3 (163)

G. P. BROOKE, Institute for Animal Health, Compton, Newbury RG20 7NN, United Kingdom (275)

T. BRYAN, Embrex, Inc., P.O. Box 13989, Research Triangle Park, North Carolina 27709 (495)

J. L. BURTON, Michigan State University, East Lansing, Michigan 48825 (61)

L. E. CARMICHAEL, James A. Baker Institute for Animal Health, College of Veterinary Medicine, Cornell University, Ithaca, New York 14853 (289)

A. CHAPMAN, Department of Veterinary Microbiology and Parasitology, School of Veterinary Medicine, Louisiana State University, Baton Rouge, Louisiana 70803 (389)

M. R. CHAPMAN, Department of Veterinary Science, Louisiana Agricultural Experiment Station, Louisiana State University Agricultural Center, Baton Rouge, Louisiana 70803 (389)

R. A. COLLINS, Institute for Animal Health, Compton, Newbury RG20 7NN, United Kingdom (275)

MARK E. COOK, Animal Sciences Department, University of Wisconsin, Madison, Wisconsin 53706 (53)

LYNETTE B. CORBEIL, Department of Pathology, University of California–San Diego, San Diego, California 92103-8416 (217)

JOSEPH F. CURLEE, JR., United Vaccines, Inc., Madison, Wisconsin 53713 (551)

PHILIPPE DESMETTRE, Merial, 254 rue Marcel Merieux, Lyon, France (359)

W. JEAN DODDS, Hemopet, 938 Stanford Street, Santa Monica, California 90403 (715)

SUSAN DUTHIE, University of Glasgow Veterinary School, Glasgow G61 1QH, United Kingdom (643)

P. DAVID ECKERSALL, University of Glasgow Veterinary School, Glasgow G61 1QH, United Kingdom (643)

J. EKSTRÖM, Department of Veterinary Microbiology, Section of Virology, Swedish University of Agricultural Sciences, Biomedical Centre Box 585, S-751 23 Uppsala, Sweden (405)

S. A. ELLIS, Institute for Animal Health, Compton, Newbury, Berkshire, United Kingdom (181)

INGE S. ERIKS, Department of Veterinary Microbiology and Pathology, College of Veterinary Medicine, Washington State University, Pullman, Washington 99164 (25)

DAVID A. ESPESETH, U.S. Department of Agriculture, Animal and Plant Health Inspection Service, Veterinary Services, Center for Veterinary Biologics, Riverdale, Maryland 20782 (585)

JAMES F. EVERMANN, Department of Veterinary Clinical Sciences and Washington Animal Disease Diagnostic Laboratory, Washington State University, Pullman, Washington 99164 (25)

F. FALIZE, Pharmaceutical Inspectorate, Ministry of Public Health, Belgium (595)

R. W. FOLSOM, Department of Veterinary Microbiology and Parasitology, School of Veterinary Medicine, Louisiana State University, Baton Rouge, Louisiana 70803 (389)

DENNIS L. FOSS, Department of Veterinary Pathobiology, University of Minnesota, St. Paul, Minnesota 55108 (83)

R. M. GADDUM, Institute for Animal Health, Compton, Newbury, Berkshire, United Kingdom (181)

DANIEL GAUDRY, Merial, Gainesville, Georgia 30501 (609)

R. GILDERSLEEVE, Embrex, Inc., P.O. Box 13989, Research Triangle Park, North Carolina 27709 (495)

LARRY T. GLICKMAN, Departments of Veterinary Pathobiology and Veterinary Clinical Sciences, Purdue University, West Lafayette, Indiana 47907 (701, 733)

JOHN R. GORHAM, Agriculture Research Service, USDA, and the Department of Veterinary Microbiology and Pathology, Washington State University, Pullman, Washington 99164 (557)

KJERSTI GRAVNINGEN, Alpharma AS, Aquatic Animal Health Division, Oslo, Norway (539)

EVAN GREGER, Alpharma NW, Aquatic Animal Health Division, Bellevue, Washington, 98005 (539)

ERIK GRUYS, Faculty of Veterinary Medicine, Utrecht University, Yalelaan 1, 3508 TD Utrecht, The Netherlands (643)

E. HADDAD, Embrex, Inc., P.O. Box 13989, Research Triangle Park, North Carolina 27709 (495)

GEORGE HAJISHENGALLIS, Department of Microbiology, University of Alabama at Birmingham, Birmingham, Alabama 35924 (105)

T. HARBIN, Merial, Limited, Athens, Georgia 30601 (571)

PETER HEEGAARD, Danish Veterinary Laboratory, 27 Bulowsvej, 1790 Frederiksburg, Denmark (643)

HARM HOGENESCH, Departments of Veterinary Pathobiology and Veterinary Clinical Sciences, Purdue University, West Lafayette, Indiana 47907 (733)

SUSAN K. HOLLINGSHEAD, Department of Microbiology, University of Alabama at Birmingham, Birmingham, Alabama 35924 (105)

D. W. HOROHOV, Department of Veterinary Microbiology and Parasitology, School of Veterinary Medicine, Louisiana State University, Baton Rouge, Louisiana 70803 (389)

MARIAN C. HORZINEK, Virology Unit, Veterinary Faculty, Utrecht University, Utrecht, The Netherlands (1)

MARGARET J. HOSIE, Department of Veterinary Pathology, University of Glasgow, Glasgow G61 1QH, United Kingdom (325)

C. J. HOWARD, Institute for Animal Health, Compton, Newbury RG20 7NN, United Kingdom (275)

K. HU, Department of Veterinary Microbiology, Section of Virology, Swedish University of Agricultural Sciences, Biomedical Centre Box 585, S-751 23 Uppsala, Sweden (405)

ROBERT J. HUSMANN, Department of Veterinary Pathobiology, University of Illinois, Urbana, Illinois 61801 (447)

DAVID R. HUSTEAD, Fort Dodge Animal Health, Overland Park, Kansas 66225 (633)

R. ILICH, Embrex, Inc., P.O. Box 13989, Research Triangle Park, North Carolina 27709 (495)

MARK W. JACKWOOD, Department of Avian Medicine, College of Veterinary Medicine, University of Georgia, Athens, Georgia 30602 (517)

OSWALD JARRETT, Department of Veterinary Pathology, University of Glasgow, Glasgow G61 1QH, United Kingdom (325)

M. E. KEHRLI, NADC-USDA-ARS, Ames, Iowa 50010 (61)

S. KING, Embrex, Inc., P.O. Box 13989, Research Triangle Park, North Carolina 27709 (495)

T. R. KLEI, Department of Veterinary Microbiology and Parasitology, School of Veterinary Medicine, Louisiana State University, and Department of Veterinary Science, Louisiana Agricultural Experiment Station, Louisiana State University Agricultural Center, Baton Rouge, Louisiana 70803 (389)

PHILLIP H. KLESIUS, United States Department of Agriculture–Agricultural Research Service, Fish Diseases and Parasites Research Laboratory, Auburn, Alabama 36830 (523)

L. S. KWONG, Institute for Animal Health, Compton, Newbury RG20 7NN, United Kingdom (275)

E. K. LEE, Iowa State University, Ames, Iowa 50011 (61)

J.-J. LETESSON, Immunology Unit, Facultes Universitaires, Notre Dame de la Paix, Namur, Belgium (275)

J. LEWIS, Veterinary Infectious Disease Organization, Saskatoon, Saskatchewan, Canada S7N 5E3 (163)

CORNELIA LIPPERHEIDE, University of Bonn, Regina-Pacis-Weg 3, Bonn D5311, Germany (643)

K. LÖVGREN-BENGTSSON, Department of Veterinary Microbiology, Section of Virology, Swedish University of Agricultural Sciences, Biomedical Centre Box 585, S-751 23 Uppsala, Sweden (405)

M. MACKOWIAK, Merial, Limited, Athens, Georgia 30601 (571)

JILLIAN F. MADDOX, Centre for Animal Biotechnology, School of Veterinary Science, The University of Melbourne, Parkville, Victoria 3052, Australia (241)

FRANCOIS MADEC, CNEVA Ploufragan, BP 53, Ploufragan 22440, France (643)

J. MAKI, Merial, Limited, Athens, Georgia 30601 (571)

BONNIE A. MALLARD, Department of Pathobiology, The Ontario Veterinary College, The University of Guelph, Guelph, Ontario, Canada N1G 2W1 (39)

STEPHEN MARTIN, Pharmacia/Upjohn, Kalamazoo, Michigan 49001 (447)

SERGE MARTINOD, Pfizer Central Research, Groton, Connecticut 06340 (657)

J. R. MCCLURE, Department of Veterinary Clinical Sciences, School of Veterinary Medicine, Louisiana State University, Baton Rouge, Louisiana 70803 (389)

ELS N. T. MEEUSEN, Centre for Animal Biotechnology, School of Veterinary Science, The University of Melbourne, Parkville, Victoria 3052, Australia (241)

SUZANNE M. MICHALEK, Department of Microbiology, University of Alabama at Birmingham, Birmingham, Alabama 35924 (105)

R. M. MOORE, Department of Veterinary Clinical Sciences, School of Veterinary Medicine, Louisiana State University, Baton Rouge, Louisiana 70803 (389)

B. MOREIN, Department of Veterinary Microbiology, Section of Virology, Swedish University of Agricultural Sciences, Biomedical Centre Box 585, S-751 23 Uppsala, Sweden (405)

PAUL S. MORLEY, Department of Veterinary Preventive Medicine, The Ohio State University, Columbus, Ohio 43210-1092 (415)

W. I. MORRISON, Institute for Animal Health, Compton, Newbury, Berkshire, United Kingdom (181)

L. MOTES-KREIMEYER, Merial, Limited, Athens, Georgia 30601 (571)

J. A. MUMFORD, Centre for Preventive Medicine, Animal Health Trust, Newmarket, Suffolk, England (379)

L. MURRAY, Embrex, Inc., P.O. Box 13989, Research Triangle Park, North Carolina 27709 (495)

MICHAEL P. MURTAUGH, Department of Veterinary Pathobiology, University of Minnesota, St. Paul, Minnesota 55108 (83)

THOMAS. J. MYERS, U.S. Department of Agriculture, Animal and Plant Health Inspection Service, Veterinary Services, Center for Veterinary Biologics, Riverdale, Maryland 20782 (585)

PETER L. NARA, Biological Mimetics Inc., Frederick, Maryland 21702 (115)

B. J. NONNECKE, NADC-USDA-ARS, Ames, Iowa 50010 (61)

CHRISTOPHER W. OLSEN, Department of Pathobiological Sciences, School of Veterinary Medicine, University of Wisconsin–Madison, 2015 Linden Drive West, Madison, Wisconsin 53706 (333)

IAN M. ORME, Department of Microbiology, College of Veterinary Medicine and Biomedical Sciences, Colorado State University, Fort Collins, Colorado 80523 (135)

K. R. PARSONS, Institute for Animal Health, Compton, Newbury RG20 7NN, United Kingdom (275)

P. P. PASTORET, Department of Immunology-Vaccinology, Faculty of Veterinary Medicine, University of Liège, Sart Tilman, B-4000 Liège, Belgium (595)

P. PHELPS, Embrex, Inc., P.O. Box 13989, Research Triangle Park, North Carolina 27709 (495)

R. POSTON, Embrex, Inc., P.O. Box 13989, Research Triangle Park, North Carolina 27709 (495)

S. S. POURCIAU, Department of Veterinary Microbiology and Parasitology, School of Veterinary Medicine, Louisiana State University, Baton Rouge, Louisiana 70803 (389)

C. A. RICKS, Embrex, Inc., P.O. Box 13989, Research Triangle Park, North Carolina 27709 (495)

JAMES A. ROTH, Institute for International Cooperation in Animal Biologics, College of Veterinary Medicine, Iowa State University, Ames, Iowa 50011 (681)

MICHAEL W. RUSSELL, Department of Microbiology, University of Alabama at Birmingham, Birmingham, Alabama 35924 (105)

LINDA J. SAIF, Food Animal Health Research Program, Ohio Agricultural Research and Development Center, The Ohio State University, Wooster, Ohio 44691 (429)

FRED W. SCOTT, Cornell Feline Health Center and the Department of Microbiology and Immunology, College of Veterinary Medicine, Cornell University, Ithaca, New York 14853 (347)

CATHARINE SCOTT-MONCRIEFF, Departments of Veterinary Pathobiology and Veterinary Clinical Sciences, Purdue University, West Lafayette, Indiana 47907 (733)

J. M. SHARMA, Veterinary PathoBiology, College of Veterinary Medicine, University of Minnesota, St. Paul, Minnesota 55108 (481)

MICHAEL SHEPPARD, Animal Health Biological Discovery, Pfizer Central Research, Groton, Connecticut 06340 (145)

CRAIG A. SHOEMAKER, United States Department of Agriculture—Agricultural Research Service, Fish Diseases and Parasites Research Laboratory, Auburn, Alabama 36830 (523)

DAVID SIEV, Center for Veterinary Biologics, U.S. Department of Agriculture, 510 South 17th Street, Ames, Iowa 50010 (749)

PAUL W. SNYDER, Departments of Veterinary Pathobiology and Veterinary Clinical Sciences, Purdue University, West Lafayette, Indiana 47907 (733)

DIANE L. SUTTON, U.S. Department of Agriculture, Animal Plant Health Inspection Service, Veterinary Services, Center for Veterinary Biologics, Licensing and Policy Development, Riverdale, Maryland 20737 (627)

C. E. SWIDERSKI, Department of Veterinary Microbiology and Parasitology, School of Veterinary Medicine, Louisiana State University, Baton Rouge, Louisiana 70803 (389)

G. TAYLOR, Institute for Animal Health, Compton, Newbury, Berkshire, United Kingdom (181)

H. W. TAYLOR, Department of Veterinary Pathology, School of Veterinary Medicine, Louisiana State University, Baton Rouge, Louisiana 70803 (389)

IAN TIZARD, Department of Veterinary Pathobiology, Texas A&M University, College Station, Texas 77843 (7)

MATHILDA J. M. TOUSSAINT, Faculty of Veterinary Medicine, Utrecht University, Yalelaan 1, 3508 TD Utrecht, The Netherlands (643)

DEOKI N. TRIPATHY, Department of Veterinary Pathobiology, University of Illinois, Urbana, Illinois 61802 (463)

S. VAN DEN HURK, Veterinary Infectious Disease Organization, Saskatoon, Saskatchewan, Canada S7N 5E3 (163)

K. VAN KAMPEN, The Van Kampen Group, Ogden, Utah (571)

JAN T. VAN OIRSCHOT, Department of Mammalian Virology, Institute of Animal Science and Health, 8200 AB Lelystad, The Netherlands (197)

P. VANNIER, Cneva Zoopôle, Les Croix, BP 53 22400 Ploufragan, France (615)

M. VILLACRÉS-ERIKSSON, Department of Veterinary Microbiology, Section of Virology, Swedish University of Agricultural Sciences, Biomedical Centre Box 585, S-751 23 Uppsala, Sweden (405)

SOMSAK VINITNANTHARAT, Alpharma NW, Aquatic Animal Health Division, Bellevue, Washington, 98005 (539)

V. WEYNANTS, Immunology Unit, Facultes Universitaires, Notre Dame de la Paix, Namur, Belgium (275)

C. WHITFILL, Embrex, Inc., P.O. Box 13989, Research Triangle Park, North Carolina 27709 (495)

BRUCE N. WILKIE, Department of Pathobiology, The Ontario Veterinary College, The University of Guelph, Guelph, Ontario, Canada N1G 2W1 (39)

C. WILLIAMS, Embrex, Inc., P.O. Box 13989, Research Triangle Park, North Carolina 27709 (495)

PETER F. WRIGHT, Canadian Food Inspection Agency, National Centre for Foreign Animal Disease, Winnipeg, Manitoba, Canada R3E 3M4 (669)

HONG-YIN WU, Department of Microbiology, University of Alabama at Birmingham, Birmingham, Alabama 35924 (105)

ROBERT J. YANCEY, JR., Animal Health Biological Discovery, Pfizer Central Research, Groton, Connecticut 06340 (257)

FEDERICO A. ZUCKERMANN, Department of Veterinary Pathobiology, University of Illinois, Urbana, Illinois 61801 (447)

PREFACE

This book presents a comprehensive view of veterinary vaccines and diagnostics—past, present, and future. The authors were all participants in the First International Veterinary Vaccines and Diagnostics Conference (IVVDC) held during the summer of 1997 at the Monona Terrace Convention Center, Madison, Wisconsin, USA.

Each session had co-chairs who selected four to six speakers. The book follows the general organization of the conference. The sessions and co-chairs were:

Vaccines and Diagnostics in Veterinary Medicine: Historical and Contemporary Perspectives, Dr. Ronald D. Schultz

Basic and Applied Immunology and Vaccinology, Dr. Pat Shewen and Dr. Chuck Czuprynski

Adjuvants and Immunomodulators, Dr. Gary Splitter and Dr. Michael P. Murtaugh

Canine Vaccines and Diagnostics, Dr. Leland E. Carmichael and Dr. Max Appel

Porcine Vaccines and Diagnostics, Dr. Linda J. Saif and Dr. Lennart Bäckström

Feline Vaccines and Diagnostics, Dr. Mary Tompkins and Dr. Fred W. Scott

Equine Vaccines and Diagnostics, Dr. Kevin Schultz and Dr. Paul Lunn

Bovine Vaccines and Diagnostics, Dr. Ivan Morrison and Dr. Manual Campos

Avian Vaccines and Diagnostics, Dr. Dick Witter and Dr. Jagdev M. Sharma

Fish, Exotic, and Wildlife Vaccines, Dr. Phillip H. Klesius and Dr. Joseph F. Curlee, Jr.

Adverse Reactions and Failures, Dr. James A. Roth and Dr. David Rosen

Standardization of Vaccines and Diagnostics, Dr. Ronald D. Schultz and Dr. Leland E. Carmichael

Regulation and Licensing, Dr. David A. Espeseth and Dr. Phillipe Vannier

Vaccinology in the 21st Century, Dr. Fred Brown and Dr. Lorne A. Babiuk

The success of the conference was made possible by the excellent oral presentations and more than 100 poster presentations and by more than 400 participants from academia, government, industry, and the practice of veterinary medicine. Major sponsors of the conference included these organizations: School of Veterinary Medicine, University of Wisconsin–Madison; American Association of Veterinary Immunologists; Bayer Animal Health; Boehringer Ingelheim Animal Health, Inc.; Ft. Dodge Laboratories; Grand Laboratories, Inc.; HESKA Corporation; IDEXX Laboratories, Inc.; Intervet; Merck Research Laboratories (now Merial Ltd.); Pfizer Animal Health; and Rhone Merieux, Inc. (now Merial Ltd.).

The topics covered herein are especially timely because of the many changes and new developments in veterinary vaccinology and diagnostics that have taken place during the past 10 years. Information on vaccines and diagnostics for virtually all the major animal species, both wild and domesticated, is included. There are many discussions on new methodologies currently being used to develop safer and more effective vaccines and for the development of rapid, effective, and simple diagnostics. In veterinary medicine, in contrast to human medicine, vaccines and associated diagnostics must be cost effective; thus, certain vaccines and diagnostics must sell for pennies if they are to be used for selected species, such as poultry. Monetary restrictions rather than any scientific/technological constraints place more significant constraints on the development of products for many of the domesticated species served by veterinary medicine. New methodologies, especially those resulting from advances in recombinant DNA technology, are making possible the development of vaccines for diseases for which there are no conventional vaccines and of replacement of conventional vaccines with safer and/or more effective vaccines as needed. New and improved diagnostics, some of which can be used with special "marker vaccines" to control or maybe some day eradicate certain diseases, are also presented. Contributors discuss methods for licensing vaccines and standardizing certain procedures and protocols worldwide to improve and simplify certain processes that are highly diverse and costly. Global distribution of many of the vaccines and diagnostics makes harmonization necessary and will help ensure the cost effectiveness of the new products.

An especially timely discussion focuses on the frequency of administration of vaccines and the adverse reactions associated with vaccines.

All the authors acknowledge the major contributions vaccines have made and will continue to make in the control of animal diseases. The major accomplishments in the improvement of animal health and wellbeing achieved through the use of many of the vaccines and diagnostics currently available are readily apparent. However, many vaccines are being given too often to animals that will benefit little or not at all from the specific vaccines. As illustrated by several authors, there are also vaccines that can cause severe adverse reactions in certain animals and there are vaccines providing little or no economic benefit for the target species; in fact, they may create an economic loss. Diagnostics that need to be improved with regard to sensitivity and/or specificity and the need for standardization and/or improved quality assurance programs are also discussed.

The frequency of vaccination is an especially common theme with respect to canine and feline vaccines. It is readily apparent to most of the authors that vaccines are designed to generate an "immunologic memory" that lasts for years or often for the life of the animal. It is also acknowledged that the current practice of annual revaccination is not necessary for many of the products used in cats and dogs and recommendations for annual vaccinations were made primarily to get pet owners to bring their pets to the veterinarian for annual physical examinations. However, the recommendation for annual revaccinations becomes less acceptable with increased adverse reactions, especially those that cause significant disease or death (for example, anaphylaxis, vaccine-associated fibrosarcomas) and as more and more vaccines become available. Adverse reactions and immune responses to self-antigens have led to the reexamination of the annual revaccination recommendation, a recommendation that has no scientific basis! For those vaccines with a long duration of immunity (for example, viral vaccines), it has been suggested that vaccination occur once every three to five years, instead of annually. Certain vaccines should not be given at all to animals, especially those at low or no risk of disease.

It is obvious that there are many safe and effective vaccines now available for many species, but even more obvious is that the future will bring many new vaccines. These new vaccines will include some that are safer and/or more effective than current vaccines and some for diseases for which no vaccines exist. Also new, totally different types of vaccines will appear on the market (for example, cancer vaccines, vaccines to prevent pregnancy). The frequency of vaccination will need to be determined for these new products, and frequency should be based on the true duration of immunity. It will be as important not to overvaccinate as it will be to ensure that as many animals as possible receive the benefit of a vaccine for diseases that cause significant morbidity and mortality. Likewise, it will be important to ensure that vaccines are not used in animals that will receive little or no benefit or in circumstances where the vaccines may cause harm.

Vaccines for prevention and treatment of cancer, for neutering pet animals and wildlife species, and for losing or gaining weight are just a few of the novel applications being developed. Vaccines that do not need to be injected but that can instead be fed or aerosolized or given in the water are required for certain species. Improved delivery methods will increase compliance for vaccines and decrease the costs associated with vaccinating food animal species and wildlife. Completely revolutionary vaccines that do not contain an antigen, only the genetic information to instruct the animal to make the antigen, are now available. These vaccines, DNA/nucleic acid vaccines, are creating a significant amount of interest because they appear to be as effective as modified live vaccines (the most effective type currently available) and safer than killed vaccines.

These are exciting times in vaccinology and diagnostic medicine, primarily because technology is providing an opportunity to make new and more effective products that can be readily delivered to a large number of animals including wildlife species when and as needed. Availability of this new technology alone should not and cannot drive product development. We must stop and ask the question, "Do we need the vaccine or the diagnostic?" We should not merely ask, "Can we make it?"

Undoubtedly the advances of the past 10 years will appear minor compared to those made during the next 10 years. We will have another opportunity to get together at the Second IVVDC (IVVDC 2000) in Oxford, England, to discuss the progress in this fast-moving field. For those unable to attend the First IVVDC, this book offers a comprehensive summary; for those who attended, it will amplify and embellish the information presented on veterinary vaccines and diagnostics.

RONALD D. SCHULTZ

I
VACCINES AND DIAGNOSTICS
Historic and Contemporary Perspectives

Vaccination: A Philosophical View

MARIAN C. HORZINEK

Virology Unit, Veterinary Faculty, Utrecht University, Utrecht, The Netherlands

I. Introduction
II. Will There Be Vaccination in the Next Millennium?
III. What Is the Future of Veterinary Vaccinology?
IV. Vaccination in the Twenty-First Century
V. Outlook

I. Introduction

Vaccination is the most successful medical and veterinary measure: More lives have been saved by immunization, more animal production safeguarded than through all other medical and veterinary activities combined. It has been possible to eradicate a disease worldwide (human smallpox), and attempts to reach the same goal for poliomyelitis and measles are viewed with optimism. But this optimism does not apply for diseases where the causative virus has a reservoir in the wild fauna (such as in the case of distemper and parvoviruses that occur in mustellids). Against these diseases vaccination will have to continue.

Vaccines were being used long before the mechanisms of immune protection became known. Today the discipline of vaccinology has acquired proper scientific status as a interdisciplinary research area that emerged from microbiology and immunology. Because vaccines make money—for the veterinarian and as a corollary also for the pharmaceutical industry—money is being invested in vaccine research. This has resulted in impressive progress. Protection can be obtained against most animal viral diseases, and it is hoped the remaining conditions (e.g., feline AIDS, feline infectious peritonitis, Aleutian disease of mink) will be controlled in the near future.

For the average small animal practice in Europe, doing vaccinations provides its financial basis; between 20 and >40% of direct and indirect income is earned from vaccinations. In view of this economy-related fact, teaching of vaccinology is clearly insufficient at most veterinary colleges. While there are chairs of immunology and vaccinology at the Utrecht Faculty of Veterinary Medicine, most of the neighboring European countries lack such chairs. The paradoxical situation exists that the average veterinary practitioner knows least about his or her most lucrative activity. This is a depressing and dangerous situation, especially in view of the future developments in veterinary medicine.

II. Will There Be Vaccination in the Next Millennium?

This is a moot question, of course, but for the virologist no prophetic talent is required to answer it in the positive sense. As long as cellular organisms populate this planet, new viruses will appear at regular intervals and infect them, at times making them ill. Consequently there will always be a reason for developing new vaccines—and there are no good alternatives. One may not expect universal antiviral chemotherapeutics, irrespective of the recent successes obtained in treating AIDS patients. (*Editor's note:* HIV drug resistant strains have already appeared to threaten the sucess of treatment.)

New viral epidemics are seen time and again (from parovirus enteritis to seal distemper to swine mystery disease), the media focus the public's attention on them, people may fall ill and die (for example, after infection with "new" influenza viruses). Research has shown that "new" viruses are the result of subtle changes in the genomes of usually well-known ones, the consequence of viral evolution.

What are these new viruses? As soon as an epidemic has been identified molecular genetic research is initiated, the results of which are almost predictable: The causative agent will be a virus that is slightly different from other, well-known viruses. The virus causing the epidemic is the result of mutations followed by selection—a process we call evolution. The driving force in the evolution of RNA viruses (among which there are numerous animal pathogens) are single nucleotide changes scattered throughout the genome, but also insertions of longer pieces of genetic information cannibalized from the cell. During transcription, viruses can acquire foreign information and incorporate it into their genome by template jumps of the transcriptase, the enzyme that copies the nucleotide sequence; if that foreign information confers a selective advantage to the "parent" virus, it will be genetically perpetuated. When the mere change of a single amino acid has dramatic effects, then the uptake of entire motives (modules) of foreign information can be expected to lead to even more dramatic alterations. Thus the mutated virus could escape preexisting immunity (to the "parent" virus) in its host, thereby causing disease; on the other hand the variant could also cross the species barrier and infect a new host.

Thus viral evolution can be tracked and analyzed in molecular terms. The vaccine industry obviously follows epidemiologic developments and is eager to market a vaccine against a newly discovered virus disease as soon as possible—this contributes to its reputation and success in the marketplace. In those cases where evolutionary trends can be predicted, where mutations can be expected in certain genes (for instance, in influenza virus), this is an efficient routine procedure: Vaccines against human influenza get a yearly update.

For a company, vaccine research, development, and registration are expensive. Assessment of the return on investment may lead to compromise, especially as far as efficiency is concerned (for instance, by accepting a marginally effective antigen concentration in a killed preparation). Placebo-controlled independent field studies to evaluate the degree of protection have been performed only in a few instances, comparative studies between preparations even more rarely. Veterinarians rely on their "experience," however objective this may be. They generally stay loyal to their manufacturers, leaving them only as a consequence of a good talk by another sales representative or of "experiences" communicated by a colleague. Undoubtedly there are differences in vaccine performance between brands, but also between viruses. The notoriously bad reputation of preparations intended to protect against feline upper respiratory disease may be due to newly evolved caliciviruses in the field against which the vaccine induces less cross-protection than years ago, when they were first developed. Research has even shown annual variation between vaccines from the same manufacturer; certain lots induced poorer immunity than others.

When discussing the bad reputation of a particular vaccine one should take a sociopsychologic phenomenon into account—that of selective observation. No veterinary practitioner or breeder will mention the thousands of healthy, well-protected vaccinees, but the single animal that falls ill (after or irrespective of vaccination) gets all the attention. When the name of the preparation is mentioned in this context, a negative image may develop as a consequence. However, this is normal—also in the statistical sense of the word: No vaccine results in 100% protection and complete safety. Any biological phenomenon follows a frequency distribution of effects, and rarely—sometimes not so rarely—one may expect no effect at all. This can bias the vet's "experience" mentioned above.

III. What Is the Future of Veterinary Vaccinology?

Van Leeuwenhoek, Beijerinck, and Kluyver are famous names in microbiology; also veterinary vaccinology has its roots in the Netherlands. When Frenkel successfully cultivated foot-and-mouth disease virus in Amsterdam using suspensions of surviving bovine tongue epithelium, the basis was laid for controlled virus growth, for the industrial production of viral antigen in cell culture. The National Veterinary Institute subsequently wrote vaccinological history: It developed the herpes virus marker vaccines and the chimeric pseudorabies/hog cholera vaccine—a molecular vaccinological highlight. Protection of dogs and mink against fatal parovirus infections using a short oligopeptide made an old immunological dream come true.

Progress has been made in both the innocuity (safety) and efficacy of veterinary vaccines. Many strategies are currently being developed to arrive at maximum possible protection with minimal side effects. Veterinary medicine leads the way: Observations can be made and experience gathered in the target species, where ethical, legal, and economical reasons prevent similar approaches in the medical environment. One of the most promising developments is vaccination with DNA that carries the information for a protection-relevant protein. The immunogenic protein is not injected into the vaccinee but the genetic information for this protein that is then synthesized by the host cell itself. Although the injected DNA does not replicate, the induced immune reaction is similar to that after modified live vaccination, with respect to MHC class I antigen presentation and the induced cellular response.

Protection is not only a function of the quality of the antigen but also of correct triggering of the immune system. To improve the immunogenicity of a preparation so-called adjuvants are added to the antigen; these are minerals such as aluminum hydroxide but also water/oil emulsions, detergents, plant glycosides, etc. The antigens are immobilized at the site of injection, and during the ensuing inflammatory reaction antigen-processing cells are attracted. Chronic local inflammations may lead to malignant transformation, however, and recent observations indicate the occurrence of fibrosarcomas at injection sites of predominantly aluminum hydroxide-containing inactivated vaccines. The principle of innocuity is stringently observed by European licensing authorities—more stringently than efficacy, one is tempted to say. It is also easier to prove. An incidence of 3 fibrosarcoma cases on

10,000 cats after leukemia or rabies vaccination is only acceptable if the disease against which it must protect carries a higher risk.

Cytokines are currently the focus of attention as immunity enhancers; they are messenger molecules of the immune system that are able to direct the response. It may be expected that they will become successors of the empirically developed adjuvants.

IV. Vaccination in the Twenty-First Century

The veterinary practitioner demands a vaccination schedule that is simple, timesaving and commercially attractive. Industry met these requirements by development of polyvalent preparations that may contain seven or more components. The small animal scene gradually adopted a yearly vaccination routine and everybody appeared satisfied: the vet, the owner, the manufacturer. One visit per client per year, one injection, one vaccine—easy enough. In biological terms this is nonsense, of course. A universal scheme cannot be expected to accommodate the immunology of the carnivore, the properties of diverse infectious agents, the changing epidemiologic situation, or the age and living environment (risks of infection) of the animal.

Injection and immunization are not synonymous. The automatic yearly application of a polyvalent preparation with disregard for the vaccinee's individual life circumstances is the dangerous development alluded to above. It may be damaging for the profession.

There is nothing wrong, in principle, with combined vaccines, if immunogenicity and protective potency have been proven for each component. Studies have shown, however, that, for example, parvo/distemper virus combinations may lead to encephalitis caused by the latter; though sporadic, these cases have received much attention and led to a more critical appraisal of combination products.

There is something wrong, however, with the unreflected yearly injection of *all* the components present in a polyvalent preparation. For some diseases this is too much, for others too little. Immunity against measles is lifelong, and I do not know of any vet who requests a yearly measles booster from her or his physician. But most small animal practitioners in Europe revaccinate dogs against distemper, which is in essence canine measles, annually. Experimentally, distemper protection has been proven for 3 years, and it may last longer. In interepidemic intervals, and this is most of the time, less frequent immunizations can therefore be defended.

The other side of the coin is vaccination against, for example, feline herpesvirus, which may be too infrequent when given once a year in a shelter situation. Crowding, immunosuppression as a consequence, infection pressure, and the notoriously poor antigenicity of herpesvirus preparations all argue in favor of more frequent applications. Experimentally, protection did not last for more than 4–6 months.

In addition to the pediatric indication—the protection of pups and kittens—there may be a geriatric indication for vaccination in the face of an aging immune system. In veterinary medicine, this is unexplored territory. In medicine, influenza vaccination has a geriatric indication.

V. Outlook

The following conclusions may be drawn from the above considerations:

- The vaccine industry should consider developing and marketing mono- and oligovalent preparations—booster vaccines containing essential immunogens and lacking superfluous ones.
- The veterinarian should continuously monitor the epidemiologic situation and adjust the booster vaccination schedule accordingly. Pups and kittens should get their first shots at the breeder's. This has a twofold advantage: the animal is immunized in its own premises, and the vet can get an idea about the management of the kennel or cattery.
- The owner should be more involved in the vaccination schedule and strategy. After all, the veterinarian is responsible for the health of a family member and should therefore be informed about familial activities and changes (vacation, birth of a baby, change of residence, acquisition of another dog or cat, etc.) to implement a made-to-measure vaccination program.
- The disciplines of veterinary microbiology should become more involved in the epidemiologic surveillance of the companion animal scene, since there is no "Veterinary Communicable Diseases Center" for the dog and cat. The appearance of new viruses, new antigenic variants, zoonotic risks, and vaccine failures need to be identified and communicated to the small animal practitioner at large, but also to the vaccine industry.

The age of empirism in vaccinology is past. To vaccinate successfully is not as easy as it seems; it requires veterinary knowledge and immunologic and microbiologic insight. It was the French poet Léon-Paul Fargue (1876–1947) who wrote: "Il n'y a pas de simplicité véritable; il n'y a que des simplifications." ("There is no real simplicity, there are only simplifications.")

Grease, Anthraxgate, and Kennel Cough: A Revisionist History of Early Veterinary Vaccines

IAN TIZARD

Department of Veterinary Pathobiology, Texas A&M University, College Station, Texas 77843

I. Introduction
II. The Origin of Vaccinia
   A. Jenner's Vaccine
   B. Dr. John Loy
   C. Grease and Horsepox
III. Anthraxgate: A Minor Nineteenth-Century Scandal
   A. Henri Toussaint
   B. Louis Pasteur
   C. William Greenfield
IV. Early Canine Distemper Vaccines
   A. Early Vaccines
   B. Ferry's Vaccine
   C. Viral Vaccines
V. Summary
References

I. Introduction

Although the general features of the historical development of both human and animal vaccines are well known and have never really been contentious, recent scholarship has produced some interesting new and perhaps controversial information on the origins and discovery of the vaccines against smallpox, anthrax, and distemper.

The birth of vaccination, and immunology in the broadest sense, stems from two key events. The recognition by Edward Jenner in 1798 that vaccinia could protect humans against smallpox and the recognition, in 1879 by Pasteur, that the general principles enshrined in Jenner's discovery were broadly applicable to a whole range of infectious diseases. The long interval between these two events was unavoidable. Only when the microbial nature of infectious disease was recognized, and the concept of spontaneous generation discredited, could the broader principles involving resistance to disease be appreciated. Once this was achieved, the principles of vaccination were rapidly applied across Western Europe and North America. Pasteur, for example reported on his fowl cholera discoveries in February 1880. In a USDA special report dated December 1880, Dr. D. E. Salmon could say the following. "At present the attention of investigators (employed by USDA) is still, for the most part turned to methods of prevention and chief among these is inoculation by means of a mitigated virus. . . ." This is not surprising since Pasteur's discovery, for the first time, provided a reasonable mechanism of controlling infectious diseases. Indeed, once the concept of microbial causes of disease became established, then the ideas of virulence and attenuation followed fairly naturally.

These two key discoveries, however, have some very interesting and controversial aspects. Thus, while we have long accepted that Edward Jenner used cowpox as his source of vaccine material, there is a considerable body of evidence to suggest that vaccinia could have originated from a now extinct horsepox virus. This could explain the mystery of the origin of vaccinia, a virus that is distinctly different from modern cowpox.

In the case of Pasteur's early discoveries, we can now see how in the years 1880 and 1881, at least three individuals, Greenfield in London and Pasteur and Toussaint in Paris, independently came to the same basic understanding of the processes of microbial attenuation and vaccination. That the priority and hence the credit was eventually accorded to Pasteur was inevitable given his very high public profile at that time. Nevertheless, Pasteur was by no means alone at the summit of the discovery of vaccination.

In effect, the limiting factor in the development of many early vaccines was the correct identification of the causal agent of a disease. The development of antidistemper vaccines demonstrates in an exemplary fashion just how this occurred. Despite the fact that the viral nature of canine distemper was clearly established as early as 1905, there was an astonishing reluctance to abandon a bacterial etiology for another 20 years. As a result, effective antidistemper vaccines were relatively late developments.

II. The Origin of Vaccinia

A. Jenner's Vaccine

Although the first documented substitution of cowpox for smallpox was performed by Benjamin Jesty in 1774, the general acceptance of vaccination dates from 1795 when Edward Jenner published his paper entitled "*An enquiry into the causes and effects of the variolae vaccinae, a disease discovered in some of the western counties of England, particularly Gloucestershire, and known by the name of The Cow Pox.*" This is a misleading title. Despite referring specifically to cowpox, the opening sentence of Jenner's paper states: "There is a disease to which the horse, from his state of domestication, is frequently subject, The Farriers have termed it the Grease" (Jenner, 1795). Jenner then proceeds to claim that the grease and cowpox were in fact the same disease. Jenner described how the grease could be transmitted from a horse to the bovine udder as a result of ". . . some particles of the infectious matter adhering to his fingers." The same disease could then, Jenner suggested, be transmitted to dairy maids. He went on to give a series of examples showing how prior infection of a human with either the grease or with cowpox could confer immunity to smallpox. Thus, in his very first example, Jenner described how Joseph Merret, an Under Gardener to the Earl of Berkeley, managed to transmit grease from his master's horses to his master's cows. He also caught it himself and was "much indisposed for several days. . . ." Merret recovered and about 25 years later was variolated against smallpox. Susceptible recipients of smallpox vaccine normally develop a lesion at the infection site. Resistant individuals do not. In the case of Joseph Merret, Jenner could never get the vaccine to take. Merret was, however, immune to smallpox and remained healthy despite the fact that other members of his family developed the disease.

Edward Jenner was convinced that the "sore" heels or "grease" of horses, which infected the farriers who shod and the grooms who attended such horses, was horsepox and could produce cowpox and that either could be used to vaccinate humans against smallpox. He continued to believe that horse-derived material was a source of vaccinia in later life. For example, in 1813, Jenner wrote to a colleague regarding vaccination and said, "for the past two months, I have been using material derived from a horse." In 1817 Jenner supplied more of this equine-derived material to the National Vaccine Establishment and it was widely used for vaccination.

B. Dr. John Loy

In 1801 Dr. John Loy, a physician from Pickering in Yorkshire obtained scab material from the hand of a patient who had been treating horses for the "grease" (Fleming, 1890; Taylor, 1993). Loy used this material to vaccinate the man's brother who developed a local lesion, which looked to Loy like a normal cowpock. Loy then went on to take more material from his original donor and use it to inoculate a cow and a child. Both developed typical lesions and the child subsequently resisted variolation. Subsequently, Loy performed a variation on this procedure. He took material from a horse with grease and used it to inoculate the udder of a cow. The cow developed "cowpox." He then took scab material from the cow and used it to inoculate a child who subsequently also resisted variolation. To reinforce these ideas, Loy went still further and vaccinated a child directly with horse-derived material and then went on to use material from this child to inoculate five additional children. All resisted variolation, a reasonably reliable indication that they had developed immunity to smallpox. Collectively these studies confirmed, at least to Dr. Loy, that there was an agent in the lesions of grease that could be transmitted to cows and that humans inoculated with this material resisted variolation.

Later in the nineteenth century there were numerous episodes where material from horses with the "grease" caused pox-like skin lesions in humans (Fleming, 1890). Two examples will suffice. In 1830, Professor Hertwig and 11 students at the Berlin Veterinary School became inoculated after handling horses affected with the grease. In 1881, a Dr. Pingaud in Paris took material from a horse with horsepox and collected the "lymph" from the pustules. He used this to inoculate 7 young soldiers of the 10th Hussars (volunteers?), 6 of whom subsequently developed characteristic vaccine vesicles. Material from these individuals was subsequently used to vaccinate an additional 64 soldiers and in 40 of these the vaccine appeared to take.

C. Grease and Horsepox

What was the "grease"? At the present time, the disease of the fetlocks of horses is considered to be due to infection with *Dermatophilus congolense*. However, Loy recognized that there were two kinds of grease: One was local and not infective to humans, the other caused a systemic febrile disease. This second form could be accompanied by eruptions elsewhere on the skin. According to Loy, only the second form was effective as a source of vaccine material. During the nineteenth century, a disease known as horsepox or equine variola was well recognized in Europe and North America. It caused skin lesions, some of which could occur on the fetlock. The disease was sporadic but was seen in Montreal in 1882, and in England at least as late as 1885. It is also of interest to note that the great Professor Chauveau himself noted that injection of vaccinia virus to horses produced lesions that ". . . differ in nothing from natural horsepox" (Fleming, 1890).

At the present time the only poxvirus that has been isolated from horses appears to be restricted to a small area of Kenya (Kaminjolo *et al.*, 1975). It is a typical orthopoxvirus that causes multiple skin lesions resembling papillomas. It is called Uasin Gishu pox after the district where the disease occurs. The virus cross-reacts in a hemagglutination inhibition assay with both vaccinia and cowpox. Because horses are a fairly recent addition to the area it has been supposed that its natural reservoir is a wild animal. It has never been identified outside the area where it was first isolated. It is, however, not inconceivable that this may actually be a relict population of the horsepox virus imported from Europe during the colonial period.

Is it possible that vaccinia may not be a cowpox but a horsepox virus? Certainly, vaccinia is distinctly different from cowpox (Fenner *et al.*, 1989). During the nineteenth century vaccinia could not, of course, be cultivated *in vitro*. It also had a relatively short storage life although attempts were made to store it on dried threads. Thus, the most effective way to ensure its efficacy was to transmit it directly between individuals by arm-to-arm inoculation. This was not always convenient and there was a constant need for new sources of vaccine. On occasion, new material could be obtained from natural cases of cowpox or, as pointed out earlier, from horsepox. When the medical authorities became aware that direct transmission was a good way to transmit other diseases, especially syphilis, they took steps to ban the procedure and made the use of calf-derived vaccine mandatory. Unfortunately, only one laboratory in the world, in Cologne, Germany, had such a vaccine. Thus, all current vaccinia stocks come from this single laboratory. Where did they get their virus from? Sadly, the records of this laboratory were lost during World War II. The origin of vaccinia is thus unknown. It is not implausible however that it did originate from the extinct disease of horses known as equine variola, horsepox or the "grease."

III. Anthraxgate: A Minor Nineteenth-Century Scandal

Around the year 1880, three investigators independently discovered how to vaccinate animals against anthrax. They were Louis Pasteur and Henri Toussaint in France and William Greenfield in England. While Louis Pasteur received all the publicity, recent scholarship suggests that the credit should at the very least be shared.

A. Henri Toussaint

Henri Toussaint was a very junior professor at the Toulouse Veterinary School. He was the one who sent a vial of blood from a cockerel dying of fowl cholera to Pasteur and enabled him to develop the first vaccine (Geison, 1995). Toussaint considered that he had actually seen the fowl cholera bacillus but although politely thanked by Pasteur, was never publically acknowledged. However, he went on to develop his own anthrax vaccine. In July 1880 he told his colleague, Henri Bouley, about his discoveries, and Bouley encouraged him to make them public. As a result, on July 12, 1880 Toussaint's initial results were presented to the National Academy of Sciences (Toussaint, 1880a). In these he described how he successfully immunized four dogs and six sheep against anthrax. Toussaint would not reveal just how he made this vaccine (he considered this somewhat premature) but on the same day deposited a sealed envelope with the secretary of the National Academy of Sciences. The members of the National Academy were very critical of Toussaint's secrecy so that on August 2, 1880, he allowed the note to be opened (Toussaint, 1880b). In it were described a method of making an anthrax vaccine for dogs and sheep. This involved heating infected defibrinated blood to 55°C for 10 minutes. Toussaint pointed out that not only were his animals protected, but there was also no local reaction at the inoculation site.

Unfortunately, soon after his method became public, Toussaint realized that this was an unpredictable method of killing the anthrax bacillus. He therefore switched to treating anthrax blood with 1–1.5% carbolic acid. Using this method, he conducted a fairly large-scale trial of this new vaccine on August 8, 1880, at the Alfort Veterinary School (Geison, 1995). In this trial, 16/20 sheep survived challenge although many got very sick indeed. At the time this was considered a partial failure and Toussaint got very discouraged. He continued his efforts but was unable to produce consistent results. Once the results of Pasteur's public experiment were publicized in 1881, Henri Toussaint appears to have given up the struggle. Late in 1881 he became mentally ill and died in 1890 at the age of 43! At his funeral, the director of the Toulouse Veterinary School declared: "His mind gave way under the weight of the great thoughts it carried."

B. Louis Pasteur

Meanwhile Louis Pasteur had turned his attention to the problem of disease prevention. Early in his studies he recognized the key features of Jennerian vaccination for what they were, namely, the use of an avirulent strain of organism to protect against a virulent strain. He also recognized that it would be important to reduce the virulence of an organism by attenuation and pointed this out to Emil Roux, his assistant, several months before he conducted his famous fowl cholera experiment in the late summer of 1879. Because he had thought all this out beforehand, at the conclusion of the experiments on fowl cholera it apparently only took him "a minute" before he exclaimed, "Well, everything explains itself—this hen has been immunized by being injected with the old culture." What has been perceived as an astonishing flash of intuition was really the culmination of long considered deliberation on the subject. Pasteur reported to the Academie des Sciences on February 9, 1880, that he had succeeded in vaccinating birds against fowl cholera. However he kept the method secret until his presentation to the Academie de Medicine in October 26, 1880 (Pasteur, 1880). The reasons for this secrecy are unclear but Pasteur probably hoped that the method could be extended to other, more significant diseases such as anthrax. This disease had been making a nuisance of itself across France, costing an estimated 20–30 million francs. (It should be noted that the relatively unknown Henri Toussaint had tried the same secrecy approach but had received very severe criticism and was obliged to reveal his methods. The renowned Louis Pasteur could not be pressured in the same way.)

Following his fowl cholera studies, Pasteur set out to make an anthrax vaccine but was surprised when in July 1880, Toussaint announced his development of an anthrax vaccine. The surprise stemmed from two concerns. One, that an unknown from the provinces might beat him to it, and two, that Toussaint had used killed organisms. Pasteur had hitherto believed that vaccination worked when the attenuated vaccine strain deprived the body of nutrients essential to bacterial growth. He could not conceive therefore how a killed vaccine might work. Fortunately for his theory, his friend Emil Roux soon showed that Toussaint's method of heating anthrax to 55°C was not lethal to all anthrax bacilli. Of course, Pasteur was basically a chemist so he sought to explain the process of attenuation by chemical laws. In fact, he considered attenuation to be chemically equivalent to oxidation. Thus, he decided that if an organism could be "oxidized" it would become avirulent. To attenuate the anthrax bacillus Pasteur decided to oxidize it by heat and so anthrax cultures were incubated at 42–43°C. This worked fairly well.

In August 1880, however, Bouley had told Pasteur and Roux about Toussaint's vaccine experiment using carbolic acid-treated anthrax blood. Just to be on the safe side, therefore, Pasteur had Roux conduct a series of studies looking at the use of chemicals to attenuate anthrax. Roux found that dilute potassium bichromate (final concentration 0.1%) added to anthrax cultures for 12–14 days was no longer lethal to sheep but effectively protected them. As a result, by January 1881 Pasteur's laboratory had two candidate anthrax vaccines available: Pasteur's heat-attenuated one and Roux's potassium bichromate-attenuated one. Studies on rabbits, suggested that the bichromate-inactivated product was more reliable and safer. Pasteur reported his studies on heat attenuation of anthrax to the Academy on February 28, 1881 (Pasteur et al., 1881a). During this presentation he took the opportunity to "knock" Toussaint's procedure emphasizing ". . . la différence qui existe entre les deux méthodes, l'incertitude de l'une, la sûreté de l'autre."

On April 27, 1881, Pasteur was goaded by the skepticism of his many critics to accept a challenge to demonstrate the effectiveness of his anthrax vaccine in public. But which one should he use in the trial? From Gerald Geison's examination of Pasteur's notebooks, it now appears that Pasteur chose to go with the bichromate-attenuated vaccine, a vaccine very similar in principle to that developed by poor Henri Toussaint. Thus on May 5, 1881, Louis Pasteur initiated a public trial of his anthrax vaccine at Pouilly-le-Fort when Roux inoculated half a flock of sheep with a bichromate-treated culture. On May 17 he gave them a second dose. On May 31 he challenged all the sheep, both vaccinated and unvaccinated. All vaccinated animals survived while the unvaccinated ones died. These results were dramatic and Pasteur rightly received international renown for this public demonstration. He officially published these results later in 1881 (Pasteur et al., 1881b). In these results, however, he claims to have used exclusively the heat-attenuated vaccine. Why Pasteur falsely claimed to have used the heat-attenuated vaccine is unclear. Certainly his later studies focused solely on the heat attenuation process and he went no further with bichromate attenuation. However, Geison, who has examined Pasteur's notebooks, suggests that Pasteur considered the heat attenuation process more reliable than the chemical attenuation process (Geison, 1995). Because Pasteur's laboratory generated a considerable income from the manufacture of anthrax vaccine, he was perhaps reluctant to draw attention to the possibility that other attenuation processes would also be effective.

C. William Greenfield

Meanwhile in London work on anthrax was under way at the Brown Animal Sanitary Institution, a small veterinary clinic operated by the University of London (Wilson, 1979). The institution not only provided veterinary care in the city but was also responsible for conducting research on animal diseases. Thus, the first superintendent, a Dr Burdon-Sanderson and his veterinary assistant, a Mr. Duguid, investigated an outbreak of anthrax in cattle in Mr. Mason's farm in February 1878. During these investigations, they found that "when the disease is transmitted by inoculation from cattle to small rodents, such as guinea-pigs, and then from them back to cattle, the character of the disease so transmitted is much milder than that of the original disease acquired in the ordinary way. The rodents die, but the bovine animals inoculated with their blood or with the pulp of their diseased spleens recover." The first experiment of this type was undertaken on March 25, 1878, when a calf was inoculated with splenic material from a guinea pig that had died of anthrax (Burdon-Sanderson, 1880). The calf got very sick, but by March 30 had completely recovered. Two yearling heifers were inoculated with guinea pig derived material on May 7 and also recovered. On May 16, all three animals were reinoculated with blood from a guinea pig that had died from anthrax. The calf got very sick again but the heifers were only mildly ill. The animals were inoculated a third time on June 10 and while two got mildly ill, one showed no symptoms whatsoever. This is clearly the first demonstration of microbial attenuation by passage through a new species.

Burdon-Sanderson resigned from the institution in 1878 and the superintendent's position was filled by Dr. William Greenfield. Greenfield followed up on these anthrax studies and reported them in the *Lancet* on April 10, 1880 and in the *Journal of the Royal Agricultural Society* that same year (Greenfield, 1880a,d). Most importantly, Greenfield saw that another key experiment was needed: "Hitherto I have had no opportunity of submitting an animal thus inoculated to the crucial test of subsequent exposure to contagion or inoculation from an original case of the disease in a bovine animal; but I have done so from a sheep which died of the artificial disease, with a favourable result." He thus described in this April report how a bovine that had been protected by prior inoculation was completely resistant to an injection of blood from a sheep that had died of anthrax. In other words, he showed that an animal was resistant not just to attenuated, guinea-pig-derived anthrax, but to the virulent forms of the disease. Greenfield then went on to predict the science of immunology thus: "Should this method of inoculation be found to succeed, even in a majority of cases, it would afford a most valuable prophylactic measure in those districts where from time to time whole herds of cattle die of the disease. It may, "however furnish another illustration of the pathological law exemplified in the relations of smallpox and vaccinia, and may serve as a basis for future more extended researches in relation to the prevention of fatal epidemic diseases." In other words, Greenfield saw, like Pasteur, that the basic principles of attenuation could be applied to give protection against other diseases.

Greenfield gave a series of lectures on his studies on anthrax at the University of London in December 1879. These were published in the *Lancet* in installments in June and July 1880 (Greenfield, 1880c,e). In the June 5 issue, he stated in a footnote: "Further experiments, completed since the delivery of this lecture, have led me to the conclusion that the poison becomes progressively less virulent in successful generations of artificial cultivation. I have thus been able to obtain a modified virus, which when inoculated produces much less severe symptoms and appears to be partially protective against future, more severe attacks."

In the July 10 issue he stated: "It has been commonly asserted that when cultivated generation after generation in indifferent media, these bacilli still retain their power of causing the disease when inoculated in susceptible animals. This I was prepared to affirm until experiment had convinced me that it was not universally true." He also said in a footnote: "Since delivering this lecture I have, I believe, established . . . that, the anthrax bacillus may be grown in activating fluids in successive generations, until it acquires an innocuous condition, though maintaining its power of germination and growth and all its morphological characters." In the same paper Greenfield described again how he passaged anthrax through a guinea pig and then took this material and gave a sublethal dose to a heifer. The heifer did not die. He therefore reinoculated the animal with a second, larger dose. The animal developed a fever, sickened but survived. Greenfield stated in his lecture, "It remains to be seen whether this modified attack will confer any future protection on this animal." In a footnote, however, he states: "These [experiments], together with still further experiments not yet published, show that protection is conferred, which is great in degree and lasts a considerable time. . . ."

In his paper in the *Royal Agricultural Society Journal* (Greenfield, 1880a), Greenfield described how he had in effect repeated Burdon-Sanderson's experiments and concluded by saying of his surviving animal: "The animal is still being kept, awaiting the opportunity of making the crucial experiment of direct inoculation from a case of anthrax in a cow." Greenfield then concludes by stating, in summary, the key requirements for all subsequent vaccine research. "It must be evident to any one who considers the matter that there are many points which must be determined by experiments of a much more extensive character than any I am able to carry out at the Brown Institution. If, as I hope, it should prove on further experiment that the earlier results are confirmed, and that the inoculation of bovine animals with the *Bacillus anthracis* cultivated artificially after transmission through guinea-pigs or some other animal serves to render bovine animals totally or partially insusceptible to the disease when transmitted by the usual channels, one great step will have been taken. But there will yet remain the questions: Is the mortality from inoculation by this method a high one, or do even a small percentage of animals die? What are the conditions under which inoculation may be best performed, and does age exercise an important influence in the fatality? And, lastly, for how long a period is protection from attack conferred? To settle these points, the inoculation of a large number of animals will be necessary, and their subsequent exposure to sources of contagion at favorable periods."

Greenfield followed up these points in a second paper published in the *Journal of the Royal Agricultural Society* of 1881 (Greenfield, 1881b). Here he demonstrated that these cattle were truly immune to anthrax. Of course, by this time the tremendous publicity generated by Pasteur's public experiment had left Dr. Greenfield well behind in the race and he went on to other areas of research. In later years he did however claim publically that it was he, not Pasteur, who first developed an anthrax vaccine.

IV. Early Canine Distemper Vaccines

A. EARLY VACCINES

Once the basic principles of vaccination had been demonstrated and publicized by Pasteur, it took very little time for the results to be applied to other domestic animals. Thus, many attempts were made to vaccinate dogs against canine distemper. Canine distemper was first described in 1580 (Whitney and Whitney, 1953). Edward Jenner was the first to differentiate between distemper and hydrophobia, the first to recognize that it was infectious, and the first to show that it was not communicable to man (Kirk, 1922). Jenner had based these conclusions on his study of an outbreak of the disease occurring among the foxhounds of Mr. Merret's employer, the Earl of Berkeley. He actually tried disinfecting the kennels by washing them, then whitewashing and then exposing them to the fumes of "marine acid."

Several individuals, including Jenner himself, believed that the skin lesions associated with distemper were related to those of cowpox and so recommended cowpox vaccination as a preventive measure. Although soon recognized as ineffective, this procedure resurfaced at intervals over the next century. In 1887, R. S. Huidekoper, the first dean at the University of Pennsylvania, explicitly pointed out that vaccination does not protect against dog distemper (Niemi, 1980). Notwithstanding this, as late as 1902, a Dr. Brown, of Cambridge, England, observed "For a number of years I inoculated puppies . . . with vaccine lymph, and with the best results. During my time I never heard of a case of distemper arising after inoculation." These observations were not widely supported. A second early method also occasionally employed was inoculation of puppies with the nasal discharge of a dog with distemper, the object being to induce a mild attack and subsequent immunity. This was not, of course, a reliable procedure and was rapidly abandoned.

In 1875, Senner was the first to discover microorganisms in association with distemper (Kirk, 1922). He isolated a micrococcus and a bacillus from the blood and lungs. Because he also found the bacillus in the spleen, liver, and kidneys he concluded that it was the causal organism. Subsequently, numerous other investigators discovered bacteria in various parts of the body of dogs with distemper and each considered that their own organism caused the disease. Some even claimed to have transmitted the disease with the organism grown in pure culture. One of these was Millais who isolated a bacillus from a case of distemper. He then heated a culture of this organism at 60°C for 10 minutes (Millais, 1890). He injected this subcutaneously into about 10 puppies. Subsequently he challenged them by intranasal inoculation and all were protected. It is unclear from his paper whether he used control, unvaccinated animals.

In 1901, a Dr. Copeman produced a vaccine by heating a broth culture of a bacillus to 60°C for 30 minutes and subsequently adding a small quantity of phenol as a preservative. This vaccine was similar to that developed for typhoid in humans at that time. Unfortunately, the reviews of Copeman's vaccine were mixed. His ungrateful countrymen were concerned that it would actually cause the disease. Others simply did not believe that it worked. Dr. Copeman himself checked into some of these unsuccessful trials and noted that in many cases the dogs failed to mount a temperature spike. He therefore concluded that ". . .

little or no immunity could be expected to ensue"—an interesting comment on the release of pyrogenic cytokines. Lest one be skeptical about Dr. Copeman's vaccine, it was approved of by no less a personage than His Grace the Duke of Beaufort. His grace vaccinated his hounds for a couple of years and declared, "If we go on with the same results, it will be the greatest boon that has ever been brought out."

In 1901, M. Chauveau reported on the use of a modified live vaccine to prevent distemper. He actually isolated his organisms not from a dog, but from a guinea pig that had died of acute septicemia. He reported that his vaccinated dogs resisted both natural and experimental infection.

About 6 months after Dr. Copeman's results were published a Dr. Physalix announced that he had isolated the causal agent of distemper (*Pasteurella canina*) and had succeeded in developing a successful vaccine (Kirk, 1922). The professor evidently had fulfilled Koch's postulates with this cocco-bacillus, claiming that it induced all the classical features of distemper when injected in pure culture. Dr. Physalix went further and developed a modified live vaccine, the organisms being attenuated by subcultivation. Two doses were administered, the first being a low dose of an avirulent culture and the second being a more virulent culture. Dr. Physalix reported that 13 out of 298 vaccinated dogs died while mortality reached 50% in controls. However, there was great reluctance to support the idea that a *Pasteurella* could cause distemper. Because of contradictory reports on the efficacy of the Physalix vaccine, a "Committee of Veterinary Surgeons" was set up in 1903 to test its merits. The committee reported thus: "The committee consider the first experiment an entire failure, and the second inconclusive but suggestive. On the other hand they regard the results of the third experiment as unimpeachable evidence that the vaccination failed to confer any immunity against distemper. . . . " The end result of the experiment was that three of the four vaccinated pups died of distemper, while only two deaths occurred among the four unvaccinated pups. Despite the disappointing results of Physalix's vaccine, Professor Lignières produced a new, polyvalent vaccine in July 1903. It contained a mixture of strains of *P. canis*, attenuated by several hundred subcultures. It appeared to be no better than its predecessor and disappeared without trace (Lignières, 1906).

B. Ferry's Vaccine

In 1912, Dr. Ferry working in the United States produced a "polyvalent polymicrobial vaccine." Ferry (1912) actually discovered *Bor-*

detella bronchiseptica and believed strongly that this was the cause of distemper. His vaccine contained several strains of the organism (six parts) together with *Staph pyogenes albus* (one part), *Staph pyogenes aureus* (one part) and *Strep pyogenes* (two parts) and was widely tested and employed. As was typical at the time he used 40 dogs to determine its efficacy. Nine received live organisms, 17 received dead organisms, and 14 were used as unvaccinated controls. Following exposure to three animals showing the typical symptoms of distemper, 8 of the control dogs died while all the vaccinated animals remained healthy. It is interesting to note that Ferry's interest stemmed from the need to provide his laboratory with healthy animals for research. The vaccine was available either as a suspension of killed organisms in saline or "in the dry state as hypodermic tablets." The vaccine was indicated for both prophylactic and therapeutic use. When used for therapeutic use it was recommended that the vaccine be administered with antidistemper serum. Ferry's vaccine was widely used and generally well regarded. However, as experience was gained, practitioners gradually became disillusioned with the product. Ferry kept faith in his project although in 1923 he suggested that the dose employed hitherto had been far too small. "As regards the efficacy of the vaccine as a prophylactic agent we have practically come to the conclusion that our dose is altogether too small, and that is the reason more favorable results have not been obtained. As pioneers in this line, perhaps we have been too conservative. We have carried on enough well controlled experiments to know that a vaccine composed of *B. bronchiseptica* will protect against distemper, and we have numerous reports from outside to the same effect; but we feel that the vaccine has not been doing itself justice and that larger doses will improve it. To condemn the specificity of the organism on an unfavorable showing of the vaccine may lead to serious results and is unfair to the organism. It was years before typhoid vaccine made a proper showing, and yet the typhoid bacillus was still considered the cause of typhoid fever." It is difficult not to feel some sympathy for Dr. Ferry since with the wisdom of hindsight his vaccine probably did have a protective effect against cases of kennel cough and against secondary infections. It is impressive to read this comment about Ferry's vaccine made in 1923: ". . . Ferry's vaccine should prove superior to others, if for no other reason than that it is polyvalent and polymicrobial, for the cocci contained may be expected to play an important role in protecting against the secondary complications with which they are so frequently proved to be associated" (Kirk, 1922). Notwithstanding this, in 1926 Hardenbergh demonstrated that a vaccine prepared against a pure culture of *"B. bronchisepticus"* was ineffective in protecting dogs against distemper (Hardenbergh, 1926).

C. Viral Vaccines

The recognition that viruses were entities distinct from bacteria was only recognized in the closing years of the nineteenth century as a direct result of increasing expertise in growing and identifying bacteria. It was in 1898 that Loeffler and Frosch showed that foot-and-mouth disease could be transmitted by material filtered through a bacteria-proof filter. Viruses were not directly observed until the development of the electron microscope around 1940. Nevertheless in 1905 Carré correctly concluded that canine distemper was caused by a filterable virus. "The specific virus or essential element of the disease is invisible, passes through the very porous meshes of the bacterial filter and is not cultivatable in various media." He found that the virus could be detected in the nasal discharge and in the pericardial effusions of affected dogs. After filtering either of these through a fine porcelain filter, 2 to 3 drops of the filtrate caused typical disease and death in susceptible dogs. The filtrate appeared to be sterile on culture. For reasons that are unclear, this discovery was ignored by distemper vaccine researchers and by the early 1920s there were two competing theories as to the cause of canine distemper. Ferry and several other investigators believed that it was caused by *"B. bronchisepticus."* Others were less certain and remembered Carré.

The first successful distemper vaccine was produced in 1923 by Dr. Puntoni in Italy. He showed that he could serially passage the disease by intracerebral inoculation and that this brain material contained no detectable bacteria. Puntoni (1923, 1924) ground up the brain of a dog with distemper, suspended it to 10% in saline, and added formalin to make a 1:10,000 solution. He kept it for two days before injecting it subcutaneously. He tried to follow this with a dose of live, attenuated vaccine. However he found the attenuation was too unreliable so he persisted with using multiple doses of the inactivated material. He apparently got significant protection. In 1927 Lebailly in France demonstrated that the spleen contains large amounts of virus and so he removed the spleen from dogs on the fourth day of visible sickness, ground it up in saline, added formalin, and produced an effective vaccine. He vaccinated 19 dogs and used 5 control animals. All vaccinated animals remained healthy, all 5 controls developed disease and 1 died. In 1926 Green demonstrated that the cause of fox distemper was a filterable agent (Whitney and Whitney, 1953), but it was not until 1928 that Laidlaw and Dunkin eventually succeeded in immunizing dogs and ferrets with formalin inactivated material derived from ferret spleen or the liver, spleen, and lymph nodes of infected dogs.

V. Summary

In conclusion, it is remarkable just how farsighted many of the early vaccine investigators were. Jenner was apparently very comfortable with contagion and even recognized that infectious agents could gradually change and adapt to a new species. Pasteur, long before his fowl cholera experiment, dreamed that attenuation could yield safe vaccines and it took him no time at all therefore to recognize the significance of that serendipitous experiment. The fact that two other investigators were also developing anthrax vaccines simultaneously is yet another example of how the times favor certain discoveries. Finally Ferry, while constrained by the fact that he had no idea that distemper was caused by a virus, recognized well the concept of secondary infection and rationalized, not unreasonably, that his vaccine might assist in controlling this.

It is also clear that we must look skeptically at the accepted historical record. Thus, it is clear that Jenner used horse-derived material as a source of vaccine material and that vaccinia may in fact be the long-lost agent of horsepox. Certainly this would not be news to many nineteenth-century investigators and veterinarians. Individuals planning to use live vaccinia in recombinant vaccines may wish to keep this in mind.

Who discovered anthrax vaccine? Burdon-Sanderson clearly recognized that he could attenuate the organism. Greenfield showed that this could protect against disease although he was far from developing an effective vaccine. Poor Henri Toussaint was probably the first to develop an effective product but did not publicize his results widely. It was left to Louis Pasteur to take the risks inherent in a high-profile public experiment and win. I believe that he richly deserves the prize.

Finally, who deserves the credit for distemper vaccine? First, Carré deserves much more credit than hitherto for discovering that distemper was caused by a virus. Second, Ferry, although misled by his identification of *B. bronchiseptica* deserves credit for realizing that his vaccine could play a role in controlling secondary infections. The true discoverer of an effective distemper vaccine was the Italian, Puntoni, but once again the publicity went to others, Laidlaw and Dunkin. Thus a pattern emerges that prior discovery matters little in the face of aggressive publicity. If nobody knows you did the experiment you might as well have never done it in the first place. Publish or perish is by no means a new phenomenon.

REFERENCES

Burdon-Sanderson, J. S. Report on experiments on anthrax conducted at the Brown Institution February 18 to June 30, 1878. *J. R. Agric. Soc.* **16**, 267–273.

Carré, H. (1905). (1905). Sur la maladie des jeunes chiens. *C. R. Hebd. Seances Acad. Sci.* **140**, 689, 1489.

Copeman, S. M. (1901). The micro-organism of distemper in the dog, and the production of a distemper vaccine. *Proc. Roy. Soc.* **64**, 459–461.

Fenner, F., Wittek, R., and Dumbell, K. R. (1989). "The Orthopoxviruses." Academic Press, San Diego, CA.

Ferry, N. S. (1912). Further studies on *Bacillus bronchisepticus,* the cause of canine distemper. *Am. Vet. Rev.* **61**, 77–79.

Fleming, G. (1890). Variola in animals and man. *Vet. J.* **31**, 73–88.

Geison, G. L. (1995). "The Private Science of Louis Pasteur." Princeton University Press, Princeton, NJ.

Greenfield, W. S. (1880a). Report on an enquiry into the nature, causes, and prevention of splenic fever, quarter evil, and allied diseases, made at the Brown Institution. *J. R. Agric. Soc.* **16**, 273–291.

Greenfield, W. S. (1880b). Preliminary note on some points in the pathology of anthrax, with especial reference to the modification of the properties of Bacillus anthracis by cultivation, and to the protective influence of inoculation with a modified virus. *Proc. R. Soc. London* **30**, 557–560.

Greenfield, W. S. (1880c). Lectures on some recent investigations into the pathology of infective and contagious diseases. Lecture III. Part I. *Lancet* **1**, 865–867.

Greenfield, W. S. (1880d). Further investigations on anthrax and allied diseases in man and animals. *Lancet* **1**, 965–967.

Greenfield, W. S. (1880e). Lectures on some recent investigations into the pathology of infective and contagious diseases. Lecture III. Part II. *Lancet* **2**, 41–44.

Greenfield, W. S. (1881a). Further investigations and allied diseases in man and animals. *Lancet* **2**, 3.

Greenfield, W. S. (1881b). Report on an experimental investigation on anthrax and allied diseases made at the Brown Institution. *J. R. Agric. Soc.* **17**, 30–39.

Hardenbergh, J. G. (1926). Present knowledge regarding the cause of canine distemper. *J. Am. Vet. Med. Assoc.* **68**, 309–320.

Jenner, E. (1795). "An enquiry into the causes and effects of the variolae vaccinae, a disease discovered in some of the western counties of England, particularly Gloucestershire, and known by the name of The Cow Pox." [*In* "Milestones in Microbiology" (T. D. Brock, ed.), pp. 121–125. Prentice-Hall, London, 1961.]

Kaminjolo, J. S., Johnson, L. W., Muhammed, S. I., and Berger, (1975). Uasin Gishu skin disease of horses in Kenya. *Bull. Anim. Health Prod. Afr.* **23**, 225–233.

Kirk, H. (1922). "Canine Distemper, its Complications, Sequelae and Treatment." Baillière, Tindall and Cox, London.

Laidlaw, P. P., and Dunkin, G. W. (1926). Studies on dog distemper. The nature of the virus. *J. Comp. Pathol. Ther.* **39**, 222–230.

Laidlaw, P. P., and Dunkin, G. W. (1928). Studies in dog distemper. The immunization of dogs. *J. Comp. Pathol. Ther.* **41**(1), 209–227.

Lebailly, C. (1927). Vaccination préventative et spécifique des chiens contra la maladies du jeune age. *C. R. Hebd. Seances Acad. Sci.* **18,** 370.
Lignières, J. (1906). Sur la maladie des chiens et le virus filtrant de Carré. *Bull. Soc. Cent. Med. Vet.* **83,** 622.
Loeffler, F., and Frosch, P. (1898). "Report of the commission for research on the foot-and-mouth disease." [*In* "Milestones in Microbiology" (T. D. Brock, ed.), pp. 149–153. 1961.] Prentice-Hall, London.
Millais, E. (1890). The pathogenic microbe of distemper in dogs and its use for protective inoculation. *Vet. J.* **30,** 313–321.
Niemi, S. M. (1980). Early veterinary biologics. *Mod. Vet. Pract.* **61,** 926–929.
Pasteur, L. (1880). De l'atténuation du virus du choléra des poules. *C. R. Hebd. Seances Acad. Sci.* **91,** 673–680.
Pasteur, L., Chamberland, C., and Roux, E. (1881a). De l'atténuation des virus et de leur retour á la virulence. Le vaccin du charbon. *C. R. Hebd. Seances Acad. Sci.* **92,** 429–435.
Pasteur, L., Chamberland, C., and Roux, E. (1881b). Le vaccin du charbon. *C. R. Hebd. Seances Acad. Sci.* **92,** 666.
Puntoni, V. (1923). Saggio di vaccinazione anticimurrosa preventiva esequita per mezzo del virus specifico. *An. Ig.* **33,** 553.
Puntoni, V. (1924). Ancora sulla vaccinazione anticimurrosa per mezzo del virus specifico. *An. Ig.* **34,** 406.
Taylor, C. E. D. (1993). Did vaccinia come from a horse? *Equine Vet. J.* **25**(1), 8–10.
Tigertt, W. D. (1980). William Smith Greenfield, M.D., F.R.C.P., Professor Superintendent, The Brown Animal Sanatory Institution (1878–81). Concerning the priority due to him for the production of the first vaccine against anthrax. *J. Hyg.* **85,** 415–420.
Toussaint, H. (1880a). De l'imunité pour le charbon, acquise à la suite d'inoculations préventives. *C. R. Hebd. Seances Acad. Sci.* **91,** 135–137.
Toussaint, H. (1880b). Procédé pour la vaccination du mouton et du jeune chien. *C. R. Hebd. Seances Acad. Sci.* **91,** 303–304.
U.S. Department of Agriculture (1880). "Contagious Diseases of Domesticated Animals," Spec. Rep. No. 22, pp. 211–214. U.S.D.A., Washington, D.C.
Whitney, L. F., and Whitney, G. D. (1953). "The Distemper Complex." Practical Science Publishing Company, Orange, CT.
Wilson, G. (1979). The Brown Animal Sanatory Institution. *J. Hyg.* **82,** 337.

Diagnostic Medicine: The Challenge of Differentiating Infection from Disease and Making Sense for the Veterinary Clinician

JAMES F. EVERMANN* AND INGE S. ERIKS†

*Department of Veterinary Clinical Sciences and Washington Animal Disease Diagnostic Laboratory, Washington State University, Pullman, Washington 99164, †Department of Veterinary Microbiology and Pathology, College of Veterinary Medicine, Washington State University, Pullman, Washington 99164

    I. Introduction
   II. Differentiating Infection Detection from Disease Diagnosis
  III. How Early Do We Want to Detect Infection?
  IV. What Are the Consequences of the Results?
   V. Where Are We Heading with Veterinary Diagnostics?
      Acknowledgments
      References

I. Introduction

Diagnostic medicine has taken on a new, broader meaning in the 1990s and reflects an expansion of clinical investigation from the diagnosis of disease to include detection of infection (Evermann, 1998). This leads to an entirely new perspective on how veterinary clinicians and diagnosticians view laboratory tests and how test results are interpreted. One must consider not only the specificity and sensitivity of the test, but the predictive value of the test, which relates directly to the clinical utility of the result (Jacobson, 1997).

    The definitive diagnosis of infectious diseases relies on a combination of clinical symptoms, history, and laboratory analyses of antemortem and/or postmortem specimens (Evermann, 1998). Disease diagnosis has customarily used diagnostic assays for early recognition of disease and rapid implementation of therapy in an individual animal basis, and when appropriate use of corrective management (segregation, culling, vaccination, etc.) on a population basis.

The detection of infection during preclinical stages has become more important as one considers the consequence of long-term infections that have prolonged incubation periods and inapparent transmission to susceptible animals in the population. This includes life-threatening diseases, such as feline infectious peritonitis (FIP), rickettsial and ehrlichial diseases and canine herpesvirus (CHV) infections. Of equal, if not more so, importance for the early detection of infection has been the increased recognition of zoonotic infections, such as rabies virus, *Salmonella typhimurium* DT104, and *Escherichia coli* O157:H7 (Evans and Davies, 1996; Slutsker *et al.,* 1997; Smith, 1996).

Together with the necessity to detect infections earlier during the preclinical stages, there has been a remarkable expansion in the availability of assays that can detect infectious microorganisms in low quantity. This increased sensitivity has been primarily through the detection of nucleic acid sequences after amplification by polymerase chain reaction (PCR, Relman and Persing, 1997). PCR can allow not only early detection of infection, but rapid speciation of organisms as well as strain typing for epidemiologic analyses (Fredricks and Relman, 1996; McDade and Anderson, 1996).

The assessment of preclinical infections allows the veterinarian the opportunity to determine the relative risk of disease occurring, and to take preventive steps to reduce or eliminate the risks depending on the consequences of the disease in the animal and/or human (if zoonotic) population.

This chapter focuses on one main issue, and that is differentiating the detection of infection from diagnosis of disease. In the course of differentiating infection from disease three questions will be addressed: (1) How early do we want to detect infection? (2) What are the consequences of the results? (3) Where are we heading with veterinary diagnostics?

II. Differentiating Infection Detection from Disease Diagnosis

Historically, the primary aim of the diagnostic laboratory was to assist the veterinarian in the diagnosis of disease. This is presented in

Fig. 1. This type of approach initially ignored the origin of the causative microorganisms and focused on the accurate diagnosis of the disease agent. An example of this type of approach was the testing of cats that were clinically ill for feline leukemia virus (FeLV). If tested positive, they were segregated or euthanized. Further examples include FIP, CHV, Johne's disease (*Mycobacterium paratuberculosis*), and the mucosal disease form of bovine viral diarrhea (BVD) virus. Expanded use of diagnostic results by the veterinarian and client allowed for some corrective management steps to be taken. These included the use of vaccination when available or segregation and culling to reduce the source of the infection in the population. Based on this latter principle, the reduction of the source of the infection, a different approach has been taken. One may consider this an epidemiologic view of the disease process (Susser and Susser, 1996a).

With a combination of more sensitive diagnostic assays, the veterinarian's concern to know the state of the preclinical infection, econom-

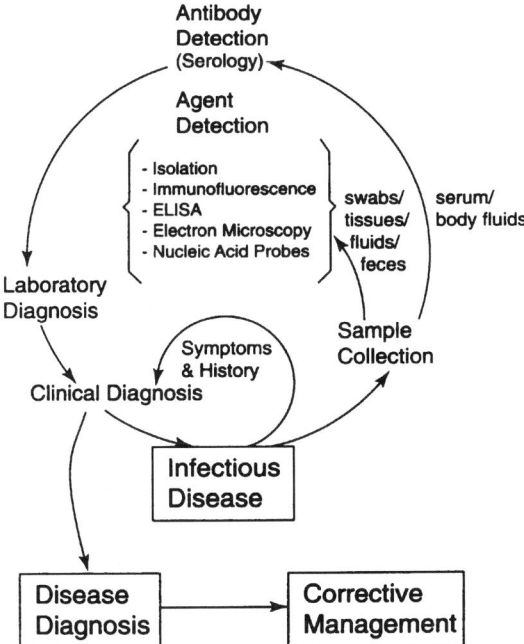

FIG. 1. Schematic depicting the historical interactions between the laboratory diagnosis of infectious disease and the steps leading to corrective management. (Modified from Evermann, 1990, with permission from W. B. Saunders Company.)

ic incentives to minimize disease by effectively controlling the infection, and concern over potential zoonotic diseases, laboratory diagnosis has taken on a different strategy. This is presented in Fig. 2. The primary emphasis in this scheme is to view animals preclinical and determine the disease risk and/or zoonotic potential of the infection. This has been the approach for some retroviral infections (Evermann and Jackson, 1997; Knowles, 1997) and bacterial infections with public health concerns, such as *E. coli* O157:H7 and *Salmonella* spp. infections (Evans and Davies, 1996; Firstenberg-Eden and Sullivan, 1997; McDonough and Simpson, 1996). The ultimate goal of the assessment of preclinical testing is to initiate a preventive management type of control. This type of approach places more emphasis on early testing and management of infected animals rather than on diseased animals.

With the shift in emphasis to preclinical testing, the knowledge of

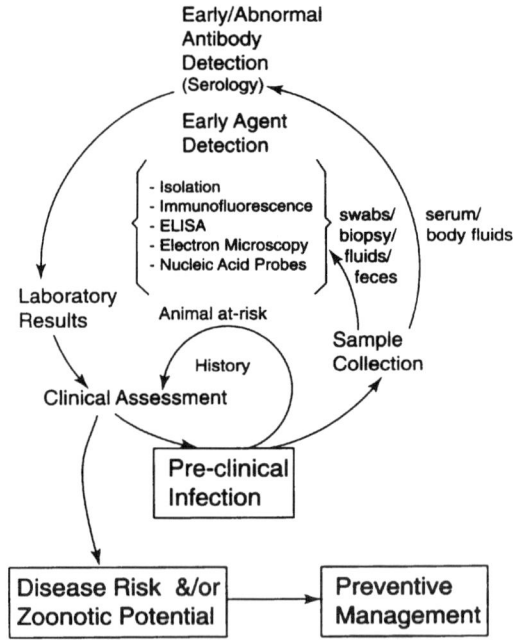

FIG. 2. Schematic depicting the current and future interactions between the laboratory testing of animals-at-risk to determine if preclinical infection has occurred, and the steps leading to preventive management. (Modified from Evermann, 1990, with permission from W. B. Saunders Company.)

the ecology of the infectious microorganism has become very important in our overall understanding of how successful the control program may be (Susser and Susser, 1996b). The control of infections with a low degree of transmissibility and narrow host range, such as caprine arthritis-encephalitis (CAE) virus, is much more realistic than the control of diseases with a wide host range, such as chlamydia and *Salmonella* spp., or those agents spread by arthropod vectors, such as arboviruses and rickettsia (Gregory and Schaffner, 1997; Hewinson *et al.*, 1997; Knowles, 1997; Raoult and Roux, 1997; Saluzzo and Dodet, 1997). The ecology of infection provides the veterinarian with vital information with which to make decisions. The ecology of six different agents is listed in Table I. The ecology of infection is divided into infection rate, attack rate (progress to become clinical), and mortality rate. It can be seen that the infection rate usually exceeds the attack rate and mortality rate in the majority of cases. Exceptions to this generalization are the mucosal disease form of BVD that occur in cattle that are tolerant to BVD and persistently infected (PI), and rabies infections in mammals (Innocent *et al.*, 1997; Smith, 1996).

Another way to view the ecology of an infection is demonstrated in Fig. 3. The schematic allows the veterinarian to readily use a graphic approach with clients to explain the differences between infection and disease. Rabies virus is used as an example of a microorganism with a low infection rate, but high mortality. (This figure would be different if one were to diagram the ecology of rabies in bats, the natural reservoir for rabies in the United States.) The CAE virus is used as an example of an infection in goats with a high infection rate, but lower attack rates, and much lower fatality rates (Fig. 4). With retroviruses, such as CAE virus, bovine leukosis virus (BLV), and equine infectious anemia (EIA), the ecology can also be subdivided into progressor (progress onto clinical signs and/or fatality) or nonprogressor (remains clinically normal, but infected and potentially contagious). With persistent bacterial infections, such as *Salmonella* and many members of the Rickettsiaciae, the ecology can be subdivided into clinical disease leading to mortality or clinical disease leading to a chronic carrier state. This chronic carrier state can then be further subdivided into inapparent infections with constant shredding and inapparent infections with intermittent shedding. With potentially zoonotic diseases, such as *Salmonella,* rickettsioses, and psittacosis (*Chlamydia psittici*), the ability to shed or transmit the organism into the environment or vectors becomes particularly relevant (Evans and Davies, 1996; Gregory and Schaffner, 1997; Raoult and Roux, 1997).

TABLE I
ECOLOGY OF INFECTION IN RELATIONSHIP TO DETECTION OF INFECTION AND DIAGNOSIS OF DISEASE

Transmissible agent	Ecology of infection			Detection of infection*	Diagnosis of disease	Vaccine available
	Infection rate (%)	Attack rate (%)	Mortality rate (%)			
Prion induced disease (scrapie)	Unknown (variable)**			IHC (biopsy) genetic typing	Clinical signs, histopath	No
Coronavirus induced disease (FIP)	85	2	99	Serology, PCR, genetic typing?	Clinical signs, histopath	Yes (40–80%)
Lentivirus induced disease (CAE)	85	30	<10	Serology, PCR, genetic typing?	Clinical signs, histopath	No
Pestivirus induced disease (BVD)	85	10	<5	Serology, PCR	Clinical signs, histopath, IHC	Yes (80–90%)
BVD-PI (immunotolerant)	1–2	50	90	PCR, VI	Clinical signs, histopath, IHC	Yes (variable)
Herpesvirus induced disease (EHV-1)	85	10	5	Serology, VI, PCR	Clinical signs, histopath, IHC	Yes (variable)

*Antemortem.
**USDA-APHIS survey in progress.

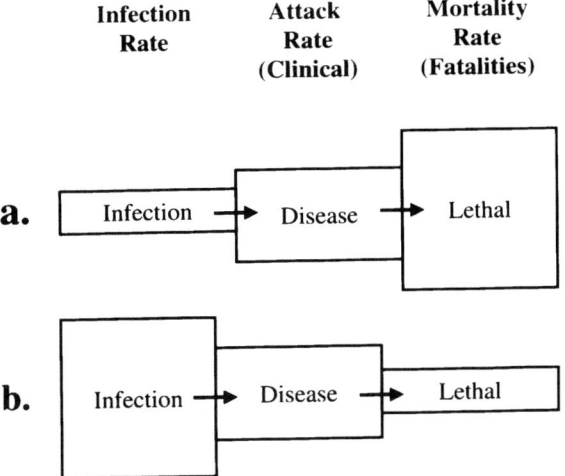

FIG. 3. Schematic depicting the conceptual view of infection rate, attack rate, and mortality rate. (a) An infection with a low infection rate, i.e., rabies virus, but a high mortality rate. (b) An infection with a high infection rate, i.e., caprine arthritis encephalitis virus, and a low mortality rate.

III. How Early Do We Want to Detect Infection?

Early detection of infection is now feasible with a number of microorganisms affecting animals. The detection may take the form of specifically identifying the nucleic acid of the infectious agent, such as bovine herpesvirus-1 in semen samples, BLV provirus is selected blood cells populations, and foodborne bacteria in dairy products (Batt, 1997; Masri *et al.*, 1996; Xie *et al.*, 1997). Although this form of early microbial detection is preclinical at this time, with further research it may be determined that these "subclinical infections" are actually causing alterations in cell structure and function leading to endocrine imbalances and decreased productivity. This form of disease has been referred to a "lesionless pathology," and will be the subject of further research (de la Torre and Oldstone, 1996).

Early detection of infection may take a "back door" approach by analyzing the host animal's genetic predisposition to infection and disease. This interesting approach has already been used in order to control the prion disease, scrapie (O'Rourke *et al.*, 1997). Sheep with a unique chromosome are highly susceptible to progress onto scrapie, an irreversible disease. Animals that are bred for genetic resistance to

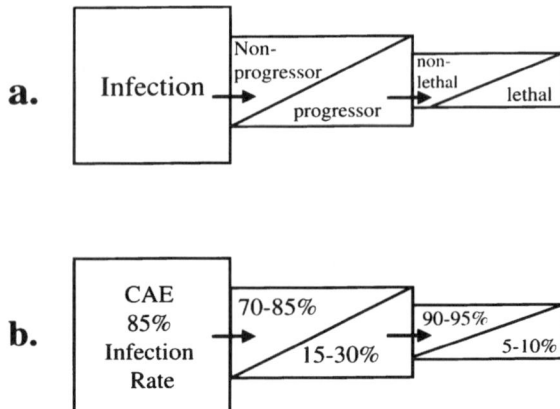

FIG. 4. Schematic depicting the conceptual view of further analyses of the attack rate into (a) progressor and nonprogressor, and the mortality rate into lethal and nonlethal. (b) How CAE virus infection occurs in this scheme.

infection and/or disease will be major factors in disease control in the future (Gavora, 1996; Malo and Skamene, 1994). The utilization of genetic testing is essential for some infections, such as the retroviruses, which serve to activate cellular oncogenes and promote disease. Identifying these cellular oncogenes would be a major step in controlling retroviral-induced diseases (Wiedemann et al., 1991).

It will be essential to clearly define what the diagnostic assay is detecting so that the veterinarian may utilize the information appropriately. Figure 5a graphically presents the use of thresholds to differentiate subclinical infection from clinical manifestations of the disease. Figure 5b shows five potential diagnostic assays, each with varying levels of sensitivity. It would be critical to understand the differences between a test with high sensitivity, which detects *subclinical infection* and a test with lesser sensitivity, but more accurately *diagnoses disease*.

IV. What Are the Consequences of the Results?

This question becomes more difficult the more one employs preclinical testing in preventive medicine programs (Clementi et al., 1995; Jacobson and Romatowski, 1996; Smith, 1995). The predictive value of a positive result may be high when an animal is clinical, such as a cat with FIP.

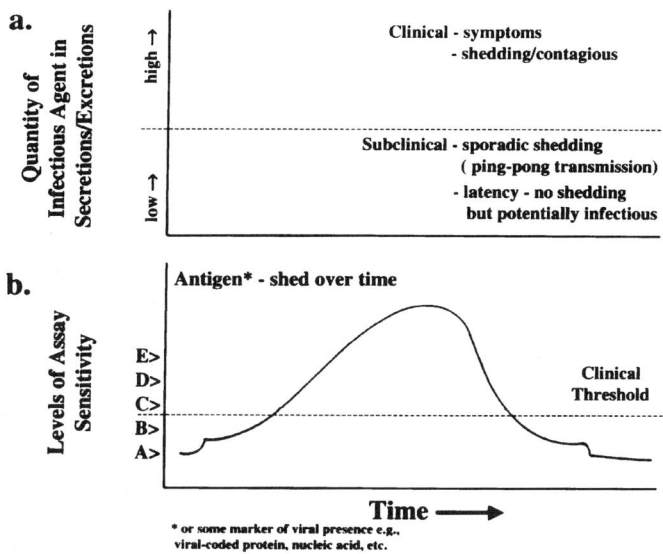

FIG. 5. (a) Schematic depicting the differences between subclinical and clinical infections where infectious agents are detected (panel a). (b) The different levels of sensitivity of detection assays, with assay A being the most sensitive and assay E being the least sensitive.

However, with early testing the problems of detecting cross-reacting viruses (feline enteric coronaviruses) increases, as does the question of whether the preclinical result accurately identifies an animal that is just infected or will progress onto disease (Evermann et al., 1995; Foley et al., 1997). In infections such as EIA, the consequences of infection are just as severe as the horse that has clinical signs of EIA. This is because the infection is regulated by the U.S. Department of Agriculture and all seropositive horses and mules are required to be reported, regardless of their health status. Assays for early detection of EIA infection have been reported to detect viral RNA in plasma samples as early as 48 hours after infection (Langemeier et al., 1996). Similarly in bovine tuberculosis, caused by Mycobacterium bovis, the consequences of a positive test result can be economically devastating due to stringent government regulations. This becomes particularly problematic because many tests currently available may cross-react with other mycobacterial species (Essey and Koller, 1994; O'Reilly and Daborn, 1995).

To determine what consequences the test results will have on the animal and the owner it is important to ask five key questions (Table II). Is the infection and/or disease of economic concern, such as EIA or *M. bovis;* is the infection and/or disease of zoonotic concern, such as *E. coli* O157:H7; where is the microbial agent when not causing disease, such as with rabies reservoirs in bat populations; what are the contributing factors to the infection and/or disease process, such as pregnancy for CHV and other herpesviruses; and what factors can animal owners/veterinarians/public health personnel control to minimize or eliminate the risk of infection and/or the disease process? Table III lists some of the consequences of the infection and/or disease process. These range from no sale, as with a goat that is CAE seropositive, to euthanasia if a horse or mule is tested EIA seropositive.

V. Where Are We Heading with Veterinary Diagnostics?

Veterinary diagnostics, like their human counterparts, are already directing efforts toward more sensitive assays, which are capable of detecting infections very early (within hours of initial infection); subclinical infections that are the result of persistent infections acquired during gestation and masked by immune tolerance; latent infections due to herpesviruses and retroviruses; and infections that pose a public health risk (Barrett *et al.,* 1997; Burr, 1996; Clarke, 1997; de la Torre and Oldstone, 1996; Rodriquez, 1997).

The evolution of diseases and the emergence of newly recognized pathogens have placed considerable pressure on new diagnostic technologies. The newer assays will assist in tracking the emerging infections, as well as linking causal association with disease to a firm cause and effect of the disease (Bryan *et al.,* 1994; Holtzman *et al.,* 1997; Hoet and Haufroid, 1997; Lipstich *et al.,* 1996; McDade and Anderson, 1996; Poland *et al.,* 1996).

TABLE II

FIVE KEY QUESTIONS TO ASK REGARDING INFECTIONS/DISEASES OF ANIMALS

1. Is the infection and/or disease of economic concern?
2. Is the infection and/or disease of zoonotic concern?
3. Where is the microbial agent when not causing disease (microbial ecology)?
4. What are the contributing factors to the infection and/or disease process?
5. What factors can animal owners/veterinarians/public health personnel control to minimize or eliminate the risk of infection/disease process?

TABLE III

THE CONSEQUENCES OF THE INFECTION/DISEASE PROCESS ON THE ANIMAL

- No sale
- Public health risk
- Early cull
- Regulatory quarantine
- Shedding of microbial agent to susceptible animals in the population
- Segregation of animal
- Euthanasia of animal(s)

The future of veterinary diagnostics is now. There are at least five directions to be pursued (Table IV), none of which is new, but continuing to evolve as the needs mandate the detection of infection earlier and the diagnosis of disease at a manageable stage (Wilson, 1994). These five directions are the development of (1) assays to monitor

TABLE IV

WHERE WE SHOULD BE GOING WITH VETERINARY DIAGNOSTICS

1. Assays to monitor immune function (immune competence)
   - Foal check
   - Calf failure of passive transfer
   - Llama/alpaca immunoglobulin status
   - CMI response
2. Assays to monitor genetic resistance/genetic susceptibility
   - Cellular receptors
   - Cellular oncogenes
   - Cellular prion proteins
3. Assays to monitor infections
   - In the environment
   - In asymptomatic vectors (potential transmissibility)
   - In asymptomatic carriers (potential shedders)
4. Assays to diagnose disease
   - Prognosis
   - Monitor response to treatment via cytokines (IL-2, IL-4, etc.)
5. Assays to track emerging infections
   - Culture *invitro*
   - Conserved PCR
   - Disease potential
   - Develop new detection assays
   - Develop new vaccines

immune function (immune competence), (2) assays to monitor genetic resistance/susceptibility, (3) assays to monitor infections, (4) assays to diagnose disease and monitor response to treatment, and (5) assays to track emerging infections. As infectious agents continue to evolve, disease expression will change, resulting in the necessity to develop new diagnostic assays (Susser and Susser, 1996b; Wilson, 1994).

Acknowledgments

The authors wish to acknowledge the technical assistance of Alison McKeirnan, John Vander Schalie, Lorraine Tanaka, Dan Bradway, Carol Petersen, and Pam Dilbeck. Additional thanks to John Gorham, Richard Ott, Tony Gallina, Terry McElwain, Colin Parrish, Niels Pedersen, Roger Spencer and Leo Bustad for scientific encouragement. Special thanks to the Biomedical Communications Group for valuable assistance with graphics. Special thanks to Danielle Bishop for invaluable assistance with the word processing.

References

Barrett, T. J., Feng, P., and Swaminathan, B. (1997). Amplification methods for detection of food-borne pathogens. In "Nucleic Acid Amplification Technologies: Application to Disease Diagnosis" (H. H. Lee, S. A. Morse, and O. Olsvik, eds.), pp. 171–181. Eaton Publishing, Natick, MA.

Batt, C. A. (1997). Molecular diagnostics for dairy-borne pathogens. J. Dairy Sci. **80**, 220–229.

Bryan, R. T., Pinner, R. W., and Berkelman, R. L. (1994). Emerging infectious diseases in the United States: improved surveillance, a requisite for prevention. Ann. N. Y. Acad. Sci. **740**, 346–361.

Burr, P. D. (1996). Detection of canine herpesvirus 1 in a wide range of tissues using polymerase chain reaction. Vet. Microbiol. **53**, 227–237.

Clarke, C. J. (1997). Paratuberculosis and molecular biology. Vet. J. **153**, 245–247.

Clementi, M., Menzo, S., Manzin, A., and Bagnarelli, P. (1995). Quantitative molecular methods in virology. Arch. Virol. **140**, 1523–1539.

de la Torre, J. C., and Oldstone, M. B. A. (1996). Anatomy of viral persistence: mechanisms of persistence and associated disease. Adv. Virus Res. **46**, 311–338.

Essey, M. A., and Koller, M. A. (1994). Status of bovine tuberculosis in North America. Vet. Microbiol. **40**, 15–22.

Evans, S., and Davies, R. (1996). Case control study of multiple-resistant *Salmonella typhimurium* DT104 infections in cattle in Great Britain. Vet. Rec. **139**, 557–558.

Evermann, J. F. (1990). Laboratory diagnosis of viral and rickettsial infections. In "Infectious Diseases of the Dog and Cat" (C. E. Greene, ed.), pp. 215–220. Saunders, Philadelphia.

Evermann, J. F. (1998). Laboratory diagnosis of viral and rickettsial infections. In "Infectious Diseases of the Dog and Cat" (C. E. Greene, ed.), 2nd ed., pp. 1–10. Saunders, Philadelphia.

Evermann, J. F., and Jackson, M. K. (1997). Laboratory diagnostic tests for retroviral infections in dairy and beef cattle. Vet. Clin. North Am.: Food Anim. Pract. **13**, 87–106.

Evermann, J. F., Henry, C. J., and Marks, S. G. (1995). Feline infectious peritonitis. *J. Am. Vet. Med. Assoc.* **206,** 1130–1134.
Firstenberg-Eden, R., and Sullivan, N. M. (1997). EZ coli rapid detection system: A rapid method for the detection of *Escherichia coli* O157 in meat and other foods. *J. Food Prot.* **60,** 219–225.
Foley, J. E., Poland, A., Carlson, J., and Pedersen, N. C. (1997). Risk factors for feline infectious peritonitis among cats in multiple-cat environments with endemic feline enteric coronavirus. *J. Am. Vet. Med. Assoc.* **210,** 1313–1318.
Fredricks, D. N., and Relman, D. A. (1996). Sequence-based identification of microbial pathogens: A reconsideration of Koch's Postulates. *Clin. Microbiol. Rev.* **9,** 18–33.
Gavora, J. S. (1996). Resistance of livestock to viruses: Mechanisms and strategies for genetic engineering. *Genet. Sel. Evol.* **28,** 385–414.
Gregory, D. W., and Schaffner, W. (1997). Psittacosis. *Semin. Respir. Infect.* **12,** 7–11.
Hewinson, R. G., Griffiths, P. C., Bevan, B. J., Kirwan, S. E., Field, M. E., Woodward, M. J., and Dawson, M. (1997). Detection of *Chlamydia psittici* DNA in avian clinical samples by polymerase chain reaction. *Vet. Microbiol.* **54,** 155–166.
Hoet, P., and Haufroid, V. (1997). Biological monitoring: State of the art. *Occup. Environ. Med.* **54,** 261–366.
Holtzman, N. A., Murphy, P. D., Watson, M. S., and Barr, P. A. (1997). Predictive genetic testing: From basic research to clinical practice. *Science* **278,** 602–604.
Innocent, G., Morrison, I., Brownlie, J., and Gettinby, G. (1997). The use of a mass-action model to validate the output from a stochastic simulation model of bovine viral diarrhea virus spread in a closed dairy herd. *Prev. Vet. Med.* **31,** 199–209.
Jacobson, R. (1997). Principles of validation of diagnostic assays for infectious diseases. In "OIE Manual of Standards for Diagnostic Tests and Vaccines" (R. Reichard, ed.), 3rd ed., pp. 8–15. Office International des Epizooties, Paris.
Jacobson, R. H., and Romatowski, J. (1996). Assessing the validity of serodiagnostic test results. *Semin. Vet. Med. Surg.* **11,** 135–143.
Knowles, D. P., Jr. (1997). Laboratory diagnostic tests for retrovirus infections of small ruminants. *Vet. Clin. North Am.: Food Anim. Pract.* **13,** 1–11.
Langemeier, J. L., Cook, S. J., Cook, R. F., Rushlow, K. E., Montelaro, R. C., and Issel, C. J. (1996). Detection of equine infectious anemia viral RNA in plasma samples from recently infected and long-term inapparent carrier animals by PCR. *J. Clin. Microbiol.* **34,** 1481–1487.
Lipstich, M., Siller, S., and Nowak, M. A. (1996). The evolution of virulence in pathogens with vertical and horizontal transmission. *Evolution* **50,** 1729–1740.
Malo, D., and Skamene, E. (1994). Genetic control of host resistance to infection. *Trends Genet.* **10,** 365–371.
Masri, S. A., Olsen, W., Nguyen, P. T., Prins, S., and Deregt, D. (1996). Rapid detection of bovine herpesvirus 1 in the semen of infected bulls by a nested polymerase chain reaction assay. *Can. J. Vet. Res.* **60,** 100–107.
McDade, J. E., and Anderson, B. E. (1996). Molecular epidemiology: Applications of nucleic acid amplification and sequence analysis. *Epidemiol. Rev.* **18,** 90–97.
McDonough, P. L., and Simpson, K. W. (1996). Diagnosing emerging bacterial infections: Salmonellosis, Campylobacteriosis, Clostridial toxicosis, and Helicobacteriosis. *Semin. Vet. Med. Surg.* **11,** 187–197.
O'Reilly, L. M., and Daborn, C. J. (1995). The epidemiology of *Mycobacterium bovis* infection in animals and man: A review. *Tuber. Lung Dis.* **76** (Suppl. 1), 1–46.
O'Rourke, K. I., Holyoak, G. R., Clark, W. W., Mickelson, J. R., Wang, S., Melco, R. P.,

Besser, T. E., and Foote, W. C. (1997). PrP genotypes and experimental scrapie in orally inoculated Suffolk sheep in the United States. *J. Gen. Virol.* **78,** 975–978.

Poland, A. M., Vennema, H., Foley, J. E., and Pedersen, N. C. (1996). Two related strains of feline infectious peritonitis virus isolated from immunocompromised cats infected with a feline enteric coronavirus. *J. Clin. Microbiol.* **34,** 3180–3184.

Raoult, D., and Roux, V. (1997). Rickettsioses as paradigms of new or emerging infectious diseases. *Clin. Microbiol.* **10,** 694–719.

Relman, D. A., and Persing, D. H. (1997). Genotypic methods for microbiol identification. *In* "PCR Protocols for Emerging Infectious Diseases" (D. H. Persing, ed.), pp. 1–32. ASM Press, Washington, DC.

Rodriguez, J. M. (1997). Detection of animal pathogens by using the polymerase chain reaction (PCR). *Vet. J.* **153,** 287–305.

Saluzzo, J. F., and Dodet, B., eds. (1997). "Factors in the Emergence of Arbivirus Diseases." Elsevier, Paris.

Slutsker, L., Ries, A. A., Greene, K. D., Wells, J. G., Hutwagner, L., and Griffin, P. M. (1997). *Escherichia coli* O157:H7 diarrhea in the United States: Clinical and epidemiologic features. *Ann. Intern. Med.* **126,** 505–513.

Smith, J. S. (1996). New aspects of rabies with emphasis on epidemiology, diagnosis and prevention of the disease in the United States. *Clin. Microbiol. Rev.* **9,** 166–176.

Smith, R. D. (1995). Evaluation of diagnostic tests. *In* "Veterinary Clinical Epidemiology. A Problem-Orientated Approach" (R. D. Smith, ed.), 2nd ed., pp. 31–52. CRC Press, Boca Raton, FL.

Susser, M., and Susser, E. (1996a). Choosing a future for epidemiology: I. Eras and paradigms. *Am. J. Public Health* **86,** 668–673.

Susser, M., and Susser, E. (1996b). Choosing a future for epidemiology: II. From black box to chinese boxes and eco-epidemiology. *Am. J. Public Health* **86,** 674–683.

Wiedemann, L. M., McCarthy, K. P., and Chang, L. C. (1991). Chromosome rearrangement, oncogene activation and other clinical events in cancer. Their use in molecular diagnostics. *J. Pathol.* **163,** 7–12.

Wilson, M. E. (1994). Disease in evolution: introduction. *Ann. N. Y. Acad. Sci.* **740,** 1–12.

Xie, B., Oyamada, T., Yoshikawa, H., Oyamada, T., and Yoshikawa, T. (1997). Detection of proviral DNA of bovine leukemia virus in cattle by a combination of *in situ* hybridization and the polymerase chain reaction. *J. Comp. Pathol.* **116,** 87–96.

II
CONCEPTS IN IMMUNOLOGY AND VACCINOLOGY

Genetic Effects on Vaccination

BRUCE N. WILKIE AND BONNIE A. MALLARD

Department of Pathobiology, The Ontario Veterinary College, The University of Guelph, Guelph, Ontario, Canada N1G 2W1

I. Introduction
II. Genetic Effects on Health and Vaccination
III. Strategies for Genetic Enhancement of Health
   A. Candidate Genes
   B. Polygenes and Quantitative Trait Loci
IV. High Immune Response Phenotype
   A. Selection for High and Low Antibody and Cell-Mediated Immune Response
   B. Response to Vaccination of Immune Response Selected Pigs
V. Discussion and Summary
   Acknowledgments
   References

I. Introduction

Vaccination is a cost-effective health maintenance tool that has enjoyed notable successes, such as the vaccine-based eradication of smallpox in 1980 (Ada, 1993). However, there are many infectious diseases for which it has not been possible to produce efficacious vaccines. Solutions may exist in increasingly refined approaches to identification of virulence factors and related immunogens and formulation of vaccines incorporating relevant genes, gene products, or synthetic immunogens together with agents that may direct immune response toward the most biologically relevant host response. However, because immunogenicity results from interaction between antigen and host, future approaches to immunization could be enhanced if the subjects of vaccination were improved for responsiveness.

In animal production, environmental influences on productivity-related traits have been studied and modified in conjunction with genetic improvement of the animals themselves. High productivity requires selection of populations most responsive to husbandry. Selection for traits such as milk production, rate of gain, or feed efficiency has dramatically altered current breeds of livestock and has been accomplished using phenotypic and genetic information largely without specific knowledge of biological variables underlying and controlling the traits. Livestock health maintenance has in contrast involved exogenous prophylactic and therapeutic procedures applied to animals selected only for production traits (Gavora and Spencer, 1983), often at the expense of health (Simianer et al., 1991; Fujji et al., 1991). If suitable criteria could be identified it may be possible to improve animal health by genetic methods.

II. Genetic Effects on Health and Vaccination

Genetic control of disease resistance is polygenic and involves quantitative trait loci (QTL) (Lie, 1990), which are additive and dictate the genetic component of variation in individual resistance to infectious disease (Fig. 1). Evidence of major gene-related resistance to infectious disease is limited to Marek's disease of chickens (Briles et al., 1977), human malaria (Hill et al., 1991), and possibly bovine leukemia (Lewin and Bernoco, 1991). The genetic influence on premature death of humans is significant for infectious diseases and exceeds environmental effects, while the opposite is true for cancer (Sørensen et al., 1988). This suggests the possibility of a genetic approach to enhanced animal health, if the same is true of livestock and contributing traits have sufficient heritability to permit genetic selection. The problem of low immune responsiveness to vaccination has been emphasized by attempts to use poorly immunogenic peptide vaccines (Outteridge, 1993). The possibility that deficient response may be due to inadequate peptide presentation by-products of the major histocompatibility gene complex (MHC) has been proposed (Outteridge, 1994). However, overall immune response phenotype is complex and may not be usefully influenced by major genes, such as the MHC, although such genes may influence responder and nonresponder phenotypes to certain antigens and are associated with quantitative differences in immune response (Mallard et al., 1989a,b) (Fig. 1).

Innate and acquired immunity are integrated and mediate resistance to infectious disease (Medzhitov et al., 1997). Selection for anti-

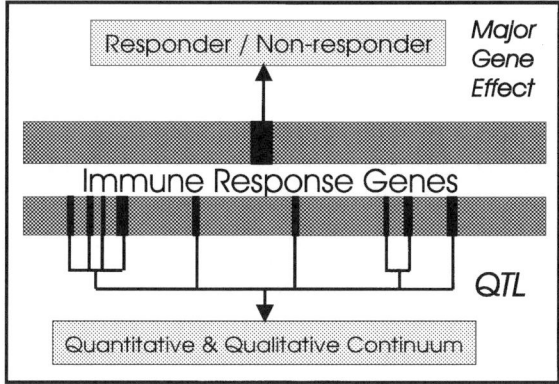

Fig. 1. Immune response and resistance to infectious disease may be influenced by genes of major effect (e.g., the major histocompatibility loci) but individual variation is mainly attributed to additive effects of multiple QTLs.

body or cell-mediated immune response has altered disease resistance and health in mice (Biozzi *et al.*, 1984; Covelli *et al.*, 1989), hence response to vaccination should vary in part as a function of host genotype. Individual variation in immune response could be exploited to derive livestock that are inherently more suitable subjects for vaccination. Variation in immune response has been recognized in various livestock species including pigs (Mallard *et al.*, 1989a,b; Edfors-Lilja *et al.*, 1994; Hessing *et al.*, 1995), cattle (Newman *et al.*, 1996), chickens (Peleg *et al.*, 1976), and fish (Strømsheim *et al.*, 1994).

III. Strategies for Genetic Enhancement of Health

Animal breeders and geneticists have contemplated a genetic approach to animal health based on observed variation in resistance and susceptibility among individuals at risk (Hutt, 1958) but breeding methods have, with few exceptions, not been applied to the problem of livestock health (Lie, 1990). Most have focused on individual diseases, such as Marek's disease of chickens (Simonsen, 1987). Heritability estimates have been made for some economically important diseases and potential benefits of genetic selection against disease have been estimated (Uribe *et al.*, 1995). Relatively low heritabilities for disease traits in dairy cattle suggest that selection could produce only small gains, possibly at the expense of productivity (Uribe *et al.*, 1995).

A genetic approach to health may be advantageous however if a suitable strategy could be described. It is unlikely that diseases can be approached individually given the risk of inverse relationships between susceptibility and resistance to individual diseases and the difficulty of altering large populations genetically. For example, cattle that are resistant to brucellosis are susceptible to extracellular bacterial pathogens, hence breeding for resistance to brucellosis would enhance susceptibility to other diseases (Templeton et al., 1990). This may reflect the complementary polarization of host resistance and microbial virulence into cellular and extracellular, broadly reflecting innate and specific cell-mediated (CMI) and antibody-mediated resistance. Although derived from models using inbred laboratory animals (Alan and Maizels, 1997) and largely unconfirmed in livestock, current concepts of polarity in host resistance-mediating events reflected in cellular and cytokine mediators are useful. The sum of events leading to resistance to intracellular pathogens (TH1) is, in part, due to quantitative and qualitative variables at the level of the antigen and its presentation regulated by host cells and cytokines which differ from those involved in generating resistance to extracellular pathogens (TH2) (Reed and Scott, 1993). The risk of genetically altering mammalian populations to evade the virulence of a specific pathogen is also high because of the genetic plasticity of pathogens relative to hosts, which allows mutation-induced avoidance of adverse conditions and rapid appearance of novel virulence attributes. Populations having a high degree of general resistance to infection and infectious disease would be expected to express TH1 and TH2 phenotypes in the appropriate contexts.

Although health could in theory be quantified by objective observation of disease prevalence, this is an impractical approach if the objective is overall resistance. Spontaneous disease is variable in incidence and husbandry is designed to limit its expression, hence masking genetic components of resistance. Artificial induction of disease to test resistance phenotype and genotype is similarly impractical. Such direct approaches are unlikely to contribute to development of improved livestock.

Genetic strategies for enhanced inherent healthiness may assume that there are genes of major effect and that such genes would be candidates for breeding schemes, such as marker-assisted selection (Beckman and Soller, 1987), designed to increase their frequency in livestock populations. Alternatively, assumptions that QTLs control inheritance of resistance lead to other schemes for optimizing frequency of resistance-related genotypes. Major genes may exist together with QTLs. These alternatives are illustrated schematically in Fig. 1.

A. CANDIDATE GENES

Genes may be proposed as candidates for genes of major effect in controlling health traits. Since the association of MHC genes with immune response in laboratory animal models (Bennaceraf, 1981) and realization that the effect was due to efficacy of binding of antigen-derived peptides to the MHC gene products in antigen presentation (Babbit *et al.*, 1985), it has become common to hypothesize major effects of MHC genes on health and immune response (Mallard *et al.*, 1989a,b). Such major genes are rarely confirmed (Briles *et al.*, 1977; Hill *et al.*, 1991) and outcomes may be due to linked genes (Lillehoj *et al.*, 1989), themselves only some of the many possible contributors to overall phenotypic variation. Nevertheless, the search for genes of major effect in health continues together with expectations of their utilization in breeding schemes based on marker-assisted selection (Beckman and Soller, 1987) and advanced gene transfection and embryologic techniques to facilitate their dissemination (Georges and Massey, 1991). The feasibility of such an approach is unconfirmed.

B. POLYGENES AND QUANTITATIVE TRAIT LOCI FOR DISEASE RESISTANCE

It is acknowledged that the phenotype, resistance to infectious disease, is polygenically controlled by QTLs and as such should be amenable to modern selection methods (van der Zijpp, 1983). Heritability of several disease-related traits is such that selection would be feasible (Uribe *et al.*, 1995); however, direct selection for health or disease is impractical due to the difficulty of objectively measuring phenotype, particularly if the objective is to obtain enhanced general disease resistance rather than improved resistance to a specific disease or response to a single vaccine (Wilkie *et al.*, 1989; Mallard *et al.*, 1998). Selection for resistance to one disease, or response to a single vaccine, has been confirmed in several instances to result in increased susceptibility to other infections and diseases (Mallard and Wilkie, 1993). Various immune response traits, including antibody and CMI response, have heritabilities in the range of 15–30% and are suitable for selection to derive high (HIR) and low (LIR) immune response populations (Mallard *et al.*, 1992, 1998). Thus immune response objectively measured by traits such as antibody production to a protein antigen, such as hen eggwhite lysozyme (HEWL), and CMI as cutaneous delayed-type hypersensitivity (DTH) to purified protein derivative (PPD) of tuberculin after immunization with bacillus Calmette–Guérin (BCG) vaccine, can be used to quantify immune response capability (Mallard *et al.*, 1992). High immune response is therefore a candidate phenotype for enhanced health that can be practically used in indirect selection for health, including improved response to vaccination (Wilkie et al., 1998).

IV. High Immune Response Phenotype

The HIR phenotype has been studied in pigs for nine generations of selection using combined phenotypic and genetic parameters expressed as estimated breeding values (EBVs) for combined antibody and CMI traits (Mallard et al., 1992, 1998). Purebred pigs selected only for production traits vary greatly as individuals in ability to produce antibody and DTH in response to immunization (Fig. 2). Both individual and breed comparisons are significantly different. This provides opportunity to alter populations of livestock to test the hypothesis that HIR is a phenotype in favorable association with enhanced health and productivity.

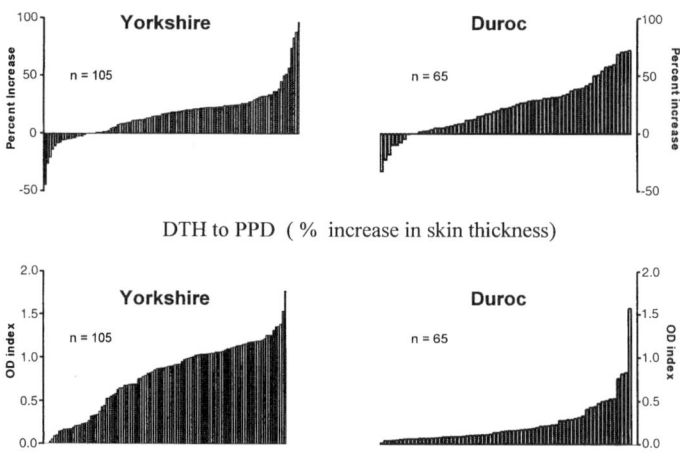

FIG. 2. Individual variation in antibody (Ab) response (day 21 serum Ab after day 0 and day 14 immunization with hen egg white lysozyme, HEWL, in quil A) and cutaneous delayed-type hypersensitivity (DTH) to purified protein derivative (PPD) after immunization with BCG vaccine. Each bar represents a purebred pig of the Yorkshire or Duroc breeds maintained in a commercial breeding herd. Antibody is expressed as OD index from an enzyme immunoassay. Cutaneous DTH is expressed as percent increase in double skinfold thickness 24 hours after intradermal injection of PPD. Methods are described in Mallard et al. (1992).

A. Selection for High and Low Antibody and Cell-Mediated Immune Response

Estimated breeding values calculated for antibody and CMI in Yorkshire pigs at generation 6 of selection for HIR and LIR are illustrated in Fig. 3. The controls (C) were not selected and reflect the original population. It is clear that livestock can be bred for overall improved immune response. The HIR and LIR pigs differ in several correlated traits. Antibody avidity is higher in the HIR animals (Appleyard *et al.*, 1992) and they produce more antibody to most and perhaps all test

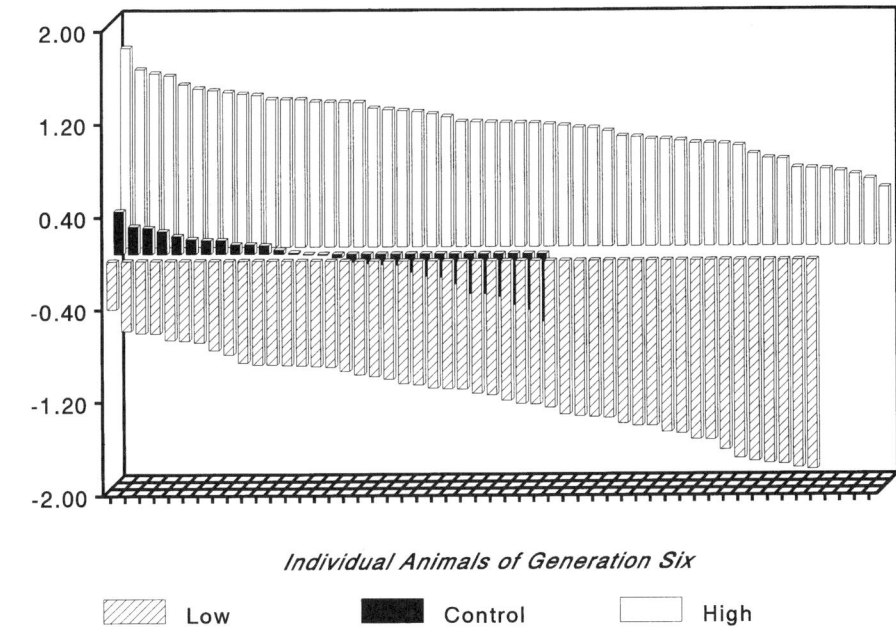

FIG. 3. Estimated breeding values (EBV) for individual Yorkshire pigs in generation 6 or selective breeding for high or low combined antibody and cell-mediated immune response. Controls were not selected for immune response. Traits incorporated in the EBV were serum antibody to secondary immunization with hen egg white lysozyme, serum IgG concentration, *in vitro* blood lymphocyte blastogenesis to Con-A, and cutaneous delayed-type hypersensitivity to PPD of tuberculin after immunization with BCG. Methods are described in Mallard *et al.* (1992).

antigens (Mallard et al., 1992, 1998). Monocyte expression of MHC II and oxygen metabolites does not vary by line (Groves et al., 1993), but NK cell numbers and function are superior in the HIR pigs (Raymond, 1997). Limited information on the disease resistance of the HIR pigs suggests that in response to experimental infection with *Mycoplasma hyorhinis* the LIR animals produce less antibody but also develop less arthritis than the HIR animals (Magnusson et al., 1998). Serositis was less in HIR than LIR pigs after infection with *M. hyorhinis*. Line-related differences in cytokine production may influence both immune response and inflammatory reactions in the HIR and LIR pigs. Cells from *M. hyorhinis*-associated arthritic joints of HIR pigs produce more interleukin 6 (IL-6) and interferon γ (IFN-γ) than those of the LIR line (J. Reddy and B. N. Wilkie, University of Guelph, personal communication, 1997) and during *M. hyorhinis* infection HIR have more serum binding proteins for the chemokines RANTES, MIP-1B, and IL-8 (Banga, 1997). The HIR pigs take significantly fewer days to reach market weight (100 kg) (Mallard et al., 1998).

B. RESPONSE TO VACCINATION OF IMMUNE RESPONSE SELECTED PIGS

Since the HIR and LIR lines differ in immune response to the index antigens used to determine phenotypic and genotypic values incorporated in the EBV used in selection, as well as in response to nonindex antigens, it was hypothesized that they would respond differently to commercial vaccines. This was tested by immunizing pigs of HIR and LIR lines with a commercial *Actinobacillus pleuropneumoniae* bacterin (Magnusson et al., 1997). Following primary and secondary vaccination of HIR and LIR pigs at generation 4 of selection, the HIR animals produced significantly ($p \leq 0.05$) more antibody to two antigens of *A. pleuropneumoniae* (carbohydrate antigen type 1, CHO 1, and lipopolysaccharide antigen type 1, LPS 1). Response to a third antigen (carbohydrate antigen type 5, CHO 5) to which HIR animals had preexisting serum antibody, indicated significantly higher HIR line response after primary but not after secondary immunization (Fig. 4).

FIG. 4. Least squares means and standard errors of serum antibody determined by enzyme immunoassay to individual antigens of *Actinobacillus pleuropnumoniae* after vaccination of pigs bred selectively for high (solid bars) and low (open bars) immune response. Vaccination occurred on days 0 and 14 (arrows). At a given time point, bars differing in letter designation differ significantly ($p \leq 0.05$) in least squares means. (Reprinted from *Vaccine*, Vol. 15, U. Magnusson et al., Antibody response to *Actinobacillus pleuropneumoniae* antigens after vaccination of pigs bred for high and low immune response, pp. 997–1000. Copyright 1997, with permission from Elsevier Science.)

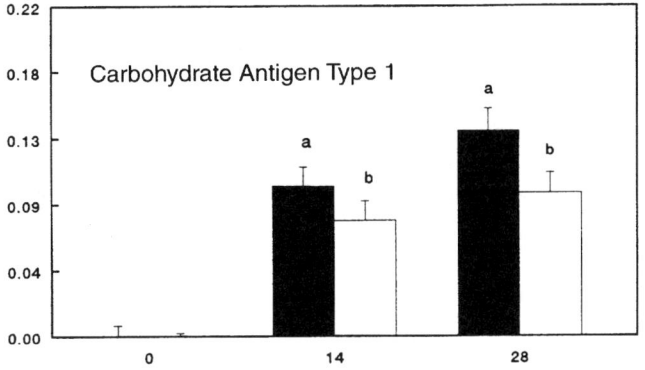
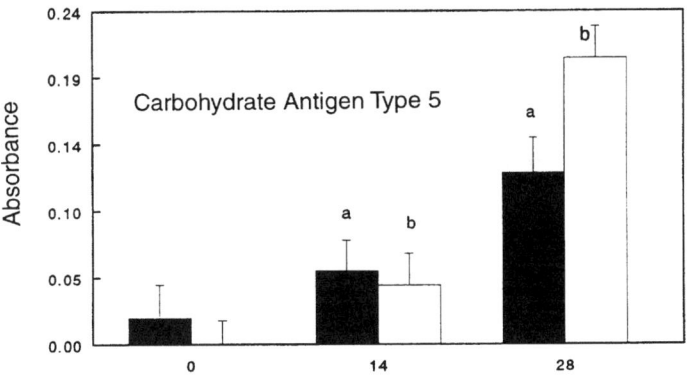
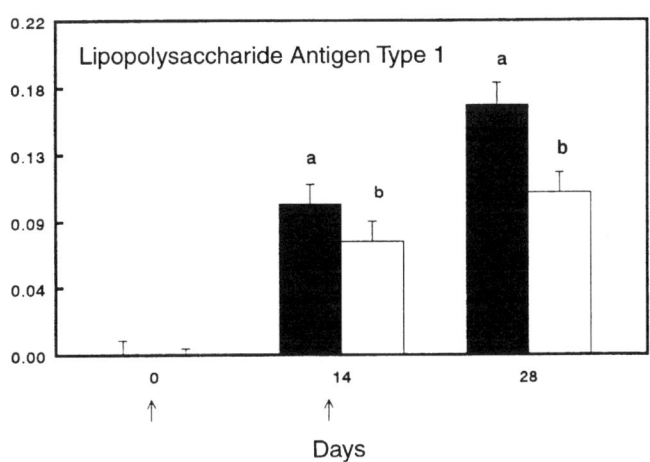

Although nonresponders were present in each line for each antigen, the percentage of these was significantly less among HIR animals for CHO 1 ($p \leq 0.01$) and LPS1 ($p \leq 0.06$) but not for CHO5 (Table I). An inactivated porcine influenza virus vaccine was used to vaccinate HIR, C, and LIR pigs of generation 9. Vaccination was performed on days 0 and 14. Serum heamagglutination inhibiting antibody was measured and at day 21 the mean HIR line response was significantly ($p \leq 0.05$) higher than the C group, which in turn was higher than the LIR (Fig. 5). In this experiment nonresponders occurred only in the LIR group where they were 38% of vaccinated animals.

V. Discussion and Summary

Immune responsiveness varies significantly among commercial pigs and other species of livestock and can be altered by selectively breeding for multiple traits reflecting ability to produce antibody and CMI (Mallard et al., 1992, 1998). Preliminary evidence indicates that HIR

TABLE I

FREQUENCY OF ANTIBODY-NEGATIVE SERA BY ENZYME IMMUNOASSAY TO PURIFIED ANTIGENS OF *ACTINOBACILLUS PLEUROPNEUMONIAE* AFTER VACCINATION OF PIGS OF HIGH AND LOW IMMUNE RESPONSE LINES[a]

	Frequency (%) of nonresponders[c]		
Test antigen[b]	High line	Low line	Value of p^d
Carbohydrate antigen type 1	7	29	0.01
Carbohydrate antigen type 5	47	48	ns[e]
Lipopolysaccharide antigen type 1	11	27	0.06

[a]Reprinted from *Vaccine*, Vol. 15, U. Magnusson et al., Antibody response to *Actinobacillus pleuropneumoniae* antigens after vaccination of pigs bred for high and low immune response, pp. 997–1000. Copyright 1997, with permission from Elsevier Science.

[b]Antigens extracted from *A. pleuropneumoniae* and used to coat microtiter plates for EIA to detect antibodies induced by immunization with a commerical vaccine.

[c]Nonresponders were defined as individuals failing to develop serum antibody as detected by having an optical denisty in EIA for the indicated antigen.

[d]Level of significance for the effect of immune response breeding line as determined by ANOVA.

[e]Not significant.

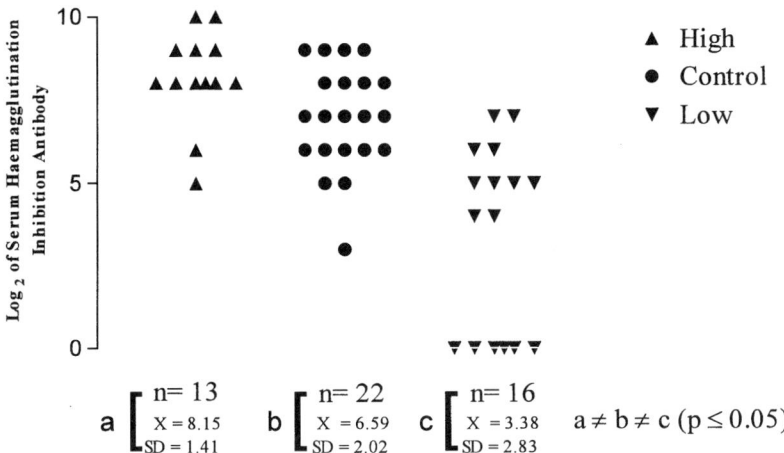

FIG. 5. Serum antibody in high, control, and low immune response Yorkshire pigs following vaccination with a commercial inactivated influenza virus vaccine. Animals were from generation 8 of selective breeding for high, low, and control antibody and cell-mediated immune response.

pigs may have advantages in productivity and in some, but not all, disease signs due to infection. More comprehensive information on the health performance of HIR and LIR pigs is presently being sought from observations on a large number of phenotyped commercial breeding pigs. Consistently with the high antibody and CMI responses of HIR pigs to index antigens, these animals have responded better to two commercial vaccines. This suggests that it may be practical to manipulate target livestock populations to improve response to vaccination. Thus the efficacy of vaccination could be enhanced not only by improved vaccines per se, but also by changing recipient animals such that the HIR phenotype and genotype become more prevalent. The method of deriving HIR and LIR lines could have further utility in vaccine development in that vaccines capable of efficaciously immunizing the LIR animals would be expected to perform well overall in unselected target populations in which most individuals would be expected to have better immune responsiveness than the LIR line. Just as improved production performance of livestock has depended on selective breeding so might improvements in general healthiness be achieved by indirect selection for candidate health-enhancing phenotype and genotype. High immune response may prove to be such a target.

Acknowledgments

Some of the data described here were derived from research conducted under grants to B. N. Wilkie from the Natural Sciences and Engineering Council of Canada and Agriculture and Agrifood Canada. Research infrastructural support was provided in part by the Ontario Ministry of Agriculture Food and Rural Affairs.

References

Ada, G. (1993). *Curr. Opin. Immunol.* **5,** 683–686.
Alan, J. E., and Maizels, R. M. (1997). *Immunol. Today* **18,** 387–392.
Appleyard, G., Wilkie, B. N., Kennedy, B. W., and Mallard, B. A. (1992). *Vet. Immunol. Immunopathol.* **31,** 229–240.
Babbit, B. A., Allen, P. M., Matseuda, G., Harber, E., and Unanue, E. M. (1985). *Nature* **317,** 359–360.
Banga, S. H. (1997). Ph.D. Thesis, University of Guelph, Guelph, Ontario, Canada.
Beckman, J. S., and Soller, M. (1987). *Bio/Technology* **5,** 673–576.
Bennaceraf, B. (1981). *Science* **212,** 1229–1238.
Biozzi, G., Mouton, D., Stiffel, C., and Bouthillier, Y. (1984). *Adv. Immunol.* **36,** 189–234.
Briles, W. E., Stone, H. A., and Cole, R. K. (1977). *Science* **195,** 193–195.
Covelli, V., Mouton, D., di Majo, V., Bouthillier, Y., Bangrazi, C., Mevel, J.-C., Rebessi, S., Doria, G., and Biozzi, G. (1989). *J. Immunol.* **142,** 1224–1234.
Edfors-Lilja, I., Wattrang, E., Magnusson, U., and Fossum, C. (1994). *Vet. Immunol. Immunopathol.* **40,** 1–16.
Fujji, J., Otsu, K., Zorzato, K., De Leon, S., Khanna, S., Weiler, J. E., O'Brien, P. J., and MacLennan, D. H. (1991). *Science* **253,** 448–451.
Gavora, J. S., and Spencer, J. L. (1983). *Anim. Blood Groups Biochem. Genet.* **14,** 159–180.
Georges, M., and Massey, J. M. (1991). *Theriogenology* **35,** 151–159.
Groves, T. C., Wilkie, B. N., Kennedy, B. W., and Mallard, B. A. (1993). *Vet. Immunol. Immunopathol.* **36,** 347–358.
Hessing, M. J. C., Coenen, G. J., Vaiman, M., and Renard, C. (1995). *Vet. Immunol. Immunopathol.* **45,** 97–113.
Hill, A. V. S., Allsopp, C. E. M., Kwiatowski, D., Anstey, N. M., Twumasi, P., Rowe, P. A., Bennett, S., Brewster, D., McMichael, A. J., and Greenwood, B. M. (1991). *Nature* **352,** 595–600.
Hutt, F. B. (1958). "Genetic Resistance to Disease in Domestic Animals." Comstock, Ithaca, NY.
Lewin, H. A., and Bernoco, D. (1991). *Anim. Genet.* **17,** 197–207.
Lie, O. (1990). *Proc. World Congr. Genet. Appl. Livest. Prod., 4th,* Vol. 26, pp. 421–425.
Lillehoj, H. S., Ruff, M. D., Bacon, L. D., Lamont, S. J., and Jeffers, T. K. (1989). *Vet. Immunol. Immunopathol.* **20,** 135–148.
Magnusson, U. L., Bossé, J., Mallard, B. A., Rosendal, S., and Wilkie, B. N. (1997). *Vaccine* **15,** 997–1000.
Magnusson, U. J., Bossé, J., Mallard, B. A., Rosendal, S., and Wilkie, B. N. (1998). *Vet. Immunol. Immunopathol.* (in press).
Mallard, B. A., and Wilkie, B. N. (1993). *Vet. Immunol. Immunopathol.* **38,** 387–394.
Mallard, B. A., Wilkie, B. N., and Kennedy, B. W. (1989a). *Vet. Immunol. Immunopathol.* **21,** 139–151.
Mallard, B. A., Wilkie, B. N., and Kennedy, B. W. (1989b). *Anim. Genet.* **20,** 167–178.

Mallard, B. A., Wilkie, B. N., Kennedy, B. W., and Quinton, M. (1992). *Anim. Biotechnol.* **3,** 257–280.
Mallard, B. A., Wilkie, B. N., Kennedy, B. W., Gibson, J. and Quinton, M. (1998). *Proc. World Congr. Genet. Appl. Livest. Prod. 5th,* Armidale, Australia, *1998.*
Medzhitov, R., Preston-Hurlbert, P., and Janeway, C. A. (1997). *Nature* **388,** 394–397.
Newman, M. J., Traux, R. E., French, D. D., Dietrich, M. A., Franke, D., and Stear, M. J. (1996). *Vet. Immunol. Immunopathol.* **50,** 43–54.
Outteridge, P. M. (1993). *Immunol. Cell Biol.* **71,** 355–366.
Outteridge, P. M. (1994). *Immunol. Cell Biol.* **72,** 256–261.
Peleg, B. A., Soller, M., Ron, N., Hornstein, K., Brody, T., and Kalmar, E. (1976). *Avian Dis.* **20,** 661–668.
Raymond, C. (1997). M.Sc. Thesis, University of Guelph, Guelph, Ontario, Canada.
Reed, S. G., and Scott, P. (1993). *Curr. Opin. Immunol.* **5,** 524–531.
Simianer, H., Solbu, H., and Schaeffer, L. R. (1991). *J. Dairy Sci.* **74,** 4358–4365.
Simonsen, M. (1987). *Vet. Immunol. Immunopathol.* **17,** 243–253.
Sørenson, T. I. A., Nielsen, G. G., Andersen, P. K., and Teasdale, T. W. (1988). *N. Engl. J. Med.* **318,** 727–732.
Strømsheim, A., Eide, D. M., Hofgaard, P. O., Larsen, H. J. S., Refstie, T., and Røed, K. H. (1994). *Vet. Immunol. Immunopathol.* **44,** 85–95.
Templeton, J. W., Estes, D. M., Price, R. E., Smith, R., and Adams, L. G. (1990). *Proc. World Congr. Genet. Appl. Livest. Prod. 4th,* Vol. 26, pp. 396–399.
Uribe, H. A., Kennedy, B. W., Martin, S. W., and Kelton, D. P. (1995). *J. Dairy Sci.* **78,** 421–430.
van der Zijpp, A. J. (1983). *World Poult. Assoc. J.* **39,** 118–131.
Wilkie, B. N., Mallard, B. A., Quinton, M., and Gibson, J. (1998). *Prog. Pig Sci. Proc. Easter Sch. Agric. Sci., 58th,* University of Nottingham, *1997.*

Nutritional Effects on Vaccination

MARK E. COOK

Animal Sciences Department, University of Wisconsin, Madison, Wisconsin 53706

I. Vaccination Cost to Performance
II. Biochemical Mechanisms in Immune-Induced Wasting
III. Conjugated Linoleic Acid
IV. Cholecystokinin and Immune-Induced Anorexia
V. Summary
   References

I. Vaccination Cost to Performance

Unquestioned is the need for immune protection against pathogenic microbial onslaught. Primed immune defense, by means of vaccination, has been an essential element in the consolidation of animals in modern forms of agriculture. All would agree that vaccinology advanced animal agriculture practices as great as nutritional and genetic developments. However, seldom is there mention of the cost associated with employment of vaccinations beyond that of the reagent and its delivery. An understanding of reduced performance associated with vaccination yields opportunity for discovery and novel therapy to improve animal performance and efficiency of production. (*Editor's note:* Ironically, the cost benefit analysis for most animal vaccines has not been reported.)

As early as the 1950s, scientists became aware of environmental factors affecting animal growth. Observation of improved animal performance in newly built animal research units demonstrated the role of the conventional microbial environment in animal performance apart from infectious disease (Hill *et al.*, 1952; Lillie *et al.*, 1952).

Animals reared in germ-free environments attained weight gains as much as 10% more than animals in contact with common environmental microbes (Lev and Forbes, 1959). Strategies to recapture reduced performance associated with conventional microbe exposure were based on antibiotic use. Antibiotics were effective in improving growth and feed efficiency in animals, but never realized the full potential achieved in germ-free environments. More recently, use of "all in all out" management practices has been effective in reducing microbial loads in growing animals and enhancing performance (Cline et al., 1992).

While the mechanism of improved performance associated with feeding antibiotics had long been in dispute (Bird, 1969; Visek, 1978), recent knowledge of immune regulation of growth now suggests that conventional microbes suppress performance by immune stimulation (Roura and Klasing, 1993). Hence, antibiotics stimulate growth by decreasing the load of immune stimulants in or on the animal.

In addition to immune stimulants in the environment, vaccination also causes a decrease in growth and poorer feed efficiency. Chamberlee and coworkers (1992) reported that "vaccination of broilers resulted in lower final body weights, poorer feed conversions, and higher 8 day and 42 day mortality (than unvaccinated broilers) . . . in the absence of overt disease."

Growth suppression following immune stimulation has been best described by Klasing et al. (1987) in poultry. Chicks immune stimulated with sheep red blood cells or endotoxin had reduced rate of gain and feed intake compared to those injected with sterile saline. The effects of immune stimulation on performance can be mimicked by directly injecting the cytokine interleukin 1 (IL-1) (Klasing et al., 1987). During the immune response, IL-1 and tumor necrosis factor are released from the macrophage. While IL-1 and related cytokines are essential for the production of interleukin 2 for lymphocyte proliferation, they also stimulate skeletal muscle degradation, and hepatic synthesis of acute phase proteins (Johnstone and Klasing, 1990; Klasing, 1988).

II. Biochemical Mechanisms in Immune-Induced Wasting

It has been hypothesized that immune-induced wasting was conserved during evolution to ensure that infected animals would fall from the herd or flock to prevent the spread of disease (Tracey et al., 1987). While immune-induced wasting may be important in maintaining wild species health, there are few benefits in modern animal production systems. Hence, the reward for controlling immune-induced weight loss in animal agriculture would be great. It was a goal to define dietary agents that could reduce immune-driven catabolism.

Research suggested that IL-1 stimulated the production of prostaglandin E_2 (PGE_2) in skeletal muscle (Goldberg et al., 1984). In turn, PGE_2 had been shown to stimulate skeletal muscle degradation and decrease net protein accretion (Rodemann and Goldberg, 1982). In addition, PGE_2 was shown to inhibit IL-1 synthesis by the macrophage (Hwang, 1989).

We hypothesized that regulation of PGE_2 production may have beneficial effects in preventing wasting associated with immune stimulation (Cook et al., 1993). Work had shown that omega-3 fatty acids were effective in reducing PGE_2 production by monocytes (Endres et al., 1989). Omega-3 fatty acids at 8% of the diet were also effective in reducing anorexia caused by immune stimulation (Hellerstein et al., 1989). However, IL-1 production by monocytes was depressed in the presence of omega-3 fatty acids. Protection against immune-induced wasting by suppressing cytokine production was not viewed as a satisfactory method in preventing immune-induced wasting. Hence, alternative nutrients were sought.

III. Conjugated Linoleic Acid

Pariza and coworkers reported on antimutagenic and anticarcinogenic activity in hamburger meat (Pariza et al., 1979, 1983; Pariza and Hargraves, 1985). The activity was purified and shown to be a mixture of geometrical and positional isomers of linoleic acid (conjugated linoleic acid) (Ha et al., 1987). Conjugated linoleic acid (CLA) (primarily cis 9-, trans 11-, and trans 10-, cis 12-octadecadienoic acid) differs from linoleic acid (cis 9-, cis 12-octadecadienoic acid) in the location of the double bonds. Since linoleic acid is the precursor of eicosanoids, it was hypothesized that it may modify cyclooxygenase metabolites and hence prevent immune-induced wasting. In addition, CLA was shown to be naturally occurring, being found in animal tissues, particularly of ruminant origin (Chin et al., 1992, 1994a).

When mice, chicks, or rats were fed CLA then injected with lipopolysaccharide (LPS), weight loss due to the immune stimulation was greatly attenuated (Cook et al., 1993; Miller et al., 1994). In addition, anorexia caused by injecting LPS was less severe in CLA fed mice. CLA prevented an increase in LPS-induced release of PGE_2 in RAW264.7

cells (Park, 1996) which mechanistically supports the original hypothesis.

Studies were immediately started to determine CLA's effect on immune function (Cook et al., 1993; Miller et al., 1994; DeVoney et al., 1997). While CLA had no effect on antibody synthesis in chicks to sheep red blood cells, CLA increased splenocyte blastogenesis and foot pad response to phytohemagglutinin. Spleen and white blood cell numbers were increased in CLA-fed animals, and the portion of CD4+ T cells, post-LPS exposure, was greater in CLA-fed animals than control fed.

Preliminary work (unpublished) also suggests that CLA feeding did not excerbate type I hypersensitivity. The trachea of guinea pigs hyperimmunized to ovalbumin did not show increase sensitivity to antigen in superfusion studies. Histamine and PGE_2 production was reduced in antigen stimulated tracheas.

The immune system has a broad array of biological activity apart from mere defense. During studies on the effects of CLA on preventing immune-induced wasting, it was discovered that CLA enhanced animal growth and feed efficiency (Chin et al., 1994b) and repartitioning of fat deposition into lean (Park et al., 1997). Also observed was a pronounced effect of CLA on increasing body and bone ash and correcting the skeletal defect in poultry known as valgus and varus leg deformities (Cook et al., 1997a). Others have reported a role of CLA in preventing atherosclerosis (Lee et al., 1994). Hence, an understanding of the role of immune regulation of physiologic processes, such as growth, yielded a number of discoveries typically viewed as unrelated to the traditional role of immune system in disease resistance.

IV. Cholecystokinin and Immune-Induced Anorexia

Immune stimulation of the host induces anorexia. During the immune reaction, cytokines induce the release of neurotransmitter peptides (i.e., cholecystokinin, CCK) which alters gastrointestinal function and stimulates satiety (Daun and McCarthy, 1993; Ohgo et al., 1992). An attempt was made to cause an autoimmune response in breeding hens to CCK. It was hypothesized that laying hens would passively transfer antibody to CCK, via the egg yolk, to the progeny, and that the CCK antibodies would tie up CCK released during immune stress. While positive results in the form of improved growth and feed efficiency were obtained, it was quickly learned that a more efficient process was to harvest and dry the anti-CCK egg yolks and feed them to growing poultry and swine. Animals fed antibodies to CCK achieved greater rates of gain and improved feed efficiency. Antibodies to other select gastrointestinal peptides have shown benefits (Cook et al., 1997b,c).

V. Summary

Immune-induced cachetic response is an example of a biological opportunity to develop technologies that ensure improved performance in animal agriculture. We have estimated that reduced performance of immune stimulated animals, whether by exposure to conventional environments or through vaccination, results in more than U.S. $500 million in reduced productivity. Nontraditional methods to alleviate the adverse effects of the immune response provide an opportunity for those skilled in the art of vaccinology and immunology to develop new technologies and feeding practices. Too often, biologists are blinded by the limits of their disciplines and rarely venture to the fringe of their field to engage in collaborations that at first glance do not seem logical. The examples of CLA and antigastrointestinal peptides suggest that new opportunities await in ensuring that the cost of the immune response is minimized and that new approaches to animal agriculture await discovery.

REFERENCES

Bird, H. R. (1969). Biological basis for the use of antibiotics in poultry feeds. In "The Use of Drugs in Animal Feeds," *Proc. Sym. Pub. 1679*, pp. 31–41. National Academy of Sciences, Washington, DC.

Chamberlee, T. N., Thompson, J. R., and Thaxton, J. P. (1992). Effects of day old vaccination on broiler performance. *Poult. Sci.* **71,** (Suppl. 1), 144 (abstr.).

Chin, S. F., Liu, W., Storkson, J. M., Ha, Y. L., and Pariza, M. W. (1992). Dietary sources of conjugated dienoic isomers of linoleic acid, a newly recognized class of anticarcinogens. *J. Food Comp. Anal.* **5,** 185–197.

Chin, S. F., Storkson, J. M., Liu, W., Albright, K. J., Cook, M. E., and Pariza, M. W. (1994a). Conjugated linoleic acid (9, 11- and 10, 12-octadecadienoic acid) is produced in conventional but not germ-free rats fed linoleic acid. *J. Nutr.* **124,** 694–701.

Chin, S. F., Storkson, J. M., Albright, K. J., Cook, M. E., and Pariza, M. W. (1994b). Conjugated linoleic acid is a growth factor for rats as shown by enhanced weight gain and improved feed efficiency. *J. Nutr.* **124,** 2344–2349.

Cline, T. R., Matrose, V. B., Scheidt, A. B., Diekman, M., Clark, K., Hurt, C., and Singleton, W. (1992). Effect of all in/all out management on the performance and health of growing-finishing pigs. *Proc. Purdue Univ. Swine Day,* pp. 0–12.

Cook, M. E., Miller, C. C., Park, Y., and Pariza, M. (1993). Immune modulation by altered nutrient metabolism: Nutritional control of immune-induced growth depression. *Poult. Sci.* **72,** 1301–1305.

Cook, M. E., Jerome, D. L., and Pariza, M. W. (1997a). Broilers fed conjugated linoleic acid had enhanced bone ash. *Poult. Sci.* **76** (Suppl. 1), 41.

Cook, M. E., Jerome, D. L., Pimentel, J., and Miller, C. (1997b). Feeding egg antibodies to cholecystokinin (CCK) improves growth and feed conversion in broilers. *Poult. Sci.* **76** (Suppl. 1), 95.

Cook, M. E., Jerome, D. L., Daley, M. J., Greenblatt, H., and Skarie, C. (1997c). Performance improvement caused by feeding broilers egg yolk antibodies to cholecystokinin is correlated with specific antibody dose not the mass of egg yolk. *Poult. Sci.* **76** (Suppl. 1), 95.

Daun, J. M., and McCarthy, D. O. (1993). The role of cholecystokinin in interleukin-1 induced anorexia. *Physiol. Behav.* **54**, 237–241.

DeVoney, D., Pariza, M. W., and Cook, M. E. (1997). Conjugated linoleic acid increases blood and splenic T-cell response post lipopolysaccharide injection. *FASEB J.* **9**, 3355.

Endres, S., Ghorbani, R., Kelley, V., Georgilis, K., Lonnemann, G., van der Meer, J. J. W. M., Cannon, J. G., Rogers, T. S., Klempner, M. S., Weber, P. C., Schaefer, E. J., Wolff, S. M., and Dinarello, C. A. (1989). The effect of dietary supplementation with n-3 polyunsaturated fatty acids on the synthesis of interleukin-1 and tumor necrosis factor by mononuclear cells. *N. Engl. J. Med.* **320**, 265–271.

Goldberg, A. L., Baracos, V., Rodemann, H. P., Waxman, L., and Dinarello, C. (1984). Control of protein degradation in muscle by prostaglandins, $Ca++$, and leukocytic pyrogen (interleukin-1). *Fed. Proc.* **43**, 1301–1306.

Ha, Y. L., Grimm, N. K., and Pariza, M. W. (1987). Anticarcinogens from fried ground beef: Heat altered derivatives of linoleic acid. *Carcinogenesis* **8**, 1881–1887.

Hellerstein, M. K., Meydani, S. N., Meydani, M., Wu, K., and Dinarello, C. A. (1989). Interleukin-1-induced anorexia in the rat. Influence of prostaglandins. *J. Clin. Invest.* **84**, 228–235.

Hill, D. C., Branion, H. D., Slinger, S. J., and Anderson, G. W. (1952). Influence of environment on the growth response of chicks to Penicillin. *Poult. Sci.* **31**, 462–466.

Hwang, D. (1989). Essential fatty acid and immune response. *FASEB J.* **3**, 2052–2061.

Johnstone, B. J., and Klasing, K. C. (1990). Nutritional aspects of leukocytic cytokines. *Nutr. Clin. Metab.* **4**, 7–29.

Klasing, K. C. (1988). Nutritional aspects of leukocytic cytokines. *J. Nutr.* **118**, 1436–1446.

Klasing, K. C., Laurin, D. E., Peng, R. K., and Fry, D. M. (1987). Immunologically mediated growth depression in chicks: Influence of feed intake, corticosterone and interleukin-1. *J. Nutr.* **117**, 1629–1637.

Lee, K. N., Kritchevsky, D., and Pariza, M. W. (1994). Conjugated linoleic acid and atherosclerosis in rabbits. *Atherosclerosis* **108**, 19–25.

Lev, M., and Forbes, M. (1959). Growth response to dietary penicillin of germ-free chicks with a defined intestinal flora. *Br. J. Nutr.* **13**, 78–84.

Lillie, R. J., Sizemore, J. R., and Bird, H. R. (1952). Environment and stimulation of growth of chicks by antibiotics. *Poult. Sci.* **31**, 466–475.

Miller, C. C., Park, Y., Pariza, M. W., and Cook, M. E. (1994). Feeding conjugated linoleic acid to animals partially overcomes catabolic response due to endotoxin injection. *Biochem. Biophys. Res. Commun.* **198**, 1107–1112.

Ohgo, S., Nakatsuro, K., Ishikawa, E., and Shigeru, M. (1992). Stimulation of cholecystokinin (CCK) release from superfused rat hypothalamo-neurohypophyseal complexes by interleukin-1 (IL-1). *Brain Res.* **593**, 25–31.

Pariza, M. W., and Hargraves, W. A. (1985). A beef-derived mutagenesis modulator inhibits initiation of mouse epidermal tumors by 7, 12-dimethylbenz[a]anthracene. *Carcinogenesis* **6,** 591–593.

Pariza, M. W., Ashoor, S. H., Chu, F. S., and Lund, D. B. (1979). Effects of temperature and time on mutagen formation in pan-fried hamburger. *Cancer Lett.* **7,** 63–69.

Pariza, M. W., Loretz, L. J., Storkson, J. M., and Holland, N. C. (1983). Mutagens and modulators of mutagenesis in fried ground beef. *Cancer Res.* **43,** 2444s–2446s.

Park, Y. (1996). Regulation of energy metabolism and the catabolic effects of immune stimulation by conjugated linoleic acid. Ph.D. Dissertation, University of Wisconsin, Madison.

Park, Y., Albright, K. J., Liu, W., Storkson, J. M., Cook, M. E., and Pariza, M. W. (1997). Effect of conjugated linoleic acid on body composition in mice. *Lipids* **32,** 853–858.

Rodemann, H. P., and Goldberg, A. L. (1982). Arachidonic acid, prostaglandin E2 and F2 influence rates of protein turnover in skeletal and cardiac muscle. *J. Biol. Chem.* **257,** 1632–1638.

Roura, E., and Klasing, K. C. (1993). Dietary antibiotics reduce immunological stress elicited by poor sanitation or consumption of excreta in broiler chicks. *Poult. Sci.* **72,** (Suppl. 1), 1 (abstr.).

Tracey, K. J., Lowry, S. F., and Cerami, A. (1987). Physiological responses to cachectin. In: tumor necrosis factor and related cytokines. *Ciba Found. Symp.* **131,** 88–108.

Visek, W. J. (1978). The mode of growth promotion by antibiotics. *J. Anim. Sci.* **46,** 1447–1469.

Effects of Stress on Leukocyte Trafficking and Immune Responses: Implications for Vaccination

M. E. KEHRLI,* J. L. BURTON,† B. J. NONNECKE,*
AND E. K. LEE**

*NADC-USDA-ARS, Ames, Iowa 50010, †Michigan State University, East Lansing, Michigan 48825, **Iowa State University, Ames, Iowa 50011

I. Introduction
II. Leukocyte Trafficking
   A. Selectins
   B. Integrins
   C. Intercellular Adhesion Molecules
   D. α_4-Integrins
III. Effects of Stress on Immunity
IV. Summary
   References

I. Introduction

A key step in host defense is leukocyte recruitment into peripheral tissues following infection by invasive microorganisms. Leukocyte trafficking through tissues is crucial to effective immune surveillance for infectious agents. Trafficking enables rapid neutrophil accumulation (innate immunity) at sites of infection or tissue injury and subsequent movement of lymphocytes through secondary lymphoid tissues for response against antigens presented in germinal centers (Adams and Shaw, 1994). The interaction of different populations of circulating leukocytes with postcapillary venule endothelial cells is essential for leukocyte emigration into tissue (Adams and Shaw, 1994). The interaction between endothelial cells and leukocytes is a dynamic process involving both cell types; the long-held view that the endothelium is a rather inert lining for blood vessels is no longer valid.

Neutrophils are a homogeneous population of rather unsophisticated effector cells whose recruitment to sites of inflammation involves the coordinated function of multiple families of adhesion molecules, cytokines, and chemoattractants. A multistep model for this process includes a transient leukocyte adhesion between the leukocyte and endothelial cells of the vessel wall mediated by members of the selectin family, followed by triggering of leukocyte activation and subsequent tight adhesion between leukocyte integrins and the intercellular adhesion molecules on endothelial cells. Selectivity in the process of leukocyte recruitment comes from the diversity of molecules capable of mediating each step.

The molecular basis of leukocyte migration to an infection site involves the combined action of multiple families of adhesion molecules and chemoattractants. The most simplistic model is that of neutrophil extravasation, initially dependent on adhesive interactions between L-selectin (CD62L) on neutrophils and E-selectin (CD62E) and P-selectin (CD62P) on endothelial cells with their respective ligands (Kishimoto, 1991; Lasky, 1992; Varki, 1994; Imhof and Dunon, 1995). These reversible selectin interactions can trigger tight adhesion or arresting of neutrophils by increasing expression and adhesiveness of another adhesion molecule, Mac-1 (CD11b/CD18), which is a member of the β_2-integrin family of leukocyte adhesion molecules (Hynes, 1992). Mac-1 on neutrophils tightly binds to its ligand on activated endothelial cells, intercellular adhesion molecule-1 (ICAM-1) (Marlin and Springer, 1987; Boyd et al., 1988; Smith et al., 1989; Diamond et al., 1991; Rothlein et al., 1991), allowing neutrophils to stop rolling and migrate along the endothelial surface to intercellular junctions of postcapillary venular endothelial cells. Neutrophils then cross the endothelial lining of vessels by migrating between endothelial cells through the intercellular junctions and enter the peripheral tissues through interaction with platelet/endothelial cell adhesion molecule 1 (PECAM-1) located in the endothelial cell junctions (Muller, 1995).

Lymphocytes have a greater diversity of effector activities and are more sophisticated in the repertoire of adhesion molecules they express. Lymphocytes emigrate from blood into tissues through the selective adhesion interactions of the α_4-integrin adhesion molecules, CD62L, cutaneous lymphocyte-associated antigen (CLA), P-selectin glycoprotein ligand (PSGL) and their endothelial cell ligands, including the mucosal addressin cell adhesion molecule 1 (MAdCAM-1), vascular cell adhesion molecule 1 (VCAM-1), the peripheral lymph node addressin (PNAd) complex, CD62E, and CD62P (reviewed in Butcher and Picker, 1996; see also Salmi and Jalkanen, 1997).

A significant dichotomy of lymphocyte trafficking exists between naive versus memory/effector lymphocytes. Naive T and B cells typically migrate through secondary lymphoid tissues (spleen, lymph nodes, Peyer's patches, and tonsil) in a pattern that ensures that their first encounter with antigen takes place in tissues where foreign antigen from tissues is presented to naive B and T cells. These lymphoid microenvironments typically promote antigen-induced differentiation while also eliminating autoreactive lymphocytes. Most lymphocytes with memory/effector functions likely also traffic through lymphoid organs, but they have the additional capacity to access and recirculate through extralymphoid sites (e.g., intestinal lamina propria, pulmonary interstitium, inflamed skin and joints) (Butcher and Picker, 1996). Moreover, the homing pattern of subsets of memory/effector lymphocytes is much more heterogeneous than that of naive lymphocytes. The specificity of these traffic patterns is dictated by the adhesion molecules expressed on the migrating lymphocyte, as well as the various receptors expressed on endothelial cells of different tissues.

Stress is an important factor that alters leukocyte trafficking, rendering animals more susceptible to infectious disease. In livestock husbandry, stress is an integral component of complex production diseases (e.g., respiratory disease in feedlot cattle and mastitis in dairy cattle). Stress can be broadly defined as situations in which homeostasis is disturbed or threatened. These perturbations can be triggered by various environmental factors to which an animal must respond physiologically to maintain equilibrium with the environment. Release of adrenocorticotropic hormone (ACTH) from the pituitary gland, which induces synthesis and secretion of glucocorticoids from the adrenal cortex, is one example of the host response to stress. Release of ACTH also occurs during a response to infection and is likely involved with the natural down-regulation of the innate immune response (De Rijk and Berkenbosch, 1994).

Studies of the effects of glucocorticoids on the immune system of cattle have identified alterations in leukocyte adhesion molecule expression that affect the normal trafficking pattern of neutrophils by causing the shedding of CD62L (Burton *et al.*, 1995). Microbicidal functions of neutrophils are suppressed by glucocorticoids (Roth and Kaeberle, 1981), as is the ability of cultured mixed populations of lymphocytes and monocytes to produce interferon γ and IgM (Doherty et al., 1995; Nonnecke et al., 1997). Glucocorticoid administration will also suppress or delay some antigen-specific immune responses to vaccines (Roth et al., 1984).

There are two major T-lymphocyte subsets: those expressing the αβ T-cell receptor for antigen and those expressing the γδ T-cell receptor. However, cattle have a much larger proportion of circulating γδ T cells than other nonruminant species (Hein and Mackay, 1991) and this has led to the general belief that γδ T cells play a particularly important role in immunity against infectious disease in cattle. Neonatal calves have a higher percentage of circulating γδ T cells (up to 70%) than mature cattle (~5%), but these values vary tremendously among individual animals of the same age (Clevers et al., 1990; Morrison and Davis, 1991). The γδ T cells preferentially traffic into epithelial areas, as opposed to αβ T cells, which preferentially traffic through secondary lymphoid tissues. Bovine γδ T cells express higher levels of the lymphocyte homing receptor, CD62L and have delayed CD62L down-regulation in response to chemoattractants compared to αβ T cells (Walcheck and Jutila, 1994). This was proposed as potentially contributing to the lack of γδ T-cell entry into secondary lymphoid tissue and, possibly, of preferential trafficking of this T-lymphocyte type to epithelial-associated tissues. Studies have shown that circulating γδ T-cell numbers decline in response to glucocorticoid administration (Burton and Kehrli, 1996) but no evidence of shedding of CD62L from γδ T cells was found. Moreover, it appears that glucocorticoid administration may actually enhance trafficking of γδ T cells into epithelial tissues.

Virtually identical changes in leukocyte trafficking patterns, cytokine and immunoglobulin secretion, and adhesion molecule expression on leukocytes have been reported in periparturient cows (Sordillo and Babiuk, 1991; Sordillo et al., 1991, 1995; Sordillo and Peel, 1992; Ishikawa et al., 1994; Detilleux et al., 1995; Lee and Kehrli, 1997). We believe that the effects of various types of stress on immune function increase an animal's susceptibility to infectious disease by altering leukocyte trafficking and functional capacities (Burton and Kehrli, 1995). This chapter summarizes the current molecular mechanisms of leukocyte trafficking and the effects of glucocorticoids on some of the molecules contributing to leukocyte egress into tissue. The recognized compromises of immune function induced by glucocorticoids released in response to stresses of castration, handling, transportation, and parturition should be considered when designing vaccination protocols for management of livestock health.

II. Leukocyte Trafficking

The immune system has a daunting task of maintaining health while facing continual exposure to the microbial world. Leukocytes routinely deal with microbes ranging from normal flora to severe life-threatening pathogens. Leukocyte trafficking is a process that ensures appropriate immune surveillance by leukocytes effecting adaptive and innate immunity. Recirculation of lymphocytes between blood and lymph via paracortical postcapillary venules of the lymph node has been recognized since 1964 (Gowans and Knight, 1964). Lymphocytes recirculate continuously, one to two times per day. Lymphocyte trafficking has also been referred to as lymphocyte homing, which alludes to the tendency of lymphocytes to preferentially recirculate through selected lymphoid tissues (e.g., the common mucosal immune system). Recognition of these recirculation patterns has also led to a segregation of research into mucosal and nonmucosal immune responses. Moreover, trafficking is not random; there are active mechanisms of leukocyte-endothelial cell recognition. Human, rodent, and porcine lymphocytes extravasate into lymph nodes through specialized paracortical postcapillary venules that are lined with high endothelium, known as high endothelial venules (HEVs) (Spalding and Heath, 1989; Harp *et al.*, 1990). In ruminants, however, these paracortical venules are smaller (i.e., morphologically not as high) and are more numerous than HEV in mice (Spalding and Heath, 1989; Harp *et al.*, 1990).

Leukocyte trafficking is regulated by various adhesion molecules on leukocytes and their ligands on postcapillary endothelial cells, involving the coordinated sequential function of several families of adhesion molecules, cytokines, and chemoattractants. Leukocyte egress is a multistep process involving a primary adhesion event that is transient and reversible in seconds. Leukocytes are then rapidly (within seconds) activated. Activation-dependent "arrest" of leukocytes to the postcapillary venule endothelial cells occurs that is stable under shear forces of blood flow, but still reversible over minutes. Diapedesis or the movement of leukocytes to an intercellular junction and migration into the tissue is the final event. Leukocyte trafficking can be regulated at any or all of the decision points. Selectivity in the process of leukocyte recruitment comes from the diversity of molecules capable of mediating each step.

A. Selectins

The first step in the process of leukocyte recruitment involves the initial contact and loose interactions required for leukocyte rolling.

The tethering of flowing leukocytes to the vessel wall and subsequent rolling is mediated by members of the selectin family (Zimmerman, 1992). Molecules of the selectin family are heavily glycosylated, single chain integral membrane proteins that include CD62P, CD62E, and CD62L (Lasky, 1992; Bevilacqua and Nelson, 1993; Tedder et al., 1995). Rolling adhesion or tethering with postcapillary venular endothelium is mediated in part by CD62L on leukocytes, a C-type lectin with affinity for sulfated, fucosylated carbohydrate determinants expressed on postcapillary venules. In mouse lymph nodes, these carbohydrate ligands are presented by a number of glycoproteins composing the peripheral node addressin (PNAd) on HEV. CD62L is concentrated on tips of leukocyte microvilli, which are the sites of initial cell–cell contact under blood flow/shear stress conditions. Other tethering receptors are expressed here too (e.g., the α_4-integrins, which bind MAdCAM-1 and VCAM-1). CD62L interaction with its ligand is reversible, which is ensured by proteolytic cleavage of CD62L near the cell membrane on cross-linking. Therefore, multivalent CD62L cross-linking is insufficient for sustained arrest of leukocytes, which requires additional reinforcing adhesion mechanisms.

CD62P is constitutively found in Weibel–Palade bodies of endothelial cells and in alpha granules of platelets (Johnston et al., 1989; Geng et al., 1990). It is mobilized to the cell surface within minutes after activation by thrombogenic and inflammatory mediators. Cell-surface expression of CD62P, by mobilization from intracellular storage granules, is short-lived (minutes) and, therefore, is likely involved in the early phase of leukocyte rolling during inflammation. By contrast, CD62E is neither synthesized constitutively nor stored within intracellular granules (Montgomery et al., 1991). Expression of CD62E by endothelial cells only occurs on stimulation by cytokines, with peak expression occurring within a few hours following inflammatory events. This indicates de novo synthesis of CD62E (Bevilacqua et al., 1987). CD62L is constitutively expressed on the surface of leukocytes and is rapidly shed by proteolytic cleavage on activation (Tedder et al., 1990; Kishimoto, 1991). CD62L appears to be crucial for recruitment of circulating neutrophils into inflamed tissue and lymphocyte homing into lymph nodes. It acts by slowing down leukocytes on contact with specific endothelial ligands prior to extravasation (Butcher, 1991; von Andrian et al., 1991). The cytoplasmic domain of CD62L appears to regulate leukocyte adhesion to endothelium independent of ligand recognition, by controlling cytoskeletal interactions (Kansas et al., 1993).

During an inflammatory response, locally produced cytokines stimulate increased expression of CD62P and CD62E on endothelial cells and these selectins interact with undefined counter ligands on leukocytes. It has been hypothesized that constitutively expressed CD62L on leukocytes initiates margination through transient interactions with carbohydrate moieties of the selectins on endothelial cells (von Andrian et al., 1993). Although ligands for the selectins have not been elucidated, it is believed that they are diverse, mucin-like, complex macromolecules that share common anionic carbohydrate moieties (Varki, 1994). Neutrophils activated by cytokines or chemoattractants shed CD62L, which is a prerequisite step for β_2-integrin-mediated tight adhesion. This activation also increases functional activity of the β_2-integrin, CD11b/CD18, leading to association of cytoskeletal proteins with the cytoplasmic tails of integrin molecules that are necessary for firm adhesion and spreading (Kishimoto et al., 1989; von Andrian et al., 1991). In a human genetic disease, termed leukocyte adhesion deficiency type 2 (LAD 2), a fucosylation failure reduces expression of all fucosylated lactosamines and is associated with markedly diminished selectin-mediated binding (Etzioni et al., 1992). Patients with LAD 2 have neutrophilia, marked defects in neutrophil motility resulting in impaired egress into diseased tissues. These patients suffer from recurrent pneumonia and bacterial infections, all of which underline the importance of the selectin-mediated adhesive interactions during inflammation (Etzioni et al., 1992).

B. Integrins

Integrins are the major family of cell surface receptors that mediate cell–cell adhesive interactions and attachment to the extracellular matrix. These integrin-mediated adhesive interactions are intimately involved in the regulation of cellular functions, including embryonic development, maintenance of tissue integrity, and leukocyte recruitment and extravasation (Albelda and Buck, 1990; Hynes, 1992; Sonnenberg, 1993). Structurally, the integrins are heterodimers composed of noncovalently linked α and β transmembrane subunits; the subunits are selected from among 16 α and 9 β glycoproteins that heterodimerize to produce more than 20 different receptors.

The β_2-integrins are mainly involved in leukocyte-endothelial cell contact. The β_2-integrin I (inverted) domain, as well as the divalent cation binding site of the α chains, are thought to be involved in ligand binding (Randi and Hogg, 1994; Stanley et al., 1994). The β_2-integrin family consists of distinct α chains, CD11a, CD11b, and CD11c, that share a common β chain (CD18). Distribution of β_2-integrins on leukocyte surfaces varies with cell type and state of activation. LFA-1

(CD11a/CD18) is expressed on all leukocytes and is the only β_2-integrin expressed on T and B lymphocytes (Arnaout, 1990). LFA-1 contributes to efficient diapedesis of all trafficking lymphocytes, although it is not absolutely required for egress. Mac-1 (CD11b/CD18) and p150,95 (CD11c/CD18) are predominantly found on granulocytes, with some expression on macrophages, and natural killer cells. They are stored in secondary and tertiary granules and lead to remarkable increases in surface expression of these molecules following cellular activation. This occurs following translocation of intracellular granules containing Mac-1 and p150,95 to the cell surface (Miller *et al.,* 1987; Arnaout, 1990).

Most activators of integrin binding also stimulate intracellular signal transduction pathways. The cytoplasmic domains of the β_2-integrins include phosphorylation sites that are accessible to these events and, thus, may mediate inside-out signaling and extracellular conformational changes (Ginsberg *et al.,* 1992). Conformational changes of LFA-1 may be brought about either by monoclonal antibody binding that activates it (Keizer *et al.,* 1988) or by intracellular elements that interact with the carboxyl terminal cytoplasmic tail of β_2-integrins (Hibbs *et al.,* 1991a,b). Conceivably, interactions of cytoskeletal proteins with the cytoplasmic domains of integrins (Pardi *et al.,* 1992), as triggered by leukocyte stimulation, might cause conformational changes and increases in integrin affinity. β_2-integrins are also involved in phagocytosis, neutrophil aggregation, superoxide production, H_2O_2 production, and intracellular killing of *Staphylococcus aureus* (Anderson *et al.,* 1986; Shappell *et al.,* 1990).

C. INTERCELLULAR ADHESION MOLECULES

Members of the immunoglobulin (Ig) superfamily of proteins on endothelial cells bind to the β_2-integrins expressed on leukocytes and play an important role in strengthening adhesive interactions between leukocytes and endothelial cells of postcapillary venules. Intercellular adhesion molecule 1 (ICAM-1) and ICAM-2 on endothelial cells are products of distinct, but homologous genes containing five and two Ig domains, respectively (Marlin and Springer, 1987; Staunton *et al.,* 1988, 1989). Immunoglobulin domains 1 and 2 of ICAM-1 are involved in binding LFA-1 (Staunton *et al.,* 1990), while Ig domain 3 of ICAM-1 mediates binding to Mac-1 (Diamond *et al.,* 1991). ICAM-1 is expressed at low levels on resting endothelial cells, but its expression is increased by various cytokines (IL-1, IFN-γ, and TNF-α). Increased expression is important for high affinity binding of leukocytes to postcapillary venules and is a prelude to subsequent egress into inflamed tissues (Dustin et al., 1986; Rothlein et al., 1988; Lo et al., 1992). Epithelial cells and stimulated leukocytes have also been shown to express ICAM-1 (Dustin et al., 1986). ICAM-2 is constitutively expressed on leukocytes and at a high level on resting endothelial cells, but its expression is not augmented by activation (de Fougerolles et al., 1991; Nortamo et al., 1991). It has been proposed that ICAM-2 plays a critical role in recirculation of resting lymphocytes (de Fougerolles et al., 1991).

Leukocytes enter the peripheral tissues through interaction with platelet/endothelial cell adhesion molecule-1 (PECAM-1 or CD31) (Muller, 1995). CD31 is another member of the Ig superfamily of glycoproteins that is involved in leukocyte endothelial cell interactions. It is expressed on the surface of monocytes, neutrophils, platelets, about 50% of circulating T cells, and natural killer cells. For neutrophils and monocytes, CD31 is known to play a major role in transmigration between endothelial cell junctions into the subendothelial matrix. This crucial role of CD31 is due to its spatial concentration at the junctions between vascular endothelial cells. Ligation of leukocyte CD31 with CD31 on endothelial cells (in the region of the tight junctions) is believed essential for endothelial transmigration of leukocytes.

D. α_4-INTEGRINS

The α_4-integrin subunit can pair with two distinct β subunits (β_1 and β_7). As heterodimers, the α_4-integrin pairs are associated with the microvillous tips of lymphocytes which initiate contact between lymphocytes and vascular endothelium (Butcher and Picker, 1996). This distribution of tethering molecules is similar to that observed for CD62L. The $\alpha_4\beta_1$-integrin heterodimer on lymphocytes binds VCAM-1 as its primary endothelial cell ligand. $\alpha_4\beta_1$ is the most abundant β_1-integrin on circulating lymphocytes (Salmi and Jalkanen, 1997) and is primarily involved with homing of memory lymphocytes into many extraintestinal inflammatory sites (Butcher and Picker, 1996). Naive lymphocytes are CD62L^{Hi}, $\alpha_4\beta_7^{Lo\text{-}Med}$, and CD11a/CD18$^+$ cells. This phenotype allows naive B or T cells to initiate their rolling contact with postcapillary venular endothelial cells primarily via CD62L, and to utilize CD11a/CD18 for eventual diapedesis into tissue. This allows CD62L to play the dominant role in controlling naive B- or T-cell trafficking through peripheral lymph nodes because naive cells are $\alpha_4\beta_7^{Lo\text{-}Med}$ (Butcher and Picker, 1996). However, naive cells also need to traffic through the Peyer's patches whose postcapillary venules express the following phenotype: CD62L ligand^{Lo}, MAdCAM-1^{Hi},

ICAM-1+, and ICAM-2+. In this instance, CD62L does not provide a tethering adhesion; instead, the high-level expression of MAdCAM-1 on the postcapillary venules of the Peyer's patches allows the $\alpha_4\beta_7$ integrins on the naive lymphocytes to mediate a tethering/rolling adhesion to MAdCAM-1, thus providing a mechanism for rolling adhesion prior to arrest and subsequent diapedesis (Butcher and Picker, 1996). Therefore, $\alpha_4\beta_7$ integrins are required for naive lymphocyte emigration into Peyer's patches but are not required for naive lymphocyte emigration into peripheral lymph nodes. Interestingly, $\alpha_4\beta_7$-integrin levels on naive cells are insufficient for binding to MAdCAM-1+, CD62L ligand- venules in the intestinal lamina propria. The CD62L^{Hi}, $\alpha_4\beta_7^{Lo-Med}$ phenotype thus ensures that naive B or T cells have access to both mucosal and peripheral secondary lymphoid tissues but not to mucosal effector sites (Butcher and Picker, 1996). However, $\alpha_4\beta_7^{Hi}$ memory lymphocytes can effectively interact with postcapillary venules in the intestinal lamina propria solely through the rolling/arresting interaction of $\alpha_4\beta_7$ and its counter endothelial ligand MAdCAM-1.

Another important dichotomy in lymphocyte trafficking is that of distinct mucosal and nonmucosal tissue traffic patterns for memory/effector lymphocytes. Memory or effector lymphocytes preferentially homing to skin or other nonmucosal sites, such as the inflamed central nervous system and heart, utilize CLA (interacting with CD62E on endothelial cells) or the $\alpha_4\beta_1$-integrin heterodimer (interacting with VCAM-1) instead of $\alpha_4\beta_7$ for rolling and arrest prior to diapedesis (Butcher and Picker, 1996).

III. Effects of Stress on Immunity

Various environmental and physiologic factors may create stress, which triggers a physiologic response to maintain equilibrium with the environment. Activation of the hypothalamic–pituitary axis including release of ACTH and secretion of glucocorticoids is one response to stress. There is a paucity of information directly comparing stress-induced changes in leukocyte trafficking patterns and immune responses to vaccination. The best examples of alterations in leukocyte trafficking and immune function associated with stress are studies of pregnancy and parturition of many species. Other examples of alterations in trafficking patterns and immune function have been demonstrated using neonates and various glucocorticoid administration models.

Periparturient and neonatal immunosuppression is suggested by the increased incidence and susceptibility of cows and calves to bacterial and viral infections during this period (Hill, 1981; Wilson, 1990). Hormonal changes during the periparturient period have been reported (Smith et al., 1973; Sasser et al., 1979; Saleem et al., 1992) and likely contribute to immunologic dysfunction. Increased plasma concentration of the endogenous opioids, β-endorphin and met-enkephalin, during the periparturient period in cows may also reduce immune function (Aurich et al., 1990). Plasma concentration of these opioids peaks at parturition and cows experiencing dystocia have significantly elevated concentrations of β-endorphin several hours postpartum compared to normal cows. Immunologic disturbances in cellular and humoral components of immune responses have been documented in cattle during the periparturient period (neutrophil chemokinesis, respiratory burst, phagocytosis, lymphocyte blastogenesis, and serum concentration of immunoglobulins) and related to the marked reduction in the ability of dairy cattle to respond to invasive microorganisms (Nagahata et al., 1988, 1992; Kehrli and Goff, 1989; Kehrli et al., 1989a,b; Saad et al., 1989; Sordillo and Babiuk, 1991; Sordillo et al., 1991, 1995; Sordillo and Peel, 1992; Cai et al., 1994; Detilleux et al., 1995).

Pregnancy has been postulated to result in suppression of cell-mediated immune function and enhancement of humoral immunity. As pregnancy progresses, Th1 cytokines [interferon γ (IFN-γ) and interleukin 2 (IL-2)] decrease and Th2 cytokines (IL-4, IL-5, IL-6, and IL-10) increase (Wegmann et al., 1993; Delassus et al., 1994). Estrogen and progesterone play roles as suppressers of cell-mediated immune responses and enhancers of humoral responses in mice (Screpanti et al., 1991). The change of immune status develops in parallel with increased plasma corticosteroid levels as well as increases in estrogen and progesterone. Corticosteroids are also known to suppress cell-mediated immune responses and enhance humoral responses by suppressing the production of Th1 cytokines such as IL-2 and IFN-γ (Vacca et al., 1992; De Rijk and Berkenbosch, 1994). Cytokine production by bovine leukocytes is also disturbed during the periparturient period in that IFN-γ and IL-2 production are reduced, suggesting suppression of Th1 immune responses (Sordillo et al., 1991; Ishikawa et al., 1994; Shafer-Weaver and Sordillo, 1997). Levels of IL-2 activity were lower and IFN-γ was not detectable in mammary gland secretions (colostrum) during the last week of gestation and at parturition when compared to levels detected 2 weeks prepartum, whereas tumor necrosis factor levels increased gradually during that period. It has been postulated that differential expression of cytokines might occur and that these changes in cytokine production might contribute to the increased incidence of disease in periparturient cows (Sordillo et al., 1992, 1995; Sordillo and Peel, 1992) and would certainly alter the host response to vaccination. Moreover, impairment of the capacity of B cells from periparturient cows to secrete IgM has been reported (Detilleux et al., 1995), which may stem from impaired IFN-γ and IL-2 production.

Leukocyte trafficking patterns change in periparturient cows; as the percentage of T cells declines [compared with cows in midlactation (~45% versus ~20%)] there is a concomitant increase in granulocytes and monocytes (Shafer-Weaver et al., 1996). Also, the proportion of $CD4^+$ cells in blood and mammary parenchyma declines postpartum, which is consistent with decreased IFN-γ secretion by lymphocytes and decreased IFN-γ in lacteal secretions of periparturient cows (Sordillo et al., 1991; Ishikawa et al., 1994; Shafer-Weaver et al., 1996; Shafer-Weaver and Sordillo, 1997; Yang et al., 1997). The immunoregulatory role of $CD8^+$ lymphocytes during the postpartum period is most likely of a suppressor nature because IL-4 mRNA was the main cytokine detectable in cultures of mononuclear cells from postpartum dairy cows while IFN-γ is the main cytokine detected from cows in middle to later stages of lactation (Shafer-Weaver and Sordillo, 1997). A neutrophilia during the immediate periparturient period is also known to be associated with reduced expression of CD62L on neutrophils in an inverse relationship with serum cortisol levels (Lee and Kehrli, 1997). These changes are consistent with the effects of glucocorticoid administration to cattle (Burton and Kehrli, 1995, 1996).

Dexamethasone causes a similar effect on IFN-γ and immunoglobulin production by mitogen-stimulated mononuclear cells, suggesting that stress induction of glucocorticoid release may impair TH1 immunity (Doherty et al., 1995; Nonnecke et al., 1997). Other studies have found that castration of bull calves causes a neutrophilia and impairment of IFN-γ production (Fisher et al., 1997a,b). Murine studies have found that B cell, NK cell, and monocyte numbers have greater stress-induced decreases than do T cells. Adrenalectomy has shown that endocrine factors (e.g., glucocorticoids) released during stress modulate this leukocyte trafficking between the blood and other immune compartments (Dhabhar et al., 1995). Cortisol or dexamethasone induce a neutrophilia in cattle without a significant increase in immature neutrophils (Burton et al., 1995), suggesting that glucocorticoids decrease the marginating pool of neutrophils, and reduce the efficiency of their egress from blood to tissues (Burton and Kehrli, 1995). Glucocorticoids also cause a gradual decline in neutrophil expression of β_2-integrins, which might contribute to some of the well-documented impairment of neutrophil complement receptor functions (Burton et al., 1995). Shedding of CD62L and concomitant neutrophilia suggests that one antiinflammatory mechanism of action of dexamethasone is to transiently preclude neutrophil egress. Studies of restraint stress in mice have also demonstrated suppressed migration of granulocytes and macrophages to inflammatory loci (Mizobe et al., 1997). Glucocorticoids in cattle also cause a dramatic decline in percentages of circulating γδ T cells (Burton and Kehrli, 1996; Nonnecke et al., 1997) and decrease in MHC class I and II molecules on remaining mononuclear cells. Since γδ T cells express high levels of CD62L (Walcheck and Jutila, 1994) and dexamethasone did not alter CD62L expression on γδ T cells (Burton and Kehrli, 1996), the putative egress of circulating γδ T cells into tissues during stress may represent a compensatory immunologic response when neutrophil egress is transiently impaired.

Where dexamethasone has been used to cause immunosuppression in calves, impairment of IgG responses to equine ferritin (delayed peak titer) and tetanus toxoid (lower peak titer) have been reported (Roth et al., 1984). Moreover, long-term administration of dexamethasone to horses can completely abrogate IgGa and IgGb responses to vaccination with nonreplicating viral antigens and yet have no effect on IgG(T) titers (Slack et al., 1997). Administration of dexamethasone and transport (800-km) stress of calves both were found to reduce expression of IL-2Rα on mononuclear cells compared to expression on mononuclear cells from control calves (Lan et al., 1995). Human patients with chronic fatigue syndrome have increased allergy problems and low production of IFN-γ and IL-2 (Rook and Zumla, 1997). In neonates, impairment of Th1 immunity has been hypothesized to result in a Th2 bias in immune responses (reviewed in Vince and Johnson, 1996) that may last in spite of subsequent boosting of immunity (Barrios et al., 1996).

Immunosuppressive and anti-inflammatory properties of glucocorticoids also include inhibition of endotoxin-stimulated cytokine production and gene expression (Hagan et al., 1992). Transcriptional regulation may be one mechanism by which glucocorticoids cause immunosuppression (Auphan et al., 1995; Scheinman et al., 1995). For example, glucocorticoids inhibit NF-κB activity by inducing transcription of the IκBα gene, thus blocking secretion of cytokines such as IL-1, IL-8, and TNF-α (Baeuerle, 1991; Grill et al., 1993). These cytokines normally up-regulate expression of β_2 integrins and down-regulate CD62L expression on leukocytes. Because plasma cortisol concentrations are known to increase during the immediate periparturient period (Smith et al., 1973; Guidry et al., 1976), and dexamethasone reduces CD62L and CD18 expression on bovine neutrophils (Burton et al., 1995), glucocorticoids could cause an increase in susceptibility to disease through their effects on adhesion molecules. Furthermore, ICAM-1 expression on endothelial cells and binding to neutrophils is reduced by dexamethasone *in vitro* (Cronstein *et al.,* 1992); this could contribute to a decline in host defense and an increased susceptibility to disease postpartum by impeding phagocyte egress. A marked reduction in expression of CD62L was found on the surface of human neonatal neutrophils, and was correlated with reduced ability of human neonatal neutrophils to adhere to endothelial cells *in vitro* (Anderson *et al.,* 1991).

IV. Summary

Increased susceptibility of animals to infectious disease during the periparturient period results in suffering and economic losses. Stress appears to delay inflammation by reducing efficiency of CD62L-mediated immune surveillance by phagocytes. It is important to note that the effects of stress are not limited to alteration of leukocyte trafficking patterns since various stressors (e.g., transport, parturition, and castration) also decrease IFN-γ secretion by lymphocytes, and may decrease antigen presentation efficiency by down-regulating class II molecule expression on antigen presenting cells, and delay or impair immune responses to vaccination. Documented immunosuppression in periparturient animals, particularly the bias toward Th2 immune responses, and also changes in general leukocyte trafficking patterns suggest that vaccination intending to elicit cell-mediated immunity may not be efficacious at this point of the production cycle. Based on findings of numerous periparturient studies on immunosuppression in cattle, waiting at least 30 days after parturition before administering routine vaccinations is recommended.

REFERENCES

Adams, D. H., and Shaw, S. (1994). Leukocyte-endothelial interactions and regulation of leukocyte migration. *Lancet* **343,** 831–836.

Albelda, S. M., and Buck, C. A. (1990). Integrins and other cell adhesion molecules. *FASEB J.* **4,** 2868–2880.

Anderson, D. C., Miller, L. J., Schmalstieg, F. C., Rothlein, R., and Springer, T. A. (1986). Contributions of the Mac-1 glycoprotein family to adherence-dependent granulocyte functions: Structure-function assessments employing subunit-specific monoclonal antibodies. *J. Immunol.* **137,** 15–27.

Anderson, D. C., Abbassi, O., Kishimoto, T. K., Koenig, J. M., McIntire, L. V., and Smith, C. W. (1991). Diminished lectin-, epidermal growth factor-, complement binding domain-cell adhesion molecule-1 on neonatal neutrophils underlies their impaired CD18-independent adhesion to endothelial cells in vitro. *J. Immunol.* **146,** 3372–3379.

Arnaout, M. A. (1990). Structure and function of the leukocyte adhesion molecules CD11/CD18. *Blood* **75,** 1037–1050.

Auphan, N., DiDonato, J. A., Rosette, C., Helmberg, A., and Karin, M. (1995). Immunosuppression by glucocorticoids: Inhibition of NF-κB activity through induction of IκB synthesis. *Science* **270,** 286–290.

Aurich, J. E., Dobrinski, I., Hoppen, H.-O., and Grunert, E. (1990). β-endorphin and met-enkephalin in plasma of cattle during pregnancy, parturition and the neonatal period. *J. Reprod. Fertil.* **89,** 605–612.

Baeuerle, P. A. (1991). The inducible transcription activator NF-κB: Regulation by distinct protein subunits. *Biochim. Biophys, Acta.* **1072,** 63–80.

Barrios, C., Brawand, P., Berney, M., Brandt, C., Lambert, P.-H., and Siegrist, C.-A. (1996). Neonatal and early life immune responses to various forms of vaccine antigens qualitatively differ from adult responses: Predominance of a Th2-biased pattern which persists after adult boosting. *Eur. J. Immunol.* **26,** 1489–1496.

Bevilacqua, M. P., and Nelson, R. M. (1993). Selectins. *J. Clin. Invest.* **91,** 379–387.

Bevilacqua, M. P., Pober, J. S., Mendrick, D. L., Cotran, R. S., and Gimbrone, M. A. J. (1987). Identification of an inducible endothelial-leukocyte adhesion molecule. *Proc. Natl. Acad. Sci. USA* **84,** 9238–9242.

Boyd, A. W., Wawryk, S. O., Burns, G. F., and Fecondo, J. V. (1988). Intercellular adhesion molecule 1 (ICAM-1) has a central role in cell-cell contact-mediated immune mechanisms. *Proc. Natl. Acad. Sci. USA* **85,** 3095–3099.

Burton, J. L., and Kehrli, M. E., Jr. (1995). Regulation of neutrophil adhesion molecules and shedding of *Staphylococcus aureus* in milk of cortisol- and dexamethasone-treated cows. *Am. J. Vet. Res.* **56,** 997–1006.

Burton, J. L., and Kehrli, M. E., Jr. (1996). Effects of dexamethasone on bovine circulating T lymphocyte populations. *J. Leukocyte Biol.* **59,** 90–99.

Burton, J. L., Kehrli, M. E., Jr., Kapil, S., and Horst, R. L. (1995). Regulation of L-selectin and CD18 on bovine neutrophils by glucocorticoids: Effects of cortisol and dexamethasone. *J. Leukocyte Biol.* **57,** 317–325.

Butcher, E. C. (1991). Leukocytes-endothelial cell recognition: three (or more) steps to specificity and diversity. *Cell* **67,** 1033–1036.

Butcher, E. C., and Picker, L. J. (1996). Lymphocyte homing and homeostasis. *Science* **272,** 60–66.

Cai, T.-Q., Weston, P. G., Lund, L. A., Brodie, B., McKenna, D. J., and Wagner, W. C. (1994). Association between neutrophil functions and periparturient disorders in cows. *Am. J. Vet. Res.* **55,** 934–943.

Clevers, H., MacHugh, N. D., Bensaid, A., Dunlap, S., Baldwin, C. L., Kaushal, A., Iams, K., Howard, C. J., and Morrison, W. I. (1990). Identification of a bovine surface antigen uniquely expressed on CD4⁻CD8⁻ T cell receptor γ/δ⁺ T lymphocytes. *Eur. J. Immunol.* **20,** 809–817.

Cronstein, B. N., Kimmel, S. C., Levin, R. I., Martiniuk, F., and Weissmann, G. (1992). A mechanism for the anti-inflammatory effects of corticosteroids: The glucocorticoid receptor regulates leukocyte adhesion to endothelial cells and expression of endothelial—leukocyte adhesion molecule 1 and intercellular adhesion molecule 1. *Proc. Natl. Acad. Sci. USA* **89,** 9991–9995.

de Fougerolles, A. R., Stacker, S. A., Schwarting, R., and Springer, T. A. (1991). Characterization of ICAM-2 and evidence for a third counter-receptor for LFA-1. *J. Exp. Med.* **174,** 253–267.

Delassus, S., Coutinho, G. C., Saucier, C., Darche, S., and Kourilsky, P. (1994). Differential cytokine expression in maternal blood and placenta during murine gestation. *J. Immunol.* **152,** 2411–2420.

De Rijk, R. H., and Berkenbosch, F. (1994). Suppressive and permissive actions of glucocorticoids: A way to control innate immunity and to facilitate specificity of adaptive immunity? *In* "Endocrinology and Metabolism: Bilateral Communication between the Endocrine and Immune Systems" (C. J. Grossman, ed.), pp. 73–93. Springer-Verlag, New York.

Detilleux, J. C., Kehrli, M. E., Jr., Stabel, J. R., Freeman, A. E., and Kelley, D. H. (1995). Study of immunological dysfunction in periparturient Holstein cattle selected for high and average milk production. *Vet. Immunol. Immunopathol.* **44,** 251–267.

Dhabhar, F. S., Miller, A. H., McEwen, B. S., and Spencer, R. L. (1995). Effects of stress on immune cell distribution: Dynamics and hormonal mechanisms. *J. Immunol.* **154,** 5511–5527.

Diamond, M. S. Staunton, D. E., Martin, S. D., and Springer, T. A. (1991). Binding of the integrin Mac-1 (CD11b/CD18) to the third immunoglobulin-like domain of ICAM-1 (CD54) and its regulation by glycosylation. *Cell* **65,** 961–971.

Doherty, M. L., Bassett, H. F., Quinn, P. J., Davis, W. C., and Monaghan, M. L. (1995). Effects of dexamethasone on cell-mediated immune responses in cattle sensitized to *Mycobacterium bovis*. *Am. J. Vet. Res.* **56,** 1300–1306.

Dustin, M. L., Rothlein, R., Bhan, A. K., Dinarello, C. A., and Springer, T. A. (1986). Induction by IL-1 and interferon-gamma: Tissue distribution, biochemistry, and function of a natural adherence molecule (ICAM-1). *J. Immunol.* **137,** 245–254.

Etzioni, A., Frydman, M., Pollack, S., Avidor, I., Phillips, M. L., Paulson, J. C., and Gershoni-Baruch, R. (1992). Brief report: Recurrent severe infections caused by a novel leukocyte adhesion deficiency. *N. Engl. J. Med.* **327,** 1789–1792.

Fisher, A. D., Crowe, M. A., O'Nuallain, E. M., Monaghan, M. L., Larkin, J. A., O'Kiely, P., and Enright, W. J. (1997a). Effects of cortisol on in vitro interferon-γ production, acute-phase proteins, growth, and feed intake in a calf castration model. *J. Anim. Sci.* **75,** 1041–1047.

Fisher, A. D., Crowe, M. A., O'Nuallain, E. M., Monaghan, M. L., Prendiville, D. J., O'Kiely, P., and Enright, W. J. (1997b). Effects of suppressing cortisol following castration of bull calves on adrenocorticotropic hormone, in vitro interferon-γ production, leukocytes, acute-phase proteins, growth, and feed intake. *J. Anim. Sci.* **75,** 1899–1908.

Geng, J.-G., Bevilacqua, M. P., Moore, K. L., McIntyre, T. M., Prescott, S. M., Kim, J. M., Bliss, G. A., Zimmerman, G. A., and McEver, R. P. (1990). Rapid neutrophil adhesion to activated endothelium mediated by GMP-140. *Nature* **343,** 757–760.

Ginsberg, M. H., Du, X., and Plow, E. F. (1992). Inside-out integrin signalling. *Curr. Opin. Cell Biol.* **4,** 766–771.

Gowans, J. L., and Knight, E. J. (1964). The route of recirculation of lymphocytes in the rat. *Proc. R. Soc. London Ser. B* **159,** 257–282.

Grilli, M., Chiu, J. S., and Leonard, M. J. (1993). NF-κB and Rel: Participants in a multiform transcriptional regulatory system. *Int. Rev. Cytol.* **143,** 1–62.

Guidry, A. J., Paape, M. J., and Pearson, R. E. (1976). Effects of parturition and lactation on blood and milk cell concentrations, corticosteroids and neutrophil phagocytosis in the cow. *Am. J. Vet. Res.* **37,** 1195–1200.

Hagan, P., Poole, S., and Bristow, A. F. (1992). Immunosuppressive activity of corticotrophin-releasing factor. *Biochem. J.* **281,** 251–254.

Harp, J. A., Pesch, B. A., and Runnels, P. L. (1990). Extravasation of lymphocytes via paracortical venules in sheep lymph nodes: Visualization using an intracellular fluorescent label. *Vet. Immunol. Immunopathol.* **24,** 159–167.

Hein, W. R., and Mackay, C. R. (1991). Prominence of γδ T cells in the ruminant immune system. *Immunol. Today* **12,** 30–34.

Hibbs, M. L., Jakes, S., Stacker, S. A., Wallace, R. W., and Springer, T. A. (1991a). The cytoplasmic domain of the integrin lymphocyte function-associated antigen 1 β subunits: Sites required for binding to intercellular adhesion molecule 1 and the phorbol ester-stimulated phosphorylation site. *J. Exp. Med.* **174,** 1227–1238.

Hibbs, M. L., Xu, H., Stacker, S. A., and Springer, T. A. (1991b). Regulation of adhesion to ICAM-1 by the cytoplasmic domain of LFA-1 integrin β subunits. *Science* **251,** 1611–1613.

Hill, A. W. (1981). Factors influencing the outcome of *Escherichia coli* mastitis in the dairy cow. *Res. Vet. Sci.* **31,** 107–112.

Hynes, R. O. (1992). Integrins: Versatility, modulation, and signaling in cell adhesion. *Cell* **69,** 11–25.

Imhof, B. A., and Dunon, D. (1995). Leukocyte migration and adhesion. *Adv. Immunol.* **85,** 345–416.

Ishikawa, H., Shirahata, T., and Hasegawa, K. (1994). Interferon-γ production of mitogen stimulated peripheral lymphocytes in perinatal cows. *J. Vet. Med. Sci.* **56,** 735–738.

Johnston, G. I., Cook, R. G., and McEver, R. P. (1989). Cloning of GMP-140, a granule membrane protein of platelets and endotheium: Sequence similarity to proteins involved in cell adhesion and inflammation. *Cell* **56,** 1033–1044.

Kansas, G. S., Ley, K., Munro, J. M., and Tedder, T. F. (1993). Regulation of leukocyte rolling adhesion to high endothelial venules through the cytoplasmic domain of L-selectin. *J. Exp. Med.* **177,** 833–838.

Kehrli, M. E., Jr., and Goff, J. P. (1989). Periparturient hypocalcemia in cows: Effects on peripheral blood neutrophil and lymphocyte function. *J. Dairy Sci.* **72,** 1188–1196.

Kehrli, M. E., Jr., Nonnecke, B. J., and Roth, J. A. (1989a). Alterations in bovine neutrophil function during the periparturient period. *Am. J. Vet. Res.* **50,** 207–214.

Kehrli, M. E., Jr., Nonnecke, B. J., and Roth, J. A. (1989b). Alterations in bovine lymphocyte function during the periparturient period. *Am. J. Vet. Res.* **50,** 215–220.

Keizer, G. D., Visser, W., Vliem, M., and Figdor, C. G. (1988). A monoclonal antibody (NKI-L16) directed against a unique epitope on the α-chain of human leukocyte function-associated antigen-1 induces homotypeic cell-cell interactions. *J. Immunol.* **140,** 1393–1400.

Kishimoto, T. K. (1991). A dynamic model for neutrophil localization to inflammatory sites. *NIH Res.* **3,** 75–77.

Kishimoto, T. K., Jutila, M. A., Berg, E. L., and Butcher, E. C. (1989). Neutrophil Mac-1 and MEL-14 adhesion proteins inversely regulated by chemotactic factors. *Science* **245,** 1238–1241.

Lan, H. C., Reddy, P. G., Chambers, M. A., Walker, G., Srivastava, K. K., and Ferguson, J. A. (1995). Effect of stress on interleukin-2 receptor expression by bovine mononuclear leukocytes. *Vet. Immunol. Immunopathol.* **49,** 241–249.

Lasky, L. A. (1992). Selectins: Interpreters of cell-specific carbohydrate information during inflammation. *Science* **258,** 964–969.

aLee, E.-K., and Kehrli, M. E., Jr. (1997). Expression of adhesion molecules on neutrophils of periparturient cows and neonatal calves. *Am. J. Vet. Res.* **59** (in press).

Lo, S. K., Everitt, J., Gu, J., and Malik, A. B. (1992). Tumor necrosis factor mediates experimental pulmonary edema by ICAM-1 and CD18-dependent mechanisms. *J. Clin. Invest.* **89,** 981–988.

Marlin, S. D., and Springer, T. A. (1987). Purified intercellular adhesion molecule-1 (ICAM-1) is a ligand for lymphocyte function-associated antigen 1 (LFA-1). *Cell* **51,** 813–819.

Miller, L. J., Bainton, D. F., Borreguard, N., and Springer, T. A. (1987). Stimulated mobilization of monocyte Mac-1 and p150,95 adhesion proteins from an intracellular vesicular compartment to the cell surface. *J. Clin. Invest.* **80,** 535–544.

Mizobe, K., Kishihara, K., Ezz-Din El-Naggar, R., Madkour, G. A., Kubo, C., and Nomoto, K. (1997). Restraint stress-induced elevation of endogenous glucocorticoid suppresses migration of granulocytes and marcophages to an inflammatory locus. *J. Neuroimmunol.* **73,** 81–89.

Montgomery, K. F., Osborn, L., Hession, C., Tizard, R., Goff, D., Vassallo, C., Tarr, P. I., Bomsztky, K., Lobb, R., Harlan, J. M., and Pohlman, T. H. (1991). Activation of endothelial-leukocyte adhesion molecule 1 (ELAM-1) gene transcription *Proc. Natl. Acad. Sci. USA* **88,** 6523–6527.

Morrison, W. I., and Davis, W. C. (1991). Differentiation antigens expressed predominantly on CD4-CD8-T lymphocytes (WC1, WC2). *Vet. Immunol. Immunopathol.* **27,** 71–76.

Muller, W. A. (1995). The role of PECAM-1 (CD31) in leukocyte emigration: Studies in vitro and vivo. *J. Leukocyte Biol.* **57,** 523–528.

Nagahata, H., Makino, S., Takeda, S., Takahashi, H., and Noda, H. (1988). Assessment of neutrophil function in the dairy cow during the perinatal period. *J. Vet. Med. B* **35,** 747–751.

Nagahata, H., Ogawa, A., Sanada, Y., Noda, H., and Yamamoto, S. (1992). Peripartum changes in antibody producing capability of lymphocytes from dairy cows. *Vet. Q.* **14,** 39–40.

Nonnecke, B. J., Burton, J. L., and Kehrli, M. E., Jr. (1997). Associations between function and composition of blood mononuclear leukocyte populations from dexamethasone-treated Holstein bulls. *J. Dairy Sci.* **80,** 2403–2410.

Nortamo, P., Salcedo, R., Timonen, T., Patarroto, M., and Gahmberg, C. G. (1991). A monoclonal antibody to the human leukocyte adhesion molecule intercellular adhesion molecule-2. Cellular distribution and molecular characterization of the antigen. *J. Immunol.* **146,** 2530–2535.

Pardi, R., Inverardi, L., Rugardi, C., and Bender, J. R. (1992). Antigen-receptor complex stimulation triggers protein kinase C-dependent CD11a/CD18-cytoskeleton association in T lymphocytes. *J. Cell Biol.* **116,** 1211–1220.

Randi, A. M., and Hogg, N. (1994). I domain of β2 integrin lymphocyte function-associated antigen-1 contains a binding site for ligand intercellular adhesion molecule-1. *J. Biol. Chem.* **269,** 12395–12398.

Rook, G. A. W., and Zumla, A. (1997). Gulf War syndrome: Is it due to a systemic shift in cytokine balance towards a Th2 profile? *Lancet* **349,** 1831–1833.

Roth, J. A., and Kaeberle, M. L. (1981). Effects of in vivo dexamethasone administration on in vitro bovine polymorphonuclear leukocyte function. *Infect. Immun.* **33,** 434–441.

Roth, J. A. Kaeberle, M. L., and Hubbard, R. D. (1984). Attempts to use thiabendazole to improve the immune response in dexamethasone-treated or stressed cattle. *Immunopharmacology* **8,** 121–128.

Rothlein, R., Czajkowski, M., O'Neil, M. M., Marlin, S. D., Mainolki, E., and Merluzzi, M. J. (1988). Induction of intercellular adhesion molecule-1 on primary and continuous cell lines by proinflammatory cytokines. Regulation by pharmcologic agents and neutralizing antibodies. *J. Immunol.* **141,** 1665–1669.

Rothlein, R., Barton, R. W., and Winquist, R. (1991). The role of intercellular adhesion molecule-1 (ICAM-1) in the inflammatory response. *In* "Cellular and Molecular Mechanisms of Inflammation" (C. Cochrane and M. Gimbrone, Jr., eds.), Vol. 1, pp. 171–180. Academic Press, San Diego, CA.

Saad, A. M., Concha, C., and ström, G. (1989). Alterations in neutrophil phagocytosis and lymphocyte blastogenesis in dairy cows around parturition. *J. Vet. Med. B* **36,** 337–345.

Saleem, M. A., Jha, P., Buckshee, K., and Farooq, A. (1992). Studies on the immunosuppressive role of steroid hormones during pregnancy. *Immunol. Invest.* **21,** 1–10.

Salmi, M., and Jalkanen, S. (1997). How do lymphocytes know where to go: Current concepts and enigmas of lymphocyte homing. *In* "Advances In Immunology" (F. J. Dixon *et al.*, eds.), pp. 139–218. Academic Press, Sand Diego, CA.

Sasser, R. G., Falk, D. E., and Ross, R. H. (1979). Estrogen in plasma of parturient paretic and normal cows. *J. Dairy Sci.* **62,** 551–556.

Scheinman, R. I., Cogswell, P. C., Lofquist, A. K., and Baldwin, A. S. J. (1995). Role of transcriptional activation of IκBα in mediation of immunosuppression by glucocorticoids. *Science* **270,** 283–286.

Screpanti, I., Meco, D., Morrone, S., Gulino, A., Mathieson, B. J., and Frati, L. (1991). In vivo modulation of the distribution of thymocyte subsets: Effects of estrogen on expression of different T cell receptor Vβ gene families in CD4-, CD8- thymocytes. *Cell Immunol.* **134,** 414–426.

Shafer-Weaver, K. A., and Sordillo, L. M. (1997). Bovine CD8+ suppressor lymphocytes alter immune responsiveness during the postpartum period. *Vet. Immunol. Immunopathol.* **56,** 53–64.

Shafer-Weaver, K. A., Pighetti, G. M., and Sordillo, L. M. (1996). Diminished mammary gland lymphocyte functions parallel shifts in trafficking patterns during the postpartum period. *Proc. Soc. Exp. Biol. Med.* **212,** 271–279.

Shappell, S. B., Toman, C., Anderson, D. C., Taylor, A. A., Entman, M. L., and Smith, C. W. (1990). Mac-1 (CD11b/CD18) mediates adherence-dependent hydrogen peroxide production by human and canine neutrophils. *J. Immunol.* **144,** 2702–2711.

Slack, J. A., Risdahl, J., Valberg, S., Murphy, M., Schram, B. R., and Lunn, D. P. (1997). Effects of corticosteroids one equine immune responses to vaccination. *Conf. Res. Workers Anim. Dis., Proc. Symp., 78th,* Chicago, p. 116.

Smith, C. W., Marlin, S. D., Rothlein, R., Toman, C., and Anderson, D. C. (1989). Cooperative interactions of LFA-1 and Mac-1 with intercellular adhesion molecule-1 in facilitating adherence and transendothelial migration of human neutrophils in vitro. *J. Clin. Invest.* **83,** 2008–2017.

Smith, V. G., Edgerton, L. A., Hafs, H. D., and Convey, E. M. (1973). Bovine serum estrogens, progestins and glucocorticoids during late pregnancy, parturition and early lactation. *J. Anim. Sci.* **36,** 391–396.

Sonnenberg, A. (1993). Integrins and their ligands. *Curr. Top. Microbiol. Immunol.* **184,** 7–35.

Sordillo, L. M., and Babiuk, L. A. (1991). Controlling acute *Escherichia coli* mastitis during the periparturient period with recombinant bovine interferon gamma. *Vet. Microbiol.* **28,** 189–198.

Sordillo, L. M., and Peel, J. E. (1992). Effects of interferon-γ on the production of tumor necrosis factor during acute *Escherichia coli* mastitis. *J. Dairy Sci.* **75,** 2119–2125.

Sordillo, L. M., Redmond, M. J., Campos, M., Warren, L., and Babiuk, L. A. (1991).

Cytokine activity in bovine mammary gland secretions during the periparturient period. *Can. J. Vet. Res.* **55,** 298–301.

Sordillo, L. M., Afseth, G., Davies, G., and Babiuk, L. A. (1992). Effects of recombinant granulocyte-macrophage colony-stimulating factor on bovine peripheral blood and mammary gland neutrophil function *in vitro. Can. J. Vet. Res.* **56,** 16–21.

Sordillo, L. M., Pighetti, G. M., and Davis, M. R. (1995). Enhanced production of bovine tumor necrosis factor-α during the periparturient period. *Vet. Immunol. Immunopathol.* **49,** 263–270.

Spalding, H. J., and Heath, T. J. (1989). Blood vessels within lymph nodes: A comparison between pigs and sheep. *Res. Vet. Sci.* **46,** 43–48.

Stanley, P., Bates, P. A., Harvey, J., Bemmett, R. I., and Hogg, N. (1994). Integrin LFA-1 α subunit contains an ICAM-1 binding site in domain V and VI. *EMBO J.* **13,** 1790–1798.

Staunton, D. E., Marlin, S. D., Stratowa, C., Dustin, M. L., and Springer, T. A. (1988). Primary structure of ICAM-1 demonstrates interaction and integrin supergene families. *Cell* **52,** 925–933.

Stauton, D. E., Dustin, M. L., and Springer, T. A. (1989). Functional cloning of ICAM-2, a cell adhesion ligand for LFA-1 homologous to ICAM-1. *Nature* **339,** 61–64.

Staunton, D. E., Dustin, M. L., Erickson, H. P., and Springer, T. A. (1990). The arrangement of the immunoglobulin-like domains of ICAM-1 and the binding sites for LFA-1 and rhinovirus. *Cell* **61,** 243–254.

Tedder, T. F., Penta, A. C., Levine, H. B., and Freedman, A. S. (1990). Expression of the human leukocyte adhesion molecule, LAM1. Identity with the TQ1 and Leu-8 differentiation antigens. *J. Immunol.* **144,** 532–540.

Tedder, T. F., Steeber, D. A., Chen, A., and Engel, P. (1995). The selectins: Vascular adhesion molecules. *FASEB J.* **9,** 866–873.

Vacca, A., Felli, M. P., Farina, A. R., Martinotti, S., Maroder, M., Screpanti, I., Meco, D., Petrangeli, E. L. F., and Gulino, A. (1992). Glucocorticoid receptor mediated suppression of the interleukin-2 gene expression through impairment of the cooperatively between nuclear factor of activated T cells and AP1 enhancer elements. *J. Exp. Med.* **175,** 637–646.

Varki, A. (1994). Selectin ligands. *Proc. Natl. Acad. Sci. USA* **91,** 7390–7397.

Vince, G. S., and Johnson, P. M. (1996). Is there a Th2 bias in human pregnancy? *J. Reprod. Immunol.* **32,** 101–104.

von Andrian, U. H., Chambers, J. D., McEnvoy, L. M., Bargatze, R. F., Arfors, K. E., and Butcher, E. C. (1991). Two-step model of leukocyte-endothelial cell interaction in inflammation: Distinct roles for LECAM-1 and the leukocyte beta-2 integrins in vivo. *Proc. Natl. Acad. Sci. USA* **88,** 7538–7542.

von Andrian, U. H., Chambers, J. D., Berg, E. L., Michie, S. A., Brown, D. A., Karolak, D., Ramezani, L., Berger, E. M., Arfors, K.-E., and Butcher, E. C. (1993). L-selectin mediates neutrophil rolling in inflammed venules through sialyl Lewis^x-dependent and -independent recognition pathways. *Blood* **82,** 182–191.

Walcheck, B., and Jutila, M. A. (1994). Bovine gamma delta T cells express high levels of functional peripheral lymph node homing receptor (L-selectin). *Int. Immunol.* **6,** 81–91.

Wegmann, T. G., Lin, H., Guilbert, L., and Mosmann, T .R. (1993). Bidirectional cytokine interactions in the maternal-fetal relationship: Is successful pregnancy a TH2 phenomenon? *Immunol. Today* **14,** 353–356.

Wilson, C. B. (1990). Developmental immunology and role of host defenses in neonatal susceptibility. *In* "Infectious Disease of the Fetus and Newborn Infant" (J. S. Remington and J. O. Klein, eds.), pp. 17–67. Saunders, Philadelphia.

Yang, T. J., Ayoub, I. A., and Rewinski, M. J. (1997). Lactation stage-dependent changes of lymphocyte subpopulations in mammary secretions: Inversion of CD4+/CD8+ T cell ratios at parturition. *Am. J. Reprod. Immunol.* **37,** 378–383.

Zimmerman, G. A. (1992). Endothelial cell interaction with granulocyte: Tethering and signaling molecules. *Immunol. Today* **13,** 93–112.

Role of Macrophage Cytokines in Mucosal Adjuvanticity

DENNIS L. FOSS AND MICHAEL P. MURTAUGH

Department of Veterinary Pathobiology, University of Minnesota, St. Paul, Minnesota 55108

I. Introduction
II. Mucosal Adjuvanticity of Cholera Toxin
   A. Structure and Function of Cholera Toxin and Related Toxins
   B. Mucosal Immunogenicity and Adjuvanticity of Cholera and Related Toxins
III. Mechanisms of Mucosal Adjuvanticity
   A. Cholera Toxin
   B. Dual-Signal Integration
   C. Role of Interleukin 1 in Mucosal Adjuvanticity
   D. Role of Interleukin 12 in Mucosal Adjuvanticity
IV. Summary
   References

I. Introduction

The mucosal immune system is in constant contact with vast numbers and varieties of antigens, both living and dead, dangerous and benign. The ability to selectively mount appropriate immunity to potentially harmful antigens and organisms, while at the same time tolerating beneficial ones, is essential to the maintenance of homeostasis. This conflict in function is especially problematic in the immune system of the intestinal tract (gut-associated lymphoid tissue or GALT), where immunity to pathogens must coexist with a need to control inflammatory responses to the lumenal flora and dietary antigens. Understanding the molecular basis of this decision will aid the rational development of mucosally delivered vaccines.

After reviewing studies of the mucosal adjuvant cholera toxin (CT), this chapter describes the ability of CT to adjuvant mucosal immunity to related antigens, such as its B subunit (CT-B), as well as unrelated antigens when administered orally or intragastrically. The second part is a brief overview of cellular and molecular mechanisms for the adjuvanticity of CT, including the role of macrophage cytokines, including interleukin 1β (IL-1β) and interleukin 12.

II. Mucosal Adjuvanticity of Cholera Toxin

Most orally administered protein antigens fail to generate detectable immunity. In contrast, some proteins such as toxins produced by enteric pathogens are able to induce robust immunity when administered orally. The best characterized of these proteins is the toxin of *Vibrio cholerae*, cholera toxin (Holmgren et al., 1996; Snider, 1995; Spangler, 1992). This unique ability to induce immunity following oral administration makes cholera toxin and related toxins potential adjuvants for oral vaccines and valuable tools for the study of mucosal immunology.

A. STRUCTURE AND FUNCTION OF CHOLERA AND RELATED TOXINS

Cholera toxin is one of a group of bacterial toxins that function via adenosine diphosphate (ADP) ribosylation of host G proteins. In addition to cholera toxin, this group consists of heat labile toxin of *Escherichia coli* and pertussis toxin (Kochnolte et al., 1996). The basic structure and function of CT and of the heat labile toxin of *E. coli* (LT) are similar, with 80% homology on the nucleotide level (Spangler, 1992). However LT tends to cause a milder disease than CT in humans (LTh-I) and causes diarrhea in swine (LTp) (Gyles and Barnum, 1969). Cholera toxin is a multimeric protein secreted by *V. cholerae* consisting of a pentamer of 11.6kDa B-subunits (CT-B) and a single 27-kDa A-subunit (CT-A). The CT-B subunit binds monosialoganglioside (G_{M1}), found on mammalian epithelial and hematopoietic cells, whereas the CT-A subunit enters the cytosol, is activated by cleavage of a disulfide bond, and catalyzes the ADP ribosylation of a G protein ($G_{s\alpha}$), increasing cyclic adenosine monophosphate (cAMP) and resulting in hypersecretion of salt and water and severe diarrhea (Spangler, 1992). This cAMP elevating activity has been the generally assumed mechanism for the adjuvant activity of CT (Snider, 1995). The importance of the CT-A induced ADP ribosylation has also been demonstrated by targeting of CT-A to B cells through an Ig binding domain (Agren et al., 1997). In this case CT-A specifically targeted to B cells had adjuvant effects very similar to those of CT. On the other hand, some recent experiments with CT or LT mutants have demonstrated adjuvant activity in the absence of toxicity (deHaan et al., 1996b; Bergquist et al., 1997). However, these adjuvants have generally been used intranasally where the requirements for adjuvanticity may be different. Signaling pathways other than cAMP have also been demonstrated for CT. For example, the ability of CT to induce IL-1 and IL-6 in human peripheral blood mononuclear cells (PBMC) was only minimally inhibited by blocking cAMP or phosphokinase (PK)A pathways, while being completely blocked by inhibition of PKC, suggesting induction of cytokines via a PKC-mediated signaling pathway (Krakauer, 1996). CT induced ADP ribosylation also has cellular effects independent of cAMP in B cells, because CT-A synergized with IL-4 or ionomycin to induce proliferation (Francis et al., 1995). Binding of CT and CT-B to the surface of cells may also induce intracellular signals regardless of the presence of CT-A. Binding of G_{M1} ganglioside induces proliferation mediated by calcium influx and nuclear expression of AP-1, a TRE binding transcription factor, but not elevated cAMP (Buckley et al., 1995).

B. MUCOSAL IMMUNOGENICITY AND ADJUVANTICITY OF CHOLERA AND RELATED TOXINS

Cholera toxin is an excellent oral immunogen in many species including mice (Elson and Ealding, 1984), rats (Pierce, 1978), and dogs (Pierce et al., 1980). Recovered human patients have detectable anti-CT IgG in serum for years after clinical cholera (Spangler, 1992).

Orally administered CT-B is weakly immunogenic. Studies showing a strong response to CT-B used purified CT-B containing small amounts of CT (Pierce, 1978). Immunization of mice with purified CT-B results in poor sIgA production, while addition of small amounts of CT markedly enhances the response (Lycke et al., 1989b). Oral administration of 25 μg of recombinant LT-B to mice results in antibody responses detected in feces and serum (Nakagawa et al., 1996). Even so, the dose required was much higher than that required for immunogenicity of CT where 2 μg is immunogenic (Lycke and Holmgren, 1986). Inclusion of 50 ng of LT with 2 μg of recombinant LT-B results in secreted IgA at distance mucosal sites including reproductive and respiratory tracts, while LT-B alone results in only intestinal IgG and IgA (deHaan et al., 1996a).

CT is an effective adjuvant for orally administered CT-B at doses much lower than are immunogenic alone. In mice as little as 20 ng of CT can act as adjuvant for 10 µg of CT-B, while as much as 500 ng of CT alone does not (Lycke and Holmgren, 1986). Likewise, as little as 2 ng of LT enhances immunity to 2 µg of LT-B (deHaan et al., 1996a). There are significant differences between species in the adjuvanticity of CT and CT-B. In humans oral CT-B is generally considered to be immunogenic. Oral cholera vaccine containing CT-B results in systemic and local (intestinal) antibody response (D. Clemens et al., 1990; J. D. Clemens et al., 1987; Svennerholm et al., 1984), but inconsistent response in distant mucosal sites including vaginal washes (Wassen et al., 1996) or in saliva (Jertborn et al., 1986) and no detectable secretory antibodies in breast milk (Clemens et al., 1990). While immunogenic orally in humans, CT-B shows no adjuvant effect for the coadministered, whole-cell portion of the cholera vaccine as measured by vibriocidal antibodies (Clemens et al., 1987). In contrast to oral administration, intranasally administered CT-B results in systemic (serum) as well as distant mucosal (vaginal) anti-CT-B responses (Bergquist et al., 1997).

While extensively studied in other species the efficacy of cholera toxin as an oral immunogen and adjuvant has not been adressed in swine. Large doses of CT (500 µg) administered directly to porcine stomach or ileum induce CT-specific antibody secreting cells in the small intestine (Loftager et al., 1995). However, the ability of CT to function as an adjuvant for CT-B or other antigens has not been defined in swine.

Since species differ in the mucosal immunogenicity and adjuvanticity of CT-B and CT we have studied the immune response to them in swine, measured as systemic and mucosal anti-CT-B IgA, both at the site of immunization and at a distant mucosal surface. Recombinant cholera toxin B-subunit, obtained from a human cholera vaccine (National Bacteriological Laboratory, Stockholm, Sweden) was administered three times at 7- to 10-day intervals at the vaccine dose of 1 mg. Cholera toxin (Sigma Chemical Co., St. Louis, MO) was administered in doses from 1 to 100 µg per animal per time point.

While most animals receiving three doses of 1 mg of CT-B alone had detectable anti-CT-B serum antibodies at 5 days after the third immunization, this response was markedly enhanced by coadministration of whole CT. For example, only animals that received at least 10 µg of CT along with 1 mg of CT-B had detectable titers of anti-CT-B serum IgA 5 days after the second administration (Fig. 1, day 15). Conversely, 10 µg of CT alone did not result in detectable anti-CT-B IgA or IgG in any animals, even after the third administration (Fig. 1, day 25). Since the adjuvant in this case (CT) also contains the antigen to which the response is being measured (CT-B), higher doses of CT alone would be expected to result in an immune response. Indeed, at the highest dose of CT (100 µg), similar responses were seen with or without the coadministration of CT-B (Fig. 2).

Mucosal anti-CT-B sIgA was also induced. Intestinal (jejunal) mucus collected 5 days after the third immunization contained anti-CT-B antibodies, with the response being primarily IgA. Animals that received three doses of 1 mg of CT-B orally had detectable anti-CT-B IgA in the jejunal mucus and the levels were increased by coadministration of 10 or 100 µg of CT (Fig. 3). Local production of this IgA was confirmed by enumeration of anti-CT-B IgA antibody secreting cells isolated from the lamina propria (Fig. 4). Anti-CT-B IgA was also detected in the saliva of immunized animals (Fig. 5). Doses of 10–100 µg of CT increased the detected anti-CT-B response, whereas the response in animals receiving 1 µg was similar to those receiving CT-B alone.

The ability of CT to act as an adjuvant to coadministered antigens has been well documented in mice. Intragastric administration of 5 mg of keyhole limpet hemocyanin (KLH) results in little or no response, while the addition of 10 µg of CT results in both local (intestinal wash) and systemic (serum) anti-KLH IgA and IgG (Lycke and Holmgren, 1986; Elson and Ealding, 1984). However, dosages for both the antigen and the adjuvant effect of CT are increased by orders of magnitude over that described above for CT-B. The adjuvant effect of CT for 10 µg of CT-B was seen with doses of 20 ng, while 10 µg was required as adjuvant for 5 mg of KLH. In contrast to CT, 10 µg of CT-B added to the KLH did not result in a detectable immune response. However, inclusion of CT (0.5 µg) with 10 µg of CT-B was as effective as 10 µg of CT in stimulating the response to KLH (Wilson et al., 1990).

Because CT has been shown to adjuvant other coadministered antigens in other species, we tested its ability to do so in swine. Serum anti-CT-B and anti-KLH IgA were determined following three weekly oral immunizations with KLH (25 mg) with or without the coadministration of CT (50 µg) and CT-B (1 mg). An additional animal also was immunized with KLH (2 mg) by a parenteral route (intramuscular). Oral CT and CT-B resulted in anti-CT-B IgA in the serum, whereas oral administration of KLH alone or with CT and CT-B did not result in detectable anti-KLH response, while parenterally administered KLH resulted in a robust response (Fig. 6). Therefore, while

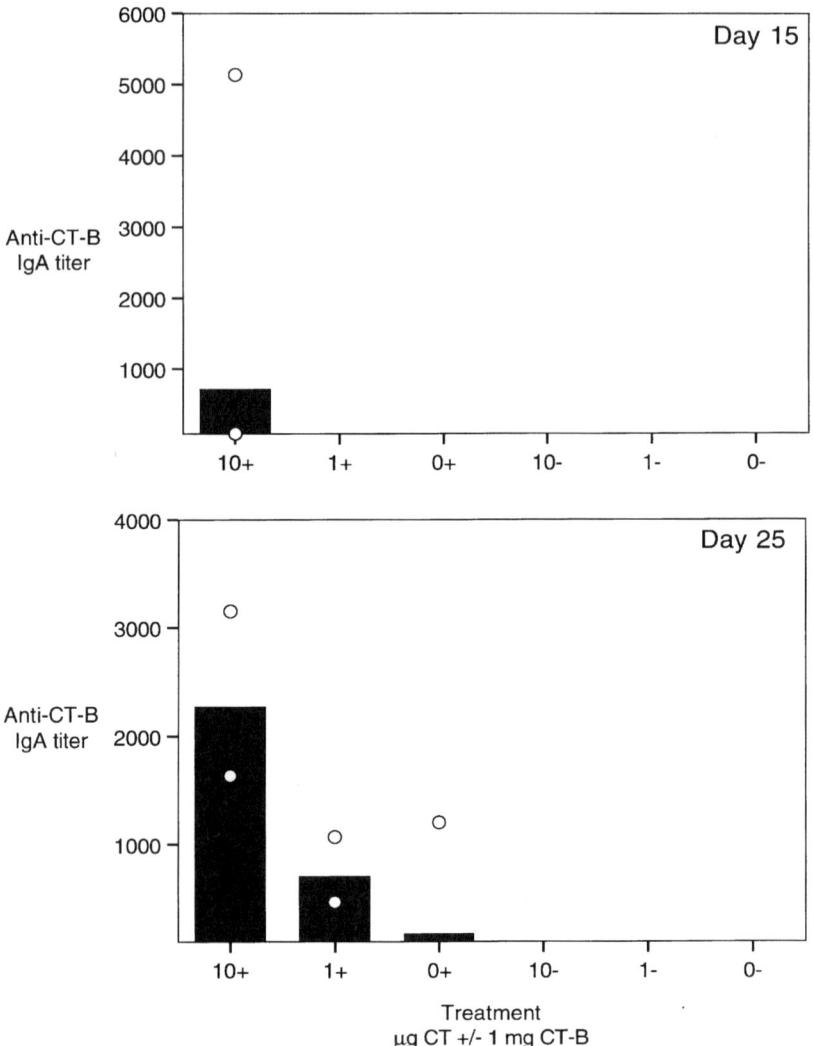

FIG. 1. Serum anti-CT-B IgA following oral administration of CT and CT-B. Treatments consisted of cholera toxin (μg) with (+) or without (−) 1 mg of CT-B, administered by gavage on days 0, 10, and 20, following neutralization of gastric acid. Serum anti-CT-B IgA was measured by enzyme-linked immunosorbent assay (ELISA) with a positive titer defined as the dilution of serum with an A_{450} of 0.5 over background. Bars represent the geometric mean titer with individual animals indicated with a circle.

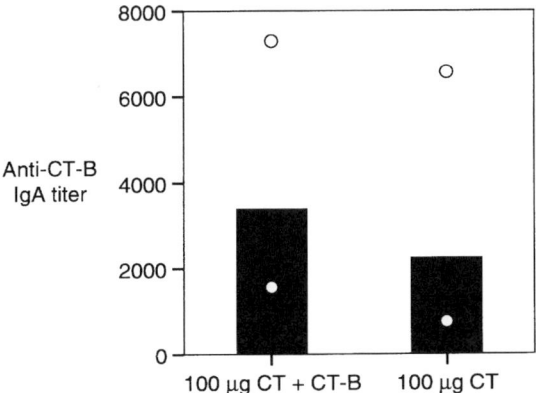

FIG. 2. Serum anti-CT-B IgA following oral administration of CT and CT-B. Treatments consisted of cholera toxin (100 μg) with or without 1 mg of CT-B, administered by gavage on days 0, 10, and 20, following neutralization of gastric acid. Serum anti-CT-B IgA was measured on day 25 by ELISA as described in Fig. 1.

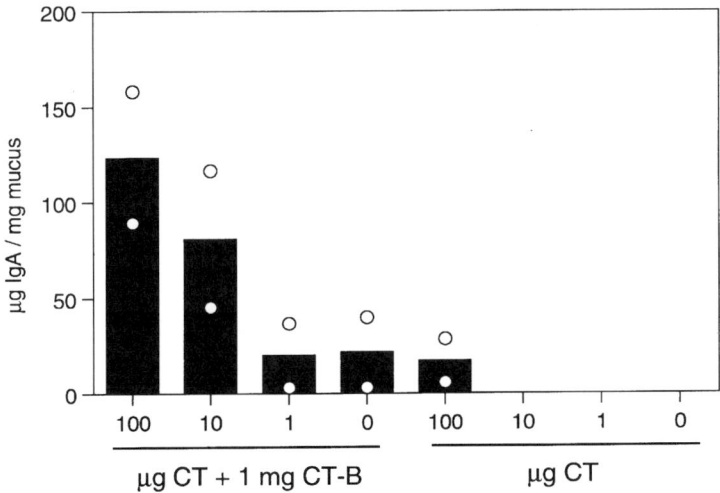

FIG. 3. Jejunal mucus anti-CT-B IgA following oral administration of CT and CT-B. Treatments were administered as described in Fig. 1. Jejunal mucus was collected by absorbent wicks; anti-CT-B IgA determined by ELISA and comparison to porcine IgA standards.

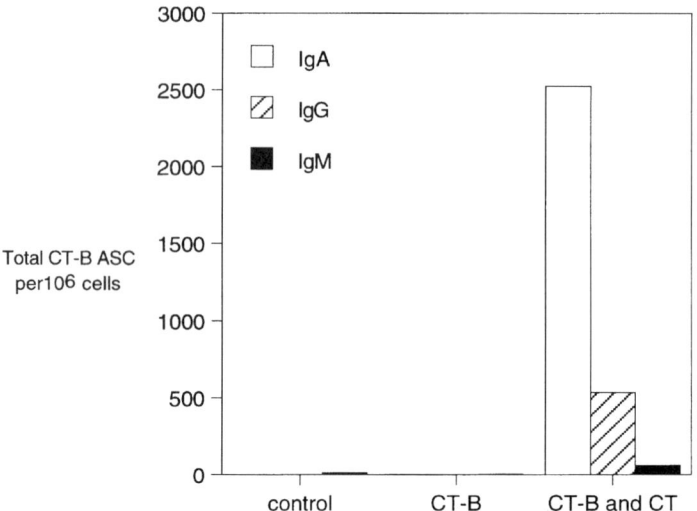

Fig. 4. Local production of anti-CT-B IgA, IgG, and IgM. Jejunal lamina propria anti-CT-B IgA, IgG, and IgM secreting cells were enumerated from the lamina propria of animals that received three doses of CT-B (1 mg) with or without CT (100 μg).

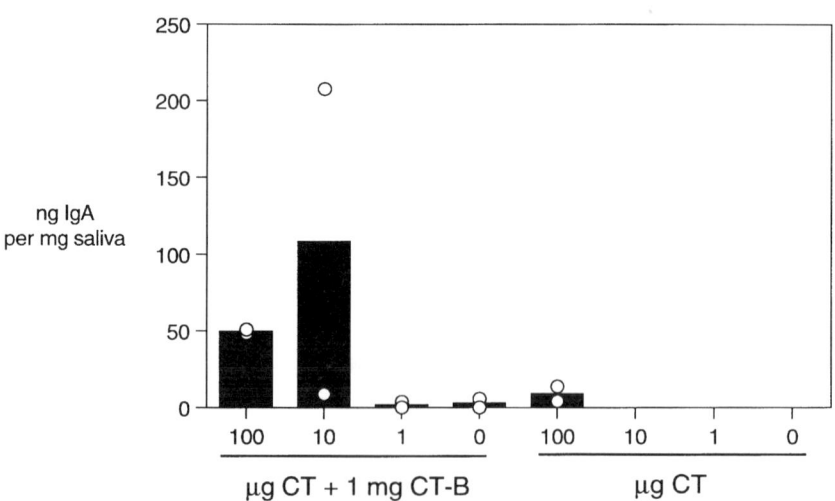

Fig. 5. Saliva anti-CT-B IgA following oral administration of CT and CT-B. Treatments were administered as described in Fig. 1. Saliva was collected by absorbent wicks; anti-CT-B IgA determined by ELISA and by comparison to porcine IgA standards.

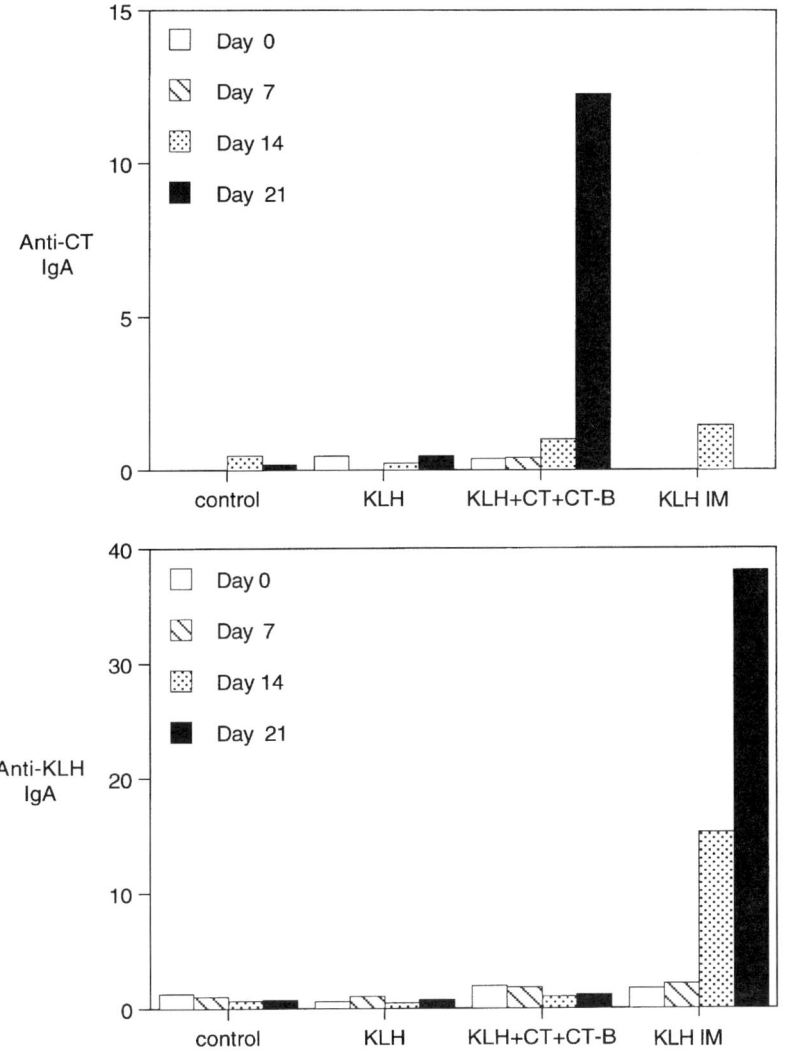

FIG. 6. Adjuvanticity of CT to coadministered KLH. Anti-CT-B and anti-KLH antibodies were measured by ELISA during and 1 week after three weekly treatments with KLH (25 mg) alone or with CT (50 μg) and CT-B (1 mg). One animal received 2 mg KLH intramuscularly. Values are expressed relative to total IgA in the serum.

having potent adjuvant properties for CT-B, no adjuvant activity was detected for coadministered KLH.

These results indicate that CT is well tolerated, is immunogenic, and has potent adjuvant properties when used with CT-B in swine. These characteristics make CT a useful tool to study mucosal immunology in swine.

III. Mechanisms of Mucosal Adjuvanticity

The ability of CT to profoundly affect the physiology of many cell types suggests many possible mechanisms for its mucosal adjuvanticity in various phases of the immune response. After providing a brief overview of these possible mechanisms, this section highlights the role of macrophage cytokines, including IL-1 and IL-12 in the mucosal adjuvanticity of CT.

A. Cholera Toxin

There are a wide variety of potential mechanisms for adjuvants to enhance mucosal immunity, many of which could be affected in some way by CT (Snider, 1995). Some examples of these mechanisms are listed:

1. *Alterations in delivery of antigen to the GALT.* Cholera toxin increases intestinal permeability to dextran (MW 3000) and also induces immune response to KLH; CT-B failed to do either (Lycke et al., 1991). Administration of CT with ovalbumin significantly increases the amount of ovalbumin in the serum within 15 min of administration (Pierre et al., 1995).

2. *Effects on antigen presentation.* The effects of CT on antigen presentation are complex. While CT inhibits intracellular processing of protein antigens by macrophages, it enhances presentation of surface MHC II molecules when administered after intracellular processing has occurred (Matousek et al., 1996). This ability to inhibit intracellular processing of proteins is dependent on the presence of whole toxin, whereas CT and CT-B both enhance expression of MHC complexes on macrophages (Matousek et al., 1996) and B cells (Nashar et al., 1997).

3. *B-cell antibody isotype switching.* CT synergizes with IL-4 to induce isotype differentiation of B cells (Lycke et al., 1990; Lycke and Strober, 1989). Induction of cAMP by CT induced the production of sterile germline γ1-RNA transcripts while binding of G_{M1} ganglioside enhanced later stages of B-cell differentiation via a calcium-dependent

mechanism (Lycke, 1993). This provides one possible mechanism for the synergy observed between CT and CT-B.

4. *Cytokine production by antigen-specific lymphocytes, induction of Th2 phenotype.* The primary immune response induced by CT administered at mucosal surfaces is secretory IgA. This response is T cell dependent and results in antigen-specific Th2 T cells in the GALT (Xuamano *et al.*, 1994; Jackson *et al.*, 1996). The ability of CT to induce a type 2 response to coadministered antigens has been well described (Marinaro *et al.*, 1997). Following oral immunization with tetanus toxiod (TT), with CT as adjuvant, the predominant T-cell response to TT is the production of IL-4 (Marinaro *et al.*, 1995). Also, the ability of CT to function as an adjuvant to coadministered KLH or ovalbumin is lost in IL-4-deficient mice (Vajdy *et al.*, 1995), although the immunogenicity of CT is maintained.

5. *Lymphocyte proliferation. In vitro,* CT inhibits IL-2 production and proliferation of T lymphocytes and B lymphocytes by a cAMP-mediated mechanism; however, the response varied with the stage of differentiation of the cells (Lycke *et al.*, 1989a). Both CT and CT-B inhibit lymphocyte proliferation *in vitro* but CT was more rapidly effective and potent (Elson *et al.*, 1995). Gavage with CT also causes a depletion of CD8+ intraepithelial lymphocytes, which have been associated with a suppressor function (Elson *et al.*, 1995).

In addition to these mechanisms, mucosal adjuvants such as CT may also mediate their effects via activation of antigen presenting cells (APC) to express costimulatory molecules and secrete cytokines. These costimulatory molecules and cytokines are, in effect, a signal from the APC to the T-cell in addition to the antigen-specific interaction of the MHC–peptide–TCR complex. These two signals are then integrated by the T cell, determining its response.

B. DUAL-SIGNAL INTEGRATION

The immune response to a given protein antigen is not inherent to the antigen but is determined by circumstances of the encounter of the antigen with the immune system. This is particularly important in the GALT where many antigens must be tolerated while immunity is generated to others. In this model of mucosal adjuvanticity, APC responds to environmental cues by producing cytokines and expressing costimulatory molecules that regulate the T-cell response to a given antigen. T cells integrate these two signals, one antigen-specific signal and the other nonspecific or innate, into the resulting immune response. The integration of antigen-specific and nonspecific signals occurs at the level of the T cell during the initiation of the immune response. The molecular basis of this integration is the two-signal model for T-cell activation (Schwartz, 1996; Jenkins *et al.*, 1991). Cognate interactions of the TCR with peptide in the context of MHC provides signal 1, which can result in activation or anergy depending on the presence of a second, nonantigen specific signal (Johnson and Jenkins, 1994). This second signal consists of surface proteins on the antigen presenting cell including CD80 and CD86 (Vella *et al.*, 1997) and proinflammatory cytokines such as IL-1β and IL-12 (Grohmann *et al.*, 1997; Pape *et al.*, 1997). Therefore, the ability of adjuvants (including CT and CT-B) to induce expression of CD80/86 and secretion of IL-1β and IL-12 may be key to their mucosal adjuvanticity.

C. Role of Interleukin-1 in Mucosal Adjuvanticity

There is clear evidence that CT induces the production of IL-1. For example, CT increases IL-1 production in a murine macrophage cell line (Lycke *et al.*, 1989a). The IL-1β promoter contains cAMP response elements (CREs), which are required for full activity of the gene (Chandra *et al.*, 1995), and CRE binding proteins are phosphorylated in response to both cAMP and LPS (Chandra *et al.*, 1995). In addition non-cAMP-dependent pathways have been described for the CT induction of IL-1, possibly involving protein kinase C (Krakauer, 1996). Besides APC, other cellular sources of IL-1 in the GALT have been described including M cells of the follicular epithelium of the Peyer's patch (Pappo and Mahlman, 1993). The ability of CT to bind specifically to epithelial cells would suggest these may be an important source of IL-1 in the GALT.

Once induced, IL-1 can augment the immune response in a number of ways. It is required for IL-12 induced IFN-γ production (Hunter *et al.*, 1995), which in turn upregulates the transcription of IL-1 mRNA (Sone *et al.*, 1994), prevents development of tolerance (Weigle *et al.*, 1987) and augments antibody production (Reed *et al.*, 1989). It also enhances the *in vivo* expansion and follicular migration of antigen stimulated T cells (Pape *et al.*, 1997). In swine, IL-1 shows adjuvant properties when administered with a *Streptococcus suis* vaccine (Blecha *et al.*, 1995).

This evidence suggests a role for IL-1 in the mucosal adjuvanticity of CT. To further understand this role, we have studied the ability of CT and CT-B to induce IL-1β in cultured porcine alveolar macrophages.

Because CT has specific adjuvant effects distinct from CT-B, we investigated the IL-1β inducing activity of CT versus that of CT-B as a potential mechanism for its mucosal adjuvanticity.

Because both the CT and CT-B contained significant quantities of endotoxin, all treatments were preincubated with polymyxin-B at a level (10 μg/ml) sufficient to block the effect of 1 μg/ml of LPS (data not shown). The specificity of the CT and CT-B effects was confirmed by specifically blocking their binding to the cell surface G_{M1} ganglioside. Preincubation with 10 μg/ml of G_{M1} ganglioside completely blocked the ability of CT or CT-B to bind to G_{M1} ganglioside in a G_{M1} ganglioside ELISA (data not shown). The pretreated CT and CT-B was then compared to untreated CT and CT-B for the ability to induce IL-1β mRNA. The pretreated CT-B induced IL-1β to levels similar to those seen with untreated CT-B (Fig. 7), indicating a non-G_{M1} ganglioside binding dependent effect. This result was in marked contrast to that seen with CT where pretreatment reduced levels of IL-1β

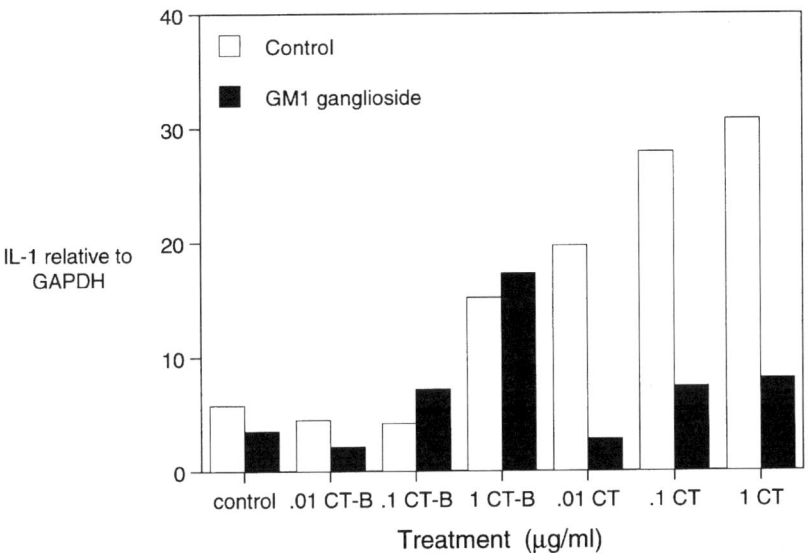

FIG. 7. Induction of IL-1β by CT in cultured porcine macrophages. Alveolar macrophages were treated with the indicated dose of CT or CT-B that had been preincubated with G_{M1} ganglioside (filled bars) or not (open bars). After 12 hours total RNA was isolated and IL-1β and GAPDH mRNA determined by northern hybridization. Values are relative phosphorimager units expressed relative to GAPDH.

mRNA to those of control (Fig. 7). These results indicate that CT specifically induces IL-1β mRNA in alveolar macrophages and CT-B does not.

D. Role of Interleukin 12 in Mucosal Adjuvanticity

Another macrophage cytokine with potent immunoregulatory functions is interleukin 12. IL-12 is a heterodimeric cytokine with glycosylated, disulfide bonded subunits of 35 and 40 kDa (Podlaski et al., 1992). The genes for the two subunits are located on separate chromosomes, 3 and 5 in human and 3 and 11 in mouse (Tone et al., 1996; Sieburth et al., 1992). IL-12 is produced by various phagocytic cells, including macrophages, and can be induced by bacteria or bacterial products. IL-12 activates NK and T cells to produce IFN-γ, which in turn activates macrophages to enhance pathogen killing as well as produce more IL-12. IL-12 acts on T cells both directly and indirectly via induced IFN-γ to have a profound impact on the acquired immune response. In this way, IL-12 may play an important role in mucosal adjuvanticity.

There are several lines of evidence for the role of IL-12 in regulation of mucosal immunity. It plays a central role in resistance to several mucosal pathogens and has been implicated in the pathogenesis of inflammatory bowel syndromes. The immune response to *Salmonella* infections is critically dependent on IL-12. Administration of *Salmonella dublin* (EL23) to mice increased IL-12, p40 mRNA expression in mesenteric lymph nodes and Peyer's patch by 6 hours after administration (Bost and Clements, 1995). Administration of exogenous IL-12 augments the protective immune response to *S. dublin,* while treatment with IL-12 neutralizing antibodies reduces survival of infected mice (Mastroeni et al., 1996; Kincycain et al., 1996). Inhibition of Substance P, which induces IL-12 in murine macrophages (Kincycain and Bost, 1997), also increases susceptibility to *Salmonella* infection in mice (Kincycain and Bost, 1996). Administration of IL-12 to either SCID or BALB/c mice prevented infection with *Cryptosporidium parvum* and neutralization of endogenous IL-12 exacerbated the infection (Urban et al., 1996). Conversely, IL-12 enhanced adult worm survival and egg production by the intestinal nematode *Nippostrongylus brasiliensis* (Finkelman et al., 1994). Lamina propria mononuclear cells from Crohn's disease patients expressed increased levels of IL-12, p35 and p40 mRNA, and secreted bioactive IL-12 (Monteleone et al., 1997).

Regulation of bioactive IL-12 secretion is complicated by the independent regulation of the two subunits and by the observation that the p40 subunit can form a homodimer that is a biological antagonist to the heterodimer (Heinzel et al., 1997). While the p40 subunit is readily induced by various stimuli, secretion of biologically active IL-12 may be better correlated by expression of the p35 subunit (Snijders et al., 1996).

The ability of CT-B binding to induce binding to AP-1 sites and the presence of AP-1 sites in the promoters for p35 and p40 subunits of IL-12 suggests the possibility of CT or CT-B inducing expression of IL-12. In addition there is also *in vivo* evidence for the induction of IL-12, p40 by LT-B in mice (Bost and Clements, 1995). The murine p40 promoter region contains recognition sequences including AP1, AP3, GATA, Pu.1, and an IFN-γ response (IRF-1) element (Tone et al., 1996). In human monocytes, treatment with IFN-γ for 16 hours primes for the production of IL-12 following subsequent LPS exposure, primarily by enhancing p40 transcription (Ma et al., 1996). In contrast to that of p40, the p35 subunit promoter contains multiple transcription start sites with no CAAT or TATA boxes, but has a GC-rich region with constitutive promoter activity in *in vitro* assays as well as potential regulatory elements such as AP1 and NF-κB sequences (Tone et al., 1996). IFN-γ directly activates the p35 promoter and has an additive effect with LPS in activating the p40 promoter (Ma et al., 1996).

To further elucidate the possible role of IL-12 in the mucosal adjuvanticity of CT, we looked for *in vivo* evidence that CT or CT-B induces IL-12 in swine. Selected GALT tissues were analyzed for IL-12, p40 by RT-PCR 5 days following the third administration of cholera toxin (100 μg per dose at 10-day intervals) or 4 days after administration of *Salmonella cholerasuis* [3.8×10^{11} colony-forming unit (cfu) avirulent strain SC54, NOBL Laboratories, Inc., Sioux Center, IA]. Both CT and SC-54 increased IL-12, p40 levels in ileal Peyer's patch and mesenteric lymph nodes (Fig. 8).

We also investigated the effect of macrophage activation on the ability of CT or CT-B to induce IL-12, p35 and p40 mRNA in cultured alveolar macrophages. The specificity of the observed responses was confirmed by performing paired experiments with and without preincubation with G_{M1} ganglioside as described in the IL-1 section. IL-12, p35 and p40 mRNA levels were determined by competitive reverse transcriptase–polymerase chain reaction (RT–PCR) (Mansfield et al., 1995) and were normalized to levels of HPRT. Results for p40 are described in Fig. 9; p35 results are similar. Activation of macrophages with IFN-γ greatly augmented the response to CT and CT-B. Both IL-12, p35 and p40 were measured at approximately 10-fold higher

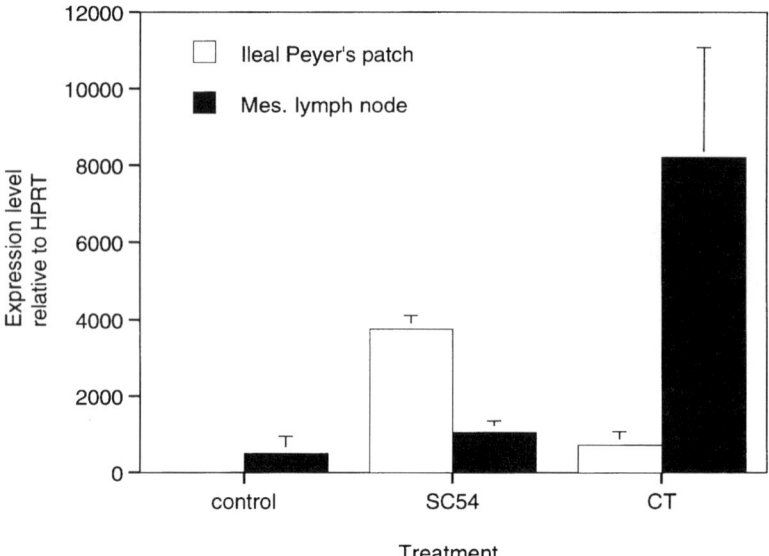

FIG. 8. Induction of IL-12, p40 in ileal Peyer's patch and mesenteric lymph nodes by CT and *S. cholerasuis* (strain SC 54). Total RNA was isolated 5 days following the third administration of cholera toxin (100 × per dose at 10-day intervals) or 4 days after administration of *S. cholerasuis* (3.8 × 10^{11} cfu) and IL-12, p40 mRNA determined by competitive RT-PCR with normalization to HPRT. Error bars are the standard deviation of three PCR reactions.

levels in activated macrophages than in nonactivated ones following treatment. In activated macrophages both CT and CT-B induced 40 to 100 fold increases in measured p35 and p40 mRNA (Fig. 9). In contrast, CT-B failed to increase p35 or p40 in nonactivated cells. However, CT increased both p35 and p40 mRNA in nonactivated cells. These effects were blocked by preincubation with G_{M1} ganglioside. These results suggest CT and CT-B may have a differential effect on IL-12 mRNA in macrophages depending on activation state.

IV. Summary

Delivery of protein antigens to the GALT can result in immunity or oral tolerance depending on the circumstances of the encounter. One mechanism by which mucosal adjuvants can affect these circumstances is by the induction of macrophage cytokines, including IL-1 and IL-12. These cytokines can directly affect the immune response by their effects on antigen-specific T cells and by the induction of IFN-γ by

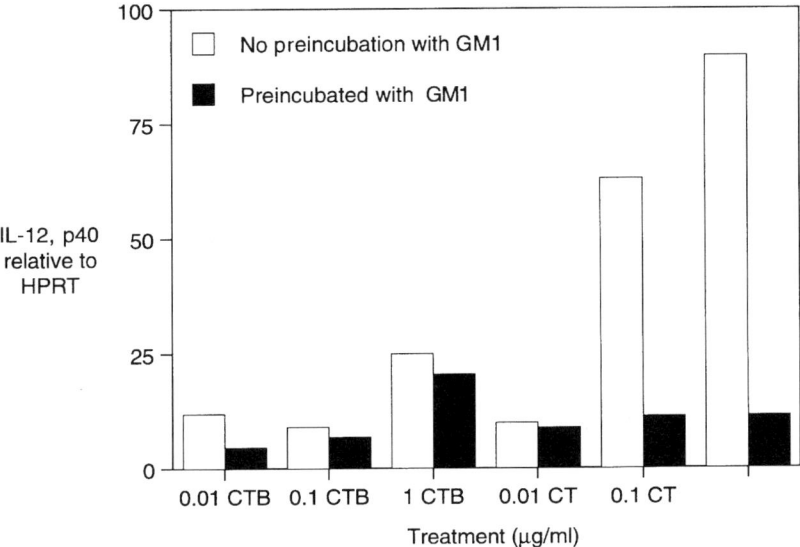

FIG. 9. Induction of IL-12, p40 in cultured alveolar macrophages by CT and CT-B. Alveolar macrophages were treated with the indicated dose of CT or CT-B that had been preincubated with G_{M1} ganglioside (filled bars) or not (open bars). After 12 hours total RNA was isolated and IL-12, p40 mRNA determined by competitive RT–PCR with normalization to HPRT.

T cells or NK cells. This IFN-γ also activates macrophages to up-regulate MHC or costimulatory molecules and by further inducing IL-1 and IL-12. In effect, mucosal adjuvants function both directly and indirectly as activators of antigen presenting cells, resulting in stimulation of the immune response to coincidental antigens.

Our studies in swine have shown CT is a potent mucosal adjuvant for CT-B. CT also increased IL-1 and IL-12 mRNA in cultured macrophages, especially after activation with IFN-γ. The effect of CT on the secretion of bioactive IL-12 protein is currently being investigated. While the mucosal adjuvanticity of CT involves a variety of mechanisms, these findings suggest a role for the induction of the macrophage cytokines IL-1 and IL-12.

REFERENCES

Agren, L. C., Ekman, L., Lowenadler, B., and Lycke, N. Y. (1997). Genetically engineered nontoxic vaccine adjuvant that combines B cell targeting with immunomodulation by cholera toxin A1 subunit. *J. Immunol.* **158,** 3936–3946.
Bergquist, C., Johansson, E. L., Lagergard, T., Holmgren, J., and Rudin, A. (1997).

Intranasal vaccination of humans with recombinant cholera toxin B subunit induces systemic and local antibody responses in the upper respiratory tract and the vagina. *Infect. Immun.* **65,** 2676–2684.

Blecha, F., Reddy, D. N., Chitko-McKown, C. G., Mcvey, D. S., Chengappa, M. M., Goodband, R. D., and Nelssen, J. L. (1995). Influence of recombinant bovine interleukin-1 beta and interleukin-2 in pigs vaccinated and challenged with *Streptococcus suis*. *Vet. Immunol. Immunopathol.* **44,** 329–346.

Bost, K. L., and Clements, J. D. (1995). In vivo induction of interleukin-12 mRNA expression after oral immunization with *Salmonella dublin* or the B subunit of *Escherichia coli* heat-labile enterotoxin. *Infect. Immun.* **63,** 1076–1083.

Buckley, N. E., Su, Y., Milstien, S., and Spiegal, S. (1995). The role of calcium influx in cellular proliferation induced by interaction of endogenous ganglioside G_{M1} with the B subunit of cholera toxin. *Biochim. Biophys. Acta* **1256,** 275–283.

Chandra, G., Cogswell, J. P., Miller, L. R., Godlevski, M. M., Stinnett, S. W., Noel, S. L., Kadwell, S. H., Kost, T. A., and Gray, J. G. (1995). Cyclic AMP signaling pathways are important in IL-1 beta transcriptional regulation. *J. Immunol.* **155,** 4535–4543.

Clemens, D., Sack, D. A., Chakraborty, J., Rao, M. R., Ahmed, F., Harris, J. R., van Loon, F. P. L., Khan, M. R., Yunis, M., and Huda, S. (1990). Field trial of oral cholera vaccines in Bangladesh: Evaluation of anti-bacterial and anti-toxin breast-milk immunity in response to ingestion of the vaccine. *Vaccine* **8,** 469–472.

Clemens, J. D., Stanton, B. F., Chakraborty, J., Sack, D. A., Khan, M. R., Huda, S., Ahmed, F., Harris, J. R., Yunus, M., Khan, M. U., Svennerholm, A. M., Jertborn, M., and Holmgren, J. (1987). B subunit-whole cell and whole cell-only vaccines against cholera: Studies on reactivity and immunogenicity. *J. Infect. Dis.* **155,** 79–85.

deHaan, L., Holtrop, M., Verweij, W. R., Agsteribbe, E., and Wilschut, J. (1996a). Mucosal immunogenicity of the *Escherichia coli* heat-labile enterotoxin: Role of the A subunit. *Vaccine* **14,** 260–266.

deHaan, L., Verweij, W. R., Feil, I. K., Lijnema, T. H., Hol, W. G. J., Agsteribbe, E., and Wilschut, J. (1996b). Mutants of the *Escherichia coli* heat-labile enterotoxin with reduced ADP-ribosylation activity or no activity retain the immunogenic properties of the native holotoxin. *Infect. Immun.* **64,** 5413–5416.

Elson, C. O., and Ealding, W. (1984). Generalized systemic and mucosal immunity in mice after mucosal stimulation with cholera toxin. *J. Immunol.* **132,** 2736–2741.

Elson, C. O., Holland, S. P., Dertzbaugh, M. T., Cuff, C. F., and Anderson, A. O. (1995). Morphologic and functional alterations of mucosal T cells by cholera toxin and its B subunit. *J. Immunol.* **154,** 1032–1040.

Finkelman, F. D., Madden, K. B., Cheever, A. W., Katona, I. M., Morris, S. C., Gately, M. K., Hubbard, B. R., Gause, W. C., and Urban, J. F. (1994). Effects of interleukin 12 on immune responses and host protection in mice infected with intestinal nematode parasites. *J. Exp. Med.* **179,** 1563–1572.

Francis, M. L., Okazaki, I., Moss, J., Kurosky, A., Pecanha, L. M. T., and Mond, J. J. (1995). cAMP-independent effects of cholera toxin on B cell activation. III. Cholera toxin a subunit-mediated ADP- ribosylation acts synergistically with ionomycin or IL-4 to induce B cell proliferation. *J. Immunol.* **154,** 4956–4964.

Grohmann, U., Bianchi, R., Ayroldi, E., Belladonna, M. L., Surace, D., Fioretti, M. C., and Puccetti, P. (1997). A tumor-associated and self antigen peptide presented by dendritic cells may induce T cell anergy in vivo, but IL-12 can prevent or revert the anergic state. *J. Immunol.* **158,** 3593–3602.

Gyles, C. L., and Barnum, D. A. (1969). A heat-labile enterotoxin from strains of *Escherichia coli* enteropathogenic for pigs. *J. Infect. Dis.* **120,** 419–426.

Heinzel, F. P., Hujer, A. M., Ahmed, F. N., and Rerko, R. M. (1997). In vivo production and function of IL-12 p40 homodimers. *J. Immunol.* **158,** 4381–4388.

Holmgren, J., Czerkinsky, C., Sun, J., and Svennerholm, A. M. (1996). Oral vaccination, mucosal immunity and oral tolerance with special reference to cholera toxin. In "Concepts in Vaccine Development" (S. T. E. Kaufmann, ed.), pp. 437–458. de Gruyter, New York.

Hunter, C. A., Chizzonite, R., and Remington, J. S. (1995). IL-1 beta is required for IL-12 to induce production of IFN-gamma by NK cells-A role for IL-1 beta in the T cell-independent mechanism of resistance against intracellular pathogens. *J. Immunol.* **155,** 4347–4354.

Jackson, R. J., Marinaro, M., VanCott, J. L., Yamamoto, M., Okahashi, N., Fujihashi, K., Kiyono, H., Chatfield, S. N., and McGhee, J. R. (1996). Mucosal immunity: Regulation by helper T cells and a novel method for detection. *J. Biotechnol.* **44,** 209–216.

Jenkins, M. K., Chen, C., Jung, G., Mueller, D. L., and Schwartz, R. H. (1991). Inhibition of antigen-specific proliferation of type 1 murine T cell clones after stimulation with immobilized anti-CD3 monclonal antibody. *J. Immunol.* **144,** 16–22.

Jertborn, M., Svennerholm, A. M., and Holmgren, J. (1986). Gut mucosal, salivary and serum antitoxic and antibacterial antibody responses in Swedes after oral immunization with B subunit-whole cell cholera vaccine. *Int. Arch. Allergy Appl. Immunol.* **75,** 38–43.

Johnson, J. G., and Jenkins, M. K. (1994). Minireview: The role of anergy in peripheral T cell unresponsiveness. *Life Sci.* **55,** 1767–1780.

Kincycain, T., and Bost, K. L. (1996). Increased susceptibility of mice to salmonella infection following *in vivo* treatment with the substance P antagonist, spantide II. *J. Immunol.* **157,** 255–264.

Kincycain, T., and Bost, K. L. (1997). Substance P-induced IL-12 production by murine macrophages. *J. Immunol.* **158,** 2334–2339.

Kincycain, T., Clements, J. D., and Bost, K. L. (1996). Endogenous and exogenous interleukin-12 augment the protective immune response in mice orally challenged with *Salmonella dublin*. *Infect. Immun.* **64,** 1437–1440.

Kochnolte, F., Haag, F., Kastelein, R., and Bazan, F. (1996). Uncovered-The family relationship of a T-cell-membrane protein and bacterial toxins. *Immunol. Today* **17,** 402–405.

Krakauer, T. (1996). Evidence for protein kinase C pathway in the response of human peripheral blood mononuclear cells to cholera toxin. *Cell. Immunol.* **172,** 224–228.

Loftager, M. K., Elmerdahl-Olsen, J., and Eriksen, L. (1995). Immune reaction of pigs following intra-gastric or ileal inoculation of attenuated *Salmonella typhi*. *Adv. Exp. Med. Biol.* **371B,** 913–917.

Lycke, N. Y. (1993). Cholera toxin promotes B cell isotype switching by two different mechanisms: cAMP induction augments germ-line Ig H-chain RNA transcripts whereas membrane ganglioside GM1-receptor binding enhances later events in differentiation. *J. Immunol.* **150,** 4810–4821.

Lycke, N., and Holmgren, J. (1986). Strong adjuvant properties of cholera toxin on gut mucosal immune responses to orally presented antigens. *Immunology* **59,** 301–308.

Lycke, N., and Strober, W. (1989). Cholera toxin promotes B cell isotype differentiation. *J. Immunol.* **142,** 3781–3787.

Lycke, N., Bromander, A., Ekman, L., Karlsson, U., and Holmgren, J. (1989a). Cellular basis of immunomodulation by cholera toxin *in vitro* with possible association to the adjuvant function *in vivo*. *J. Immunol.* **142,** 20–27.

Lycke, N., Bromander, A. K., and Holmgren, J. (1989b). Role of local IgA antitoxin-producing cells for intestinal protection against cholera toxin challenge. *Int. Arch. Allergy Appl. Immunol.* **88,** 273–279.

Lycke, N., Severinson, E., and Strober, W. (1990). Cholera toxin acts synergistically with IL-4 to promote IgG1 switch differentiation. *J. Immunol.* **145,** 3316–3324.

Lycke, N., Karlsson, U., Sjolander, A., and Magnusson, K. E. (1991). The adjuvant action of cholera toxin is associated with an increased intestinal permeability for luminal antigens. *Scand. J. Immunol.* **33,** 691–698.

Ma, X. J., Chow, J. M., Gri, G., Carra, G., Gerosa, F., Wolf, S. E., Dzialo, R., and Trinchieri, G. (1996). The interleukin 12 p40 gene promoter is primed by interferon gamma in monocytic cells. *J. Exp. Med.* **183,** 147–157.

Mansfield, L. S., Urban, J. F., Murtaugh, M. P., Zarlenga, D. S., Foss, D. L., and Lunney, J. K. (1995). Development of IL-10, IL-12 and HPRT internal cDNA competetors for measuring swine cytokine gene expression by quantitative RT-PCR. *Conf. Res. Workers Anim. Dis. Proc.* **76,** 131.

Marinaro, M., Staats, H. F., Hiroi, T., Jackson, R. J., Coste, M., Boyaka, P. N., Okahashi, N., Yamamoto, M., Kiyono, H., Bluethmann, H., Fujihashi, K., and McGhee, J. R. (1995). Mucosal adjuvant effect of cholera toxin in mice results from induction of T helper 2 (Th2) cells and IL-4. *J. Immunol.* **155,** 4621–4629.

Marinaro, M., Kiyono, H., van Cott, J. L., Okahashi, N., van Ginkel, F. W., Pascual, D. W., Ban, E., Jackson, R. J., Staats, H. F., and McGhee, J. R. (1997). Vaccines for selective induction of Th-1 and Th2-cell responses and their roles in mucosal immunity. *In* "Essentials of Mucosal Immunology" (M. F. Kagnoff and H. Kiyono, eds.), pp. 461–475. Academic Press, San Diego, CA.

Mastroeni, P., Harrison, J. A., Chabalgoity, J. A., and Hormaeche, C. E. (1996). Effect of interleukin 12 neutralization on host resistance and gamma interferon production in mouse typhoid. *Infect. Immun.* **64,** 189–196.

Matousek, M. P., Nedrud, J. G., and Harding, C. V. (1996). Distinct effects of recombinant cholera toxin B subunit and holotoxin on different stages of class II MHC antigen processing and presentation by macrophages. *J. Immunol.* **156,** 4137–4145.

Monteleone, G., Biancone, L., Marasco, R., Morrone, G., Marasco, O., Luzza, F., and Pallone, F. (1997). Interleukin 12 is expressed and actively released by Crohns disease intestinal lamina propria mononuclear cells. *Gastroenterology* **112,** 1169–1178.

Nakagawa, I., Takahashi, I., Kiyono, H., McGhee, J. R., and Hamada, S. (1996). Oral immunization with the B subunit of the heat-labile enterotoxin of *Escherichia coli* induces early Th1 and late Th2 cytokine expression in Peyer's patches. *J. Infect. Dis.* **173,** 1428–1436.

Nashar, T. O., Hirst, T. R., and Williams, N. A. (1997). Modulation of B-cell activation by the B subunit of *Escherichia coli* enterotoxin—Receptor interaction up-regulates MHC class II, B7, CD40, CD25 and ICAM-1. *Immunology* **91,** 572–578.

Pape, K. A., Khoruts, A., Mondino, A., and Jenkins, M. K. (1997). Inflammatory cytokines enhance the in vivo clonal expansion and differentiation of antigen-activated CD4(+) T cells. *J. Immunol.* **159,** 591–598.

Pappo, J., and Mahlman, R. T. (1993). Follicle epithelial M cells are a source of interleukin-1 in Peyer's patches. *Immunology* **78,** 505–507.

Pierce, N. F. (1978). The role of antigen form and function in the primary and secondary intestinal immune response to cholera toxin and toxoid in rats. *J. Exp. Med.* **148,** 195–200.

Pierce, N. F., Cray, W. C., and Engel, P. F. (1980). Antitoxic immunity to cholera in dogs immunized orally with cholera toxin. *Infect. Immun.* **27,** 632–637.

Pierre, P. G., Mbongolo Mbella, E. G., and Vaerman, J. P. (1995). Effect of cholera toxin and its B subunit on intestinal permeability for ovalbumin. *Adv. Exp. Med. Biol.* **371B,** 1519–1521.
Podlaski, F. J., Nanduri, V. B., Hulmes, J. D., Pan, Y. C., Levin, W., Danho, W., Chizzonite, R., Gately, M. K., and Stern, A. S. (1992). Molecular characterization of interleukin 12. *Arch. Biochem. Biophys.* **294,** 230–237.
Reed, S. G., Pihl, D. L., Conlon, P. J., and Grabstein, K. H. (1989). IL-1 as adjuvant. Role of T cells in the augmentation of specific antibody production by recombinant human IL-1 alpha. *J. Immunol.* **142,** 3129–3133.
Schwartz, R. H. (1996). Models of T cell anergy—Is there a common molecular mechanism. *J. Exp. Med.* **184,** 1–8.
Sieburth, D., Jabs, E. W., Warrington, J. A., Li, X., Lasota, J., LaForgia, S., Kelleher, K., Huebner, K., Wasmuth, J. J. and Wolf, S. F. (1992). Assignment of genes encoding a unique cytokine (IL-12) composed of two unrelated subunits to chromosomes 3 and 5. *Genomics* **14,** 59–62.
Snider, D. P. (1995). The mucosal adjuvant activities of ADP-ribosylating bacterial enterotoxins. *Crit. Rev. Immunol.* **15,** 317–348.
Snijders, A., Hilkens, C. M. U., Kraan, T. C. T. M. V., Engel, M., Aarden, L. A., and Kapsenberg, M. L. (1996). Regulation of bioactive IL-12 production in lipopolysaccharide-stimulated human monocytes is determined by the expression of the p35 subunit. *J. Immunol.* **156,** 1207–1212.
Sone, S., Orino, E., Mizuno, K., Yano, S., Nishioka, Y., Haku, T., Nii, A., and Ogura, T. (1994). Production of IL-1 and its receptor antagonist is regulated differently by IFN gamma and IL-4 in human monocytes and alveolar macrophages. *Eur. Respir. J.* **7,** 657–663.
Spangler, B. D. (1992). Structure and function of cholera toxin and the related *Escherichia coli* heat-labile enterotoxin. *Microbiol. Rev.* **56,** 622–647.
Svennerholm, A. M., Jertborn, M., Gothefors, L., Karim, A. M. M. M., Sack, D. A., and Holmgren, J. (1984). Mucosal antitoxic and antibacterial immunity after cholera disease and after immunization with a combined B subunit-whole cell vaccine. *J. Infect. Dis.* **149,** 884–893.
Tone, Y., Thompson, S. A. J., Babik, J. M., Nolan, K. F., Tone, M., Raven, C., and Waldmann, H. (1996). Structure and chromosomal location of the mouse interleukin-12, p35 and p40 subunit genes. *Eur. J. Immunol.* **26,** 1222–1227.
Urban, J. F., Fayer, R., Chen, S. J., Gause, W. C., Gately, M. K., and Finkelman, F. D. (1996). IL-12 protects immunocompetent and immunodeficient neonatal mice against infection with *Cryptosporidium parvum*. *J. Immunol.* **156,** 263–268.
Vajdy, M., Kosco-Vilbois, M. H., Kopf, M., Kohler, G., and Lycke, N. (1995). Impaired mucosal immune responses in interleukin 4-targeted mice. *J. Exp. Med.* **181,** 41–53.
Vella, A. T., Mitchell, T., Groth, B., Linsley, P. S., Green, J. M., Thompson, C. B., Kappler, J. W., and Marrack, P. (1997). CD28 engagement and proinflammatory cytokines contribute to T cell expansion and long-term survival in vivo. *J. Immunol.* **158,** 4714–4720.
Wassen, L., Schon, K., Holmgren, J., Jertborn, M., and Lycke, N. (1996). Local intravaginal vaccination of the female genital tract. *Scand. J. Immunol.* **44,** 408–414.
Weigle, W. O., Scheuer, W. V., Hobbs, M. V., Morgan, E. L., and Parks, D. E. (1987). Modulation of the induction and circumvention of immunological tolerance to human gamma-globulin by interleukin 1. *J. Immunol.* **138,** 2069–2074.
Wilson, A. D., Clark, C. J., and Stokes, C. R. (1990). Whole cholera toxin and B subunit

act synergistically as an adjuvant for the mucosal immune response in mice to keyhole limpet haemocyanin. *Scand. J. Immunol.* **31,** 443–451.

Xuamano, J. C., Jackson, R. J., Fujihashi, K., Kiyono, H., Staats, H. F., and McGhee, J. R. (1994). Helper Th1 and Th2 cell responses following mucosal or systemic immunization with cholera toxin. *Vaccine* **12,** 903–911.

Cholera Toxin B Subunit as an Immunomodulator for Mucosal Vaccine Delivery

MICHAEL W. RUSSELL, HONG-YIN WU,
GEORGE HAJISHENGALLIS, SUSAN K. HOLLINGSHEAD,
AND SUZANNE M. MICHALEK

Department of Microbiology, University of Alabama at Birmingham, Birmingham, Alabama 35924

I. Introduction
II. Responses to Mucosal Immunization with SBR-CTA2/B
III. Responses to Mucosal Immunization with *Salmonella* Expressing SBR-CTA2/B
IV. Discussion and Summary
   Acknowledgments
   References

I. Introduction

Efforts to exploit oral immunization for the development of vaccines against the great majority of infections, which directly afflict or invade through the mucosal surfaces, were formerly limited by the modest responses usually generated by the mucosal immune system. In part, this is probably because its functions include the maintenance of homeostasis despite being continuously challenged by large quantities of mainly harmless environmental antigens and commensal microorganisms. In addition, most nonviable immunogens have no affinity for the M cells overlying mucosal inductive sites, such as the intestinal Peyer's patches, and are readily digested within the gut lumen. Thus, oral immunization with conventional nonviable immunogens typically requires the repeated administration of large doses, and yields modest responses that do not persist. However, numerous developments during the past decade have addressed these problems. Notably, the exceptionally potent mucosal immunogenic and adjuvant properties of cholera toxin (CT) and related enterotoxins (Elson, 1989) have been exploited by coupling other antigens to the nontoxic B subunit of CT (Czerkinsky *et al.*, 1989), or by seeking to detoxify the heat-labile toxin (LT) of *Escherichia coli* while preserving its adjuvanticity (Di Tommaso *et al.*, 1996; Dickinson and Clements, 1995). Furthermore, the intranasal (i.n.) route of vaccine administration has been found to be more effective than the intragastric (i.g.) route, and it has the advantage that CTB then displays adjuvant effects that are not evident i.g. (Wu and Russell, 1993).

The adjuvant activity of CT coadministered i.g. with antigens appears to depend on the adenine diphospho-ribosylating activity of the toxic A1 subunit, whereas CT-B subunit lacks adjuvant activity by this route (Holmgren *et al.*, 1993). However, CT-B can serve as a carrier of other antigens chemically coupled to it, but effective immunization with such conjugates by the i.g. route still depends on the adjuvant activity of intact CT coadministered with it (Russell and Wu, 1991). Moreover, preparation of chemical conjugates is cumbersome and yields an incompletely defined product. While straightforward genetic coupling of peptide antigens to CT-B as fusion proteins has been successfully accomplished (Jagusztyn-Krynicka *et al.*, 1993), fusion of large peptides and immunogenic proteins tends to disrupt the folding and assembly of CT-B into pentamers, with consequent loss of crucial G_{M1}-binding activity (Dertzbaugh and Elson, 1993). The crystallographic resolution of the molecular structure of LT (Sixma *et al.*, 1991), which closely resembles CT, suggested another approach: Protein antigen segments might be genetically fused to the A2 subunit of CT (replacing the toxic A1 subunit), coexpressed with CT-B, and assembled into chimeric immunogens resembling the composition of CT but completely lacking its toxicity (Hajishengallis *et al.*, 1995). The plasmid encoding this antigen-CTA2/B chimeric protein may also be expressed in an attenuated live carrier such as *Salmonella typhimurium*, thereby eliminating the need for purifying the product (Hajishengallis *et al.*, 1996a).

II. Responses to Mucosal Immunization with SBR-CTA2/B

The surface protein adhesin, AgI/II, of *Streptococcus mutans*, and its 42-kDa saliva-binding region (SBR) (Hajishengallis *et al.*, 1994) were used to demonstrate the concept, and the cloning strategy was designed to be readily applicable to other protein antigen sequences of interest. Thus the A1 sequence in the gene for CT was replaced with DNA encoding SBR so that an SBR-CTA2 fusion protein was coexpressed in *E. coli* with CT-B (Hajishengallis *et al.*, 1995). On transport into the periplasm, the polypeptides were assembled into a chimeric protein of the composition SBR-CTA2.CTB$_5$, as shown by gel electrophoretic analysis and Western blotting, and by G_{M1} enzyme-linked immunosorbent assay (ELISA) showing that the binding activity of CT-B was retained and coupled with antigenically identifiable SBR. The chimeric protein was purified from periplasmic lysates by ammonium sulfate precipitation and by chromatography on anion-exchange and molecular size-exclusion media (Hajishengallis *et al.*, 1995).

When mice were immunized i.g. with this protein, salivary IgA and serum IgG and IgA antibodies to the parent AgI/II were induced, and the addition of CT enhanced the responses (Table I). Antibodies to CT were also induced, especially when CT was included (Hajishengallis *et al.*, 1995). Intranasal immunization was also effective, particularly for the generation of salivary IgA antibodies even in the absence of added CT and with a lower dose of chimeric protein (Table I). Subsequent experiments have indicated that one or two doses given i.n. may be sufficient to generate responses by 3–4 weeks after immunization (H.-Y. Wu, unpublished data). Furthermore, continued monitoring of responses showed that both salivary IgA and serum IgG antibodies could be sustained for up to at least 11 months (Hajishengallis *et al.*, 1996b).

TABLE I

SERUM AND SALIVARY ANTIBODY RESPONSES TO *S. MUTANS* AgI/II INDUCED IN MICE BY I.G. OR I.N. IMMUNIZATION WITH SBR-CTA2/B CHIMERIC PROTEIN GIVEN THREE TIMES AT 10-DAY INTERVALS

		Antibody response to AgI/II 7 days after last dose[a]		
Immunogen and dose	Route	Serum IgG (μg/ml)	Serum IgA (μg/ml)	Salivary IgA (% Ab/Ig)[b]
SBR-CTA2/B (100μg)	i.g.	88 $\times/\div$4.0	0.5 $\times/\div$3.8	0.4 $\times/\div$2.8
SBR-CTA2/B (100μg) + CT (5 μg)	i.g.	363 $\times/\div$1.8	3.7 $\times/\div$2.1	4.7 $\times/\div$1.5
SBR-CTA2/B (25 μg)	i.n.	38 $\times/\div$1.6	1.5 $\times/\div$1.7	2.8 $\times/\div$1.6

[a]Geometric mean $\times/\div$SD (N = 4 or 5). No antibodies to AgI/II were detectable prior to immunization or in unimmunized control mice.

[b]Salivary antibodies expressed as percent of total salivary IgA concentration.

Examination of cells isolated from Peyer's patches and mesenteric lymph nodes of i.g. immunized mice for proliferation in response to culture with AgI/II revealed the presence of AgI/II-specific T cells (Table II), and the proliferative response increased with the number of immunogen doses administered (Toida *et al.*, 1997). In addition, SBR-CTA2/B induced stronger T-cell responses than SBR alone, and the use of CT adjuvant further enhanced the proliferative response to AgI/II (Table II). AgI/II-responsive T cells can also be found in the cervical lymph nodes after i.n. immunization with AgI/II-CT-B conjugates (Wu *et al.*, 1997). Furthermore, both type 1 and 2 cytokines were expressed by antigen-specific T cells from both the inductive sites and the draining lymph nodes corresponding to the route of administration (Wu *et al.*, 1996, 1997).

III. Responses to Mucosal Immunization with *Salmonella* Expressing SBR-CTA2/B

Several strains of *S. typhimurium* have been attenuated by gene deletion to serve as live carriers for mucosal immunization because of their ability to colonize the gut-associated lymphoid tissues (GALT).

TABLE II

PROLIFERATIVE RESPONSES OF CELLS FROM PEYER'S PATCH (PP), MESENTERIC LYMPH NODE (MLN), AND SPLEEN CELLS FROM MICE IMMUNIZED I.G. WITH SBR (40 μG), SBR-CTA2/B (100 μG), OR SBR-CTA2/B (100 μG) PLUS CT (5 μG)

Organ	Mean stimulation index[a] ±SD in cells from mice immunized with:			
	Controls	SBR	SBR-CTB	SBR-CTB + CT
PP	1.31	2.62	5.44	6.14
	±0.06	±0.48	±0.49	±1.34
MLN	1.16	2.39	3.00	4.57
	±0.12	±0.23	±0.30	±0.60
Spleen	1.50	1.80	2.06	2.05
	±0.33	±0.23	±0.14	±0.09

[a]Groups of five mice were killed 10 days after immunization with three doses (at 10-day intervals) of immunogens shown, and cells were cultured in triplicate for 5 days with 0.5 μg/ml of AgI/II and assessed for proliferation by [^{3}H]thymidine uptake. Results are shown as stimulation indices calculated as cpm (stimulated culture)/mean cpm (unstimulated culture).

An $aroA^-/aroD^-$ strain (BRD509) that lacks genes essential for aromatic amino acid synthesis was transformed with the pSBR-CT$^{\Delta A1}$ plasmid that expresses SBR-CTA2/B together with temperature-regulated plasmid GP1-2, which encodes T7 RNA polymerase (Hajishengallis *et al.*, 1996a). On growth at 30°C, expression of GP1-2 and hence of SBR-CTA2/B (which is dependent on a T7 promoter) is suppressed, but at 37°C *in vitro,* and presumably therefore *in vivo,* large quantities of SBR-CTA2/B are produced at the expense of continued growth. After i.g. immunization with 10^9 live bacteria, mice developed salivary IgA and serum IgG antibodies to AgI/II (Table III), as well as to CT and to *Salmonella* (Hajishengallis *et al.*, 1996a), which were further elevated after a second dose of 10^9 cells (Table III). Both salivary IgA and serum IgG responses persisted for at least 7 months after the administration of a second dose (Table III). Intranasal administration of 10^8 recombinant *Salmonella* also induced antibody responses of comparable magnitude. When mice were immunized i.g. with another recombinant strain of *Salmonella* containing a plasmid that expressed SBR alone (as well as plasmid GP1-2), and therefore producing SBR but not CTA2/B, the antibody responses to AgI/II were lower (Hajishengallis *et al.*, 1996a). Furthermore, analysis of the IgG subclasses (IgG1 and IgG2a) of serum antibodies to AgI/II suggested a slight shift toward a Th1 pattern of T-cell regulation (Harokopakis *et al.*, 1997), as typically occurs with *Salmonella,* in contrast to the adjuvant effects of CT, which tends to drive Th2-regulated responses.

TABLE III

SERUM IgG AND SALIVARY IgA ANTIBODY RESPONSES TO *S. MUTANS* AgI/II INDUCED BY I.G. IMMUNIZATION OF MICE WITH 10^9 CFU OF *S. TYPHIMURIUM* EXPRESSING SBR-CTA2/B CHIMERIC PROTEIN

		Geom. mean $\times/\div$ SD (N = 4 or 5)	
Immunization	Tested after:	Serum IgG (μg/ml)	Salivary IgA (%Ab/Ig)[a]
Salmonella/SBR-CTA2/B × 1	7 days	2.4 $\times/\div$ 1.5	1.0 $\times/\div$ 1.7
Salmonella/SBR-CTA2/B × 2[b]	7 days	39 $\times/\div$ 1.7	0.8 $\times/\div$ 1.7
Salmonella/SBR-CTA2/B × 2[c]	7 months	8.1 $\times/\div$ 1.3	0.8 $\times/\div$ 1.6

[a]Salivary antibodies expressed as percent of total salivary IgA concentration.
[b]Second immunization (10^9 cfu) given after 7 weeks.
[c]Second immunization (10^{10} cfu) given after 15 weeks.

IV. Discussion and Summary

Among numerous approaches to the mucosal application of antigens that have been found to induce high levels of IgA antibodies in secretions (Mestecky et al., 1997), the use of CT-B as a carrier that promotes uptake of coupled antigens by the inductive sites of the mucosal immune system has been particularly successful. The activity may also depend on CT-B functioning as an adjuvant that stimulates antigen-presenting cells, T cells, and B cells (Snider, 1995). Although the addition of CT, which has more potent adjuvant properties than CT-B, enhances responses to chimeric Ag-CTA2/B immunogens given by the i.g. route in rodents, use of this in humans is precluded by its extreme toxicity. However, the i.n. route in mice, monkeys, and humans appears to be considerably more effective, and not dependent on the use of intact CT adjuvant (Bergquist et al., 1997; Russell et al., 1996; Wu and Russell, 1993). Indeed, recombinant CT-B totally lacking contaminating toxin serves as an adjuvant i.n. in mice (Wu and Russell, 1998), and CT or LT mutants lacking toxic activity hold great promise as mucosal adjuvants, particularly by the i.n. route (Di Tommaso et al., 1996; Dickinson and Clements, 1995; Douce et al., 1995; Yamamoto et al., 1997). Reasons for the efficacy of i.n. immunization undoubtedly include the lack of exposure of immunogens to acid and digestive enzymes in the gastrointestinal tract, as well as more immediate contact with the site of antigen uptake, which in rodents is thought to be the nasal lymphoid tissue (NALT). NALT is considered to be the functional equivalent of Waldeyer's ring (Kuper et al., 1992), the pharyngeal ring of tonsils, adenoids, and related lymphoid structures in humans and other primates. Although anatomically quite distinct, both have accumulations of B and T lymphocytes, and accessory cells necessary to mount an immune response, surmounted by an epithelium that includes M-like cells which probably take up antigens and pass them to underlying immunocompetent cells. The binding of CT-B to intestinal Peyer's patch M cells has been demonstrated (Neutra et al., 1996). We have shown that i.n. immunization leads to the development of antigen-responsive T cells and antibody-secreting cells in NALT, the draining cervical lymph nodes, and salivary glands of mice (Wu et al., 1996, 1997; Wu and Russell, 1993, 1997).

Live carrier organisms that colonize the GALT are particularly attractive as delivery vehicles for mucosal vaccines, because an appropriate recombinant organism, rendered avirulent by multiple mutations, and expressing several different vaccine antigens should be easy and cheap to produce and administer (Curtiss et al., 1989). Furthermore, organisms such as *Salmonella* induce a different pattern of immune responses than CT (or CT-B) and its homologs, as shown by the relative balance of cytokines associated with Th1 or Th2 cells, and the consequent proportions of IgG antibody subclasses (Klimpel *et al.*, 1995). Selection of delivery technologies for mucosal vaccines, or combinations thereof, may allow responses to be tailored to the most appropriate pattern for combatting a particular infection. Limitations of recombinant carrier organisms, however, include the difficulty of obtaining an adequate level of expression, and surface expression in particular, of the cloned protein antigen, though this will probably be solved by genetic manipulation. More problematic is the fact that the recipient animal mounts a strong response to the carrier itself, often exceeding the response to the desired antigen. Although this may not be undesirable if the carrier itself is an attenuated pathogen against which protection is useful, it may effectively preclude repeated use of the same carrier organism for subsequent booster or other immunizations, because the immune response to the carrier interferes with its colonization of the gastrointestinal tract. Remarkably, however, CT-B does not appear to suffer this disadvantage (Wu and Russell, 1994), possibly because the affinity of CT-B for G_{M1} ganglioside is considerably higher than any antibody affinity, so that CT-B is therefore able to bind to its receptor even in the presence of antibody.

The mucosal immune system appears to respond at different levels, probably according to the intensity and duration of the stimulus. Food and other such nonoffensive environmental antigens that are always present induce secretory IgA antibodies but only at low levels, and immunization with a bolus of most novel, but nonviable, antigens such as foreign proteins induces a modest response that does not persist for long after the cessation of antigen application. Such findings led to the notion that memory was not a feature of mucosal immunity, whereas it is now becoming clear that, with appropriate stimulation, mucosal immune responses can persist for a long time, and may display anamnestic characteristics. Similarly, it seems that nonthreatening commensal microorganisms also induce only modest responses. In these instances, the S-IgA antibodies can inhibit absorption of the antigen and adequately restrain the commensals. The noninflammatory properties of IgA antibodies are probably important in maintaining the integrity of the mucosal epithelium which is itself a defensive barrier. At the same time, the mechanisms that manifest themselves as oral tolerance suppress systemic immune responses, including IgG antibodies and cell-mediated immunity, thereby limiting potentially damaging and unnecessary inflammatory responses. It may also be envisaged that this moderation of responses avoids overstimulation of the immune system to stimuli that are constantly present but not threatening, yet maintains an unobtrusive check on the commensal microbiota. In contrast, an aggressive challenge, represented by an invasive pathogen or a noxious toxin, induces a vigorous response from the mucosal immune system, resulting in high levels of S-IgA antibodies, and also serum IgG and cell-mediated immunity. Furthermore, the continued production of S-IgA antibodies at the mucosal surface, as well as its precursor IgA in the submucosa, may serve anti-inflammatory functions that limit tissue damage and maintain mucosal integrity. Exploitation of the properties of toxins such as CT and its nontoxic B subunit, as well as genetically engineered avirulent microbial vectors, should enable us to construct effective novel mucosal vaccines for protection against the majority of infections that begin at mucosal surfaces.

Acknowledgments

These studies were supported by U.S. Public Health Service (National Institutes of Health) grants DE06746, DE08182, and DE09081.

References

Bergquist, C., Johansson, E. L. Lagergård, T., Holmgren, J., and Rudin, A. (1997). Intranasal vaccination of humans with recombinant cholera toxin B subunit induces systemic and local antibody responses in the upper respiratory tract and the vagina. *Infect. Immun.* **65,** 2676–2684.

Curtiss, R., Kelly, S. M., Gulig, P. A., and Nakayama, K. (1989). Selective delivery of antigens by recombinant bacteria. *Curr. Top. Microbiol. Immunol.* **146,** 35–49.

Czerkinsky, C., Russell, M. W., Lycke, N., Lindblad, M., and Holmgren, J. (1989). Oral administration of a streptococcal antigen coupled to cholera toxin B subunit evokes strong antibody responses in salivary glands and extramucosal tissues. *Infect. Immun.* **57,** 1072–1077.

Dertzbaugh, M. T., and Elson, C. O. (1993). Reduction in oral immunogenicity of cholera toxin B subunit by N-terminal peptide addition. *Infect. Immun.* **61,** 384–390.

Dickinson, B. L., and Clements, J. D. (1995). Dissociation of *Escherichia coli* heat-labile enterotoxin adjuvanticity from ADP-ribosyltransferase activity. *Infect. Immun.* **63,** 1617–1623.

Di Tommaso, A., Saletti, G., Pizza, M., Rappuoli, R., Dougan, G., Abrignani, S., Douce, G., and De Magistris, M. T. (1996). Induction of antigen-specific antibodies in vaginal secretions by using a nontoxic mutant of heat-labile enterotoxin as a mucosal adjuvant. *Infect. Immun.* **64,** 974–979.

Douce, G., Turcotte, C., Cropley, I., Roberts, M., Pizza, M., Domenghini, M., Rappuoli, R., and Dougan, G. (1995). Mutants of *Escherichia coli* heat-labile toxin lacking ADP-ribosyltransferase activity act as nontoxic, mucosal adjuvants. *Proc. Natl. Acad. Sci. USA* **92,** 1644–1648.

Elson, C. O. (1989). Cholera toxin and its subunits as potential oral adjuvants. *Curr. Top. Microbiol. Immunol.* **146,** 29–33.
Hajishengallis, G., Koga, T., and Russell, M. W. (1994). Affinity and specificity of the interactions between *Streptococcus mutans* antigen I/II and salivary components. *J. Dent. Res.* **73,** 1493–1502.
Hajishengallis, G., Hollingshead, S. K., Koga, T., and Russell, M. W. (1995). Mucosal immunization with a bacterial protein antigen genetically coupled to cholera toxin A2/B subunits. *J. Immunol.* **154,** 4322–4332.
Hajishengallis, G., Harokopakis, E., Hollingshead, S. K., Russell, M. W., and Michalek, S. M. (1996a). Construction and oral immunogenicity of a *Salmonella typhimurium* strain expressing a streptococcal adhesin linked to the A2/B subunits of cholera toxin. *Vaccine* **14,** 1545–1548.
Hajishengallis, G., Michalek, S. M., and Russell, M. W. (1996b). Persistence of serum and salivary antibody responses after oral immunization with a bacterial protein antigen genetically linked to the A2/B subunits of cholera toxin. *Infect. Immun.* **64,** 665–667.
Harokopakis, E., Hajishengallis, G., Greenway, T. E., Russell, M. W., and Michalek, S. M. (1997). Mucosal immunogenicity of a recombinant *Salmonella typhimurium*-cloned heterologous antigen in the absence or presence of co-expressed cholera toxin A2/B subunits. *Infect. Immun.* **65,** 1445–1454.
Holmgren, J., Lycke, N., and Czerkinsky, C. (1993). Cholera toxin and cholera B subunit as oral-mucosal adjuvant and antigen vector systems. *Vaccine* **11,** 1179–1184.
Jagusztyn-Krynicka, E. K., Clark-Curtiss, J. E., and Curtiss, R. (1993). *Escherichia coli* heat-labile toxin subunit B fusions with *Streptococcus sobrinus* antigens expressed by *Salmonella typhimurium* oral vaccine strains: Importance of the linker for antigenicity and biological activities of the hybrid proteins. *Infect. Immun.* **61,** 1004–1015.
Klimpel, G. R., Asuncion, M., Haithcoat, J., and Niesel, D. W. (1995). Cholera toxin and *Salmonella typhimurium* induce different cytokine profiles in the gastrointestinal tract. *Infect. Immun.* **63,** 1134–1137.
Kuper, C. F., Koornstra, P. J., Hameleers, D. M. H., Biewenga, J., Spit, B. J., Duijvestijn, A. M., van Breda Vriesman, P. J. C., and Sminia, T. (1992). The role of nasopharyngeal lymphoid tissue. *Immunol. Today* **13,** 219–224.
Mestecky, J., Michalek, S. M., Moldoveanu, Z., and Russell, M. W. (1997). Routes of immunization and antigen delivery systems for optimal mucosal immune responses in humans. *Behring Inst. Mitt.* **98,** 33–43.
Neutra, M. R., Frey, A., and Kraehenbuhl, J. P. (1996). Epithelial M cells: Gateways for mucosal infection and immunization. *Cell* **86,** 345–348.
Russell, M. W., and Wu, H.-Y. (1991). Distribution, persistence, and recall of serum and salivary antibody responses to peroral immunization with protein antigen I/II of *Streptococcus mutans* coupled to the cholera toxin B subunit. *Infect. Immun.* **59,** 4061–4070.
Russell, M. W., Moldoveanu, Z., White, P. L., Sibert, G. J., Mestecky, J., and Michalek, S. M. (1996). Salivary, nasal, genital, and systemic antibody responses in monkeys immunized intranasally with a bacterial protein antigen and cholera toxin B subunit. *Infect. Immun.* **64,** 1272–1283.
Sixma, T. K., Pronk, S. E., Kalk, K. H., Wartna, E. S., van Zanten, B. A. M., Witholt, B., and Hol, W. G. J. (1991). Crystal structure of a cholera toxin-related heat-labile enterotoxin from *E. coli. Nature* **351,** 371–377.
Snider, D. P. (1995). The mucosal adjuvant activities of ADP-ribosylating bacterial enterotoxins. *Crit. Rev. Immunol.* **15,** 317–348.
Toida, N., Hajishengallis, G., Wu, H.-Y., and Russell, M. W. (1997). Oral immunization

with the saliva-binding region of *Streptococcus mutans* AgI/II genetically coupled to the cholera toxin B subunit elicits T helper cell responses in gut-associated lymphoid tissues. *Infect. Immun.* **65,** 909–915.

Wu, H.-Y., and Russell, M. W. (1993). Induction of mucosal immunity by intranasal application of a streptococcal surface protein antigen with the cholera toxin B subunit. *Infect. Immun.* **61,** 314–322.

Wu, H.-Y., and Russell, M. W. (1994). Comparison of systemic and mucosal priming for mucosal immune responses to a bacterial protein antigen given with or coupled to cholera toxin (CT) B subunit, and effects of pre-existing anti-CT immunity. *Vaccine* **12,** 215–222.

Wu, H.-Y., and Russell, M. W. (1997). Nasal lymphoid tissue, intranasal immunization, and compartmentalization of the common mucosal immune system. *Immunol. Res.* **16,** 187–201.

Wu, H.-Y., and Russell, M. W. (1998). Induction of mucosal and systemic immune responses by intranasal immunization using recombinant cholera toxin B subunit as an adjuvant. *Vaccine* **16,** 286–292.

Wu, H.-Y., Nikolova, E. B., Beagley, K. W., and Russell, M. W. (1996). Induction of antibody-secreting cells and T helper and memory cells in murine nasal lymphoid tissue. *Immunology* **88,** 493–500.

Wu, H.-Y., Nikolova, E. B., Beagley, K. W., Eldridge, J. H., and Russell, M. W. (1997). Development of antibody-secreting cells and antigen-specific T cells in cervical lymph nodes after intranasal immunization. *Infect. Immun.* **65,** 225–235.

Yamamoto, S., Kiyono, H., Yamamoto, M., Imaoka, K., Yamamoto, M., Fujihashi, K., Van Ginkel, F. W., Noda, M., Takeda, Y., and McGhee, J. R. (1997). A nontoxic mutant of cholera toxin elicits Th2-type responses for enhanced mucosal immunity. *Proc. Natl. Acad. Sci. USA* **94,** 5267–5272.

Deceptive Imprinting: Insights into Mechanisms of Immune Evasion and Vaccine Development

PETER L. NARA

Biological Mimetics Inc., Frederick, Maryland 21702

I. Introduction and Background
II. Deceptive Imprinting
    A. "Original Antigenic Sin" Gone Awry?
    B. Hyperactive B-Cell Ig-Mediated Control of T Cells?
    C. A Fleeting Memory?
III. Refocusing the Immune Response by Masking Epitopes Responsible for Deceptive Imprinting: Novel Approach to Vaccination
References

I. Introduction and Background

Introduction of foreign antigens into the host either by vaccination or infection in many cases leads to the production of specific antibody (Briles and Davie, 1980). Depending on the nature of the immunogen and the various pathways leading to B-cell activation, the clonality of the response is ultimately evoked by a given epitope. In most cases, antigens represent a broad array of epitopes, and consequently the antibody response is chemically heterogeneous and antigenically specific. Some multideterminant antigens (i.e., albumin, hen egg white lysosyme, HBV epitope, Cro Lac-gp41, etc.) result in an unequal response to some epitopes (reviewed in Benjamin *et al.*, 1984). Given the potentially large B-cell repertoire in any given host, selective recognition of specific epitope by a limited population of B cells needs explanation to determine whether this restriction is due to the immunogen itself or to host factors involved in its selection. Some antigens induce a less variable response, for example, certain carbohydrate antigens,

antigens with structural–functional homology to self (Table I), and other antigens displaying a limited, highly ordered, or redundant number of immunodominant cross-reactive sites, more often with an unique steric presentation (i.e., streptococcal group A-variant cell wall) (Briles and Davie, 1980). For example, an animal immunized with one of these antigenic determinants and later exposed to a different, but structurally similar, determinant responds to this second determinant by initially producing antibody (sometimes with a higher affinity) to the original antigen, a phenomena referred to as OAS (original antigenic sin). In many cases, depending on the antigen concentration, continued boosting with the first or second Ag also elicits a normal primary response to the second antigen (Briles and Davie, 1980).

As mentioned previously this OAS-like phenomena, was first described for an influenza viral infection of humans by Francis in 1953. Since then, OAS-like phenomena have been recognized, to influenza, in other virus families such as toga-, paramyxo-, and enteroviruses (Fenner *et al.,* 1974). Previously, this phenomenon has been considered (on a population level) a beneficial immune response of the host (Angelova and Shvartsman, 1982). Providing various members of the population long-term, protective anamnestic immune responses to different strains of the virus would ensure that some members of the population are immune to viral agents constantly undergoing antigenic drift and shift. In general, the OAS phenomenon challenges the dogma of immunologic response specificity and the induction of its memory in a host where there may be sequential infections over time with two different, but antigenically related, strains of virus; or, as it appears is the case with human immunodeficiency virus type 1 (HIV-1) (Table I), with

TABLE I

B-Cell Abnormalities during HIV-1 Infection[a]

1. Infection with HIV-1 triggers a strong/sustained antibody response; approximately 10–50% of peripheral IgG-secreting B cells are HIV-1 specific.
2. Evidence for high-frequency of preexisting HIV-1 reacting B cells in normal seronegative individuals; 45–50% in both adults (1/16,200–49,000) and newborns (1/11,800–26,600).
3. Evidence for abnormal selection of the B-cell repertoire: (1) 26% show overt IgG monoclonal populations demonstrating cross-reactivity with p66 pol and p55 gag.
4. Skewing of K/L ratios, shared idiotypes to different HIV-1 components.
5. Gp120 acting as a natural ligand for VHIII gene product; B-cell superantigen.
6. Structural similarities of gp120 and IgH variable and FW region segments.

[a]Reviewed in Nara (1996).

the generation of such viruses during a "single" persistent viral infection. Evidence is mounting that a number of microbial pathogens including HIV-1 and other animal lentiviruses may have evolved to use it in a diversionary way (Table II). By providing immunodominant epitopes capable of undergoing antigenic variation to the immune system these pathogens appear to limit or fix the humoral and cell-mediated responses to the initial resident pathogen and thus the host seems unable to eliminate and/or control the agent. This type of host response is somewhat reminiscent of those described in the fields of parsitology and tumor immunology some 25 years ago. The terms *concomitant immunity, premunition,* and *heterotypic immunity* were coined to describe an immune response in a host whereby the resident pathogens or tumor cells were tolerated but pathogenic varients in the surrounding "transmission community" were susceptible. This prevented them from establishing a new infection in the same host (reviewed in Mitchell, 1991).

II. Deceptive Imprinting

A. "Original Antigenic Sin" Gone Awry

As mentioned before, the immunologic phenomenon of OAS was considered in the literature to be a type of protective host response in humans and rats infected with strains of influenza. Recently, however, Nara and colleagues have proposed data and a model of OAS that represents an immunopathogenic immune-diverting and evading strategy termed *deceptive imprinting* (Kohler *et al.,* 1994). The nature of this phenomenon as modeled in HIV-1-infected chimpanzees is presented below based on previously published data.

TABLE II

Proposed Mechanisms for Immunodominant Epitope-Mediated Suppression/Restriction Consumption/Exhaustion of "Critical" Cells or Factors

---

1. Induction of nonantigenic specific regulatory cells (e.g., NK cells)
2. Antibody-dependent epitope masking
3. Carrier-induced epitopic suppression (effector cell or idiotypic network)
4. Ig-mediated TGF-B CTL suppression
5. B-cell-driven T-cell diversification (superantigen-like effects)
6. "Unconventional" B-cell antigen/superantigens (leading to VH expansion)

Chimpanzees previously immunized with given immunoaffinity purified homologous gp120 and challenged within weeks with the same virus stock elicited a rather robust anamnestic response capable of neutralizing at high titer both the homologous strain as well as those of divergent strains (~18% sequence difference in the gp120). Viruses were re-isolated from the immunized animals PBMC's, and found to be completely neutralization resistant. Interestingly, when the genetic sequences of these viruses were determined they were found to be only 1 to 5% divergent in the glycoprotein 120 (gp120) and completely identical in the V3 region to which the neutralizing response was directed (Nara et al., 1990). The mechanism(s) by which this polyclonal antibody response was raised, to effectively neutralize a widely divergent strain (not previously seen by the animals immune system), but incapable of neutralizing a genetically near identical varient (with identical V3 sequence), is unknown and extremely important if one is to understand how to develop effective humoral immunogens against HIV-1. One potential explanation is that cross-reactive neutralizing antibodies are elicited that recognize epitopes present on these divergent strains. Why they are also not capable of neutralizing the closely related reisolated viruses remains unknown at this time. This phenomenon may be due to the virus since specific distant-site mutations (to the neutralization epitope) were found to occur in both the gp120 and gp41 molecule, which conveyed some structural and/or functional alteration to the neutralization epitope itself thus making the neutralizing antibody ineffective (Back et al., 1993; Nara et al., 1990). In addition, the kinetics of the subsequent neutralizing response were determined for the early neutralization escape viruses and generally found to be slow to develop and of low titer, while the neutralizing titers to both the parental and divergent strains continued to increase to a plateau level after 6 months to 1 year. The explanation and name given to this immunologic phenomena, briefly mentioned earlier in the chapter, was initially described as a form of "original antigenic sin" (Nara and Goudsmit, 1991). It has subsequently been explained in terms of "clonal dominance" (Kohler et al., 1992) and later refined to include other aspects related to the phenomena and distinguish it as an immunopathogenic strategy. It is now termed *deceptive imprinting* (Kohler et al., 1994).

In general, the model and the data suggest that antigenic variation of the virus envelope secondary to random mutation/selection theory provide for the continued presentation of either identical and/or cross-reactive pseudo-neutralizing epitopes present on "escape variants." Continued clonal expansion of a limited functional B-cell repertoire restricts the subsequent complete development and/or functional maturation of the humoral response to other less immunodominant epitopes (for a review, see Muller et al., 1992; Nara and Goudsmit, 1991). The initial immune clonal expansion seems to be initiated through a viral clonal expansion in both chimpanzees (Nara et al., 1990) and humans (Zhang et al., 1993; Zhu et al., 1993; McNearney et al., 1992; Pang et al., 1992; Wolfs et al., 1992) with saturation of follicular dendritic cells of the germinal centers with a genetically homogenous HIV-1. The resulting immune response from the deceptive imprinting model predicts that an overt clonality for specific and/or all structural viral antigens should exist. This immunologic mechanism effectively restricts or reduces the available T- and B-cell immune repertoire, which functionally limits the polyclonal nature of the response while expanding a founder population of cross-reactive B and T cells. The signature of such an immune response is observed as a stable oligoclonal population of antibody (reviewed in Muller et al., 1992) and clonality of the T-cell response to viral encoded antigens. Recently, evidence has been reported which supports this model at the T-cell level. Both limited Vβ usage (Pantaleo et al., 1994) during ARS and unusual oligoclonal TCR usage/expansions in later stage patients of specific cytotoxic T cells over long periods of time have been reported during HIV-1 infection (Kalams et al., 1994).

B. HYPERACTIVE B-CELL IG-MEDIATED CONTROL OF T CELLS?

1. *The B-Cell Problem*

During HIV-1 infection there is a generalized abnormality of the B-cell compartment that persists during the transition from early HIV-1 infection to AIDS (Schnittman et al., 1986). Circulating B cells are at a high level of activation, and *in vitro* mononuclear cell cultures have demonstrated both polyclonal activation and production of anti-HIV-1 antibodies (Martinez-Maza et al., 1987; Briault et al., 1988) and have also demonstrated that 26% of sera from HIV-infected individuals contains monoclonal IgG populations. McGrath and coworkers (Ng et al., 1988) purified an electrophoretic spike of oligoclonal Ig origin from the sera from an ARC patient that had high titer binding activity (>1:100,000) to a variety of HIV epitopes (p66^{pol}, p55^{gag}, p53^{pol}, p41^{gag}, and p24^{gag}). Later IgG oligoclonal bands in HIV-1-infected patients were shown to be directed against HIV-1 specific determinants (Amadori et al., 1990, reviewed in Amadori and Chieco-Bianchi, 1992). The striking finding regarding this oligoclonal activation is the spontaneous *in vitro* production of anti-HIV-1 antibody occurring in unstimulated PBMC, bone marrow, lymph nodes and cerebrospinal fluid cultures from HIV-1-infected patients. In general, the antibody is directed against the gp120; however, antibody against gag and pol were reported (Amadori and Chieco-Bianchi, 1992). Paradoxically, this spontaneous B-cell activation is associated with a poor B-cell response to mitogens and antigens both in terms of proliferation and antibody response to recall antigens (Lane *et al.*, 1983).

Isoelectric focusing (IEF) was used to evaluate the clonal diversity of B-cell responses, because this method is capable of identifying single clone products, i.e., spectrotypic patterns reflect the number of actively secreting specific B-cell clones. This approach has been used to study the antibody spectrotypes in the sera and in the cerebrospinal fluid of certain HIV-1-infected patients, and the detection of oligoclonal spectrotypes (characterized by a few clusters of bands) has been interpreted as evidence that during infection there are expansions of a limited number of B-cell clones. We have also employed IEF to evaluate the antibody responses of HIV-1-infected patients (reviewed in Muller *et al.*, 1992). In a longitudinal study, each patient (during 12–36 months of evaluation) maintained a stable characteristic spectrotype of anti-gp120 antibodies. Each spectrotype is a form of immune fingerprinting and indicates a rather continuous production over time. Therefore, clonal dominance or polyclonal restriction was consistently present over the course of the disease, regardless of clinical stage or the development of ARC and AIDS.

Studies in VH gene usage stress the qualitative abnormality of the humoral immune responses in seropositive patients. Maturation of VH3L genes, the largest in the B cell of genes, was found in some studies to be selectively depleted, suggesting that B cells from HIV-1-infected patients present a maturational arrest at the level of the germinal center. Further evidence for this was demonstrated more directly when it was observed that gp120 had Fab surface receptor-binding capacity for B cells that were similar to that of Staphylococcal protein A (Berberian, 1993). This binding demonstrated that between 20 and 40% of human peripheral B cells bind gp120 and lead to the induction and synthesis of Ig enriched in VH3 IgM *in vitro*. Although circumstantial, the gp120 molecule may behave like those of SpA, which acts as a B-cell superantigen through specific interactions with B cells expressing VH3 heavy chain rearrangements (Silverman and Kohler, 1992). For mechanisms on how both expansion and clonal depletion of VH3 occurs in HIV-1-infected people the reader is referred to a previous review (Muller *et al.*, 1992). Although both protein and carbohydrate type antigens can elicit a clonally expanded population of B cells, molecular and serologic approaches have shown their selection of VH subfamily specificity can be very different. Some antigens, termed *conventional* such as *Haemophilus influenzae,* type B polysaccharide, use only two types of VH3 H chains preferentially. So-called *unconventional* antigens, however, such as staphlococcal protein A, have been found to bind both the Fcγ-binding site of the Ig framework structure and the Fab receptor on some IgM, IgG, IgA, and IgE expressing B-cell surface molecules (Langone, 1982; Harboe, 1974; Inganas, 1981). These antigens do not appear to be limited to a small number of germ line gene elements within the large VH3 family and elicit a B-cell superantigenic-like effect (reviewed in Silverman 1994). A direct mechanism of B-cell activation via carbohydrate and/or protein on the gp120 could lead to the types of B cell clonal expansions previously described above.

Mechanism(s) to reinforce the preferential expansion of B-cell clones just described, for a single antigen administration, would be that which, on a second administration with a similar antigen, suppresses and/or limits the antibody response. This type of phenomenon exists and is well known as *antigenic competition* (Albright et al., 1970; Moller and Sjöberg, 1970; Waterson, 1970; Schecter, 1968; Radovick and Talmadge, 1967). Various mechanisms for this phenomena have been demonstrated and/or proposed (Table II) to occur with various antigens from numerous sources (i.e., bacteria, plant and animal virus, animal proteins, tumor antigens, etc.). In general, antigens capable of eliciting these types of responses share various immunochemical characteristics such as high charge, hydrophilicity, and mobility/freedom of rotation; they tend to be proline-rich and often have proposed loop-like structures with repeated sequences and steric/spatial distribution on the pathogens surface. Recently the phenomenon of antigenic competition with regard to MHC class II presentation has been found to be dependent on both the specificity of the internalization and subsequent presentation. Kittlesen *et al.* (1993) have suggested that the B cell processing pathway for an endogenous antigen which is recognized by MHC class II-restricted T cells is different from that for exogenous antigen internalized nonspecifically, the latter being resistant to protein synthesis inhibitors and sensitive to antigenic competition (Kakiuchi *et al.,* 1990). Also Lorenze *et al.* (1990) demonstrated that presentation of antigen internalized through the mannose receptor into macrophages is resistant to competition by self proteins. Thus antigens internalized through both the mannose receptor and a specific antigen receptor seem to be processed via similar pathways. The biochemical, biophysical, and immunochemical properties of shed and virion associated gp120 make it a very good candidate for eliciting a direct B-cell clonal dominance and/or clonal dominance through one or more mechanisms associated with the phenomenon of antigenic competition.

2. The T-Cell Problem

What mechanism could help contribute to suppression or limit the polyclonal nature of the cytolytic T-cell response? Recently Rowley *et al.* (1993; Rowley and Stach, 1993) have demonstrated that IgG and TGF-B form complexes with macrophages through Fc receptors that localize at antigenic sites and play important roles in homeostasis of immunity by augmenting proliferation of already activated dominant lymphocyte clones (Coffman *et al.*, 1989), promoting isotope switching (Lin and Stavnezer, 1992; Kuruvilla *et al.*, 1991; Coffman *et al.*, 1989; Sonoda *et al.*, 1989), and suppressing activation/proliferation of new specific antigen-reactive clones that may arise during ongoing immunity (Kuruvilla *et al.*, 1991). It seems remarkable that a mechanism of nonantigenic-specific IgG-mediated suppression and/or regulation of CTL should exist. If it provides some protective mechanism in the immunity of pregnancy, as has been speculated, it would have evolved as an important and conserved mechanism (Rowley and Stach, 1993). On the other hand, failure to develop both CTL-mediated and B-cell immunity to particular protective epitopes while expanding the response to other antigens expressed sequentially after a first or dominant immunization should be detrimental to individuals bearing immunogenic tumors or infected with organisms that give rise to variants expressing new or cross-reactive epitopes. The aforementioned discussions of newly discovered basic immunologic networks regulating the presentation of antigen to the immune system are providing a wealth of scientific opportunities to piece together an old phenomenon that may provide insights into the immunity of pregnancy, immunologic memory, how to develop new ways for promoting allograft survival in transplant recipients, and, last, a way for the induction of more effective immunity to the class of currently "vaccine-resistant" microbes and tumor cells.

C. A FLEETING MEMORY?

Classically, substances that elicit B-cell responses are divided into thymus-independent (TI) and thymus-dependent (TD) antigens depending on whether they require substantial cooperating mature T cells to generate an antibody response. Because the final consequence of antigenic stimulation is the same for both types of substances, that is, proliferation, differentiation and maturation of B cells to antibody products, this implies that TI antigens can circumvent or substitute for signals delivered by T lymphocytes for TD antigens. Traditionally, the TI antigens, both TI-1 and TI-2, immunochemically consist of bacterial polysaccharides, such as lipopolysaccharide of gram-negative bacteria and pneumococcal saccharides, as well as sheep red blood cells and human gamma globulin (Veenhoff and Seijen, 1982). The primary immune response to TI antigens is primarily restricted to antibodies of the IgM isotype, although interestingly, a few examples exist in the literature of primary responses of the IgG isotype, mainly IgG3 (Sharon et al., 1975; Kunkle and Klaus, 1981; Mosier et al., 1974; Rude et al., 1976; Humbert et al., 1979).

Previous studies of B-cell responses during HIV-1 infection provide evidence suggesting that an abnormal activation of the B-cell compartment occurs during HIV-1 infection (Table I). The nature of some of these B-cell abnormalities suggests that immunodominant antigen(s) elicit strong yet nonfunctional immune responses. These antigens have cross-reactive properties and homology residing at the protein, glycoprotein, and carbohydrate residues level (Kohler et al., 1994). There has been a long-standing belief that bacterial antigens residing on or within the host as normal flora, or those species attempting to colonize the host from outside, play an important role in the evolution of the B-cell immune response (Colle et al., 1990). Since they are mostly found on the membrane and/or cell wall of the microorganisms, they are naturally present in the environment and may contribute to the permanent stimulation of the immune system.

1. Evidence for a Primary IgG Response to the HIV-1 Env in Experimentally Infected Chimpanzees and Natural Infections of Humans

Due to previous reports of a clonally dominant humoral response and immunochemical evidence for homology of carbohydrate of HIV-1 and bacterial polysaccharides we investigated the early Ig isotype responses to a native, unprocessed, oligomeric form of gp160 secreted from human T-cell line (HUT78) (Kalyanaraman et al., 1988, 1990; Van Cott et al., 1995) in experimentally HIV-1-infected chimpanzees ($n = 5$) and preseroconversion serum panels (Boston Biomedical, Inc.; Serologics, Inc.) from HIV-1-infected humans ($n = 13$). In addition, serologic responses to gp120 and p24 by RIP and neutralization were studied and compared to the serologic response to the oligomeric form of the protein.

Preliminary serologic studies utilized the controlled experimental inoculation of varying 10-fold dilutions of HIV-1 IIIB in chimpanzees. All infected animals mounted a serologic response that correlated with a persistent HIV-1 infection as determined by reisolation of virus from their peripheral blood cells over a period of weeks to months to years (Arthur et al., 1989). Interesting, however, were the different kinetics and/or sensitivities of antibody responses to various structural proteins of the virus. Sampling every 2 weeks following infection revealed that the earliest antibody detectable by any of the serologic assays was detected with the oligomeric HIV-1 envelope gp160$_{451}$ antigen (Table III). The IgG gp160$_{451}$ ELISA was positive at dilutions of 1:100–1:1600 at 2 weeks (animals 856 and 923) and 4–6 weeks earlier in animals 854 and 851 than that detected by RIP to the immunoaffinity-purified gp120 HIV-1 IIIB and/or the detection of neutralizing antibody to the homologous virus. Samples for one animal, 911, were not available for testing on days 40 and 54; however, by day 70 a titer of 1:6400 was observed. Interestingly, no IgM was detected against the oligomeric gp160$_{451}$ from any of the samples taken from these animals. Thus it appears that a very early and rapid humoral response to infection with HIV-1 results from a population of antibody recognizing full-length, oligomeric, glycosylated HIV-1 gp160. There is a possibility that sampling every 2 weeks may have missed a short-lived response. Previous work (data not shown) demonstrated that IgM to the p24 was detectable in these samples, however, was not detected until 1–2 weeks later. Prior to this finding, the detection of low-level neutralizing antibody to the homologous virus was the earliest antibody/serologic assay capable of detecting a specific antiviral response (Nara et al., 1990).

Analysis of the human acute seroconversion panels ($n = 6$ of 13) are presented in Table IV. These major Ig isotypic profiles were observed. One patient exhibited a primary IgG in the absence of any IgM and/or in association with a weak and low titered IgM response (panel 2211 and C). The second pattern was that of a more simultaneous rise in both IgM and IgG (2215, I, and 2214). In general (4/6), the serologic reactivity to the oligo gp160$_{451}$ was detected prior to, or as an indeterminant reaction of, the EIA and WB. However, in one patient (panel A), the IgM response preceded the IgG response.

2. Summary and Conclusions

These findings are consistent with a earlier report in which the earliest detectable HIV-1 specific antibody, as measured by live cell immunofluorescence of infected T cells (Race et al., 1991), was found to be of the IgG isotype. In addition, these antibodies were found to cross-react with many different isolates. Other studies also demonstrated

INSIGHTS INTO IMMUNE INVASION AND VACCINE DEVELOPMENT 125

TABLE III

Ig Isotype Kinetics and Serologic Profiles for HIV-1 Experimentally Infected Chimpanzees

Animal	Dose (TCID$_{50}$)	Day	ELISA Oligo gp160 IgM	ELISA Oligo gp160 IgG	RIP p24	RIP gp120	Neutralization
856	4000	−7	—	—	—	—	—
		0	—	—	—	—	—
		12	—	—	—	—	—
		26	—	1:100	—	—	—
		40	—	1:1600 1:50	—	—	1:4–8
923	400	−7	—	—	—	—	—
		0	—	—	—	—	—
		12	—	—	—	—	—
		26	—	—	—	—	—
		40	—	1:100	—	—	—
		54	—	1:800	1:50	—	1:4–8
		70	—	1:1600 1:50	—	—	1:8
		84	—	1:3200 1:50	1:250	—	1:32
911	400	−7	—	—	—	—	—
		0	—	—	—	—	—
		12	—	—	—	—	—
		26	—	—	—	—	—
		40	—	ND	1:50	T/50	1:8
		54	—	ND	1:50	—	1:8–16
		70	—	1:6400 1:50	1:250	—	1:16
854	40	−7	—	—	—	—	—
		0	—	—	—	—	—
		12	—	—	—	—	—
		26	—	—	—	—	—
		40	—	1:400	—	—	—
		54	—	1:6400 1:50	—	—	—
		70	—	>1:6400	1:50	T/50	1:4–8
851	4	−7	—	—	—	—	—
		0	—	—	—	—	—
		12	—	—	—	—	—
		26	—	—	—	—	—
		40	—	1:50	—	—	—
		54	—	1:3200 T/50	—	—	—
		70	—	>1:6400	50	—	—
		84	—	>1:6400	50	1:250	1:4–8

the presence of such antibody in both natural infections of humans (Moore *et al.*, 1994) and experimental infections of chimpanzees (Nara *et al.*, 1990). Recently, this oligo gp160 preparation was found to be

TABLE IV

Ig Isotype Kinetics and Serologic Profiles during Primary HIV-1 Infection of Humans

| Panel No. | Day | ELISA Oligo gp160* | | Serology | | Virus |
		IgM	IgC	EIA	WB	p24
2211	7	0.00	0.01	−	−	nd
	10	0.00	0.01	−	−	nd
	21	0.12	0.86	+	+	nd
	24	0.13	1.23	+	+	nd
	28	0.14	1.41	+	+	nd
	34	0.05	1.64	+	+	nd
	41	0.04	1.99	+	+	nd
	81	0.01	2.37	+	+	nd
C	0	0.00	0.00	−	−	−
	7	0.00	0.00	−	−	+
	9	0.03	0.01	−	−	+
	14	0.13	0.14	−	−	−
	16	0.04	0.13	+/−	−	−
	21	0.03	0.36	+/−	−	−
	23	0.02	0.39	+/−	−	−
	28	0.01	0.32	+	−	−
	30	0.01	1.06	+	+	−
	72	0.00	1.62	+	+	−
2215	0	0.01	0.01	−	−	nd
	2	0.01	0.01	−	−	nd
	7	1.09	0.98	+	+	nd
	9	1.16	1.46	+	+	nd
	14	0.95	1.72	+	+	nd
	63	0.02	2.22	+	+	nd
I	0	0.02	0.02	−	+	+
	7	0.11	0.07	−	+	+
	14	0.44	0.38	+	+	+
	16	0.63	0.96	+	+	+
	21	1.08	1.28	+	+	+
	23	0.99	1.31	+	+	+
	28	0.55	1.50	+	+	−
	30	0.27	1.69	+	+	−
2214	0	0.01	0.05	−	−	nd

more sensitive than the other commercial diagnostic EIA and WB tests (Van Cott et al., 1995; Nair et al., 1994) in detecting an early antibody response following infection. Furthermore, this recognition may require a higher ordered conformation involving an oligomeric form of the protein (Van Cott et al., 1995); however, more experiments need to be performed.

The presence of an IgG isotype occurring alone or in combination with a simultaneous IgM response is evidence of nonconventional primary response. Assuming no technical or logistical limitations (i.e., sampling period) are to blame, the results of these studies suggest that a somewhat unconventional primary antibody response to an oligomeric form of the HIV-1 envelope may be initiated very early following HIV-1 infection. In explanation of these results one could consider that this is either a specific or cross-reactive anamnestic recall response or an as-yet unappreciated IgG-specific primary response as mentioned earlier. At least three papers that have been published to date suggest the response may be of the former type (i.e., may have a preexisting antibody repertoire to the viral envelope). The first, by Zubler et al. (1991), demonstrates that a high percentage (40–50%) of a rather limited number of subjects, both HIV-1 negative adults ($n = 9$) and newborns ($n = 6$), had relatively high frequencies of HIV-1 envelope-specific and reactive B cells, 1:16,200–49,000 and 1:11,800–26,600, respectively. Work by Davis et al. (1990) describes specific and cross-reactive antibody in normal laboratory volunteers, all who recognized the same two major immunodominant peptides (amino acids 601–615 and 771–785) in gp41 recognized by HIV-1 infected sera. Additional data from that paper demonstrated that amino acid sequences similar to those found in the gp41 of HIV-1 were shared and common among a number of common viruses infecting humans, that is, rhinovirus, poliovirus, herpesvirus, Coxsackie virus, and paramyxoviruses (Gonzalez-Scarano et al., 1987). In addition, they exhibited the same IgG2 subclass restriction to rhinovirus type 2. Similar Ig class restriction(s) have been reported by other authors for epitopes in gp41 (Sundqvist et al., 1986; Mathiesen et al., 1988, 1989). Additional cross-reactivity to the gp41 (amino acids 705–752) has also been seen with increasing frequency using the new and more sensitive EIAs (Sayre et al., 1996).

Evidence for homology of the HIV-1 envelope to bacterial polysaccharides comes from several areas. First, gp120 is capable of binding with high affinity to mannose binding protein, a plasma acute phase c-type lectin binding protein whose specificity is for trimeric CHO moieties found through a Ca^{2+}-dependent interaction (Ezekowitz et al., 1989). In addition, naturally occurring carbohydrate-specific antibodies (Tomiyama et al., 1991) as well as murine monoclonal antibodies directed at specific N- and O-linked carbohydrate epitopes (Hansen et al., 1990) have been reported to recognize HIV-1 envelope glycoproteins. Putting these findings into context would suggest that the HIV-1 envelope may have evolved an antigenic makeup consisting of epitopes that are "naturally immunodominant" yet paradoxically nonprotective. This type of phenomenon has been observed before for many multideterminant experimental antigens and chronic-active pathogens (for a review, see Nara, 1996). Immunologically speaking, these epitopes would exist due to their inherent ability to cross-react with preexisting B-cell clones (i.e., protein or carbohydrate epitopes), which when stimulated would rapidly proliferate and outcompete other clones and clonally dominate the primary response (Fig. 1). Other antigens such as Staph A and polylactosamine-1 have been shown to interact with antibody binding pocket of the Ig molecule in unconventional ways. These antigens and others appear to bind more preferentially to one chain or the other or outside such as is the case with Staph A. This is reminiscent of the T-cell superantigens which bind the MHC receptor outside the conventional binding pocket. A large body of literature exists suggesting that antibodies regulate both B- and T-cell responses (Stach and Rowely, 1993). More relevant to this discussion and vaccination, however, is how a primary dominant antibody response to an immunodominant epitope subsequently shapes the responses to other less immunodominant epitopes in the multideterminant antigen. This has been most recently and elegantly shown for the surface Ag protein for hepatitis B virus (Vijayakrishanan *et al.*, 1994). In this study it was clearly demonstrated that the presence of a rapid and early antibody response to an immunodominant epitope on that viral antigen prevented, delayed, and selectively down-regulated the maturation of epitope-specific B/T-cell-specific responses to a less immunodominant epitope. The previously reported preexisting antibodies cross-reactive with the HIV-1 gp120/41, along with our findings of an unconventional primary Ig isotype response to the HIV-1 envelope, may be evidence for the evolution and selection of a virus envelope which is immunochemically proved for anamnestic recall to other dominant and established immune networks. Thus the strategy known as *deceptive imprinting* may have evolved as strategy to decoy and dysregulate the immune responses toward nonprotective epitopes while attenuating or preventing the recognition of other more conserved and possibly protective epitopes. Further defining the epitopes and mechanism(s) involved in this deceptive imprinting and/or recall response(s) will be important for future vaccine design involving the viral envelope. We are currently investigating novel strategies to mask or eliminate these epitopes from the native protein while preserving the relevant and protective epitopes.

III. Refocusing the Immune Response by Masking Epitopes Responsible for Deceptive Imprinting: Novel Approach to Vaccination

A striking disparity exists between the small number of diseases that are currently prevented by available vaccines and the many infectious diseases for which either no vaccine exists or the vaccine is of such low efficacy that no real protection is provided in the vaccinated outbred populations of man or animals. A recent National Institutes of Health Strategic Panel Report on Immunology and Vaccines pointed out that one of the principal reasons for vaccine failure is the the inability of current vaccine technologies to overcome antigenic variation and the lack of knowledge concerning the mechanism(s) employed by the pathogens to survive despite an ongoing immune response. As discussed earlier, it appears that one somewhat cosmopolitin strategy is the initial presentation of nonprotective and/or type restricted immunodominant epitopes to the host immune system, resulting in a deceptive response from which the host does not ever fully regain. Again it is important to note that these "deceptive responses" may result in type restricted protection and semiprotective immunity and does not have to result in the complete absence of any microbe controlling immunity.

Briefly, our previous work on masking these epitopes through site-directed technology resulted in the immune masking through either the reduction of charge and/or the introduction of N-linked carbohydrate to amino acid triplet sequones (N-X-T/S), where X is any amino acid other than P or D (Garrity et al., 1997). A set of four site-specific N-linked mutations was enough to immune dampen a 32-amino-acid determinant in the third hypervariable domain known to contain epitopes responsible for deceptive imprinting in the HIV-1 envelope. Two of the four sites were glycoslated while two sites were not; however, the combination of reductuion of charge and N-linked glycosylation resulted in this determinant becoming immune masked while retaining the complex conformation of the envelope molecule as determined by recombinant binding studies to the CD4 molecule and the resulting viable infectious virus when these were reintroduced into recombinant viruses. The resulting immune responses demonstrated antibodies to new conserved domains while at the same time shifting the neutralizing antibody response to a more broadly *in vitro* protective pool. Interestingly, single site-specific N-linked mutants where also observed to refocus the antibody response to adjacent epitopes, both N and COOH terminal to the N-linked site within the domain.

Dampening immunodominant epitopes such as the V3 domain of the HIV-1 envelope can lead to a shift or refocusing of the antibody response to an otherwise silent, qualitatively different, second-order epitope. Whether this strategy can be used to reveal as-yet more potent and broader neutralizing epitopes awaits ongoing research at this time. In closing, however, it is important to note that a large number of persistent and annually recurring pathogens in addition to HIV-1 complicate host immunity through the presentation of immunodominant, hypervariable, or repeated epitopes; and many if not most of these pathogens including cancer cells have resisted conventional vaccine strategies. These epitopes appear to play little or no role in protection. If the evolution of these epitopes is part of a clever pathogenic strategy to evade immune clearance, then the approach of masking these epitopes and refocusing the immune response away from them may have a broad application in the development of more broadly protective second-generation vaccines and therapeutics.

REFERENCES

Albright, J. F., Omer, T. F., and Deitchman, J. W. (1970). Antigenic competition: Antigens compete for a cell occurring with limited frequency. *Science* **167,** 196.

Amadori, A., and Chieco-Bianchi, L. (1992). B cell activation and HIV infection: Protective or potentially detrimental response? *Int. Rev. Immunol.* **9,** 15–24.

Angelova, L. A., and Shvartsman, Y. S. (1982). Original antigenic sin to influenza in rats. *Immunology* **46,** 183–188.

Arthur, L. A., Bess, J. W., Jr., Waters, D. J., Pyle, S. W., Kelliher, J. C., Nara, P. L., Krohn, K., Robey, W. G., Langlois, A. J., Gallo, R. C., and Fischinger, P. J. (1989). Challenge of chimpanze (*Pan troglodytes*) immunized with human immunodeficiency virus envelope glycoprotein, gp120. *J. Virol.* **63,** 5046–5053.

Back, N. K. T., Smit, L., Schutten, M., Nara, P. L., Tersmette, M., and Goudsmit, J. (1993). Mutations in human immundeficiency virus type 1 gp41 after sensivity to neutralization by gp120 antibodies, *J. Virol.* **67,** 6897–6902.

Benjamin, D. C., Berzofksky, J. A., East, I. J., Gurd, F. N., Hannum, C., Leach, S. J., Margoliash, E., Michael, J. G., Miller, A., Prager, E. M., Reichlin, M., Sercarz, E. E., Smith-Gill, S. J., Todd, P. E., and Wilson, A. C. (1984). The antigenic structure of proteins: A reprasial. *Annu. Rev. Immunol.* **2,** 67–101.

Berberian, L. (1993). Immunoglobulin VH3 gene products: Natural ligands for HIV gp120. *Science* **261,** 1588–1591.

Briault, S., Courtois-Capella, M., Duarter, F., Aucouturier, P., and Preud'Homme, J. L. (1988). Isotypy of serum monoclonal immunoglobulins in human immundeficiency virus-infected adults, *Clin. Exp. Immunol.* **74**(2), 182–184.

Briles, D. E., and Davie, J. M. (1980). Clonal nature of the immune response. II. The effect of immunization on clonal commitment. *J. Exp. Med.* **152,** 151–160.

Coffman, R. L., Lebman, D. A., and Shrader, B. (1989). Transforming growth factor beta specifically enhances IgA production by lipopolysaccharide-stimulated murine B lymphocytes. *J. Exp. Med.* **170**(3), 1039–1044.

Colle, J.-H., LeMoal, M. A., and Truffa-Bachi, P. (1990). Immunological memory. *Crit. Rev. Immunol.* **10**, 270–288.

Davis, D., Chaudhri, B., Stephens, D. M., Carne, C. A., Willers, C., and Lachmann, P. J. (1990). The immunodominance of epitopes within the transmembrane protein (gp41) of human immunodeficiency virus type 1 may be determined by the hosts previous exposure to similar epitopes on unrelated antigens. *J. Gen. Virol.* **71**, 1975–1983.

Ezekowitz, R. A. B., Kuhlman, M., Groopman, J. E., and Byrn, R. A. (1989). A human serum mannose-binding protein inhibits *in vitro* infection by the human immunodeficiency virus. *J. Exp. Med.* **169**, 185–196.

Fenner, F., McAuslan, B. R., Mims, C. A., Sambrook, J., and White, D. O., eds. (1974). "The Biology of Animal Viruses," 2nd ed., pp. 417–418. Academic Press, London.

Garrity, R., Rimmelzwaan, G., Minassin, A., Tsai, W.-P., Lin, G., de Jong, J.-J., Goudsmit, J., and Nara, P. L. (1997). Refocusing neutralizing antibody response by targeted dampening of an immunodominant epitope. *J. Immunol.* **159**, 279–289.

Gonzalez-Scarano, F., Waxham, M. N., Ross, A. M., and Hoxie, J. A. (1987). Sequence similarities between human immunodeficiency virus gp41 and paramyxovirus fusion proteins. *AIDS Res. Hum. Retroviruses* **3**, 242–252.

Hansen, J.-E.S., Clausen, H., Nielsen, C., Teglbjaerg, L. S., Hansen, L. L., Nielsen, C. M., Dabelsteen, E., Mathiesen, L., Hakomori, S.-I., and Nielsen, J. O. (1990). Inhibition of human immunodeficiency virus (HIV0 infection *in vitro* by anticarbohydrate monoclonal antibodies: Peripheral blycosylation of HIV envelope blycoprotein gp120 may be a target for virus neutralization. *J. Virol.* **64**, 2833–2840.

Harboe, M. (1974). Recognition of two distinct groups of human IgM and IgA based on different binding to staphylococci. *Scand. J. Immunol.* **3**(4), 471–482.

Humbert, J., Motta, I., and Truffa-Bachi, P. (1979). TNP-LPS induces an IgG anti-TNP immune response in mice. *Cell. Immunol.* **47**, 211.

Inganas, M. (1981). Comparison of mechanisms of interaction between protein A from Staphylococcus aureus and human monoclonal IgG, IgA and IgM in relation to the classical FC gamma and the alternative F(ab')2 epsilon protein A interactions. *Scand. J. Immunol.* **13**(4), 343–352.

Kakiuchi, T., Watanabe, M., Hozumi, N., and Nariuchi, H. (1990). Differential sensitivity of specific and nonspecific antigen-presentation by B cells to a protein synthesis inhibitor, *J. Immunol.* **145**(6), 1653–1658.

Kalams, S. A., Johnson, R. P., Trocha, A. K., Dynan, M. J., Ngo, H. S., D'Aquila, R. T., Kurnick, J. T., and Walker, B. D. (1994). Longitudinal analysis of the T cell receptor (TCR) gene usage by humnan immunodeficiency virus 1 envelope-speecific cytotoxic T lymphocyte clones reveals a limited TCR repertoire. *J. Exp. Med.* **179**, 1261–1271.

Kalyanaraman, V. S., Pal, R., Gallo, R. C., and Sarngadharan, G. (1988). A unique human immunodeficiency virus culture secreting soluble gp160. *AIDS Res. Hum. Retroviruses* **4**, 319–329.

Kalyanaraman, V. S., Rodriguez, V., Veronese, F., Rahman, R., Lusso, P., DeVico, A. L., Copeland, T., Oroszlan, S., Gallo, R. C., and Sarngadharan, M. G. (1990). Characterization of the secreted, native gp120 and gp160 of the human immunodeficiency virus type 1. *AIDS Res. Hum. Retroviruses* **6**, 371–380.

Kittlesen, D. J., Brown, L. R., Braciale, V. L., Sambrook, J. P., Gething, M.-J., and Braciale, T. J. (1993). Presentation of newly synthesized glycoproteins to CD4+ T lymphocytes. An analysis using influenza hemagglutinin transport mutants. *J. Exp. Med.* **177**, 1021–1030.

Kohler, H., Goudsmit, J., and Nara, P. L. (1992). Clonal antibody dominance in HIV-1

infection: Cause for a limited and failing immune response to HIV-1 infection and vaccination. *J. Acquired Immune Defic. Syndr.* **5,** 1158–1168.
Kohler, H., Muller, S., and Nara, P. (1994). Deceptive imprinting in the immune response against HIV-1 *Immunol. Today* **13,** 475–478.
Kunkl, A., and Klaus, G. G. B. (1981). The generation of memory cells. V. Preferential priming of IgG1 B memory cells by immunization with antigen IgG2 antibody complexes. *Immunology* **44,** 163.
Kuruvilla, A. P., Shah, R., Hochwald, G. M., Liggitt, H. D., Palladino, M. A., and Thorbecke, G. J. (1991). Protective effect of transforming growth factor beta 1 on experimental autoimmune diseases in mice. *Proc. Natl. Acad. Sci. USA* **88**(7), 2918–2921.
Langone, J. J. (1982). Protein A of *Staphylococcus aureus* and related immunoglobulin receptors produced by streptococci and pneumonococci. *Adv. Immunol.* **32,** 157–252.
Lin, Y. A., and Stavnezer, J. (1992). Regulation of transcription of the germ-line Ig alpha constant region by an ATF element and by novel transforming growth factor-beta 1 responsive elements. *J. Immunol.* **149,** 2914.
Lorenz, R. G., Blum, J. S., and Allen, P. M. (1990). Constitutive competition by self proteins for antigen presentation can be overcome by receptor-enhanced uptake. *J. Immunol.* **144,** 1600–1608.
Martinez-Maza, O., Crabb, E., Mitsuyasu, R. T., Fahey, J. L., and Giorgi, J. V. (1987). Infection with human immundeficiency virus (HIV) is associated with an increase in B lymphocyte activation and immaturity. *J. Immunol.* **138,** 3720–3724.
Mathiesen, T., Sonnerborg, A., von Sydow, M., Gaines, H., and Wahren, B. (1988). IgG subclass reactivity against human immunodeficiency virus (HIV) and cytomegalovirus in cerebrospinal fluid and serum from HIV-infected patients. *J. Med. Virol.* **25,** 17–26.
Mathiesen, T., Chiodl, F., Broliden, P.-A., Houghten, A. J., Utter, R. A., Wahren, B., and Norrby, E. (1989). Analysis of a subclass-restricted HIV-1 gp41 epitope by omission peptides. *Immunology* **67,** 1–7.
Mitchell, G. F. (1991). Co-evolution of parasites and adaptive immune responses. *Immunol. Today* **12**(3), A2–A5.
Moller, G., and Sjöberg, O. (1970). Effect of antigenic competition on antigen-sensitive cells and on adoptively transferre immunocompetent cells. *Cell. Immunol.* **1**(1), 110–121.
Moore, J. P., Cao, Y., Ho, D. D., and Koup, R. A. (1994). Development of the anti-gp120 antibod response during seroconversion to human immunodeficiency virus type 1. *J. Virol.* **68,** 5142–5155.
Mosier, D. E., Johnson, B. M., Paul, W. E., and McMaster, P. R. B. (1974). Cellular requirements for the primary *in vitro* antibody response to DNP-Ficoll. *J. Exp. Med.* **139,** 1354.
Muller, S., Nara, P., D'Amelio, R., Biselli, R., Gold, D., Wang, H., and Silverman, G. J. (1992). Clonal patterns in the human immune response to HIV-1 infection. *Int. Rev. Immunol.* **9,** 1–13.
Nair, B. C., Ford, G., Kalyanaraman, V. S., Zafari, M., Fang, C., and Sarngadharan, M. G. (1994). Enzyme immunoassay using negative envelope glycoprotein (gp160) for detection of human immunodeficiency virus type 1 antibodies. *J. Clin. Microbiol.* **32**(6), 1449–1456.
Nara, P. L. (1996). Humoral immunity to HIV-1: Lethal force or trojan horse? *In* "Immunology of HIV Infection" (S. Gupta, ed.), Chapter 12. Plenum, New York.
Nara, P. L., and Goudsmit, J. (1991). Clonal dominance of the neutralizing response to the HIV-1 V3 epitope: Evidence for "Originak Antigenic Sin" during vaccination and

infection in animals, including humans. In "Vaccines 91" (R. A. Lerner, H. Ginsberg, R. M., Chanock, and F. Brown, eds.), pp. 51–58. Cold Spring Harbor Lab. Press, Cold Spring Harbor, NY.

Nara, P. L., Smit, L., Dunlop, N., Natch, W., Merges, M., Waters, D., Kelliher, J., Gallo, R. C., Fischinger, P. J., and Goudsmit, J. (1990). Emergence of viruses resistant to neutralization by V3-specific antibodies in experimental human immunodeficiency virus type 1 IIIB infection of chimpanzees. *J. Virol.* **64,** 3779–3791.

Ng, V. L., Hwang, K. M., Reyes, G. R., Kaplan, L. D., Khayam-Bashi, H., Hadely, W. K., and McGrath, M. (1988). High titer anti-HIV antibody reactivity associated with a paraprotein spike in a homosexual male with AIDS related complex. *Blood* **71,** 1397–1401.

Pang, S., Sclesinger, Y., Darr, E. S., Moudgil, T., Ho, D. D., and Chen, I. S. (1992). Rapid generation of sequence variation during primary infection. *AIDS* **6,** 453–460.

Pantalero, G., Demarest, J. F., Soudeyns, H., Grazioso, C., Denis, F., Adelsberger, J. W., Borrow, P., Saag, M. S., Shaw, G. M., Sekaly, R. P., and Fauci, A. S. (1994). Major expansion of CD8+ T cells with a predominant VB usuage during the primary immune response to HIV. *Nature* **370,** 463–467.

Race, E. M., Ramsey, K. M., Lucia, H. L., and Cloyd, M. W. (1991). Human immunodeficiency virus elicits antibody not detected by standard tests: Implications for diagnostics and viral immunology. *Virology* **184,** 716–722.

Radovick, J., and Talmadge, D. W. (1967). Antigenic competition: Cellular or humoral. *Science* **158,** 512.

Rowley, D. A., and Stach, R. M. (1993). A first or dominant immunization. I. Suppression of simultaneous cytolytic T cell responses to unrelated alloantigens. *J. Exp. Med.* **178**(3), 835–840.

Rude, E., Wrede, J., and Gundelach, M. L. (1976). Production of IgG antibodies and enhanced response of nude mice to DNP-AE-Dextran. *J. Immunol.* **116,** 527.

Sayre, K. R., Dodd, R. Y., Tegtmeier, G., Layug, L., Alexander, S. S., and Busch, M. P. (1996). False-positive human immunodeficiency virus type 1 Western blot tests in noninfected blood donors. *Transfusion* **36,** 45–52.

Schechter, I. (1968). Antigenic competition between polypeptidyl determinants in normal and tolerant rabbits. *J. Exp. Med.* **127,** 237–250.

Schnittman, S., Lane, H., Higgins, S., and Folks, T. (1986). Direct polyclonal activation of human B lymphocytes by the acquired immune deficiency syndrome virus. *Science* **233,** 1084–1088.

Sharon, R., McMaster, P. R. B., Kask, A. M., Owens, J. D., and Paul, W. E. (1975). DNP-Lys-Ficol: A T-independent antigen which elicits both IgM and IgG anti-DNP antibody-secreting cells. *J. Immunol.* **114,** 1585.

Silverman, G. J. (1994). Superantigens and the spectrum of unconventional B-cell antigens. *Immunologist* **2**(2), 51–57.

Silverman, G. J., and Kohler, H. (1992). Clonal restriction in human antibody responses to unfections. *Int. Rev. Immunol.* **9**(1), 1–57.

Sonoda, E., Matsumoto, R., Hitoshi, Y., Ishii, T., Sugimoto, M., Araki, S., Tominaga, A., Yamaguchi, N., and Takatsu, K. (1989). Transforming growth factor beta induces IgA production and acts additively with interleukin 5 for IgA production. *J. Exp. Med.* **170**(4), 1415–1420.

Stach, R. M., and Rowley, D. A. (1993). A first or dominant immunization. II. Induced immunoglobulin carries transforming growth factor β and suppresses cytolytic T cell responses to unrelated alloantigens. *J. Exp. Med.* **178,** 841–852.

Sundqvist, V.-A., Linde, A., Kurtu, R., Werner, A., Helm, E. B., Popovic, M., Gallo, R. C., and Wahren, B. (1986). Restricted IgG subclass responses to HTLV III/LAV and to cytomegalovirus in patients with AIDS and lymphadenopathy syndrome. *J. Infect. Dis.* **153,** 970–973.

Tomiyama, T., Lake, D., Masuho, Y., and Hersh, E. M. (1991). Recognition of human immunodeficiency virus glycoproteins by natural anti-carbohydrate antibodies in human serum. *Biochem. Biophys. Res. Commun.* **177,** 279–285.

Van Cott, T. C., Veit, S. C. D., Kalyanaraman, V., Earl, P., and Birx, D. L. (1995). Characterization of a soluble, oligomeric HIV-1 gp160 protein as a potential immunogen. *J. Immunol. Methods* **183,** 103–117.

Veenhoff, E., and Seijen, H. G. (1982). Plasma cells and their precursors I. Preexistent antibody-forming cell precursors dominate the anti-sheep red blood cell and anti-human gamma globulin IgG antibody response. *Immunobiology* **162,** 165–174.

Vijayakrishnan, L., Kumar, V., Agrewala, J. N., Mishra, G. C., and Rao, K. V. S. (1994). Antigen-specific early primary humoral responses modulate immunodominance of B cell epitopes. *J. Immunol.* **153,** 1613.

Waterson, R. H. (1970). Antigenic competition: A paradox. *Science* **170,** 1108.

Wolfs, T. F., Zwart, G., Bahher, M., and Goudsmit, J. (1992). Hiv-1 genomic RNA diversification following sexual and parental transmission. *Virology* **189,** 103–110.

Zhang, L. Q., MacKenzie, P., Cleland, A., Holmes, E. C., Leigh-Brown, A. J., and Simmonds, P. (1993). Selection for specific sequences in the external envelope protein of human immunodeficiency virus type 1 upon primary infection. *J. Virol.* **67,** 3345–3356.

Zhu, T., Mo, H., Wang, N., Nam, D. S., Cho, Y., Kuup, R. A., and Ho, D. D. (1993). Genotypic and phenotypic characterization of HIV-1 in patients with primary infection. *Science* **261,** 1179–1181.

Zubler, R. H., Perrin, L. H., Doucet, A., Zhang, X., Huang, Y.-P., and Miescher, P. A. (1991). Frequencies of HIV-reactive B cells in seropositive and seronegative individuals. *Clin. Exp. Immunol.* **87,** 31–36.

Vaccination against Tuberculosis: Recent Progress

IAN M. ORME

Department of Microbiology, College of Veterinary Medicine and Biomedical Sciences, Colorado State University, Fort Collins, Colorado 80523

I. Introduction
II. Acquired Immunity and Memory Immunity to Tuberculosis Infection
III. Types of Vaccines
    A. Auxotrophic and Recombinant Vaccines
    B. Subunit Vaccines
    C. DNA Vaccines
IV. Can We Increase Herd Resistance to Bovine Tuberculosis?
    Acknowledgment
    References

I. Introduction

The World Health Organization (WHO) recently declared a global emergency for tuberculosis. The mortality rate from this disease had remained stable for the past decade at about 2.5 to 2.9 million deaths per year, but recently, primarily because of the human immunodeficiency virus (HIV) epidemic, this rate appears to have surged.

The bacille Calmette-Guérin (BCG) vaccine for tuberculosis was developed more than 70 years ago, and has been used throughout the world, with the exception of the United States. Unfortunately, despite the fact that the vaccine is safe and cheap to make, in many controlled field trials the actual efficacy of vaccination has been low or close to zero. In the Poster Child for BCG vaccination, a massive field conducted in India in the 1980s, you were actually better off being in the placebo group (Hitze, 1980).

As a result, the National Institutes of Health (NIH) has established programs designed to stimulate new vaccine research, and the initial response has been very encouraging with multiple laboratories coming up with highly innovative approaches to the problem. These are still at the animal model testing stage, but provide optimism that new strategies will emerge (Orme, 1995). These may not replace BCG vaccination, given its high rate of efficacy in young children, but should certainly be considered as boosters for vaccinated people in young adulthood where the efficacy of BCG vaccination seems to wane.

There are certain parallels between human tuberculosis (TB) and cattle TB. Both diseases are controlled well in certain countries (United States, Australia), but are rife in others (Africa, Asia) (Stanford et al., 1991). In some countries (Ireland, England, New Zealand) the incidence is low due to good control measures, but the disease is far from eradicated and hence continues to take a significant toll on the economy of these countries (Tweddle and Livingstone, 1994). To date there has been no serious attempt at vaccination as a control measure, but the time may now have come to consider this option.

This brief review describes the different approaches now being taken, and progress to date in terms of testing in mouse and guinea models of infection (see Table I).

II. Acquired Immunity and Memory Immunity to Tuberculosis Infection

The purpose of vaccination is to establish a state of memory immunity, which in the context of tuberculosis consists of a long-lived population of recirculating CD4 T cells that can secrete interferon (IFN) on recognition of specific antigens and hence mediate an accelerated recall of specific acquired resistance. It remains unknown precisely how the tuberculosis bacillus is killed following macrophage activation by IFN, but a combination of a drop in phagosomal pH combined with the production of noxious compounds such as peroxynitrite are probably the most important factors.

Our current picture of the response in the lungs to tuberculosis infection involves the interaction of a bacillus that has managed to carry through the tidal airspace and is engulfed on the alveolar or peribronchial surface by an alveolar macrophage. The bacillus begins to replicate inside this cell, provoking it to spread and tightly adhere to the alveolar endothelial surface. The most probable consequence is that the cell is able to kill the bacilli, but if this does not occur the likelihood increases that the cell will lyse and bacilli will then erode into the thin basement membrane between the alveolus and the adjacent capillary.

TABLE I

Vaccine Development: Recent Progress

Vaccine type	Animal	Activity	Notes
CFP in incomplete Freund's	Mouse	Yes	Initially as good as BCG; then activity waned
CFP in DDA adjuvant	Mouse	Yes	Good long-term protection
CFP-derived proteins in Synthex	GP	Yes	Reduction in bacterial load; survival >10 weeks
CFP in MPL plus IL-2	GP	Yes	Protection >30 weeks; no necrosis in lungs
M. vaccae	Mouse	No	
M. vaccae	GP	No	
M. microti	GP	No	
M. vaccae expressing 19 kDa	Mouse	No	Infection made worse
rBCG expressing Osp-A	Mouse	Yes	BCG efficacy not compromised
rBCG expressing human/murine cytokines	Mouse	Yes	BCG efficacy not compromised
rBCG expressing Osp-A	GP	Yes	BCG efficacy not compromised
rBCG expressing human/murine cytokines	GP	Yes	BCG efficacy not compromised
Low-dose BCG versus high-dose	GP	Yes	Equally effective
Low-dose BCG versus high-dose	Mouse	Yes	Equally effective
Merck DNA-Ag85	Mouse	Yes	Protective
Merck DNA-Ag85	GP	Yes	Protects against caseous necrosis
TB and BCG auxotrophs	Mouse	Yes	Slowly cleared
TB and BCG auxotrophs	GP	Yes	Not fully resolved

This seems a likely event for two reasons:

1. It explains the long-known phenomenon that soon after initial infection some bacilli "reseed" to other areas of the lung where the V/Q ratio is more favorable to the aerobic organism. Obviously this involves moving via the blood (the mouse has almost no lymphatic drainage to the alveolar region of the lung).
2. I do not think it feasible that T-cell sensitization occurs at the initial infectious site. Instead I believe a few bacilli escape to the liver and/or spleen where they are engulfed by "professional" antigen presenting cells that in turn present "filtrate" proteins to CD4 T cells (Orme et al., 1993a).

The infection is then controlled and contained by acquired cell-mediated immunity. The CMI response consists of two components; protective immunity, mediated by T cells and driven by the Th1 cytokine axis [interleukin 12 (IL-12) driving IFN secretion by T cells] (Orme et al.,

1993b; Cooper et al., 1997a) and delayed-type hypersensitivity, in which tumor necrosis factor (TNF)-stimulated local tissue cells release chemokines that attract the influx of monocytes from the blood to wall off any potential further dissemination of the infection (Fig. 1).

After the expression of protective immunity by short-lived CD4 T cells this population is gradually replaced by a second subset of CD4 cells that are long lived and are retained even after the animal is given chemotherapy to destroy remaining bacilli. We have some evidence that this population is phenotypically different ($CD44^{hi}$ $CD45RB^{lo/neg}$ $CD49e^+$) but other than that, this population is poorly understood (Griffin and Orme, 1994).

III. Types of Vaccines

A. AUXOTROPHIC AND RECOMBINANT VACCINES

A major area of research during the past several years has been in the use of mycobacteria as genetic vectors, in which recombinant bacilli can be used to overexpress species-specific or foreign genes (Jacobs and Bloom, 1992; Cirillo et al., 1995; Stover et al., 1996). Progress has been slow because mycobacteria are particularly difficult to manipu-

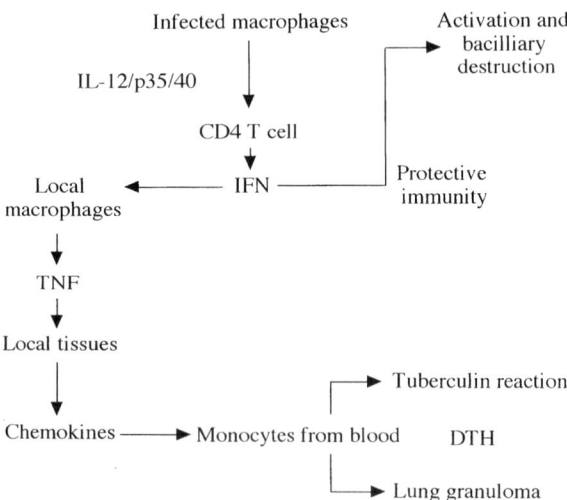

FIG. 1. Cell-mediated immunity to tuberculosis; a combination of protective immunity and delayed type hypersensitivity. The former is driven by the cytokines IL-12 and IFN, the latter by TNF and locally produced chemokines.

late in this manner, but gradually the tools have become available for this endeavor.

One such advance has been in the production of mycobacterial auxotrophs (McAdam et al., 1995). In that study transposons were transposed into the BCG genome in a relatively random fashion. Three auxotrophs, two for leucine and one for methionine, were isolated from the library of transposon insertions in BCG. They were characterized by sequencing and these data suggested homology to the leuD gene of *Escherichia coli* and a sulfate-binding protein of cyanobacteria, respectively. When inoculated intravenously into C57BL/6 mice, the leucine auxotrophs, in contrast to the parent BCG strain or the methionine auxotroph, showed an inability to grow *in vivo* and were cleared from the lungs and spleen over several weeks.

Another approach involved the technique of allelic exchange, a process that is relatively easy in most bacteria but only recently achieved in mycobacteria, to produce leucine auxotrophic mutants by homologous recombination for both the Erdman and H37Rv strains of *Mycobacterium tuberculosis* (Balasubramanian et al., 1996).

A possible application for such bacilli could be in vaccination of HIV-positive individuals. To demonstrate safety of these auxotrophs, they were injected into severe combined immunodeficient (SCID) mice and it was then shown that viable BCG could no longer be detected in control mice receiving the auxotrophs after 16–32 weeks, and that infected SCID mice survived for at least 230 days. In contrast, all SCID mice died within 8 weeks after being given BCG vaccine. In addition, several of the auxotrophs produced comparable protection against intravenous and intratracheal challenges with *M. tuberculosis,* suggesting that auxotrophic strains of BCG may represent a potentially safe and useful vaccine against tuberculosis for populations at risk for HIV (Guleria et al., 1996; Bange et al., 1996).

Turning to recombinant mycobacteria, a major advance has been in the characterization of phages that infect these bacteria. The best known is L5, which efficiently transforms slow-growing mycobacteria. The ability to easily generate stable recombinants in these slow-growing mycobacteria without the requirement for continual selection has since led to the construction of recombinant BCG vaccines and the isolation and characterization of mycobacterial pathogenic determinants in animal model systems (Lee et al., 1991; Stover et al., 1992).

B. Subunit Vaccines

The potential use of subunit vaccines has generated a lot of activity during the past decade. These range from selecting one particular major antigen, or a few antigens of high immunogenicity, to whole subcellular fractions. Of these, the one that has received the most attention is the culture filtrate fraction (CFP) (Orme et al., 1993a; Andersen, 1994).

This could be the subject of a whole review by itself, but to summarize, unfractionated CFP, or fractions thereof, delivered in a number of different adjuvants, can increase the resistance of mice and guinea pigs to M. tuberculosis infection (Pal and Horwitz, 1992; Horwitz et al., 1995; Roberts et al., 1995; Andersen, 1997). The results are not dramatic, and only two approaches (CFP plus IL-2, and Ag85 DNA, see below) have been shown to prevent caseous necrosis in the guinea pig model over the long term, but this progress leaves room for optimism.

In fact, our approach has been somewhat different to that of others. Instead of using strong adjuvants that induce strong DTH, we have combined CFP with a milder adjuvant (MPL; Ribi ImmunoChem) and then "nudged" the Th1 response using IL-2 (Chiron). This formulation does not induce initial protection in the form of reduced bacterial load as BCG does, but it does lead to prolonged survival of guinea pigs with (1) prevention of caseous necrosis and (2) it does not sensitize pigs to PPD (Huygen et al., 1998). We have had a similar result with a DNA vaccine (see next section).

C. DNA Vaccines

When the first results with "naked DNA" became available the results were very surprising to most of us, and yet multiple examples of using DNA vaccines have now arisen.

Immunization of mice with DNA encoding the Ag85A mycolyl transferase antigen of M. tuberculosis induces a small but significant protection (Huygen et al., 1996). No such protection is seen in guinea pigs, but good long-term survival with a lymphocytic response in the lungs and no necrosis has been achieved (Baldwin et al., 1998).

Similar protection in mice has been observed using DNA encoding stress proteins (Silva et al., 1994, 1996; Tascon et al., 1996; Ragno et al., 1997). Characterization of T-cell clones obtained after this immunization procedure showed that both CD4 and CD8 T cells were generated. In tests of antimycobacterial activity against M. tuberculosis, both in infected macrophages in vitro and by adoptive transfer of protection with T-cell clones injected into irradiated mice, the most effective clones were the most cytotoxic and secretion of IFN made only a secondary contribution. [Such findings, however, are contrary to literature that shows that secretion of IFN is essential to protection (Cooper *et al.*, 1993; Flynn *et al.*, 1993) and that cytolysis is not a protective mechanism (Laochumroonvorapong *et al.*, 1997; Cooper *et al.*, 1997b.]

IV. Can We Increase Herd Resistance to Bovine Tuberculosis?

It has often been said that tuberculosis in humans is detectable, preventable, and treatable. And yet, given the global incidence of the disease, it is clear that we stand on the edge of a potential worldwide disaster. The situation in terms of bovine tuberculosis is certainly much better (as the Australians will attest, having apparently eradicated the disease), but in certain areas of the world control involves constant test and slaughter surveillance, while in other countries (Mexico, Latin America) only rudimentary control measures exist.

It is not feasible to treat bovine tuberculosis, and efforts to control vectors (badgers, possum) are both ineffective and offensive to certain animal rights organizations (who, in this case at least, have a reasonable point) (Cowan, 1996). The only other major option is vaccination, and this has been stymied by the lack of a good candidate and the obvious fear factor generated by any potential use of the BCG vaccine. Despite this, there is good evidence that BCG might be effective in cattle (Hancox, 1994; Clifton-Hadley, 1995; Newell and Hewinson, 1995; Adwell *et al.*, 1995; Buddle *et al.*, 1995; Hughes *et al.*, 1996).

With new vaccines becoming available, however, particularly those that do not disable the skin test, the time may be coming whereby wholesale herd vaccination could be feasible. Certainly, in areas of the world where BCG vaccination in humans has been effective, the incidence of tuberculosis has fallen dramatically. Of course, the flip side is that BCG has been totally ineffective in other areas of the world, and it is quite possible that BCG in cattle may cause more overall harm than good; we simply do not know. However, with trade agreements such as NAFTA likely to increase the flow of cattle from one country to another, this may be a good time to find out.

Acknowledgment

This work was supported in part by grant AI-75320 from the NIAID, NIH, and by USDA Training Grant 96-38420-3039.

References

Aldwell, F. E., Pfeffer, A., DeLisle, G. W., Jowett, G., Heslop, J., Keen, D., Thomson, A., and Buddle, B. M. (1995). Effectiveness of BCG vaccination in protecting possums against bovine tuberculosis. *Res. Vet. Sci.* **58,** 90–95.

Andersen, P. (1994). The T cell response to secreted antigens of *Mycobacterium tuberculosis*. *Immunobiology* **191,** 537–547.
Andersen, P. (1997). Host responses and antigens involved in protective immunity to *Mycobacterium tuberculosis*. *Scand. J. Immunol.* **45,** 115–131.
Balasubramanian, V., Pavelka, M. S., Jr., Bardarov, S. S., Martin, J., Weisbrod, T. R., McAdam, R. A., Bloom, B. R., and Jacobs, W. R., Jr. (1996). Allelic exchange in *Mycobacterium tuberculosis* with long linear recombination substrates. *J. Bacteriol.* **178,** 273–279.
Bange, F. C., Brown, A. M., and Jacobs, W. R., Jr. (1996). Leucine auxotrophy restricts growth of *Mycobacterium bovis* BCG inmacrophages. *Infect. Immun.* **64,** 1794–1799.
Buddle, B. M., Keen, D., Thomson, A., Jowett, G., McCarthy, A. R., Heslop, J., De Lisle, G. W., Stanford, J. L., and Aldwell, F. E. (1995). Protection of cattle from bovine tuberculosis by vaccination with BCG by the respiratory or subcutaneous route, but not by vaccination with killed *Mycobacterium vaccae*. *Res. Vet. Sci.* **59,** 10–16.
Cirillo, J. D., Stover, C. K., Bloom, B. R., Jacobs, W. R., Jr., and Barletta, R. G. (1995). Bacterial vaccine vectors and bacillus Calmette-Guérin. *Clin. Infect. Dis.* **20,** 1001–1009.
Clifton-Hadley, R. S. (1995). Badgers, bovine tuberculosis and the age of reason. *Br. Vet. J.* **152,** 243–246.
Cooper, A. M., Dalton, D. K., Stewart, T. A., Griffin, J. P., Russell, D. G., and Orme, I. M. (1993). Disseminated tuberculosis in interferon gamma gene-disrupted mice. *J. Exp. Med.* **178,** 2243–2247.
Cooper, A. M., D'Souza, C., Frank, A. A., and Orme, I. M. (1997a). The course of *Mycobacterium tuberculosis* infection in the lungs of mice lacking expression of either perforin- or granzyme-mediated cytolytic mechanisms. *Infect. Immun.* **65,** 1317–1320.
Cooper, A. M., Magram, J., Ferrante, J., and Orme, I. M. (1997b). Interleukin 12 (IL-12) is crucial to the development of protective immunity in mice intravenously infected with *Mycobacterium tuberculosis*. *J. Exp. Med.* **186,** 39–45.
Cowan, P. E. (1996). Possum biocontrol: Prospects for fertility regulation. *Reprod. Fertil. Dev.* **8,** 655–660.
Flynn, J. L., Chan, J., Triebold, K. J., Dalton, D. K., Stewart, T. A., and Bloom, B. R. (1993). An essential role for interferon gamma in resistance to *Mycobacterium tuberculosis* infection. *J. Exp. Med.* **178,** 2249–2254.
Griffin, J. P., and Orme, I. M. (1994). Evolution of CD4 T-cell subsets following infection of naive and memory immune mice with *Mycobacterium tuberculosis*. *Infect. Immun.* **62,** 1683–1690.
Guleria, I., Teitelbaum, R., McAdam, R. A., Kalpan, G., Jacobs, W. R., Jr., and Bloom, B. R. (1996). Auxotrophic vaccines for tuberculosis. *Nat. Med.* **2,** 334–337.
Hancox, M. (1994). Bovine tuberculosis: Does badger culling help? *Trends Microbiol.* **2,** 337–338.
Hitze, K. (1980). Results of the controlled trial on BCG conducted in the district of Chingleput in southern India. *Bull. Int. Union Against Tuberc.* **55,** 13–14.
Horwitz, M. A., Lee, B. W., Dillon, B. J., and Harth G. (1995). Protective immunity against tuberculosis induced by vaccination with major extracellular proteins of *Mycobacterium tuberculosis*. *Proc. Natl. Acad. Sci. USA* **92,** 1530–1534.
Hughes, M. S., Neill, S. D., and Rogers, M. S. (1996). Vaccination of the badger (*Meles meles*) against *Mycobacterium bovis*. *Vet. Microbiol.* **51,** 363–379.
Huygen, K., Content, J., Deni, O., Montgomery, D. L., Yawman, A. M., Deck, R. R., DeWitt, C. M., Orme, I. M., Baldwin, S., D'Souza, C., Drowart, A., Lozes, E., Vanden-

bussche, P., Van Vooren, J. P., Liu, M. A., and Ulmer, J. B. (1996). Immunogenicity and protective efficacy of a tuberculosis DNA vaccine. *Nat. Med.* **2,** 893–898.

Jacobs, W. R., Jr., and Bloom, B. R. (1992). Use of recombinant BCG as a vaccine delivery vehicle. *Adv. Exp. Med. Biol.* **327,** 175–182.

Laochumroonvorapong, P., Wang, J., Liu, C. C., Ye, W., Moreira, A. L., Elkon, K. B., Freedman, V. H., and Kaplan, G. (1997). Perforin, a cytotoxic molecule which mediates cell necrosis, is not required for the early control of mycobacterial infection in mice. *Infect. Immun.* **65,** 127–132.

Lee, M. H., Pascopella, L., Jacobs, W. R., Jr., and Hatfull, G. F. (1991). Site-specific integration of mycobacteriophage L5: Integration-proficient vectors for *Mycobacterium smegmatis, Mycobacterium tuberculosis,* and bacille Calmette-Guerin. *Proc. Natl. Acad. Sci. USA* **88,** 3111–3115.

McAdam, R. A., Weisbrod, T. R., Martin, J., Scuderi, J. D., Brown, A. M., Cirillo, J. D., Bloom, B. R., and Jacobs, W. R., Jr. (1995). *In vivo* growth characteristics of leucine and methionine auxotrophic mutants of *Mycobacterium bovis* BCG generated by transposon mutagenesis. *Infect. Immun.* **63,** 1004–1012.

Newell, D. G., and Hewinson, R. G. (1995). Control of bovine tuberculosis by vaccination. *Vet. Rec.* **136,** 459–463.

Orme, I. M. (1995). Prospects for new vaccines against tuberculosis. *Trends Microbiol.* **3,** 401–404.

Orme, I. M., Andersen, P., and Boom, W. H. (1993a). T cell response to *Mycobacterium tuberculosis. J. Infect. Dis.* **167,** 1481–1497.

Orme, I. M., Roberts, A. D., Griffin, J. P., and Abrams, J. S. (1993b). Cytokine secretion by CD4 T lymphocytes acquired in response to *Mycobacterium tuberculosis* infection. *J. Immunol.* **151,** 518–525.

Pal, P. G., and Horwitz, M. A. (1992). Immunization with extracellular proteins of *Mycobacterium tuberculosis* induces cell-mediated immune responses and substantial protective immunity in a guinea pig model of pulmonary tuberculosis. *Infect. Immun.* **60,** 4781–4792.

Ragno, S., Colston, M. J., Lowrie, D. B., Winrow, W. R., Blake, D. R., and Tascon, R. (1997). Protection of rats from adjuvant arthritis by immunization with naked DNA encoding for mycobacterial heat shock protein 65. *Arthritis Rheum.* **40,** 277–283.

Roberts, A. D., Sonnenberg, M. G., Ordway, D. J., Furney, S. K., Brennan, P. J., Belisle, J. T., and Orme, I. M. (1995). Characteristics of protective immunity engendered by vaccination of mice with purified culture filtrate protein antigens of *Mycobacterium tuberculosis. Immunology* **85,** 502–508.

Silva, C. L., Silva, M. F., Pietro, R. C., and Lowrie, D. B. (1994). Protection against tuberculosis by passive transfer with T-cell clones recognizing mycobacterial heat-shock protein 65. *Immunology* **83,** 341–346.

Silva, C. L., Silva, M. F., Pietro, R. C., and Lowrie, D. B. (1996). Characterization of T cells that confer a high degree of protective immunity against tuberculosis in mice after vaccination with tumor cells expressing mycobacterial hsp65. *Infect. Immun.* **64,** 2400–2407.

Stanford, J. L., Grange, J. M., and Pozniak, A. (1991). Is Africa lost? *Lancet* **338,** 557–558.

Stover, C. K., de la Cruz, V. F., Bansal, G. P., Hanson, M. S., Fuerst, T. R., Tascon, R. E., Colston, M. J., Ragno, S., Stavropoulos, E., Gregory, D., and Lowrie, D. B. (1996). Vaccination against tuberculosis by DNA injection. *Nat. Med.* **2,** 888–892.

Tweddle, N. E., and Livingstone, P. (1994). Bovine tuberculosis control and eradication programs in Australia and New Zealand. *Vet. Microbiol.* **40,** 23–39.

Viral Vectors for Veterinary Vaccines

MICHAEL SHEPPARD

Animal Health Biological Discovery, Pfizer Central Research, Groton, Connecticut 06340

I. Introduction and Background
II. Viral Vector Construction
III. Advantages and Disadvantages of Viral Vectors for Vaccine Delivery
IV. Construction of Safer Viral Vectors for Vaccine Delivery
   A. Deletion of Nonessential Genes
   B. Deletion of Essential Genes
   C. Replication Limited Virus
V. Examples of Reported Viral Veterinary Vaccine Vectors
VI. Commercially Available Viral Vaccine Vectors for Veterinary Use
VII. Summary
    References

I. Introduction and Background

Numerous reviews have described the use of viral vectors for possible vaccine delivery (e.g., Cavanagh, 1985; Sheppard and Fahey, 1989; Wray and Woodward, 1990; Graham and Prevec, 1992; Boyle and Heine, 1993; Hilleman, 1994; Martin, 1994; Dorner, 1995; Babiuk *et al.*, 1996; Perkus and Paoletti, 1996). However, in this review I will focus solely on the use of viral vectors for delivery of veterinary vaccines. It is without question that vaccination plays an essential role in veterinary medicine, providing the major and often the only prophylatic approach for the control of infectious diseases. In spite of the vast array of currently available vaccines veterinarians and the livestock producers continue to express the need for vaccines that not only maintain the best features of killed or subunit vaccines (such as safety) as well as the best features of conventional modified live vaccines (such as efficacy) but improve on them. As well as the need for continual improvement of vaccines there exists a need for new vaccines either to new diseases (e.g., chicken anemia virus or porcine reproductive and respiratory syndrome virus) or to old diseases for which vaccines are not available or no longer meet the requirements of the end user (e.g., bovine virus diarrhea virus vaccines). As well as new vaccines there is also need for vaccines with special features that allow potential customers to design disease control programs that suit their specific needs on top of offering greater safety and improved protection. The design and construction of these new veterinary vaccines is a major challenge facing the field of vaccinology. With the continued demand of improving vaccines and producing new ones it is easier for potential vaccine candidates to fail to meet the increased level of requirements that are expected. The failure of some vaccines can result from problems associated with delivery, such as insufficient or no induction of the appropriate protective immune response. The development of delivery systems to produce vaccines that are more effective, offer greater safety, are convenient to administer, and are compatible with customer practices is part of the challenge for vaccinologists. The development of safe and convenient live viral vectors for the delivery of veterinary vaccines is *one* possible way of meeting some of these challenges. Recombinant DNA technology has allowed more detailed characterization of the genetic organization of many viruses to such an extent that regions suitable for insertion of foreign genetic material have been identified. This has resulted in the development of numerous types of viral vectors from a wide variety of viral families. Some of these viral vectors have been developed with the potential for delivering and expressing gene(s) from a foreign pathogen and so act as a vaccine vector (Table I). The viral vector is often genetically attenuated or cannot complete its replication cycle in the animal to be immunized, and thus produces no clinical disease. Although initially the majority of viral vector development centered around poxviruses, especially vaccinia (Panicali and Paoletti, 1982; Macket *et al.*, 1982), it was not long before viral vector development witnessed a virtual explosion in the types of viruses developed into vectors. These included herpeviruses (Post *et al.*, 1982), adenoviruses (Berkner and Sharp, 1982), retroviruses (Wei *et al.*, 1981), papoviruses (Southern and Berg, 1982), polyoma virus (Fried and Ruley, 1982), picornaviruses (Kitson *et al.*, 1991), Semliki Forest virus (SFV; Zhou *et al.*, 1994), Sindbis virus (Pugachev *et al.*, 1995), and even some plant viruses (Jagadish *et al.*, 1996; Dalsgaard *et al.*, 1997).

TABLE I

Characteristics of the More Common Virus Groups Used as Vectors

Characteristics	Pox viruses	Adenoviruses	Herpes viruses	Retroviruses
Genome	180–300 kb	30–45 kb	150–200 kb	9.2 kb
Max. Insert	>30 kb	>5 kb	30 kb	8 kb
Max. Titer	10^7–10^9	10^8–10^{11}	10^6–10^8	10^6–10^9
Administration	Scarification/injection	Injection/aerosol/oral	Injection/water	Injection
Safety	Problems with immunosuppressed	Inflammation	Latency	Genomic insertion
Background expression by vector	Yes	Yes	Yes	No

II. Viral Vector Construction

Greater understanding of the structure and function of a wide range of viruses at the genetic level has opened up ways of designing novel viral vaccine vectors which should improve the quality and effectiveness of some future vaccines as major prophylatic tools. Viral vaccine vectors have really developed from a greater technological understanding of viruses at the genetic level, where today they have become a viable alternative strategy as one method for the delivery of vaccines. The concept of viral vectors was first highlighted by Bernard Moss and others in the early 1980s (Mackett *et al.*, 1982; Panicali and Paoletti, 1982), where they showed that vaccinia virus could be engineered to carry and express foreign genes (Panicali and Paoletti, 1982; Mackett *et al.*, 1982). From the time when Moss and others first demonstrated that vaccinia virus could be developed as a vector for the expression of foreign genes, the technology has been exploited to apply to a variety of virus families as well as a variety of foreign genes including those that encode antigens from pathogens. As a result both DNA and RNA viruses have been developed as viral vaccine vectors (Table I).

To produce viral vaccine vectors it is first necessary to study the genome of the vector to a stage of understanding where at least one region suitable for insertion of foreign genetic material has been identified. Second, genes from pathogens that encode proteins that will induce an appropriate protective immune response and can be stably integrated into the vector's genome and expressed need to be identified. Finally, it is necessary to insert the foreign gene(s) in such a way as to ensure the correct and sufficient expression of the foreign gene(s).

The ideal viral vaccine vector would have all or at the very least some of the following features:

- Safe and nonpathogenic for the vaccinate
- Evoke the appropriate protective immune response
- Single host or limited host range
- Stable genome
- No integration into the host genome
- Readily accessible region(s) for insertion of foreign genetic material
- Able to tolerate well insertion of foreign genetic material and expression of foreign gene(s)
- Convenient to deliver and fits with management practices
- Relatively simple and cost effective to produce
- Limited background gene expression by the vector

III. Advantages and Disadvantages of Viral Vectors for Vaccine Delivery

Live viral vectors offer several advantages for vaccine delivery compared to killed, subunit, or conventional modified live vaccines. First, because of the possibility of delivering divalent or even perhaps multivalent vaccines, using a single type of vector can result in a single manufacturing process rather than several and possibly even a single vaccination rather than several. Therefore, vectored vaccines have the potential to be less expensive to the manufacturer and the end user. Because the foreign gene is being expressed in the cells of its natural host, it is expected that any post-translational modifications required will be correct and produce an authentic antigen, as opposed to *Escherichia coli* or baculovirus systems (among others) that do not always produce authentic foreign proteins. Depending on the vector selected it may be possible to deliver the vectored vaccine more conveniently to the mammal or bird by spray or water or some other means rather than by needle injection. Such a mass administration approach may be particularly relevant to the poultry industry. The vector could also be constructed to deliver simultaneously an immunomodulator (e.g., gamma interferon), which could modify the type or magnitude of the immune response to allow the vaccine to be successful or more successful than it would be otherwise. The vector only expresses the antigens from the pathogen that are required to elicit a protective immune response and therefore reduces or eliminates the chance of disease by being exposed to the whole pathogen as with a killed or modified live vaccine. Finally, the appropriate viral vectors will induce both cell-mediated and humoral immune responses and in some cases are particularly suitable for inducing a local immune response in the mucosal surfaces.

One of the main disadvantages of using viral vectors for vaccine delivery is that like subunit vaccines each vector can only deliver one or a relatively small number of foreign antigens to the host animal and therefore rely on those being able to elicit a completely protective immune response. Also the only antigens that can be delivered are those that are encoded by nucleic acid. Thus such things as lipopolysaccharides are not deliverable. With any vector, regardless of type, only a limited amount of foreign genetic material can be inserted into the vector's genome stably and expressed appropriately. One must always be wary of altered tissue tropisms due to the expression of the foreign gene(s). Of course the effectiveness of a viral vector is limited by preexisting immune response in the animal from prior exposure to the virus used to construct the vector. Finally, as with all live vaccines there is the problem of shelf life and compatibility with other vaccine preparations.

IV. Construction of Safer Viral Vectors for Vaccine Delivery

To produce viral vaccine vectors successfully it is necessary to ensure that the vector itself does not pose any disease threat to the animal that receives the vaccination or to the person delivering the vaccine to the animal. Most often this is achieved by attenuating the viral vector in some way. Until recent times the means of generating a live attenuated virus had been entirely empirical. This process usually involved the passaging of the virus in cell culture or animals that were not the natural host, followed by testing of the resulting viruses for decreased virulence in the natural host. The basis for attenuation is most often unknown, and may be a result from as minor as a single base change, and thus the chance of reversion back to virulence is always a possibility. This type of traditional method for generating a live attenuated virus is not necessarily the most attractive method for generating a viral vaccine vector. With the advent of molecular biology and our improved knowledge of viruses at the genetic level it is now possible to generate live attenuated viruses with precise genetic changes, improving their safety and thus make them more suitable as vectors for vaccine delivery.

A. Deletion of Nonessential Genes

A good example is the deletion of the thymidine kinase (TK) gene. While the deletion of the TK gene has little or no effect on virus growth in cell culture, TK deleted viruses can be significantly attenuated *in vivo* ((Buller *et al.*, 1985; Kit *et al.*, 1985, 1986; Becker *et al.*, 1986). This feature has been exploited successfully for the development of live attenuated herpesvirus vaccines (McGregor *et al.*, 1985; Kit *et al.*, 1985; Marchioli *et al.*, 1987; Moorman *et al.*, 1990) as well as safer herpesvirus and poxvirus vectors (e.g., Buller *et al.*, 1985; Bayliss *et al.*, 1991; Mulder *et al.*, 1994; Hu *et al.*, 1997).

B. Deletion of Essential Genes

If an essenential gene is deleted from a virus, the virus can only grow if the gene or gene product is provided in trans. This virus is phenotypically normal but genotypically defective and cannot replicate in the host because the deleted gene product is not available. This type of virus can replicate *in vitro* with the help of a genetically engineered supporting cell line that expresses the deleted gene product. The stage of the virus life cycle of which the gene product is required will govern how far through the replication cycle a virus will proceed. In some cases (e.g., if the essential deleted gene is required for virus penetration of the cell) the virus will complete a single round of replication in the host but the progeny viruses will not be able to invade any other cell. (Farrell *et al.*, 1994; McLean *et al.*, 1994; Peeters *et al.*, 1994). However, if the deleted essential gene is an early gene that is required to activate other viral genes, then the number of viral proteins synthesized may be limited and the viral genome may not be able to complete even a single cycle of replication (Chen and Knipe, 1996; Brehm *et al.*, 1997; Da Costa *et al.*, 1997).

C. Replication Limited Virus

A third alternative, which has been exploited successfully, is to use a virus that can only completely replicate in one species as a vector in another species, where it cannot complete an entire cycle of replication but can commence a replication cycle sufficiently to allow expression of the foreign gene (Tartaglia et al., 1992). The canarypox virus (CPV) vector, termed *ALVAC*, has successfully been exploited to the degree of commercial success. The CPV is restricted to avian cells only for productive replication but can be used to vaccinate mammals where it can elicit an immune response to the foreign gene product without completing an entire cycle of replication (Tartaglia et al., 1992, 1993; Taylor et al., 1995). The human adenovirus type 5 (HAV-5) has also been exploited in a similar fashion to the CPV (Table III) but has the disadvantage that this virus is a human pathogen and so has yet to be exploited commercially.

Several other strategies are also available and in some cases have been exploited successfully in order to generate safe viral vectors for vaccine delivery. Table V (next section) provides a summary of some of these possible approaches.

V. Examples of Reported Viral Veterinary Vaccine Vectors

Even though there are a great many examples of viral vectors reported in the literature since they were first described in 1982, the number of publications reporting the use of viral vectors for veterinary vaccine delivery is not that large. After searching for published papers that describe viral vectors with veterinary vaccine applications, especially those that could be described as purposely developed for veterinary use, the obvious conclusion was that even though this research was first described in 1982 the veterinary side is still in its infancy. Publications describing viral vectors for veterinary vaccine delivery can be divided into several groups, which are represented in the following tables: Table II, poxvirus vectors; Table III, adenovirus vectors; Table IV, herpesvirus vectors; and Table V, other virus vectors. Although these four tables probably do not include every single publication describing viral vectors for veterinary vaccine delivery they do describe the majority of published papers and present the reader with an idea of the limited amount of research that has occurred in this field during the last 15 years.

VI. Commercially Available Viral Vaccine Vectors for Veterinary Use

At the time of writing this review only three viral vectored vaccines for use in the veterinary field have been licensed for release. All three are based on poxvirus vectors and the three vectors represent the

TABLE II
Poxvirus Vectors for the Delivery of Veterinary Vaccines

Vector	Pathogen	Antigen	Test animal	Reference
CPV	RHDV	Capsid	Rabbit	Fischer et al., (1997)
RPV	FPV/rabies	VP2/G	Cat	Hu et al., (1997)
CPV/VV	CDV	F/HA	Ferret	Stephensen et al., (1997)
FPV	NDV	F/HN	Chicken	Taylor et al., (1996)
Myxoma	Influenza	HA	Rabbit	Kerr and Jackson (1995)
SPV	PrV	gp50/gp63	Swine	van der Leek et al., (1994)
CPV	FeLV	env/gag	Cat	Tartaglia et al., (1993)
VV	Rabies	G	Fox	Brochier et al., (1991)
PPV	NDV	F	Chicken	Latellier et al., (1991)
FPV	NDV	HA/NA	Chicken	Boursnell et al., (1990a)
FPV	NDV	HN/F	Chicken	Boursnell et al., (1990b)
VV	BLV	env	Rabbit	Ohishi et al., (1990)
VV	EHV-1	gp13	Mouse	Guo et al., (1989)
VV	PrV	gp50/63/I/X	Mouse	Kost et al., (1989)
FPV	Rabies	G	Dog/cat	Taylor et al., (1988)
VV	FeLV	env	Cat	Gilbert et al., (1987)
VV	Rabies	G	Fox	Blancou et al., (1986)
VV	Rabies	G	Mouse	Kieny et al., (1984)

Key: VV, vaccinia virus; FPV, fowl poxvirus; PPV, pigeon poxvirus; SPV swine poxvirus; CPV, canary poxvirus; RHDV, rabbit hemorrhagic disease virus; CDV, canine distemper virus; FPV, feline parvovirus; PrV, pseudorabies virus; FeLV, feline leukemia virus; NDV, Newcastle disease virus; BLV, bovine leukosis virus; EHV, equine herpes virus.

TABLE III
Adenovirus Vectors for the Delivery of Veterinary Vaccines

Vector	Pathogen	Antigen	Test animal	Reference
OAV	*Tinea ovis*	45W	Sheep	Rothel et al., (1997)
HAV-5	PRCV	Spike	Swine	Callebaut et al., (1996)
HAV-5	TGE	Spike	Swine	Torres Iet al., (1996)
HAV-5	Rabies	G	Skunk	Yarosh et al., (1996)
HAV-5	BCV	HEG	Cotton rat	Bacca-Estrada et al., (1995)
HAV-5	FIV	env	Cat	Gonin et al., (1995)
HAV-5	PRCV	Spike	Swine	Callebaut et al., (1994)
HAV-5	PrV	gD	Swine	Adam et al., (1994)
HAV-5	Rabies	G	Dog	Prevec et al., (1990)
HAV-5	PrV	gp50	Rabbit/mouse	Eloit et al., (1990)

Key: OAV, ovine adenovirus; HAV-5, human adenovirus type 5; PRCV, porcine respiratory corona virus; TGE, transmissible gastroenteritis virus; BCV, bovine corona virus; PrV, pseudorabies virus.

TABLE IV

HERPES VIRUS VECTORS FOR THE DELIVERY OF VETERINARY VACCINES

Vector	Pathogen	Antigen	Test animal	Reference
HVT	NDV	HN/F	Chicken	Reddy et al., (1996)
HVT	MDV	gpAB	Chicken	Reddy et al., (1996)
FHV-1	FeLV	env	Cat	Willemse et al., (1996)
PrV	HCV	gpE1	Swine	Mulder et al., (1994)
HVT	MDV	gpB	Chicken	Ross et al., (1993)
BHV-1	PrV	gpC	Swine	Kit et al., (1992)
FHV-1	FeLV	gag/env	Cat	Wardley et al., (1992)
BHV-1	FMDV	cp-epitopes	Cattle	M. Kit et al., (1991)
BHV-1	FMDV	cp-epitopes	Cattle	S. Kit et al., (1991)
PrV	HCV	gpE1	Swine	van Zijl et al., (1991)

Key: HTV, herpes virus of turkeys; FHV, feline herpes virus; BHV, bovine herpes virus; PrV, pseudorabies virus; NDV, Newcastle disease virus; MDV, Marek's disease virus; FMDV, foot-and-mouth disease virus; HCV, hog cholera virus.

evolution in poxvirus vector development. The first vector approved was the vaccinia virus vector carrying the rabies G glycoprotein gene (e.g., Kieny et al., 1984; Blancou et al., 1986; Brochier et al., 1991). In terms of complying with the characteristics of a desirable vector for vaccine delivery in the veterinary setting, this vector has the greatest number of undesirable characteristics. However, it satisfied an unmet need and as a result was released in various parts of the world. The second vector to be licensed for release was the fowlpox virus vector. This vector delivers the Newcastle disease virus HN and F genes and is designed to vaccinate poultry (e.g., Boursnell et al., 1990a,b; Taylor et al., 1996). While this vector has the desirable characteristic of only replicating in poultry it also has some limitations that affect its use in the field. The third vector licensed is the canarypox virus vector and represents the state-of-the-art poxvirus vector. This vector was developed to deliver the HA and F genes of canine distemper virus and is the most recently available of the three vector vaccines (e.g., Stephensen et al., 1997).

VII. Summary

Whatever strategy is adopted for the development of viral vectors for delivery of veterinary vaccines there are several key points to consider: (1) Will the vectored vaccine give a delivery advantage compared to

TABLE V
OTHER VIRUS VECTORS FOR THE DELIVERY OF VETERINARY VACCINES

Vector	Pathogen	Antigen	Test animal	Reference
CPMV	MEV	VP2 epitope	Mink	Dalsgaard et al. (1997)
Poliovirus	FMDV	Epitopes	Guinea pig	Kitson et al. (1991)
Retrovirus	NDV	HN	Chicken	Morrison et al. (1990)
Retrovirus	Influenza	HA	Chicken	Hunt et al. (1988)

OTHER ALTERNATIVE VIRAL VECTORS THAT HAVE THE POTENTIAL FOR VETERINARY VACCINE DELIVERY

Amplicons	VLPs	SFV	Sinbis	Bacteriophage
Frenkel et al. (1994)	Jagadish et al. (1996)	Atkins et al. (1996)	Pugachev et al. (1995)	Bastien et al. (1997)
Smith et al. (1995)	Porter et al. (1996)	Mossman et al. (1996)		
Fink et al. (1996)	Roy (1996)	Zhou et al. (1995)		
Pechan et al. (1996)	Schodel et al. (1994a)	Zhou et al. (1994)		
Starr et al. (1996)	Schodel et al. (1994b)			

Key: CPMV, cowpea mosaic virus; MEV, mink enteritis virus.

what's already available? (2) Will the vectored vaccine give a manufacturing advantage compared to what's already available? (3) Will the vectored vaccine provide improved safety compared to what's already available? (5) Will the vectored vaccine increase the duration of immunity compared to what's already available? (6) Will the vectored vaccine be more convenient to store compared to what's already available? (7) Is the vectored vaccine compatible with other vaccines? If there is no other alternative available then the answer to these questions is easy. However, if there are alternative vaccines available then the answers to these questions become very important because the answers will determine whether a vectored vaccine is merely a good laboratory idea or a successful vaccine.

REFERENCES

Adam, M., Lepottier, M. F., and Eloit, M. (1994). Vaccination of pigs with replication-defective adenovirus vectored vaccines: The example of pseudorabies. *Vet. Microbiol.* **42,** 205–215.

Atkins, G. J., Sheahan, B. J., and Liljestrom, P. (1996). Manipulation of the Semliki Forest virus genome and its potential for vaccine construction. *Mol. Biotechnol.* **5,** 33–38.

Babiuk, L. A., van Drunen Littel-van den Hurk, S., Tikoo, S. K., Lewis, P. J., and Liang, X. (1996). Novel viral vaccines for livestock. *Vet. Immunol. Immunopathol.* **54,** 355–363.

Bacca-Estrada, M. E., Liang, X., Babiuk, L. A., and Yoo, D. (1995). Induction of mucosal immunity in cotton rats to haemagglutinin-esterase glycoprotein of bovine coronavirus by recombinant adenovirus. *Immunology* **86,** 134–140.

Bastien, N., Trudel, M., and Simard, C. (1997). Protective immune responses induced by the immunization of mice with a recombinant bacteriophage displaying an epitope of the human respiratory syncytial virus. *Virology* **234,** 118–122.

Bayliss, C. D., Peters, R. W., Cook, J. K. A., Reece, R. L., Howes, K., Binns, M. M., and Boursnell, M. E. G. (1991). A recombinant fowlpox virus that expresses the VP2 antigen of infectious bursal disease virus induces protection against mortality caused in the virus. *Arch. Virol.* **120,** 193–205.

Becker, Y., Hadar, J., Taylor, E., Ben-Hur, T., Raibstein, I., Rosen, A., and Darai, G. (1986). A sequence in HpaI P fragment of herpes simplex virus 1 DNA determines intraperitoneal virulence in mice. *Virology* **149,** 255–259.

Berkner, K. L., and Sharp, P. A. (1982). Preparation of adenovirus recombinants using plasmids of viral DNA. *In* "Eukaryotic Viral Vectors" (Y. Gluzman, ed.), pp. 193–198. Cold Spring Harbor Lab., Cold Spring Harbor, NY.

Blancou, J., Kieny, M. P., Lathe, R., Lecocq, J. P., Pastoret, P. P., Soulebot, J. P., and Desmettre, P. (1986). Oral vaccination of the fox against rabies using a live recombinant vaccinia virus. *Nature* **322,** 373–375.

Boursnell, M. E. G., Green, P. F., Campbell, J. I. A., Deuter, A., Peters, R. W., Tomley, F. M., Samson, A. C. R., Emmerson, P. T., and Binns, M. M. (1990a). A fowlpox virus vaccine vector within insertion sites in the terminal repeats: Demonstration of the efficacy using the fusion gene of Newcastle disease virus. *Vet. Microbiol.* **23,** 305–316.

Boursnell, M. E. G., Green, P. F., Samson, A. C. R., Campbell, J. I. A. Deuter, A., Peters, R. W., Millar, N. S., Emmerson, P. T., and Binns, M. M. (1990b). A recombinant fowlpox virus expressing the hemagglutinin-neuraminidase gene of Newcastle disease virus (NDV). Protects chickens against by NDV. *Virology* **178,** 297–300.

Boyle, D. B., and Heine, H. G. (1993). Recombinant fowlpox virus vaccines for poultry. *Immunol. Cell Biol.* **71,** 391–397.

Brehm, M. A., Bonneau, R. N., Knipe, D. M., and Tevethia, S. S. (1997). Immunization with a replication-deficient mutant of herpes simplex virus type 1 (HSV-1) induces a CD8+ cytotoxic T-lymphocyte response and confers a level of protection comparable to that of wild-type HSV-1. *J. Virol.* **71,** 3534–3544.

Brochier, B., Kieny, M. P., Costy, F., Coppens, P., Bauduin, B., Lecocq, J. P., Languet, B., Chappuis, G., Desmettre, P., Afiademanyo, K., Libois, R., and Pastoret, P. P. (1991). Large-scale eradication of rabies using recombinant vaccinia-rabies vaccine. *Nature* **354,** 520–522.

Buller, R. M. L., Smith, G. L., Cremer, K., Notkins, A., and Moss, B. (1985). Decreased virulence of recombinant vaccinia virus expressed vectors is associated with a thymidine kinase negative phenotype. *Nature* **317,** 813–815.

Callebaut, P., Pensaert, M., and Enjuanes, L. (1994). Construction of a recombinant adenovirus for the expression of the glycoprotein S antigen of porcine respiratory coronavirus. *In* "Coronaviruses" (H. Lande and J. F. Vantherot, eds.), pp. 469–470. Plenum, New York.

Callebaut, P., Enjuanes, L., and Pensaert, M. (1996). An adenovirus recombinant expressing the spike glycoprotein of porcine respiratory coronavirus is immunogenic in swine. *J. Gen. Virol.* **77,** 309–313.

Cavanagh, D. (1985). Viral and bacterial vectors of immunogens. *Vaccine* **3,** 45–48.

Chen, Y. M., and Knipe, D. M. (1996). A dominant mutant form of the herpes simplex virus ICP8 protein decreases viral late gene transcription. *Virology* **221,** 281–290.

Da Costa, X. J., Bourne, N., Stanberry, L. R., and Knipe, D. M. (1997). Construction and characterization of a replication-defective herpes simplex virus 2 ICP8 mutant strain and its use in immunization studies in a guinea pig model of genital disease. *Virology* **232,** 1–12.

Dalsgaard, K., Uttenthal, A., Jones, T. D., Xu, F., Merryweather, A., Hamilton, W. D. O., Langeveld, J. P. M., Boshuizen, R. S., Kamstrup, S., Lomonossoff, G. P., Porta, C., Vela, C., Casal, J. I., Meloen, R. H., and Rodgers, P. B. (1997). Plant-delivered vaccine protects target animals against a viral disease. *Nat. Biotechnol.* **15,** 248–252.

Dorner, F. (1995). An overview of vaccine vectors. *Dev. Biol. Stand.* **84,** 23–32.

Eloit, M., Gilardi- Hebenstreit, P., Toma, B., and Perricaudet, M. (1990). Construction of a defective adenovirus vector expressing the pseudorabies virus glycoprotein gp50 and its use as a live vaccine. *J. Gen. Virol.* **71,** 2425–2431.

Farrell, H. E., McClean, C. S., Harley, C., Efstathiou, S., Inglis, S., and Minson, A. C. (1994). Vaccine potential of a herpes simplex virus type 1 mutant with an essential gene deleted. *J. Virol.* **68,** 927–932.

Fink, D. J., DeLuca, N. A., Goins, W. F., and Glorioso, J. C. (1990). Gene transfer to neurons using herpes simplex virus based vectors. *Annu. Rev. Neurosci.* **19,** 265–287.

Fischer, L., LeGros, F.-X., Mason, P. W., and Paoletti, E. (1997). A recombinant canarypox virus protects rabbits against a lethal rabbit hemorrhagic disease virus (RHDV) challenge. *Vaccine* **15,** 90–96.

Frenkel, N., Singer, O., and Kwong, A. D. (1994). Minireview: The herpes simplex virus amplicon—a versatile defective virus vector. *Gene Ther.* **1,** Suppl. 1, S40–S46.

Fried, M., and Ruley, E. (1982). Use of polyoma virus vector. In "Eukaryotic Viral Vectors" (Y. Gluzmann, ed.), pp. 67–70. Cold Spring Harbor Lab., Cold Spring Harbor, NY.
Gilbert, J. H., Pedersen, N. C., and Nunberg, J. H. (1987). Feline leukemia virus envelope protein expression encoded by a recombinant vaccinia virus: Apparent lack of immunogenicity in vaccinated animals. Virus Res. **7,** 49–67.
Gonin, P., Fournier, A., Oualikene, W., Moraillon, A., and Eloit, M. (1995). Immunization trial of cats with a replication defective adenovirus type 5 expressing the ENV gene of feline immunodeficiency virus. Vet. Microbiol. **45,** 393–401.
Graham, F. L., and Prevec, L. (1992). Adenovirus-based expression vectors and recombinant vaccines. In "Vaccines: New Approaches to Immunological Problems" (R. W. Ellis, ed.), pp. 363–430. Butterworth-Heinemann, Boston.
Guo, P. X., Goebel, S., Davis, S., Perkus, M. E., Languet, B., Desmettre, P., Allen, G., and Paoletti, E. (1989). Expression in recombinant vaccinia virus of the equine herpesvirus 1 gene encoding glycoprotein 13 and protection of immunized animals. J. Virol. **63,** 4189–4198.
Hilleman, M. R. (1994). Recombinant vector vaccines in vaccinology. Dev. Biol. Stand. **82,** 3–20.
Hu, L., Ngichabe, C., Trimarchi, C. V., Esposito, J. J., and Scott, F. W. (1997). Raccoon poxvirus live recombinant feline panleukopenia virus VP2 and rabies virus glycoprotein bivalent vaccine. Vaccine **15,** 1466–1472.
Hunt, L. A., Brown, D. W., Robinson, H. L., Naeve, C. W., and Webster, R. G. (1988). Retrovirus-expressed hemagglutinin protects against lethal influenza virus infections. J. Virol. **62,** 3014–3019.
Jagadish, M. N., Edwards, S. J., Hayden, M. B., Grusovin, J., Vandenberg, K., Schoofs, P., Hamilton, R. C., Shukla, D. D., Kalnins, H., McNarmara, M., Haynes, J., Nisbet, I. T., Ward, C. W., and Pye, D. (1996). Chimeric potyvirus-like particles as vaccine carriers. Intervirology **39,** 85–92.
Kerr, P. J., and Jackson, R. J. (1995). Myxoma virus as a vaccine vector for rabbits—antibody levels to influenza virus hemagglutinin presented by a recombinant myxoma virus. Vaccine **13,** 1722–1726.
Kieny, M. P., Lathe, R., Drillen, R., Spehner, D., Skory, S., Schmitt, D., Wiktor, T., Koprowski, H., and Lecocq, J. P. (1984). Expression of rabies virus glycoprotein from a recombinant vaccinia virus. Nature **312,** 163–166.
Kit, M., Kit, S., Little, S. P., DiMarchi, R. D., and Gale, C. (1991). Bovine herpesvirus-1 (infectious bovine rhinotracheitis virus)-based viral vectors which expresses foot-and-mouth disease epitopes. Vaccine **9,** 564–572.
Kit, S., Kit, M., and Pirtle, E. C. (1985). Attenuated properties of thymidine kinase-negative deletion mutant of pseudorabies virus. Am. J. Vet. Res. **46,** 1359–1367.
Kit, S., Kit, M., and McConnell, S. (1986). Intramuscular and intravaginal vaccination of pregnant cows with thymidine kinase-negative, temperature-resistant infectious bovine rhinotracheitis virus (bovine herpesvirus 1). Vaccine **4,** 55–61.
Kit, S., Kit, M., DiMarchi, R. D., Little, S. P., and Gale, C. (1991). Modified-live infectious bovine rhinotracheitis virus vaccine expressing monomer and dimer forms of foot-and-mouth disease cupsid protein epitopes on surface of hybrid virus particles. Arch. Virol. **120,** 1–17.
Kit, S., Otsuka, H., and Kit, M. (1992). Expression of porcine pseudorabies virus genes by a bovine herpesvirus-1 (infectious bovine rhinotracheitis virus) vector. Arch. Virol. **124,** 1–20.

Kitson, J. D. A., Burke, K. L., Pullen, L. A., Belsham, G., and Almond, J. W. (1991). Chimeric polioviruses that include sequences derived from two independent antigenic sites of foot-and-mouth disease virus (FMDV) induce neutralizing antibodies against FMDV in guinea pigs. *J. Virol.* **65,** 3068–3075.

Kost, T. A., Jones, E. V., Smith, K. M., Reed, A. P., Brown, A. L., and Miller, T. J. (1989). Biological evaluation of glycoproteins mapping to two distinct mRNAs within the BamHI fragment 7 of pseudorabies virusi expression of the coding regions by vaccinia virus. *Virology* **171,** 365–376.

Letellier, C., Burny, A., and Meulemans, G. (1991). Construction of a pigeonpox virus recombinant: expression of the Newcastle disease virus (NDV) fusion glycoprotein and protection of chickens against NDV challenge. *Arch. Virol.* **118,** 43–56.

Mackett, M., Smith, G. L., and Moss, B. (1982). Vaccinia virus: A selectable eukaryotic cloning and expression vector. *Proc. Natl. Acad. Sci. USA* **79,** 7415–7419.

Marchioli, C. C., Yancey, R. J., Wardley, R. C., Thomsen, D. R., and Post, L. E. (1987). A vaccine strain of pseudorabies virus with deletions in the thymidine kinase and glycoprotein X genes. *Am. J. Vet. Res.* **48,** 1577–1583.

Martin, S. J. (1994). Vaccine design: Future possibilities and potentials. *Biotechnol. Adv.* **12,** 619–624.

McGregor, S., Easterday, B. C., Kaplan, A. S., and Ben-Porat, T. (1985). Vaccination of swine with thymidine kinase-deficient mutants of pseudorabies virus. *Am. J. Vet. Res.* **46,** 1494–1497.

McLean, C. S., Erturk, M., Jennings, R., Ni Challanain, D., Minson, A. C., Duncan, I., Boursnell, M. E. G., and Inglis, S. C. (1994). Protective vaccination against primary and recurrent disease caused by herpes simplex virus (HSV) type 2 using a genetically disabled HSV-1. *J. Infect. Dis.* **170,** 1100–1109.

Moorman, R. J. M., deRover, T., Briaire, J., Peters, B. P. H., Gielkens, A. L. J., and van Oirschot, J. T. (1990). Inactivation of the thymidine kinase gene of a gI deletion mutant of pseudorabies virus generates a safe but still highly immunogenic vaccine strain. *J. Gen. Virol.* **71,** 1591.

Morrison, T., Hinshaw, V. S., Sheerer, M., Cooley, A. J., Brown, D., McQuain, C., and McGinnes, L. (1990). Retroviral expressed hemagglutinin-neuraminidase protein protects chickens from Newcastle disease virus induced disease. *Microbiol. Pathol.* **9,** 387–396.

Mossman, S. P., Bex, F., Berglund, P., Arthos, J., O'Neil, S. P., Riley, D., Maul, D. H., Bruck, C., Momin, P., Burny, A., Fultz, P. N., Mullins, J. I., Liljestrom, P., and Hoover, E. A. (1996). Protection against lethal simian immunodeficiency virus SIVsmmPBj14 disease by a recombinant Semliki Forest virus gp160 vaccine and by a gp120 subunit vaccine. *J. Virol.* **70,** 1953–1960.

Mulder, W. A. M., Priem, J., Glazenburg, K. L., Wagenaar, F., Gruys, E., Gielkens, A. L. J., Pol, J. M. A., and Kimman, T. G. (1994). Virulence and pathogenesis of non-virulent and virulent strains of pseudorabies virus expressing envelope glycoprotein E1 of hog cholera virus. *J. Gen. Virol.* **75,** 117–124.

Ohishi, K., Suzuki, H., Maruyama, T., Yamamoto, T., Funahashi, S., Miki, K., Ikawa, Y., and Sugimoto, M. (1990). Induction of neutralizing antibodies against bovine leukosis virus in rabbits by vaccination with recombinant vaccinia virus expressing bovine leukosis virus envelope glycoprotein. *Am. J. Vet. Res.* **51,** 1170–1173.

Panicali, D., and Paoletti, E. (1982). Construction of poxviruses as cloning vectors: Insertion of the thymidine kinase gene from herpes simplex virus into the DNA of infectious vaccinia virus. *Proc. Natl. Acad. Sci. USA* **79,** 4927–4931.

Pechan, P. A., Fotaki, M., Thompson, R. L., Dunn, R., Chase, M., Chiocca, E. A., and Breakefield, X. O. (1996). A novel 'piggyback' packaging system for herpes simplex virus amplicon vectors. *Hum. Gene Ther.* **7,** 2003–2013.

Peeters, B., Bouma, A., deBruin, T., Moorman, R., Gielkens, A., and Kimman, T. (1994). Non-transmissible pseudorabies virus gp50 mutants: A new generation of safe live vaccines. *Vaccine* **12,** 375–380.

Perkus, M. E., and Paoletti, E. (1996). Recombinant virus as vaccination carrier of heterologous antigens. *In* "Concepts in Vaccine Development" (S. H. E. Kaufman, ed.), pp. 379–422. de Gruyter, Berlin.

Porter, D. C., Melsen, L. R., Compans, R. W., and Morrow, C. D. (1996). Release of virus-like particles from cells infected with poliovirus replicons which express human immunodeficiency virus type 1 Gag. *J. Virol.* **70,** 2643–2649.

Post, L. E., Norrild, B., Simpson, T., and Roizman, B. (1982). Chicken ovalbumin gene fused to a herpes simplex virus a promoter and linked to a thymidine kinase gene is regulated like a viral gene. *Mol. Cell. Biol.* **2,** 233–240.

Prevec, L., Campbell, J. B., Christie, B. S., Belbeck, L., and Graham, F. L. (1990). A recombinant human adenovirus vaccine against rabies. *J. Infect. Dis.* **161,** 27–30.

Pugachev, K. V., Mason, P. W., Shope, R. E., and Frey, T. K. (1995). Double-subgenomic sindbis virus recombinants expressing immunogenic proteins of Japanese encephalitis virus induce significant protection in mice against lethal JEV infection. *Virology,* **212,** 587–594.

Reddy, S. K., Sharma, J. M., Ahmad, J., Reddy, D. N., McMillen, J. K., Cook, S. M., Wild, M. A., and Schwartz, R. D. (1996). Protective efficacy at a recombinant herpesvirus of turkeys as an *in ovo* vaccine against Newcastle and Marek's disease in specific-pathogen-free chickens. *Vaccine* **14,** 469–477.

Ross, L. J., Binns, M. M., Typers, P., Pastorek, J., Zelnik, V., and Scott, S. (1993). Construction and properties of a turkey herpesvirus recombinant expressing the Marek's disease virus homologue of glycoprotein B of herpes simplex virus. *J. Gen. Virol.* **74,** 371–377.

Rothel, J. S., Boyle, D. B., Both, G. W., Pye, A. D., Waterkeyn, J. G., Wood, P. R., and Lightowlers, M. W. (1997). Sequential nucleic acid and recombinant adenovirus vaccination induces host-protective immune responses against *Taenia ovis* infection in sheep. *Parasite Immunol.* **19,** 221–227.

Roy, P. (1996). Genetically engineered particulate virus-like structures and their use as vaccine delivery systems. *Intervirology* **39,** 62–71.

Schodel, F., Peterson, D., Hughes, J., and Milich, D. (1994a). Hepatitis B virus core particles as a vaccine carrier moiety. *Int. Rev. Immunol.* **11,** 153–165.

Schodel, F., Wirtz, R., Peterson, D., Hughes, J., Warren, R., Sudoff, J., and Milich, D. (1994b). Immunity to malaria elicited by hybrid hepatitis B virus core particles carrying circumsporozoite protein epitopes. *J. Exp. Med.* **180,** 1037–1046.

Sheppard, M., and Fahey, K. J. (1989). Herpesviruses and adenoviruses as potential vectors for the poultry industry. *Aust. Vet. J.* **66,** 421–423.

Smith, R. L., Geller, A. I., Escudero, K. W., and Wilcox, C. L. (1995). Long-term expression in sensory neurons in tissue culture from herpes simplex virus type 1 (HSV-1) promoters in an HSV-1-derived vector. *J. Virol.* **69,** 4593–4599.

Southern, P., and Berg, P. (1982). Mammalian cell transformation with SV40 vector. *In* "Eukaryotic Viral Vectors" (Y. Gluzman, ed.), pp. 41–45. Cold Spring Harbor Lab., Cold Spring Harbor, NY.

Starr, P. A., Lim, F., Grant, F. D., Trask, L., Lang, P., Yu, L., and Geller, A. I. (1996). Long-term persistence of defective HSV-1 vectors in the rat brain is demonstrated by reactivation of vector gene expression. *Gene Ther.* **3,** 615–623.

Stephensen, C. B., Welter, J., Thaker, S. R., Taylor, J., Tartaglia, J., and Paoletti, E. (1997). Canine distemper virus (CDV) infection of ferrets as a model for testing Morbillivirus vaccine strategies: NYVAC- and ALVAC-based CDV recombinants protect against symptomatic infection. *J. Virol.* **71,** 1506–1513.

Tartaglia, J., Perkus, M. E., Taylor, J., Norton, E. K., Audonnet, J.-C., Cox, W. I., Davis, S. W., VanderHoeven, J., Meignier, B., Riviere, M., Languet, B., and Paoletti, E. (1992). NYVACP: A highly attenuated strain of vaccinia virus. *Virology* **188,** 217–232.

Tartaglia, J., Jarrett, O., Neil, J. C., Desmettre, P., and Paoletti, E. (1993). Protection of cats against feline leukemia virus by vaccination with a canarypox virus recombinant, ALVAC-FL. *J. Virol.* **67,** 2370–2375.

Taylor, J., Weinberg, R., Languet, B., Desmettre, P., and Paoletti, E. (1988). Recombinant fowlpox virus inducing protective immunity in non-avian species. *Vaccine* **6,** 497–503.

Taylor, J., Meignier, B., Tartaglia, J., Languet, B., VanderHoeven, J., Franchini, G., Trimarchi, C., and Paoletti, E. (1995). Biological and immunogenic properties of a canarypox-rabies recombinant, ALVAC-RG (vCP65) in non-avian species. *Vaccine* **13,** 539–549.

Taylor, J., Christensen, L., Gettig, R., Goebel, J., Bouquet, J. F., Mickle, T. R., and Paoletti, E. (1996). Efficacy of a recombinant fowlpox-based Newcastle disease virus vaccine candidate against velogenic and respiratory challenge. *Avian Dis.* **40,** 173–180.

Torres, J. M., Alonso, C., Ortega, A., Mittal, S., Graham, F., and Enjuanes, L. (1996). Tropism of human adenovirus type 5-based vectors in swine and their ability to protect against transmissible gastroenteritis coronavirus. *J. Virol.* **70,** 3770–3780.

van der Leek, M. L., Feller, J. A., Sorensen, G., Isaacson, W., Adams, C. L., Borde, D. J., Pfeiffer, N., Tran, T., Moyer, R. W., and Gibbs, E. P. (1994). Evaluation of swinepox virus as a vaccine vector in pigs using Aujeszky's disease (pseudorabies) virus gene insert coding for glycoproteins gp50 and gp63. *Vet. Rec.* **134,** 13–18.

van Zijl, M., Wensvoort, G., deKluyver, E., Hulst, M., van der Gulden, H., Gielkens, A., Berns, A., and Moormann, R. (1991). Live attenuated pseudorabies virus expressing envelope glycoprotein E1 of hog cholera virus protects swine against both pseudorabies and hog cholera. *J. Virol.* **65,** 2761–2765.

Wardley, R. C., Berlinski, P. J., Thomsen, D. R., Meyer, A. L., and Post, L. E. (1992). The use of feline herpesvirus and baculovirus as vaccine vectors for the gag and env genes of feline leukemia virus. *J. Gen. Virol.* **73,** 1811–1818.

Wei, C.-M., Gibson, M., Spear, P. G., and Scolnick, E. M. (1981). Construction and isolation of a transmissible retrovirus containing the src gene of Harvey murine sarcoma virus and the thymidine kinase gene of herpes simplex virus type 1. *J. Virol.* **39,** 935–944.

Willemse, M. J., van Schooneveld, S. H. B., Chalmers, W. S. K., and Sondermeijer, P. J. A. (1996). Vaccination against feline leukemia using a new feline herpesvirus type I vector. *Vaccine* **14,** 1511–1516.

Wray, C., and Woodward, M. S. (1990). Biotechnology and veterinary science: Production of veterinary vaccines. *Rev. Sci. Tech. Off. Int. Epizoot.* **9,** 779–794.

Yarosh, O. K., Wandeler, A. I., Graham, F. L., Campbell, J. B., and Prevec, L. (1996). Human adenovirus type 5 vectors expressing rabies glycoprotein. *Vaccine* **14,** 1257–1264.

Zhou, X., Berglund, P., Rhodes, G., Parker, S. E., Jondal, M., and Liljestrom, P. (1994).

Self replicating Semliki Forest virus RNA as recombinant vaccine. *Vaccine* **12,** 1510–1514.

Zhou, X., Berglund, P. Zhao, H., Lilijestrom, P., and Jondal, M. (1995). Generation of cytotoxic and humoral immune responses by nonreplicative recombinant Semliki Forest virus. *Proc. Natl. Acad. Sci. USA* **92,** 3009–3013.

DNA Immunization: Present and Future

L. A. BABIUK, J. LEWIS, S. VAN DEN HURK, AND R. BRAUN

Veterinary Infectious Disease Organization, Saskatoon, Saskatchewan, S7N 5E3, Canada

I. Introduction
II. Universality of DNA Immunization
III. Induction of Immunity
IV. Role of Different Antibodies in Clearing Viruses
V. Vaccine Delivery
VI. DNA Immunization in the Face of Passive Antibody
VII. Regulatory
VIII. Epilogue
    Acknowledgments
    References

I. Introduction

Vaccination continues to be the most cost-effective means by which economic losses and animal suffering from infectious diseases can be prevented. To date the majority of the preventive measures directed against controlling infectious diseases include immunization with live attenuated or inactivated vaccines. Although these approaches have been at least partially successful, the quest for better and safer methods of immunization continues. One possible approach to improve both safety and efficacy of vaccination and especially duration of immunity is by employing DNA immunization. One of the first indications that this approach could possibly be effective was the report by Wolff *et al.* (1990) that introduction of a plasmid encoding a reporter gene into an animal resulted in expression of that gene *in vivo*. As a result of these initial observations, it became clear that injected DNA could persist for extended periods of time in muscle tissues. Whether the DNA integrates or not is still an issue of debate with most reports suggesting that it does not, but there are also reports that under certain conditions it may integrate (Schubbert et al., 1997). Determination of whether integration occurs or not is critical if we hope to use this approach for immunization of companion or food producing animals.

These initial studies were quickly extended to demonstrate that not only was the plasmid capable of expressing protein, the protein produced by the plasmid DNA also induced an immune response in animals (Robinson et al., 1993; Cox et al., 1993; Ulmer et al., 1993). Furthermore, the immune response induced by the plasmids introduced intramuscularly or intradermally resembled the responses induced by a natural infection, with antigen being processed through both exogenous and endogenous pathways and able to produce both a Th1 and Th2 type response depending on the antigen and route of immunization (Michel et al., 1995; Pertmer et al., 1996; Feltquate et al., 1997).

The past 5 years have seen a rapid evolution in our understanding of DNA immunization and this approach is rapidly being adopted by academics and biological companies as a potentially valuable adjunct to current vaccine strategies. This review attempts to summarize some of the lessons learned as well as some outstanding questions that need to be addressed before this technique is widely embraced as the "third generation" of vaccination.

II. Universality of DNA Immunization

Following the early observation that plasmids encoding human growth hormone and human alpha-1 antitrypsin (Tang et al., 1992) could produce humoral responses, numerous groups have investigated the possible application of this approach to protect individuals from various infectious agents. Reports employing genes from various viral, bacterial, and parasitic agents as well as cancers have clearly demonstrated the broad application of DNA immunization for induction of immunity (Manickan et al., 1997; Whalen and Davies, 1995; Donnelly et al., 1994). Furthermore, this phenomenon appears to work in many species including those of veterinary importance such as mammals, birds, and aquatic species (Manickan et al., 1997; Anderson et al., 1996; Cox et al., 1993; Davis et al., 1993; Donnelly et al., 1993; Fynan et al., 1993; Hoffman et al., 1994; Lowrie et al., 1994; Robinson et al., 1993; Tang et al., 1992; Williams et al., 1991; Xiang et al., 1994; Xu and Liew, 1994). Probably the most significant observation employing DNA immunization is that not only is an immune response induced, but animals are often protected from challenge. Initially some groups expressed frustration in not being able to detect antibody responses to the specific gene product introduced by plasmid immunization and even questioned whether the early reports were as universally applicable as originally thought. However, it rapidly became apparent that antibody levels may not be the best readout to measure immune responses induced by DNA immunization since animals with no or very low levels of antibody were protected from disease (Yokonama et al., 1995; R. Schultz, personal communications). This was apparent even in diseases where it was thought that antibody was primarily responsible for protection. As a result, we have begun to reassess the role of antibody and cell-mediated immune responses in recovery from disease. (*Editor's note:* We have also begun to better understand that the mechanisms responsible for protective immunity in vaccinated animals can be quite different than immunity in immunologically mature animals.) However, even with the recognition that plasmids generally induce stronger cell-mediated immune responses than humoral responses, considerable investigation still needs to be conducted with respect to the following questions: (1) What are the different types of antigens and their forms, such as whether they are compartmentalized in the cell, expressed on the cell surface, or excreted from the cell? (2) What are the biological activity and toxicity of the specific protein in question? If the protein is extremely toxic, the level of expression will need to be below the toxic threshold or the protein will need to be excreted extracellularly to prevent rapid death of the transfected cell. If death occurs too early, the quantity of antigen may be insufficient to induce an immune response. (3) What are the genotype, size, and age of the animals used for primary immunization? Presently, it is very easy to induce immune responses in mice, but it is much more difficult to repeat the same experiments, even with identical gene constructs, in large animals. Unfortunately, no one has systematically investigated the inherent differences between large animals and rodents to determine which factors are critical for induction of immune responses. With respect to genotype of the animal, experiments in mice clearly indicate that the immune responses are not the same in different strains. (Barry et al., 1995). (4) What are the route of administration and method of gene delivery? Even if the plasmids are putatively delivered to the same site, the method of delivery may influence the response. For example, intradermal immunization with needles induces different immune responses than introduction of the same plasmids by a gene gun (Feltquate et al., 1997). Whether this is related to the quantity of DNA or the specific cells that take up the DNA is unknown. Similarly, immune responses with the same genes introduced intradermally versus intramuscularly are not identical (Pertmer et al., 1996). Thus, a considerable amount of research still needs to be conducted with respect to the presentation of antigens expressed by the plasmids to the immune system. Because the quantity of DNA introduced by different routes of administration may qualitatively influence the immune response, especially since DNA itself has immunostimulatory properties, investigations of these parameters and routes must be very carefully controlled before hypotheses regarding these factors are proposed. Thus, even though it is being widely accepted that DNA immunization is a real phenomenon, considerable investigations are still required to optimize this method of immunization before it replaces conventional vaccines used in veterinary medicine today. Indeed, it is possible that even if we answer all of these questions and fully understand the parameters involved in DNA immunization, some of the very effective conventional vaccines, which induce sterile immunity, may never be supplanted by this new technique. However, the vaccine companies, practitioners, and producers will have a repertoire of different approaches to control infectious diseases. This is the goal that we are all striving for, in an attempt to provide the highest level of protection, yet meeting different management needs of producers.

III. Induction of Immunity

It is clear that various routes of administration can result in the induction of immunity; however, the transfected cells that are the most relevant in antigen presentation are still unclear. Indeed, there are a number of possible ways to induce immunity by plasmids depending on which cells are transfected. Presently, evidence exists which indicates that transfected muscle cells may play a role, possibly indirect, in induction of immunity. Following transfection of myoblasts *in vitro* and transplantation of these transfected myoblasts *in vivo*, Ulmer et al. (1993) clearly showed that these transfected myoblasts could induce immunity and that the myoblasts needed to be viable for induction of immunity to occur. These results suggest that the myocytes must continue to produce antigen if immunity is expected to develop. Furthermore, it was shown that the CTL responses in F1 mice (H-2^d × H-2^k) were restricted to the MHC-1 haplotype of the bone marrow cells, suggesting that antigen released from myocytes is taken up by antigen presenting cells to prime the CTL response. These results show that myocytes may be involved in inducing immunity following intramuscular immunization, and that plasmid uptake by nonmyocytic cell types is not critical for induction of immunity. However, these results do not exclude the possibility that other cells can also take up plasmids and participate in the induction of immunity. Clearly, all of the injected DNA does not stay at the injection site. This was eloquently demonstrated by S. A. Johnston (personal communications) who administered DNA into the ear of the mouse and then removed the ear shortly after. A similar experiment was performed by Torres et al. (1997) who demonstrated that excision of an injected muscle bundle within 10 minutes of DNA injection did not affect the magnitude or longevity of the immune response. In these instances the immune response was equivalent to that induced in intact animals. Studies have shown that professional antigen presenting cells such as dendritic cells are very efficient in uptake and antigen presentation (Bos and Kapsenberg, 1993; Ertl et al., 1995). Thus, it is proposed that these cells may take up the DNA and migrate to the regional lymph nodes where they induce immunity following production of extremely low concentrations of protein. Indeed, it has been shown that in vivo transfection of dermal Langerhan cells is followed by mobilization and trafficking to the draining lymph node (Condon et al., 1996). Although it is difficult "to quantitate" the amount of antigen produced in vivo following DNA immunization, it seems plausible that this continuous stimulation with protein by antigen presenting cells could induce immunity with 100–1000 times less antigen than is required by subunit vaccines. Thus, as long as antigen continues to be presented, lymphocyte maturation occurs, resulting in both immunity and memory.

A considerable number of investigations have focused on the type of immune response induced following DNA immunization and have generally suggested that the method of delivery plays a role in deviating the immune response toward a Th1 or Th2 type response (Feltquate et al., 1997; Pertmer et al., 1996). Generally, intramuscular and to a lesser extent intradermal administration in saline favors a Th1 type response whereas administration by gene gun may favor a Th2 response. Although this may occur in specific cases, we should be cautious in accepting this as dogma. Thus, the quantity of DNA administered may also influence the type of immune response as well as the specific antigens and form of antigen seen by the immune system itself (Roman et al., 1997; Cardoso et al., 1996). For example, using measles hemagglutinin, the preferential induction of a Th1 or Th2 type immune response was dependent on whether the H protein was membrane anchored or secreted.

To investigate the role of antigenic form in induction of immunity we constructed different plasmids encoding the epitopes of a single protein: BHV-1 glycoprotein gD. These plasmids encoded three different forms of antigen including the authentic gD with a transmembrane anchor intact; a form that had the transmembrane anchor removed, thereby resulting in secretion of the antigen; and finally an intracellular form with signal and transmembrane domain sequences removed. Using this model, we demonstrated that although different antigenic forms induced both humoral and cell-mediated immunity, the character of the immune response varied. The cytosolic and membrane-anchored forms clearly favored antibody of the IgG_{2a} isotype (81%) and the secreted form favored the IgG_1 isotype (94%). These results demonstrate that the cellular compartment in which the antigen originates can, depending on the antigen, be important in determining the type of immune response induced. Thus, it is imperative to understand the pathogenesis of each pathogen being investigated and the type of immune response required for protection before embarking on a vaccination regime.

Because most bacterial antigens would not generally be expressed on the mammalian cell membrane, one could introduce transmembrane sequences from viruses to have these antigens presented on the surface. One would, however, need to ensure that this did not alter the protein by glycosylation, etc. Thus, introduction of specific upstream sequences could ensure that the protein was compartmentalized in a manner appropriate for induction of the most effective immune response.

In addition to construction of plasmids expressing antigens in specific cell compartments, the immune response can also be redirected by inducing plasmids encoding cytokines, or costimulatory molecules delivered as coadministered plasmids, bicistronic plasmids or plasmids encoding genetic fusions between antigens and costimulatory molecules (Conry et al., 1996; Kim et al., 1997; Levitsky, 1997). Inclusion of these immune regulatory molecules and immunizing protocols may not only enhance the level of immunity, but may redirect the immune response to the desired balance. For example, coadministration of a cytokine encoding interferon-γ and TNF-α altered the character of the humoral immune response from one showing predominantly IgG_1 antibody isotypes to a more balanced profile as a result of significant increases in the means IgG_{2a} antibody levels. In contrast, moderate doses of plasmids encoding GM-CSF appeared to exaggerate the IgG_1 response (P. J. Lewis, unpublished results).

IV. Role of Different Antibodies in Clearing Viruses

If neutralization of a virus is critical and this is enhanced by complement it would be important to induce the specific antibody isotype that binds complement. Because this will be different in different species it will be critical to design the antigens for each species. Murine immunoglobulin isotype IgG_{2a} possesses effector functions that occur to a lesser extent, or not at all, in isotypes IgG_{2b}, IgG_1, and IgG_3. IgG_{2a} is more effective at fixing complement and also facilitates antibody-dependent cell-mediated cytotoxicity (Neuberger and Rajewsky, 1981; Ravetch and Kinet, 1991). These effector functions have been demonstrated to contribute significantly to the efficacy of protection against several different viral pathogens, although the correlation between isotype and protection is not always clearly evident (Ishizaka et al., 1995; Corbeil et al., 1996). DNA-based vaccines, when injected intramuscularly or intradermally into mice, often result in serum immunoglobulin responses that show a predominance of IgG_{2a} (Feltquate et al., 1997). Current conventional recombinant or killed vaccines typically drive humoral responses that are characterized by high titers of IgG_1 with little, if any, IgG_{2a}. The only potential mechanism, other than DNA-based vaccines, to elevate serum levels of IgG_{2a} would involve formulating recombinant antigens with adjuvants such as saponins, monophosphoryl lipid A (MPL), or lipophilic muramyl dipeptides (MDP) which deviate responses toward a potent Th1 type of immunity (Cox and Coulter, 1997). Although the advantages of DNA-based vaccines seem obvious, it has recently been suggested that above a certain serum immunoglobulin threshold concentration such things as functional affinity and isotype are irrelevant with regard to protection (Bachmann et al., 1997). This observation suggests that IgG_{2a} effector functions may only be relevant at subthreshold serum concentration ranges. It may also simply reflect observations based on a single viral disease model in mice and may not be a universal phenomenon in a murine model of viral infection.

V. Vaccine Delivery

One of the major concerns in veterinary medicine today is the development of injection site reaction following administration of conventionally killed vaccines. These reactions are primarily due to the adjuvants used in the vaccine. In the case of DNA immunization, many of these concerns should be eliminated or dramatically reduced. Furthermore, this method of immunization provides opportunities for focusing the immune response to the desired site. For example, if mucosal immunity is desired the vaccines can be delivered to the mucosal site either as free DNA (Kuklin et al., 1997), by a gene gun (Keller et al.,

1996), in various lipofectins (Caplen et al., 1995; Stribling et al., 1992), or by bacteria (Sizemore et al., 1995, 1997). In these cases the plasmid would induce immunity, not only at the site of administration but also at other mucosal sites due to the homing of lymphocytes to the common mucosal immune system (Croitoru and Bienenstock, 1994). However, it is also possible to induce mucosal immunity by administration of the vaccine into the inner surface of the ear. The lymph nodes draining the ear appear to be the same ones draining the upper respiratory tract/oral cavity mucosal surfaces (Gao et al., 1995). Results have clearly shown that administration of DNA vaccines intradermally into the ear provides excellent immunity, possibly due to the rich environment of antigen presenting cells in the dermis/epidermis of the ear (Bos and Kapsenberg, 1993). The ear provides a real advantage for immunization of large animals due to the minimal economic value of this part of the animal as well as the ease of administration of the vaccine.

The main advantage of gene gun delivery is the quantity of DNA that is required to induce immunity. In most reports, 100 times less DNA is required to induce immunity by the gene gun versus intramuscular administration. The main reason for this is the relative inefficiency of delivery of the DNA by intramuscular administration. Following intramuscular administration it appears that the majority of the DNA is placed extracellularly and therefore is subject to degradation by nucleases. Indeed, we obtained better results by multiple injection versus a single injection intramuscularly. Thus, more injection sites with the same amount of DNA will enhance the chances of administering some of the DNA into the muscle fiber. It is not known whether the muscle fiber of large animals is more resilient than that of mice, but it is possible that this is the reason why larger quantities of DNA are required for large animals than for mice. The increased quantity of DNA used in immunization of large animals may also qualitatively influence the immune response generated. This is probably due to the immunomodulatory effects of DNA (Roman et al., 1997). Note that the level of immunity and efficiency of various routes of administration may also be related to the specific antigen and genotype of the animal. Thus, every antigen would be processed and their epitopes expressed differently, which may influence the immune response. This is an obvious area which requires more investigation. Previously, we described the qualitative differences between immune responses induced by subunit or conventional antigens and those induced by DNA immunization. However, we reemphasize that in addition to the compartmentalization of antigen, the route of delivery and the quantity of DNA used in immunization may also influence the type of immune response.

Presently, many individuals are attempting to optimize delivery with respect to the route, dose, and delivery vehicles to not only improve the efficacy of immunization, but also to reduce the cost of such an approach. As we continue to develop more efficient methods of delivery, hopefully we should also learn more about the factors that are critical in redirecting the immune response to that required for most efficient control of the specific pathogen. Furthermore, it will be critical to determine whether sterile immunity (protection from infection) or protective immunity, as demonstrated by the absence of disease, is the most important. These two types of protective immunity are critically different and must be taken into consideration in any overall vaccine strategy.

VI. DNA Immunization in the Face of Passive Antibody

One of the greatest impediments to effective disease control in veterinary medicine is the inability to immunize animals in the presence of passive antibody. Because the level of passive antibody acquired at birth varies, individual animals need to be tested to determine when the antibody decays to a level when vaccination could be effective. Unfortunately, this is expensive and will rarely be implemented into common management practices for most species. As a result, animals are either overvaccinated, to protect those few animals whose antibody levels have decayed, or, alternatively, vaccination is delayed until all animals are seronegative. Thus, if it was possible to immunize animals at birth, in the presence of significant levels of antibody, such that they can develop active immunity, then at no time in the animal's life would there be a window of susceptibility to disease. The observation that neonatal animals can respond to nucleic acid vaccination is encouraging in this regard (Hassett et al., 1997). Indeed, in some cases there is an inverse relationship between age and induction of immune responses with younger animals being better responders (Barry and Johnston, 1997). In addition to overcoming maternal antibody, it appears that DNA immunization can even overcome immune tolerance as demonstrated with hepatitis B transgenic mice (Davis et al., 1997). Although tolerance is not the same as passive immunity, it does demonstrate that nucleic acid immunization can overcome many barriers to immunization experienced by conventional vaccines.

At present there is still uncertainty as to how universal and under what conditions DNA immunization can overcome maternal antibody or immune tolerance. Our laboratory has passively immunized 6- to

7-week-old mice with polyclonal mouse antibody to BHV-1 gD and showed that 90% of the mice did seroconvert following DNA immunization. Similarly, Bruce Smith (personal communication) has observed that immunity to canine parvovirus can be induced in the presence of passive antibody. The possible reason for this ability to overcome passive immunity is that the antigens are produced endogenously and are presented directly to the immune system.

It is possible that specific isotypes of antibody acquired passively may influence the ease of induction of active immunity. Results in our laboratory indicate that IgG_1 may be more efficient in blocking immune responses than IgG_2 or IgM (L. A. Babiuk et al., unpublished results). Clearly, more experiments need to be performed to demonstrate the parameters that may affect the development of immunity in the presence of antibody. Regardless, even if a specific isotype limits the development of active immunity, the persistence of plasmids for an extended period of time might result in induction of immunity once the antibody decays to low levels. If this did indeed occur, then animals would respond as their passive immunity waned and would not experience an extended period of susceptibility between the time when antibody disappeared and they were vaccinated. This would greatly increase the flexibility producers would have in animal management.

VII. Regulatory

Regulatory agencies have had extensive experience in developing regulations for conventional vaccines and have a successful track record in ensuring the safety of these vaccines prior to licensing. Unfortunately, there is limited experience with this new generation of vaccines. As a result, many questions are being asked regarding the safety of these vaccines. These concerns include (1) whether plasmids integrate into the chromosomes, (2) the distribution of cell types harboring the plasmid, especially germ cells, (3) the length of persistence and whether continued stimulation of the immune system may lead to immune tolerance or other immunopathologic effects, and (4) whether anti-DNA antibodies are formed following injection of DNA. Because of possibilities that some of these concerns may be realized, the regulatory agencies are approaching the risk–benefit aspects of nucleic acid immunization in an extremely cautious manner.

To help alleviate the concerns of regulatory agencies, many researchers are actively pursuing answers to these questions. To date there is minimal evidence for DNA integration even after numerous attempts to detect integration and association of nucleic acid immunization with enhanced tumor formation (Nichols et al., 1995). However, attempts are being made to further reduce the risk of homologous recombination and integration by constructing plasmids with sequences having no, or limited homology, with mammalian genomes. In the case of the concerns over the development of anti-DNA antibodies experience indicates that in normal individuals induction of anti-DNA antibodies is extremely difficult to achieve (Madaio et al., 1984). In those cases where there has been development of anti-DNA antibodies, the DNA was modified or was injected in the presence of Freund's complete adjuvant (Gilkeson et al., 1995; Puccetti et al., 1995). Since most DNA vaccines will not be incorporated into Freund's adjuvant this should be of minimal concern. However, as the trend toward multicomponent vaccines continues, manufacturers should be cautious not to combine nucleic acid vaccines with adjuvanted subunit or conventional vaccines since the risk of inducing anti-DNA antibodies may increase.

Persistence of DNA and extended stimulation of an immune response does not presently appear to induce any immunopathologic events, nor is tolerance to the protein established (Davis et al., 1996). Whether this will ever occur with a specific protein is sheer speculation. Hopefully, as we gain experience with these vaccines, the regulatory agencies should adopt broad-based regulations, not requiring researchers to conduct all of the safety tests with every single nucleic acid-based vaccine in the future. For example, if the specific plasmid backbone has been shown not to integrate, not to cause induction of anti-DNA antibodies, etc., it seems excessive to repeat all of the timeconsuming expensive experiments for every new gene introduced into the plasmid and in every potential species which may be immunized with nucleic acid vaccines.

Although the above concerns are legitimate, others appear to be based on less than scientific parameters. For example, is the injected plasmid shed in the urine, feces, or other body secretions and then transmitted to other hosts? Secondly, can the presence of a few remaining plasmids (greater than 90% of the plasmid is eliminated in the first day) in the meat of food animals be a health risk to individuals consuming the meat? Because of the presence of DNAases in the environment and the gastrointestinal tract this risk should be considered almost nonexistent. More importantly, it is impossible to evaluate this risk effectively.

VIII. Epilogue

The rapid evolution of DNA immunization has been made possible by the availability of biotechnological tools as well as our understanding of pathogenesis and the individual proteins/glycoproteins involved in inducing protective immunity to many pathogens. Furthermore, our understanding of the role of different immune responses in protecting animals from disease or enhancing infection has increased dramatically. Thus, there is a much better appreciation that an immune response in itself may not be protective, which means that we must ensure that the correct immune response is induced to the appropriate antigens or epitopes. Furthermore, our long history describing impediments to effective immunization in real-world situations is allowing us to address these shortcomings quickly. Based on these insights, DNA immunization protocols must be designed to address these perceived impediments to effective immunization. The first challenge is the potential to overcome the suppression of active immune responses in neonates to administered vaccines by maternally derived passive immunity. Although this is still controversial, a number of recent reports indicate that DNA immunization may overcome, at least in part, the suppressive effects of maternal antibodies (Lewis et al., 1997; Hassett et al., 1997). If this is indeed the case, the ability to immunize animals, or at least prime animals' immune response at an early age, could dramatically improve productivity by narrowing the window of susceptibility to disease. Furthermore, such an approach will be easy to introduce into present-day management practices. Management conditions must always be in the forefront of any new approaches if adoption of this strategy is ever going to occur. Because animals are often exposed to multiple pathogens simultaneously, simultaneous immunization to all these pathogens needs to be considered. Indeed, many vaccines today contain multiple antigens. Unfortunately, in some cases, these cocktail vaccines are not as effective as we want them to be because some components interfere with immune responses to other components. Although there are limited reports of coadministration of multiple plasmids, preliminary evidence suggests that in at least the cases tested there was minimal interference (R. Braun, unpublished results). Thus, it is theoretically possible to combine numerous plasmids into a single vaccine. This will, however, require improved efficacy of transfection so that the quantity of DNA introduced in one injection is reasonable. From a practical point of view, administration of multiple plasmids should provide both economical sense as well as reduce stress on the animals. The ability to combine plasmids encoding for vaccine antigens and immune modulators should further enhance the efficacy of these vaccines.

By combining the appropriate plasmids we should be able to alter the vaccine architecture in such a way as to greatly enhance the spectrum of protection as well as improve compliance.

In addition to administering multiple plasmids, administration of these vaccines should not introduce significant tissue reactions even though some CTL activity may be generated at the injection site. The issue of tissue site reactions and vaccination residues is one of the greatest concerns today with regards to meat quality. Presently, inactivated vaccines are introduced with adjuvants which induce injection site reactions and leave residues in the meat. Since DNA immunization uses no additional adjuvants and can be delivered to sites such as the ear, where damage to the carcass is minimal or nonexistent this approach appears very attractive to producers of food animals. This will also be extremely important for companion animals where owners are concerned about their animal's welfare. Preliminary studies suggest that delivery to mucosal sites can induce mucosal immunity, thereby, preventing or limiting infection as well as reducing disease. However, for this to become routine and fit into current management practices, better delivery and transfection efficiencies are required.

For veterinary vaccines, economics is always an important consideration. For example, in poultry, vaccines must be produced for $0.01–0.05/dose. In cattle, this increases by an order of magnitude. In contrast, in humans, vaccines can sell for 100–500 times the cost of a poultry vaccine. To make these vaccines economical for food producing animals, it will require improvement of delivery to enhance transfection efficiency, enhanced expression of the antigens, and economical purification of the plasmid.

Possibly one of the greatest advantages of DNA immunization is the ability to protect animals from disease as well as develop companion diagnostics to differentiate vaccinated animals from natural infected ones. Because many diseases can establish carrier states, this approach can easily be adopted to current management systems and allow culling of carriers. Thus, countries can vaccinate against specific diseases and eventually eradicate the disease. This will be especially attractive following the introduction of exotic diseases into a country. Thus, following accidental introduction it should be possible to rapidly eliminate the disease at a much reduced economic cost. Once an exotic disease is eradicated they can continue to test animals in the surrounding area and prove that they are all free of the disease. This would return a country to a disease-free status at much lower costs than the present approaches used today.

ACKNOWLEDGMENTS

Work in the authors' laboratory is supported by the Medical Research Council and the Natural Sciences and Engineering Research Council of Canada. We thank Michelle Balaski and Joyce Sander for typing the manuscript.

REFERENCES

Anderson, E. D., Mourich, D. V., Fahrenkrug, S. C., La Patra, S., Shepherd, J., and Leong, J. A. (1996). Genetic immunization of rainbow trout (Oncorhynchus mykiss) against infectious hematopoietic necrosis virus. *Mol. Mar. Biol. Biotechnol.* **5,** 114–122.

Bachmann, M. F., Kalinke, U., Althage, A., Freer, G., Burkhart, C., Roost, H.-P., Aguet, M., Hengartner, H., and Zinkernagel, R. M. (1997). The role of antibody concentration and avidity in antiviral protection. *Science* **276,** 2024–2027.

Barry, M. A., Lai, W. C., and Johnston, S. A. (1995). Protection against mycoplasma infection using expression-library immunization. *Nature* **377,** 632–635.

Barry, M. M., and Johnston, S. A. (1997). Biological features of genetic immunization. *Vaccine* **15,** 788–791.

Bos, J. D., and Kapsenberg, M. L. (1993). The skin immune system: Progress in cutaneous biology. *Immunol. Today* **14,** 75–78.

Caplen, N. J., Alton, E. W. F. W., Middleton, P. G., Dorin, J. R., Stevenson, B. J., Gao, X., Durham, S. R., Jeffery, P. K., Hodson, M. E., Coutelle, C., Huang, L., Porteous, D. J., Williamson, R., and Geddes, D. M. (1995). Liposome-mediated *CFTR* gene transfer to the nasal epithelium of patients with cystic fibrosis. *Nat. Med.* **1**(1), 39–46.

Cardoso, A. I., Blixenkrone-Moller, M., Fayolle, J., Liu, M., Buckland, R., and Wild, T. F. (1996). Immunization with plasmid DNA encoding for the measles virus haemagglutinin and nucleoprotein leads to humoral and cell-mediated immunity. *Virology* **225,** 293–299.

Condon, C., Watkins, S. C., Celluzi, C. M., Thompson, K., and Falo, L. D. J. (1996). DNA-based immunization by *in vivo* transfection of dendritic cells. *Nature Med.* **2,** 1122–1128.

Conry, R. M., Widera, G., LoBuglio, A. F., Fuller, J. T., Moore, S. E., Barlow, D. L., Turner, J., Yang, N.-S., and Curiel, D. T. (1996). Selective strategies to augment polynucleotide immunization. *Gene Ther.* **3,** 67–74.

Corbeil, S., Seguin, C., and Trudel, M. (1996). Involvement of the complement system in the protection of mice from challenge with respiratory syncytial virus Long strain following passive immunization with monoclonal antibody 18A2B2. *Vaccine* **14**(6), 521–525.

Cox, G. J. M., Zamb, T. J., and Babiuk, L. A. (1993). Bovine Herpesvirus 1: Immune responses in mice and cattle injected with plasmid DNA. *J. Virol.* **67,** 5664–5667.

Cox, J. C., and Coulter, A. R. (1997). Adjuvants—a classification and review of their modes of action. *Vaccine* **15**(3), 248–256.

Croitoru, K., and Bienenstock, J. (1994). Characteristics and functions of mucosa-associated lymphoid tissue. *Handb. Mucosal Immunol.*, pp. 141–150.

Davis, H. L., Michel, M. L., and Whalen, R. G. (1993). DNA based immunization induces continuous secretion of hepatitis B surface antigen and high levels of circulating antibody. *Hum. Mol. Genet.* **2**, 1847–1851.

Davis, H. L., Mancini, M., Michel, M.-L., and Whalen, R. G. (1996). DNA-mediated immunization to hepatitis B surface antigen: Longevity of primary response and effect of boost. *Vaccine* **14**(9), 910–915.

Davis, H. L., Brazolot-Millan, C. L., and Watkins, S. C. (1997). Immune-mediated destruction of transfected muscle fibers after direct gene transfer with antigen-expressing plasmid DNA. *Gene Ther.* **4**, 181–188.

Donnelly, J. J., Ulmer, J. B., and Liu, M. A. (1993). Immunization with polynucleotides: A novel approach to vaccination. *Immunologist* **2**, 20–26.

Donnelly, J. J., Ulmer, J. B., and Liu, M. A. (1994). Immunization with polynucleotides: A novel approach to vaccination. *Immunologist* **2**, 20–26.

Ertl, H. C., Verma, P., He, Z., and Xiang, Z. Q. (1995). Plasmid vectors as antiviral vaccines. *Ann. N.Y. Acad. Sci.* **772**, 77–97.

Feltquate, D. M., Heaney, S., Webster, R. G., and Robinson, H. L. (1997). Different T helper cell types and antibody isotypes generated by saline and gene gun DNA immunization. *J. Immunol.* **158**, 2278–2284.

Fynan, E. F., Webster, R. G., Fuller, D. H., Haynes, J. R., Santoro, J. S., and Robinson, H. L. (1993). DNA vaccines: Protective immunizations by parenteral, mucosal, and gene gun inoculations. *Proc. Natl. Acad. Sci. USA* **90**, 11478–11482.

Gao, Y., Daley, M. J., and Splitter, G. A. (1995). BHV-1 glycoprotein 1 and recombinant interleukin 1B efficiently elicit mucosal IgA response. *Vaccine* **13**, 871.

Gilkeson, G. S., Pippen, A. M., and Pisetsky, D. S. (1995). Induction of cross-reactive anti-dsDNA antibodies in pre-autoimmune NZB/NZW mice by immunization with bacterial DNA. *J. Clin. Invest.* **95**, 1398–1402.

Hassett, D. E., Zhang, J., and Whitton, L. (1997). Neonatal DNA Immunization with a plasmid encoding an internal viral protein is effective in the presence of maternal antibodies and protects against subsequent viral challenge. *J. Virol.* **71**(10), 7881–7888.

Hoffman, S. L., Sedegah, M., and Hedstrom, R. C. (1994). Protection against malaria by immunization with a plasmodium yoelei circums parazoite protein nucleic acid vaccine. *Vaccine* **12**, 1529–1533.

Ishizaka, S. T., Piacente, P., Silva, J., and Mishkin, E. M. (1995). IgG subtype is correlated with efficiency of passive protection and effector function of anti-herpes simplex virus glycoprotein D monoclonal antibodies. *J. Infect. Dis.* **172**, 1108–1111.

Keller, E. T., Burkholder, J. K., Shi, F., Pugh, T. D., McCabe, D., Malter, J. S., MacEwen, E. G., Yang, N.-S., and Ershler, W. B. (1996). In vivo particle-mediated cytokine gene transfer into canine oral mucosa and epidermis. *Cancer Gene Ther.* **3**(3), 186–191.

Kim, J. J., Ayyavoo, V., Bagarazzi, M. L., Chattergoon, M. A., Dang, K., Wang, B., Boyer, J. D., and Weiner, D. B. (1997). In vivo engineering of a cellular immune response by co-administration of IL-12 expression vector with a DNA immunogen. *J. Immunol.* **158**, 816–826.

Kuklin, N., Daheshia, M., Karem, K., Manickan, E., and Rouse, B. T. (1997). Induction of mucosal immunity against herpes simplex virus by plasmid DNA immunization. *J. Virol.* **71**, 3138–3145.

Levitsky, H. (1997). Accessories for naked DNA vaccines. *Nat. Biotechnol.* **15**, 619–620.

Lewis, P. J., Cox, G. J. M., van Drunen Littel-van den Hurk, S., and Babiuk, L. (1997). Polynucleotide vaccines in animals: Enhancing and modulating responses. *Vaccine* **15**(8), 861–864.

Lowrie, D. B., Tascon, R. E., Colston, M. J., and Silva, C. L. (1994). Towards a DNA vaccine against tuberculosis. *Vaccine* **356,** 152–154.
Madaio, M. P., Hodder, S., Schwartz, R. S., and Stollar, B. D. (1984). Responsiveness of autoimmune and normal mice to nucleic acid antigens. *J. Immunol.* **132,** 872–876.
Manickan, E., Karem, K. L., and Rouse, B. T. (1997). DNA vaccines—A modern gimmick or a boon to vaccinology? *Crit. Rev. Immunol.* **17,** 139–154.
Michel, M.-L., Davis, H. L., Schleef, M., Mancini, M., Tiollais, P., and Whalen, R. G. (1995). DNA-mediated immunization to the hepatitis B surface antigen in mice: Aspects of the humoral response mimic hepatitis B viral infection in humans. *Proc. Natl. Acad. Sci. USA* **92,** 5307–5311.
Neuberger, M. S., and Rajewsky, K. (1981). Activation of mouse complement by mouse monoclonal antibodies. *Eur. J. Immunol.* **11,** 1012–1016.
Nichols, W. W., Ledwith, B. J., Manam, S. V., and Troilo, P. J. (1995). Potential DNA vaccine integration into host cell genome. *Ann. N.Y. Acad. Sci.* **772,** 30–39.
Pertmer, T. M., Roberts, T. R., and Haynes, J. R. (1996). Influenza virus nucleoprotein-specific immunoglobulin G subclass and cytokine responses elicited by DNA vaccination are dependent on the route of vector DNA delivery. *J. Virol.* **70,** 6119–6125.
Puccetti, M. B., Dolcher, M. P., Madaio, M. P., and Migliorini, P. (1995). Induction of anti-DNA antibodies in non-autoimmune mice by immunization with DNA-DNAase I complex. *Clin. Exp. Rheumatol.* **13,** 7–10.
Ravetch, J. V., and Kinet, J. P. (1991). Fc receptors. *Annu. Rev. Immunol.* **9,** 457–492.
Robinson, H. L., Hunt, L. A., and Webster, R. G. (1993). Protection against a lethal influenza virus challenge by immunization with a haemagglutinin-expressing plasmid DNA. *Vaccine* **11**(9), 957–960.
Roman, M., Martin-Orozco, E., Goodman, J. S., Nguyen, M.-D., Sato, Y., Ronaghy, A., Kornbluth, R. S., Richman, D. D., Carson, D. A., and Raz, E. (1997). Immunostimulatory DNA sequences function as T helper-1-promoting adjuvants. *Nat. Med.* **3**(8), 849–854.
Schubbert, R., Renz, D., Schmitz, B., and Doerfler, W. (1997). Foreign (M13) DNA ingested by mice reaches peripheral leukocytes, spleen, and liver via the intestinal wall mucosa and can be covalently linked to mouse DNA. *Proc. Natl. Acad. Sci. USA* **94,** 961–966.
Sizemore, D. R., Branstrom, A. A., and Sadoff, J. C. (1995). Attenuated *Shigella* as a DNA delivery vehicle for DNA-mediated immunization. *Science* **270,** 299–302.
Sizemore, D. R., Branstrom, A. A., and Sadoff, J. C. (1997). Attenuated bacteria as a DNA delivery vehicle for DNA-mediated immunization. *Vaccine* **15,** 804–807.
Stribling, R., Brunette, E., Liggitt, D., Gaensler, K., and Debs, R. (1992). Aerosol gene delivery *in vivo*. *Proc. Natl. Acad. Sci. USA* **89,** 11277–11281.
Tang, D.-C., DeVit, M., and Johnston, S. A. (1992). Genetic immunization is a simple method for eliciting an immune response. *Nature* **356,** 152–154.
Torres, A. T., Iwasaki, A., Barber, B. H., and Robinson, H. L. (1997). Differential dependence on target site tissue for gene gun and intramuscular DNA immunizations. *J. Immunol.* **158,** 4529–4532.
Ulmer, J. B., Donnelly, J. J., Parker, S. E., Rhodes, G. H., Felgner, P. L., Dwarki, V. J., Gromkowski, S. H., Deck, R. R., DeWitt, C. M., Friedman, A., Hawe, L. A., Leander, K. R., Martinez, D., Perry, H. C., Shiver, J. W., Montgomery, D. L., and Liu, M. (1993). Heterologous protection against influenza by injection of DNA encoding a viral protein. *Science* **259,** 1745–1749.
Whalen, R. G., and Davies, H. L. (1995). DNA-mediated immunization and the energetic

immune response to hepatitis B surface antigen. *Clin. Immunol. Immunopathol.* **75,** 1–2.

Williams, R. S., Johnston, S. A., Reidy, M., Devit, M. J., McElligott, S. C., and Sanford, J. C. (1991). Introduction of foreign genes into tissues of living mice by DNA coated microprojectiles. *Proc. Natl. Acad. Sci. USA* **88,** 2726–2730.

Wolff, J. A., Malone, R. W., Williams, P., Ascadi, G., Jani, A., and Felgner, P. L. (1990). Direct gene transfer into mouse muscle *in vivo. Science* **247,** 1465–1468.

Xiang, Z. Q., Spitalnik, S., Tran, M., Wunner, W. H., Cheng, J., and Ertle, H. C. (1994). Vaccination with a plasmid vector carrying the rabies virus glycoprotein gene induces protective immunity against rabies virus. *Virology* **199,** 132–140.

Xu, D., and Liew, F. Y. (1994). Genetic vaccination against Leishmaniasis. *Vaccine* **12,** 1534–1536.

Yokonama, M., Zhang, J., and Whitton, J. L. (1995). DNA immunization confers protection against lethal lymphocytic choriomeningitis virus infection. *J. Virol.* **69,** 2684–2688.

Contribution of Advances in Immunology to Vaccine Development

W. I. MORRISON, G. TAYLOR, R. M. GADDUM, AND S. A. ELLIS

Institute for Animal Health, Compton, Newbury, Berkshire, United Kingdom

I. Introduction
II. Advances in Immunology Relevant to Vaccine Development
   A. Recognition of Antigens by Antibodies and T Cells
   B. Two Intracellular Pathways of Antigen Processing
   C. Different Cytokine Profiles of Responding T Cells
   D. Identification of Protective Antigens
III. Mechanisms of Immune Protection against Bovine Respiratory Syncytial Virus
   A. Disease Caused by RSV
   B. Role of Antibody and T-Cell Responses in Protection
   C. Identification of Expressed Bovine Class I MHC Genes
   D. Use of Cloned Class I Genes for Functional Studies
IV. Summary
   References

I. Introduction

The early pioneers of vaccination established that attenuated or killed organisms could be used to stimulate protective immune responses. These approaches to vaccine development have remained essentially unchanged for much of the twentieth century, progress having been made largely as a consequence of technical advances in methods for culturing microorganisms, identification of adjuvants, and refinement of inactivation and attenuation procedures. Although these developments have resulted in the production of vaccines that have had a major impact on the control of livestock diseases, there remain many important diseases for which either vaccines are not available or the current vaccines are only partially effective.

During the last two decades, developments in recombinant DNA technology have created new opportunities for vaccine production both through genetic manipulation of pathogens and by enabling the identification of defined antigenic subunits that induce protective immune responses. That the application of this new technology has been slow to yield vaccine products is due in part to the fact that, until recently, there was limited knowledge of the immunology of many of the target diseases and of how antigens are processed and recognized by the immune system. Recent advances in immunology, coupled to further developments in the application of DNA technology, now provide a strong conceptual framework for the rational development of new vaccines.

In this chapter, we consider recent developments in immunology that are particularly relevant to vaccination and discuss how studies of the bovine immune system are contributing to vaccine development in cattle, focusing particularly on our work on MHC and T-cell responses to viral infections.

II. Advances in Immunology Relevant to Vaccine Development

A. Recognition of Antigens by Antibodies and T Cells

Although it has been known for more than 20 years that T lymphocytes recognize antigen that has been processed and presented in association with major histocompatibility complex (MHC) molecules on the surface of antigen presenting cells, the precise nature of the antigenic epitopes and how they are recognized was only elucidated following the determination of the molecular structure of the MHC and T-cell receptor molecules. Structural studies have demonstrated that processed peptides bind within a groove in the membrane-distal portion of class I and class II MHC molecules, formed by two parallel α helices and a floor of β-pleated sheets (Engelhard, 1994; Stern et al., 1994). Class I and class II molecules bind peptides of different length, 8–10 amino acids and 13–17 amino acids, respectively. In both cases the peptide-binding groove contains pockets that accommodate the side chains of amino acids at particular positions on the bound peptide; for most class I molecules, these binding interactions involve residues at position 2 and at the C terminus of the peptide. Because the binding residues tend to be conserved for any given class I molecule, they represent a motif for that molecule, which can be exploited when screening protein sequences for potential epitopes (Rammensee et al., 1993). Such motifs are less well defined for class II molecules (Hammer et al., 1993). Antigen recognition by the T-cell receptor involves interaction both with the α helices of the MHC molecule and with residues in the bound peptide that have upward-facing side chains (i.e., different residues from those involved in MHC binding).

Much of the polymorphism in MHC molecules occurs in and around the peptide-binding groove and although this variation does not affect the overall structure of the groove it results in subtle differences that influence the nature of the peptides that each molecule will bind (Matsumura et al., 1992; Rammensee, 1995). Consequently, T cells from animals expressing different MHC molecules usually recognize different epitopes from the same pathogen and in some cases these epitopes may be on different proteins. This effect is particularly pronounced with $CD8^+$ cytotoxic T lymphocyte (CTL) responses, which in individual animals are often focused on only a few epitopes, and, when pathogen strains are antigenically polymorphic, can lead to variation between animals in the strain specificity of the CTL response (Vitiello and Sherman, 1983). MHC-determined variation in strain specificity of CTL responses has been observed in responses of cattle to infection with the protozoan parasite *Theileria parva* (Goddeeris et al., 1990; Taracha et al., 1995). Such variation has important implications for vaccination strategies utilizing individual proteins, because the potential T-cell epitopes within such proteins may not be recognized by all individuals in an outbred population.

In contrast to the situation with T-cell epitopes which are generated by denaturation and degradation of proteins, most epitopes recognized by antibodies are dependent to varying degrees on the native conformation of the target molecules. In the early 1980s, a series of studies demonstrated that antibodies raised against intact proteins recognized short peptide fragments of the proteins (Geyson et al., 1984). These observations encouraged the belief that it would be possible to use synthetic peptides for vaccination. Numerous studies aimed at stimulating immunity with peptides were undertaken, in which animals were immunized with synthetic peptides, representing B-cell epitopes, conjugated either to other polypeptides from the same pathogen or to unrelated proteins, to provide the necessary T helper cell epitopes. With a few exceptions, these attempts at immunization were unsuccessful. In many instances, the synthetic peptide failed completely to induce antibodies against the parent protein or organism or resulted in antibodies that only recognized denatured antigen, while in other cases strong antibody responses did occur but were at best only partially effective in mediating protection.

Subsequent studies of protein structure have highlighted the fact that so-called linear peptides exhibit a degree of conformation and that they tend to be conformationally flexible such that they can adopt structures that differ from that of the parent molecule. The process of conjugating a peptide to a carrier molecule may also affect the conformation of the peptide, resulting in antibodies of low avidity for the pathogen in question. There are, in addition, many B-cell epitopes that have a complex conformational structure dependent on the tertiary structure of the proteins.

Another potential disadvantage of using peptides representing single antigenic sites to stimulate protective antibody responses is the possibility of selecting for antigenic mutations in the pathogen. The risk of selecting such mutations is highest for RNA viruses, because of the inherent high error rate in RNA replication, and would be of particular concern in the case of viruses, such as foot-and-mouth disease virus, which already have a propensity to undergo antigenic changes and for which antibody responses have an important role in protection.

B. Two Intracellular Pathways of Antigen Processing

In the early 1980s it was discovered that $CD4^+$ T cells recognized antigen presented by class II MHC molecules whereas $CD8^+$ T cells recognized antigen presented by class I. The functional significance of this finding was subsequently clarified by the discovery that class I and class II molecules bind peptides generated within different subcellular compartments of antigen-presenting cells. Antigens derived from organisms that replicate in the cytoplasm of cells were shown to be degraded by proteases within the cytosol and the resultant peptides translocated by specialized transporter for antigen presentation (TAP) molecules into the endoplasmic reticulum where they associate with newly synthesized class I molecules destined for the cell surface (Williams et al., 1996). By contrast, organisms or proteins taken into antigen presenting cells by phagocytosis or endocytosis were shown to undergo enzymatic degradation within endosomes and associate within an endosomal compartment with newly synthesized class II molecules before being expressed on the cell surface (Wubbolts et al., 1997). Binding of peptides to class II molecules in the endoplasmic reticulum is prevented by association with an invariant polypeptide, part of which lies in the peptide-binding groove. This invariant chain also facilitates transport of the newly synthesized class II molecules from the Golgi into endosomes, where it is removed by enzymatic degradation allowing peptide loading (Teyton and Peterson, 1992).

These alternative routes of antigen processing are known as the endogenous (class I) and exogenous (class II) pathways, respectively. The processing of antigens by the endogenous pathway and presentation by class I is confined mainly to organisms, such as viruses and some bacteria and protozoa, that replicate intracellularly either in the cytosol or within endocytic vacuoles that allow entry of polypeptides into the cytosol. This requirement for endogenous processing to induce CD8+ T-cell responses imposes constraints on the types of antigen delivery systems that can be used when considering the development of subunit vaccines against pathogens for which a CD8+ T-cell response is required to mediate immunity. Immunization of animals with killed organisms or their component proteins generally fails to induce CD8+ T-cell responses, whereas virus vectors, liposomes, and naked DNA have all been shown to be effective at stimulating these responses.

C. Different Cytokine Profiles of Responding T Cells

In 1986 Mosmann and colleagues, working with mouse T-cell clones, described two types of CD4+ T cell, termed Th1 and Th2, that were distinguished by the cytokines they produced. This and subsequent studies established that activated Th1 cells secrete interleukin-2 (IL-2) and interferon-γ (IFN-γ) but not IL-4, IL-5, IL-10, and IL-13, whereas the converse applies to activated Th2 cells (Mosmann and Coffman, 1989). A similar dichotomy has been reported for human T cells (Weiranga et al., 1990). Strong cross-regulation between Th1 and Th2 cells has been demonstrated: IFN-γ produced by Th1 cells inhibits the induction of Th2 responses and both IL-4 and IL-10 have inhibitory effects on the induction of Th1 responses (Mosmann et al., 1991; Fitch et al., 1993). In mice, Th1 responses result in activation of macrophages, production of antibody of the IgG_{2a} isotype and delayed-type hypersensitivity reactions, while Th2 responses give rise to eosinophilia and production of antibody of the IgG_1 and IgE isotypes. Studies of murine models of parasitic infections have proved invaluable in elucidating the biological significance of the difference in Th1 and Th2 cells. Infection of mice with *Leishmania major*, or immunization with *Leishmania* antigens, can induce either Th1 or Th2 T-cell responses, depending on the strain of mouse and route of immunization. Th1 responses result in control of infection and immunity, whereas Th2 responses lead to enhanced disease (Liew, 1990). Conversely, Th2 responses have been shown to be essential for immunity to the murine intestinal nematode *Trichuris muris* (Else and Grencis, 1991). These are just two of a growing number of examples in which the outcome of infection is strongly influenced by the cytokine profile of the responding T cells (Abbas et al., 1996). Although Th1 and Th2 responses have not been particularly well defined for cattle, preliminary studies in several laboratories indicate that similar pathways of T-cell differentiation do occur (Brown et al., 1994; Estes et al., 1995).

The mechanisms that determine the bias in T-cell responses are not well understood but there is evidence that modulation of the function of antigen presenting cells is involved, by direct or indirect effects of pathogens or antigenic components thereof on their expression of cytokines or costimulatory molecules (Fitch et al., 1993; Carter and Dutton, 1996). These phenomena have obvious relevance to vaccine design. Clearly, it is important to have an understanding of whether the protective responses against target pathogens involve Th1 or Th2 CD4$^+$ T-cell responses and whether there is the potential for one or other of these responses to potentiate disease. There is also the potential to incorporate into subunit vaccines cytokines or other costimulatory molecules that have the capacity to bias the T-cell response to a Th1 or Th2 cytokine profile.

D. IDENTIFICATION OF PROTECTIVE ANTIGENS

Advances in our understanding of how the immune response operates to control pathogenic organisms, coupled with the development of reagents and techniques to identify functionally important cells and molecules of the immune system, have opened the way for studies to define the immune responses that mediate protection against specific pathogens. Such information can, in turn, be exploited to develop strategies for identification of protective antigens, based on the use of specific antibodies or T-cell lines known to have protective activity. Techniques for screening biochemically fractionated antigen preparations or antigens expressed by cDNA libraries to identify molecules recognized by antibodies or CD4$^+$ T cells are well established. The identification of antigens recognized by CD8$^+$ T cells presents a more difficult challenge because of the need for antigen to be processed by the endogenous route for association with class I MHC. In the case of simple viruses for which the genome sequence is known, the individual genes can be expressed in target cells either by transfection or by incorporation into a virus vector, and the cells screened for susceptibility to killing by specific CD8+ T cells. Recently a system for identifying CD8+ T-cell target antigens by screening complex cDNA libraries expressed in COS-7 cells has been described (Coulie et al., 1994). This system involves several rounds of screening pools of cDNA clones based on the ability of the expressed product to stimulate tumor necrosis factor (TNF) production in the specific T-cell line. However, this approach requires expression of the restricting class I molecule in the COS cells and this is only feasible where cloned genes for the appropriate class I molecule are available.

The development of biochemical techniques for analysis of MHC-bound peptides is also providing new approaches to the identification of antigens recognized by T lymphocytes. As already discussed, class I MHC peptide-binding motifs can be used to identify potential T-cell epitopes in organisms of known genome sequence (Rammensee et al., 1993). The mixture of peptides eluted from a particular MHC molecule can also be subjected to more rigorous fractionation procedures, utilizing high-pressure liquid chromatography or mass spectrometry, which allow amino acid sequencing of individual peptides (Hunt et al., 1992). Although, at present, this approach requires large amounts of starting material and has been used successfully only in a few laboratories, it is likely that further refinement of the methodologies will result in improved sensitivity and reproducibility. Once this has been achieved, the approach will find wide application, particularly in the identification of T-cell epitopes in complex organisms such as bacteria and parasites.

III. Mechanisms of Immune Protection against Bovine Respiratory Syncytial Virus

A. Disease Caused by RSV

Bovine respiratory syncytial virus (RSV) is an important cause of pneumonia in calves. The disease occurs as annual winter outbreaks in housed calves during the first 6 months of life. The virus infects respiratory epithelium in both the upper and lower respiratory tract, the latter affecting predominantly bronchioles, causing a severe bronchiolitis. A closely related virus in man, human RSV, causes a disease in infants with similar pathologic and epidemiologic features. Unlike bovine RSV, the human virus infects mice and, although the disease produced in mice differs in a number of respects from that in the natural hosts, this model has been used extensively to study the immunology of RSV. Bovine RSV tends to be difficult to grow in tissue culture and generally loses virulence following passage. Nevertheless, disease can be reproduced experimentally with some isolates that have undergone limited passage in culture, although, because of the ubiquitous nature of the virus, such studies need to be carried out in specific pathogen-free animals.

An important problem encountered in the development of some RSV vaccines has been the occurrence of enhanced disease following challenge with the virus. This problem is particularly well documented for a human vaccine that utilized formalin-inactivated virus incorporated in alhydrogel (Fulginiti et al., 1969). The need for a detailed understanding of the immunology of RSV is, therefore, widely recognized not only to identify those immune responses that mediate protection but also to highlight the responses that result in enhanced disease.

B. Role of Antibody and T-Cell Responses in Protection

The available evidence indicates that antibody and T-cell responses are both involved in immunity to RSV. Animals that have recovered from infection produce antibodies specific for a number of viral proteins including the attachment (G) and fusion (F) envelope glycoproteins. Studies in which monoclonal antibodies (MAb) raised against different viral proteins of human RSV were tested for their ability to protect mice against infection with the virus showed that some of the MAb specific for the F and G glycoproteins were protective (Taylor et al., 1984). The F-specific MAbs exhibited particularly potent protective activity, which correlated with the ability to inhibit cell fusion in infected cell cultures and was associated with specificity for either of two sites on the F protein (Taylor et al., 1992). These MAbs not only protected when administered prior to infection with RSV but also resulted in rapid clearance of virus when given to mice with established infections. A more limited study in calves with bovine MAb specific for the F protein, produced from heterohybridomas, confirmed that such antibodies could also protect against bovine RSV (Thomas et al., 1998).

Experiments in which $CD4^+$, $CD8^+$, or γ/δ T-cell subsets were transiently depleted by administration of specific MAb, in calves infected with RSV, have provided evidence that $CD8^+$ T cells are required for prompt resolution of infection with RSV (Taylor et al., 1995; Thomas et al., 1996). Calves depleted of $CD8^+$ T cells exhibited prolonged nasal shedding of virus and more extensive replication of virus in the lungs in comparison to calves given a control MAb. By contrast, these parameters were unaffected by depletion of $CD4^+$ or γ/δ T cells. These results differ from findings with the mouse model, in which the course of infection was unaffected by depletion of CD4+ or CD8+ T cells, and prolonged infection only occurred when both subsets were depleted (Graham et al., 1991).

Immunohistological analyses of lung tissue from RSV-infected calves have revealed an influx of T lymphocytes of all three subsets during infection, with CD8+ T cells representing the predominant population in most animals (Taylor et al., 1995; Thomas et al., 1996). Moreover, the T-lymphocyte infiltrate in bronchiolar epithelium was composed largely of CD8+ cells. Lymphocytes isolated from lung tissue of calves 10 days after infection (i.e., 2–3 days after clearance of virus) exhibited cytotoxic activity for autologous RSV-infected cells (Gaddum et al., 1996a). Similar, but lower, levels of cytotoxicity were detected in peripheral blood mononuclear cells between 7 and 10 days after infection. The cells mediating this cytotoxicity were shown to reside in the CD8+ T-cell population, were specific for RSV-infected cells, and were class I MHC restricted. They also killed target cells infected with strains of bovine RSV which differ predominantly in the antigenicity of the G protein. These data coupled with the *in vivo* T-cell depletion studies indicate that virus-specific CD8+ CTL play an important role in the resolution of primary infections with RSV in cattle. Current studies are directed toward identification of the RSV proteins recognized by the CTL using target cells infected with recombinant vaccinia viruses expressing individual viral proteins.

Given the previous evidence for a protective role of antibodies specific for the F and G glycoproteins, the absence of an effect on infection of CD4+ T-cell depletion was somewhat surprising and raises questions concerning the quality of the early antibody response to primary infection with RSV in calves.

Studies of the immune responses of mice inoculated with recombinant vaccinia viruses expressing the F or G glycoproteins of human RSV have revealed a marked difference in the CD4+ T cell responses stimulated by the two antigens (Openshaw et al., 1992; Alwan and Openshaw, 1993; Alwan et al., 1993). The F glycoprotein induced CD8+ CTL and a typical Th1-dominated CD4 response, with production of high levels of IL-2 and IFN-γ, whereas the G protein induced T cells that produced abundant IL-4 and IL-5 but low levels of IL-2 and IFN-γ. Significantly, following challenge with RSV, mice immunized with the G glycoprotein exhibited more severe lung pathology, a prominent feature of which was infiltration of eosinophils. Also, immunization of mice with formalin-inactivated RSV prior to challenge with RSV has been shown to result in enhanced pulmonary pathology, which could be abrogated by depletion of CD4+ T cells or treatment with MAbs specific for IL-4 or IL-10 (Connors et al., 1992, 1994). On the basis of these findings, it has been proposed that the previously described enhanced disease associated with certain inactivated vaccines may have been due to preferential induction of a Th2 CD4 T-cell response. However there is as yet little evidence to support or refute this hypothesis from studies of the disease in the natural hosts.

C. IDENTIFICATION OF EXPRESSED BOVINE CLASS I MHC GENES

As discussed earlier, access to animals of defined MHC genotype for studies of cytotoxic T-cell responses provides the opportunity to dissect the fine specificity of the response and to determine the influence of the MHC on viral antigen specificity. Until recently characterization of class I MHC antigens in cattle relied on typing with alloantisera. More than 50 class I serologic specificities are defined by available antisera, the majority of which are reported to represent the products of a single locus (Davies et al., 1993). However, initial molecular analyses of class I in a *Bos indicus* animal identified two class I genes expressed on one MHC haplotype (Bensaid et al., 1991). Our subsequent studies, carried out mainly in Friesian/Holstein (*B. taurus*) animals, have confirmed that cattle have more than one expressed polymorphic class I gene (Ellis et al., 1996; S. A. Ellis, unpublished data). This conclusion is based on sequencing of full-length cDNA clones isolated from animals carrying different haplotypes and biochemical analyses of the expressed class I gene products in transfected cell lines in comparison with class I molecules expressed in the animals of origin and in animals bred as homozygous for the haplotypes in question. Fourteen class I genes expressed by 9 different haplotypes have been analyzed in this way. Of 6 haplotypes for which we believe that all of the expressed classical class I genes have been identified, five have two expressed genes, whereas one (serologic specificity A18) has only one expressed gene. Although these data indicate that two class I genes are expressed on most haplotypes, comparative analyses of the gene sequences have so far not allowed assignment of the individual genes to particular loci. Indeed, preliminary phylogenetic analyses indicate that there may be four or more potentially expressable class I genes with different patterns of gene expression by different haplotypes (S. A. Ellis and E. Holmes, unpublished data).

D. USE OF CLONED CLASS I GENES FOR FUNCTIONAL STUDIES

Transfected cell lines expressing bovine class I MHC molecules are being exploited in two ways to study the fine specificity of CTL responses to RSV. First, transfected cells infected with recombinant vaccinia viruses expressing different RSV proteins are being used as target cells to identify the viral proteins recognized by CTL in the context of the individual class I molecules. This approach is being pursued, initially, using animals homozygous for the A18 class I specificity as donors of cytotoxic T cells and transfected cells expressing the A18 molecule as targets. Analysis of this particular MHC haplotype is simplified by the fact that it has only one expressed class I gene. Preliminary findings indicate that the nucleoprotein and matrix 2 (M2) protein of RSV are recognized by RSV-specific CTL when presented by the A18 class I molecule (R. Gaddum, unpublished data). Further studies are under way to confirm this finding and to examine the specificity of the response on different MHC backgrounds.

The second way in which the transfected cell lines are being used to study the specificity of CTL responses is by defining the peptide binding motifs of the expressed class I molecules. This involves affinity purification of the bovine class I molecules from detergent lysates of the transfected cells using a MAb specific for bovine class I heavy chain (IL-A88), elution, and purification of bound peptides by reverse phase–high-pressure liquid chromatography and amino acid sequencing of the purified peptide pools. The resultant sequence information identifies conserved residues in the bound peptides, that is, anchor residues, as well as amino acids that are preferentially represented at other sites in the peptides. We have identified detailed peptide binding motifs for six of the cloned class I molecules (Hegde et al., 1995; Gaddum et al., 1996b; S. Ellis, R. M. Gaddum, and A. C. Willis, unpublished data). The characteristics of the motifs are generally similar to those identified for murine and human class I molecules, in that there is usually an anchor at position 2 and at the C terminus, although in some instances a primary anchor could not be identified at the latter site.

The peptide-biding motif data can be used to identify potential T-cell epitopes in pathogens of known sequence. Synthetic peptides representing potential epitopes can then be used to prime target cells of the appropriate MHC phenotype to test for recognition by CTL. Experience from analyses of human and murine CTL indicates that approximately 75% of known epitopes contain predicted binding motifs. This approach will be applied in the analysis of bovine CTL responses to RSV, by screening peptides with predicted binding motifs from those viral proteins that have been shown to be recognized by CTL.

These studies of CTL responses to RSV, when completed, will provide information on the viral antigens and the epitopes therein that are recognized by CTL, and the extent to which MHC phenotype results in variation between animals in the antigenic specificity of the response.

IV. Summary

During the last 10 years, investigation of the bovine immune system has generated knowledge and reagents that can now be applied to study the mechanisms of immunity to disease and the identity of antigens recognized by protective immune responses. Such studies can indicate which antigens are likely to be effective in subunit vaccines and also highlight the type of antigen delivery system that will be required for a vaccine to induce a protective immune response. In the case of bovine RSV, studies of immune responses in the target host have demonstrated that both antibody and CTL responses play an important role in immunity. Both the F and G glycoproteins have been identified as targets of protective antibodies, and systems have been established that will allow the identification of the viral antigens recognized by CTL. Further studies of CD4$^+$ T-cell responses to the virus are required to determine whether or not components of the response have the potential to enhance disease and, therefore, need to be avoided in vaccination strategies.

References

Abbas, A. K., Murphy, K. M., and Sher, A. (1996). Functional diversity of helper T lymphocytes. *Nature* **383,** 787–793.

Alwan, W. H., and Openshaw, P. J. (1993). Distinct patterns of T- and B-cell immunity to respiratory syncytial virus induced by individual viral proteins. *Vaccine* **11,** 431–437.

Alwan, W. H., Record, F. M., and Openshaw, P. J. M. (1993). Phenotypic and functional characterisation of T cell lines specific for individual respiratory syncytial virus proteins. *J. Immunol.* **150,** 5211–5218.

Bensaid, A., Kaushal, A., Baldwin, C. L., Clevers, H., Young, J. R., Kemp, S. J., MacHugh, N. D., Toye, P. G., and Teale, A. J. (1991). Identification of expressed bovine class I MHC genes at two loci and demonstration of physical linkage. *Immunogenetics* **33,** 247–254.

Brown, W. C., Woods, V. M., Chitko-McKown, C. G., Hash, S. M., and Rice-Ficht, A. C. (1994). Interleukin-10 is expressed by bovine type 1 helper, type 2 helper and unrestricted parasite-specific T-cell clones and inhibits proliferation of all three subsets in an accessory-cell-dependent manner. *Infect. Immun.* **62,** 4697–4708.

Carter, L. L., and Dutton, R. W. (1996). Type 1 and type 2: a fundamental dichotomy for all T-cell subsets. *Curr. Opin. Immunol.* **8,** 336–342.

Connors, M., Kulkarni, A. B., Firestone, C.-Y., Holmes, K. L., Morse, H. C., III, Sotnikov, A. V., and Murphy, B. R. (1992). Pulmonary histopathology induced by RSV challenge of formalin-inactivated RSV-immunised BALB/c mice is abrogated by depletion of CD4$^+$ T cells. *J. Virol.* **66,** 7444–7451.

Connors, M., Giese, N. A., Kulkarni, A. B., Firestone, C.-Y., Morse, H. C., III, and Murphy, B. R. (1994). Enhanced pulmonary histopathology induced by respiratory syncytial virus (RSV) challenge of formalin-inactivated RSV-immunised BALB/c mice is abrogated by depletion of interleukin 4 (IL-4) and IL-10. *J. Virol.* **68,** 5321–5325.

Coulie, P. G., Brichard, V., Van Pel, A., Wolfel, T., Schneider, J., Traversari, C., Mattei, S., De Plaen, E., Lurquin, C., Szikora, J.-P., Renauld, J. C., and Boon, T. (1994). A new gene coding for a differentiation antigen recognised by autologous cytotoxic T lymphocytes on HLA-A2 melanomas. *J. Exp. Med.* **180,** 35–42.

Davies, C. J., Joosten, I., Bernoco, D., Arriens, M. A., Bester, J., Ceriotti, G., Ellis, S., Hensen, E. J., Hines, H. C., Horin, P., Kristensen, B., Lewin, H. A., Meggiolaro, D., Morgan, A. L. G., Nilsson, P. R., Oliver, R. A., Orlova, A., Ostergard, H., Park, C. A., Schuberth, H.-J., Simon, M., Spooner, R. L., and Stewart, J. A. (1993). Polymorphism of bovine MHC class I genes. Joint Report of the Fifth International Bovine Lymphocyte Antigen (BoLA) Workshop, Interlaken, Switzerland, 1 August 1993. *Eur. J. Immunogenet.* **21,** 239.

Ellis, S. A., Staines, K. A., and Morrison, W. I. (1996). cDNA sequence of cattle MHC class I genes transcribed in serologically defined haplotypes A18 and A31. *Immunogenetics* **43,** 156–159.

Else, K. J., and Grencis, R. K. (1991). Cellular immune responses to the murine nematode parasite *Trichuris muris*. I. Differential cytokine production during acute or chronic infection. *Immunology* **72,** 508–513.

Engelhard, V. H. (1994). Structure of peptides associated with class I and class II MHC molecules. *Annu. Rev. Immunol.* **12,** 181–207.

Estes, D. M., Hirano, A., Heussler, V. T., Dobbelaere, D. A. E., and Brown, W. C. (1995). Expression and biological activities of bovine interleukin 4: Effects of recombinant bovine interleukin 4 on T cell proliferation and B cell differentiation and proliferation in vitro. *Cell. Immunol.* **163,** 268–279.

Fitch, F. W., McKisic, M. D., Lancki, D. W., and Gajewski, T. F. (1993). Differential regulation of murine lymphocyte subsets. *Annu. Rev. Immunol.* **11,** 29–48.

Fulginiti, V. A., Eller, J. J., Sieber, O. F., Joyner, J. W., Minamitani, M., and Meiklejohn, G. (1969). Respiratory virus vaccines; an aqueous trivalent parainfluenza virus vaccine and an alum precipitated respiratory syncytial virus vaccine. *Am. J. Epidemiol.* **89,** 435–448.

Gaddum, R. M., Cook, R. S., Thomas, L. H., and Taylor G. (1996a). Primary cytotoxic T-cell responses to bovine respiratory syncytial virus in calves. *Immunology* **88,** 421–427.

Gaddum, R. M., Willis, A. C., and Ellis, S. A. (1996b). Peptide motifs from three cattle MHC (BoLA) class I antigens. *Immunogenetics* **43,** 238–239.

Geyson, H. M., Meloen, R. H., and Barteling, S. J. (1984). Use of peptide synthesis to probe viral antigens for epitopes to a resolution of a single amino acid. *Proc. Natl. Acad. Sci. USA* **81,** 3998–4002.

Goddeeris, B. M., Morrison, W. I., Toye, P. G., and Bishop, R. (1990). Strain specificity of bovine *Theileria parva*-specific cytotoxic T cells is determined by the phenotype of the restricting class I MHC. *Immunology* **69,** 38–44.

Graham, B. S., Bunton, L. A., Wright, P. F., and Karzon, D. T. (1991). Role of T lymphocyte subsets in the pathogenesis of primary infection and rechallenge with respiratory syncytial virus in mice. *J. Clin. Invest.* **88,** 1026–1033.

Hammer, J., Valsasnini, P., Tolba, K., Bolin, D., Higelin, J., Takacs, B., and Sinigaglia, F. (1993). Promiscuous and allele-specific anchors in HLA-DR-binding peptides. *Cell* **74,** 197–203.

Hegde, N. R., Ellis, S. A., Gaddum, R. M., Tregaskes, C. A., Sarath, G., and Srikumaran, S. (1995). Peptide motif of a cattle MHC class I antigen BoLA-A11. *Immunogenetics* **42,** 302–303.

Hunt, D. F., Henderson, R. A., Shabanowitz, J., Sakaguchi, K., Michel, H., Sevilir, N.,

Cox, A. L., Appella, E., and Engelhard, V. H. (1992). Characterisation of peptides bound to the class I MHC molecule HLA-A2.1 by mass spectrometry. *Science* **255,** 1261–1263.

Liew, F. Y. (1990). Regulation of cell-mediated immunity to leishmaniasis. *Curr. Top. Microbiol. Immunol.* **155,** 63–84.

Matsumura, M., Fremont, D. H., Peterson, P. A., and Wilson, I. A. (1992). Emerging principles for the recognition of peptide antigens by MHC class I molecules. *Science* **257,** 927–934.

Mosmann, T. R., and Coffman, R. L. (1989). Th1 or Th2 cells: Different patterns of lymphokine secretion lead to different functional properties. *Annu. Rev. Immunol.* **7,** 145–173.

Mosmann, T. R., Cherwinski, H., Bond, M. W., Giedlin, M. A., and Coffmann, R. L. (1986). Two types of murine helper T cell clone. I. Definition according to profiles of lymphokine activities and secreted proteins. *J. Immunol.* **136,** 2348–2357.

Mosmann, T. R., Schumacher, J. H., Street, N. F., Budd, R., O'Garra, A., Fong, T. A. T., Bond, M. W., Moore, K. W. M., Sher, A., and Fiorentino, D. F. (1991). Diversity of cytokine synthesis and function of mouse $CD4^+$ T cells. *Immunol. Rev.* **123,** 209–229.

Openshaw, P. J. M., Clarke, S. L., and Record, F. M. (1992). Pulmonary eosinophilic response to respiratory syncytial virus infection in mice sensitized to the major surface glycoprotein G. *Int. Immunol.* **4,** 493–500.

Rammensee, H.-G. (1995). Chemistry of peptides associated with MHC class I and class II molecules. *Curr. Opin. Immunol.* **7,** 85–96.

Rammensee, H.-G., Falk, K., and Rotzchse, O. (1993). Peptides naturally presented by MHC class I molecules. *Annu. Rev. Immunol.* **11,** 213–244.

Stern, L. J., Brown, J. H., Jardetzky, T. S., Gorga, J. C., Urban, R. G., Strominger, J. L., and Wiley, D. C. (1994). Crystal structure of the human class II MHC protein HLA-DR1 complexed with an influenza virus peptide. *Nature* **368,** 215–221.

Taracha, E. L. N., Goddeeris, B. M., Teale, A. J., Kemp, S. J., and Morrison, W. I. (1995). Parasite strain specificity of bovine cytotoxic T cell responses to *Theileria parva* is determined primarily by immunodominance. *J. Immunol.* **155,** 4854–4860.

Taylor, G., Stott, E. J., Bew, M., Fernie, B. F., Cote, P. J., Collins, A. P., Hughes, M., and Jebbett, J. (1984). Monoclonal antibodies protect against respiratory syncytial virus infection in mice. *Immunology* **52,** 137–142.

Taylor, G., Stott, E. J., Furze, J., Ford, J., and Sopp, P. (1992). Protective epitopes on the fusion protein of respiratory syncytial virus recognised by murine and bovine monoclonal antibodies. *J. Gen. Virol.* **73,** 2217–2223.

Taylor, G., Thomas, L. H., Wyld, S. G., Furze, J., Sopp, P., and Howard C. J. (1995). Role of T-lymphocyte subsets in recovery from respiratory syncytial virus infection in calves. *J. Virol.* **69,** 6658–6664.

Teyton, L., and Peterson, P. A. (1992). Invariant chain—a regulator of antigen presentation. *Trends Cell Biol.* **2,** 52–56.

Thomas, L. H., Cook, R. S., Howard, C. J., Gaddum, R. M., and Taylor, G. (1996). Influence of selective T-lymphocyte depletion on the lung pathology of gnotobiotic calves and the distribution of different T-lymphocyte subsets following challenge with bovine respiratory syncytial virus. *Res. Vet. Sci.* **61,** 38–44.

Thomas, L. H., Cook, R. S., Wyld, S. G., Furze, J. M., and Taylor, G. (1998). Passive protection of gnotobiotic calves using monoclonal antibodies directed at different epitopes on the fusion protein of bovine respiratory syncytial virus. *J. Infect. Dis.* **177,** 874–880.

Vitiello, A., and Sherman, L. A. (1983). Recognition of influenza-infected cells by cytolytic

T lymphocyte clones: Determinant selection by class I restriction elements. *J. Immunol.* **131,** 1635–1640.

Weiranga, E. A., Snook, M., DeGroot, C., Chrétien, I., Bos, J., Jansen, H., and Kapsenberg, M. (1990). Evidence for compartmentalisation of functional subsets of CD4 T lymphocytes in atopic patients. *J. Immunol.* **144,** 4651–4656.

Williams, D. B., Vassilakos, A., and Suh, W.-K. (1996). Peptide presentation by MHC class I molecules. *Trends Cell Biol.* **6,** 267–273.

Wubbolts, R., Fernandez-Borja, M., and Neefjes, J. (1997). MHC class II molecules: Transport pathways for antigen presentation. *Trends Cell Biol.* **7,** 115–118.

III
BOVINE VACCINES AND DIAGNOSTICS

Bovine Viral Vaccines, Diagnostics, and Eradication: Past, Present, and Future

JAN T. VAN OIRSCHOT

Department of Mammalian Virology, Institute of Animal Science and Health, 8200 AB Lelystad, The Netherlands

I. Introduction
II. Foot-and-Mouth Disease
   A. Vaccines
   B. Diagnostics
   C. Eradication
III. Infectious Bovine Rhinotracheitis
   A. Vaccines
   B. Diagnostics
   C. Eradication
IV. Bovine Virus Diarrhea
   A. Vaccines
   B. Diagnostics
   C. Eradication
V. Posteradication Period
References

I. Introduction

Foot-and-mouth disease (FMD) virus (FMDV) has been eradicated from the Western world. In the recent past, various European countries have become free of bovine herpesvirus 1 (BHV1), which causes infectious bovine rhinotracheitis (IBR) and pustular vulvovaginitis (IPV), and other countries have eradication schemes implemented for this virus. Recently, in the Scandinavian countries, programs have been started with the aim of eradicating bovine virus diarrhea (BVD) virus (BVDV) and in the future other European countries may follow this example. These three bovine viral diseases will be used to illustrate the evolution in vaccines and diagnostics in relation to eradication.

II. Foot-and-Mouth Disease

FMD is among the most contagious viral diseases of cloven-hoofed animals and is causing enormous economic losses. The virus belongs to the genus Aphtovirus of the family *Picornaviridae* and consists of seven serotypes, A, O, C, SAT 1, SAT 2, SAT 3, and Asia 1, that do not cross-protect. Typically, vesicles are found in the mouth and on the feet. The virus is prevalent on the South American, African, Asian, and Eastern European continents, and has been eradicated from the Western world.

A. Vaccines

Only conventional killed vaccines are available on the market. These vaccines usually contain a high amount of antigen that is rendered noninfectious with ethyleneimine or other aziridine compounds or formalin and is formulated with aluminum hydroxide and saponin (Quil a) or mineral oil as adjuvant. Most vaccines are multivalent, that is, they contain more than one serotype. The basis for these conventional killed vaccines was laid down decades ago.

Soon after the discovery of FMDV, attempts were made, by trial and error, to develop a vaccine. First, living cattle were used to produce the virus. The first efficacious vaccines consisted of lymph from tongue lesions, which was inactivated with formaldehyde and adsorbed to alum (Waldmann *et al.*, 1937). However, the amount of antigen produced by this procedure was too low: vesicular material from one cow yielded virus for only a few hundred doses of vaccine. Nevertheless, this type of vaccine was used in the field, until the beginning of the 1950s. At that time, the so-called Frenkel vaccine became available. Frenkel was the first to grow virus in tissue culture on a scale large enough to start systematic annual vaccination. He cultured fragments of epithelium from cattle tongues collected at slaughterhouses *in vitro*, and could do so because penicillin that prevented bacterial contamination had become available (Frenkel, 1947). The first comprehensive vaccination program against FMD started in the Netherlands in 1952. It dramatically reduced the number of outbreaks, and other countries soon followed the Dutch example (Brown, 1989). The development of large-scale suspension cultures using the baby hamster kidney cell line BHK21 enabled the production of huge quantities of vaccine (Capstick *et al.,* 1965), which are still required today.

The current conventional killed FMD vaccines have some disadvantages: (1) The antigenic variation among virus isolates within one serotype; hence, vaccination with one subtype may protect only partially against challenge with another subtype, which makes the choice of vaccine strains highly critical. (2) To maintain a high level of (herd) immunity, annual or biannual vaccination is necessary. (3) Last but not least, it has been demonstrated that many outbreaks in Europe have originated from virus escaping from laboratories or vaccine production plants or to the use of improperly inactivated vaccines (Beck and Strohmaier, 1987).

These problems have stimulated research on modern approaches of FMD vaccine development. The viral protein 1 (VP1) has been expressed in *Escherichia coli* (Kleid *et al.,* 1981), synthetic peptides have been designed (Bittle *et al.,* 1982), peptides have been linked to hepatitis B virus core antigen using vaccinia virus as a vector (Clarke *et al.,* 1987), and a peptide has been expressed in BHV1 (Kit *et al.,* 1991). Recently, a genetically altered mutant of FMDV has been constructed that lacks the coding region for a proteinase. This mutant was shown to be attenuated and considered to be a vaccine candidate (Mason *et al.,* 1997). Use of a part of the viral particle or a protein in a vaccine has the advantage that infected or convalescent animals can be differentiated from vaccinated animals. For this purpose, a protein-specific or epitope-specific enzyme-linked immunosorbent assay (ELISA) to detect antibodies needs to be developed. The existing conventional killed vaccines, depending on their grade of purification, contain low to negligible amounts of nonstructural viral proteins. Consequently, antibodies against these proteins may be absent after vaccination, whereas these are induced after infection. Tests that detect antibodies against nonstructural proteins may allow infected animals to be distinguished from vaccinated ones (Berger *et al.,* 1990; Bergmann *et al.,* 1993). Hence, modern FMDV vaccines and perhaps also existing FMDV vaccines can serve as marker vaccines.

1. Efficacy and Safety

In Europe, the efficacy must be assessed in compliance with the European Pharmacopoeia. Three groups of five cattle are injected with three dilutions of the vaccine and challenged 2–4 weeks later with virulent virus of the same type as the vaccine strain. A 50% protective dose of 3 or more is required. Vaccinal immunity is induced in cattle in a few days (Doel *et al.*, 1994) and may last for several years after the last vaccination (Terpstra *et al.*, 1990).

The European Pharmacopeia and the Manual of Standards of the Office International des Epizooties (1996) also prescribes safety tests. It is, of course, imperative for killed FMDV vaccines that the antigen be inactivated so that no residual infectivity exists.

B. Diagnostics

Diagnostic techniques are extensively described in the OIE manual (1996). Briefly, the preferred method for the detection of FMDV antigen and serotype is an indirect sandwich ELISA using sera against the seven serotypes as "capture" sera. Virus isolation and subsequent identification, or polymerase chain reactions (PCRs) can also be used. For the detection of antibody, virus neutralization tests and ELISAs are mostly used. However, vaccination interferes with serodiagnosis and, therefore, in vaccinating countries the diagnosis is focused on detection of the virus in samples of animals suspected of FMD.

C. Eradication

Control is directed toward easing the impact of disease, and eradication is focused on elimination of the agent (Schnurrenberger *et al.*, 1987). Eradication could be considered as the ultimate aim of animal disease control programs. Foot-and-mouth disease virus has been eliminated in the Western world. In most countries, the systematic vaccination with vaccines of high quality has significantly reduced the number of outbreaks. It is noteworthy that the conventional killed vaccines used for the yearly vaccinations were primarily developed, not to confer herd immunity, but to prevent disease upon an infection with field virus. The awareness of farmers and veterinarians of this severe infectious disease, coupled with rapid and sensitive laboratory diagnosis and rigorous control measures, were also essential in the eradication of FMD. In 1992, the EU implemented a nonvaccination policy against FMD, which was justified because of the absence of the virus on its territory, and the estimated economic benefits related to it. Since then, only a few outbreaks in Italy and Greece have been recorded. Many countries still suffer from FMD, and thus constitute a threat for the reintroduction of the virus in fully susceptible populations. Consequently, countries free from FMDV have instituted surveillance, importation control, rapid diagnosis, and emergency programs. Part of the emergency control is the accessibility to FMDV vaccine banks installed in various countries. It would be appropriate if a worldwide concerted effort could begin to eradicate FMDV globally, as has been done for smallpox and is now under way for polio in humans.

III. Infectious Bovine Rhinotracheitis

The causative virus (BHV1) belongs to the genus Varicellovirus from the subfamily of alphaherpesvirinae. It infects bovines and causes, apart from rhinotracheitis, vulvovaginitis and balanoposthitis. After a primary infection, viral DNA remains latent in sensory ganglia of the host for life. Some countries have eradicated BHV1, but the virus is still prevalent on all continents.

A. VACCINES

To reduce the severity of clinical illness due to BHV1 infection, conventional live or killed vaccines have been widely used. Although these vaccines indeed have been shown to be efficacious under experimental conditions, BHV1 seems to persist in bovine populations at varying levels in different countries and continents. This is probably also the result of nonsystematic and *ad hoc* vaccination strategies. A disadvantage of vaccination with conventional vaccines is the interference with serodiagnosis and determination of the prevalence of infection in herds, regions, and countries.

To circumvent this problem, marker vaccines have been developed and some of these have been on the market since 1995. A marker vaccine can be defined as a vaccine based on deletion mutants or containing one or more microbial proteins, which allows the distinction between infected and vaccinated individuals based on the respective antibody responses. (*Editor's note:* Marker vaccines would be used to identify vaccinated/infected cattle since vaccination is unable to prevent infection; thus the vaccinated animal will be latently infected with the virluent virus.)

1. Types of Marker Vaccines

Deletion mutants of BHV1 have been compared for virulence and immunogenicity. It was found that mutants lacking glycoprotein G (gG) or gE were the most immunogenic and were sufficiently attenuated to serve as candidates for a marker vaccine. The gC deletion mutant had too high a residual virulence to be suitable as vaccine strain (Kaashoek *et al.*, 1998). However, when a gC deletion was combined with an inactivated thymidine kinase gene a candidate vaccine strain arose (Kit, 1988; Flores et al., 1993). A subunit vaccine containing only gD was shown to be more promising than a vaccine based on gB or gC (van Drunen Littel-van den Hurk et al., 1993).

A BHV1 field strain underwent multiple passages over cell cultures and consequently formed some small plaques. After such a small plaque was biologically cloned three times, it yielded a virus deleted of the entire gE gene and the downstream located US9 gene (Rijsewijk et al., 1993). This virus provided the basis for a live and a killed BHV1 marker vaccine (Kaashoek et al., 1994, 1995; Strube et al., 1995, 1996). An experimental gD subunit vaccine has been developed by inserting the gD gene lacking the transmembrane anchor into Madin–Darby bovine kidney cells resulting in a cell line that constitutively secretes gD into its supernatant (Kowalski et al., 1993). This supernatant was then formulated with an adjuvant (van Drunen Littel-van den Hurk, et al., 1994).

2. Efficacy

With both gE-negative vaccines and the experimental gD subunit vaccine a series of experiments have been performed to assess their efficacy.

a. Vaccination-Challenge Experiment. The efficacy of these vaccines as assessed in separate vaccination-challenge experiments has been reported previously (Kaashoek et al., 1994, 1995; van Drunen Littel-van den Hurk et al., 1994). In other studies, the most potent adjuvant for the killed gE-negative vaccine was selected, and minimum vaccine dose, vaccination regimen, and duration of immunity were established. Both the live and killed gE-negative vaccines were found to induce immunity for at least half a year (Strube et al., 1996). From these studies, however, we cannot determine which is the most efficacious vaccine, because all were performed under different conditions in different laboratories. Therefore, we tested their comparative efficacy in three successive identical experiments, each of which combined a vaccination-challenge experiment with a transmission experiment. The experimental gD-subunit vaccine was produced by Pfizer Animal Health group, USA, according to the principle described by van Drunen Littel-van den Hurk et al. (1994). The gE-negative vaccines were based on the conventionally obtained gE-negative strain Difivac (Kaashoek et al., 1994) and produced by Bayer AG, Germany. In each of the three experiments 30 BHV1 seronegative yearling cattle were randomly assigned to three groups of 10; each group was housed separately in identically conditioned isolation stables. Two groups were vaccinated twice, each with a different vaccine and the third group served as unvaccinated control. Four weeks after the second vaccination, 5 randomly assigned cattle from each group were placed in another stable. The 5 remaining cattle in each group were challenged intranasally with the virulent Lam strain of BHV1 (Kaashoek et al., 1996a). After 24 hours, the 5 cattle that were not inoculated were placed back into their original stable. In the first experiment the live gE-negative and the killed gE-negative vaccine were tested, in the second the killed gE-negative and the gD-subunit vaccine, and in the third the live gE-negative and the gD-subunit vaccine. The live vaccine was found more efficacious than the killed and the gD-subunit vaccine, as evidenced by less severe clinical signs and much lower virus titres in nasal swabs after challenge. The gD-subunit vaccine induced hardly any protection; it only shortened the duration of virus shedding (Bosch et al., 1996).

 b. *Transmission Experiment.* The 5 in-contact yearlings in each group of the three above successive experiments were used to assess whether these three vaccines lowered the transmission of challenge virus. They were examined to detect whether they became infected, as indicated by virus presence in nasal swabs or development of antibodies against gE (van Oirschot et al., 1997). All 15 in-contact control cattle, and all 10 cattle vaccinated with either the killed vaccine or the gD-subunit vaccine became infected. All gD-subunit vaccinated in-contact cattle shed virus and 7 out of 10 cattle given the killed vaccine shed virus from their nasal fluids. Four out of 10 yearlings administered the live vaccine became infected. The reproduction ratio R, defined as the average number of infections caused by one infectious individual, was 0.9 in the group given the live vaccine, which was significantly less than that in the control group. Hence, only the live gE-negative vaccine induced herd immunity in these small populations under these experimental conditions (Bosch et al., 1997a).

 c. *Reactivation Experiment.* The 30 unvaccinated infected cattle of the above described three successive experiments were used for three "reactivation" experiments, to assess whether vaccination after infection could reduce the rate of reactivation and subsequent virus shedding. In each experiment, comprising 10 cattle, one of the three above-described vaccines was used. Vaccinations were carried out 4 and 8 weeks after infection in 5 randomly chosen cattle. The other 5 served as unvaccinated controls. From 2 weeks after the second vaccination all cattle were subjected to a commonly applied dexamethasone treatment to reactive putative latent BHV1. None of the vaccines could prevent reactivation and its subsequent virus shedding, because all 30 cattle had virus in their nasal swabs after dexamethasone treatment. However, the amount, but not the duration, of virus shed was significantly lower in the cattle vaccinated with the killed gE-negative vaccine or the gD-subunit vaccine than in the cattle given the live vaccine. Surprisingly, the live vaccine elicited no antiviral effect at all (Bosch et al., 1997b). Hence, to lower the chance that latently infected cattle may act as a source of BHV1 transmission the killed vaccines can better be used than the live one. However, mixing the live vaccine in an adjuvant, as is common in pseudorabies live gE-negative vaccines, may enhance its efficacy in this respect.

d. Field Trial. The killed gE-negative and the gD-subunit vaccines were also evaluated for efficacy in a field trial, which may be considered the ultimate test for a vaccine. A randomized, double-blind, placebo-controlled field trial that comprised 130 farms and approximately 16,000 head of cattle was performed. The advantage of the use of marker vaccines is that their efficacy can be measured relatively easily, because they allow monitoring of the incidence of infections in vaccinated populations. The herds were randomly assigned to one of three groups: two vaccinated groups and one placebo-treated group. In the beginning of the trial all cattle above 3 months were bled and vaccinated twice, at a 4-week interval. A third vaccination was applied about 6 months after the start. The trial lasted 13 months and at the end all cattle above 3 months were bled again. The number of cattle that turned gE-seropositive during the trial was determined. A herd was considered to have had an "outbreak" when at least one animal had produced antibodies against gE. The gD-subunit and the killed gE-negative vaccine reduced the number of herds with an outbreak, but only in case of the gD-subunit vaccine was it statistically significant. The BHV1 transmission within the infected herds was significantly reduced by both vaccines. However, the reproduction ratios R were significantly higher than 1, indicating that major outbreaks may sometimes still occur in vaccinated herds (Bosch et al., 1997c). A field trial with the live gE-negative vaccine is under way.

3. Safety

The safety of the live gE-negative vaccine has been extensively studied. After intranasal administration a slight serous nasal discharge was observed and high virus titers were detected in nasal swabs (Kaashoek et al., 1994; Strube et al., 1995, 1996). However, sentinel calves did not develop antibodies against BHV1, indicating that the vaccine virus was not transmitted after intranasal inoculation. A few calves of 2 weeks of age shed virus at very low titers after intramuscular vaccination, but virus shedding was not detected in 3-month-old intramuscularly vaccinated calves (Strube et al., 1996). Under field conditions, spread of gE-negative vaccine virus could not be detected either after intramuscular vaccination (Van der Poel and Hage, 1998). The vaccine virus was safe for breeding bulls and did not infect the fetus, after being administered intravenously. The gE-negative vaccine virus was found to remain latent in trigeminal ganglia after intranasal administration, but not after intramuscular vaccination (F. A. C. Van Engelenburg *et al.*, unpublished observations), and could be reactivated by corticosteroid treatment by some (Straub, 1996), but not other research workers (Kaashoek *et al.*, 1994, 1996b; Bosch *et al.*, 1997b).

The killed gE-negative and the gD-subunit vaccines were not associated with any adverse effect in the field trial described above, wherein about 16,000 cattle were vaccinated three times. In a separate experiment, the killed gE-negative vaccine was found to induce a slight rise in temperature and a slight decrease in milk production. Both minor effects were considered to be acceptable (Bosch *et al.*, 1997d).

B. Diagnostics

1. Identification of the Agent

Virus isolation from nasal swabs during an acute clinical infection, in conjunction with antibody detection in paired serum samples is still the most common method to be used for diagnosis. After growth in cell culture, the virus can be identified by neutralization with a monospecific antiserum or with the use of monoclonal antibodies. Monoclonal antibodies have been developed that can differentiate between BHV1 subtype 1.1 (IBR-like) and subtype 1.2 (IPV-like) strains (Wyler *et al.*, 1989; Rijsewijk *et al.*, 1997). Further characterization is possible by DNA restriction enzyme analysis.

Swab material put onto coverslips and tissues that are collected postmortem can be examined for the presence of BHV1 antigen by standard immunofluorescence or immunoperoxidase tests. ELISAs to capture antigen have also been described (Edwards and Gitao, 1987).

Nucleic acid detection has not as yet found a widespread application in routine diagnostics. Particularly, the PCR has the advantage of being more sensitive and quicker than virus isolation. However, the disadvantage is that it is prone to contamination, and therefore many precautions must be taken to avoid false-positive results. A PCR has been developed for detecting BHV1 DNA in semen (Van Engelenburg *et al.*, 1993).

2. Antibody Detection

The neutralization test is still the "gold standard" for detection of antibodies against BHV1. However, ELISAs are being used more and more. Neutralization tests with a prolonged (up to 24 hours) virus-serum incubation period are generally more sensitive than those with a shorter incubation period. Some ELISAs are more or as sensitive as 24-hour neutralization tests, whereas others are less sensitive (Perrin *et al.*, 1993; Kramps *et al.*, 1994). In general, ELISAs using the blocking principle appeared to be the most sensitive (Perrin *et al.*, 1993). Based on a study in the European Union (EU), wherein 49 laboratories participated a strong-positive, a weak-positive, and a negative serum were selected as EU reference sera (Perrin *et al.*, 1994). These sera are presently also recognized as such by the Office International des Epizooties. It was found that 24-hour neutralization tests and blocking ELISAs were the most reliable in scoring the weak-positive EU serum as positive, and that about half of the commercial ELISAs scored this serum as negative (Kramps *et al.*, 1996).

Milk is increasingly used for the detection of antibodies (von Forschner *et al.*, 1986). Herds with a BHV1 seroprevalence of 10–20% are usually scored positive when their bulk milk is tested in an ELISA (Hartman *et al.*, 1997). Bulk milk is not suitable to detect the introduction of a single seropositive animal (Frankena *et al.*, 1997).

Companion diagnostic tests are applied along with marker vaccines. A gE-ELISA has been developed that is based on the use of two different monoclonal antibodies against the gE of BHV1. Thus, this ELISA detects antibodies against two different epitopes on gE (van Oirschot *et al.*, 1997). A commercial gE-ELISA has been launched based on the use of one monoclonal antibody, which has the same specificity as one of the monoclonal antibodies used in the above gE-ELISA. This commercial ELISA had a specificity and a sensitivity of 99% compared with the noncommercial gE-ELISA (Kaashoek, 1995). The commercial gE-ELISA is also suitable for detecting antibodies in milk and bulk milk samples (Wellenberg *et al.*, 1998a,b). These or other tests, for example, a gB-ELISA or gC-ELISA, can be used as companion diagnostics tests for a gD-subunit vaccine.

C. Eradication

Switzerland, Denmark, and other European countries have eradicated BHV1 by test and removal strategies. Cattle with antibodies against BHV1 were detected by ELISA tests and removed from the population. By following this procedure for several years these countries, which obviously did not vaccinate, eventually succeeded in eliminating the virus. This scheme can only be adopted when the initial prevalence is low. However, in countries with a moderate to high prevalence of BHV1, marker vaccines may be systematically and continuously used to eradicate the virus or to lower the prevalence in order to eliminate the last few percentages of infected cattle. Such a program is now in force in the Netherlands, where a half-yearly vaccination of cattle above 3 months of age in infected herds has been made obligatory. Based on the above findings live and killed gE-negative vaccines may be used, but the use of the live vaccine is recommended. The Dutch eradication program foresees cessation of vaccination around the year 2005 and the absence of BHV1-seropositive cattle in the country around 2010.

IV. Bovine Virus Diarrhea

Bovine virus diarrhea virus is a member of the genus Pestivirus in the family of *Flaviridae*. It gives rise to high economic losses due to the birth of weak or malformed calves, the development of mucosal disease, severe postnatal infections, and the immunosuppression it induces. The virus is prevalent worldwide and conventional vaccines are applied in attempts to minimize the damage. The first eradication programs are in force in the Scandinavian countries.

A. Vaccines

Since the beginning of the 1960s many conventional live and killed vaccines, comprising a cytopathic strain of BVDV, have been developed and marketed. The first vaccines were probably primarily aimed at preventing the most severe clinical form of BVDV, namely, mucosal disease. It is, however, not surprising that cases of mucosal disease still occurred after vaccination. At that time, it was not known that mucosal disease only develops in persistently infected immunotolerant (PI) cattle, which are the result of congenital infection between 30 and 120 days of gestation.

Since the mid-1970s, it has become clear that vaccines should be developed that primarily prevent congenital BVDV infection. The first partially successful attempts to create such a vaccine were made by McClurkin *et al.* (1975). The need to develop vaccines for ameliorating disease after postnatal infection decreased, because it was recognized that most infections ran a subclinical course. Recently, however, the emergence of more virulent BVDV strains (Corapi *et al.*, 1989) restimulated interest in vaccines to prevent the severe postnatal disease they can cause.

Because BVDV can be involved in the pathogenesis of bovine respiratory disease, for example, by inducing immunosuppression, BVDV vaccines are often incorporated in multivalent vaccines for the prevention of respiratory disease.

A complication for the development of BVDV vaccines is the fact that BVDV is antigenically diverse. Two antigenic groups of strains have been identified and there is a tendency to divide group I into at least two subgroups (Van Rijn et al., 1997). It may be expected that more antigenic variants will be recognized. The impact of this antigenic diversity is not well established yet, but the first data suggest that it indeed has consequences for vaccine development (Bolin and Ridpath, 1996; Bruschke et al., 1997). On the other hand, a live BVDV type I vaccine strain protected calves against the severe clinical signs of a BVDV type II infection (Cortese et al., 1996).

1. Efficacy

The efficacy is primarily evaluated in a vaccination-challenge experiment in the target host. A few of these experiments to test BVDV vaccines for their efficacy to prevent congenital infection have been performed (McClurkin et al., 1975; Harkness et al., 1987; Meyling et al., 1987; Brownlie et al., 1995; Brock and Grooms, 1996; Zimmer et al., 1996). None of the vaccines tested induced complete protection against fetal infection. However, one conventional killed vaccine was found to protect all fetuses of vaccinated cows, but the fetus of one of the six unvaccinated control cows was not infected (Brownlie et al., 1995). No data are available on this vaccine's duration of immunity and ability to induce protection against BVDV from another subgroup. Pregnant ewes have also been used to assess the efficacy of BVDV vaccines and again none of the tested vaccines was 100% efficacious (Carlsson et al., 1991; Bruschke et al., 1997, 1998). The vaccines tested for their efficacy against postnatal infection were shown to give variable results. The most promising vaccine for the prevention of congenital infections induced, after three vaccinations, complete protection when calves were challenged 2 weeks after the last vaccination (Howard et al., 1994). Cortese et al. (1996) found considerable clinical protection in calves given a conventional live combination vaccine based on a type I strain, that were challenged with a type strain. Ideally, vaccines should be evaluated for efficacy in the field. However, well-designed field trials have not been reported that demonstrate BVDV vaccine to be effective in the field, but there is clinical evidence that most modified live BVDV vaccines protect cattle against severe disease and death in the face of natural challenge.

2. Safety

Apart from being efficacious, vaccines must also be safe. Live BVDV vaccines have always been associated with a variety of adverse effects, such as induction of mucosal disease, fetal infection, and immunosuppression (van Oirschot et al., 1998). It is obvious that killed vaccines are safe in this respect.

In view of the above, it can be concluded that there is ample room for improving the efficacy as well as safety of BVDV vaccines.

B. Diagnostics

1. Identification of the Agent

A postnatal clinical infection can best be diagnosed by virus isolation from nasal swabs or blood along with the demonstration of a fourfold rise in antibody titer in paired serum samples. An *in utero* infection can be diagnosed by examining precolostral serum for virus and antibodies. A persistently infected immunotolerant calf can be identified by detecting the virus in blood or leukocytes twice at a 3-week interval. The diagnoses "mucosal disease" is well established if both noncytopathic and cytopathic BVDV can be isolated from the same severely affected animal. Because BVDV is a common contaminant of fetal bovine serum and of primary bovine cells, cell cultures must be regularly checked for absence of BVDV to avoid a false-positive result.

Various antigen detection tests have been reported for the detection of BVDV in blood leukocytes of PI cattle (Shannon et al., 1991; Entrican et al., 1995).

Detection of BVDV RNA by reverse PCR is increasingly used for diagnosing BVD. An advantage is that the amplified cDNA product can be directly sequenced, which may result in a further (sub)typing of the virus (Hofmann et al., 1994; Pellerin et al., 1994).

Methods that detect antigen or nucleic acid must detect the full range of antigenic or genetic diversity of BVDV.

2. Antibody Detection

Enzyme-linked immunosorbent assays and neutralization tests are most commonly used for the detection of antibodies against BVDV (Edwards, 1990; Paton et al., 1991; Kramps et al., 1998). The use of different BVDV strains in neutralization tests can result in large differences in antibody titers (Dekker et al., 1995). To determine the BVDV antibody status of a herd a bulk milk ELISA can be useful (Niskanen, 1993). A high level of antibodies may indicate the presence of one or more PI cattle in a herd (Bitsch and Roensholt, 1995).

C. ERADICATION

Norway was the first country to make BVD a notifiable disease and to start an eradication program. Other Scandinavian countries followed this example. These programs are essentially based on (1) identification of free herds, (2) prevention of infection of these herds, and (3) reduction of the number of infected herds. A decrease in infected herds may be achieved by detection and subsequent removal of PI cattle from herds (Bitsch and Roensholt, 1995). Caution should be taken in generalizing the latter observation, because it has been reported that after removal of the PI cattle BVDV can still circulate in herds for months or years (Barber and Nettleton, 1993; Moerman *et al.*, 1993). In addition, a bull has recently been described that probably had been postnatally infected, had high neutralizing antibody titers, and shed BVDV in semen intermittently (Voges, 1997). The significance of these sources of infection in the framework of eradication programs needs to be determined.

An efficacious marker vaccine that prevents the birth of persistently infected calves would be of benefit in the eradication of BVDV, particularly in countries with large cattle herds.

V. Posteradication Period

After a virus disease has been eradicated, vaccination against this disease should be stopped. An intensive surveillance program should start to demonstrate continuously that the virus has not been reintroduced. In case of clinical suspicion of an exotic disease, it is of great importance to very rapidly diagnose these infections. The laboratory diagnosis for FMD can be made in 1 day, but routine diagnosis for IBR and BVD usually takes longer, because in most laboratories virus isolation is still the method of choice. It would be worthwhile to develop animal side tests (which are already available for diseases in pets) for the above diseases in order to accelerate the diagnosis. Promising new biosensors have been described for possible use as animal side tests (Lin *et al.*, 1997; Holtz and Asher, 1997). Once the diagnosis has been made a variety of zoosanitary measures can then be taken to prevent further dissemination of the virus. One of these measures could be an emergency ring vaccination. In this regard, it is of importance to have marker vaccines available that induce early herd immunity, that is, prevent or reduce further spread of the virus within and between herds in a couple of days. Although FMDV vaccines induced clinical immunity in 3–4 days (Doel *et al.*, 1994), it is not known whether they can give rise to herd immunity in such short a time. In the case of IBR, it has been demonstrated that calves vaccinated with a live gE-negative marker vaccine 2 days before a contact infection did shed significantly less virus than unvaccinated calves (Kaashoek and van Oirschot, 1996), thus were less infectious. Hence, an emergency vaccination may substantially contribute to reducing the spread of BHV1 after an outbreak. The onset of immunity after a BVDV vaccination is not well known. Theoretically, there are various approaches to develop more rapidly acting emergency marker vaccines, that is, the use of special adjuvants and cytokines.

REFERENCES

Barber, D. M. L., and Nettleton, P. F. (1993). Investigations into bovine viral diarrhoea virus in a dairy herd. *Vet. Rec.* **133,** 549–550.
Beck, E., and Strohmaier, K. (1987). Subtyping of European foot-and-mouth disease virus strains by nucleotide sequence determination. *J. Virol.* **61,** 1621–1629.
Berger, H.-G., Straub, O. C., Ahl, R., Tesar, M., and Marquardt, O. (1990). Identification of foot-and-mouth disease virus replication in vaccinated cattle by antibodies to nonstructural virus proteins. *Vaccine* **8,** 213–216.
Bergmann, I. E., De Mello, P. A. Neitzert, E., Beck, E., and Gomes, I. (1993). Diagnosis of persistent aphtovirus infection and its differentiation from vaccination response in cattle by use of enzyme-linked immunoelectrotransfer blot analysis with bioengineered nonstructural viral antigens. *Am. J. Vet. Res.* **54,** 825–831.
Bitsch, V., and Roensholt, L. (1995). Control of bovine viral diarrhea virus infection without vaccines. *Vet. Clin. North Am.* **11,** 627–640.
Bittle, J. L., Houghten, R. A., Alexander, H., Shinnick, T. M., Sutcliffe, J. G., Lerner, R. A., Rowlands, D. J., and Brown, F. (1982). Protection against foot-and-mouth disease by immunization with a chemically synthesized peptide predicted from the viral nucleotide sequence. *Nature* **298,** 30–33.
Bolin, S. R., and Ridpath, J. F. (1996). Glycoprotein E2 of bovine viral dairrhea virus expressed in insect cells provides calves limited protection from systemic infection and disease. *Arch. Virol.* **141,** 1463–1477.
Bosch, J. C., Kaashoek, M. J., Kroese, A. H., and van Oirschot, J. T. (1996). An attenuated bovine herpesvirus 1 marker vaccine induces a better protection than two inactivated marker vaccines. *Vet. Microbiol.* **52,** 223–234.
Bosch, J. C., De Jong, M. C. M., Van Bree, J., and van Oirschot, J. T. (1997a). Quantification of transmission of bovine herpesvirus 1 in cattle vaccinated with marker vaccines. Ph.D. Thesis, pp. 53–66. J. C. Bosch, Utrecht University.
Bosch, J. C., Kaashoek, M. J., and van Oirschot, J. T. (1997b). Inactivated bovine herpesvirus 1 marker vaccines are more efficacious in reducing virus excretion after reactivation than a live marker vaccine. *Vaccine* **15,** 1512–1517.
Bosch, J. C., De Jong, M. C. M., Franken, P., Frankena, K., Hage, J. J., Kaashoek, M. J., Maris-Veldhuis, M. A., Noordhuizen, J. P. T. M., Van der Poel, W. H. M., Verhoeff, J., Weerdmeester, K., Zimmer, G. M., and van Oirschot, J. T. (1998). An inactivated gE-negative marker vaccine and an experimental gD-subunit vaccine reduce the incidence of bovine herpesvirus 1 infections in the field. *Vaccine* **16,** 265–271.

Bosch, J. C., Frankena, K., and van Oirschot, J. T. (1997d). Effect on milk production of vaccination with a bovine herpesvirus 1 gene-deleted vaccine. *Vet. Rec.* **140,** 196–199.
Brock, K. V., and Grooms D. L. (1996). Evaluation of a modified-live bovine viral diarrhoea virus vaccine by fetal challenge. *Proc. European Society for Veterinary Virology Symp. Pestivirus Infect., 3rd, 1996,* Lelystad, The Netherlands, pp. 177–179.
Brown, F. (1989). The development of chemically synthesized vaccines. *Adv. Vet. Sci. Comp. Med.* **33,** 173–193.
Brownlie, J., Clarke, M. C., Hooper, L. B., and Bell, G. D. (1995). Protection of bovine fetus from bovine viral diarrhoea virus by means of a new inactivated vaccine. *Vet. Rec.* **137,** 58–62.
Bruschke, C. J. M., Moormann, R. J. M., van Oirschot, J. T., and Van Rijn, P. A. (1997). A subunit vaccine based on glycoprotein E2 of bovine virus diarrhea virus induces fetal protection in sheep against homologous challenge. *Vaccine* **15,** 1940–1945.
Bruschke, C. J. M., van Oirschot, J. T., and Van Rijn, P. A. (1998). Two conventional inactivated BVDV vaccines do not prevent congenital infection. In preparation.
Capstick, P. B., Garland, A. J. M., Chapman, W. G., and Masters, R. C. (1965). Production of foot-and-mouth disease virus antigen from BHK 21 clone 13 cells grown and infected in deep suspension cultures. *Nature* **205,** 1135–1136.
Carlsson, U., Alenius, S., and Sundquist, B. (1991). Protective effect of an ISCOM bovine virus diarrhoea virus (BVDV) vaccine against an experimental BVDV infection in vaccinated and non-vaccinated pregnant ewes. *Vaccine* **9,** 577–580.
Clarke, B. E., Newton, S. E., Carroll, A. R., Francis, M. J., Appleyard, G., Syred, A. D., Highfield, P. E., Rowlands, D. J., and Brown, F. (1987). Improved immunogenicity of a peptide epitope after fusion to hepatitis B core protein. *Nature* **330,** 381–384.
Corapi, W. V., French, T. W., and Dubovi, E. J. (1989). Severe thrombocytopenia in young calves experimentally infected with noncytopathic bovine viral diarrhea virus. *J. Virol.* **63,** 3934–3943.
Cortese, V. S., West, K., and Ellis, J. (1996). Clinical and immunologic responses to type 2 BVDV challenge in vaccinated and unvaccinated calves. *Proc. World Buiatrics Congr., 19th,* Edinburgh, *1996,* pp. 610–613.
Dekker, A., Wensvoort, G., and Terpstra, C. (1995). Six antigenic groups within the genus pestivirus as identified by cross neutralization assays. *Vet. Microbiol.* **47,** 317–329.
Doel, T. R., Williams, L., and Barnett, P. V. (1994). Emergency vaccination against foot-and-mouth disease: Rate of development of immunity and its implications of the carrier state. *Vaccine* **12,** 592–600.
Edwards, S. (1990). The diagnosis of bovine virus diarrhoea-mucosal disease in cattle. *Rev. Sci. Tech. Off. Int. Epizoot.* **9,** 115–130.
Edwards, S., and Gitao, G. C. (1987). Highly sensitive antigen detection procedures for the diagnosis of infectious bovine rhinotracheitis: Amplified ELISA and reverse passive haemagglutination. *Vet. Microbiol.* **13,** 135–141.
Entrican, G., Dand, A., and Nettleton, P. F. (1995). A double monoclonal antibody ELISA for detecting pestivirus antigen in the blood of viraemic cattle and sheep. *Vet. Microbiol.* **43,** 65–74.
Flores, E. F., Osorio, F. A., Zanella, E. L., Kit, S., and Kit, M. (1993). Efficacy of a deletion mutant bovine herpesvirus-1 (BHV-1) vaccine that allows serologic differentiation of vaccinated from naturally infected animals. *J. Vet. Diagn. Invest.* **5,** 534–540.
Frankena, K., Franken, P., Vandehoek, J., Koskamp, G., and Kramps, J. A. (1997). Probability of detecting antibodies to bovine herpesvirus 1 in bulk milk after the introduction of a positive animal on to a negative farm. *Vet. Rec.* **140,** 90–92.

Frenkel, H. S. (1947). La culture du virus de la fièvre aphteuse sur l'epithelium de la langue des bovides. *Bull. Off. Int. Epizoot.* **28,** 155–162.
Harkness, J. W., Roeder, P. L., Drew, T., Wood, L., and Jeffrey, M. (1987). The efficacy of an experimental inactivated BVD-MD vaccine. *Proc. CEC Semin. Res. Anim. Husb., 1985,* pp. 233–250.
Hartman, A., Van Wuijckhuise, L., Frankena, K., Franken, P., Wever, P., De Wit, J., and Kramps, J. (1997). Within-herd BHV-1 prevalence prediction from an ELISA on bulk milk. *Vet. Rec.* **140,** 484–485.
Hofmann, M. A., Brechtbühl, K., and Stäuber, N. (1994). Rapid characterization of new pestivirus strains by direct sequencing of PCR-amplified cDNA from the 5' noncoding region. *Arch. Virol.* **139,** 217–229.
Holtz, J. H., and Asher, S. A. (1997). Polymerized colloidal crystal hydrogel films as intelligent chemical sensing materials. *Nature* **389,** 829–832.
Howard, C. J., Clarke, M. C., Sopp, P., and Brownlie, J. (1994). Systemic vaccination with inactivated bovine virus diarrhoea virus protects against respiratory challenge. *Vet. Microbiol.* **42,** 171–179.
Kaashoek, M. J. (1995). Marker vaccines against bovine herpesvirus 1 infections. Ph.D. Thesis, Utrecht University.
Kaashoek, M. J., and van Oirschot, J. T., (1996). Early immunity induced by a live gE-negative bovine herpesvirus 1 marker vaccine. *Vet. Microbiol.* **53,** 191–197.
Kaashoek, M. J., Moerman, A., Madic, J., Rijsewijk, F. A. M., Quak, J., Gielkens, A. L. J., and van Oirschot, J. T. (1994). A conventionally attenuated glycoprotein E-negative strain of bovine herpesvirus type 1 is an efficacious and safe vaccine. *Vaccine* **12,** 493–444.
Kaashoek, M. J., Moerman, A., Madic, J., Weerdmeester, K., Maris-Veldhuis, M., Rijsewijk, F. A. M., and van Oirschot, J. T. (1995). An inactivated vaccine based on a glycoprotein E-negative strain of bovine herpesvirus 1 induces protective immunity and allows serological differentiation. *Vaccine* **13,** 342–346.
Kaashoek, M. J., Straver, P. H., Van Rooij, E. M. A., Quak, J., and van Oirschot, J. T. (1996a). Virulence, immunogenicity and reactivation of seven bovine herpesvirus 1.1. strains: clinical and virological aspects. *Vet. Rec.* **139,** 416–421.
Kaashoek, M. J., Rijsewijk, F. A. M., and van Oirschot, J. T. (1996b). Persistence of antibodies against bovine herpesvirus 1 and virus reactivation two to three years after infection. *Vet. Microbiol.* **53,** 103–110.
Kaashoek, M. J., Rijsewijk, F. A. M., Ruuls, R. C., Keil, G. M., Thiry, E., Pastoret, P.-P., and van Oirschot, J. T. (1998). Virulence, immunogenicity, and reactivation of bovine herpesvirus 1 mutants with a deletion in the gC, gG, gI, gE, or gI and gE gene. *Vaccine* **16,** 802–809.
Kit, S. (1988). Genetically engineered pseudorabies and infectious bovine rhinotracheitis virus vaccine. In "Technological Advances in Vaccine Development," pp. 183–195. Alan R. Liss, New York.
Kit, M., Kit, S., Little, S. P., DiMarchi, R. D., and Gale, C. (1991). Bovine herpesvirus-1 (infectious bovine rhinotracheitis virus)-based viral vector which expresses foot and mouth disease epitopes. *Vaccine* **9,** 564–572.
Kleid, D. G., Yansura, D., Small, B., Dowbenko, D., Moore, D. M., Grubman, M. J., McKercher, P. D., Morgan, D. O., Robertson, B. H., and Bachrach, H. L. (1981). Cloned viral protein vaccine for foot-and-mouth disease: Responses in cattle and swine. *Science* **214,** 1125–1129.
Kowalski, J., Gilbert, S. A., Van Drunen Littel-Van den Hurk, S., Van den Hurk, J., Babiuk, L. A., and Zamb, T. J. (1993). Heat-shock promoter-driven synthesis of

secreted bovine herpesvirus glycoproteins in transfected cells. *Vaccine* **11,** 1100–1107.

Kramps, J. A., Magdalena, J., Quak, J., Weerdmeester, K., Kaashoek, M. J., Maris-Veldhuis, M. A., Rijsewijk, F. A. M., Keil, G., and van Oirschot, J. T. (1994). A simple, specific, and highly sensitive blocking enzyme-linked immunosorbent assay for detection of antibodies to bovine herpesvirus 1. *J. Clin. Microbiol.* **32,** 2175–2181.

Kramps, J. A., Perrin, B., Edwards, S., and van Oirschot, J. T. (1996). A European interlaboratory trial to evaluate the reliability of serological diagnosis of bovine herpesvirus 1 infections. *Vet. Microbiol.* **53,** 153–161.

Kramps, J. A., Van Maanen, C. A., Van de Wetering, G., Stienstra, G., Quak, J., Brinkhof, J., Roensholt, L., and Nylin, B. (1998). A simple, rapid and reliable enzyme-linked immunosorbent assay for the detection of bovine virus diarrhoea virus (BVDV) specific antibodies in cattle serum, plasma and bulk milk. *Vet. Microbiol.* (in press).

Lin, V. S.-Y., Motesharei, K., Dancil, K.-P. S., Sailor, M. J., and Ghadiri, M. R. (1997). A porous silicon-based optical interferometric biosensor. *Science* **278,** 840–843.

Mason, P. W., Piccone, M. E., McKenna, T.St.-C., Chinsangaram, J., and Grubman, M. J. (1997). Evaluation of a live-attenuated foot-and-mouth disease virus as a vaccine candidate. *Virology* **227,** 96–102.

McClurkin, A. W., Coria, M. F., and Smith, R. L. (1975). Evaluation of acetylethyleneimine-killed bovine viral diarrhea-mucosal disease virus (BVD) vaccine for prevention of BVD infection of the fetus. *Proc. U.S. Anim. Health Assoc.* **79,** 114–123.

Meyling, A., Ronsholt, L., Dalsgaard, K., and Jensen, A. M. (1987). Experimental exposure of vaccinated and non-vaccinated pregnant cattle to isolates of bovine viral diarrhoea virus (BVDV). *Proc. CEC Semin. Res. Anim. Husb. 1985,* pp. 225–231.

Moerman, A., Straver, P. J., De Jong, M. C. M., Quak, J., Baanvinger, T., and van Oirschot, J. T. (1993). A long term epidemiological study of bovine viral diarrhoea infections in a large herd of dairy cattle. *Vet. Rec.* **132,** 622–626.

Niskanen, R. (1993). Relationship between levels of antibody to bovine viral diarrhoea virus in bulk tank milk and the prevalence of cows exposed to the virus. *Vet. Rec.* **133,** 341–344.

"OIE Manual of Standards for Diagnostic Tests and Vaccines," 3rd ed. Office International des Epizooties, Paris, 1996.

Paton, D. J., Ibata, G., Edwards, S., and Wensvoort, G. (1991). An ELISA detecting antibody to conserved pestivirus epitopes. *J. Virol. Methods* **31,** 315–324.

Pellerin, C., Van den Hurk, J., Lecomte, J., and Tijssen, P. (1994). Identification of a new group of bovine viral diarrhea virus strains associated with severe outbreaks and high mortalities. *Virology* **203,** 260–268.

Perrin, B., Bitsch, V., Cordioli, P., Edwards, S., Eloit, M., Guérin, B., Lenihan, P., Perrin, M., Ronsholt, L., van Oirschot, J. T., Vanopdenbosch, E., Wellemans, G., Wizigmann, G., and Thibier, M. (1993). A European comparative study of serological methods for the diagnosis of infectious bovine rhinotracheitis. *Rev. Sci. Tech. Off. Int. Epizoot.* **12,** 969–984.

Perrin, B., Calvo, T., Cardioli, P., Coudert, M., Edwards, S., Eloit, M., Guérin, B., Kramps, J. A., Lenihan, P., Paschaleri, E., Perrin, M., Schon, J., van Oirschot, J. T., Vanopdenbosch, E., Wellemans, G., Wizigmann, G., and Thibier, M. (1994). Selection of European Union standard reference sera for use in the serological diagnosis of infectious bovine rhinotracheitis. *Rev. Sci. Tech. Off. Int. Epizoot.* **13,** 947–960.

Rijsewijk, F. A. M., Kaashoek, M. J., Madic, J., Paal, H., Ruuls, R., Gielkens, A. L. J., and van Oirschot, J. T. (1993). Characterization of a DNA rearrangement found in the

unique short region of the Za strain of bovine herpesvirus type 1 and the vaccine properties of this strain. *Proc. Int. Herpesvirus Workshop, 18th,* Pittsburgh, p. C-67.
Rijsewijk, F. A. M., Kaashoek, M. J., Langeveld, J. P. M., Meloen, R. M., Judek, J., Bienkowska-Szewczyk, K., Maris-Veldhuis, M. A., and van Oirschot, J. T. (1997). Identification of epitopes on glycoprotein C of bovine herpesvirus 1 that allow differentiation between 'IBR-like' and 'IPV-like' strains. *Proc. 22nd Int. Herpesvirus Workshop,* San Diego, California, Aug. 2–8, 1997, p. 288.
Schnurrenberger, P. R., Sharman, R. S., and Wise, G. H. (1987). "Attacking Animal Disease. Concepts and Strategies for Control and Eradication." Iowa State University Press, Ames.
Shannon, A. D., Richards, S. G., Kirkland, P. D., and Moyle, A. (1991). An antigen capture ELISA detects pestivirus antigens in blood and tissues of immunotolerant carrier cattle. *J. Virol. Methods* **34,** 1–12.
Straub, O. C. (1996). Experiences with two BHV1 marker vaccines. *Proc. World Buiatrics Congr., 19th,* Edinburgh, *1996,* pp. 38–41.
Strube, W., Abar, B., Bergle, R. D., Block, W., Heinen, E., Kretzdorn, D., Rodenbach, C., and Schmeer, N. (1995). Safety aspects in the development of an infectious bovine rhinotracheitis marker vaccine. *Dev. Biol. Stand.* **84,** 75–81.
Strube, W., Auer, S., Block, W., Heinen, E., Kretzdorn, D., Rodenbach, C., and Schmeer, N. (1996). A gE deleted infectious bovine rhinotracheitis marker vaccine for use in improved bovine herpesvirus 1 control programs. *Vet. Microbiol.* **53,** 181–189.
Terpstra, C., Van Maanen, C., and Van Bekkum, J. G. (1990). Endurance of immunity against foot-and-mouth disease in cattle after three consecutive annual vaccinations. *Res. Vet. Sci.* **49,** 236–242.
Van der Poel, W. H. M., and Hage, J. J. (1998). Onderzoek naar spreiding van een intramusculair toegediend levend gE-negatief BHV1 markervaccin op twee rundveebedrijven. *Tijdschr. Diergeneeskd.* **123,** 109–111.
van Drunen Littel-van den Hurk, S., Tikoo, S. K., Liang, X., and Babiuk, L. A. (1993). Bovine herpesvirus-1 vaccines. *Immunol. Cell Biol.* **71,** 405–420.
van Drunen Littel-van den Hurk, S., Van Donkersgoed, J., Kowalski, J., Van den Hurk, J. V., Harland, R., Babiuk, L. A., and Zamb, T. J. (1994). A subunit gIV vaccine, produced by transfected mammalian cells in culture, induces mucosal immunity against bovine herpesvirus-1 in cattle. *Vaccine* **12,** 1295–1302.
Van Engelenburg, F. A. C., Maes, R. K., van Oirschot, J. T., and Rijsewijk, F. A. M. (1993). Rapid and sensitive detection of bovine herpesvirus type 1 in bovine semen by a polymerase chain reaction based assay. *J. Clin. Microbiol.* **31,** 3129–3135.
van Oirschot, J. T., Kaashoek, M. J., Maris-Veldhuis, M. A., Weerdmeester, K., and Rijsewijk, F. A. M. (1997). An enzyme-linked immunosorbent assay to detect antibodies against glycoprotein gE of bovine herpesvirus 1 allows differentiation between infected and vaccinated cattle. *J. Virol. Methods* **67,** 23–34.
van Oirschot, J. T., Bruschke, C. J. M., and Van Rijn, P. A. (1998). Vaccination of cattle against bovine viral diarrhoea. *Vet. Microbiol.* (in press).
Van Rijn, P. A., Van Gennip, H. G. P., Leendertse, C. H., Bruschke, C. J. M., Paton, D. J., Moormann, R. J. M., and van Oirschot, J. T. (1997). Subdivision of the *Pestivirus* genus based on envelope glycoprotein E2. *Virology* **237,** 337–348.
Voges, H. (1997). A non-viraemic AI bull persistently shedding BVDV in his semen—a previously unidentified source of infection. *Proc. Eur. Symp. Control BVD Virus Cattle,* Lillehammer, *1997,* p. 45.
von Forschner, E., Bünger, I., Küttler, D., and Mehrkens, L. (1986). IBR/IPV—Serodiagnostik mit ELISA-methoden an blut-, einzelmilch- und tankmilchproben; kontroll-

massnahmen zur erhaltung unverdächtiger rinderherden; sanierungswege unter berücksichtigung der impfung. *Dtsch. Tierärztl. Wochenschr.* **93,** 281–344.

Waldmann, O., Kobe, Z., and Pyl, G. (1937). Die aktive Immunisierung des Rindes gegen Maul- und Klauenseuche. *Zentralbl. Bakteriol., Parasitenkd., Infektionskr. Hyg., Abt. I: Orig.* **138,** 401–412.

Wellenberg, G. J., Verstraten, E. R. A. M., Mars, M. H., and van Oirschot, J. T. (1998a). Detection of bovine herpesvirus 1 glycoprotein E antibodies in individual milk samples by enzyme-linked immunosorbent assays. *J. Clin. Microbiol.* **36,** 909–913.

Wellenberg, G. J., Verstraten, E. R. A. M., Mars, M. H., and van Oirschot, J. T., (1998b). Detection of antibodies to glycoprotein E of bovine herpesvirus 1 in bulk samples by an ELISA. *Vet. Rec.* **192,** 219–220.

Wyler, R., Engels, M., and Schwyzer, M. (1989). Infectious bovine rhinotracheitis/vulvovaginitis. *In* "Herpesvirus Diseases of Cattle, Horses and Pigs" (G. Wittmann, ed.), pp. 1–72. Kluwer Academic Publishers, Boston, Dordrecht, and London.

Zimmer, G. M., Wentink, G. H., Brinkhof, J., Bruschke, C. J. M., Westenbrink, F., Crauwels, A. P. P., and De Goey, I. (1996). Model for testing efficacy of bovine viral diarrhoea virus vaccines against intrauterine infection. *Proc. World Buiatrics Congr., 19th* Edinburgh, *1996,* pp. 217–220.

Immunization and Diagnosis in Bovine Reproductive Tract Infections

LYNETTE B. CORBEIL

Department of Pathology, University of California–San Diego, San Diego, California 92103-8416

I. Introduction and Background
II. *Brucella abortus* Infection
III. *Haemophilus somnus* Infection
   A. Syndrome
   B. Protective Antigens/Virulence Factors and Evasion of Defense
   C. Immunity
   D. Diagnosis of *H. somnus* Infection
IV. *Campylobacter fetus* subsp. *venerealis* Infection
   A. Syndrome
   B. Immunity
   C. Evasion of Host Defense
   D. Diagnosis of Venereal Campylobacteriosis
V. *Tritrichomonas foetus* Infection
   A. Syndrome
   B. Prevalence
   C. Protective Antigens
   D. Immunity
   E. Cellular Evidence of Genital Mucosal Inductive Sites
   F. Immunity to Trichomoniasis in Bulls
   G. Evasion of Immune Responses to *T. foetus*
   H. Diagnosis of Trichomoniasis
VI. Summary and Future Directions
   Acknowledgments
   References

I. Introduction and Background

Bovine genital tract infections that result in pregnancy loss are large economic problems in the cattle industry. Such infections can occur by either a hematogenous or an ascending route. Because the diseases acquired by the former route are primarily septicemic whereas local ascending infections are primarily sexually transmitted diseases (STDs), diagnosis and control may differ between these two categories. Systemic infection with localization in the gravid uterus occurs in leptospirosis, systemic campylobacteriosis (caused by *Campylobacter fetus* subsp. *fetus*), listeriosis, brucellosis, neosporosis, and *Haemophilus sommus* infection. In these syndromes, systemic immunity and serologic diagnosis are often reliable. Brucellosis is a good example of hematogenous infection with localization in the gravid uterus since systemic vaccination and serologic assays for diagnosis have been practiced for decades. *Haemophilus sommus* infection is also presented because it is a septicemic disease causing abortion and many other sequelae, thus immunity is complex. Two examples of STDs, venereal campylobacteriosis and trichomoniasis, are also discussed. The emphasis will be on trichomoniasis since this is a current worldwide problem.

II. *Brucella abortus* Infection

In North America, cattle have been immunized against *Brucella abortus* induced abortion by systemic inoculation with attenuated live strain 19 vaccine for many years. The use of a live attenuated strain is considered to be critical in stimulating cell-mediated immunity (CMI). Because *B. abortus* is a facultative intracellular parasite, this arm of the immune response may be most protective. With this vaccine, a strong antibody response to the O polysaccharide side chains of the "smooth" lipopolysaccharide (LPS) also results (Table I). The antibody response probably plays some role in protection because several investigators have shown that antibody will passively protect mice against *B. abortus* infection (Pardon, 1977; Araya and Winter, 1990; Plommet

TABLE I

VACCINE STRAINS OF *BRUCELLA ABORTUS*

Vaccine	Form	LPS	O Polysaccharide	Seroconversion
Strain 19	Live	Smooth	Yes	Yes
Strain 45/20	Killed	Rough	No	No
Strain RB51	Live	Rough	No	No

and Plommet, 1983; Limét et al., 1987; Montaraz et al., 1986). The systemic antibody response to LPS O polysaccharide is also the basis of the serologic assays for diagnosis of brucellosis. In countries where control of brucellosis is achieved by serologic diagnosis and euthanasia of positive animals as well as by vaccination to prevent disease, the antibody response to the O polysaccharide of strain 19 can interfere with diagnosis of animals infected by wild-type *B. abortus*. This problem has been largely avoided by only immunizing female calves between 4 and 8 months of age so that antibody titers have fallen below diagnostic levels by the time of serologic testing. Another approach to circumvent this problem has been use of killed *B. abortus* strain 45/20, which has rough LPS (i.e., no O polysaccharide side chains). This does avoid false-positive serologic assays (Table I) but killed vaccines are not usually as efficient in stimulating CMI, and strain 45/20 is said to be somewhat unstable in producing only rough LPS. Recently a new stable rough mutant of *B. abortus* (strain RB51) was developed (Schurig et al., 1991, 1996) that has no O side chains and is killed by bovine complement (Corbeil et al., 1988). Live strain RB51 protects mice (Jimenez de Bagues et al., 1994) and cattle (Cheville et al., 1993) against brucellosis without stimulating antibodies to LPS O polysaccharide (Table I) and causing seroconversion (Stevens et al., 1994). This vaccine is now licensed for use in the United States and is being introduced in other countries.

III. *Haemophilus somnus* Infection

A. SYNDROME

Septicemia with vasculitis is the hallmark of *H. somnus* infection (Harris and Janzen, 1989; Humphrey and Stephens, 1983). In the pregnant cow, the organism localizes in the gravid uterus to cause abortion (Miller et al., 1983). We were able to reproduce hematogenous abortion by inoculation of *H. somnus* into the jugular vein or intrabronchially (Widders et al., 1986). Other sequelae of septicemia include thrombotic meningoencephalitis, arthritis, and myocarditis (Harris and Janzen, 1989; Humphrey and Stephens, 1983). In addition, *H. somnus* may cause local infections of the respiratory or reproductive tracts with resulting pneumonia or infertility (Harris and Janzen, 1989; Humphrey and Stephens, 1983; Kwiecien and Little, 1991). Even with this array of syndromes, it is likely that the most common result of colonization by *H. somnus* is the asymptomatic genital carrier state

(Humphrey et al., 1985; Ward and Corbeil, 1983). Thus pathogenesis is complex.

B. PROTECTIVE ANTIGENS/VIRULENCE FACTORS AND EVASION OF DEFENSE

Haemophilus somnus is a gram-negative bacterium, so it is not surprising that endotoxin is a virulence factor. This has been characterized as lipooligosaccharide (LOS) without O polysaccharide side chains (Inzana et al., 1988). The LOS undergoes antigenic variation during the course of infection (Inzana et al., 1992), which may explain, in part, the chronic nature of the respiratory infection (Gogolewski et al., 1989) and perhaps other syndromes. Adherence to several cell types (Corbeil et al., 1995) is a second virulence factor which may be mediated by a recently discovered surface fibrillar network (Corbeil et al., 1997a). This network and a 76-kDa surface antigen are also associated with serum resistance (Widders et al., 1988, 1989a; Cole et al., 1992, 1993) and with Fc binding of IgG_2 to the surface of *H. somnus* (Yarnall et al., 1988a,b; Cole et al., 1992; Corbeil et al., 1997a). Both serum resistance and Fc binding could result in evasion of host defense. Others have shown that *H. somnus* suppresses PMN function (Chiang et al., 1986; Czuprynski and Hamilton, 1985; Pfeifer et al., 1992), which would allow evasion of effector cells of the immune response. Iron acquisition via binding of transferrin to the surface of *H. somnus* (Yu et al., 1992) allows evasion of the host iron sequestration mechanisms and may account for species specificity (Ogunnariwo et al., 1990). Lastly, several outer membrane proteins have been characterized and may serve as virulence factors (Tagawa et al., 1993a-d). In particular, a 40-kDa OMP appears to be protective (Gogolewski et al., 1988). This could be the same as one of two 40-kDa lipoprotein genes cloned and characterized by Theisen et al. (1992, 1993). The expressed protein of one of these has some protective capacity (Rioux et al., 1994).

C. IMMUNITY

Gogolewski *et al.* (1987a) showed that convalescent phase serum was passively protective against experimental pneumonia in calves, as would be expected for an extracellular pathogen. Monospecific bovine antibody to a 40-kDa OMP was also passively protective (Gogolewski et al., 1988). This protective antibody was shown to be specific for the 40-kDa OMP, not the closely associated 41-kDa major OMP or a 39-kDa OMP (Corbeil et al., 1991b). Protection was associated with IgG2 antibodies in these experiments as well as in experimental abortion (Corbeil et al., 1997b; Gogolewski et al., 1988; Widders et al., 1989b). This is characteristic of other pyogenic infections also (Nansen, 1972). Later studies showed that protective IgG_2 antibodies may have recognized more epitopes of the 40-kDa OMP than IgG_1 antibodies (Corbeil et al., 1997b). Because there are two allotypes of IgG_2 in cattle, we then evaluated functions of these two allotypes which may be associated with differences in protective ability. Thus far, bovine IgG_2A1 has been shown to be less effective in activating bovine C than IgG_2A2 (Bastida-Corcuera and Corbeil, 1995). Because C appears to be important in resistance to *H. somnus* infection (Nielsen et al., 1981), this difference in function of the IgG_2 allotypes may reflect genetic differences in resistance to infection. Although vaccines have been available for *H. somnus* infection for some time, it is unclear whether they protect against reproductive tract infection. New vaccines that stimulate IgG_2 responses to the most protective antigens may be most effective. (*Editor's note:* A graduate student in my laboratory found the predominant isotype of antibody to a commercial *H. somnus* [Pfizer] was IgG_1 not IgG_2.)

D. Diagnosis of *H. somnus* Infection

Currently, culture is the best method used in diagnosis of *H. somnus* infection. This is slightly problematic, however, since many cattle are asymptomatic carriers (Humphrey et al., 1985; Ward and Corbeil, 1983), suggesting that a positive culture may not be diagnostic of the cause of disease. The diagnosis of *H. somnus* disease can be confirmed by demonstration of the organism in the tissue lesions by immunohistochemistry (Gogolewski et al., 1987b) but this takes time. Serologic assays have not been very helpful since many *H. somnus* antigens are cross-reactive with antigens of other bovine pathogens (Corbeil, 1990; Corbeil et al., 1991b; Kania et al., 1990). We showed, however, that the IgG_2 response increases most after disease (Widders et al., 1989b) and that an IgG_2 response to a purified 270-kDa antigen is diagnostic (Yarnall and Corbeil, 1989). Because this serologic assay distinguished between asymptomatic carriers and animals with *H. somnus* pneumonia or abortion, and was rapid, it may be helpful in the field.

IV. *Campylobacter fetus* subsp. *venerealis* Infection

A. Syndrome

Campylobacteriosis (previously vibriosis) caused by the extracellular gram-negative bacterium *Campylobacter fetus* subsp. *venerealis,* is a STD. An ascending infection of heifers or cows with these extracellular gram-negative bacteria causes mild to moderate endometritis and infertility or abortion. Although the infection is usually self-limiting, heifers and cows may remain infected for many months (Corbeil et al., 1981). After clearance, females are resistant to reinfection for a short period of time. Then vaginal infection may be detected but ascension to the uterus and resulting endometritis is prevented for a longer period of time. Thus the individual cow does not abort or show signs of infertility but the infection can be spred by the bull to susceptible heifers or cows in the herd. This also suggests a difference between vaginal and uterine immunity. The bull, on the other hand, is a long-term asymptomatic carrier. Young bulls may clear the infection but older bulls are often carriers for life. This age difference may be due to the fact that older bulls have deeper preputial epithelial crypts, providing a good niche for these microaerophilic bacteria (Samuelson and Winter, 1966).

B. Immunity

A reliable vaccine has been available for campylobacteriosis for many years (Corbeil et al., 1981). This may be the first STD in any species for which an efficacious vaccine has been developed so the principles of protection are well worth noting. Passive protection by antibody has been demonstrated (Berg et al., 1979), illustrating the principle that antibody is most important in protection against extracellular pathogens. Local immunity clears the infection slowly with the uterus becoming free of infection before the vagina (Corbeil et al., 1981). Convalescent animals are partially protected. This immunity wanes over a couple of years. The antibody response to *C. fetus* infection is predominantly IgA in the vagina but predominantly IgG in the uterus (Table II). When kinetics of the vaginal antibody response to

TABLE II

Immunity to *C. fetus* Subsp. *venerealis*

	Predominant Ig class		
	Uterus	Vagina	Clearance[a]
Systemic immunization	IgG	IgG	Rapid-Ut and Vag
Convalescent (local) immunity	IgG	IgA	Slower-Ut then Vag

[a]Ut, uterus; Vag, vagina.

infection were evaluated in weekly samples of vaginal secretions, the IgM response was detected first, followed by the IgA response, the IgG_1 response, and lastly the IgG_2 response. The duration of the IgA response was greatest (average > 56 weeks) followed by the IgG_1 response (average 42 weeks), the IgG_2 response (average 28 weeks), and lastly the IgM response (average 3 weeks) (Corbeil et al., 1974b, 1981). Prophylactic systemic immunization, with killed cells in oil adjuvant, results in high levels of IgG antibody in serum, uterine secretions, and vaginal secretions (Corbeil et al., 1974a,b). This results in both protection against subsequent infection (Corbeil and Winter, 1978) and cure of chronic infection of the female reproductive tract (Schurig et al., 1975). Infection of bulls with C. fetus subsp. venerealis results in long-term preputial infection but systemic vaccination with whole killed cells in oil adjuvant is both prophylactic and therapeutic (Bouter et al., 1973). This dramatic protection against an extracellular noninvasive pathogen of the genital mucosa by systemic immunization is a breakthrough in understanding defense of the reproductive tract. It has also resulted in essentially controlling this infection where vaccination is practiced. The ability to cure a chronic STD in both males and females by systemic immunization is even more unusual.

C. Evasion of Host Defense

In nonimmunized animals, C. fetus infection of the vagina can be very persistent (> 74 weeks in one study; Corbeil et al., 1981) in the face of a vaginal antibody response. This suggested that C. fetus had developed effective means of evading local immune responses. These included (1) resistance of virulent strains to killing by bovine C (Corbeil et al., 1974a), (2) coating with vaginal IgA antibodies that are not opsonic in a competition with opsonic IgG antibodies (Corbeil et al., 1974a, 1975a), (3) variation of surface antigens in the face of the local immune response (Corbeil et al., 1975b). Even so, in systemically immunized animals, IgG antibodies transudated to the genital secretions are able to prevent (Corbeil and Winter, 1978; Corbeil et al., 1981) and even cure (Schurig et al., 1975) infection. This illustrates that the dynamic interaction between host and parasite can be manipulated to favor the host even when the parasite has multiple strategies to evade host protective responses.

D. Diagnosis of Venereal Campylobacteriosis

Diagnosis of C. fetus subsp. venerealis infection is primarily by culture under microaerophilic conditions on selective media. The vaginal mucus agglutination test has also been useful for diagnosis in heifers or cows and an indirect fluorescent antibody assay has been used with preputial scrapings in bulls.

V. *Tritrichomonas foetus* Infection

A. Syndrome

Bovine trichomoniasis is clinically similar to bovine venereal campylobacteriosis. Both are chronic STDs caused by extracellular noninvasive pathogens (the former due to protozoa and the latter to gram-negative bacteria). Both infections result in reproductive failure, with serious economic consequences (Corbeil *et al.*, 1981; Clark *et al.*, 1983a; Goodger and Skirrow, 1986; Rae, 1989). Trichomoniasis is essentially asymptomatic in bulls but is manifest as endometritis, salpingitis, placentitis, and fetal loss in cows (BonDurant, 1985; Skirrow and BonDurant, 1988). In the only published study of pathogenesis where heifers were infected by breeding to an infected bull, no inflammation or fetal loss was detected prior to 60 days of infection (Parsonson *et al.*, 1976). Inflammation and fetal loss occurred primarily between 60 and 90 days (Parsonson *et al.*, 1976). Reports of fetal loss from field cases have been from days 16–17 through 7 months of gestation (BonDurant, 1985; Rhyan *et al.*, 1988; Skirrow and BonDurant, 1988). After pregnancy loss, cattle may remain infertile for weeks to months but fertility is usually regained in time. The few animals with pyometra may have more severe sequelae (BonDurant, 1985; Skirrow and BonDurant, 1988).

B. Prevalence

Trichomoniasis is prevalent in areas of the world where natural breeding is used. Herd infection rates in the United States have been reported to be as high as 44% of herds in Nevada (Kvasnicka *et al.*, 1989), and 16% of herds in California (BonDurant *et al.*, 1990). In one study in Australia, 40% of herds were infected (Dennett *et al.*, 1974) and in another study in Costa Rica 7–16% of herds were infected (Perez *et al.*, 1992). When individual animals in a herd are compared, the prevalence rate in bulls is often higher than that in cows (Dennett *et al.*, 1974; Akinboade, 1980). This difference could be due to differences in duration of infection in cows and bulls. The infection is self-limiting in cows, although the duration is quite variable (BonDurant,

1985; Skirrow and BonDurant, 1988). Clearance may occur between 12 and 22 weeks (Skirrow and BonDurant, 1988). A few cows may carry the infection right through the gestation period (Skirrow, 1987). Experimental infections have lasted from an average of 8 weeks (BonDurant et al., 1993) to 19.5 weeks (Skirrow and BonDurant, 1990a). In these studies, the duration may have been related to dose since animals with the lowest doses cleared earlier than those with the highest doses. The duration of infection in bulls is much longer than in cows. Bulls greater than 4 years of age at infection generally remain asymptomatic preputial carriers for life whereas 2- to 4-year-old bulls carry the infection for shorter times (Clark et al., 1974; Skirrow and BonDurant, 1988). These differences between bulls and cows may be related to differences in local immune responses of the male and female genital mucosa.

C. Protective Antigens

To better understand mechanisms of protective immunity, it is necessary to identify protective antigens and the most appropriate immune responses. Many trichomonad antigens are recognized by the immune system when cattle are immunized systemically with *T. foetus* (Hall et al., 1986; Corbeil et al., 1989). Burgess used monoclonal antibodies to identify a 150-kDa surface antigen that was highly conserved in various *T. foetus* isolates (Burgess, 1986, 1988). Various monoclonal antibodies (mAbs) were shown to mediate complement lysis, act as opsonins or inhibit adhesion and cytotoxicity (Burgess, 1986; Burgess and McDonald, 1992). We also prepared a bank of murine mAbs to *T. foetus* to determine which antigens may be most protective. Those with potentially protective functions were selected for further study. Of four mAbs that reacted strongly with *T. foetus* in ELISA, two (TF1.17 and TF1.15) were most promising (Hodgson et al., 1990). These two mAbs immobilized and agglutinated *T. foetus,* mediated complement lysis, and prevented adherence of *T. foetus* to bovine vaginal epithelial cells (Hodgson et al., 1990). Later we showed that the two mAbs recognized different epitopes of the same surface antigen (Ikeda et al., 1993) and that this antigen was conserved in 50 isolates from Costa Rica, Argentina, and various areas of the United States (Ikeda et al., 1993; BonDurant et al., 1993; Corbeil, 1994). Other studies showed that the antigen ran as a broad band in SDS–PAGE due to heterologous glycosylation as demonstrated by thymol staining (Hodgson et al., 1990). The target surface antigen was immunoaffinity purified using mAb TF1.17. The immunoaffinity purified antigen also ran as a broad band between ~45 and 75 kDa, which reacted with both the mAb and with bovine antiserum to whole *T. foetus* cells in Western blots (Fig. 1). This bovine antiserum reacted with many bands in whole-cell lysates but only the wide band in the purified preparation (Fig. 1), demonstrating the relative antigenic purity of the immunoaffinity purified TF1.17 antigen.

D. IMMUNITY

Immune responses to *T. foetus* infection have been studied by several groups. Convalescent cows are usually resistant to reinfection but this immunity wanes over approximately 2 years (Clark *et al.*, 1983a). Skirrow and BonDurant (1990b) showed that infected heifers mounted minimal systemic antibody responses but strong IgA and IgG1 vaginal, cervical, and uterine responses by 7–12 weeks after intravaginal inoculation of *T. foetus*. The IgG_2 response was slightly later and was more

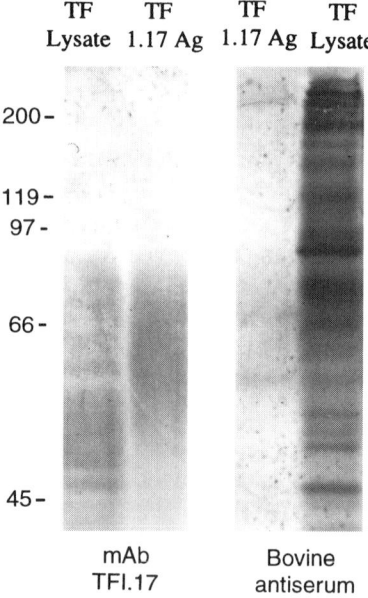

FIG. 1. Western blot of SDS–PAGE gel wells loaded with *T. foetus* (TF) lysate or immunoaffinity purified TF1.17 antigen (Ag). The panel on the left was reacted with TF1.17 mAb and the panel on the right with antiserum from a cow (#64) immunized intramuscularly with live *T. foetus* (see Corbeil *et al.*, 1989, and Hodgson *et al.*, 1990, for methods of antibody production). Molecular weight standards on the left.

transient. The duration of the IgA response was greatest (Skirrow and BonDurant, 1990b). Our subsequent studies of experimentally infected heifers also showed vaginal (BonDurant et al., 1993) and uterine (Anderson et al., 1996) antibody responses to infection. These observations of resistance of convalescent cows to reinfection, and demonstration of mucosal antibody responses, encouraged development of vaccines. Currently, killed cell vaccines are available and are at least partially protective (Kvasnicka et al., 1989, 1992).

To investigate protection with specific antigens, we immunized virgin heifers with immunoaffinity purified TF1.17 antigen mixed with incomplete Freund's adjuvant (IFA). A second group was immunized intramuscularly with TF1.17 cross-linked with glutaraldehyde and mixed with IFA plus dextran sulfate. The two immunized groups and an adjuvant control group were challenged with 10^6 *T. foetus* strain D1, 2 weeks after the last (third) immunizing dose (BonDurant et al., 1993). Both immunized groups cleared the infection more quickly than the adjuvant controls ($p < 0.005$). Importantly, most of the immunized animals cleared the infection by 7 weeks, which was the time that Parsonson et al. (1976) had shown was just before the beginning of inflammation and fetal loss. Most of the controls cleared after this time. By ELISA with whole-cell *T. foetus* antigen, it was shown that cervicovaginal mucus IgA and IgG_1 responses peaked at about the time of clearance of infection in vaccinated animals (BonDurant et al., 1993). This raised the question of whether IgA of IgG antibodies were more protective in this infection. To attempt to determine this, we first immunized mice with TF1.17 antigen plus several different adjuvants by different routes (Corbeil et al., 1998). A group of mice immunized twice subcutaneously with TF1.17 antigen and Quil A adjuvant, followed by an intravaginal immunization with killed *T. foetus* in Quil A, had greatest genital mucosal response. Thus heifers were immunized twice subcutaneously with TF1.17 antigen and Quil A, followed by killed *T. foetus* given either subcutaneously or intravaginally. These two groups and a group of unimmunized controls were challenged intravaginally with 10^6 *T. foetus*, 3 weeks after the last immunization. Again, immunized animals cleared the infection faster than the controls ($p - 0.0225$). By 7 weeks postchallenge, only 25% of vaccinated animals were still culture positive whereas 60% of controls were positive (Corbeil et al., 1998). There was no statistically significant difference in clearance rate between immunized groups even though the intravaginally boosted group had much higher vaginal IgA responses to TF1.17 antigen and the systemically boosted group had higher vaginal IgG anti-TF1.17 antigen levels (Corbeil et al., 1998). In parallel to our previous studies with *C. fetus* and *T. foetus* immunity, however, the vaginal IgA response was of greater duration than the IgG response (Corbeil *et al.*, 1981, 1998). This may be important in the field, when not all animals are challenged naturally within 3 weeks of the last immunizing dose.

Clearance data in vaccinated and in controls from both of these immunization experiments (BonDurant *et al.*, 1993; Corbeil *et al.*, 1998) were then averaged since there were no significant differences between groups. It was clear that the majority of controls did not clear the infection before 7 weeks and that less than 25% of immunized animals were still positive by 7 weeks (Fig. 2). This is true even though false negative cultures sometimes showed more animals to be positive on one week than on the week before (eg week 7 vs 8, Fig. 2). This data shows that with three different adjuvant formulations and two different combinations of routes, TF1.17 antigen was still protective. The clearance of immunized animals generally before 7 weeks should prevent the development of inflammation and fetal loss, if the results of the studies of Parsonson *et al.*, (1976), as described under Section V,A, can be generalized.

E. CELLULAR EVIDENCE OF GENITAL MUCOSAL INDUCTIVE SITES

Studies of genital immunity in both venereal campylobacteriosis and trichomoniasis address the question of the role of the female reproduc-

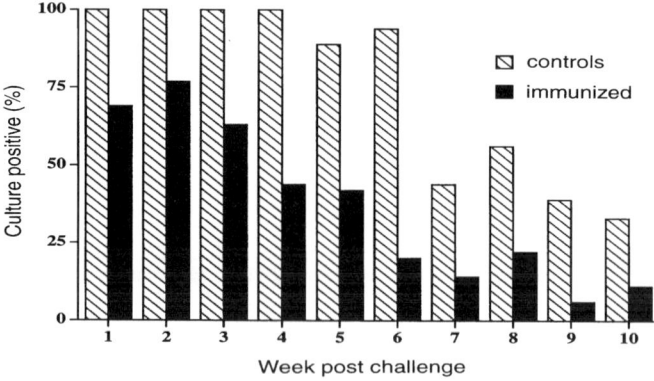

FIG. 2. Clearance of *T. foetus* after immunization of heifers with TF1.17 antigen and challenge with *T. foetus* strain D1 (data averaged from experiments reported by BonDurant *et al.*, 1993, and Corbeil *et al.*, 1998). In each of two experiments there were two immunized groups and one control group for totals of 36 and 18 animals, respectively.

tive tract as an inductive site for local immune responses. Immunization of the vagina with nonreplicating or soluble antigens has often resulted in poor vaginal IgA responses when compared with immunization via the gastrointestinal (GI) or respiratory tracts. Because of this, others have suggested that immunization via the GI tract (Haneberg *et al.*, 1994) or the respiratory tract (DiTomaso *et al.*, 1996) may be best for stimulating vaginal IgA responses. Whether the genital mucosa is an inductive site for local immune responses has been a controversial issue due to lack of organized lymphoid tissue in the female reproductive tract (Haneberg *et al.*, 1994). Yet infections of the female reproductive tract by extracellular noninvasive pathogens such as *C. fetus* and *T. foetus* induce local IgA responses in the virtual absence of systemic antibody responses (Corbeil *et al.*, 1974a,b; BonDurant *et al.*, 1993; Skirrow and BonDurant, 1990b). To investigate this, vaginal, cervical, uterine, and uterine tube (or oviduct) tissues of *T. foetus* TF1.17 immunized and control heifers were studied at necropsy 10 weeks after infection with *T. foetus* (Anderson *et al.*, 1996). Animals that cleared infection within the first 2 weeks (or never became infected) had little organized lymphoid tissue at any site. However, 5 of 8 nonimmunized infected heifers had focal aggregates of nodular lymphoid tissue in the endometrium and 4 of 16 systemically immunized animals had lymphoid nodules in the endometrium (Anderson *et al.*, 1996). Some of these lymphoid aggregates appeared to have germinal centers similar to secondary lymphoid follicles. Sequential uterine secretions were available from a few of these heifers as well as some heifers from a separate study. Analysis of IgA antibody reactivity to TF1.17 antigen by ELISA showed that most infected heifers had uterine IgA responses by 6 weeks postchallenge, which increased at weeks 8 and 10 (Anderson *et al.*, 1996). Noninfected heifers had essentially no IgA response to TF1.17 antigen. These observations suggested that organized lymphoid tissue, in the endometrium at least, was stimulated by the presence of infection and that a specific IgA response in uterine secretions resulted. Subsequent studies were done with heifers immunized to enhance local IgG or IgA anti-TF1.17 levels (systemic priming with TF1.17 antigen followed by systemic or local boosting with killed *T. foetus* as described earlier). These studies revealed lymphoid nodules in both the vaginal and uterine superficial submucosa of nonimmunized infected heifers, as well as systemically primed, vaginally boosted heifers (Corbeil *et al.*, 1998). These were the groups with the highest vaginal IgA response. The systemically primed and boosted group did not have detectable lymphoid nodules in the vaginal mucosa and had predominantly IgG_1 antibodies in the vaginal secretions. Presumably this was transferred from serum since serum IgG_1 antibody levels were highest in this group. The presence of many eosinophils within and below the vaginal epithelial cells may have been related to leakage of IgG across the epithelium (Corbeil et al., 1998). Many plasma cells and large mononuclear cells were detected just below the epithelium of all groups, suggesting local antibody synthesis was occurring. In other immunohistochemical studies of heifers infected with *T. foetus*, we have shown that *T. foetus* antigen is taken up by epithelial cells and is present in large mononuclear cells below the basement membrane (J. C. Rhyan, M. L. Anderson, R. H. BonDurant, and L. B. Corbeil, unpublished data). Because vaginal and uterine epithelial cells are known to present antigen (Wira and Rossol, 1995a,b) and other antigen-presenting cell types with morphology of dendritic cells or macrophages were seen to take up *T. foetus* antigen, it appears that antigen uptake and presentation is occurring in the bovine female genital tract. Therefore, all the requirements for inductive sites (antigen uptake, antigen presentation, and organized lymphoid tissue) as well as local antibody synthesis appear to be present in the infected bovine vagina and uterus.

F. Immunity to Trichomoniasis in Bulls

Experimental immunization of bulls against trichomoniasis has been reported to be successful for both prophylaxis and therapy (Clark et al., 1983b, 1984), as for bovine venereal campylobacteriosis (Bouter et al., 1973). Clark et al. (1983b) first showed that systemic vaccination with killed *T. foetus* in oil adjuvant could prevent or eliminate infection of bulls up to 5 years of age but was not protective or therapeutic in bulls older than 5.5 years old. Subsequently the same research group showed that systemic immunization with *T. foetus* membranes and oil adjuvant could clear bulls of infection. In two experiments with a total of 8 immunized and 8 control bulls between 5 and 8 years of age, 6 immunized bulls were free of infection by 2 weeks after the third immunization but only 1 control bull cleared the infection (Clark et al., 1984). This suggests that immunization may be useful in control of infection in bulls. Studies in our laboratories showed that infected bulls have significantly more IgG_1, IgA, IgM, and IgG_2 antibodies to TF1.17 antigen in preputial secretions than culturally negative bulls (J. C. Rhyan, R. H. BonDurant, G. A. K. Mutwiri, and L. B. Corbeil, unpublished data). The changes in isotypic antibody levels in preputial secretions of immunized bulls may provide insight into mechanisms of immune protection of the preputial surface.

G. Evasion of Immune Responses to *T. foetus*

As in *C. fetus* infection, however, *T. foetus* has strategies to evade the antibody response. Although *T. foetus* can be killed by bovine C (Hodgson *et al.*, 1990) or polymorphonuclear leukocytes (Aydintug *et al.*, 1990, 1993) and specific IgG antibody, it has developed means of avoiding these defenses. First, the parasite releases extracellular (Thomford *et al.*, 1996) cysteine proteinases that digest bovine IgG_1 and IgG_2 (Talbot *et al.*, 1991). Second, *T. foetus* nonspecifically binds bovine IgG via the Fc or hinge region to its surface, presumably masking antigens for specific antigen–antibody reactions but not initiating the complement cascade or opsonization (Corbeil *et al.*, 1991a). As for *C. fetus*, these evasive mechanisms appear to contribute to the chronic nature of the infection but systemic or local immunization with appropriate antigens and adjuvants can shift the dynamic interaction in favor of host defense.

H. Diagnosis of Trichomoniasis

The gold standard for diagnosis of trichomoniasis is culture of vaginal or preputial secretions (Abbitt and Ball, 1978; BonDurant, 1985). Microscopic examination of wet mounts is also used but is less sensitive than culture (BonDurant, 1985). Even though culture is more sensitive than wet mount observation, three successive negative cultures are necessary to rule out trichomoniasis in bulls (BonDurant, 1985; Skirrow and BonDurant, 1988). The situation is more complex in females because the number of trichomonads in vaginal mucus varies with the estrus cycle. Numbers of organisms in vaginal mucus also decrease as the infection progresses. False-negative cultures occur when numbers fall below the threshold of detectability but the next week animals may be culturally positive again, as we have often observed in experimental trichomoniasis (Anderson *et al.*, 1996; Corbeil *et al.*, 1998). Reported sensitivities of diagnosis by culture were 81–97% for bulls and 58.7% for cows (Goodger and Skirrow, 1986; Skirrow and BonDurant, 1988). With even less sensitivity by direct microscopic examination, better diagnostic methods are a priority. DNA probes have been offered (Speer and White, 1991), but sensitivity and specificity were controversial (Appell *et al.*, 1993). Later, others showed that with a different probe and polymerase chain reaction (PCR)-based amplification, as low as 10 trichomonads could be detected in reproductive tract secretions (Ho *et al.*, 1994). With this method 47 of 52 (90.4%) of positive samples were detected and there were no false positives. Culture, on the other hand, detected 44 of 52 positive samples (84.6%). This method is promising. Another approach has involved antibody based tests. Serology was not helpful in early studies (Skirrow and BonDurant, 1988) because little systemic immune response is stimulated by this mucosal pathogen (BonDurant et al., 1993; Skirrow and BonDurant, 1990b). However, we recently showed that ELISA based on TF1.17 antigen and the IgA antibody responses in cervicovaginal secretions was diagnostic by 6 weeks after experimental infection. If whole *T. foetus* cells were used as antigen, cross-reactive antibodies in vaginal secretions were a problem. If IgG_2 or IgG_1 antibody against TF1.17 antigen was assayed, then not all infected animals were detected by 6 weeks postinfection (Ikeda et al., 1995). Vaginal IgA responses to TF1.17 antigen above preinfection levels were detected in all infected animals (Ikeda et al., 1995), the antigen is conserved (Ikeda et al., 1993), and the infection is a chronic herd problem. Thus, detection by 6 weeks after infection of the first heifer or cow would be quite useful in initiating control procedures for the herd. This is especially true given that little fetal loss is thought to occur until after 7 weeks. Perhaps even more promising is a more sensitive serologic assay than was available in the past. BonDurant et al. (1996) have recently developed a sensitive hemolytic assay based on antigen shed from *T. foetus* and adsorbed to bovine erythrocytes. Serum antibody which mediated C lysis of coated erythrocytes was detected by 2 weeks postinfection of heifers and remained high throughout the 10-week observation period (BonDurant et al., 1996). Sensitivity was 94% and specificity was 95.6%. Thus, although culture is still the gold standard, new PCR assays, vaginal IgA ELISA, and passive hemolysis assays for serum antibody have possible applications.

VI. Summary and Future Directions

Immunoprophylaxis, immunotherapy, and immunodiagnostics for bovine reproductive tract infections have been very successful. A new systemic vaccine for brucellosis protects without seroconversion. Thus the serologic assays based on antibody to *B. abortus* LPS oligosaccharide side chains can be used with less concern about interpretation due to interference by immune responses to vaccine. For abortion following *H. somnus* septicemia, commercial vaccines are available but it is unclear whether they protect against *H. somnus* abortion. Diagnosis is also problematic since asymptomatic carriers are common and diagnosis is primarily by culture. However, virulence factors and protective antigens have been identified. Also a diagnostic ELISA using purified antigen and conjugate detecting IgG_2 has been presented. Therefore, new immunodiagnostic and immunoprophylactic measures are on the horizon. There are two commercially available vaccines for STDs in cattle, which provides hope for development of vaccines for STDs in other species. Even though both *C. fetus* and *T. foetus* have developed effective strategies to evade immune protection, vaccines have tipped the balance in favor of host defense. New information on protective antigens of *T. foetus* and the characteristics of protective responses, such as kinetics of isotypic antibody responses on mucosal surfaces, should permit even more effective second-generation vaccines. New diagnostic techniques for trichomoniasis including a hemolytic assay, vaginal mucus IgA ELISA, and a nucleic acid probe are promising.

ACKNOWLEDGMENTS

The research from our laboratories was supported in part by NIH grant number A133540 and USDA grant numbers 93-37204-9332, 94-37204-0852, and 97-35204-4771. Special thanks to Sharon McFarlin for preparing the manuscript and Karen Arnold for preparing the figures.

REFERENCES

Abbitt, B., and Ball, L. (1978). Diagnosis of trichomoniasis in pregnant cows by culture of cervico-vaginal mucus. *Theriogenology* **9,** 267–270.
Akinboade, O. A. (1980). Incidence of bovine trichomoniasis in Nigeria. *Rev. Elev. Med. Vet. Pays Trop.* **33,** 381–384.
Anderson, M. L., BonDurant, R. H., Corbeil, R. R., and Corbeil, L. B. (1996). Immune and inflammatory responses to reproductive tract infection with *Tritrichomonas foetus* in immunized and control heifers. *J. Parasitol.* **82,** 594–600.
Appell, L. H., Mickelsen, W. D., Thomas, M. W., and Haromon, W. M. (1993). A comparison of techniques used for the diagnosis of *Tritrichomonas foetus* infections in beef bulls. *Agri-Practice* **14,** 30–34.
Araya, L. N., and Winter, A. J. (1990). Comparative protection of mice against virulent and attenuated strains of *Brucella abortus* by passive transfer of immune T cells or serum. *Infect. Immun.* **58,** 254–256.
Aydintug, M. K., Leid, R. W., and Widders, P. R. (1990). Antibody enhances killing of *Tritrichomonas foetus* by the alternative pathway. *Infect. Immun.* **58,** 944–948.
Aydintug, M. K., Widders, P. R., and Leid, R. W. (1993). Bovine polymorphonuclear leukocyte killing of *Tritrichomonas foetus*. *Infect. Immun.* **61,** 2995–3002.
Bastida-Corcuera, F., and Corbeil, L. B. (1995). Complement activation of bovine IgG2 A1 and A2 allotypes. *Int. Vet. Immunol. Symp., 4th,* p. 297.
Berg, R. L., Firehammer, B. D., Border, M., and Myers, L. L. (1979). Effects of passively and actively acquired antibody on bovine campylobacteriosis (vibriosis). *Am. J. Vet. Res.* **40,** 21–25.
BonDurant, R. H. (1985). Diagnosis, treatment and control of bovine trichomoniasis. *Compend. Contin. Educ. Pract. Vet.* **7,** S179–S188.

BonDurant, R. H., Anderson, M. L., Blanchard, P., Hird, D., Danaye-Elmi, C., Palmer, C., Sischo, W. M., Suther, D., Utterback, W., and Weigler, B. J. (1990). Prevalence of trichomoniasis among California beef herds. *J. Am. Vet. Med. Assoc.* **196,** 1590–1593.

BonDurant, R. H., Corbeil, R. R., and Corbeil, L. B. (1993). Immunization of virgin cows with surface antigen TF1.17 of *Tritrichomonas foetus*. *Infect. Immun.* **61,** 1385–1394.

BonDurant, R. H., Van Hoosear, K. A., Corbeil, L. B., and Bernoco, D. (1996). Serological responses to in vitro-shed antigen(s) of *Tritrichomonas foetus* in cattle. *Clin. Diagn. Lab. Immunol.* **3,** 432–437.

Bouter, R., Dekeyser, J., Vandeplassche, M., VanAert, A., Brone, E., and Bonte, P. (1973). *Vibrio fetus* infection in bulls: Curative and preventive vaccination. *Br. Vet. J.* **129,** 52–57.

Burgess, D. E. (1986). *Tritrichomonas foetus*: Preparation of monoclonal antibodies with effector function. *Exp. Parasitol.* **62,** 266–274.

Burgess, D. E. (1988). Clonal and geographic distribution of a surface antigen of *Tritrichomonas foetus*. *J. Protozool.* **35,** 119–122.

Burgess, D. E., and McDonald, C. M. (1992). Analysis of adhesion and cytotoxicity of *Tritrichomonas foetus* to mammalian cells by use of monoclonal antibodies. *Infect. Immun.* **60,** 4253–4259.

Cheville, N. F., Stevens, M. G., Jensen, A. E., Tatum, F. M., and Halling, S. M. (1993). Immune responses and protection against infection and abortion in cattle experimentally vaccinated with mutant strains of *Brucella abortus*. *Am. J. Vet. Res.* **54,** 1591–1597.

Chiang, Y. W., Kaeberle, M. L., and Roth, J. A. (1986). Identification of suppressive components in *Haemophilus somnus* fractions which inhibit bovine polymorphonuclear leucocyte function. *Infect. Immun.* **52,** 792–797.

Clark, B. L., Parsonson, I. M., and Dufty, J. H. (1974). Experimental infection of bulls with *Tritrichomonas foetus*. *Aust. Vet. J.* **50,** 189–191.

Clark, B. L., Dufty, J. H., and Parsonson, I. M. (1983a). The effect of *Tritrichomonas foetus* infection on calving rates in beef cattle. *Aust. Vet. J.* **60,** 71–74.

Clark, B. L., Dufty, J. H., and Parsonson, I. M. (1983b). Immunisation of bulls against trichmoniasis. *Aust. Vet. J.* **60,** 178–179.

Clark, B. L., Emery, D. L., and Dufty, J. H. (1984). Therapeutic immunisation of bulls with the membranes and glycoproteins of *Tritrichomonas foetus* var. Brisbane. *Aust. Vet. J.* **61,** 65–66.

Cole, S. P., Guiney, D. G., and Corbeil, L. B. (1992). Two linked genes for outer membrane proteins are absent in four non-disease strains of *Haemophilus somnus*. *Mol. Microbiol.* **6,** 1895–1902.

Cole, S. P., Guiney, D. G., and Corbeil, L. B. (1993). Molecular analysis of a gene encoding serum-resistance-associated 76 kDa surface antigen of *Haemophilus somnus*. *J. Gen. Microbiol.* **139,** 2135–2143.

Corbeil, L. B. (1990). Molecular aspects of some virulence factors of *Haemophilus somnus*. *Can. J. Vet. Res.* **54,** S57-S62.

Corbeil, L. b. (1994). Vaccination strategies against *Tritrichomonas foetus*. *Parasitol. Today* **10,** 103–106.

Corbeil, L. B., and Winter, A. J. (1978). Animal model for the study of genital secretory immune mechanisms: Venereal vibriosis in cattle. *In* "Immunobiology of *Neisseria gonorrhoeae*" (G. F. Brooks, E. C. Gotschlick, K. K. Holmes, W. D. Sawyer, and F. E. Young, eds.), pp. 293–299. Am. Soc. Microbiol., Washington, DC.

Corbeil, L. B., Schurig, G. D., Duncan, J. R., Corbeil, R. R., and Winter, A. J. (1974a).

Immunoglobulin classes and biological functions of *Campylobacter* (*Vibrio*) antibodies in serum and cervicovaginal mucus. *Infect. Immun.* **10,** 422–429.

Corbeil, L. B., Duncan, J. R., Schurig, G. G. D., Hall, C. E., and Winter, A. J. (1974b). Bovine venereal vibriosis variations in immunoglobulin class of antibodies in genital secretions and serum. *Infect. Immun.* **10,** 1084–1090.

Corbeil, L. B., Corbeil, R. R., and Winter, A. J. (1975a). Bovine venereal vibriosis: Activity of inflammatory cells in protective immunity. *Am. J. Vet. Res.* **36,** 403–406.

Corbeil, L. B., Schurig, G. G. D., Bier, P. J., and Winter, A. J. (1975b). Bovine venereal vibriosis: Antigenic variation of the bacterium during infection. *Infect. Immun.* **3,** 854–856.

Corbeil, L. B., Schurig, G. G., Duncan, J. R., Wilkie, B. N., and Winter, A. J. (1981). Immunity in the female reproductive tract based on the response to *Campylobacter fetus*. *Adv. Exp. Med. Biol.* **137,** 729–743.

Corbeil, L. B., Blau, K., Inzana, T. J., Nielsen, K. H., Jacobson, R. H., Corbeil, R. R., and Winter, A. J. (1988). Killing of *Brucella abortus* by bovine serum. *Infect. Immun.* **56,** 3251–3261.

Corbeil, L. B., Hodgson, J. L., Jones, D. W., Corbeil, R. R., Widders, P. R., and Stephens, L. R. (1989). Adherence of *Tritrichomonas foetus* to bovine vaginal epithelial cells. *Infect. Immun.* **57,** 2158–2165.

Corbeil, L. B., Hodgson, J. L., and Widders, P. R. (1991a). Immunoglobulin binding by *Tritrichomonas foetus*. *J. Clin. Microbiol.* **29,** 2710–2714.

Corbeil, L. B., Kania, S. A., and Gogolewski, R. P. (1991b). Characterization of immunodominant surface antigens of *Haemophilus somnus*. *Infect. Immun.* **59,** 4295–4301.

Corbeil, L. B., Gogolewski, R. P., Stephens, L. R., and Inzana, T. J. (1995). *Haemophilus somnus*: Antigen analysis and immune responses *In* "Haemophilus, Actinobacillus, and Pasteurella" (W. Donachie *et al.,* eds.), pp. 63–73 Plenum, New York.

Corbeil, L. B., Bastida-Corcuera, F. D., and Beveridge, T. J. (1997a). *Haemophilus somnus* immunoglobulin binding proteins and surface fibrils. *Infect. Immun.* **65,** 4250–4257.

Corbeil, L. B., Gogolewski, R. P., Kacskovics, I., Nielsen, K. H., Corbeil, R. R., Morrill, J. L., Greenwood, R., and Butler, J. E. (1997b). Bovine IgG2a antibodies to *Haemophilus somnus* and allotype expression. *Can. J. Vet. Res.* **61,** 207–213.

Corbeil, L. B., Anderson, M. L., Corbeil, R. R., Eddow, J. M., and BonDurant, R. H. (1998). Female reproductive tract immunity in bovine trichomoniasis. *Am. J. Reprod. Immunol.* **39,** 189–198.

Czuprynski, C. J., and Hamilton, H. L. (1985). Bovine neutrophils ingest but do not kill *Haemophilus somnus* in vitro. *Infect. Immun.* **50,** 431–436.

Dennett, D. P., Reece, R. L., Barasa, J. O., and Johnson, R. H. (1974). Observations on the incidence and distribution of serotypes of *Tritrichomonas foetus* in beef cattle in North-Eastern Australia. *Aust. Vet. J.* **50,** 427–431.

Di Tomaso, A., Saletti, G., Pizza, M., Rappuoli, R., Dougan, G., Abrignani, S., Douce, G., and de Magistris, M. (1996). Induction of antigen-specific antibodies in vaginal secretions by using a nontoxic mutant of heat-labile enterotoxin as a mucosal adjuvant. *Infect. Immun.* **64,** 974–979.

Gogolewski, R. P., Kania, S. A., Inzana, T. J., Widders, P. R., Liggitt, H. D., and Corbeil, L. B. (1987a). Protective ability and specificity of a convalescent serum from *Haemophilus somnus* pneumonia. *Infect. Immun.* **55,** 1403–1411.

Gogolewski, R. P., Leathers, C. W., Liggitt, H. D., and Corbeil, L. B. (1987b). Experimental *Haemophilus somnus* pneumonia in calves and immunoperoxidase localization of bacteria. *Vet. Pathol.* **24,** 250–256.

Gogolewski, R. P., Kania, S. A., Liggitt, H. D., and Corbeil, L. B. (1988). Protective ability of monospecific sera against 78 kDa and 40-kDa outer membrane antigens of *"Haemophilus somnus."* *Infect. Immun.* **56,** 2301–2316.

Gogolewski, R. P., Schaefer, D. C., Wasson, S. K., Corbeil, R. R., and Corbeil, L. B. (1989). Pulmonary persistence of *Haemophilus somnus* in presence of specific antibody. *J. Clin. Microbiol.* **27,** 1767–1774.

Goodger, W. J., and Skirrow, S. Z. (1986). Epidemiologic and economic analyses of an unusually long epizootic of trichomoniasis in a large California dairy herd. *J. Am. Vet. Med. Assoc.* **189,** 772–776.

Hall, M. R., Huang, J.-C., Ota, R., Redelman, D., Hanks, D., and Taylor, R. E. L. (1986). Characterization of *Tritrichomonas foetus* antigens, using bovine antiserum. *Am. J. Vet. Res.* **47,** 2549–2553.

Haneberg, B., Kendall, D., Amerongen, H. M., Apter, F. M., Kraehenbuhl, J.-P., and Neutra, M. R. (1994). Induction of specific immunoglobulin A in the small intestine, colon-rectum, and vagina measured by a new method for collection of secretions from local mucosal surfaces. *Infect. Immun.* **62,** 15–23.

Harris, F. W., and Janzen, E. D. (1989). The *Haemophilus somnus* disease complex (Hemophilosis): A review. *Can. Vet. J.* **30,** 816–822.

Ho, M. S. Y., Conrad, P. A., Conrad, P. J., LeFebvre, R. B., Perez, E., and BonDurant, R. H. (1994). Detection of bovine trichomoniasis with a specific DNA probe and PCR amplification system. *J. Clin. Microbiol.* **32,** 98–104.

Hodgson, J. L., Jones, D. W., Widders, P. R., and Corbeil, L. B. (1990). Characterization of *Tritrichomonas foetus* antigens by use of monoclonal antibodies. *Infect. Immun.* **58,** 3078–3083.

Humphrey, J., and Stephens, L. (1983). *Haemophilus somnus:* A review. *Vet. Bull.* **53,** 987–1004.

Humphrey, J. D., Little, P. B., Stephens, L. R., Barnum, D. A., Doig, P. A., and Thorsen, J. (1985). Prevalence and distribution of *Haemophilus somnus* in the male bovine reproductive tract. *Am. J. Vet. Res.* **43,** 792–795.

Ikeda, J. S., BonDurant, R. H., Campero, C. M., and Corbeil, L. B. (1993). Conservation of a protective surface antigen of *Tritrichomonas foetus*. *J. Clin. Microbiol.* **31,** 3289–3295.

Ikeda, J. S., BonDurant, R. H., and Corbeil, L. B. (1995). Bovine vaginal antibody responses to immunoaffinity-purified surface antigen of *Tritrichomonas foetus*. *J. Clin. Microbiol.* **33,** 1158–1163.

Inzana, T. J., Iritani, B., Gogolewski, R. P., Kania, S. A., and Corbeil, L. B. (1988). Purification and characterization of lipooligosaccharides from four strains of *Haemo philus somnus*. *Infect. Immun.* **56,** 2830–2837.

Inzana, T. J., Gogolewski, R. P., and Corbeil, L. B. (1992). Phenotypic phase variation in *Haemophilus somnus* lipooligosaccharide during bovine pneumonia and after *in vitro* passage. *Infect. Immun.* **60,** 2943–2951.

Jimenez de Bagues, M. P., Elzer, P. H., Jones, S. M., Blasco, J. M., Enright, F. M., Schurig, G. G., and Winter, A. J. (1994). Vaccination with *Brucella abortus* mutant RB51 protects BALB/c mice against virulent strains of *B. abortus, melitensis* and *ovis*. *Infect. Immun.* **62,** 4990–4996.

Kania, S. A., Gogolewski, R. P., and Corbeil, L. B. (1990). Characterization of a 78-kilodalton outer membrane protein of *Haemophilus somnus*. *Infect. Immun.* **58,** 237–244.

Kvasnicka, W. B., Taylor, R. E. L., Huang, J.-C., Hanks, D., Tronstad, R. J., Bosomworth, A., and Hall, M. R. (1989). Investigations of the incidence of bovine trichomoniasis in

Nevada and of the efficacy of immunizing cattle with vaccine containing *Tritrichomonas foetus*. *Theriogenology* **31,** 963–971.

Kvasnicka, W. G., Hanks, D., Haung, J.-C., Hall, M. R., Sandblom, D., Chu, H.-J., Chavez, L., and Acree, W. M. (1992). Clinical evaluation of the efficacy of inoculating cattle with a vaccine containing *Tritrichomonas foetus*. *Am. J. Vet. Res.* **53,** 2023–2027.

Kwiecien, J. M., and Little, P. B. (1991). *Haemophilus somnus* and reproductive disease in the cow. *Can. Vet. J.* **32,** 595–601.

Limét, J., Plommet, A.-M., Dubray, G., and Plommet, M. (1987). Immunity conferred upon mice by anti-LPS monoclonal antibodies in murine brucellosis. *Ann. Inst. Pasteur Immunol.* **138,** 417–424.

Miller, R. B., Lein, D. H., McEntee, K. E., Hall, C. E., and Shaw, S. (1983). *Haemophilus somnus* infection of the reproductive tract of cattle: A review. *J. Am. Vet. Med. Assoc.* **182,** 1390–1392.

Montaraz, J. A., Winter, A. J., Hunter, D. M., Sowa, B. A., and Adams, L. G. (1986). Protection against *Brucella abortus* in mice with O-polysaccharide-specific monoclonal antibodies. *Infect. Immun.* **51,** 961–963.

Nansen, P. (1972). Selective immunoglobulin deficiency in cattle and susceptibility to infection. *Acta. Pathol. Microbiol. Scand. Sect.* **80,** 49–54.

Nielsen, K., Duncan, J. R., Stemshorn, B., and Ruckerbauer, G. (1981). Relationship of humoral factors (antibody and complement) to immune responsiveness, resistance and diagnostic serology. *Adv. Exp. Med.* **137,** 367–389.

Ogunnariwo, J. A., Cheng, C. Y., Ford, J. A., and Schryvers, A. B. (1990). Response of *Haemophilus somnus* to iron limitation; expression and identification of a bovine-specific transferrin receptor. *Microb. Pathog.* **9,** 397–406.

Pardon, P. (1977). Resistance against a subcutaneous *Brucella* challenge of mice immunized with living or dead *Brucella* or by transfer of immune serum. *Ann. Immunol. Paris* **128C,** 1025–1037.

Parsonson, I. M., Clark, B. L., and Dufty, J. H. (1976). Early pathogenesis and pathobiology of *Tritrichomonas foetus* infection in virgin heifers. *J. Comp. Pathol.* **86,** 59–66.

Perez, E., Conrad, P. A., Hird, D., Ortuno, A., Chacon, J., BonDurant, R., and Noordhuizen, J. (1992). Prevalence and risk factors for *Trichomonas foetus* infection in cattle in northeastern Costa Rica. *Prev. Vet. Med.* **14,** 155–165.

Pfeifer, C. G., Campos, M., Beskorwayne, T., Babiuk, L. A., and Potter, A. A. (1992). Effect of *Haemophilus somnus* on phagocytosis and hydrogen peroxide production by bovine polymorphonuclear leukocytes. *Microb. Pathog.* **13,** 191–202.

Plommet, M., and Plommet, A.-M. (1983). Immune serum-mediated effects on brucellosis evolution in mice. *Infect. Immun.* **41,** 97–105.

Rae, D. O. (1989). Impact of trichomoniasis on the cow-calf producer's profitability. *J. Am. Vet. Med. Assoc.* **49,** 42–45.

Rhyan, J. C., Stackhouse, L. L., and Quinn, W. J. (1988). Fetal and placental lesions in bovine abortion due to *Tritrichomonas foetus*. *Vet. Pathol.* **25,** 350–355.

Rioux, C. R., Harland, R. J., Theisen, M., Rawlyk, N. H., and Potter, A. A. (1994). Protective capacity of the antigenic 40-kilodalton LppB lipoprotein of *Haemophilus somnus* against experimental infection in cattle. *Haemophilus, Actinobacillus Pasteurella Int. Conf. Abstr.* T6, p. 35.

Samuelson, J. D., and Winter, A. J. (1966). Bovine vibriosis: The nature of the carrier state in the bull. *J. Infect. Dis.* **116,** 581–592.

Schurig, G. D., Hall, C. E., Corbeil, L. B., Duncan, J. R., and Winter, A. J. (1975). Bovine venereal vibriosis: Cure of infection in females by systemic immunization. *Infect. Immun.* **11,** 245–251.

Schurig, G. G., Roop, R. M., Bagchi, T., Buhrman, D., Boyle, S. M., and Sriranganathan, N. (1991). Biological properties of RB51: A stable rough strain of *Brucella abortus*. *Vet. Microbiol.* **28,** 171–188.

Schurig, G. G., Boyle, S. M., and Sriranganathan, N. (1996). *Brucella abortus* vaccine strain RB51: A brief review. *Arch. Vet. Med.* **23,** 19–22.

Skirrow, S. (1987). Identification of trichomonad-carrier cows. *J. Am. Vet. Med. Assoc.* **191,** 553–554.

Skirrow, S. Z., and BonDurant, R. H. (1988). Bovine trichomoniasis. *Vet. Bull.* **58,** 591–603.

Skirrow, S. Z., and BonDurant, R. H. (1990a). Induced *Tritrichomonas foetus* infection in beef heifers. *J. Am. Vet. Med. Assoc.* **196,** 885–889.

Skirrow, S. Z., and BonDurant, R. H. (1990b). Immunoglobulin isotype of specific antibodies in reproductive tract secretions and sera in *Tritrichomonas foetus*-infected heifers. *Am. J. Vet. Res.* **51,** 645–653.

Speer, C. A., and White, M. W. (1991). Bovine trichomoniasis. *Large Anim. Vet.* **46,** 18–20.

Stevens, M. G., Hennager, D. G., Olsen, S. C., and Cheville, N. F. (1994). Serological response in diagnostic test for brucellosis in cattle vaccinated with *Brucella abortus* or RB51. *J. Clin. Microbiol.* **32,** 1065–1066.

Tagawa, Y., Ishikawa, H., and Yuasa, N. (1993a). Purification and partial characterization of the major outer membrane protein of *Haemophilus somnus*. *Infect. Immun.* **61,** 91–96.

Tagawa, Y., Haritani, M., Ishikawa, H., and Yuasa, N. (1993b). Characterization of a heat-modifiable outer membrane protein of *Haemophilus somnus*. *Infect. Immun.* **61,** 1750–1755.

Tagawa, Y., Haritani, M., Ishikawa, H., and Yuasa, N. (1993c). Antigenic analysis of the major outer membrane protein of *Haemophilus somnus* with monoclonal antibodies. *Infect. Immun.* **61,** 2257–2259.

Tagawa, Y., Haritani, M., and Yuasa, N. (1993d). Characterization of an immunoreactive 17.5-kilodalton outer membrane protein of *Haemophilus somnus* by using a monoclonal antibody. *Infect. Immun.* **61,** 4153–4157.

Talbot, J. A., Neilsen, K., and Corbeil, L. B. (1991). Cleavage of proteins of reproductive secretions by extracellular proteinases of *Tritrichomonas foetus*. *Can. J. Microbiol.* **37,** 384–390.

Theisen, M., Rioux, C. R., and Potter, A. A. (1992). Molecular cloning, nucleotide sequence, and characterization of a 40,000-molecular-weight lipoprotein of *Haemophilus somnus*. *Infect. Immun.* **60,** 826–831.

Theisen, M., Rioux, C. R., and Potter, A. A. (1993). Molecular cloning, nucleotide sequence, and characterization of *LppB*, encoding an antigenic 40-kilodalton lipoprotein of *Haemophilus somnus*. *Infect. Immun.* **61,** 1793–1798.

Thomford, J. W., Talbot, J. A., Ikeda, J. S., and Corbeil, L. B. (1996). Characterization of extracellular proteinases of *Tritrichomonas foetus*. *J. Parasitol.* **82,** 112–117.

Ward, A. C. S., and Corbeil, L. B. (1983). A selective medium for Gram-negative pathogens from bovine respiratory and reproductive tracts. *Am. Assoc. Vet. Lab. Diagn.* **26,** 103–112.

Widders, P. R., Paisley, L. G., Gogolewski, R. P., Evermann, J. F., Smith, J. W., and Corbeil, L. B. (1986). Experimental abortion and the systemic immune response in cattle to *Haemophilus somnus*. *Infect. Immun.* **54,** 555–560.

Widders, P. R., Smith, J. W., Yarnall, M., Mcguire, T. C., and Corbeil, L. B. (1988). Nonimmune immunoglobulin binding of *Haemophilus somnus*. *J. Med. Microbiol.* **26,** 307–311.

Widders, P. R., Dorrance, L. A., Yarnall, M., and Corbeil, L. B. (1989a). Immunoglobulin-binding activity among pathogenic and carrier isolates of *H. somnus*. *Infect. Immun.* **57,** 639–642.

Widders, P. R., Dowling, S. C., Gogolewski, R. P., Smith, J. W., and Corbeil, L. B. (1989b). Isotypic antibody responses in cattle infected with *Haemophilus somnus*. *Res. Vet. Sci.* **46,** 212–217.

Wira, C. R., and Rossol, R. M. (1995a). Antigen-presenting cells in the female reproductive tract: Influence of sex hormones on antigen presentation in the vagina. *Immunology* **84,** 505–508.

Wira, C. R., and Rossol, R. M. (1995b). Antigen-presenting cells in the female reproductive tract: Influence of the estrous cycle on antigen presentation by uterine epithelial and stromal cells. *Endocrinology* **136,** 4526–4534.

Yarnall, M., and Corbeil, L. B. (1989). Antibody response to an *Haemophilus somnus* Fc receptor. *J. Clin. Microbiol.* **27,** 111–117.

Yarnall, M., Gogolewski, R. P., and Corbeil, L. B. (1988a). Characterization of two *Haemophilus somnus* Fc receptors. *J. Gen. Microbiol.* **134,** 1993–1999.

Yarnall, M., Widders, P. R., and Corbeil, L. B. (1988b). Isolation and characterization of Fc receptors from *Haemophilus somnus*. *Scand. J. Immunol.* **28,** 129–137.

Yu, S., Gray-Owen, S. D., Ogunnariwo, J., and Schryvers, A. B. (1992). Interaction of ruminant transferrins with transferrin receptors in bovine isolates of *Pasteurella haemolytica* and *Haemophilus somnus*. *Infect. Immun.* **60,** 2992–2994.

Progress and Expectations for Helminth Vaccines

ELS N. T. MEEUSEN AND JILLIAN F. MADDOX

Centre for Animal Biotechnology, School of Veterinary Science, The University of Melbourne, Parkville, Victoria 3052 Australia

I. Introduction
II. Vaccination Using Defined Parasite Antigens
III. Vaccine-Induced Immune Responses
IV. Simulation Models for Host–Parasite Population Dynamics
V. Host Immunity and Population Dynamics of Gastrointestinal Nematode Infections
VI. Development of a Simple Model for Vaccination against *Haemonchus*
VII. Summary
References

I. Introduction

During the last 10 years, our knowledge of the immune responses generated during parasite infections has increased enormously and many studies have been published showing induction, or manipulation, of protection against parasites in experimental animal models. These advances, however, have not yet been translated into effective vaccines for the important internal parasites of livestock. This can in part be attributed to the fact that different immune rejection and immune evasion mechanisms may exist for each host–parasite system. It is therefore important, for the development of effective parasite vaccines, that the major parasite infections are identified for which vaccination would be a commercially acceptable alternative to existing control strategies, and that the vaccination studies are performed in these host–parasite systems. In the United States, dairy and beef cattle are of far greater importance than sheep, while the latter still form a substantial part of the livestock industry in countries such as Australia, New Zealand, the United Kingdom, and South Africa. Internal parasites of major economic importance in the dairy and beef cattle industry are liver fluke (*Fasciola hepatica*) and the gastrointestinal trichostrongylid nematodes, of which *Ostertagia ostertagi* and *Haemonchus placei* are the most pathogenic (Reinemeyer, 1994). The same liver fluke species also infects sheep, and similar trichostrongylid nematodes are major pathogens in the ovine gastrointestinal tract, including *Ostertagia circumcincta* (now reclassified as *Teladorsagia circumcincta*), *Haemonchus contortus,* and *Trichostrongylus* spp. More protection studies against internal parasites have been performed in sheep than in cattle, probably due to the lower cost of sheep as experimental animals, but it would be expected that, due to the close evolutionary relationship between both the sheep and cattle hosts and the parasite species, many of the findings in sheep would be transferable to cattle. However, even between these closely related ruminants, there can be significantly different effects as exemplified by the different responses of sheep and cattle to infection with the same liver fluke species (see below). While there is some evidence that mounting an immune response may impair the productivity of an animal, this may also depend on the mechanism of immunity that occurs in a specific host–parasite interaction because conflicting results have been reported (Kimambo *et al.,* 1988; Bisset *et al.,* 1997).

This article concentrates on several aspects of vaccine development for these major internal parasites of sheep and cattle. Rather than being an exhaustive overview of work published over the years, which has been dealt with in other papers (Newton, 1995; Emery, 1996; Munn, 1997) we aim to highlight some of the special characteristics of the host immune response and the parasite's molecular makeup that we consider to have important implications for vaccine development. In addition, we speculate on the predicted outcomes that the availability of vaccines may have on parasite population dynamics.

II. Vaccination Using Defined Parasite Antigens

Because large-scale culturing of parasitic helminths for the production of live or modified live vaccines is not economically feasible for the majority of helminth infections, it has long been recognized that helminth vaccines will require the production of defined, subunit parasite material produced in a commercially acceptable manner such as by recombinant DNA methods or chemical synthesis. These methods, however, require the identification of the crucial parasite molecules that, together with appropriate adjuvant systems, will induce useful levels of protection in the host animal. Vaccination with whole parasite homogenates has generally been unsuccessful in protecting against helminth infections. Several previous livestock vaccination studies have used various semipurified parasite extracts for experimental trials. The resulting protection levels, if any, have generally been variable, and are not easily reproducible in other laboratories because of the biochemically ill-defined nature of the antigen preparations. While the efforts involved in identifying and purifying particular parasite molecules can be substantial, it is highly desirable for further studies to be performed with well-defined parasite antigens. Identification of the target, protective antigen(s) is the major initial hurdle to vaccine development. Conventional or "natural" antigens are recognized by the host's immune system during natural infection. Only a few of these antigens, however, are likely to play a crucial role in inducing protective immunity. Attempts to identify these "protective" antigens have generally been based on the reactivity of parasite molecules with antibodies circulating in the serum of infected animals. The most recent report in the literature of successful vaccination of sheep against a gastrointestinal nematode used a 31-kDa antigen of *T. circumcincta* identified through its reactivity with immune serum of genetically resistant sheep (McGillivery *et al.*, 1992). However, further extensive studies from the same laboratory have been unable to reproduce these results and the antigen has not been further characterized (Morton *et al.*, 1995). Recently, the search for "natural" protective parasite antigens has been aided by the development of a new antibody tool for antigen identification (Meeusen and Brandon, 1994a,b). Instead of relying on parasite-specific antibodies circulating in the serum of infected animals, this technique isolates parasite-specific antibodies directly from the antibody-secreting B cells (ASCs) induced in the lymph nodes draining the site of parasite rejection. These "ASC probes" provide a much more restricted antibody recognition profile than that found with serum antibodies, reflecting the limited number of parasite antigens present at the specific time and place where rejection is taking place. Using ASC probe technology, a 70- to 90-kDa *H. contortus* antigen (*Hc-sL3*) has been identified as being specifically recognized during a rejection response in *H. contortus* immune sheep (Bowles *et al.*, 1994). The *Hc-sL3* antigen was shown to be present on the surface of L3 larvae and is expressed in a stage-specific manner (Ashman *et al.*, 1995; Raleigh and Meeusen, 1996). Vaccination trials using small amounts of purified *Hc-sL3* antigen have shown significant levels of protection to *H. contortus* infection in vaccinated sheep resulting in 50 and 65% reductions in worm burden and egg counts, respectively (Meeusen, 1996; Jacobs et al., 1998). This technology has also been applied to liver fluke infections using a rat model, because these animals develop a pronounced immunity to reinfection with *F. hepatica.* Three distinct antigens were identified using ASC probes during rejection of liver fluke infection in immune rats, one of which was present on the surface of the infective juvenile fluke (Meeusen and Brandon, 1994a; Tkalcevic et al., 1996). Monoclonal antibodies against these antigens have been generated but, so far, none of the antigens has been tested in vaccine trials. One interesting characteristic common to all the parasite antigens so far identified during the early stages of infection is that each of the antigens is only expressed for a limited time (2–3 days) by the parasite and is absent from subsequent infection stages. Protection induced by these antigens would therefore be expected to act on the early infective stages of the parasite, before they reach the liver in the case of *F. hepatica* or moult to the blood-sucking adult stage in *H. contortus* infection. Antigens identified as potential vaccine candidates using ASC probes have so far been shown to be protective in helminth (Jacobs et al., 1998), ectoparasite (Bowles et al., 1996), and bacterial (Walker et al., 1994) infections.

An alternative approach to conventional vaccination using "natural" protective antigens, extensively pursued in recent years, is to identify and isolate parasite molecules that may be crucial to parasite infectivity or survival in the host and use these molecules to immunize the host. The resulting immune response is then expected to neutralize the activity of these molecules, thereby killing or severely affecting the viability of the parasite. The choice of these molecules, which are generally enzymes, is more based on their presumed functional activity rather than on their interaction with the host's immune response. In many cases, no immune response is generated against these molecules during natural infection in which case they are also referred to as "novel" or "hidden" antigens. One such antigen for which no or very poor antibody responses are generated during natural infection is glutathione *S*-transferase (GST) of liver fluke, thought to be important in detoxification reactions. Vaccination of sheep with liver fluke GST has been shown to cause a reduction in parasite burden in approximately half of the immunized sheep (Sexton et al., 1990), while significant reductions in both fluke numbers and weight were obtained in most vaccinated cattle (Morrison et al., 1996). Another target for liver fluke vaccines is the cysteine proteases. Cysteine proteases are an ubiquitous group of enzymes in many parasites contributing to parasite migration into host tissues, digestion of proteins for nutrients, and immune evasion strategies. Vaccination with purified liver fluke cysteine proteases resulted in significant reductions in worm fertility in sheep (Wijffels *et al.*, 1994), while in cattle both worm numbers and egg viability were significantly affected by vaccination, especially when the cysteine protease was administered in combination with liver fluke hemoglobin (Dalton *et al.*, 1996). The cysteine proteases used in these vaccination trials are immunogenic during natural infection and antibody levels are boosted after infection of vaccinated animals (Dalton *et al.*, 1996).

For nematode parasites, the most pronounced reductions in worm burdens have been achieved using gut-derived molecules as vaccines, with the best documented example being the aminopeptidase, H11, from adult *H. contortus* worms (Newton, 1995; Munn, 1997). Immunization with native H11 has been shown to confer between 60 and 96% reductions in worm burden in different breeds of sheep. Analogous aminopeptidase molecules are also present in *Trichostrongylus* and *Ostertagia* parasites but protection studies in these species have not been published. Because these two nematode species are not predominant bloodfeeders, host antibodies will have less access to gut antigens, and protection of this type of vaccination is expected to be less efficient than for *H. contortus*. H11 and other gut-derived antigens are generally considered to be "hidden" antigens, although a slight boosting of the antibody response has been reported after infection of vaccinated sheep.

III. Vaccine-Induced Immune Responses

A revolution in our understanding of the initiation and control of various types of immune responses has come about with the identification of the various leukocyte regulatory molecules or cytokines and their differential production during infection (Sherr and Coffman, 1992; Mosmann and Sad, 1996). While knowledge of this area is at present still expanding dramatically, major implications for vaccine research, especially with respect to parasite vaccines have become apparent (Meeusen, 1996). It had long been recognized that helminth-induced immune responses are distinct from those of other infectious agents, especially in their induction of distinct effector cells (mast cells and eosinophils) and antibody isotypes (IgE). While the exact mechanisms of parasite rejection have not been elucidated, it is likely that some of these specialized components are involved in this process. The induction of these parasite-specific mechanisms has now been found to be under the control of distinct cytokines including interleukin 4 (IL-4) and IL-5, generally referred to as Th2-type cytokines. In the case of intestinal parasites, a clear correlation between induction of Th2-type cytokines and protection has been established in mouse models (Grencis, 1993). The induction of Th2-type cytokines through vaccination is therefore an important consideration for vaccine research that aims to mimic the natural protective response through immunization with purified antigens (Meeusen, 1996). The particular adjuvant used in the vaccine preparation plays a critical role in directing the type of immune response induced against the vaccine antigen (Cox and Coulter, 1997). The importance of adjuvants for the induction of immunity against helminths through vaccination is exemplified by the vaccine trials with the *Hc-sL3 H. contortus* antigen identified during a natural rejection response in sheep. In this case, protection was only achieved when aluminium hydroxide was used as the adjuvant and no protection was observed when Freund's complete adjuvant (FCA) or Quil A were used in the vaccine preparation (Jacobs *et al.*, 1998; Turnbull *et al.*, 1992). In the past FCA has been used mainly as the experimental adjuvant due to its strong induction of both cellular and humoral immune responses. The aluminum hydroxide adjuvant, on the other hand, typically induces much lower antibody responses than Quil A or FCA but has been reported to be the best inducer of Th2-type cytokine responses including eosinophilia (Cox and Coulter, 1997). It has been postulated that eosinophil-dependent killing, mediated by antibodies specific to the *Hc-sL3* antigen, may be an important mechanism in resistance during naturally induced and vaccine-induced immunity to larval challenges (Rainbird *et al.*, 1998).

In contrast to vaccination with natural antigens, the immune response induced against functional of "hidden" antigens does not need to have any relevance to natural immunity. The levels of protection against *H. contortus* afforded by vaccination with native H11 gut antigen have been reported to correlate significantly with antibody levels and inhibition of enzyme activity by the antibodies (Newton, 1995; Munn, 1997). So far, all of the published trial data in this system have been derived from experiments performed with FCA as the adjuvant. FCA has also been the adjuvant used in most of the liver fluke trials; however, no correlation between protection and antibody levels induced by vaccination were found in either the GST or cysteine protease immunization trials (Sexton *et al.*, 1990; Dalton *et al.*, 1996). It is possible that, in the liver fluke system, the partial weakening of the worms by vaccination against functional molecules of the fluke may allow the natural immune response of the host to become effective.

This would explain why protection with similar molecules seems to be more protective in cattle than in sheep, because cattle demonstrate increased resistance to natural infection compared to sheep. In more recent liver fluke vaccine studies, different adjuvants have been tested in vaccination trials in cattle using GST from *F. hepatica* (Morrison et al., 1996). Significant protection has been reported with either Quil A of poly-D-lactide coglycolide microspheres when used in combination with squalene Montanide 80. It is clear that, for commercial application, FCA would be unacceptable due to its known side effects, and more studies using commercially acceptable adjuvant are required to be able to fully evaluate the effect of vaccination with functional or "hidden" antigens. Aluminium hydroxide-based adjuvants have been widely used in human and veterinary vaccines but are unlikely to generate the type and level of response required for "functional" antigen vaccines.

IV. Simulation Models for Host–Parasite Population Dynamics

One way to examine the possible effects of vaccines that differ in their targets and efficacy on host–parasite population dynamics is by the development of simulation models. In general, there are two main uses for simulation models. The first is to gain a better understanding of the system being modeled by testing the assumptions on which the model is based, and the second is for predicting the consequences of different scenarios. A good model can be used to elicit theoretical outcomes rapidly and cheaply for a variety of hypothetical control strategies under different conditions. A simple overview of the processes involved in building simulation models is given next.

Prior to building a simulation model the assumptions and constraints underlying the system being modeled need to be stipulated. This leads to the development of influence diagrams and flowcharts to show the relationships between the various elements of the system that are considered to be important. The next step involves the development of mathematical formulas that describe the various functions within the model. Often these formulas are derived from other studies or, if such information does not exist, from intuition based on experience of the system being modeled. Many biological systems can be extremely complex, hence it is important for the model to focus on key issues rather than utilizing all the information that is available for the various aspects of the system. Therefore, when developing a host–parasite model to compare the outcomes of different parasite control regimes based on host treatment, such as drenching or vaccinating, minimal attention should be paid to parts of the system relating to parasite stages external to the host such as pasture management regimes or alterations in climate. It is assumed that these external stages will behave in a consistent fashion for all of the host treatment regimes. However, if the model will also be required to compare combinations of host treatment and external control measures then variables representing the external control measures will also need to be incorporated. Once a model has been built, it is tested with real data, if such data are available, or simulated data to check that it produces sensible outcomes. The model is then modified in an iterative fashion, by the inclusion of additional parameters and the modification of formulas, to increase the success of the model in explaining real situations or expected outcomes. Finally the model is used to examine theoretical systems and to test the impact on the dynamics of the system of parameter variation. At this point a number of real trials should be performed to determine how well the model predicts reality.

In the past there have been two major barriers to building detailed simulation models for vaccination of grazing livestock against helminth parasites. The first of these was that the tools needed to perform such modeling were not widely available or easy to use. This problem has largely been overcome by the enormous increase in affordable computing power, together with the development of user-friendly, powerful simulation software on a variety of hardware platforms. In addition, courses on modeling systems dynamics are now taught at many universities. The major obstacle that now prevents the construction of good models for vaccination of livestock against helminth parasites is the lack of knowledge about some aspects of livestock-parasite interactions. The main targets of the models that have been developed to date have pertained to (1) examining the effects of drenching as a control measure, both from the perspective of controlling parasite numbers in the host and development of resistance to anthelmintics by the parasite, and (2) variation in the number of infective larvae on pasture in response to changing pasture conditions and climate (Barnes et al., 1988; Gettinby et al., 1989; Barnes and Dobson, 1990; Echevarria et al., 1993; Coyne and Smith, 1994). In contrast, few models have been developed to examine the effects of vaccination of livestock against helminth parasites. A number of models have been generated to examine the development of natural host immunity to gastrointestinal nematodes (Coyne and Smith, 1994; Smith, 1988; Dobson et al., 1990; Barnes and Dobson, 1993) and these can be used as a basis for modeling different vaccination strategies. One drawback of many of the older models is that they use seasonal data on the levels of pasture contamination as inputs rather than deriving the pasture contamination from the model itself. Some of the more recent models do incorporate this link (Leathwick *et al.*, 1992). The consequences of the reduction in pasture contamination as a result of vaccination will be of importance when investigating the effects of vaccination.

Although many factors influence host–parasite relationships, the underlying feature of many of the important gastrointestinal nematode parasite infections of grazing livestock is basically a positive feedback loop. The host ingests the infective stage of the parasite from the pasture; this stage matures and then produces new infective forms on the pasture that are subsequently ingested by the host (Fig. 1). Without intervention this would lead to an exponential increase in the parasite burden of an animal. However, this cycle is normally con-

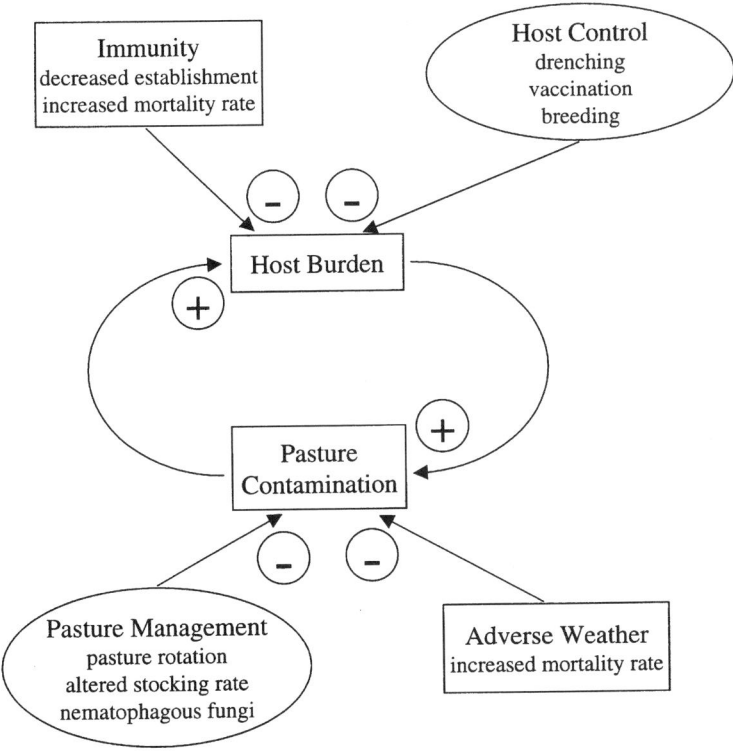

FIG. 1. Diagram of influences on helminth infections in livestock.

trolled, and can be manipulated, by a number of factors affecting both the parasite stages that are within the host and those that are on the pasture. These include the development of immunity to the parasite by the host, the treatment of the host with drenches that kill virtually all parasites within the host at the time of drenching, and the effects of both climate and pasture management on the pasture stages of the parasite.

The population dynamics of both host and parasite populations need to be considered when developing models to assess the effects of vaccinating against helminth parasites of livestock. The duration of immunity needed will differ for livestock kept for different purposes, with only relatively short-term immunity needed for meat lambs as compared to the longer term immunity that will be needed for breeding and wool-producing animals. The effects of combining vaccination with other control strategies such as drenching or pasture management will also need to be considered, particularly as some drenches have been shown to have immunosuppressive effects (Stankiewicz *et al.,* 1996). Hence different models will need to be constructed for different purposes. Because we are focusing on highlighting special characteristics of the host's immune response to parasites, only those factors pertinent to the effects of host immunity on parasite population dynamics are discussed below.

V. Host-Immunity and Population Dynamics of Gastrointestinal Nematode Infections

A large number of studies have shown that the immunity of the host is influenced by the host's age, genotype, and sex, by the number of parasites that it has been exposed to, and by its nutritional status. Two major host population subsets are most susceptible to gastrointestinal nematode parasites. These are young animals that have not yet developed an acquired immunity against the parasites, in particular young male animals, and periparturient females whose immunity to the parasite is reduced. The periparturient rise in helminth parasite eggs shed by the dams is often a major source of infection for young animals (Donald and Waller, 1973) and is also the principal means for the major livestock nematode parasites to survive during adverse weather conditions. Although young animals are separated from their dams at weaning and placed on relatively clean pasture, they have usually commenced pasture consumption before this time and hence have already acquired worm burdens. These burdens result in contamination of the "clean" pasture and consequently are an important source of infection for the weaned animals.

A common property of helminth infections is that of overdispersion of the parasite within the host population. That is, the majority of the host population parasite burden is carried by only a small proportion of hosts. This situation is characterized by a high variance to mean ratio, and is commonly represented by a negative binomial function defined by two parameters, a mean and an exponent k. The hosts within a population with higher burdens are assumed to be those that have genotypes that render them more susceptible to the parasite. Another feature of livestock helminth infections is that a threshold of exposure is needed before immunity develops. It is not known whether this exposure threshold is higher in the livestock that are genetically susceptible, or whether these animals just mount a poorer immune response. It has also been found that animals classified as resistant in response to infection at a young age are also more resistant in the periparturient period (Woolaston, 1992; Barger, 1993).

In addition to the general properties of host–parasite relationships for gastrointestinal nematodes described above, each host–parasite interaction is characterized by a number of features that are particular to the species involved. It is likely that the relative importance of different host immune mechanisms will vary for different parasites, and the methods by which the parasite attempts to evade the host immune response will also vary between species. For example, the relative impacts of hypobiosis of L4 larvae and the dramatic effects of "self-cure" differ for *Ostertagia* and *Haemonchus* infections of sheep.

VI. Development of a Simple Model for Vaccination against *Haemonchus*

A simple model for predicting the effects of vaccination against *H. contortus* has been adapted from the model of Coyne and Smith (1994) for *H. contortus* infection in grazing lambs. This model contains the following assumptions: (1) constant fecundity of female worms; (2) decreases in parasite establishment (immune exclusion) and changes in the morality rate of established worms as a consequence of the host's exposure to infection; (3) no arrested larvae; and (4) constant mortality and maturation rates for free-living stages. Simulation is started by setting both an initial pasture contamination level and an initial host parasite burden. The subsequent contamination level of the pasture is mainly determined by the number of eggs deposited onto the pasture by the host. This in turn determines the infection rate of the host. Control treatments such as vaccination, using vaccines of differing efficacy, and drenching can be superimposed onto the model to investigate possible outcomes (Fig. 2). The vaccination data incorporated in the model are based on the known effects of experimental vaccines, with L3 larvae being targeted by the *Hc-sL3* vaccine and adult worms by the H11 vaccine, and on hypotheses about the level and duration of host immunity invoked by the prototype vaccine that can be tested in

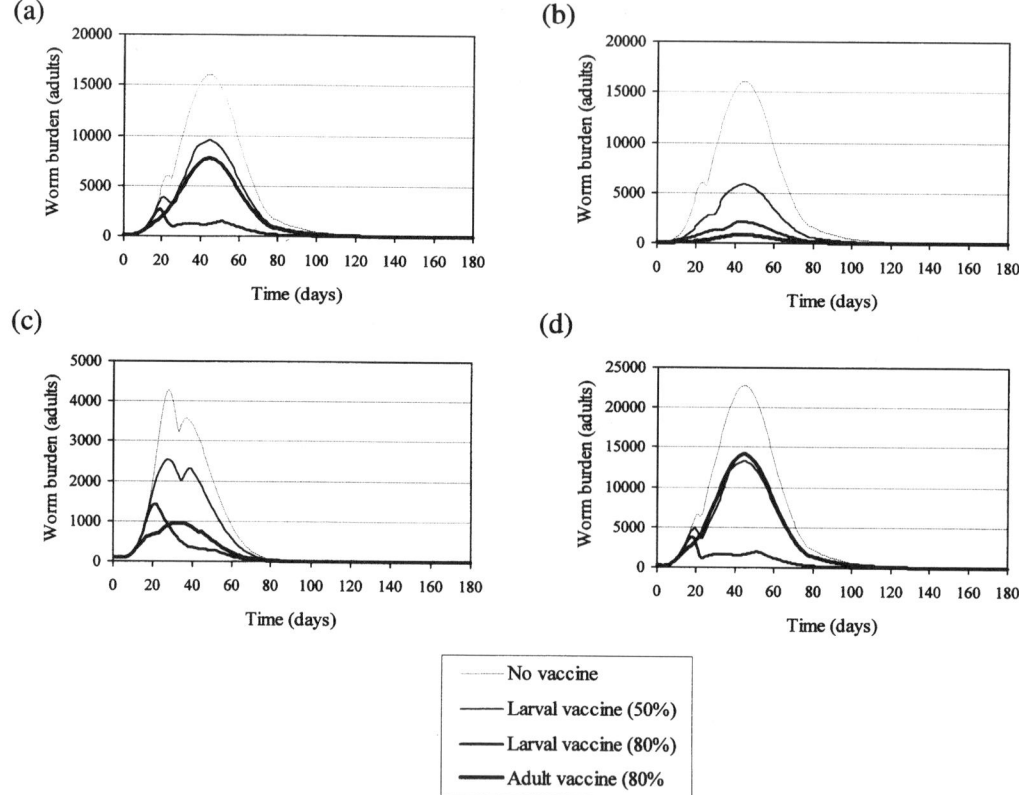

FIG. 2. Simulation of effects of vaccination against *H. contortus* in a grazing population of lambs. A larval vaccine giving 50 or 80% protection against larval forms and an adult vaccine giving 80% protection against adult forms are incorporated into the model: (a) initial worm burden = 100, vaccination after weaning, low pasture adversity; (b) initial worm burden = 100, vaccination prior to weaning, low pasture adversity; (c) initial worm burden = 100, vaccination after weaning, moderate pasture adversity; and (d) initial worm burden = 200, vaccination after weaning, low pasture adversity.

further vaccination trials. It is assumed that the immunity induced through a combination of vaccination and natural exposure lasts throughout the grazing season for lambs. This is likely to be true when a natural antigen vaccine, such as *Hc-sL3*, is used where the response induced by vaccination is boosted by natural infection. It is less likely to be true for "hidden" antigen vaccines where no significant boosting is expected. The use of "hidden" antigen vaccines may therefore require additional strategies such as slow-release devices for antigen delivery to maintain high antibody responses throughout this time period. As expected, the model predicts that the use of vaccines that confer higher levels of immunity will have greater effects than those conferring lower levels of immunity. It is also clear from this simple model that the effect of vaccination procedures against larvae and adults will have different dynamics on worm populations. Further models could test the effect of a combined larval/adult vaccine and of combining vaccination with different drenching or pasture rotation strategies. Note that the behavior of this model is constrained by the formulas it uses that relate both immune exclusion (modeled as a declining sigmoidal function) and parasite mortality (modeled as a linear function) to cumulative host experience of infective larvae and that these formulas are likely to need modification. The usefulness of this model could be increased by incorporating elements representing a number of other features of host–parasite interactions such as aggregated parasite distributions, hypobiosis, altered duration of immunity, and the possibility of "self-cure."

VII. Summary

The large amount of scientific progress made in the last 5 years has allowed a more rational approach to the design of nematode vaccines to develop. Successful experimental trials have been published using two different approaches, one aiming to boost acquired host immunity through vaccination with natural immunogens, the other affecting parasite viability by targeting parasite molecules crucial for nutrition or survival in the host. The individual or combined action of these two vaccination procedures will need to be evaluated with respect to their potential effects on animal health and productivity in the field. To this effect, more data are required concerning the level and duration of immunity of the vaccine-induced protection using acceptable adjuvant systems. In addition, the age at which vaccination is effective and the effect of vaccination on highly susceptible or temporarily immunosupressed individuals will need to be considered. In the case of gastrointestinal nematodes, the level of pasture contamination with infective larvae is dependent on the worm burdens in the host animal and, in turn, affects the buildup of natural resistance in the host. An appreciation of these complex interactive factors is best achieved through computer simulation models using the powerful simulation software that has recently become available. Further animal trials will need to be performed to establish the necessary data to incorporate into the models and to adapt the model outcomes to the trial results. These epidemiologic and simulation studies should be pursued in parallel with vaccine development so that a better appreciation is gained of the requirements of a successful commercial vaccine.

REFERENCES

Ashman, K., Mather, J., Wiltshire, C., Jacobs, H. J., and Meeusen, E. (1995). Isolation of a larval surface glycoprotein from *Haemonchus contortus* and its role in immune evasion. *Mol. Biochem. Parasitol.* **70,** 175–179.

Barger, I. A. (1993). Influence of sex and reproductive status on susceptibility of ruminants to nematode parasitism. *Int. J. Parasitol.* **23,** 463–469.

Barnes, E. H., and Dobson, R. J. (1990). Population dynamics of *Trichostrongylus colubriformis* in sheep: Computer model to simulate grazing systems and the evolution of anthelmintic resistance. *Int. J. Parasitol.* **20,** 823–831.

Barnes, E. H., and Dobson, R. J. (1993). Persistence of acquired immunity to *Trichostrongylus colubriformis* in sheep after termination of infection. *Int. J. Parasitol.* **23,** 1019–1026.

Barnes, E. H., Dobson, R. J., Donald, A. D., and Waller, P. J. (1988). Predicting populations of *Trichostrongylus colubriformis* infective larvae on pasture from metereological data. *Int. J. Parasitol.* **18,** 767–774.

Bisset, S. A., Vlassoff, A., West, C. J., and Morrison, L. (1997). Epidemiology of nematodosis in Romney lambs selectively bred for resistance or susceptibility to nematode infection. *Vet. Parasitol.* **70,** 255–269.

Bowles, V. M., Meeusen, E. N. T., Chandler, K., Verhagen, A., Nash, A. D., and Brandon, M. R. (1994). The immune response of sheep infected with larvae of the sheep blowfly *Lucilia cuprina* monitored via efferent lymph. *Vet. Immunol. Immunopathol.* **172,** 71–76.

Bowles, V. M., Meeusen, E. N. T., Young, A. R., Nash, A. D., Andrews, A. E., and Brandon, M. R. (1996). Vaccination of sheep against larvae of the sheep blowfly (*Lucilia cuprina*). *Vaccine* **14,** 1347–1352.

Cox, J. C., and Coulter, A. R. (1997). Adjuvants—a classification and review of their modes of action. *Vaccine* **15,** 248–256.

Coyne, M. J., and Smith, G. (1994). *In* "Parasitic and Infectious Diseases: Epidemiology and Ecology" (M. E. Scott and G. Smith, eds.), pp. 235–247. Academic Press, San Diego, CA.

Dalton, J. P., McGonigle, S., Rolph, T. P., and Andrews, S. J. (1996). Induction of protective immunity in cattle against infection with *Fasciola hepatica* by vaccination with cathepsin L proteinases and with hemoglobin. *Infect. Immun.* **64,** 5066–5074.

Dobson, R. J., Waller, P. J., and Donald, A. D. (1990). Population dynamics of *Trichostrongylus colubriformis* in sheep: Model to predict the worm population over time as a function of infection rate and host age. *Int. J. Parasitol.* **20,** 365–373.

Donald, A. D., and Waller, P. J. (1973). Gastro-intestinal nematode parasite populations in ewes and lambs and the origin and time course of infective larval availability in pastures. *Int. J. Parasitol.* **3,** 219–233.

Echevarria, F. A. M., Gettinby, G., and Hazelwood, S. (1993). Model predictions for anthelmintic resistance among *Haemonchus contortus* populations in southern Brazil. *Vet. Parasitol.* **47,** 315–325.

Emery, D. L. (1996). Vaccination against worm parasites of animals. *Vet. Parasitol.* **64,** 31–45.

Gettinby, G., Soutar, A., Armour, J., and Evans, P. (1989). Anthelmintic resistance and the control of ovine ostertagiasis: A drug action model for genetic selection. *Int. J. Parasitol.* **19,** 369–376.

Grencis, R. K. (1993). Cytokine-mediated regulation of intestinal helminth infections: The *Trichuris muris* model. *Ann. Trop. Med. Parasitol.* **87,** 643–647.

Jacobs, H. J., Wiltshire, C., Ashman, K., and Meeusen, E. N. T. (1998). Vaccination against the gastrointestinal nematode, *Haemonchus contortus,* using a purified larval surface antigen. *Vaccine* (in press).

Kimambo, A. E., MacRae, J. C., and Dewey, P. J. S. (1988). The effect of daily challenge with *Trichostrongylus colubriformis* larvae on the nutrition and performance of immunologically resistant sheep. *Vet. Parasitol.* **28,** 205–212.

Leathwick, D. M., Barlow, N. D., and Vlassoff, A. (1992). A model for nematodiasis in New Zealand lambs. *Int. J. Parasitol.* **22,** 789–799.

McGillivery, D. J., Yong, W. K., Adler, B., and Rifkin, G. G. (1992). A purified stage-specific 31 kDa antigen as a potential protective antigen against *Ostertagia circumcincta* infection in lambs. *Vaccine* **10,** 607–613.

Meeusen, E., and Brandon, M. R. (1994a). The use of antibody-secreting cell probes to reveal tissue-restricted immune responses during infection. *Eur. J. Immunol.* **24,** 469–474.

Meeusen, E. N. T., and Brandon, M. R. (1994b). Antibody secreting cells as specific probes for antigen identification. *J. Immunol. Methods* **172,** 71–76.

Meeusen, E. N. T. (1996). Rational design of nematode vaccines: Natural antigens. *Int. J. Parasitol.* **26,** 813–818.

Morrison, C. A., Colin, T., Sexton, J. L., Bowen, F., Wicker, J., Friedel, T., and Spithill, T. W. (1996). Protection of cattle against *Fasciola hepatica* infection by vaccination with glutathione S-transferase. *Vaccine* **14,** 1603–1612.

Morton, R. E., Yong, W. K., Riffkin, G. G., Bozas, S. E., Spithill, T. W., Adler, B., and Parsons, J. C. (1995). Inability to reproduce protection against *Teladorsagia circumcincta* in sheep with a purified stage-specific 31 kDa antigen complex. *Vaccine* **13,** 1482.

Mosmann, T. R., and Sad, S. (1996). The expanding universe of T-cell subsets: Th1, Th2 and more. *Immunol. Today* **17,** 138–146.

Munn, E. A. (1997). Rational design of nematode vaccines–hidden antigens. *Int. J. Parasitol.* **27,** 359–356.

Newton, S. E. (1995). Progress on vaccination against *Haemonchus contortus*. *Int. J. Parasitol.* **25,** 1281–1289.

Rainbird, M. A., Macmillan, D., and Meeusen, E. N. T. (1998). Eosinophil-mediated killing of Haemonchus contortus larvae: Effect of eosinophil activation and role of antibody, complement and IL-5. *Parasite Immunol.* **20,** 93–103.

Raleigh, J. M., and Meeusen, E. N. T. (1996). Developmental expression of *Haemonchus contortus* antigens. *Int. J. Parasitol.* **26,** 673–675.

Reinemeyer, C. R. (1994). Parasitisms of dairy and beef cattle in the United States. *J. Am. Vet. Med. Assoc.* **205,** 670–680.

Sexton, J. L., Milner, A. R., Panaccio, M., Waddington, J. Wijffels, G., Chandler, D., Thompson, C., Wilson, L., Spithill, T. W., Mitchell, G. F., and Campbell, N. J. (1990). Glutathione S-transferase. Novel vaccine against *Fasciola hepatica* infection in sheep. *J. Immunol.* **145,** 3905–3910.

Sherr, A. F., and Coffman, R. L. (1992). Regulation of immunity to parasites by T-cells and T-cell-derived cytokines. *Annu. Rev. Immunol.* **10,** 385–409.

Smith, G. (1988). The population biology of the parasitic stages of *Haemonchus contortus*. *Parasitology* **96,** 185–190.

Stankiewicz, M., Cabaj, W., Pernthaner, A., Jonas, W., and Rabel, B. (1996). Drug-abbreviated infections and development of immunity against *Trichostrongylus colubriformis* in sheep. *Int. J. Parasitol.* **26,** 97–103.

Tkalcevic, J., Brandon, M. R., and Meeusen, E. N. T. (1996). *Fasciola hepatica:* Rapid switching of stage-specific antigen expression after infection. *Parasite Immunol.* **18,** 139–147.

Turnbull, I. F., Bowles, V. M., Wiltshire, C. J., Brandon, M. R., and Meeusen, E. N. T. (1992). Immunization of sheep with surface antigens from *Haemonchus contortus* larvae. *Int. J. Parasitol.* **22,** 537.

Walker, J., Jackson, H., Eggleton, D., Meeusen, E. N. T., and Brandon, M. R. (1994). Identification of a novel antigen from *Corynebacterium pseudotuberculosis* that protects sheep against caseous lymphadenitis. *Infect. Immun.* **62,** 2562–2567.

Wijffels, G. L., Salvatore, L., Dosen, M., Waddington, J. Wilson, L., Thompson, C., Campbell, N., Sexton, J., Wicker, J., Bowen, F., Friedel, T., and Spithill, T. W. (1994). Vaccination of sheep with purified cysteine proteinases of *Fasciola hepatica* decreases worm fecundity. *Exp. Parasitol.* **78,** 132–148.

Woolaston, R. R. (1992). Selection of merino sheep for increased and decreased resistance to *Haemonchus contortus:* Peri-parturient effect on faecal egg counts. *Int. J. Parasitol.* **22,** 947–953.

Vaccines and Diagnostic Methods for Bovine Mastitis: Fact and Fiction

ROBERT J. YANCEY, JR.

Animal Health Biological Discovery, Pfizer Central Research, Groton, Connecticut 06340

I. Introduction
II. Vaccines for Contagious Pathogens
   A. *Staphylococcus aureus*
   B. *Streptococcus agalactiae*
   C. Other Contagious Pathogens
III. Vaccines for Environmental Mastitis Pathogens
   A. *Streptococcus uberis*
   B. Clinical Coliform Mastitis
IV. Diagnostic Methods
V. Summary
   References

I. Introduction

Bovine mastitis is the most costly disease to animal agriculture in the United States and much of the world (Bramley et al., 1996). The U.S. economic loss is estimated to be approximately $185/cow annually and the overall yearly cost of mastitis has been estimated in the billions of dollars. These losses are primarily due to lower milk yields, reduced milk quality, and higher production costs.

A number of problems are uniquely associated with vaccination of dairy cows for mastitis (Anderson, 1978; Colditz and Watson, 1985; Yancey, 1993). One of these is that mastitis, an inflammation of the mammary gland, is usually an immune response of the gland to invading microorganisms; that is, the disease is the immune response. Therefore, specific enhancement of the immune response through vaccination might also exacerbate the disease. In addition, due to the large volume of the udder there is dilution of the immune components such as specific immunoglobulin, lymphocytes, and phagocytes. Furthermore, the enormous surface area of the mammary epithelium makes immune surveillance difficult. Also, it is well documented that milk components have an inhibitory effect on phagocytic cells, making them less able to kill invading bacteria. Moreover, for many bacteria, with the notable exception of *Streptococcus uberis* (discussed later) the milk is an excellent growth medium. Another of these problems is that the number of mastitis pathogens are numerous and heterogeneous. Watts (1988) estimated that there are more than 135 agents of bovine mastitis. Bovine mastitis is almost always caused by bacteria.

Regardless of the numbers of different etiologic agents of mastitis, the majority of infections are caused by staphylococci, streptococci, and gram-negative bacilli. These are categorized as either contagious pathogens (e.g., *Staphylococcus aureus, Streptococcus agalactiae,* and *Streptococcus dysgalactiae*) or environmental pathogens (e.g., the coliforms and *S. uberis*). Vaccine efforts for bovine mastitis have concentrated mainly on these more common contagious and environmental pathogens.

In addition to the problems associated with vaccination, the success of a mastitis vaccine is difficult to define. Should success be based on reduction in severity or frequency of clinical symptoms, prevention of new intramammary infections (IMIs), or elimination of existing IMIs (spontaneous cure rates)? Ideally, a successful vaccine would do all of these, but even the most successful mastitis vaccine, the core antigen vaccine for clinical coliform mastitis, only accomplishes a reduction in severity and frequency of mastitis. These and other issues are discussed in this review.

II. Vaccines for Contagious Pathogens

A. *Staphylococcus aureus*

Although *S. aureus* bacterins have been commercially available for many years, efficacy has not been proven in large, independent field trials, the data from which have been published in peer-reviewed journals. One such commercial bacterin composed of a lysate of five isolates of different bacteriophage types was found to be effective in challenge models. Williams *et al.,* (1975) found that vaccinated cows had lower clinical scores, lower somatic cell counts (SCCs), and developed fewer cases of chronic staphylococcal mastitis than did nonvaccinated cows.

TABLE I

EFFICACY OF S. AUREUS BACTERINS IN AN EXPERIMENTAL CHALLENGE MODEL[a]

Treatment	Number			Spontaneous cures (%)
	Cows	Quarters	New IMI	
Bacterin	10	40	33	24 (73)[b]
Protein A	10	40	29	24 (83)[b]
Nonvaccinated	10	37	30	14 (47)

[a]Data from Pankey et al., 1985.
[b]$p < 0.025$ compared to nonvaccinated controls.

With this same vaccine, there was a significant increase in the number of spontaneous cures in challenge trials by Pankey et al., (1985), although the rate of clinical mastitis and new IMIs was unaffected through three lactations (Table I). In addition, in this trial the SCC was significantly lower for the vaccinated group compared to the control group, even though milk production was not affected.

Although not published in a peer-reviewed journal, Pankey et al., (1983) also found this same bacterin was effective at increasing spontaneous cure rates compared to nonvaccinated controls in a field trial in three herds in New Zealand (Table II). Again, there was no difference

TABLE II

EFFICACY OF VARIOUS S. AUREUS VACCINES IN FIELD TRIALS

Study (reference)	Number of cows/herds	Statistically significant ($p \leq 0.05$) reduction in:			
		Mastitis	New IMI	Old IMI	SCC
Pankey et al. (1983)	201/3	N	N	Y	NR
Warson and Schwartzkoff (1990)	582/5	Y	Y	NR	NR
Watson et al. (1996)	1819/7	N(Y)[a]	NR	NR	NR
Nordhaug et al. (1994a)	108/16	N	N	NR	N
Giraudo et al. (1997)	30/1	Y	Y	NR	N
Calzolari et al. (1997)	164/2	Y	Y	NR	Y

Key: N, no; Y, yes; NR, not reported.
[a]Significant reduction in the high prevalence herd.

in new IMIs. The results of these studies suggest that while bacterins may not be efficacious when clinical mastitis or a new IMI is the primary criterion, vaccination with certain bacterins may provide a management tool to reduce the frequency of subclinical infections within a herd.

Inclusion of staphylococcal toxoids and capsular or pseudocapsular materials in *S. aureus* vaccines has been a common strategy for a number of years. Watson and his colleagues in Australia were the first to demonstrate the importance of inducing opsonic IgG_2 subclass antibody in protection against *S. aureus*. Watson showed that a vaccine composed of killed *S. aureus* that had been cultured under conditions that simulated *in vivo* growth and induced a pseudocapsule, combined with staphylococcal toxoids, and a mineral oil/dextran sulfate adjuvant, provided statistically significant protection from clinical mastitis in dairy heifers experimentally challenged with different strains of *S. aureus* (Watson, 1992). Nickerson *et al.*, (1993), confirmed that Watson's vaccine was effective in reducing new IMI in experimental challenge studies. When this vaccine was tested in a blinded trial involving 582 cows from five commercial dairies in Australia, the incidence of clinical mastitis and new IMI due to *S. aureus* was significantly reduced (Watson and Schwartzkoff, 1990; Table II). In a more recent, larger field trial (1819 cows, 7 herds), Watson's group found that, although the number of clinical and subclinical mastitis cases due to *S. aureus* was lower in the vaccinated groups (45 cases in vaccinates, 67 in controls), these differences were not statistically significant across all herds (Watson *et al.*, 1996; Table II). In the herd with the highest prevalence of staphylococcal mastitis, however, vaccination did significantly reduce the incidence of clinical and subclinical mastitis due to *S. aureus* (Watson *et al.*, 1996).

In another recent placebo-controlled, multicenter field trial involving 108 heifers, a vaccine was used that contained killed whole cells with pseudocapsule and the α- and β-toxoids in a mineral oil adjuvant (Nordhaug *et al.*, 1994a; Table II). While the incidence of clinical and subclinical *S. aureus* mastitis was numerically lower in the vaccinated group, 8.6 versus 16.0% for the control group, these differences were not statistically significant. The rates of *S. aureus* IMI were also similar among groups. While high concentrations of IgG_2 isotype antibody to the pseudocapsule were obtained in the serum of these cows, the milk IgG_2 concentrations were not significantly different from nonvaccinates. α-Toxin neutralizing IgG_1 isotype antibodies were found in both the milk and serum. The IgG_1 neutralizing titer to β-toxin was significantly greater in the serum but not the milk, compared to nonvaccinates (Nordhaug et al., 1994b).

A "combination vaccine" was recently evaluated in three commercial dairies in Argentina with a high incidence of staphylococcal mastitis (Calzolari et al., 1997; Giraudo et al., 1997; Table 2). This vaccine was composed of two different strains of S. aureus, a crude capsular extract of S. aureus, and one strain each of S. uberis and S. agalactiae, in an aluminum hydroxide adjuvant. The vaccine provided statistically significant protection from clinical and subclinical mastitis due to S. aureus in both heifers and cows compared to adjuvant-vaccinated controls. In addition, there was a significant reduction in IMI due to S. aureus. There was no impact on IMI or mastitis due to Streptococcus spp. provided by this vaccine. It will be interesting to determine whether this vaccine will be effective against S. aureus in larger field trials.

Vaccines for S. aureus based on chemically linked conjugates of the type 5 capsular polysaccharide (CP5) or surface proteins such as the fibronectin-binding proteins and protein A have also been the subject of active investigation (Yancey, 1993). A conjugate of CP5 with ovalbumin administered in Freund's incomplete adjuvant to cows resulted in production of IgG_2 isotype antibodies to CP5 (Gilbert et al., 1994). This is the antibody subtype in cows which has been reported to be important for opsonic activity. In a recent report the authors chemically conjugated CP5 to α-toxin (Herbelin et al., 1997). Lactating cows immunized SC with α-toxin, α-toxin and CP5, or α-toxin coupled to CP5, in Freund's complete adjuvant, showed enhanced immune recruitment of activated neutrophils into the mammary gland. Vaccination of dairy cows with protein A in Freund's complete adjuvant was found to increase the rate of spontaneous cures and to reduce the SCC (Table I), although no significant differences were noted in new IMI, milk production, or clinical mastitis (Pankey et al., 1985). Although fibronectin has never been conclusively shown to have a role in colonization of the mammary gland by S. aureus, the fibronectin-binding receptor on the surface of S. aureus has been investigated as a potential vaccine for mastitis (Flock, 1992; Mamo et al., 1994; Nelson et al., 1991). Recombinantly produced fusion proteins of the fibronectin-binding receptor reduced the severity of S. aureus-induced mastitis in the mouse model and the incidence of clinical mastitis in a small number of dairy cows. While these results with capsule and surface receptors are exciting, none of these experimental vaccines have been shown to provide protection in field trials.

It is not obvious why the most recently tested bacterin (Calzolari et *al.*, 1997; Giraudo *et al.*, 1997) should be more effective than the pseudocapsule/toxoid-containing bacterins or experimental vaccines discussed earlier. It may be simply that all of these *S. aureus* vaccines are more effective statistically, in herds with a high prevalence of *S. aureus* infection. It might also be that selection of the strains of *S. aureus* or the adjuvants used with these vaccines were important variables in these studies. Sutra and Poutrel (1994) have suggested that a vaccine for *S. aureus*-induced mastitis should contain capsular polysaccharide of the predominant bovine serotypes (5 and 8) coupled to purified surface proteins (e.g., the fibronectin receptor or protein A) and/or α- and β-toxin. Based on additional studies (Colditz and Watson, 1985; Herbelin *et al.*, 1997), these antigens should be provided in an adjuvant and/or by a route of administration that induces IgG_2 opsonic antibody and provides for rapid recruitment of immune-activated neutrophils. Nonetheless, it appears that *S. aureus* vaccines require more work before they are ready for large-scale application.

B. *Streptococcus agalactiae*

For *S. agalactiae*, vaccine research has been less prodigious. Immunization of cows with formalin-killed *S. agalactiae* cells has not been protective (Calzolari *et al.*, 1997; Giraudo *et al.*, 1997; Mackie *et al.*, 1983). Poutrel, Rainard, and their colleagues have studied a surface antigen that is present on many bovine streptococci, the X-protein, as a potential protective immunogen (Rainard and Poutrel, 1991; Rainard *et al.*, 1994, 1995). They found that antibody raised in cows to this antigen was opsonic for *S. agalactiae* (Rainard *et al.*, 1994, 1995) and enhanced phagocytosis of *S. agalactiae*. Also, these researchers found that the group B capsular polysaccharide, when conjugated to a protein such as ovalbumin, induced opsonic antibody (Rainard, 1992). X-protein may also be coupled with the group B capsular polysaccharide to enhance the protective response to both immunogens (Rainard and Poutrel, 1991). However, no efficacy trial data in cows have been published yet with those antigens as vaccines.

C. Other Contagious Pathogens

Little success has been reported with vaccination against other contagious agents such as *S. dysgalactiae* or *Mycoplasma* spp. While *S. dysgalactiae* is usually classified as an environmental pathogen, effective control of this agent by teat disinfection and dry cow therapy makes it more properly classed as a contagious pathogen (Bramley,

1997; Oliver et al., 1997). While a number of potential virulence factors of this streptococcal species have been identified that might serve as protective antigens (Oliver et al., 1997), few vaccination studies have been published (Stark and Norcross, 1970). A formalin-killed vaccine of *Mycoplasma bovis* was found not to prevent infection or milk production changes by *M. bovis* upon experimental challenge of lactating cows. The quarters of the vaccinated animals did resolve infections sooner than the control cows, although the SCC of these cows remained high and milk production was severely reduced in the challenged quarters (Boothby et al., 1986a,b).

III. Vaccines for Environmental Mastitis Pathogens

A. *Streptococcus uberis*

With *S. uberis*, previous exposure does provide resistance to infection, at least with the homologous strain (Hill, 1988). Nonetheless, simple bacterins fail to provide protection in the field against this important environmental pathogen (Calzolari et al., 1997; Giraudo et al., 1997).

Studies at the Institute for Animal Health in Compton, United Kingdom, have shown that both killed and especially live vaccines provide protection against experimental challenge; however, protection occurred against challenge only with the homologous strain (Finch et al., 1994, 1997; Leigh, 1997). The investigators concluded that protection did not correlate with the presence of either milk or serum opsonizing antibody and that the appearance of neutrophils in the milk (increased SCC) did not reduce the numbers of *S. uberis* in a quarter as it does for coliform mastitis. Leigh (1997) hypothesized that interference with bacterial colonization or growth within the gland may have provided the protection from clinical mastitis seen in these studies. To further evaluate this hypothesis, the Compton group has investigated the role of the plasminogen activator, PauA, as a protective antigen for *S. uberis*-induced mastitis. PauA is an extracellularly secreted enzyme of approximately 30,000 kDa which is specific for bovine and ovine plasminogen. The production of this enzyme and activation of plasminogen to plasmin in the milk may facilitate growth of *S. uberis* by providing to this auxotrophic organism necessary amino acids and peptides. (Leigh, 1994; Fig. 1).

Leigh (1997) recently reported challenge-infection experiments in a small number of lactating cows using 100 µg of partially purified PauA

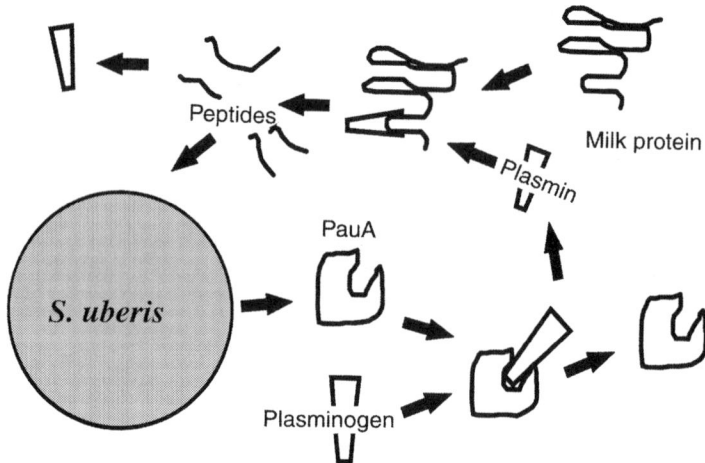

FIG. 1. Possible role of PauA for *S. uberis* in the milk. PauA, which is released from *S. uberis* into the milk as a soluble enzyme, activates bovine plasminogen to plasmin. Plasmin cleaves milk proteins such as casein providing peptides and amino acids necessary for growth of *S. uberis*.

as the antigen (Table III). PauA was administered with either incomplete Freud's adjuvant (IFA) or an experimental adjuvant containing an oil-in-water emulsion (SB62). Cows were vaccinated subcutaneously (SC) either five times (IFA) or two times (SB62) and challenged a few weeks after final vaccination with virulent *S. uberis* in one or two of four quarters of each cow. In the SB62 adjuvanted group, significant protection against clinical mastitis was obtained and the majority of quarters cleared the infections. Since the PauA was isolated from a different strain than the challenge strain, the partially purified PauA provided heterologous strain protection in this trial. It was recently determined that the *pauA* genes from two strains of *S. uberis*, one from the United States and one from the United Kingdom, had over 99% sequence identity, suggesting that this antigen is highly conserved across strains (E. L. Rosey, R. J. Yancey, and J. A. Leigh, unpublished data, 1997). Whether highly purified PauA will provide protection in natural infection trials remains to be determined.

B. Clinical Coliform Mastitis

One of the major advances in the control of environmental mastitis in the United States has been the introduction of core antigen vaccines

TABLE III

EFFECT OF VACCINATION WITH PARTIALLY PURIFIED PauA ON *S. UBERIS*-INDUCED CLINICAL MASTITIS IN MID-LACTATION DAIRY COWS[a]

Group	Adjuvant	Number of cows	Quarters with: Clinical mastitis	Quarters with: Spontaneous cures	Percent protection
Control	None	5	9/9	0/9	0
PauA	IFA[b]	4	5/8	3/8	37.5
PauA	SB62[c]	4	3/8	5/8	62.5[d]

[a]Data from Leigh, 1997.
[b]IFA, incomplete Freund's adjuvant, 100 μg PauA per dose, five SC immunizations.
[c]SB62, experimental adjuvant, 100 μg PauA per dose, two SC immunizations.
[d]$p \leq 0.05$ based on reduction of clinical mastitis compared to controls.

for clinical coliform mastitis (CCM). The core antigen vaccines are bacterins composed of *Escherichia coli* or *Salmonella* spp. strains which have lesions in their ability to produce complete lipopolysaccharide (LPS), resulting in rough (R) mutants. The most studied of these strains is *E. coli* J5, a genetically stable Rc mutant of *E. coli* O111:B4, and Re-17, a rough mutant of *Salmonella typhimurium*. These organisms have an exposed LPS core region that induces antibody that is cross-reactive with most gram-negative bacteria, especially during periods of active growth (Tyler *et al.*, 1990). Core antigen bacterins provide protection against the multiple serotypes of *E. coli* as well as other genera and species of gram-negative bacteria which cause CCM. Several field studies have shown these core antigen bacterins to be effective in reducing the number of cases of CCM in herds (Cullor, 1991, 1993; Hogan *et al.*, 1990, 1992a; González *et al.*, 1989, 1995; McClure *et al.*, 1994; Table IV). The mean percentage reduction of CCM from five field trials using J5 bacterins and one trial using a *S. typhimurium* Re-17 bacterin was 55% (Table IV). If the two-dose *Salmonella* bacterin study was eliminated from this analysis, the mean reduction provided by J5 bacterins was 65%. Whether the Re-17 core vaccine is less efficacious than the J5 bacterins has yet to be shown in published trial data.

It was observed that while the core antigen vaccines do not always reduce the incidence of new coliform IMI at calving (there was a reduction in new coliform IMI in the first year, but not year 2 of the study), the percentage of quarters which develop CCM when IMI occur was significantly lower in the vaccinated cows (Hogan *et al.*, 1990, 1992a).

TABLE IV

EFFICACY OF CORE ANTIGEN VACCINES AT REDUCING CLINICAL COLIFORM MASTITIS (CCM) IN FIELD TRIALS

Study (reference)	Vaccine type	Group	No. of doses	No. of cows	No. of cases of CCM	Incidence (%)	Reduction (%)
Gonzáles et al. (1989)	J5[a]	Control	0	227	29	12.8	80
		Vaccinates	3	233	6	2.6	
Cullor (1991)	J5	Control	3	229	25	10.9	70
		Vaccinates	3	212	7	3.3	
Cullor (1993)	J5	Control	3	421	48	11.4	65
		Vaccinates	3	424	17	4.0	
Hogan et al. (1990, 1992a)	J5	Control	0	112	9	8.0	78
		Vaccinates	3	113	2	1.8	
Gonzáles et al. (1995)	J5	Control	0	60	27	45.0	72
		Vaccinates	3	180	23	12.8	
McClure et al. (1994)	Re-17[b]	Control	0	646	90	13.9	41
		Vaccinates	2	646	53	8.2	
MEAN	Core	Control	1	283	38	13.4	55
		Vaccinates	2.8	301	18	6.0	

[a]*E. coli* strain J5.
[b]*Salmonella typhimurium* strain Re-17.

The exact mechanism whereby vaccination with core antigen vaccines protects dairy cows has yet to be determined.

A partial budget analysis of vaccinating cows against CCM with the J5 vaccine concluded that when >1% of cow lactations resulted in CCM, increased profits of $57/cow lactation could be obtained (De-Graves and Fetrow, 1991). This resulted in a 1700% return on investment from the J5 vaccination program as used in this model.

In contrast to the results of field trials, challenge trials have not been as successful in demonstrating the efficacy of J5 vaccines. Hill (1991) found that vaccination with J5 would not protect cows from a virulent strain of *E. coli*. Others have since demonstrated that challenge models can be used to demonstrate differences between vaccinated and nonvaccinated animals, but the differences between groups are usually small (Hogan et al., 1992b, 1995). It was also found in one study that changes in management practices can minimize differences afforded by J5 vaccination (Hogan et al., 1992a). Additionally, a recent study showed that vaccination of cows during lactation can cause a significant, short-term reduction in milk production (Musser and Anderson,

1996). Most of the manufacturers of core antigen vaccines, however, recommend vaccinating cows during the dry period, a time when milk production would not be affected. While the core antigen vaccines are not a magic bullet for CCM, they do provide the dairy producer with a valuable tool.

IV. Diagnostic Methods

For diagnosis of clinical mastitis, the eyes and hands of the milker are a sensitive enough tool to assess milk quality and udder appearance. However, for subclinical mastitis, the mainstay diagnostic methods are assessment of the SCC and bacterial culture (Bramley et al., 1996; Kitchen, 1981; Mackie, 1994).

The SCC is assessed at the bulk tank and at the cow level. At cow side, the SCC can be estimated by detergent-based tests such as the California Mastitis Test or the Wisconsin Mastitis Test. More quantitative SCCs are commonly obtained by automated electronic cell counting methods such as those employed by the Dairy Herd Improvement Association (DHIA) (Bramley et al., 1996; Kitchen, 1981). Detection of subclinical mastitis and early detection of clinical mastitis has been facilitated by measuring changes in electrical conductivity of the foremilk. While the methodology and equipment are not yet ready for commercial application, this early detection procedure has been shown to enhance cure rates and reduce the time required for return to normal milk when coupled with appropriate antimicrobial therapy (Milner et al., 1996, 1997). Various other assays have been used to identify subclinical mastitis and the NAGase test (Kitchen, 1981; Mattila et al., 1986) and chloride ion concentration tests have been the more commonly used of these assays (Bramley et al., 1996; Kitchen, 1981; Mackie, 1994). However, these assays at present are primarily suited to a research setting.

Pathogen diagnosis can rely on standard procedures (Harmon et al., 1990) or a commercial identification system (Watts and Yancey, 1994). To rapidly detect coliform mastitis the chromogenic limulus amebocyte lysate test, sold commercially as the Limast® test, has been used primarily in Europe (Hakogi et al., 1989; Keefe and Leslie, 1997). Recently, a rapid bacteriologic test system, the HyMast® test, was introduced to rapidly allow decisions for clinical mastitis therapy (Jansen et al., 1997; Keefe and Leslie, 1997). This test consists of a vial with a screw-top cap to which is attached a paddle with media selective for coliforms or gram-positive bacteria embedded on either side. The vial is filled with an aseptically collected milk sample, the cap with its paddle is replaced, and the vial is inverted to inoculate the media. The milk is discarded and the vial with its inoculated paddle is incubated for 8–12 hours and again for 24 hours, at which times it is bacteriologically evaluated. Growth on the gram-positive selective side indicates that the quarter should be treated immediately with an appropriate gram-positive active, commercial antibiotic therapy. Growth on the coliform selective side or no growth indicates no antimicrobial therapy should be used (Jensen et al., 1997).

An ELISA assay, the Prostaph 1™ test has been used for diagnosis of staphylococcal mastitis through detection of S. aureus-specific antibody. The sensitivity and specificity of the test has been determined to be 60–73% and 61–97%, respectively, depending on the study (Hicks et al., 1994; Snoep et al., 1995; Watts et al., 1992). While the test is convenient to screen for S. aureus in a herd, it is most accurate at detecting uninfected rather than infected animals (Watts et al., 1992).

Various typing systems have been used to identify isolates. These systems have been used as valuable tools in epidemiologic studies for mastitis pathogens. Biotyping, phage typing, DNA fingerprinting, and 16S rRNA probes have been used to differentiate or follow strains of mastitis pathogens (Aarestrup and Jensen, 1996; Hill and Brady, 1989; Hill and Leigh, 1989; Hillerton, 1997; Jayarao et al., 1991). However, again, these assays at present are primarily suited to a research setting.

V. Summary

A number of problems are uniquely associated with vaccination of dairy cows for mastitis. One of these is that the number of mastitis pathogens is numerous and heterogeneous. Vaccine efforts have concentrated mainly on the major mastitis pathogens. While at least one S. aureus bacterin has been commercially available for a number of years, no large-scale, independent field trials have been published in refereed journals which support the efficacy of this vaccine. Experimental vaccines for S. aureus composed of pseudocapsule-enriched bacterins supplemented with α- and/or β-toxoids appear promising, but none of these has been commercialized. With S. uberis, some protection against homologous strain challenges was reported recently with a live strain and a bacterin, but other data from the same laboratory showed this vaccine would not protect against heterologous challenge strains. At this time there is only one highly effective vaccine for mastitis, the core-antigen vaccine for coliform mastitis. All of the commercially available vaccines for this indication are bacterins of rough mutants of *E. coli* strain J5 or *Salmonella* spp. Preliminary success with an experimental vaccine based on the plasminogen activator of *S. uberis* is a very different approach for a mastitis vaccine. Little success has been reported with vaccination against other mastitis pathogens. For diagnostic methods, the high somatic cell count, as measured by direct count or indirect assays, remains the cornerstone of mastitis diagnosis. However, for subclinical mastitis, bacterial cell culture is a reliable diagnostic method. Pathogen identification may rely on older biochemical testing methods or newer commercial identification systems, depending on the laboratory budget. ELISA assays also have been used to assess herd infection status. Epidemiologic studies have used DNA fingerprinting and ribotyping, but none of these methods has yet produced an easily utilized commercial format. Within the next decade, additional efficacious vaccines for several of the most common agents for bovine mastitis are likely. A review written at that time then can be more fact than fiction.

References

Aarestrup, F. M., and Jensen, N. E. (1996). Genotypic and phenotypic diversity of *Streptococcus dysgalactiae* strains isolated from clinical and subclinical cases of bovine mastitis. *Vet. Microbiol.* **53**, 315–323.

Anderson, J. C. (1978). The problem of immunization against staphylococcal mastitis. *Br. Vet. J.* **134**, 412–420.

Boothby, J. T., Jasper, D. E., and Thomas, C. B. (1986a). Experimental intramammary inoculation with *Mycoplasma bovis* in vaccinated and unvaccinated cows: Effect on milk production and milk quality. *Can. J. Vet. Res.* **50**, 200–204.

Boothby, J. T., Jasper, D. E., and Thomas, C. B. (1986b). Experimental intramammary inoculation with *Mycoplasma bovis* in vaccinated and unvaccinated cows: Effect on the mycoplasmal infection and cellular inflammatory response. *Cornell Vet.* **76**, 188–197.

Bramley, A. J. (1997). Environmental streptococci: summary and issues. In "Udder Heath Management for Environmental Streptococci," pp. 95–103. National Mastitis Council, Madison, WI.

Bramley, A. J., Cullor, J. S., Erskine, R. J., Fox, L. K., Harmon, R. J., Hogan, J. S., Nickerson, S. C., Oliver, S. P., Smith, K. L., and Sordillo, L. M. (1996). "Current Concepts of Bovine Mastitis," 4th ed. National Mastitis Council, Madison, WI.

Calzolari, A., Giraudo, J. E., Rampone, H., Odierno, L., Giraudo, A. T., Frigerio, C., Bettera, S., Raspanti, C., Hernández, J., Wehbe, M., Mattea, M., Ferrari, M., Larriestra, A., and Nagel, R. (1997). Field trials of a vaccine against bovine mastitis. 2. Evaluation in two commercial dairy herds. *J. Dairy Sci.* **80**, 854–858.

Colditz, I. G., and Watson, D. L. (1985). The immunophysiological basis for vaccinating ruminants against mastitis. *Aust. Vet. J.* **62**, 145–153.

Cullor, J. S. (1991). The *Escherichia coli* J5 vaccine: Investigating a new tool to combat coliform mastitis. *Vet. Med.* **86,** 836–844.
Cullor, J. S. (1993). J5 *Escherichia coli:* A core antigen vaccine for coliform mastitis. *In* "1993 Coliform Mastitis Symposium," Fox, L. K., and Gay, C. C., eds., pp. 30–35. Veterinary Learning Systems, Washington State University, Pullman.
DeGraves, F. J., and Fetrow, J. (1991). Partial budget analysis of vaccinating dairy cattle against coliform mastitis with an *Escherichia coli* J5 vaccine. *J. Am. Vet. Med. Assoc.* **199,** 451–455.
Finch, J. M., Hill, A. W., Field, T. R., and Leigh, J. A. (1994). Local vaccination with killed *Streptococcus uberis* protects the bovine mammary gland against experimental intramammary challenge with the homologous strain. *Infect. Immun.* **62,** 3599–3603.
Finch, J. M., Winter, A. Walton, A. W., and Leigh, J. A. (1997). Further studies on the efficacy of a live vaccine against mastitis caused by *Streptococcus uberis*. *Vaccine* **15,** 1138–1143.
Flock, J.-I. (1992). Vaccination against *S. aureus* infections based on a fibronectin binding protein. *Abstr., Int. Sympo. Staphylococci Staphylococcal Infect. 7th.* p. 58.
Gilbert, F. B., Poutrel, B., and Sutra, L. (1994). Immunogenicity in cows of *Staphylococcus aureus* type 5 capsular polysaccharide-ovalbumin conjugate. *Vaccine* **12,** 369–374.
Giraudo, J. E., Calzolari, A., Rampone, H., Rampone, A., Giraudo, A. T., Bogni, C., Larriestra, A., and Nagel, R. (1997). Field trials of a vaccine against bovine mastitis. 1. Evaluation in heifers. *J. Dairy Sci.* **80,** 845–853.
González, R. N., Cullor, J. S., Jasper, D. E., Farver, T. B., Bushnell, R. B., and Oliver, N. N. (1989). Prevention of clinical coliform mastitis in dairy cows by a mutant *Escherichia coli* vaccine. *Can. J. Vet. Res.* **53,** 301–305.
González, R. N., Wilson, D. J., and Sears, P. M. (1995). Efficacy of an *Escherichia coli* J5 bacterin to prevent clinical coliform mastitis under three immunization schedules. *In* "Third IDF International Mastitis Seminar," S-4, Saren, A., and Soback, S., eds., pp. 64–68. M. Lachmann Printers Ltd., Haifa, Israel.
Hakogi, E., Tamura, H., Tanaka, S., Kohata, A., Shimada, Y., and Tabuchi, K. (1989). Endotoxin levels in milk and plasma of mastitis-affected cows measured with a chromogenic limulus test. *Vet. Microbiol.* **20,** 267–274.
Harmon, R. J., Eberhart, R. J., Jasper, D. E., Langlois, B. E., and Wilson, R. A. (1990). "Microbiological Procedures for the Diagnosis of Bovine Udder Infection," 3rd ed. National Mastitis Council, Arlington, VA.
Herbelin, C., Poutrel, B., Gilbert, F. B., and Rainard, P. (1997). Immune recruitment and bactericidal activity of neutrophils in milk of cows vaccinated with staphylococcal α-toxin. *J. Dairy Sci.* **80,** 2025–2034.
Hicks, C. R., Eberhart, R. J., and Sischo, W. M. (1994). Comparison of microbiologic culture, an enzyme-linked immunosorbent assay, and determination of somatic cell count for diagnosing *Staphylococcus aureus* mastitis in dairy cows. *J. Am. Vet. Med. Assoc.* **204,** 255–260.
Hill, A. W. (1988). Protective effect of previous intramammary infection with *Streptococcus uberis* against subsequent clinical mastitis in the cow. *Res. Vet. Sci.* **44,** 386–387.
Hill, A. W. (1991). Vaccination of cows with rough *Escherichia coli* mutants fails to protect against experimental intramammary bacterial challenge. *Vet. Res. Commun.* **15,** 7–16.
Hill, A. W., and Brady, C. A. (1989). A note on the isolation and propagation of lytic phages from *Streptococcus uberis* and their potential for strain typing. *J. Appl. Bacteriol.* **67,** 425–431.

Hill, A. W., and Leigh, J. A. (1989). DNA fingerprinting of *Streptococcus uberis:* A useful tool for epidemiology of bovine mastitis. *Epidemiol. Infect.* **103,** 165–171.
Hillerton, J. E. (1997). Detection and identification of environmental streptococcal infections. In "Udder Health Management for Environmental Streptococci," pp. 51–58. National Mastitis Council, Madison, WI.
Hogan, J. S., Smith, K. L., Todhunter, D. A., and Schoenberger, P. S. (1990). Efficacy of *Escherichia coli* J5 vaccine for preventing coliform mastitis. In "International Symposium on Bovine Mastitis," pp. 200–204. National Mastitis Council, Arlington, VA.
Hogan, J. S., Smith, K. L., Todhunter, D. A., and Schoenberger, P. S. (1992a). Field trial to determine efficacy of an *Escherichia coli* J5 mastitis vaccine. *J. Dairy Sci.* **75,** 78–84.
Hogan, J. S., Weiss, W. P., Todhunter, D. A., Smith, K. L., and Schoenberger, P. S. (1992b). Efficacy of an *Escherichia coli* J5 mastitis vaccine in an experimental challenge trial. *J. Dairy Sci.* **75,** 415–422.
Hogan, J. S., Weiss, W. P., Smith, K. L., Todhunter, D. A., Schoenberger, P. S., and Sordillo, L. M. (1995). Effects of an *Escherichia coli* J5 vaccine on mild clinical coliform mastitis. *J. Dairy Sci.* **78,** 285–290.
Jensen, J., Leslie, K., and Kelton, D. (1997). Utilizing and evaluating the Hymast test on dairy farms. In "1997 Regional Meeting of the National Mastitis Council," pp. 1–9. National Mastitis Council, Madison, WI.
Jayarao, B. M., Oliver, S. P., Tagg, J. R., and Matthews, K. R. (1991). Genotypic and phenotypic analysis of *Streptococcus uberis* isolated from bovine mammary secretions. *Epidemiol. Infect.* **107,** 543–555.
Keefe, G., and Leslie, K. (1997). Therapy protocols for environmental streptococcal mastitis. In "Udder Health Management for Environmental Streptococci," pp. 75–86. National Mastitis Council, Madison, WI.
Kitchen, B. J. (1981). Review of the progress of dairy science: Bovine mastitis: Milk compositional changes and related diagnostic tests. *J. Dairy Sci.* **48,** 167–188.
Leigh, J. A. (1994). Purification of a plasminogen activator from *Streptococcus uberis*. *FEMS Microbiol. Lett.* **118,** 153–158.
Leigh, J. A. (1997). Progress in the development of vaccines against environmental streptococcal mastitis. In "Udder Health Management for Environmental Streptococci," pp. 59–74. National Mastitis Council, Madison, WI.
Mackie, D. P. (1994). Monoclonal antibody technology: applications in veterinary science. *Biotechnol. Adv.* **12,** 703–710.
Mackie, D. P., Pollock, D. A., Meneely, D. J., and Logan, E. F. (1983). Clinical features of consecutive intramammary infections with *Streptococcus agalactiae* in vaccinated and non-vaccinated heifers. *Vet. Rec.* **112,** 472–476.
Mamo, W., Jonsson, P., Flock, J.-I., Lindberg, M., Müller, H.-P., Wadstrom, T., and Nelson, L. (1994). Vaccination against *Staphylococcus aureus* mastitis: Immunological response of mice vaccinated with fibronectin binding protein (FnBP-A) to challenge with *S. aureus*. *Vaccine* **12,** 988–992.
Mattila, T., Pyörälä, S., and Sandholm, M. (1986). Comparison of milk antirypsin, albumin, N-acetyl-β-D-glucosaminidase, somatic cells and bacteriological analysis as indicators of bovine subclinical mastitis. *Vet. Res. Commun.* **10,** 113–124.
McClure, A. M., Christopher, E. E., Wolff, W. A., Fales, W. H., Krause, G. F., and Miramonti, J. (1994). Effect of Re-17 mutant *Salmonella typhimurium* bacterin toxoid on clinical coliform mastitis. *J. Dairy Sci.* **77,** 2272–2280.
Milner, P., Page, K. L., Walton, A. W., and Hillerton, J. E. (1996). Detection of clinical

mastitis by changes in electrical conductivity of foremilk before visible changes in milk. *J. Dairy Sci.* **79**, 83–86.

Milner, P., Page, K. L., and Hillerton, J. E. (1997). The effects of early antibiotic treatment following diagnosis of mastitis detected by a change in the electrical conductivity of milk. *J. Dairy Sci.* **80**, 859–863.

Musser, J. M. B., and Anderson, K. (1996). Effect of vaccination with an *Escherichia coli* bacterin-toxoid on milk production in dairy cattle. *J. Am. Vet. Med. Assoc.* **209**, 1291–1293.

Nelson, L., Flock, J.-I., Höök, M., Lindberg, M., Müller, H. P., and Wadstrom, T. (1991). Adhesions in staphylococcal mastitis as vaccine components. *Flem. Vet. J.* **62**, (Suppl. 1), 111–125.

Nickerson, S. C., Owens, W. E., and Boddie, R. L. (1993). Effect of a *Staphylococcus aureus* bacterin on serum antibody, new infection, and mammary histology in nonlactating dairy cows. *J. Dairy Sci.* **76**, 1290–1297.

Nordhaug, M. L., Nesse, L. L., Norcross, N. L., and Gudding, R. (1994a). A field trial with an experimental vaccine against *Staphylococcus aureus* mastitis in cattle: 1. Clinical parameters. *J. Dairy Sci.* **77**, 1267–1275.

Nordhaug, M. L., Nesse, L. L., Norcross, N. L., and Gudding, R. (1994b). A field trial with an experimental vaccine against *Staphylococcus aureus* mastitis in cattle: 2. Antibody response. *J. Dairy Sci.* **77**, 1276–1284.

Oliver, S. P., Calvinho, L. F., and Almeida, R. A. (1997). Characteristics of environmental *Streptococcus* species involved in mastitis. In "Udder Health Management for Environmental Streptococci," pp. 1–35. National Mastitis Council, Madison, WI.

Pankey, J. W., Duirs, G., Murray, G., and Twomey, A. (1983). Evaluation of a commercial bacterin against *Staphylococcus aureus* in New Zealand. In "Dairy Research Report 1983," pp. 157–161. Hill Farm Research Station, Homer, LA.

Pankey, J. W., Boddie, N. T., Watts, J. L., and Nickerson, S. C. (1985). Evaluation of protein A and a commercial bacterin as vaccines against *Staphylococcus aureus* mastitis by experimental challenge. *J. Dairy Sci.* **68**, 726–731.

Rainard, P. (1992). Isotype antibody response in cows to *Streptococcus agalactiae* group B polysaccharide-ovalbumin conjugate. *J. Clin. Microbiol.* **30**, 1856–1862.

Rainard, P., and Poutrel, B. (1991). Immunization against mastitis: A practical goal? *Flem. Vet. J.* **62**, (Suppl. 1), 141–149.

Rainard, P., Sarradin, P., and Poutrel, B. (1994). Phenotypic variability of X-protein expression by mastitis-causing *Streptococcus agalactiae* of serotype NT/X and opsonic activities of specific antibodies. *Microb. Pathog.* **16**, 359–372.

Rainard, P., Sarradin, P., and Poutrel, B. (1995). Bovine antibodies to X-protein promote phagocytic killing of *Streptococcus agalactiae*. In Third IDF International Mastitis Seminar," S-1, Saren, A., and Soback, S., eds., pp. 82–83. M. Lachmann Printers Ltd. Haifa, Israel.

Snoep, J. J., Sampimon, O. C., Bloemert, J., and van der Mei, J. (1995). Evaluation of an ELISA test to detect *Staphylococcus aureus* antibodies in bovine milk samples. In Third IDF International Mastitis Seminar," S-2, Saren, A., and Soback, S., eds., pp. 28–32. M. Lachmann Printers, Ltd. Haifa, Israel.

Stark, D. M., and Norcross, N. L. (1970). Response of bovine immunization against *Streptococcus dysgalactiae*. *Cornell Vet.* **60**, 604–612.

Sutra, L., and Poutrel, B. (1994). Virulence factors involved in pathogenesis of bovine intramammary infections due to *Staphylococcus aureus*. *J. Med. Microbiol.* **40**, 79–89.

Tyler, J. W., Cullor, J. S., Spier, S. J., and Smith, B. P. (1990). Immunity targeting common core antigens of gram-negative bacteria. *J. Vet. Intern. Med.* **4**, 17–25.

Watson, D. L. (1992). Vaccination against experimental staphylococcal mastitis in dairy heifers. *Res. Vet. Sci.* **53,** 346–353.

Watson, D. L., and Schwartzkoff, C. L. (1990). A field trial to test the efficacy of a staphylococcal mastitis vaccine in commercial dairies in Australia. *In* "International Symposium on Bovine Mastitis," pp. 73–76. National Mastitis Council, Arlington, VA.

Watson, D. L., McColl, M. L., and Davies, H. I. (1996). Field trial of a staphylococcal mastitis vaccine in dairy herds: Clinical, subclinical and microbiological assessments. *Aust. Vet. J.* **74,** 447–450.

Watts, J. L. (1988). Etiologic agents of bovine mastitis. *Vet. Microbiol.* **16,** 41–66.

Watts, J. L., and Yancey, R. J., Jr. (1994). Identification of veterinary pathogens by the use of commercial identification systems and new trends in antimicrobial susceptibility testing of veterinary pathogens. *Clin. Microbiol. Rev.* **7,** 346–356.

Watts, J. L., Owens, W. E., Ray, C. H., and Washburn, P. J. (1992). Evaluation of the Prostaph 1™ test for detection of *Staphylococcal aureus* intramammary infections in dairy cattle. *Agri-Practice* **13,** 31–34.

Williams, J. M., Shipley, G. R., Smith, G. L., and Gerber, D. L. (1975). A clinical evaluation of *Staphylococcus aureus* bacterin in the control of staphylococcal mastitis in cows. *Vet. Med. / Small Anim. Clin.* May, pp. 587–594.

Yancey, R. J. (1993). Recent advances in bovine vaccine technology. *J. Dairy Sci.* **76,** 2418–2436.

T-Cell Responses and the Influence of Dendritic Cells in Cattle

C. J. HOWARD,* R. A. COLLINS,* P. SOPP,* G. P. BROOKE,* L. S. KWONG,* K. R. PARSONS,* V. WEYNANTS,† J.-J. LETESSON,† AND G. P. BEMBRIDGE*

Institute for Animal Health, Compton, Newbury RG20 7NN, United Kingdom, †Immunology Unit, Facultes Universitaires, Notre Dame de la Paix, Namur, Belgium

I. Introduction
II. Identification of the Major T-Cell Populations in Cattle
III. Role of Different T-Cell Populations *in Vivo*
IV. Identification of Subpopulations of CD4 and CD8 T Cells that Differ in Function
V. Activation Requirements and Function of γδ T Cells
VI. T-Cell Responses Induced by Dendritic Cells
References

I. Introduction

If an effective strategy for immunization is to be developed, we must develop an in-depth knowledge of the immune system of the target species and the means to analyze the responses at the cellular and molecular level. The importance of the cellular collaboration between particular antigen-presenting cells (APCs) in the initiation of naive T-cell responses and the subsequent direction and regulation of the specific T-cell response is appreciated, if not fully understood, at the molecular level. Specific T-cell responses in naive animals are induced by the presentation of processed antigen presented in the context of MHC class I or class II stimulating $CD8^+$ and $CD4^+$ T cells. The cytokines produced by the responding T cells have differing effects and the bias of the T-cell cytokine response can be affected by interactions between the microbe, its antigens, the local microenvironment, the APC, and other cells not acting in a cognitive manner. The purpose of this paper is to summarize some of the information that relates to cattle T-cell function *in vitro* and *in vivo* and the properties of dendritic cells, since these cells are pivotal in the stimulation of naive T cells.

II. Identification of the Major T-Cell Populations in Cattle

The major T-cell populations in cattle have been identified with monoclonal antibodies (mAbs) to differentiation antigens expressed by the cells produced and characterized in the laboratories of origin or within a series of international workshops that have been held. The nomenclature used follows the human CD nomenclature where there is sufficient evidence to conclude that the homologous molecule is being identified in cattle and humans. If human CD homologs are not evident WC (workshop cluster) numbers were assigned to the mAb and molecules recognized. Thus cattle leukocytes are defined in terms of the CD or WC antigens expressed (Howard *et al.*, 1991a; Howard and Naessens, 1993; Naessens and Hopkins, 1996; Naessens *et al.*, 1996, 1997).

In blood three major populations of T lymphocytes are identified as being either $\alpha\beta$ TCR$^+$, CD3$^+$, CD2$^+$, CD6$^+$ CD4$^+$ (MHC class II restricted) or $\alpha\beta$ TCR$^+$, CD3$^+$ CD2$^+$, CD6$^+$, CD8$^+$ (MHC class I restricted) or $\gamma\delta$ TCR$^+$, CD3$^+$, CD2$^-$, CD6$^-$, CD4$^-$, CD8$^-$, WC1$^+$. Thus, the major T-cell populations can be identified by expression of CD4 or CD8 or WC1. However, further minor populations are evident within peripheral blood mononuclear cells (PBMCs) that do not fall within this simple categorization and this simple scheme does not apply to T cells in tissues. For example, CD8$^+$ T cells that are $\gamma\delta$ TCR$^+$ are evident as a minor subset of CD8$^+$ T cells in blood but a major subset in the intestinal epithelium; similarly, the WC1 antigen is not present on the major $\gamma\delta$ TCR$^+$ population in the spleen of calves (Clevers *et al.*, 1990; Hein and Mackay, 1991; Wyatt *et al.*, 1994, 1996). This is clearly important when assessing changes in T-cell populations in (1) tissues (2) following activation (3) in response to infection.

III. Role of Different T-Cell Populations *in Vivo*

A question, asked in relation to specific infections, that is central to vaccine design is whether a particular T-cell population is primarily responsible for recovery from infection or immunity to reinfection. Differences between different infections would influence the appropriate strategy selected for immunization. One way in which it is possible to analyze this in cattle is to use mAb to specifically deplete particular T-cell populations and to determine the effect on immune responses and infection. Our investigations have shown that depletion of $CD8^+$ cells but not $CD4^+$ or $WC1^+$ cells has a profound effect on the ability of calves to recover from respiratory syncytial virus (RSV) infection. This indicates a central role for MHC class I restricted $CD8^+$ T cells, and by analogy with other species' cytotoxic T-cell responses, in recovery (Taylor *et al.*, 1995). In contrast, depletion of $CD4^+$ T cells resulted in a prolonged viremia and nasal shedding following challenge with bovine virus diarrhea virus (BVDV) (Howard *et al.*, 1992). This was even more pronounced if the dose of CD4 mAb was increased (Fig. 1), resulting, in the example shown, in a viremia lasting 28 days and nasal shedding 53 days postinfection compared to the 14 or 7 days noted in normal gnotobiotic calves of the same age (Howard *et al.*, 1992). The appearance of neutralizing antibody was delayed following CD4 depletion and its appearance in serum coincided with the disappearance of viremia. However, nasal shedding persisted after the appearance of antibody (Fig. 1). Thus, neutralizing antibody dependent on T-cell help, activation of other effector mechanisms, or a direct antiviral effect of the $CD4^+$ T cells could be playing greater or lesser roles at different sites within the animal. A lesser effect of depletion noted with rotavirus

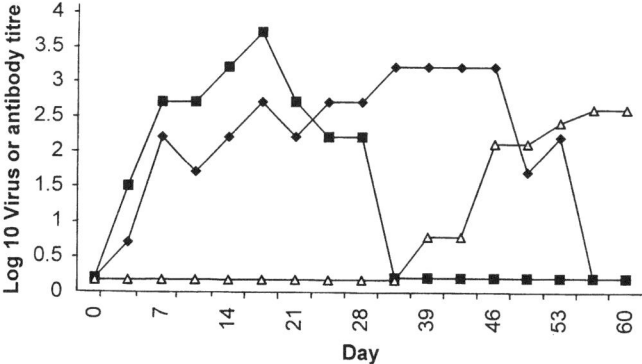

FIG. 1. Prolonged viremia and nasal shedding of BVDV after *in vivo* depletion of CD4 T cells by intravenous inoculation of mAb. Inoculations of about 25 mg of mAb CC8 were given daily for first 8 days. Intranasal inoculation with BVDV strain Pe515nc. ■, virus titer in blood; ♦, virus in nasal swabs; △, neutralizing antibody titer.

(Oldham *et al.*, 1993) may have resulted from incomplete depletion in the gut epithelium compared to the blood. Studies of *Trypanosoma congolense* (Sileghem and Naessens, 1995) indicated that in contrast to the results from a mouse model, CD8$^+$ T cells did not play a major role in immunity in cattle.

IV. Identification of Subpopulations of CD4 and CD8 T Cells that Differ in Function

CD4$^+$ and CD8 T$^+$ populations in blood and other tissues are not homogeneous and differences in expression of surface molecules between T-cell subsets is related to their functional differences. The objective of an immunization strategy should be to produce an appropriate type of T-cell response, a CD4 T-cell response that is biased toward a Th1- or Th2-type cytokine pattern depending on what is appropriate for combatting a particular pathogen.

Both the CD4 and CD8 T-cell populations can be divided into subsets based on expression of different isoforms of the leukocyte common antigen, CD45. Three-color flow cytometry (FCM) showed that most CD4 and CD8 cells express either the high molecular weight isoform (CD45R) or the low molecular weight isoform (CD45RO) and a few expressed both (Fig. 2). *In vitro* proliferation assays with CD4$^+$ T cells

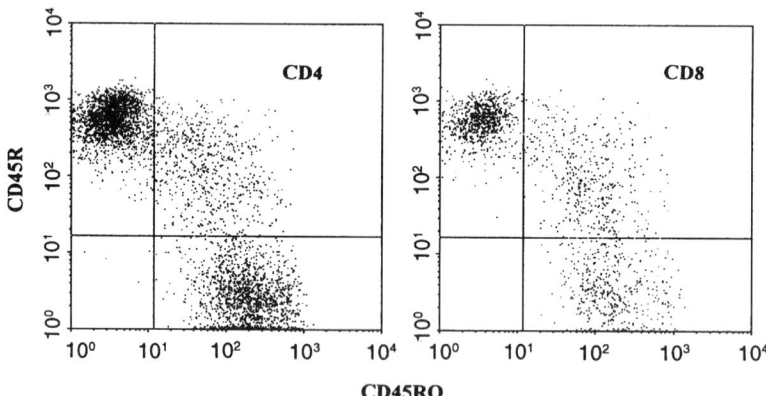

FIG. 2. Dot plots showing three distinct subpopulations of CD4$^+$ and CD8$^+$ T cells. Bovine (*Bos taurus*) peripheral blood mononuclear cells (PBMCs) were stained with mAb IL-A116 (CD45RO), CC76 (CD45R) and CC30 (CD4), or CC58 (CD8) in three-color flow cytometry. The dot plots show gated CD4 cells (left) and gated CD8 cells (right) and demonstrate three distinct subpopulations—those that are CD45RO$^-$, CD45R$^+$ (upper left quadrant), CD45RO$^+$, CD45R$^+$ (upper right), and CD45RO$^+$, CD45R$^-$ (lower right).

involving *Trypanosoma brucei,* RSV, or ovalbumin showed that resting memory T cells were within the CD45R⁻ and CD45RO⁺ subset (Bembridge *et al.,* 1995; Howard *et al.,* 1991b). However, when the CD4⁺, CD45RO⁺, CD45R⁺ population as well as the CD4⁺, CD45RO⁺, CD45R⁻ and the CD4⁺, CD45RO⁻, CD45R⁺ subsets were assessed, both of the CD45RO⁺ populations responded (unpublished). Thus memory CD4 T cells are CD45RO⁺, CD45R⁻ and CD45RO⁺, CD45R⁺ while naive CD4 T cells are CD45RO⁻, CD45R⁺.

These differences in CD45 isoform expression are related to cytokine synthesis. An analysis of the CD4⁺, CD45RO⁺ and CD4⁺, CD45RO⁻ populations by reverse transcriptase polymerase chain reaction (RT–PCR) showed that there were transcripts for interleukin 2 (IL-2) in both the RO⁺ and RO⁻ subsets but that transcripts for interferon γ (IFN-γ) and IL-4 were only in the RO⁺ subset (Bembridge *et al.,* 1995). Thus, resting memory CD4 T cells are capable of rapidly synthesizing cytokines that are not produced by naive cells in the same time. An analysis of cytokine production by cytoplasmic staining with mAb revealed further differences between the subsets (Fig. 3). Following activation of cattle PBMC with PMA/ionomycin, in a manner similar to that described for mice (Openshaw *et al.,* 1995), 18% of CD4 cells expressed cytoplasmic IFN-γ. These cells were within the CD45RO⁺ and CD45R⁺ populations. However, only 1–2% of the CD4⁺ T cells stained with mAb to IL-4, a percentage similar to that seen with human PBMC. All IL-4⁺ cells were within the CD45RO⁺ population (Fig. 4). Thus, it appears that IFN-γ is produced mainly by the CD45RO⁺, CD45R⁺ and CD45RO⁺, CD45R⁻ CD4⁺ T cells but IL-4 only by the CD45RO⁺, CD45R⁻ CD4⁺ T cells. Salmon *et al.* (1994) showed that with more rounds of stimulation naive T cells initially produced IFN-γ and subsequently IL-4 as they ceased synthesis of IL-2 and became susceptible to apoptosis. This may relate to these observations although it does not fit the concept that the different cytokine bias is a result of the type of stimulation. The findings of Powrie *et al.* (1994) may also be of relevance. In this case CD45RB⁺ and CD45RB⁻ T cells in mice differed in production of IL-4 and this was related to pathology or immunity in passive transfer experiments.

Although subsets of CD8 T cells were also evident based on CD45 isoform expression, cytolytic precursors evident in PBMC after immunization with *Theileria parva* were not in a particular subset (Bembridge *et al.,* 1995; Howard *et al.,* 1991b). These cells are clearly the progeny of cells that have experienced antigen and can be considered as memory cells. Consequently, for CD8 T cells memory does not relate to CD45 isoform expression. However, an analysis by cytoplasmic staining with mAb showed that after activation 19% stained for IFN-γ

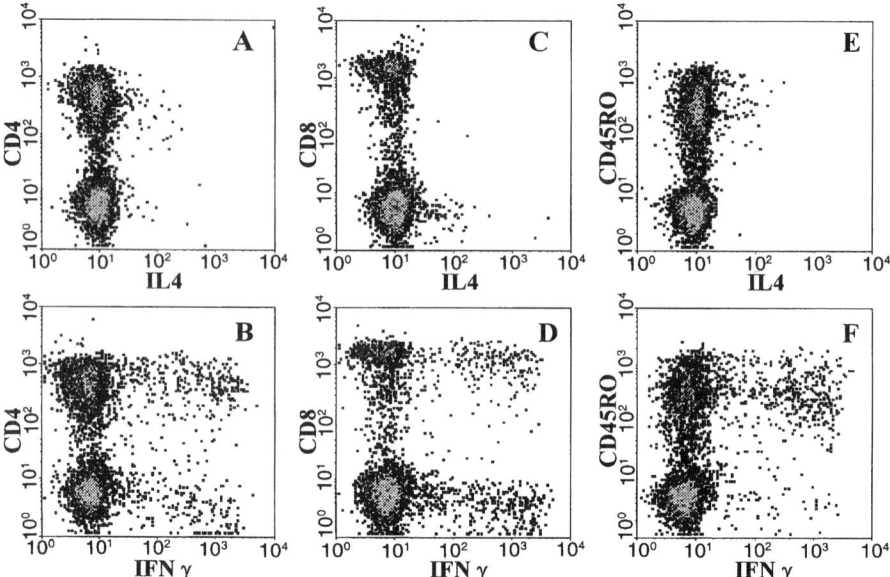

FIG. 3. Density plots showing cytokine staining in subpopulations of bovine peripheral blood mononuclear cells (PBMCs). PBMCs were cultured in the presence of 50 ng/ml PMA (phorbol 12-myristate 13-acetate), 1 μg/ml ionomycin, and 10 μg/ml brefeldin A for 5 hours. The cells were stained with mAb CC8 (CD4), CC58 (CD8), or IL-A116 (CD45RO), fixed in phosphate-buffered saline (PBS) containing 1% w/v paraformaldehyde, permeabilized in Permeabilisation Solution™ (Becton Dickinson) then stained with mAb 11C12 (IL-4) or 6H5 (IFN-γ). The two-color staining demonstrates that most IL-4+ cells were CD4+ and that no CD8+ cells produced IL-4 (A and C). IFN-γ, however, was synthesized by both CD4+ and CD8+ cells (B and D). IFN-γ was produced mainly by cells expressing CD45RO but production of IL-4 was entirely within the CD45RO+ cells (E and F).

(Fig. 4). Because most of the IFN-γ producing cells were CD45RO+ it is likely that early synthesis of IFN-γ after stimulation is predominantly by the CD8 RO+ cells.

An explanation of these observations is that following activation naive CD4 and CD8 T cells change from being CD45RO− to CDRO+. Thus, this change relates to activation of naive T cells. Memory CD4+ T cells in PBMC remain CD45RO+ while some CD8+ T cells return to being CD45RO−. If expression of CD45RO is regarded as relating to the activation state of the cell, then an interpretation is that CD4 cells after experiencing antigenic stimulation remain partially activated but not all CD8 T cells do. Hence CD45RO expression coincidentally

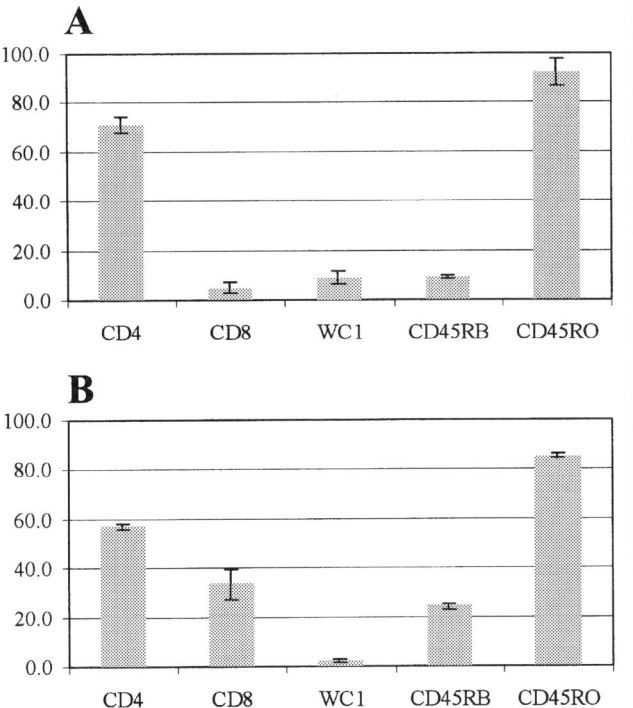

FIG. 4. Histograms showing the phenotype of (A) IL-4 and (B) IFN-γ positive cells in PBMCs. The y axis indicates the percentage of (A) IL-4 or (B) IFN-γ positive cells that also stain for the cell surface marker indicated on the x axis. (Cell culture and staining methods are summarized in the Fig. 3 legend.)

relates to memory for CD4 T cells but not CD8 T cells. In fact, evidence has been put forward indicating that some memory CD4 T cells are CD45RO$^-$ and require a higher level of stimulation to induce proliferation than do CD45RO$^+$ memory cells (Pilling *et al.*, 1996).

The innovative studies on the recirculation of T cells in sheep (Mackay, 1991; Mackay *et al.*, 1990, 1992) showed that CD4 and CD8 T cells in PBMC were CD45R$^+$ and CD45R$^-$ (a mAb reactive with CD45RO was not available at that time). For CD4 T cells the CD45R$^+$ subset was shown to be naive based on proliferation assays. Efferent lymph contained predominantly CD45R$^+$ cells and afferent lymph CD45R$^-$ cells. The selective recirculation of memory T cells to tissues was proposed and substantiated in a number of publications in sheep and mice. A comparison of blood, pseudoafferent lymph, and gut-derived lymphocytes of cattle (Table I) similarly shows selective recirculation of CD45RO+ CD4 and CD8 T cells to the body surfaces.

V. Activation Requirements and Function of γδ T Cells

As has been noted in a number of publications, ruminant blood, in particular when taken from young animals, contains a high proportion of γδ T lymphocytes (Table I) (Clevers et al., 1990; Hein and Mackay, 1991; Mackay and Hein, 1991; Wyatt et al., 1994). Most of these γδ T cells in blood express the WC1 antigen and the majority of functional studies have been with this WC1+ population, which appears to selec-

TABLE I

THREE-COLOR FLOW CYTOMETRIC ANALYSIS OF CD45 ISOFORM EXPRESSION BY CATTLE T CELLS[a]

T-cell subset	Percentage of cells expressing the high and low Mr isoforms of CD45			Cell population examined
	CD45RO− CD45R+	CD45RO+ CD45R+	CD45RO+ CD45R−	
CD4	61±3	10±2	29±5	2- to 5-Month-old calves PBMC
	31±23	9±1	60±22	Afferent lymph
	39±13	12±3	49±15	Adult cattle PBMC
	1±1	2±1	96±1	Adult cattle gut epithelium
CD8	70±6	12±3	18±7	2- to 5-Month-old calves PBMC
	50±36	16±3	34±35	Afferent lymph
	55±7	20±4	25±4	Adult cattle PBMC
	9±3	19±7	72±10	Adult cattle gut epithelium
WC1	<1	5±3	95±3	2- to 5-Month-old calves PBMC
	<1	3±2	97±2	Afferent lymph
	1±1	3±1	97±1	Adult cattle PBMC
	ng	ng	ng	Adult cattle gut epithelium

[a]The figures are the means of three animals ± SD and are the percentage of cells within the CD4, CD8, or WC1 populations that were CD45RO−, CD45R+, or CD45RO+, CD45R+, or CD45RO+, CD45R−. ng, not given, the percentage of WC1+ cells in the intestinal epithelium from adult cattle was too low to get meaningful values. PBMC from 2- to 5-month-old calves and adult cattle >3 years old were examined as well as pseudoafferent lymph from 2- to 5-month old calves and gut epithelium from adult cattle. Three-color immunofluorescent staining was with mAb to CD4, CD8, or WC1 together with mAb to the high Mr isoform, CD45R; mAb CC76, and the low Mr isoform, CD45RO; mAb IL-A116 (Bembridge et al., 1995; Howard et al., 1991b, 1997).

tively recirculate from blood to skin (Mackay *et al.*, 1990, 1992) and has been shown to be thymus dependent (Hein *et al.*, 1990). A cDNA encoding the WC1 antigen was cloned and sequenced and the molecule shown to be a member of the scavenger receptor family with no clear human or rodent homolog (Clevers *et al.*, 1990). More recent studies in ruminants (Kirkham *et al.*, 1997; Takamatsu *et al.*, 1997) have reported that the molecule provides a negative signal that reduces proliferation. This, together with the finding that the WC1 antigen is actually a family of molecules produced by a cluster of up to 50 genes that are differentially expressed on WC1$^+$ cells (Walker *et al.*, 1994; Wijngaard *et al.*, 1992-1994) implies some specificity in ligand binding and possibly control of expansion or effector function by these cells.

The function of WC1$^+$ $\gamma\delta$ T cells has not been established and it is possible that the high proportion present in ruminant blood is a consequence of a biochemical event occurring around gestation rather than indicating a selective advantage. However, the high numbers of these cells in PBMC has enabled resting cells from blood to be isolated and their activation requirement and possible role in immunity investigated. The WC1 cells do not synthesize CD28 transcripts, which is considered to be a major costimulatory molecule on $\alpha\beta$ TCR$^+$ cells. Thus, the established mechanism of stimulating T cells involving one signal through the TCR and one through CD28 cannot be used (Howard *et al.*, 1996). However, CD25 is expressed by WC1$^+$ T cells in PBMC and an *in vitro* model using parasite transformed monocytes has shown that the WC1 cells will proliferate to two signals, one from the surface of a transformed monocyte and one from a cytokine mediated by CD25 (Collins *et al.*, 1996). That it is the TCR providing one signal has not been established in this model but since Concanavalin-A and IL-2 will stimulate proliferation (Clevers *et al.*, 1990) it is likely. Investigations using the autologous MLR indicate that the proliferative response is polyclonal and not MHC restricted (Hanby Flarida *et al.*, 1996; Okragly *et al.*, 1996). On initial stimulation of blood-derived WC1$^+$ cells, few cytokine transcripts are evident (*Collins et al., 1996*); hence the requirement for the provision of IL-2 by other T cells. But, WC1$^+$ T-cell lines produce a variety of cytokine transcripts after more prolonged activation (R. A. Collins unpublished; Brown *et al.*, 1994) which indicates a role in influencing some immune responses. More recent investigations have shown these cells to be capable of presenting native protein antigens to CD4$^+$ memory T cells (Collins *et al.*, 1998) and to express FcγRIII (CD16) which mediate specific recognition of antigen (Collins *et al.*, 1997). Investigations *in vivo,* in which the WC1$^+$ T cells were specifically depleted with mAbs, indicated a possible role in the regulation of B-cell responses (Howard *et al.*, 1989). Although depletion studies *in vivo* have not provided evidence for a role in immunity to RSV or BVDV infection (Howard *et al.*, 1992; Taylor *et al.*, 1995) they did confirm a possible modulatory role on antibody responses. Furthermore, changes in the number of circulating WC1+ cells in cattle after infection with *Mycobacterium tuberculosis* have been noted (Pollock *et al.*, 1996) perhaps indicating a role in the immune response to this bacterium.

VI. T-Cell Responses Induced by Dendritic Cells

Dendritic cells are the most potent of the professional APCs and utilize a number of mechanisms of antigen uptake for presentation to T cells (Lanzavecchia, 1996). Dendritic cells are the only APCs recognized as capable of stimulating naive T cells (Steinman, 1991). At the body surfaces they form a network of cells that effectively take up antigen and then migrate to the draining lymph node where processed antigen is presented to T cells. Associated with this migration to the draining lymph node is a down-regulation of capacity to take up antigen and up-regulation of capacity to stimulate T cells. Dendritic cells are present in low numbers in blood and body tissues but in relatively high numbers in afferent lymph draining the skin. Thus, in ruminants afferent lymph veiled cells (ALVCs), the dendritic cells in afferent lymph, can be isolated by cannulation and used for investigations of cells that have not been subjected to prolonged isolation procedures involving a variety of culture conditions which may result in changes in their properties (Emery *et al.*, 1987; McKeever *et al.*, 1991). ALVCs from cattle are phenotypically distinguishable from monocytes according to expression of a number of surface molecules. Thus, monocytes express CD11b, CD14, and the IgG_2 binding Fcγ2R that are not expressed by ALVCs while ALVCs express the WC6 molecule at a high level and monocytes do not (Howard *et al.*, 1997). Furthermore, cattle ALVCs have been shown to be much more effective than monocytes in ability to stimulate specific T-cell responses *in vitro* (McKeever *et al.*, 1991) and after pulsing *in vitro* with antigen able to prime naive T cells *in vivo* (McKeever *et al.*, 1992).

ALVCs are not a homogeneous population (Howard *et al.*, 1997 McKeever *et al.*, 1991) and two major populations defined with a panel of mAb have been shown to vary in their ability to stimulate CD4 and CD8 T-cell responses. Thus, one ALVC population is CD11a−, Myd-1+, and the other CD5+, CD11a+, MyD-1−, CC81+. Both populations present native antigen, ovalbumin, to CD4 T cells and both stimulate proliferation of allogeneic CD4 T cells. But, the CD11a⁻ population is more effective at stimulating allogeneic CD8 T cells as well as in presenting RSV antigen to resting memory T cells or VSG from *Trypanosoma brucei* to a T-cell clone (Howard *et al.*, 1997; McKeever *et al.*, 1991). The molecular basis for the differences in function of the ALVC subsets has not been defined. Binding of CTLA4-Ig or CD40L fusion proteins was similar for both, indicating similar levels of expression of CD80/CD86 and CD40. Uptake of markers of macropinocytosis and mannose receptor mediated uptake did not obviously distinguish either subset. One difference between the subsets is expression of a molecule that we have recently cloned from cattle and named MyD-1 that mediates binding of T cells (Brooke and Howard, 1996).

It is clear that our knowledge of the ruminant immune system has progressed during recent years but that much still remains unknown. Much progress has been attributed to the application of modern technologies to answering old questions. There is a pressing need to address the importance of various leukocyte populations in the development of immunity. In view of the central role of dendritic cells in the initiation of primary T-cell responses, the findings related to the different function of ALVC subsets may be of significance for the induction of immunity in naive animals. Immunization strategies designed to target these cells are likely to be beneficial particularly when vectors are constructed to contain cytokine genes that will bias an appropriate immune response, as has been shown to be possible with cattle cytokine genes *in vitro* (Kuhnle *et al.*, 1996).

References

Bembridge, G. P., MacHugh, N. D., McKeever, D., Awino, E., Sopp, P., Collins, R. A., Gelder, K. I., and Howard, C. J. (1995). CD45RO expression on bovine T cells: Relation to biological function. *Immunology* **86**, 537–544.

Brooke, G. P., and Howard, C. J. (1996). Characterisation, cloning and sequencing of a novel molecule, MyD-1, expressed on cattle and human antigen presenting cells. *Immunology* **89**, 40.

Brown, W. C., Davis, W. C., Choi, S. H., Dobbelaere, D. A., and Splitter, G. A. (1994). Functional and phenotypic characterization of WC1⁺ γδ T cells isolated from Babesia bovis-stimulated T cell lines. *Cell. Immunol.* **153**, 9–27.

Clevers, H., MacHugh, N. D., Bensaid, A., Dunlap, S., Baldwin, C. L., Kaushal, A., Iams, K., Howard, C. J., and Morrison, W. I. (1990). Identification of a bovine surface antigen uniquely expressed on CD4-CD8⁻ T cell receptor γδ T lymphocytes. *Eur. J. Immunol.* **20**, 809–817.

Collins, R. A., Sopp, P., Gelder, K. I., Morrison, W. I., and Howard, C. J. (1996). Bovine γδ TCR⁺ T lymphocytes are stimulated to proliferate by autologous *Theileria annulata*-infected cells in the presence of interleukin-2. *Scand. J. Immunol.* **44**, 444–452.

Collins, R. A., Gelder, K. I., and Howard, C. J. (1997). Nucleotide sequence of cattle FcGRIII: Its identification in GD T cells. *Immunogenetics* **45**, 440–443.

Collins, R. A., *et al.* (1998). Submitted for publication.

Collins, R. A., Werling, D., Duggan, S. E., Bland, A. P., Parsons, K. R., and Howard, C. J. (1998). γδ T cells present antigen to CD4+ αβ T cells. *J. Leukocyte Biol.* **63**, 707–714.

Emery, D. L., MacHugh, N. D., and Ellis, J. A. (1987). The properties and functional activity of non-lymphoid cells from bovine afferent (peripheral) lymph. *Immunology* **62**, 177–183.

Hanby Flarida, M. D., Okragly, A. J., and Baldwin, C. L. (1996). Autologous mixed leucocyte reaction and the polyclonal activation of bovine γδ T cells. *Res. Vet. Sci.* **61**, 65–71.

Hein, W. R., and Mackay, C. R. (1991). Prominence of γδ T cells in the ruminant immune system. *Immunol. Today* **12**, 30–34.

Hein, W. R., Dudler, L., and Morris, B. (1990). Differential peripheral expansion and in vovo antigen reactivity of αβ and γδ T cells emigrating from the early lamb thymus. *Eur. J. Immunol.* **20**, 1805–1813.

Howard, C. J., and Naessens, J. (1993). Summary of workshop findings for cattle. *Vet. Immunol. Immunopathol.* **39**, 25–47.

Howard, C. J., Sopp, P., Parsons, K. R., and Finch, J. (1989). In vivo depletion of BoT4 (CD4) and of non-T4/T8 lymphocyte subsets in cattle with monoclonal antibodies. *Eur. J. Immunol.* **19**, 757–764.

Howard, C. J., Morrison, W. I., Bensaid, A., Davis, W., Eskra, L., Gerdes, J., Hadam, M., Hurley, D., Leibold, W., Letesson, J. J., *et al.* (1991a). Summary of workshop findings for leukocyte antigens of cattle. *Vet. Immunol. Immunopathol.* **27**, 21–27.

Howard, C. J., Sopp, P., Parsons, K. R., McKeever, D. J., Taracha, E. L., Jones, B. V., MacHugh, N. D., and Morrison, W. I. (1991b). Distinction of naive and memory BoCD4 lymphocytes in calves with a monoclonal antibody, CC76, to a restricted determinant of the bovine leukocyte-common antigen, CD45. *Eur. J. Immunol.* **21**, 2219–2226.

Howard, C. J., Clarke, M. C., Sopp, P., and Brownlie, J. (1992). Immunity to bovine virus diarrhoea virus in calves: The role of different T-cell subpopulations analysed by specific depletion in vivo with monoclonal antibodies. *Vet. Immunol. Immunopathol.* **32**, 303–314.

Howard, C. J., Sopp, P., Brownlie, J., Parsons, K. R., Kwong, L. S., and Collins, R. A. (1996). Afferent lymph veiled cells stimulate proliferative responses in allogeneic CD4+ and CD8+ T cells but not γδ TCR+ T cells. *Immunology* **88**, 558–564.

Howard, C. J., Sopp, P., Brownlie, J., Kwong, L. S., Parsons, K. R., and Taylor, G. (1997). Identification of two distinct populations of dendritic cells in afferent lymph that vary in their ability to stimulate T cells. *J. Immunol.* **159**, 5372–5382.

Kirkham, P. A., Takamatsu, H. H., and Parkhouse, R. M. E. (1997). Growth arrest of γδ T cells induced by monoclonal antibody against WC1 correlates with activation of multiple tyrosine phosphatases and dephosphorylation of a MAP kinase erk2. *Eur. J. Immunol.* **27**, 717–725.

Kuhnle, G., Collins, R. A., Scott, J. E., and Keil, G. M. (1996). Bovine interleukins 2 and 4 expressed in recombinant bovine herpesvirus 1 are biologically active secreted glycoproteins. *J. Virol.* **77**, 2231–2240.

Lanzavecchia, A. (1996). Mechanisms of antigen uptake for presentation. *Curr. Opin. Immunol.* **8**, 348–354.

Mackay, C. R. (1991). T-cell memory: The connection between function, phenotype and migration pathways. *Immunol. Today* **12**, 189–192.

Mackay, C. R., and Hein, W. R. (1991). Marked variation in γδ T cell numbers and

distribution throughout the life of sheep. *Curr. Top. Microbiol. Immunol.* **173,** 107–111.
Mackay, C. R., Marston, W. L., and Dudler, L. (1990). Naive and memory T cells show distinct pathways of lymphocyte recirculation. *J. Exp. Med.* **171,** 801–817.
Mackay, C. R., Marston, W. L., Dudler, L., Spertini, O., Tedder, T. F., and Hein, W. R. (1992). Tissue-specific migration pathways by phenotypically distinct subpopulations of memory T cells, *Eur. J. Immunol.* **22,** 887–895.
McKeever, D. J., MacHugh, N. D., Goddeeris, B. M., Awino, E., and Morrison, W. I. (1991). Bovine afferent lymph veiled cells differ from blood monocytes in phenotype and accessory function. *J. Immunol.* **147,** 3703–3709.
McKeever, D. J., Awino, E., and Morrison, W. I. (1992). Afferent lymph veiled cells prime CD4$^+$ T cell responses in vivo. *Eur. J. Immunol.* **22,** 3057–3061.
Naessens, J., and Hopkins, J. (1996). Introduction and summary of workshop findings. *Vet. Immunol. Immunopathol.* **52,** 213–235.
Naessens, J., Howard, C. J., and Hopkins, J. (1996). Nomenclature for clusters of differentiation of ruminant leukocyte antigens. *Immunologist* **4,** 207–210.
Naessens, J., Howard, C. J., and Hopkins, J. (1997). Nomenclature and characterization of leukocyte differentiation antigens in ruminants. *Immunol. Today* **18,** 365–368.
Okragly, A. J., Hanby Flarida, M., Mann, D., and Baldwin, C. L. (1996). Bovine $\gamma\delta$ T-cell proliferation is associated with self-derived molecules constitutively expressed in vivo on mononuclear phagocytes. *Immunology* **87,** 71–79.
Oldham, G., Bridger, J. C., Howard, C. J., and Parsons, K. R. (1993). In vivo role of lymphocyte subpopulations in the control of virus excretion and mucosal antibody responses of cattle infected with rotavirus. *J. Virol.* **67,** 5012–5019.
Openshaw, P., Murphy, E. E., Hosken, N. A., Maino, V., Davis, K., Murphy, K., and O'Garra, A. (1995). Heterogeneity of intracellular cytokine synthesis at the single-cell level in polarized T helper 1 and T helper 2 populations. *J. Exp. Med.* **182,** 1357–1367.
Pilling, D., Akbar, A. N., Bacon, P. A., and Salmon, M. (1996). CD4$^+$ CD45RA$^+$ T cells from adults respond to recall antigens after CD28 ligation. *Int. Immunol.* **8,** 1737–1742.
Pollock, J. M., Pollock, D. A., Campbell, D. G., Girvin, R. M., Crockard, A. D., Neill, S. D., and Mackie, D. P. (1996). Dynamic changes in circulating and antigen-responsive T-cell subpopulations post-Mycobacterium bovis infection in cattle. *Immunology* **87,** 236–241.
Powrie, F., Correa Oliveira, R., Mauze, S., and Coffman, R. L. (1994). Regulatory interactions between CD45RBhigh and CD45RBlow CD4$^+$ T cells are important for the balance between protective and pathogenic cell-mediated immunity. *J. Exp. Med.* **179,** 589–600.
Salmon, M., Piling, D., Borthwick, N. J., Viner, N., Janossy, G., Bacon, P. A., and Akbar, A. N. (1994). The progressive differentiation of primed T cells is associated with an increasing susceptibility to apoptosis. *Eur. J. Immunol.* **24,** 892–899.
Sileghem, M., and Naessens, J. (1995). Are CD8 T cells involved in control of African trypanosomiasis in a natural host environment? *Eur. J. Immunol.* **25,** 1965–1971.
Steinman, R. M. (1991). The dendritic cell system and its role in immunogenicity. *Annu. Rev. Immunol.* **9,** 271–296.
Takamatsu, H. H., Kirkham, P. A., and Parkhouse, R. M. E. (1997). A $\gamma\delta$ cell specific surface receptor (WC1) signalling G0/G1 cell cycle arrest. *Eur. J. Immunol.* **27,** 105–110.
Taylor, G., Thomas, L. H., Wyld, S. G., Furze, J., Sopp, P., and Howard, C. J. (1995). Role of lymphocyte subsets in recovery from respiratory syncytial virus infection of calves. *J. Virol.* **69,** 6658–6664.

Walker, I. D., Glew, M. D., Ma, O. K., Metcalfe, S. A., Clevers, H. C., Wijngaard, P. L., Adams, T. E., and Hein, W. R. (1994). A novel multi-gene family of sheep γδ T cells. *Immunology* **83,** 517–523.

Wijngaard, P. L., Metzelaar, M. J., MacHugh, N. D., Morrison, W. I., and Clevers, H. C. (1992). Molecular characterization of the WC1 antigen expressed specifically on bovine CD4-CD8- gamma delta T lymphocytes. *J. Immunol.* **149,** 3273–3277.

Wijngaard, P. L., MacHugh, N. d., Metzelaar, M. J., Romberg, S., Bensaid, A., Pepin, L., Davis, W. C., and Clevers, H. C. (1994). Members of the novel WC1 gene family are differentially expressed on subsets of bovine CD4-CD8- γδ T lymphocytes. *J. Immunol.* **152,** 3476–3482.

Wyatt, C. R., Madruga, C., Cluff, C., Parish, S., Hamilton, M. J., Goff, W., and Davis, W. C. (1994). Differential distribution of γδ T-cell receptor lymphocyte subpopulations in blood and spleen of young and adult cattle. *Vet. Immunol. Immunopathol.* **40,** 187–199.

Wyatt, C. R., Brackett, E. J., Perryman, L. E., and Davis, W. C. (1996). Identification of γδ T lymphocyte subsets that populate calf ileal mucosa after birth. *Vet. Immunol. Immunopathol.* **52,** 91–103.

IV
CANINE AND FELINE VACCINES

Canine Viral Vaccines at a Turning Point— A Personal Perspective

L. E. CARMICHAEL

James A. Baker Institute for Animal Health, College of Veterinary Medicine, Cornell University, Ithaca, New York 14853

I. General Remarks
II. Veterinary Vaccines
III. Comments on Selected Vaccines
  A. Canine Distemper
  B. Canine Parvovirus Type 2
  C. Canine Coronavirus
IV. Summary
  References

I. General Remarks

Widely divergent views regarding vaccines and their use have appeared recently in the human, veterinary, and lay literature (Carmichael, 1983; Dodds, 1991; Holmes, 1996; Pitcairn, 1995; Priest, 1996; Smith, 1995; Starita-Mehan, 1997; Tizzard, 1990; Yarnall, 1995). Strong opinions have been voiced by many individuals and an increasingly wide public desires to know "the facts" that underlie vaccine use and the basis of immunization regimens. Impassioned, sometimes uninformed, concerns have been expressed regarding vaccine efficacy and safety, the need for certain vaccines, the frequency with which vaccines are given, the need for annual vaccination—indeed, whether vaccines should be used at all! The questions are not new, for they have been raised since Mithridates VI, an ancient Greek king of Pontus (first century B.C.), attempted to protect himself against poisoning by repeatedly taking small amounts of noxious substances, in honey (*theriaca*)—a practice not dissimilar to certain contemporary holistic rituals. Since the introduction of variolation from the Near East in the early eighteenth century and the use of cowpox virus against *Variola,* there has been public concern about the safety and efficacy of vaccines. The concept of "safety" has changed with time, for reactions that were common when the risks of a serious disease were great are no longer acceptable. Although often exaggerated by individual passions, or groups who champion their own doctrines, several of today's concerns seem justified, especially when recognized problems with vaccines are not addressed in the light of existing knowledge or experience.

Rapidly changing attitudes toward pets, their value, and their health care have provoked vigorous and widespread discussion of the use of vaccines in small animal practice (Priest, 1996; Schultz, 1995; Smith, 1995; Tizzard, 1990). This is illustrated by the full-day session devoted to vaccination practices at the July 1996 annual meeting of the AVMA at Cornell's Feline Practitioner's Seminar and at a recent symposium, European Symposium on Pet Vaccinology, held in France in September 1996. In an ideal world, real or presumed problems with companion animal vaccines would be addressed quickly and responsibly by industry, government regulatory officials, and the veterinary profession as soon as they are identified. Unfortunately, problems often are neglected or avoided. This circumstance contributes to confusion and the creation of myths, which often are enhanced by differing views of "experts" who have sometimes formed their own conclusions with limited data or biased sampling designs.

Misfortunes with vaccines are well documented in the literature (Appel and Gillespie, 1972; Carmichael, 1983, 1997; Martin, 1985; Rikula *et al.,* 1995; Tizzard, 1990; Wilbur *et al.,* 1994; Wilson *et al.,* 1986). They have often become elevated to catastrophes, however, especially by those who advocate a radical philosophy but ignore the benefits provided by vaccines. Concerns have sometimes led to the senseless conclusion that all vaccines are dangerous and are a direct or indirect cause of chronic illness ("vaccinosis"). "Vaccinoses" are claimed to range from "devastated immune systems, laziness, bowel disease, bloat, stained teeth, ulcers, chronic gastroenteritis, autoimmune hemolytic anaemia, and seizures," to list but a few conditions that have been cited (Duval and Giger, 1996; Pitcairn, 1995; Priest, 1996; Starita-Mehan, 1997; Yarnall, 1995). However, there are truths between the passion of some and indifference of others. Advocates of holistic vaccination practices, as inane as they may seem, may actually be doing a service to pet fanciers by bringing issues to the fore which have largely been ignored. Unfortunately, legitimate safety or efficacy problems have sometimes been disregarded until major misfortunes occurred. Nevertheless, those who experienced the rampant distemper outbreaks prior to the mid-1960s are amazed by the arguments presented by some critics, especially holistic believers. Within 2–3 years of the advent of efficacious distemper vaccines, the disease practically disappeared in vaccinated populations, but it has reappeared whenever vaccination had diminished. Undeniable progress has been made in the suppression of canine distemper and infectious hepatitis and, more recently, in controlling the canine parvovirus pandemic in a remarkably short period of time. However, the recent outbreaks of distemper in Scandinavia, and this year's epizootic in Alaska and northern Quebec, illustrate what may occur when distemper vaccine efficacy, or vaccine use, diminishes. Veterinarians and the public have become more sophisticated—and litigious; they want to know the facts about treatments they use. Unfortunately, many essential facts regarding vaccines are lacking and myths continue to flourish.

Questions commonly asked by dog owners/breeders and veterinarians are usually complex: Are all vaccines available for dogs necessary? Are vaccines safe in very young pups? How effective are they in preventing disease? Do both live and inactivated vaccines produce a sterilizing immunity so as to interrupt transmission? How soon does immunity occur after vaccination and how long does it endure? Why do vaccines continue to be developed against diseases that are still poorly understood? Are too many agents packaged as multicomponent vaccines, and what are the consequences? It has been well established that the immune system can respond normally to several different antigens—an issue that seem to persist; however, some combined vaccines that had inadequate field trial data prior to release have given rise to serious consequences in regard to safety. Unfortunately, answers to the questions above often reflect individual experiences, vested interests, or a disinclination to state that true answers are not known.

It has been estimated that more than 50% of office visits to veterinarians are associated with vaccination. Several vaccines for dogs (and cats) have been licensed that have poor or questionable efficacy; yet they continue to be produced and promoted, for example, *Leptospira* bacterins, some canine coronavirus (CCV) vaccines and, in the recent past, several canine parvovirus type 2 (CPV-2) vaccines.

New or "improved" vaccines are introduced almost yearly, yet even perfunctory examination reveals a sparse amount of data that often overstates claims for a particular product. On the other hand, questions posed by veterinarians, dog owners, or by those who oppose vaccination on philosophical grounds often defy factual answers because of the paucity of published results. Questions are often based on the perception that valid data are available. Also, many individuals do not accept the reality that vaccination, as other medical practices, sustain some risk. To a large extent, problems in standardizing veterinary vaccines resist solution because of the complexity inherent in the number of different vaccines and viral strains available for pet animals, most of which are poorly characterized.

I share the belief that expectations for vaccines are at a turning point. In this article I outline some personal views and experiences, note unsettled problems, and point out the difficulties in resolving some of the commonly asked questions. Notwithstanding, I am aware that my remarks will have little impact unless veterinarians concerned with dog vaccines show the same concerns as those raised by the Feline Practitioner's Association and act to gain a better understanding of vaccines, how they work, a realistic appreciation of the problems that can occur, and how they might be remedied.

II. Veterinary Vaccines

Most veterinary vaccines continue to be developed empirically. With the technology now available, new vaccines will doubtless continue to be developed, including subunit vaccines, vectored recombinant vaccines, deletion mutants, nucleic acid (plasmid DNA) vaccines and, perhaps, even "recombinant nosodes" (sic). When made available, however, their merits should be evaluated against presently used products, not merely for the sake of novelty. Some recombinant vaccines, for example, vaccinia-vectored rabies for wildlife, a recently licensed canary pox-vectored distemper vaccine, and a Lyme disease vaccine, have shown merit in their utility, safety, or, in some cases, superior efficacy.

With few exceptions, modified live virus (MLV) vaccines are the most common products used worldwide (Appel, 1987; Carmichael, 1997). Most vaccines comprise virus strains which were selected as spontaneous mutants that emerged from the native viral populations during repeated passage in cell cultures or other laboratory hosts. But, the majority of vaccines consist of viral populations that contain multiple mutations and few canine vaccinal strains have been biologically cloned so as to suppress the generation of nonimmunizing mutants during laboratory passage to vaccine.

Mutants that grow in the intended host, yet are replication restricted in critical tissues, constitute vaccines with different degrees of loss of natural virulence ("attenuated virus") Nonimmunizing mutants also may emerge during laboratory passage. Such variants may fail to grow in the natural host, yet proliferate luxuriantly in tissue cultures or chick embryos. Because "attenuation" means reduction, not absolute loss of the capacity to produce disease, safety problems may not be revealed until extensive field tests have been conducted; unfortunately, this has occurred after a product has been licensed and marketed. A conspicuous example of such failure was the large number of dogs that died or suffered serious illness following the introduction of a live canine coronavirus vaccine in 1983 (Martin, 1985; Wilson et al., 1986). Also a vaccine judged harmless for one species may provoke illness in another one (Appel, 1987; Carmichael, 1997; Tizzard, 1990). Because of the uncertainty of absolute safety with certain vaccines, for example, distemper vaccinal strains propagated in canine cell cultures, live viral vaccines are not recommended for most wildlife species, pregnant animals, unweaned pups, or pups that are ill. Yet, breeders and some veterinarians continue to vaccinate pregnant dams, pups as early as 2 weeks of age, or use vaccines for pet species where safety information in limited (e.g., ferrets).

Efficacy problems persist with certain "primary" vaccines, such as some canine parvovirus vaccines and certain canine distemper products (Appel and Gillespie, 1972; Carmichael, 1983, 1989, 1997; Schultz, 1995, 1996). However, the recent improvements in several canine vaccines, especially parvovirus vaccines that previously had poor or marginal efficacy, have been greatly improved, and they now appear to provoke good immune responses. Whether the improvements will be sustained depends in large measure on the care taken by vaccine producers in selecting and conserving their seed stock.

III. Comments on Selected Vaccines

A. Canine Distemper

Virtually all licensed canine distemper (CD) vaccines consist of living attenuated viral strains (Appel, 1987; Appel and Gillespie, 1972; Carmichael, 1997). The majority are produced from the egg-adapted or avian cell culture-adapted Onderstepoort strain or the "Rockborn strain," which is propagated in canine cell cultures (Rockborn et al., 1965). The Rockborn strain is produced legitimately only by the European company authorized by Professor Gunnar Rockborn (G. Rockborn, personal communication, 1996). That virus had undergone ≤56 passages in cell cultures, whereas virus in several U.S. vaccines is used at lower passage levels. Thus, the designation "Rockborn CD vaccine strain" has been misrepresented by several authors in the past, including myself. Certain products also are claimed to contain the attenuated "Snyder Hill" strain, also grown in canine cell cultures, or a ferret-origin strain cultivated in avian cell cultures. It is difficult to determine the origin of those viral strains in a CD vaccine because some strains have been given novel designations by manufacturers and there are few genetic markers. Regardless of the viral strain employed, attenuated CD vaccines have proved highly effective when administered to dogs lacking maternal immunity, but they are variably effective in dogs with low levels of maternal antibodies.

As noted earlier, a recognized problem with certain CD vaccines, especially those propagated in canine cell cultures, is the variable occurrence of postvaccinal (PV) encephalitis, but actual risks are unknown (Appel, 1978, 1987; Appel and Gillespie, 1972; Carmichael, 1983). Some CD vaccines are virulent for several zoo or wildlife species, some of which are now considered pets (e.g., ferrets, skunks, raccoons). Also, reversion to virulence of the attenuated Rockborn CD strain was demonstrated after serial passage in dogs, or in dog lung macrophages (Appel, 1978). The canine cell-adapted vaccines are not recommended for pups less than 6 weeks of age, or for wildlife species, because of the greater risk of postvaccinal encephalitis. Field experience has demonstrated enhanced virulence of CD vaccines produced in canine cell cultures when administered in combination with certain other viruses. The most conspicuous have been canine adenovirus type 1 (CAV-1, ICH) and live CCV vaccines (Carmichael, 1983; Martin, 1985; Wilson *et al.*, 1986). Notwithstanding the risks noted earlier, the Rockborn strain (at passage level ~55–60) has been used as our laboratory's principal experimental vaccine for >40 years. The vaccine is adjusted to ~10^3 TCD_{50}/dose, since the minimal immunizing dose (MID) of the "Rockborn strain" is ~20 TCD_{50}. In such instances, no cases of postvaccinal encephalitis have been observed in field use. However, in laboratory experiments, when the vaccine dose was more than $10^{5.5}$ TCD_{50}/ml, or when vaccine was given together with live CAV-1, encephalitis was a frequent occurrence about 10–12 days postvaccination. Field reports indicated that the frequency of postvaccinal CD encephalitis diminished greatly after the substitution of CAV-2 for CAV-1 in combined vaccines. This was one reason for advocating the substitution of CAV-2 for CAV-1 in canine vaccines, in addition to the marked, but not total, reduction in postvaccinal "blue eyes" that occurred frequently after vaccination with CAV-1. Manufacturers who utilize canine cell-grown CDV should, therefore, determine optimal safe doses. My personal view is that each vaccine should have the MID indicated on the package insert; this seems important with both CD and CPV-2 vaccines, but for different reasons (see below). If CD vaccinal titers were kept low ($\leq 10^{3.0}$/dose), the excellent immunity provided by CD vaccines grown in canine cells would probably be attended by a more acceptable risk of PV reactions.

Duration of immunity data for most commercial distemper vaccines are limited. In one study at the Baker Institute (L. E. Carmichael, unpublished results, 1980), nine beagles were vaccinated with the Rockborn CD strain and maintained in strict isolation. All dogs had high levels of neutralizing antibodies >6 years later. Also, we have recently confirmed 6.5-year immunity (SN titers $\geq$1:80) in male dogs that were vaccinated with a commercial (multiple) vaccine and kept as breeding stock in a kennel that maintains strict isolation. Nevertheless, the rates of immunity following vaccination differ between CD vaccines (Appel, 1987; Appel and Gillespie, 1972; Carmichael, 1977; Rikula et al., 1995). As with other canine vaccines, maternal antibodies interfere with immunization. Recently, substantial differences were reported in the ability of CD vaccines to immunize pups with similar levels of maternal antibodies at 6–7 weeks of age (Schultz, 1996).

Early studies on duration of antibody persistence at levels that were estimated to ensure immunity (neutralizing antibody titers $\geq$1:100) waned within 1 year in 33% of dogs vaccinated with the chick embryo-adapted "Lederle low passage" CD strain; 2 years later another 33% had antibody titers <1:100 (Baker et al., 1962, 1962). Those limited data appear to be the basis for the common practice of annual revaccination. Whether an SN titer of 1:100, by the tests done then, is required for protection is uncertain, for it has been stated that SN titers of 1:20 are protective (Appel and Gillespie, 1972; Gorham, 1966). Neutralizing antibodies to the low egg passage Onderstepoort strain also have been reported to last from 3 to 6 years in almost 90% of dogs kept in isolation (Prydie, 1966). Since distemper vaccine efficacy has generally improved in recent years, it now seems reasonable, without being radical, to discontinue recommending annual vaccination after the first year of life, and to limit vaccinations to 3- to 5-year intervals. Notwithstanding, most veterinarians and dog breeders will likely continue annual vaccinations for pecuniary, or other, reasons.

1. Comments

- Minimum immunizing doses for each canine vaccinal strain should be determined. Egg-adapted CD strains appear to vary somewhat in efficacy, while canine-cell adapted strains vary in their capacity to provoke PV encephalitis. The "Rockborn-type" strains should probably contain about 500 MIDs, unless safety has been ensured.
- Duration of immunity data are needed. Some vaccines, especially those propagated in the chick embryo or Vero cells, appear to provoke shorter durations of immunity than do other vaccines. However, data are scant. Such data are essential to the formulation of rational recommendations.
- Safety of canine-cell grown ("Rockborn-type" including "Snyder Hill" strains) should be more rigorously studied, especially if used in combination with other agents.
- There is a need for an effective nonliving CD vaccine especially for wildlife species. Promising experiments with a recombinant (canarypox) distemper product that protected dogs against challenge with virulent distemper virus suggests the possibility for success of such vaccines (Taylor et al., 1994). One recombinant CD product has recently been licensed in the United States, but unequivocal recommendation should be withheld until field studies have demonstrated its efficacy and duration of immunity.

B. Canine Parvovirus Type 2

Several vaccines have been developed for CPV-2 infection, but immune response data on most CPV-2, or CPV-2a, -b, strains are limited, except for brief periods (2–3 weeks) following vaccination. Immunity to CPV-2 is believed to be antibody mediated and hemagglutination-inhibiting (HI) titers $\geq 1:80$ are considered protective (Carmichael, 1983, 1994, 1997; Carmichael et al., 1983; Pollock and Carmichael, 1990). However, serologic tests are not standardized and comparison of antibody titers from different laboratories is not too meaningful (Luff et al., 1987).

Inactivated and MLV vaccines are available in most countries for immunization of dogs. Although inactivated vaccines for CPV-2 provide only limited protection against infection, dogs may be exempt from disease for several months (Carmichael, 1983; Pollock and Carmichael, 1990). Like distemper, reports of the actual duration of immunity to inactivated CPV-2 vaccines are very limited. It is not known whether immunologic memory provides immunity beyond the period when antibody has declined below detectable levels; it also is not known whether all killed vaccines perform in a similar manner since the magnitude of the antibody responses is related to the amount of viral antigen administered. Because inactivated vaccines do not interrupt transmission of virulent virus, except for belief periods of time (~2–3 months), they are not recommended where large numbers of dogs are raised, that is, breeding kennels, pet shops, and animal shelters or where dogs are at high risk of exposure, such as at shows or field trials. It should be obvious that inactivated vaccines should not be followed by MLV vaccines, or the reverse, because antibodies engendered by the killed vaccine will neutralize the live virus; in the latter instance, the killed vaccine would be wasted if the MLV vaccine had immunized.

Efficacious modified-live CPV-2 vaccines have been highly successful in preventing parvovirus infection when administered to seronegative pups, or to dogs with very low antibody titers. They normally engender rapid and enduring immunity, and it is probable that immunity persists for several years. HI antibody titers >1:320 persisted for periods as long as 6 years in 13 dogs vaccinated with one strain (Cornell LP strain 780916). In recent tests, 5 male dogs that had received a commercial product (combined vaccine), and were maintained in a commercial specific pathogen-free colony, had titers >1:320 more than 6.5 years later. Similar studies with other CPV-2 vaccines have not been published, but tests done in our laboratory in 1987–1990 revealed that serum HI antibody titers in dogs that had received certain commercial vaccines had declined to ≤1:10 within 2–2.5 years. Thus, differences have been observed between vaccines, but several of the ones tested earlier have now been replaced by "new generation" products.

As with CD, a principal cause of vaccination failures in pups is maternal antibody interference, which has been amply exploited by biologics producers in promoting "new vaccines" that claim to immunize pups earlier than do competing products. The reality is that live virus vaccines differ in their capacity to evade low levels of antibodies and no vaccine has been shown to immunize pups at the time when they have maternal antibody levels that prevent infection with virulent virus. The concept of the "critical period" (or "window of vulnerability") was developed to describe that period of time when pups become susceptible to infection with virulent virus, but respond unpredictably to vaccines (Carmichael, 1989; Pollock and Carmichael, 1990). The critical period has been shown to range from 2–5 weeks, but it is briefer with some vaccines than with others; that is some vaccines may immunize pups earlier than to others, regardless of age (Carmichael, 1989, 1997; Hoskins et al., 1995; Schultz, 1995). Failures to respond to efficacious vaccines relate to prevaccination antibody titers, but not age.

No modified live CPV-2 vaccine has been reported to cause adverse reactions, and the myth of "immunosuppression" by virulent CPV-2, or vaccine virus, has been discredited (Brunner and Swango, 1985; Phillips and Schultz, 1987). Indeed, a recent study in Japan indicated that modified live CPV-2 vaccines enhance cellular immune responses; when vaccine was given to dogs prior to surgery, it prevented the postsurgical immunosuppression attending the use of halothane anaesthesia (Taura et al., 1995).

Despite the general benefit derived from CPV-2 vaccines, consistent efficacy has been a recurring problem. Several commercial (MLV) vaccines that were studied in our laboratory, and found effective at the time they were launched, later had poor efficacy. This is likely due to genetic heterogeneity of the seed stock. Such occurrences have prompted new products, including vaccines prepared from isolates that represent variants (CPV-2a,-b) of the original CPV-2. However, it is evident that vaccines prepared from the more recent CPV-2 types have no discernible advantage over efficacious vaccines prepared from the original isolates.

The USDA's "master seed principle" does not appear to function well with CPV-2 vaccines. The principle may be sound, but it hasn't always worked in practice. Reasons are not documented for most vaccine strains, but mutant viruses that fail to provoke immunity often predominate after several passages in cell cultures (Fig. 1). Vaccines that we have examined, with two exceptions, consisted of mixed viral populations. Manufacturers should, therefore, prepare seed virus from biologically cloned stock, selecting those clones shown to immunize and which are stable during subsequent cell culture passage from seed stock to vaccine.

The term *high titered vaccine* has been promoted in advertising, but the term has very little meaning if the minimal immunizing dose is not revealed. A few years ago, we tested two widely used commercial vaccines that had viral infectivity titers of $>10^{5.5}$/dose, yet they provoked only low antibody responses in SPF dogs. Those products have been supplanted by "new generation vaccines" and it appears that, during the past 2 years, most CPV-2 vaccines have improved substantially. Most CPV-2 cases/outbreaks now are reported in unvaccinated dogs, breeding kennels or animal shelters. Animals vaccinated with ineffective vaccines, pups that had interfering levels of maternal antibodies at the time of vaccination, and puppies in contaminated kennels are at

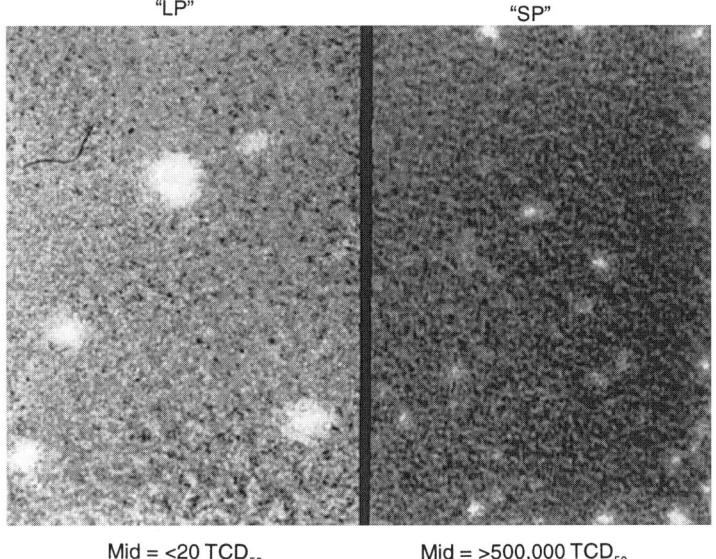

FIG. 1. Plaque variants and immunogenicity of clones from a CPV-2 vaccine (1989). The original vaccinal virus population was predominantly "small plaque" (SP), with approximately 2% "large plaques" (LP). The vaccine immunized pups at $> 10^{5.5}$ TCD$_{50}$ virus, but not with 10^2 TCD$_{50}$. Selected "SP" variants failed to provoke HI antibody responses in pups, even at doses $> 10^6$. In contrast, the "LP" variants produced strong HI antibody responses (1:5120–1:10,240) within 10 days of vaccination.

greatest risk of infection, especially where stringent hygiene is not practiced.

Claims that a vaccine will "immunize pups more efficiently at a particular age" are misleading. Failures to respond to good vaccines are related to prevaccinal antibody levels, not age. Also, we have never observed the failure of a susceptible dog to respond to efficacious CPV-2 or CD vaccines, regardless of breed (e.g., Rottweilers). Studies in our laboratory of nearly 1000 field sera from breeding-age dogs indicated that approximately 25% would not be expected to respond until after 12 weeks of age. On the other hand, studies on vaccine response-versus-age have reported a higher success rate at 12 weeks of age in pups from kennels where the dams' antibody titers were probably low as a result of vaccination rather than infection (Hoskins et al., 1995; Larson and Schultz, 1996; Schultz, 1995). Control of CPV-2 during the initial 3 months of a pup's life should be based on stringent management and prudent vaccination—pups should be isolated as much as possible and kept in a sanitary environment. The availability of dependable good vaccines is essential.

The foregoing remarks notwithstanding, the general success of vaccines in controlling canine parvovirus infections has been remarkable and the improved vaccine efficacy during the past 2 years inspires confidence that parvoviral infections will continue to be uncommon in vaccinated dogs reared in hygienic environments. Inactivated, MLV, and heterotypic (feline parvovirus) vaccines are currently available, but homologous MLV vaccines are recommended for most dogs because they interrupt virulent CPV-2 transmission. Although all MLV vaccines have not been the same with regard to their efficacy there have been no documented safety problems with any parvovirus vaccine in the 15 years since introduction. Note also that attempts to "boost" low antibody titer (i.e., HI titers $\geq 1:40$) are ineffectual.

1. Comments

- Vaccinal seed stock strains should be biologically cloned to provide more uniform and stable viral populations in order to ensure more constant efficacy.
- It would seem beneficial to consider "primary" vaccines for pups less than 3 months of age that contain only CD and CPV-2 components. Multiple vaccines are suggested at 12 weeks of age, unless pups are at high risk for respiratory infections (e.g., animal shelters, pet shops, etc.).
- Because efficacious CPV-2 and distemper vaccines have been shown to provide immunity for at least 5 years, revaccination at 3–5 years, after the first year, seems a conservative strategy.
- Parvovirus vaccines are exceptionally safe. Dogs that develop signs and symptoms of parvovirus infection within 5 days of vaccination should be considered as infected with virulent virus prior to, or at the time of, vaccination. This is still a common occurrence where parvovirus is more likely to be present in the environment (e.g., "puppy mills," pet shops, dog shows, animal shelters, veterinary clinics).

C. CANINE CORONAVIRUS

Canine infections caused by CCV, a virus that may infect both cats and dogs, occur as sporadic cases or kennel outbreaks of mild to explosive (Appel, 1987; Binn *et al.,* 1975; Pastoret, 1984; Pollock and Carmichael, 1990; Tennant *et al.,* 1993). Although CCV is frequently observed by electron microscopy (EM) in the feces of both normal and diarrheic dogs, the true role that CCV plays in canine enteric illness, or the need for vaccines, has yet to be agreed on; however, millions of doses have been sold. Disease associated with CCV is usually attended by low mortality, but occasional deaths occur in young pups. Dogs under stress of intensive training or crowding and those who shelter additional enteric pathogens seem to be at greater risk of illness.

The biology of CCV, and its close relatives in cats and pigs, is still unclear. Serologic cross-reactions have been demonstrated between CCV, feline infectious peritonitis/feline enteric coronavirus (FECV), and transmissible gastroenteritis of swine, but cross-protection has been reported only between CCV and FECV (Coyne and May, 1995). Most reports on CCV have been case reports or epizootiologic studies, where CCV particles in diarrheic feces have ranged from <1% of normal stools to about 75% prevalence in rescue kennels (Rimmelzwaan, 1990; Tennant et al., 1993; Vieler and Herbst, 1995). A controlled study in the Netherlands detected CCV by ELISA tests in 7% of normal stools and in 11% of diarrheic stools (Rimmelzwaan, 1990). Cases are rarely reported since they are usually mild, with the exception of infrequent outbreaks with fatal cases, usually in young pups.

Laboratory studies have confirmed that mixed infections by CCV and CPV-2 result in more severe disease than that caused by either virus alone (Appel, 1988), an argument commonly used to justify the use of CCV vaccines. However, it has not been reported that vaccination of dogs with CCV vaccine prevents the severe manifestations of concurrent, or closely spaced, infections with both viruses. One study of possible benefit by an inactivated CCV vaccine to prevent the serious consequences attending infection by CPV-2 and CCV failed to demonstrate protection (M. Appel, unpublished results, 1985). Also, dual infections now appear to be rare in vaccinated dogs as the result of the extensive use of CPV-2 vaccines.

Both inactivated and live CCV vaccines are available (Carmichael, 1997; Coyne and may, 1995; Edwards et al., 1985; Fulker et al., 1995). The history of CCV vaccines is convoluted and not without misfortune. The first licensed modified live CCV vaccine was rescinded shortly after its introduction in 1983 because of severe adverse reactions with lesions that resembled those of FIP (Martin, 1985). Those reactions occurred in an estimated 5% of vaccinated pups, generally ones <12 weeks of age. An inactivated CCV vaccine that had been licensed also was withdrawn from the market shortly after it was issued because of inadequate efficacy. In addition, a second licensed modified live CCV product which was combined with a canine cell-grown CD vaccine was withdrawn because of a high frequency of post vaccinal CD encephalitis. That vaccine has since been reformulated to exclude the distemper component, which appears to have contributed to the problem. Interestingly, the latter CCV vaccine strain had been licensed for use in California, where it had been marketed for more than 10 years in combination with CPV-2 and distemper vaccine, and the manufacturer affirmed that there had been no adverse reactions. The most recently licensed CCV vaccine comprises a killed FECV product, but information on that vaccine, as well as with most others, is limited mainly to promotional information.

The status of CCV infection is controversial since authenticated cases or outbreaks are seldom reported. Notwithstanding, in January 1997 we made several isolations of a CCV from an outbreak of mild enteric disease in a kennel in New Jersey that breeds and trains dogs for the blind. Of interest was the finding that the recent isolates differed from previous isolates in their failure to grow in feline cell cultures and its lack of affinity for the aminopeptidase-N cell receptor, typical of other coronaviruses from cats, dogs, pigs, and humans which were studied (Tresnan et al., 1996; D. Tresnan, personal communication, 1997).

It would seem, therefore, that the development and distribution of CCV vaccines was mainly the result of marketing decisions, not clearly demonstrated need. On the other hand, factual information on CCV disease is limited. Presently, there seems inadequate medical justification for recommending the use of coronavirus vaccines in dogs until further research results are available. A lesson from the experiences with CCV vaccines is that veterinarians should be cautious when administering new products, especially when little data are available other than that required for product licensing.

IV. Summary

The most important canine viral infections are distemper and CPV-2. Problems of variable CD vaccine safety and efficacy persist, but CD vaccines have greatly reduced the prevalence of disease and cases in vaccinated dogs are now rare. Canine hepatitis (ICH, CAV-1 infection) also has been controlled well by vaccines for more than 35 years and it is now rare; the sporadic cases seen in the 1990s have usually occurred in unvaccinated dogs. CAV-2 vaccines should, therefore, continue to be given since they have proved to be safe and effective, and prevent hepatitis as well as adenoviral tracheobronchitis. Failure to vaccinate would likely result in increase in cases of ICH, a serious disease, but never as significant as distemper and CPV infection.

"Are we vaccinating too often?" The question is complex, but the dominant opinion is "yes" (Smith, 1995). The question cannot be responded to unequivocally, however, since manufacturers employ different strains that vary in their immunizing capacity and, probably, duration of immunity. This question was frequent with distemper in the 1960s. At that time, many veterinarians tested batches of the vaccine they used by providing pre- and postvaccinal sera to competent diagnostic laboratories. That practice appeared to benefit veterinarians and dogs, as well as the quality of vaccines.

Unfortunately, many owners and some veterinarians seem to hold the view that infectious diseases such as parvovirus infection can be controlled by frequent vaccination alone. The common practice of dog breeders of vaccinating their animals several times each year is senseless.

Revaccination for distemper and parvovirus infection is suggested at 1 year of age, but recommendations regarding the frequency of most vaccinations given after that time are unclear. Since most distemper and CPV-2 vaccines probably provide immunity that endures several years, vaccination at 3- to 5-year intervals, after the first year, seems a reasonable practice until more data on duration of immunity become available.

"Are too many kinds of vaccines being promoted for dogs?" Distemper and parvovirus vaccines are essential; canine adenovirus vaccines are recommended since the few cases brought to our attention in recent years have been in unvaccinated dogs. Vaccination against respiratory infections is recommended for most dogs, especially those in kennels, or if they are to be boarded. Need has not been clearly established for coronavirus vaccines; Lyme disease vaccines (see below) are useful in preventing illness in areas where the disease exists, but are unnecessary elsewhere since dogs respond rapidly to appropriate antibiotics; current *Leptospira* bacterins are without benefit since they contain serovars that fail to protect in most areas (noted below).

Lyme disease (LD) was not considered here, but newer recombinant (OspA) vaccines are now available that appear to be safe and effective for at least 1 year and they have not caused vaccine-induced postvaccinal lameness, which has been documented with certain whole-cell Lyme disease bacterins. Lyme disease vaccines should be restricted to dogs in, or entering, endemic areas where infested ticks reside. More than 85% of LD cases occur in the mid-Atlantic and Northeastern

States, about 10% in six Midwestern states (Michigan, Minnesota, and Wisconsin), and a smaller percentage in restricted areas of northern California and the Pacific Northwest.

Leptospirosis also was not discussed here, but vaccines are commonly reported as a cause of anaphylaxis and current vaccines do not contain the serovars prevalent in most regions. The vast majority of cases diagnosed at the New York State Diagnostic Lab at Cornell are *grippotyphosa* and *pomona* serovars and there have been no recent cases caused by *canicola* or *icterohemorrhagiae* serovars. Because leptospirosis is an important disease of dogs, there is an urgent need for more research and the development of safer vaccines that contain the prevalent serovars. In Mexico, dogs may be infected with several serovars and some canine vaccines contain 8–10 serovars.

The *conditio sine qua non* is the availability of consistently good vaccines. Without standardization of vaccines, it seems difficult to formulate general vaccine recommendations. Effort should be directed to improving and standardizing the important vaccines in current use, not the development of new products, unless need is demonstrated.

The public is becoming increasingly aware of vaccine problems, perhaps even more so than the benefits of vaccination. The reality that all vaccines carry some risk is not fully perceived by many owners and veterinarians. Alternative veterinary medicine is now a growing reality; such practices are being taught in some veterinary colleges and questions pertaining to vaccine safety and efficacy will continue to vex veterinarians, vaccinologists, and vaccine producers. They will have to be addressed. There is a need for better appreciation of the risk of adverse reactions (Duval and Giger, 1996).

Finally, the issues that have been discussed, or recommendations that might be made, will have little influence unless biologics manufacturers and regulatory officials exercise greater responsibility in controlling vaccine quality. This could be encouraged by the appointment of a committee of unbiased experts to review vaccines for each disease and provide recommendations based on available evidence. This view has been discussed at meetings on several occasions during the past 30 years, but it has been largely neglected because of considerations that involve industry interests, indifferent or overburdened government authorities, and the trust by veterinarians and dog owners in advertising. Vaccines and vaccination guidelines for physicians are supervised by the American Academy of Pediatric's Committee on Infectious Diseases and the Advisory Committee on Immunization Practices who advise the medical profession and regulatory authorities (Holmes, 1996). Until the veterinary profession insists on a responsible advisory council, concerns and questions regarding vaccines will continue to be met by conflicting opinions and open the door to "Nosodes" and "Thuja"—whose benefits seem to be understood only by those who use and profit from them.

REFERENCES

Appel, M. (1978). Reversion to virulence of attenuated canine distemper virus *in vivo* and *in vitro*. *J. Gen. Virol.* **41,** 385–390.
Appel, M. (1987). "Virus Infections of Vertebrates," (M. C. Horzinek, ser. ed.), Vol. 1, pp. 29–160. Elsevier, Amsterdam.
Appel, M. J. G. (1988). Does canine coronavirus augment the effects of subsequent parvovirus infection? *Vet. Med.* **83,** 360–366.
Appel, M. J. G., and Gillespie, J. H. (1972). Canine distemper virus. *Virol. Monogr.* **11,** 27–48.
Baker, J. A., Robson, D. S., Carmichael, L. E., Gillespie, J. H., and Hildreth, B. (1961). Control procedures for infectious diseases of dogs. *Proc. Anim. Care Panel* **11,** 234–244.
Baker, J. A., Robson, D. S., Hildreth, B., and Pakkala, B. (1962). Breed response to distemper vaccination. *Proc. Anim. Care Panel* **12,** 157–162.
Binn, L. N., Lazar, E. C., Keenan, *et al.* (1975). Recovery and characterization of a coronavirus from military dogs with diarrhea. *Proc. 78th Annu. Meet., U.S. Anim. Health Assoc.,* Roanoke, VA, *1974,* pp. 359–366.
Brunner, C. J., and Swango, L. J. (1985). Canine parvovirus infection: Effects on the immune system and factors that predispose to severe disease. *Compend. Contin. Educ. Vet. Pract.* **7,** 979–988.
Carmichael, L. E. (1983). Immunization strategies in puppies—why failures? *Compend. Contin. Educ. Pract. Vet.* **5,** 1043–1051.
Carmichael, L. E. (1989). Canine parvovirus immunization: "Myths and realities." *Am. Kennel Club Gaz.,* Decembers, pp. 94–102.
Carmichael, L. E. (1994). Canine parvovirus type-2: An evolving pathogen of dogs. *Ann. Med. Vét.* **138,** 459–464.
Carmichael, L. E. (1997). Vaccines for dogs. *In* "Veterinary Vaccinology" (P.-P. Pastoret, J. Balncou, P. Vannier, and C. Vereschueren, eds.), pp. 327–331. Elsevier, Amsterdam.
Carmichael, L. E., Pollock, R. H. V., and Joubert, J. C. (1983). A modified live canine parvovirus strain with novel plaque characteristics. II: Immune response. *Cornell Vet.* **71,** 13–29.
Coyne, M. J., and May, S. W. (1995). Considerations in using a canine coronavirus vaccine. *Top. Vet. Med.* **6,** 32–34.
Dodds, W. J. (1991). Vaccine safety and efficacy. *Kennel Hotline* **8,** 2–4.
Duval, D., and Giger, U. (1996). Vaccine-associated immune-mediated hemolytic anaemia in the dog. *J. Vet. Intern. Med.* **10,** 290–295.
Edwards, B. G., Fulker, R. H., and Acree, W. M. (1985). Evaluating a canine coronavirus vaccine through antigen extinction and challenge studies. *Vet. Med.* **80,** 28–33.
Fulker, R., Wasmoen, T., Atchison, H.-J., and Acree, W. (1995). Efficacy of an inactivated vaccine against clinical disease caused by canine coronavirus. *In* "Corona- and Related Viruses" (P. J. Talbot and G. A. Levy, eds.). Plenum, New York.
Gorham, J. R. (1966). Duration of vaccination immunity and the influence on subsequent prophylaxis. *J. Am. Vet. Med. Assoc.* **149,** 699–704.

Holmes, S. J. (1996). Review of recommendations of the Advisory Committee on Immunization Practices. Centers for Disease Control and Prevention. *J. Infect. Dis.* **174,** S342–S344.

Hoskins, J. D., Taylor, H. W., and Gourley, K. R. (1995). Challenge trial of a new attenuated canine parvovirus vaccine. *Proc. Annu. Vet. Med. Forum Am. Coll. Vet. Intern. Med.* Vol. 13, p. 1012.

Larson, L. J., and Schultz, R. D. (1996). High-titer canine parvovirus vaccine: Serologic response and challenge-of-immunity study. *Vet. Med.,* March, pp. 1–5.

Luff, P. R., Wood, G. W., and Thornton, P. H. (1987). Canine parvovirus serology: Collaborative assay. *Vet. Rec.* **120,** 270–273.

Martin, M. L. (1985). Canine coronavirus enteritis and a recent outbreak following modified live virus vaccination. *Compend. Contin. Educ. Pract. Vet.* **7,** 1012–1017.

Pastoret, P.-P. (1984). Les infections digestives d'origin viral chez le chien. *Ann. Med. Vet.* **128,** 473–483.

Phillips, T. R., and Schultz, R. D. (1987). Failure of vaccine or virulent strains of canine parvovirus to induce immunosuppressive effects on the immune system of the dog. *Viral Immunol.* **1,** 135–144.

Pitcairn, R. H. (1995). "Dr. Pitcairn's Complete Guide to Natural Health for Dogs & Cats," 2nd ed. Rodale Press, Emmaus, PA.

Pollock, R. H. V., and Carmichael, L. E. (1990). Enteric viruses. *In* "Infectious Diseases of the Dog and Cat" (C. E. Greene, ed.), pp. 226–283. Saunders, Philadelphia.

Priest, S. A. (1996). Holistic remedies are getting a shot in the arm. Homeopathic nosodes are being examined by some vets as an option to yearly vaccines. *Dog World,* January, pp. 24–30.

Prydie, J. (1966). Persistence of antibodies following vaccination against canine distemper and effect of revaccination. *Vet. Rec.* **78,** 486–488.

Rikula, U., Sihvonen, L., Voipio, H. M. *et al.* (1995). Serum antibody response to canine distemper virus vaccines in beagle dogs. *Front. Lab. Anim. Sci.,* p. 199.

Rimmelzwaan, G. (1990). Application of enzyme-linked immunosorbant assay systems for the serology and antigen detection in parvovirus, coronavirus and rotavirus infections in dogs in The Netherlands. Canine Parvovirus Infection: Novel Approaches to Diagnosis and Immune Prophylaxis, pp. 39–56. Thesis, State University of Utrecht, The Netherlands.

Rockborn, G., Lannek, N., and Norby, E. (1965). Comparison between the immunizing effect in dogs and ferrets of living distemper vaccines attenuated in dog tissue culture and embryonated eggs. *Res. Vet. Sci.* **6,** 423.

Schultz, R. D. (1995). Emerging issue: Vaccines strategies for canine viral enteritis. *Proc. Int. Gastroenter. Symp. North Am. Vet. Conf.,* pp. 19–24.

Schultz, R. D. (1996). Canine distemper: Comparison of the leading multi-component commercial vaccines. *Infect. Dis. Bull.,* Millsboro, DE, Intervet, pp. 1–2.

Smith, C. A. (1995). Current concepts: Are we vaccinating too much? *J. Am. Vet. Med. Assoc.* **207,** 421–425.

Starita-Mehan, D. (1997). "The Dangers of Vaccinations, and the Advantages of Nosodes for Disease Prevention. Nosode Vaccination." County Way Veterinary Care, Boring, OR.

Taura, Y., Ishi, K., Nagami, M. *et al.* (1995). Changes in lymphoproliferation and DTH responses after vaccination immediately before surgery in puppies. *J. Vet. Med. Sci.* **57,** 899–904.

Taylor, J., Tartaglia, J., Riviere, *et al.* (1994). Applications of canarypox (ALVAC) vectors in human and veterinary vaccines. *Dev. Biol. Stand.* **82,** 131–135.

Tennant, B. J., Gaskell, R. M., Jones, R. C. *et al.* (1993). Studies on the epizootiology of canine coronavirus. *Vet. Rec.* **132,** 7–11.

Tizzard, I. (1990). Risks associated with the use of live vaccines. *J. Am. Vet. Med. Assoc.* **196,** 1851–1858.

Tresnan, D. B., Levis, R., and Holmes, K. V. (1996). Feline aminopeptidase N serves as a receptor for feline, canine, porcine and human coronavirus in serogrup I. *J. Virol.* 8669–8674.

Vieler, E., and Herbst, W. (1995). electron microscopic determination of viruses in feces of dogs with diarrhea. *Tieraerztl. Prax.* **23,** 66–69.

Wilbur, L. A. *et al.* (1994). Abortion and death in pregnant bitches associated with a canine vaccine contaminated with bluetongue virus. *J. Am. Vet. Med. Assoc.* **204,** 1762–1765.

Wilson, R. B., Holladay, J. A., and Cave, J. A. (1986). A neurologic syndrome associated with use of a canine coronavirus-parvovirus vaccine in dogs. *Compend. Contin. Educ. Pract. Vet.* **8,** 117–124.

Yarnall, C. (1995). "Cat Care Naturally." Tuttle, Rutland, VT.

Editor's note: See also two very recent publications on this subject:

Schultz, R. D. (1998). Current and future canine and feline vaccination programs. *Vet. Med.* **93**(3), 233–254.

Schultz, R. D. (1998). Vaccine immunity challenges for the 21st century. *Suppl. Compend. Contin. Educ. Pract. Vet.* **20**(5B), 5–18.

Forty Years of Canine Vaccination

M. J. G. APPEL

James A. Baker Institute for Animal Health, College of Veterinary Medicine, Cornell University, Ithaca, New York 14853

I. Introduction
II. Rabies Virus
III. Canine Distemper Virus
IV. Canine Parvovirus
V. Canine Coronavirus
VI. Canine Adenovirus Type 1 (Infectious Canine Hepatitis Virus)
VII. Canine Adenovirus Type 2
VIII. Canine Parainfluenza Virus
IX. *Bordetella bronchiseptica*
X. *Borrelia burgdorferi*
XI. Leptospirosis
XII. Summary
References

I. Introduction

With the exception of rabies vaccines, which were introduced earlier this century, efficacious canine vaccines for the protection from infectious diseases were developed during the past 40 years. Research and development during this time period has focused on controlling fatal infectious diseases like canine distemper, infectious canine hepatitis, canine parvovirus infections, or leptospirosis. Later developments addressed the need to control nonfatal diseases such as kennel cough or Lyme disease.

Modified live virus (MLV) vaccines became the products of choice to control fatal virus infections in dogs. They induce rapid and prolonged cellular as well as humoral immune responses after a single inoculation in susceptible animals. Historically, inactivated (killed) virus vaccines did not sufficiently control disease induced by canine distemper virus (CDV) or canine parvovirus (CPV).

Inactivated virus vaccines were successfully developed to prevent rabies virus infections. Inactivated bacterial products (bacterins) protected dogs from certain strains of leptospirosis, *Bordetella bronchiseptica*, and *Borrelia burgdorferi* (Lyme disease).

The induction of mucosal immunity by intranasal inoculation of modified live products provided better protection against infectious agents that cause kennel cough than parenteral inoculation with inactivated products.

A recombinant vaccine for canine distemper that was introduced recently may be the beginning of a new era in vaccine production. It was intended to increase the safety level of vaccination. However, the efficacy of this vaccine is probably not comparable to MLV vaccines. More recombinant vaccines can be expected to appear on the market. Undoubtedly, DNA vaccines will be introduced for some of the canine infectious diseases which may have a similar effect. In most cases in vaccine production, the enhancement of one factor comes at the sacrifice of the other. There is presently a tendency to produce safer products. The question remains whether the safer products sufficiently control disease outbreaks.

It appears that canine infectious diseases are presently controlled well by vaccination. This may be the time for some fine tuning to address lesser problems such as the possible autoimmune responses in some breeds after multiple vaccinations. The question has been raised: "Are we vaccinating too much?" and the answer is probably yes. Limited data are available for the duration of a vaccine-induced immunity against CDV, CPV, and canine adenovirus (CAV) in dogs kept in isolation. Immunity against these diseases lasts for several years and annual revaccinations may be unwarranted. In addition, if reliable and affordable quick tests for levels of maternal antibody were to become available, the multiple puppy vaccinations could be reduced to one or two inoculations. The use of oro/nasal vaccinations could be developed for more products, perhaps combined with newly developed vectors, which may reduce the risk of abnormal reactions after needle inoculations. There will be additional innovations to reduce the risk of vaccination but to maintain the protection of the animals.

The purpose of this paper is to give a brief, historical review of canine vaccine development during the past 40 years.

II. Rabies Virus

Rabies in dogs has been known since the fifth century B.C. and the dog has long been known to be a principal transmitter of rabies.

The first rabies vaccine was developed by Pasteur in the early 1880s when he adapted "street" virus to rabbits by serial intracerebral passage (Pasteur, 1885). The Pasteur vaccine was predominantly used for human vaccination. Chloroform or ether inactivated virus vaccines for dogs prepared from infected brain suspensions became available in the 1920s (Kelser, 1930). The development of live attenuated rabies virus, vaccines in low egg passage (LEP) and high egg passage (HEP) (Koprowski, 1954) led to effective vaccination of dogs (Tierkel et al., 1953; Sikes, 1975). However, on rare occasions the live attenuated vaccines caused rabies-like disease in dogs. They are no longer available. Greatly improved inactivated virus vaccines prepared from rabies virus grown in diploid cell culture are now commonly used in dogs (Pastoret et al., 1997). Although antigenic differences between virus strains were found by monoclonal antibody, the vaccines cross-protect against different strains (Wiktor and Koprowski, 1980). Most of the inactivated rabies virus vaccines on the market today induce immunity in dogs that lasts for 3 years.

Promising results have been reported with rabies ISCOMES (Osterhaus et al., 1986). In addition, newer developments include viral vectors expressing rabies virus G protein. The vectors are nonreplicating in mammalian hosts (avipox viruses). Both fowlpox and canarypox recombinant rabies vaccines induced protective immunity in dogs and proved to be safe (Taylor et al., 1988, 1991).

In several countries around the world, canine rabies was greatly reduced by mass immunization of dogs (Bögel et al., 1982). A great reduction in wildlife rabies, the source for dog rabies, has been accomplished by the introduction of live attenuated rabies virus by the oral route (Baer et al., 1971). Initially a MLV vaccine was applied that was later replaced by a vaccinia recombinant rabies vaccine (Pastoret et al., 1997). Results with oral vaccination of dogs remain inconclusive.

III. Canine Distemper Virus

Canine distemper (CD) is caused by a morbillivirus closely related to measles and rinderpest viruses. The first vaccine against CD was made by Puntoni (1923) from formalin inactivated brain tissue from dogs with distemper encephalitis. Inactivated CDV vaccines, which were used earlier this century, have not been able to control the disease and are no longer commercially available in the United States. They include a limited protection against disease and no protection against infection. Another approach was attempted by Laidlaw and Dunkin in the late 1920s (1928) by simultaneous inoculation of virulent virus and antiserum. Results were not satisfactory. Vaccination with the heterotypic measles virus (MV) induces protection from disease but not from infection with CDV (Appel et al., 1984). It has the advantage of inducing partial immunity in pups with maternal antibody (reviewed by Appel and Gillespie, 1972).

The first MLV vaccine for CD was developed by Green and Carlson (1945) by passaging virus 50 times in ferrets. The vaccine was widely used. Unfortunately, clinical signs and death frequently occurred after vaccination.

The MLV vaccines for CD that have controlled the disease and that are still in use today were developed in the late 1950s. The virus was adapted to embryonating hen eggs by Cabasso and Cox (1952, Lederle strain) and by Haig (1956, Onderstepoort strain). Both strains were later adapted to tissue culture. Rockborn (1960) introduced a canine kidney cell culture adapted CDV vaccine that is still used worldwide today. There are advantages and disadvantages in both types of vaccines. The Rockborn vaccine induces complete immunity in virtually 100% of susceptible dogs. However, products from some companies induce postvaccinal encephalitis (PVE), which has not been seen in 35 years of the original and authentic product from Behringwerke (Hoechst) in Germany. The Onderstepoort strain does not induce PVE, however, the seroconversion rate of this product in general is lower, and its H glycoprotein profile differs from the H protein of field isolates (Harder et al., 1996).

A promising approach was taken by De Vries et al. (1988). They incorporated the CDV-H and -F proteins into immune stimulating complexes (ISCOMES), which protected dogs from CDV infection. Because both the H and the F proteins are important in producing immunity against CDV, any future recombinant or DNA vaccine should incorporate both. In 1997, a recombinant CDV vaccine containing the H and F genes in a canarypox virus carrier was introduced (Stephensen et al., 1997).

IV. Canine Parvovirus

A new enteric disease of dogs that resembled panleukopenia of cats and mink enteritis appeared in 1978 in North America, Europe, and

Australia. A parvovirus was isolated (Appel *et al.,* 1979) and was tentatively classified as canine parvovirus type 2 (CPV-2) (Carmichael and Binn, 1981). This was in contrast to the "minute virus of canines" that was isolated in 1967 and was referred to as canine parvovirus type 1 (Binn *et al.,* 1970). CPV-2 is believed to be a mutant of feline panleukopenia virus (FPV) or mink enteritis virus; however, the origin of this "new" virus remains unknown.

Soon after the detection of CPV-2 inactivated and heterotypic (FPV) vaccines were introduced that controlled the disease only to a limited extent (Appel *et al.,* 1979; Pollock and Carmichael, 1982). A ML-CPV-2 vaccine became available in 1980 (Carmichael *et al.,* 1981, 1983; Pollock and Carmichael, 1983) that was safe and more efficacious than the inactivated or heterotypic vaccine. The vaccine protects dogs from infection as well as from disease. Antibody titers of ≥1:80 tested by HI are considered to be protective. Dogs vaccinated with ML-CPV-2 vaccine and kept in isolation thereafter had protective antibody titers at least 5 years after vaccination (L. E. Carmichael, personal communication).

CPV-2 has further mutated into CPV-2a and CPV-2b and new vaccines have been introduced (Parrish, 1991). However, the original ML-CPV-2 vaccine protects dogs against present field strains of CPV-2 (Appel and Carmichael, 1987).

Vaccination failures are frequently found when maternal antibody interferes with immunization. Pups become susceptible to virulent CPV-2 before they are susceptible to vaccination. This "window" of susceptibility may last from 2 to 5 weeks (Pollock, 1984). Although claimed by vaccine producers, none of the presently available vaccines eliminates this "window" entirely.

V. Canine Coronavirus

Canine coronavirus (CCV) causes a mild gastroenteritis in dogs (Appel *et al.,* 1980; Carmichael and Binn, 1981). However, it may enhance the pathogenicity of CPV-2 infection (Appel, 1988). The virus was first isolated from sentry dogs with diarrhea in 1971 (Binn *et al.,* 1975). The distribution of the virus in dogs appears to be worldwide (Pensaert and Callebaut, 1978; Rimmelzwaan, 1990).

Inactivated CCV vaccines were introduced in the 1980s (Edwards *et al.,* 1985). The vaccine protects dogs from disease but not from infection. Because of the mild nature of the disease and the limited protection by the killed vaccine its use in dogs is debatable.

In 1983 a ML-CCV vaccine was introduced in combination with other canine vaccines including CPV and CDV. The vaccine was withdrawn from the market 2 months later because adverse reactions were seen in more than 900 dogs with central nervous signs or death in more than 300 dogs (Martin 1985; Wilson et al., 1986).

A different strain of ML-CCV was recently licensed in the United States, which by itself appears to be safe and efficacious. However, in combination with ML-CDV (Rockborn strain), it produced PVE in a large number of dogs. The combination was withdrawn from the market and replaced with the same ML-CCV, but a canarypox vectored CDV that is incapable of causing PVE. A ML feline enteric coronavirus vaccine antigenically related to CCV also became available recently.

VI. Canine Adenovirus Type 1 (Infectious Canine Hepatitis Virus)

Infectious canine hepatitis (ICH) or hepatitis contagious canis (HCC), formerly known as "epizootic fox encephalitis," is caused by canine adenovirus type 1 (CAV-1). A comprehensive report about the disease in dogs was made by Rubarth in 1947. Besides acute hepatitis, CAV-1 is known to be responsible for other diseases (e.g., encephalopathy, neonatal disease, respiratory disease, chronic hepatitis, interstitial nephritis, and ocular lesions) (reviewed by Koptopoulos and Cornwell, 1981). The virus was isolated and production of a MLV vaccine followed after serial passage in dog or swine cells (Cabasso et al., 1954, 1958). The safety of this product was limited; the vaccine induced "blue eyes" in some dogs, was shed in urine, and produced kidney lesions. The CAV-1 vaccine for the control of ICH was replaced in the 1970s by CAV-2 vaccines, which induce protection from ICH virus infection without the undesirable side effects of CAV-1 vaccines (Appel et al., 1975).

Inactivated vaccines for CAV-1 are not on the market in the United States. (*Editor's note:* A vaccine now manufactured by Bayer Animal Health [previously BioCor, previously Tech America] still contains an inactivated CAV-1 component according to the package insert. It also contains inactivated CAV-2.) They are available in other countries and have been found to be safe and efficacious for limited time periods (Miller et al., 1980).

VII. Canine Adenovirus Type 2

In 1961 a virus designated Toronto A26/61 was isolated by Ditchfield *et al.* (1962) from dogs in Canada suffering from laryngotracheitis and kennel cough. The virus is one of the agents causing severe kennel cough in nonvaccinated puppies in pet shop situations that may simulate canine distemper (Appel, 1981). It was found to be antigenically related to CAV-1; however, the tissue tropism of both viruses is entirely different (Appel *et al.*, 1973). The virus was later classified as CAV-2 (Hamelin *et al.*, 1984; Marusyk *et al.*, 1970). The attenuated CAV-2 proved to protect dogs against infection with both CAV-1 and CAV-2 (Appel *et al.*, 1975). Because ML-CAV-2 vaccine is safer than CAV-1 vaccine, the former replaced the latter in the 1970s. Intranasal vaccine is now available in combination with *B. bronchiseptica* and canine parainfluenza virus to protect dogs from kennel cough. It has the advantage over parenteral injection by inducing immunity in pups with maternal antibody (Appel *et al.*, 1975) and, therefore, only one inoculation is needed.

VIII. Canine Parainfluenza Virus

Canine parainfluenza virus (CPIV) is one of the main causes of canine infectious tracheobronchitis or "kennel cough"and has a worldwide distribution (Appel and Percy, 1970; Binn and Lazar, 1970). The virus was first isolated from laboratory dogs with respiratory disease (Binn *et al.*, 1967). CPIV is closely related to simian virus 5 (SV5) (Binn *et al.*, 1967; Crandell *et al.*, 1968), and to human parainfluenza 2 (Hsiung, 1972).

Attenuated CPIV vaccines were introduced in the 1970s in combination with *B. bronchiseptica* in two forms: One with inactivated *B. bronchiseptica* for parenteral inoculation (Chladek *et al.*, 1981; Emery *et al.*, 1976) and one with ML *B. bronchiseptica* for intranasal inoculation (Glickman and Appel, 1981; Kontor *et al.*, 1981). Because protection from infection by both agents depends on mucosal immunity with IgA production, the latter protects from infection and disease while the former protects only from disease. In addition, maternal antibody does not interfere with the intranasal application. More recently a ML-CAV-2 component has been added to the intranasal vaccine. CAV-2 is also involved in the kennel cough complex. A genome analysis of virulent and attenuated strains of CPIV was made by Yonezawa (1985).

Although CPIV vaccine-induced immunity probably lasts longer than 1 year, annual revaccination with the combined vaccine is recommended because immunity to *B. bronchiseptica* is limited.

IX. *Bordetella bronchiseptica*

Bordetella bronchiseptica is the main cause of canine infectious tracheobronchitis or "kennel cough," a highly contagious respiratory disease of dogs (Bemis et al., 1977a; Binn et al., 1968; Wagener et al., 1984; Wright et al., 1973). Infection with B. bronchiseptica is not restricted to dogs. A variety of other species become infected with the agent including pigs, cats, and rodents. Although B. bronchiseptica is highly susceptible to antibiotics in vitro, the in vivo effect is limited because the organisms attach to the cilia of trachea and bronchi (Bemis and Appel, 1977; Bemis et al., 1977b).

The immune response to B. bronchiseptica in dogs is slow. Although dogs become resistant to reinfection and clearance is initiated by 3 weeks after infection, total clearance of the bronchial tree takes about 3 months (Bemis et al., 1977b). The mucosal immunity resulting from infection or intranasal vaccination lasts for about 1 year (Bemis et al., 1977b). Muscosal immunity with IgA production is essential for protection from infection.

As with CPIV, two forms of vaccine have been developed: one inactivated bacterin in adjuvant for parenteral inoculation (Chladek et al., 1981; McCandlish and Thompson, 1978) and one ML in combination with CPIV for intranasal installation (Bey et al., 1981; Glickman and Appel, 1981; Kontor et al., 1981; Shade and Goodnow, 1979). The latter protects from infection with virulent B. bronchiseptica and from disease while the former protects only from disease. Two inoculations of susceptible pups are needed to induce protection. Maternal antibody does not interfere with intranasal vaccination and only one inoculation is needed. In addition, parenteral inoculation with killed organisms and adjuvant may cause undesired local reactions.

X. Borrelia burgdorferi

Lyme disease or Lyme borreliosis is caused by the spirochete Borrelia burgdorferi (Barbour, 1984). The agent is transmitted by hard shell ticks (Ixodes species) (Spach et al., 1993; Appel, 1990). Lyme disease is seen in humans (Steere, 1989), dogs (Appel et al., 1993; Levy et al., 1993), cats (May et al., 1994), horses, and cattle (Parker and White, 1992) after natural infection. The disease in humans on the North American continent was first described by Steere et al. (1978) and in dogs by Lissman et al., 1984).

For vaccination strategies it has to be taken into consideration that Lyme disease in the United States is caused by B. burgdorferi sensu stricto. Additional strains of B. garinii, B. afzelii, and B. japonica are known to occur worldwide. Cross-protection between strains is limited (Lovrich et al., 1994). In addition, Lyme disease is prominent in endemic areas. Ninety percent of Lyme disease in the United States was found in the Northeast, with the upper Mississippi region and Northern California following in frequency. Vaccination of dogs should only be recommended in endemic areas and in dogs exposed to ticks. It has been commented that only 5% of seropositive dogs develop clinical lameness that responds well to antibiotic treatment. However, a high percentage of exposed nonlame dogs have a subclinical chronic polyarthritis (Appel *et al.*, 1993) and antibiotic treatment does not entirely eliminate persistent infection (Straubinger *et al.*, 1997a,b). In addition, a fatal renal syndrome has been observed in a limited number of dogs (Dambach *et al.*, 1997).

A whole-cell bacterin for dogs was introduced in 1992 by Chu *et al.* A single protein vaccine for dogs prepared from recombinant outer surface protein A of *B. burgdorferi* became available in 1996 (Ma *et al.*, 1996; Chang *et al.*, 1995). The OspA vaccine has the advantage that it induces a specific borreliacidal antibody in dogs (or other species) that prevent transmission of *B. burgdorferi* from ticks to dogs (Straubinger *et al.*, 1995). OspA vaccines for human Lyme disease are in testing stages. (*Editor's note:* Extensive research attempting to demonstrate infection and disease from *B. burgdorferi* in cattle, calves, fetuses *in utero*, and adult pregnant cows by Schultz and others in 1993 showed that cattle are highly resistant to infection and no clinical disease was produced in any aged animal, including young fetuses, when multiple isolates of *B. burgdorferi* were given at low or high doses.)

XI. Leptospirosis

Leptospirosis is a zoonotic disease of worldwide distribution. The disease may be fatal in dogs if left untreated (Hartman *et al.*, 1986). Dogs may recover clinically after antibiotic treatment but may die from kidney failure and uremia several months or years later. The disease is caused by infection with antigenically distinct serovars of *Leptospira*. The most common serovars isolated from dogs used to be *L. icterohaemorrhagiae, L. canicola, L. pomona*, and *L. grippotyphosa*. However, in recent years outbreaks of leptospirosis in dogs infected with different serovars have been reported from Canada and Long Island. Raccoons, opossums, deer, rodents, and domestic livestock are reservoirs for most *L. serovars* with the exception of *L. canicola*, which is transmitted from dog to dog.

Bivalent bacterins for dogs that contain *L. canicola* and *L. icterohaemorrhagiae* have been on the market since the 1950s (Hartman *et al.*, 1984a,b). They are prepared from chemically inactivated whole cells, which make them relatively allergenic. Because immunity after vaccination is highly serovar specific, immunized dogs are not protected from other types that are common in many areas and that may infect dogs. They may also suppress the immune response in young puppies and vaccination of pups less than 9 or 10 weeks of age should not be recommended. In addition, the vaccine-induced immunity in dogs is often less than 6 months, and repeated vaccination in endemic areas would be essential for protection (Broughton and Scarnell, 1985). It would be highly desirable to have specific outer surface (envelope) proteins for the immunization of dogs (Bey and Johnson, 1982) like the OspA vaccine in Lyme disease, another spirochetal disease, to reduce the risk of anaphylactic shock and other vaccine-related disorders. In addition, more serovars should be included in leptospira vaccines that correlate with the serovars in endemic areas.

Public health considerations: Leptospira-contaminated urine is highly infectious for people. Persistent infection in healthy vaccinated dogs with leptospiruria has been found with resulting development of the disease in people (Feigin *et al.*, 1973).

XII. Summary

During the last 40 years vaccines have been developed that have greatly reduced the incidence of infectious diseases of dogs. In general, modified live products have been superior to inactivated vaccines for dogs. It can be expected that recombinant and/or DNA vaccines may dominate the market in the future.

Although most vaccines on the market are safe and efficacious, there have been exceptions where disease was induced by vaccination or dogs were not protected. The failure of protection may in part be due to variations in individual vaccine batches. Only potency tests but not efficacy tests are required, which may not be sufficient. For example, a virus titer in a vaccine may be meaningless if the minimum protective dose is not known. Overattenuated virus (e.g., CDV-Ond or parvovirus in cat cells) may have a high titer in tissue culture but is not immunogenic.

The question of frequency of vaccination of dogs should be addressed. Annual revaccinations for CDV, CPV, and CAV are probably not needed. However, it would be desirable to collect more data to support less frequent vaccinations. Annual immunization for bacterial diseases such as kennel cough, Lyme disease, and leptospirosis should continue. It also would be desirable to develop more oro/nasal vaccines, perhaps combined with newly developed vectors that are less likely to induce undesirable side effects that may be seen after parenteral vaccination.

Finally a word of warning against homeopathic "nosodes" to replace tested canine vaccines. They will appear highly effective as long as the majority of dogs remain vaccinated. As soon as a nonvaccinated dog population is large enough to allow virulent agents to spread, disease outbreaks will occur and we will be back where we began 40 years ago.

REFERENCES

Appel, M. J. G. (1981). Canine infectious tracheobronchitis (kennel cough): A status report. *Compend. Contin. Educ. Pract. Vet.* **3,** 70–79.
Appel, M. J. G. (1988). Does canine coronavirus augment the effects of subsequent parvovirus infection? *Vet. Med.,* April, pp. 360–366.
Appel, M. J. G. (1990). Lyme disease in dogs and cats. *Compend. Contin. Educ. Pract. Vet.* **12,** 617–626.
Appel, M. J. G., and Carmichael, L. E. (1987). Can a commercial vaccine protect pups against a recent field isolate of parvovirus? *Vet. Med.,* October, pp. 1091–1094.
Appel, M. J. G., and Gillespie, J. H. (1972). Canine distemper virus. "Handbook of Virus Research" (S. Gard, C. Hallauer, and K. F. Meyer, eds.), Vol. 11, pp. 1–96. Springer-Verlag, New York.
Appel, M. J. G., and Percy, D. H. (1970). SV5-like parainfluenza virus in dogs. *J. Am. Vet. Med. Assoc.* **156,** 1778–1781.
Appel, M. J. G., Bistner, S. E., Menegus, M., Albert, D. A., and Carmichael, L. E. (1973). Pathogenicity of low-virulence strains of two canine adenovirus types. *Am. J. Vet. Res.* **34,** 543–550.
Appel, M. J. G., Carmichael, L. E., and Robson, D. S. (1975). Caine adenovirus type 2—induced immunity to two canine adenoviruses in pups with maternal antibody. *Am. J. Vet. Res.* **36,** 1199–1202.
Appel, M. J. G., Scott, F. W., and Carmichael, L. E. (1979). Isolation and immunisation studies of a canine parvo-like virus from dogs with haemorrhagic enteritis. *Vet. Rec.* **105,** 156–159.
Appel, M. J. G., Meunier, P., Pollock, R., Greisen, H., Carmichael, L., and Glickman, L. (1980). Canine viral enteritis. A report to practitioners. *Canine Pract.* **7,** 22–23.
Appel, M. J. G., Shek, W. R., Shesberadaran, H., and Norrby, E. (1984). Measles virus and inactivated canine distemper virus induce incomplete immunity to canine distemper. *Arch. Virol.* **82,** 73–82.
Appel, M. J. G., Allan, S., Jacobson, R. H., Lauderdale, T.-L., Chang, Y.-F., Shin, S. J., Thomford, J., Todhunter, R. J., and Summers, B. A. (1993). Experimental Lyme disease in dogs produces arthritis and persistent infection. *J. Infect. Dis.* **167,** 651–664.
Baer, G. M., Abelseth, M. K., and Debbie, J. G. (1971). Oral vaccination of foxes against rabies. *Am. J. Epidemiol.* **93,** 487–490.
Barbour, A. G. (1984). Isolation and cultivation of Lyme disease spirochetes. *Yale J. Biol. Med.* **57,** 521–525.
Bemis, D. A., and Appel, M. J. G. (1977). Aerosol, parenteral and oral antibiotic treatment of *Bordetella bronchiseptica* infections in dogs. *J. Am. Vet. Med. Assoc.* **170,** 1082–1086.
Bemis, D. A., Carmichael, L. E., and Appel, M. J. G. (1977a). Naturally occurring respiratory disease in a kennel caused by *Bordetella bronchiseptica*. *Cornell Vet.* **67,** 282–293.
Bemis, D. A., Greisen, H. A., and Appel, M. J. G. (1977b). Pathogenesis of canine bordetellosis. *J. Infect. Dis.* **135,** 753–761.

Bey, R. F., and Johnson, R. C. (1982). Leptospiral vaccines in dogs: Immunogenicity of whole cell and outer envelope vaccines prepared in protein-free medium. *Am. J. Vet. Res.* **43,** 831–834.

Bey, R. F., Shade, F. J., Goodnow, R. A., and Johnson, R. C. (1981). Intranasal vaccination of dogs with live avirulent *Bordetella bronchiseptica:* A correlation of serum agglutination titer and the formation of secretory IgA with protection against experimentally-induced infectious tracheobronchitis. *Am. J. Vet. Res.* **42,** 1131–1132.

Binn, L. N., and Lazar, E. C. (1970). Comments on epizootiology of parainfluenza SV5 in dogs. *J. Am. Vet. Med. Assoc.* **156,** 1774–1777.

Binn, L. N., Eddy, G. A., Lazar, E. C., Helms, J., and Murnane, T. (1967). Viruses recovered from laboratory dogs with respiratory disease. *Proc. Soc. Exp. Biol. Med.* **126,** 140–145.

Binn, L. N., Lazar, E. C., Rogul, M., Shepler, V. M., Swango, L. J., Claypoole, T., Hubbard, D. W., Asbill, S. G., and Alexander, A. D. (1968). Upper respiratory disease in military dogs: Bacterial mycoplasma and viral studies. *Am. J. Vet. Res.* **29,** 1809–1815.

Binn, L. N., Lazar, E. C., Eddy, G. A., and Kajima, A. (1970). Recovery and characterization of a minute virus of canines. *Infect. Immun.* **1,** 503–508.

Binn, L. N., Lazar, E. C., Keenan, K. P., Huxsoll, D. L., Marchwicki, B. S., and Strano, A. J. (1975). Recovery and characterization of a coronavirus from military dogs with diarrhea. *Proc. 78th Annu. Meet., U.S. Anim. Health Assoc., 1974,* pp. 359–366.

Bögel, K., Andral, L., Beran, G., Schneider, L. G., and Wandeler, A. (1982). Dog rabies elimination. *Int. J. Zoonoses* **9,** 97–112.

Broughton, E. S., and Scarnell, J. (1985). Prevention of renal carriage of leptospirosis in dogs by vaccination. *Vet. Rec.* **117,** 307–311.

Cabasso, V. J., and Cox, H. R. (1952). A distemper-like strain of virus derived from a case of canine encephalitis: Its adaptation to the chick embryo and subsequent modification. *Cornell Vet.* **42,** 96–107.

Cabasso, V. J., Stebbins, M. R., Norton, T. W., and Cox, H. R. (1954). Propagation of infectious canine hepatitis virus in tissue culture. *Proc. Soc. Exp. Biol. Med.* **85,** 239–245.

Cabasso, V. J., Stebbins, M. R., and Avampato, J. M. (1958). A bivalent live virus vaccine against canine distemper (CD) and infectious canine hepatitis (ICH). *Proc. Soc. Exp. Biol. Med.* **99,** 46–51.

Carmichael, L. E., and Binn, L. N. (1981). New enteric viruses in the dog. *Adv. Vet. Sci. Comp. Med.* **25,** 1–37.

Carmichael, L. E., Joubert, J. C., and Pollock, R. V. H. (1981). A modified live canine parvovirus strain with novel plaque characteristics. I. Viral attenuation and dog response. *Cornell Vet.* **71,** 408–427.

Carmichael, L. E., Joubert, J. C., and Pollock, R. V. H. (1983). A modified live canine parvovirus vaccine. II. Immune response. *Cornell Vet.* **73,** 13–29.

Chang, Y.-F., Appel, M. J. G., Jacobson, R. H., Shin, S. J., Harpending, P. Straubinger, R., Patrican, L. A., Mohammed, H., and Summers, B. A. (1995). Recombinant OspA protects dogs against infection and disease caused by *Borrelia burgdorferi. Infect. Immun.* **63,** 3543–3549.

Chladek, D. W., Williams, J. M., Gerber, D. L., Harris, L. L., and Murdok, F. M. (1981). A canine parainfluenza—*Bordetella bronchiseptica* vaccine: Part I. Immunogenicity. *Am. J. Vet. Res.* **12,** 266–270.

Chu, H.-J., Chavez, L. G., Blumer, B. M., Sebring, R. W., Wasmoen, T. L., and Acree, W. M. (1992). Immunogenicity and efficacy study of a commercial *Borrelia burgdorferi* bacterin. *J. Am. Vet. Med. Assoc.* **201,** 403–411.

Crandell, R. A., Brumlow, W. B., and Davison, V. E. (1968). Isolation of a parainfluenza virus from sentry dogs with upper respiratory disease. *Am. J. Vet. Res.* **29,** 2141–2147.

Dambach, D. M., Smith, C. A., Lewis, R. M., and Van Winkle, T. J. (1997). Morphologic, immunohistochemical, and ultrastructural characterization of a distinctive renal lesion in dogs putatively associated with *Borrelia burgdorferi* infection: 49 cases (1987–1992). *Vet. Pathol.* **34,** 85–96.

De Vries, P., Uytdehaag, F. G. C., and Osterhaus, A. D. M. E. (1988). Canine distemper virus (CDV) immune-stimulating complexes (iscoms), but not measles virus iscoms, protect dogs against CDV infection. *J. Gen. Virol.* **69,** 2071–2083.

Ditchfield, J., MacPherson, L. W., and Zbitnew, A. (1962). Association of a canine adenovirus (Toronto A26/61) with a outbreak of laryngotracheitis (kennel cough). A preliminary report. *Can. Vet. J.* **3,** 238–247.

Edwards, B. G., Fulker, R. H., and Acree, W. M. (1985). Evaluating a canine coronavirus vaccine through antigen extinction and challenge studies. *Vet. Med.* **80,** 28–33.

Emery, J. B., House, J. A., Bittle, J. L., and Spotts, A. M. (1976). A canine parainfluenza viral vaccine: Immunogenicity and safety *Am. J. Vet. Res.* **37,** 1323–1327.

Feigin, R. D., Lobes, L. A., Anderson, D., and Pickering, L. (1973). Human leptospirosis from immunized dogs. *Ann. Intern. Med.* **79,** 777–785.

Glickman, L. T., and Appel, M. J. (1981). Intranasal vaccine trail for canine infectious tracheobronchitis (kennel cough). *Lab. Anim. Sci.* **31,** 397–399.

Green, R. G., and Carlson, W. E. (1945). The immunization of foxes and dogs to distemper with ferret-passage virus. *J. Am. Vet. Med. Assoc.* **107,** 131–142.

Haig, D. A. (1956). Canine distemper-immunization with avianised virus. *Onderstepoort J. Vet. Res.* **27,** 19–53.

Hamelin, C., Marsolais, G., and Assaf, R. (1984). Interspecific differences between the DNA restriction profiles of canine adenoviruses. *Experientia* **40,** 482.

Harder, T. C., Kenter, M., Vos, H., Siebelink, K., Huisman, W., van Amerongen, G., Örvell, C., Barrett, T., Appel, M. J. G., and Osterhaus, A. D. M. E. (1996). Canine distemper virus from diseased large felids: Biological properties and phylogenetic relationships. *J. Gen. Virol.* **77,** 397–405.

Hartman, E. G., van Houten, M., van der Donk, J. A., and Frik, J. F. (1984a). Serodiagnosis of canine leptospirosis by solid-phase enzyme-linked immunosorbent assay. *Vet. Immunol. Immunopathol.* **7,** 33–42.

Hartman, E. G., van Houten, M., van der Donk, J. A., and Frik, J. F. (1984b). Determination of specific anti-leptospiral immunoglobulins M and G in sera of experimentally infected dogs by solid-phase enzyme-linked immunosorbent assay. *Vet. Immunol. Immunopathol.* **7,** 43–51.

Hartman, E. G., van den Ingh, T. S. G. A. M., and Rothuizen, J. (1986). Clinical, pathological and serological features of spontaneous canine leptospirosis. An evaluation of the IgM- and IgG- specific ELISA. *Vet. Immunol. Immunopathol.* **13,** 261–271.

Hsiung, G. D. (1972). Parainfluenza 5 virus infection in man and animals. *Prog. Med. Virol.* **14,** 241–274.

Kelser, R. A. (1930). Rabies vaccine chloroform-treated. *J. Am. Vet. Med. Assoc.* **77,** 595–603.

Kontor, E. J., Wegrzyn, R. J., and Goodnow, R. A. (1981). Canine infectious tracheobronchitis: effects of an intranasal live canine parainfluenza—*Bordetella bronchiseptica* vaccine on viral shedding and clinical tracheobronchitis (kennel cough). *Am. J. Vet. Res.* **42,** 1694–1698.

Koprowski, H. (1954). Biological modification of rabies virus as a result of its adaptation to chicks and developing chick embryos. *Bull. WHO* **10,** 709–724.

Koptopoulos, G., and Cornwell, H. J. C. (1981). Canine adenoviruses: A review. *Vet. Bull.* **51,** 135–142.

Laidlaw, P. P., and Dunkin, G. W. (1928). Studies in dog distemper. V. The immunization of dogs. *J. Comp. Pathol.* **41,** 209–227.

Levy, S. A., Barthold, S. W., Dambach, D. M., and Wasmoen, T. L. (1993). Canine Lyme borreliosis. *Compend. Contin. Educ. Pract. Vet.* **15,** 833–848.

Lissman, B. A., Bosler, E. M., Camay, H., Ormiston, B. G., and Benach, J. L. (1984). Spirochete-associated arthritis (Lyme disease) in a dog. *J. Am. Vet. Med. Assoc.* **185,** 219–220.

Lovrich, S. D., Callister, S. M., Lim, L. C. L., DuChateau, B. K., and Schell, R. F. (1994). Seroprotective groups of Lyme borreliosis spirochetes from North American and Europe. *J. Infect. Dis.* **170,** 115–121.

Ma, J., Hine, P. M., Clough, E. R., Fish, D., Coughlin, R. T., Beltz, G. A., and Shew, M. G. (1996). Safety, efficacy, and immunogenicity of a recombinant Osp subunit canine Lyme disease vaccine. *Vaccine* **14,** 1366–1374.

Martin, M. L. (1985). Canine coronavirus enteritis and a recent outbreak following modified live virus vaccination. *Compend. Contin. Educ. Pract. Vet.* **7,** 1013–1017.

Marusyk, R. G., Norrby, E., and Lundqvist, U. (1970). Biophysical comparison of two canine adenoviruses. *J. Virol.* **5,** 507–512.

May, C., Carter, S. D., Barnes, A. McLean, C., Bennett, D., Coutts, A., and Grant, C. K. (1994). *Borrelia burgdorferi* infection in cats in the UK. *J. Small Anim. Pract.* **35,** 517–520.

McCandlish, I. A. P., and Thompson, H. (1978). Vaccination against canine bordetellosis using an aluminium hydroxide adjuvant vaccine. *Res. Vet. Sci.* **25,** 51–57.

Miller, A. S. H., Curtis, R., and Furminger, I. G. S. (1980). Persistence of immunity to infectious canine hepatitis using a killed vaccine. *Vet. Rec.* **106,** 343–344.

Osterhaus, A. D. M. E., Sundquist, B., Morein, B., and Steenis, G. (1986). Comparison of an experimental rabies iscom subunit vaccine with inactivated dog kidney cell vaccine. *Proc. Int. Vet. Immunol. Symp., 1st,* Guelph, Canada.

Parker, J. L., and White, K. K. (1992). Lyme borreliosis in cattle and horses. *Cornell Vet.* **82,** 253–274.

Parrish, C. R. (1991). Mapping specific functions in the capsid structure of canine parvovirus and feline panleukopenia virus suing infectious plasmid clones. *Virology* **183,** 195–205.

Pasteur, L. (1885). Méthodes pour prévenir la rage après morsure. *C. R. Hebd. Séances. Acad. Sci.* **101,** 765–772.

Pastoret, P.-P., Brochier, B., Aguilar-Setién, A., and Blancou, J. (1997). Part 2. Vaccination against rabies. *In* "Veterinary Vaccinology" (P.-P. Pastoret, J. Blancou, P. Vannier, and C. Verschueren, eds.), Chapter 18, pp. 616–628. Elsevier, Amsterdam.

Pensaert, M., and Callebaut, P. (1978). The coronaviruses: Clinical and structural aspects with practical implications. *Ann. Méd. Vét.* **122,** 301–322.

Pollock, R. V. H. (1984). The parvoviruses. Part II. Canine parvovirus. *Compend. Contin. Educ. Pract. Vet.* **6,** 653–656.

Pollock, R. V. H., and Carmichael, L. E. (1982). Dog response to inactivated canine parvovirus and feline panleukopenia vaccines. *Cornell Vet.* **72,** 16–35.

Pollock, R. V. H., and Carmichael, L. E. (1983). Use of modified feline panleukopenia virus vaccine to immunize dogs against canine parvovirus. *Am. J. Vet. Res.* **44,** 169–175.

Puntoni, V. (1923). Saggio di vaccinazione anticimurrosa preventiva eseguita per mezzo del virus specifico. *Ann. Igiene* **33,** 553.

Rimmelzwaan, G. (1990). Application of enzyme-linked immunosorbent assay systems for the serology and antigen detection in parvovirus, coronavirus and rotavirus infections in dogs in The Netherlands. Canine Parvovirus Infection: Novel Approaches to Diagnosis and Immune Prophylaxis. Theses; pp. 39–56. State University of Utrecht, The Netherlands.

Rockborn, G. (1960). A preliminary report on efforts to produce a living distemper vaccine in tissue culture. *J. Small Anim. Pract.* **1**, 53.

Rubarth, S. (1947). An acute virus disease with liver lesions in dogs (hepatitis contagiosa canis). A pathologico-anatomical and etiologic investigation. *Acta Pathol. Microbiol. Scand.* **24**, Suppl. 69.

Shade, F. J., and Goodnow, R. A. (1979). Intranasal immunization of dogs against *Bordetella bronchiseptica*-induced tracheobronchitis (kennel cough) with modified live *Bordetella bronchiseptica* vaccine. *Am. J. Vet. Res.* **40**, 1241–1243.

Sikes, R. K. (1975). Canine and feline vaccines—past and present. *In* "The Natural History of Rabies" (G. M. Baer, ed.), Vol. 2, pp. 177–187. Academic Press, New York.

Spach, D. H., Liles, W. C., Campbell, G. L., Quick, R. E., Anderson, D. E., Jr., and Fritsche, T. R. (1993). Tick-borne diseases in the United States. *N. Engl. J. Med.* **329**, 936–947.

Steere, A. C. (1989). Lyme disease. *N. Engl. J. Med.* **321**, 586–596.

Steere, A. C., Broderick, T. F., and Malawista, S. E. (1978). Erythema chronicum migrans and Lyme arthritis: Epidemiologic evidence for a tick vector. *Am. J. Epidemiol.* **108**, 312–321.

Stephensen, C. B., Welter, J., Thaker, S. R., Taylor, J., Tartaglia, J., and Paoletti, E. (1997). Canine distemper virus (CDV) infection of ferrets as a model for testing *Morbillivirus* vaccine strategies: NYVAC- and ALVAC-based CDV recombinants protect against symptomatic infection. *J. Virol.* **71**, 1506–1513.

Straubinger, R. K., Staubinger, A. F., Jacobson, R. H., Chang, Y.-F., Summers, B. A., Erb, H. N., and Appel, M. J. G. (1997a). Two lessons from the canine model of Lyme disease: Migration of *Borrelia burgdorferi* in tissues and persistence after antibiotic treatment. *J. Spirochete Tick-borne Dis.* **4**, 24–31.

Straubinger, R. K., Summers, B. A., Chang, Y.-F., and Appel, M. J. G. (1997b). Persistence of *Borrelia burgdorferi* in experimentally infected dogs after antibiotic treatment. *J. Clin. Microbiol.* **35**, 111–116.

Straubinger, R. K., Chang, Y.-F., Jacobson, R. H., and Appel, M. J. G. (1995). Sera from OspA-vaccinated dogs, but not those from tick-infected dogs, inhibit in vitro growth of *Borrelia burgdorferi*. *J. Clin. Microbiol.* **33**, 2745–2751.

Taylor, J., Weinberg, R., Languet, B., Desmettre, P., and Paoletti, E. (1988). Recombinant fowlpox virus inducing protective immunity in nonavian species. *Vaccine* **6**, 497–503.

Taylor, J., Trimarchi, C., Weinberg, R., Languet, B., Guillemin, F., Desmettre, P., and Paoletti, E. (1991). Efficacy studies on a canarypox rabies recombinant virus. *Vaccine* **9**, 190–193.

Tierkel, E. S., Kissling, R. E., Eidson, M., and Habel, K. (1953). A brief survey and progress report of controlled comparative experiments in canine rabies immunization. *Proc. Book, 90th Annu. Meet., Am. Vet. Med. Assoc.,* Toronto, Canada, *1953,* pp. 443–445.

Wagener, J. S., Sobonya, R., Minnich, L., and Taussig, L. M. (1984). Role of canine parainfluenza virus and *Bordetella bronchiseptica* in kennel cough. *Am. J. Vet. Res.* **45**, 1862–1866.

Wiktor, T. J., and Koprowski, H. (1980). Antigenic variants of rabies virus. *J. Exp. Med.* **152**, 99–112.

Wilson, R. B., Holladay, J. A., and Cave, J. S. (1986). A neurologic syndrome associated with use of a canine coronavirus-parvovirus vaccine in dogs. *Compend. Contin. Educ. Pract. Vet.* **8,** 117–123.

Wright, N., Thompson, H., Taylor, D., and Cornwell, H. (1973). *Bordetella bronchiseptica:* a reassessment of its role in canine respiratory disease. *Vet. Rec.* **91,** 486–487.

Yonezawa, Y. (1985). Genome analysis of virulent and attenuated strains of canine parainfluenza virus. *Br. Vet. J.* **141,** 192–194.

Analysis of the Protective Immunity Induced by Feline Immunodeficiency Virus Vaccination

MARGARET J. HOSIE AND OSWALD JARRETT

Department of Veterinary Pathology, University of Glasgow, Bearsden Road, Glasgow G61 1QH, United Kingdom

I. Introduction and Background
II. Whole Inactivated Virus Vaccines
III. Subunit Vaccines
IV. DNA Vaccination
References

I. Introduction and Background

Since its discovery in 1986 (Pedersen *et al.*, 1987), feline immunodeficiency virus (FIV) infection has been shown to result in an immunodeficiency in cats that is similar to AIDS in human beings. The virus is now recognized as a long-established and important feline pathogen (Brown *et al.*, 1994) and an appropriate animal model for human immunodeficiency virus (HIV) infection, playing a key role in the development of vaccines against HIV (Willett *et al.*, 1997). Furthermore, because FIV induces significant disease in cats, the development of an effective FIV vaccine is of great veterinary interest.

Following infection with FIV, cats display multiple clinical signs and frequently develop opportunistic infections (Pedersen *et al.*, 1987; Yamamoto *et al.*, 1989), immune dysfunction (Siebelink *et al.*, 1990; Hoffman-Fezer *et al.*, 1992), hematologic changes (Yamamoto *et al.*, 1989; Shelton *et al.*, 1990b), and neoplasia (Shelton *et al.*, 1990a). Abnormalities of $CD4^+$ and $CD8^+$ subpopulations develop, similar to those found in HIV-infected human beings. Decreased CD4:CD8 ratios have been documented in infected cats due to decreased CD4+ (Novotney et al., 1990; Hoffman-Fezer et al., 1992) and increased CD8+ cell numbers (Willett et al., 1993).

II. Whole Inactivated Virus Vaccines

Protection against FIV infection has been achieved by immunizing cats with a whole inactivated virus (WIV) vaccine (Yamamoto et al., 1991b, 1993; Hosie et al., 1995) produced from the FL4 feline lymphoblastoid cell line, which is persistently infected with the Petaluma isolate of FIV (FIV/PET) (Yamamoto et al., 1991a). This cell line appears to be unique, since it produces large amounts of FIV/PET rich in envelope glycoprotein (Env) that is well preserved during purification. In contrast, FIV grown in other cell lines contains relatively little or no Env protein (Hosie, 1994).

However, the protective immunity induced by this WIV vaccine does not extend to the heterologous isolate FIV/Glasgow-8 (FIV/GL-8) (Hosie, et al., 1995), which is antigenically distinct from FIV/PET by virus neutralization (Osborne et al., 1994) and established a higher viral load following infection compared to FIV/PET (M. J. Hosie and D. Klein, unpublished data). Thus, this vaccine system provides an excellent model in which to analyze the viral and host determinants contributing to the protection afforded by lentiviral vaccines.

The protection that we observed with the inactivated FIV vaccines was associated with a type-specific neutralizing antibody response and was retained when the challenge virus was propagated in an unrelated cell line (Hosie et al., 1995), indicating that the response was virus, and not cell, specific. As shown in Fig. 1a, cats immunized with the WIV vaccine prepared from FL4 cells developed significantly higher titers of virus neutralizing antibodies (VNAs) against the homologous FIV/PET isolate compared to the FIV/GL-8 isolate, suggesting that VNAs played a significant role in vaccine-induced protection.

To define more clearly the immunologic correlates of protection by the WIV vaccine, we immunized cats by a regime that ensured that only a proportion of the vaccinates would be protected and then compared the immune responses in protected and nonprotected cats (Hosie and Flynn, 1996). As shown in Fig. 1b, the protected cats developed significantly higher titers of VNA compared to the unprotected cats following immunization. When the protected cats were rechallenged 8 months later, the VNA levels had declined in some cats and there was

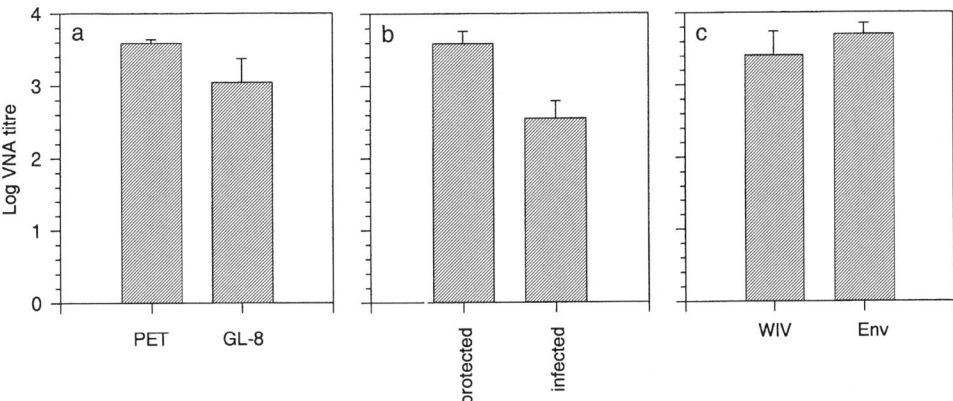

FIG. 1. Log titers of VNAs in protected compared to infected whole inactivated virus (WIV) vaccinates. (a) Log titers of VNAs generated by WIV vaccine against FIV/PET compared to FIV/GL-8. (b) Log titers of VNAs induced by WIV compared to Env subunit vaccine. (c) The mean log titers of VNA are shown ±2 standard errors of the mean (SEM).

no correlation between VNA levels and protection. It was observed, however, that the cats resisting this second challenge were those that possessed significant levels of cytotoxic T cells (CTLs) to Env proteins immediately following vaccination, 8 months previously. We concluded from these results that the high levels of VNA induced by the WIV vaccine may block infection with cell-free virus and prevent the establishment of a persistent infection, while CTLs are responsible for long-term protection and the maintenance of immunologic memory (Hosie and Flynn, 1996; Flynn et al., 1996).

The protective effect of antibodies has also been shown by the passive transfer of resistance to challenge with plasma from cats either immunized with WIV vaccines or infected with FIV/PET (Hohdatsu et al., 1993). Because passive transfer of plasma from cats immunized with uninfected T cells in which the vaccine virus was grown did not protect cats, anti-FIV antibodies and not anticellular antibodies appeared to be responsible for vaccine protection. Similarly, neonatal kittens receiving high levels of maternal antiviral antibodies from either vaccinated or infected queens were protected from FIV infection, whereas kittens that received low levels of maternal antibody became infected (Pu et al., 1997).

III. Subunit Vaccines

Although WIV vaccines have proved effective against FIV infection and may be valuable in identifying critical determinants of protective immunity against lentiviruses, the potential hazards associated with WIV vaccines have led to the investigation of alternative vaccines. Immunization with affinity-purified Env induced high titers of VNA as shown in Fig. 1c but gave only partial protection, suppressing the virus load in the blood mononuclear and lymph node cells of vaccinates (Hosie *et al.*, 1996). These results indicated that high titers of VNA are not sufficient for complete protection.

Despite equally high levels of VNA being elicited by both WIV and affinity-purified Env, the cats immunized with the affinity-purified Env displayed a more potent antibody response to determinants that were neither conformational nor glycosylation dependent, implying that the purified Env preparation was partially denatured. It is probable that antibodies directed to nonconformational epitopes of the surface Env protein neutralize FIV *in vitro* by a mechanism that does not operated against virus *in vivo*. Indeed, these antibodies may enhance infection *in vivo*. Evidence that supports this possibility comes from experiments in which immunization with peptides derived from the V3 loop of FIV gp120 also elicited VNA (Lombardi *et al.*, 1994) and yet the peptides not only failed to protect cats from challenge with the homologous virus but appeared to enhance viral replication (Lombardi *et al.*, 1994).

In other experiments, cats immunized with ISCOMs containing recombinant vaccinia virus (rVV)-expressed gp120 or gp160 displayed evidence of enhanced infection (Siebelink *et al.*, 1995). This enhancement of infection could be transferred to naive cats with plasma, indicating that the enhancement was mediated by FIV envelope-specific antibodies. Although there are at present no reliable systems for the detection of FIV-enhancing antibodies, complement and Fc receptor-mediated mechanisms of antibody-dependent enhancement have been described for HIV and SIV infection (Boyer *et al.*, 1991; Takeda *et al.*, 1988). It is possible that the affinity and specificity of the anti-Env antibodies induced by vaccination determines whether protection or enhancement result. Thus the rVV-expressed Env ISCOM vaccine may have induced Env-specific antibodies of different affinity and specificity to those induced by the WIV vaccine (Siebelink *et al.*, 1995). However, enhancement of infection has also been observed following immunization of cats with a recombinant FIV p24 ISCOM vaccine which did not induce anti-FIV envelope antibodies (Hosie et al., 1992), indicating that enhancement may arise by several different mechanisms.

IV. DNA Vaccination

Because vaccination with subunit vaccines has proved less successful than WIV immunization, the alternative strategy of DNA vaccination has been investigated. Having demonstrated that an infectious molecular clone of FIV/PET is fully expressed in vivo, giving rise to persistent infections in cats following intramuscular inoculation (Rigby et al., 1997), we proceeded to develop a novel DNA vaccine. The infectious molecular clone was rendered defective by engineering a 33 codon deletion in the pol gene. The defective clone, designated △RT, was administered by intramuscular inoculation at four different sites, each receiving 100 μg DNA on weeks 0, 10, and 23. The cats were challenged intraperitoneally 3 weeks after the third immunization with 25 ID_{50} of the homologous FIV/F-14 molecular clone and were monitored for 12 weeks following challenge.

No virus could be isolated from samples of peripheral blood mononuclear cells taken on the day of challenge, indicating that there was no reversion to virulence of the mutant provirus prior to challenge. Following challenge, 4/10 cats inoculated with △RT DNA remained virus-free whereas all 10 control cats became infected. Thus, there was significant protection in the 10 cats immunized with the △RT compared to the controls. Furthermore, the △RT vaccinates developed significantly lower clinical scores following challenge compared to the controls (Fig. 2, Table I). It was concluded that protective immunity to lentivirus infection may be induced by DNA immunization in the absence of a VNA response (Hosie et al., 1998).

These studies have shown that the use of DNA vaccines may be exploited in future trials to further define the mechanism of protection against FIV infection by manipulating the immune response in favor of inducing either CTL or humoral responses by coadministering type I or type II cytokine genes, respectively. If successful, these experiments will provide extremely useful information for guiding the choice of candidate vaccines for use against HIV infection in human beings.

Acknowledgments

Work in the authors' laboratory was funded by the U.K. Medical Research Council and by the European Commision's Concerted Action on FIV vaccination.

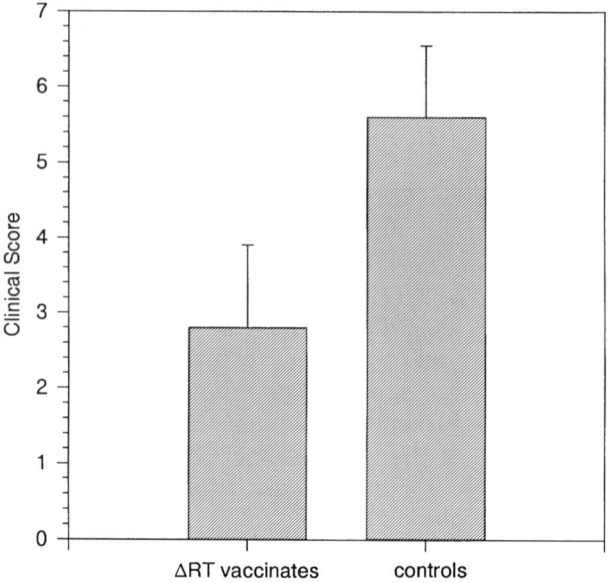

FIG. 2. Comparison of clinical scores (±2 SEM) following challenge of △RT DNA vaccinates and controls.

TABLE I

CLINICAL SCORING CRITERIA

Virus isolation	
Positive at 3 weeks pc	1
Positive at 6 weeks pc	1
Immunoblot analysis of plasma pc	
Positive at 6 weeks pc	1
Positive at 9 weeks pc	1
Viral load quantitation	
Virus isolated from 2×10^6 PBMC	1
Virus isolated from 2×10^5 PBMC	1
Virus isolated from 2×10^4 PBMC	1
Possible maximum score	7

REFERENCES

Boyer, V., Desgranges, C., Trabaud, M.-A., Fischer, E., and Kazatchkine, M. D. (1991). *J. Exp. Med.* **173,** 1151–1158.
Brown, E. W., Yuhki, N., Packer, C., and O'Brien, S. J. (1994). *J. Virol.* **68,** 5953–5968.
Flynn, J. N., Keating, P., Hosie, M. J., Mackett, M., Stephens, E. B., Beatty, J. A., Neil, J. C., and Jarrett, O. (1996). *J. Immunol.* **157,** 3658–3665.
Hoffman-Fezer, G., Thum, J., Ackley, C., Herbold, M., Mysliwietz, J., Thefeld, S., Hartmann, K., and Kraft, W. (1992). *J. Virol.* **66,** 1484–1488.
Hohdatsu, T., Pu, R., Torres, B. A., Trujillo, S., Gardner, M. B., and Yamamoto, J. K. (1993). *J. Virol.* **67,** 2344–2348.
Hosie, M. J. (1994). *Br. Vet. J.* **150,** 25–39.
Hosie, M. J., and Flynn, J. N. (1996). *J. Virol.* **70,** 7561–7568.
Hosie, M. J., Osborne, R., Reid, G., Neil, J. C., and Jarrett, O. (1992). *Vet. Immunol. Immunopathol.* **35,** 191–198.
Hosie, M., Osborne, R., Yamamoto, J. K., Neil, J. C., and Jarrett, O. (1995). *J. Virol.* **69,** 1253–1255.
Hosie, M. J., Dunsford, T. H., de Ronde, A., Willett, B. J., Cannon, C. A., Neil, J. C., and Jarrett, O. (1996). *Vaccine* **14,** 405–411.
Hosie, M. J., Flynn, N., Rigby, M., Cannon, C., Dunsford, T., Mackay, N., Argyle, D., Willett, B. J., Miyazawa, T., Onions, D., Jarrett, O., and Neil, J. C. (1998). *J. Virol.* **72** (in press).
Lombardi, S., Garzelli, C., Pistello, M., Massi, C., Matteucci, D., Baldinotti, F., Cammaroto, G., Da Prato, L., Bandecchi, P., Tozzini, F., and Bendinelli, M. (1994). *J. Virol.* **68,** 8374–8379.
Novotney, C., English, R. V., Housman, J., Davidson, M. G., Naisse, M. P., Jeng, C., Davis, W. C., and Tompkins, M. B. (1990). *AIDS* **4,** 1213–1218.
Osborne, R., Rigby, M., Siebelink, K., Neil, J. C., and Jarrett, O. (1994). *J. Gen. Virol.* **75,** 3641–3645.
Pedersen, N. C., Ho, E. W., Brown, M. L., and Yamamoto, J. K. (1987). *Science* **235,** 790–793.
Pu, R., Okada, S., Little, E. R., Xu, B., Stoffs, W. V., and Yamamoto, J. K. (1997). *AIDS* **9,** 235–242.
Rigby, M. A., Hosie, M. J., Willett, B. J., Mackay, N., McDonald, M., Cannon, C., Dunsford, T., Jarrett, O., and Neil, J. C. (1997). *AIDS Res. Hum. Retroviruses* **13,** 405–412.
Shelton, G. H., Grant, C. K., Cotter, S. M., Gardner, M. B., Hardy, W. D. J., and DiGiacomo, R. F. (1990a). *J. AIDS* **3,** 623–630.
Shelton, G. H., Linenberger, M. L., Grant, C. K., and Abkowitz, J. L. (1990b). *Blood* **76,** 1104–1109.
Siebelink, K. H. J., Chu, I., Rimmelzwaan, G. F., Wiejer, K., van Herwijnen, R., Knell, P., Egberink, H. F., Bosch, M., and Osterhaus, A. D. M. E. (1990). *AIDS Res. Hum. Retroviruses* **6,** 1373–1378.
Siebelink, K. H. J., Tijhaar, E., Huisman, R. C., Huisman, W., de Ronde, A., Darby, I. H., Francis, M. J., Rimmelzwaan, G. F., and Osterhaus, A. D. M. E. (1995). *J. Virol.* **69,** 3704–3711.
Takeda, A., Tuazon, C. V., and Ennis, F. A. (1988). *Science* **242,** 580–583.
Willett, B. J., Hosie, M. J., Neil, J. C., and Jarrett, O. (1993). *Immunology* **78,** 1–6.
Willett, B. J., Flynn, J. N., and Hosie, M. J. (1997). *Immunol. Today* **18,** 182–189.

Yamamoto, J. K., Hansen, H., Ho, E. W., Morishita, T. Y., Okuda, T., Saura, T. R., Nakamura, R. M., and Pedersen, N. C. (1989). *J. Am. Vet. Med. Assoc.* **194,** 213–220.

Yamamoto, J. K., Ackley, C. D., Zochlinski, H., Louie, H., Pembroke, E., Torten, M., Hansen, H., Munn, R., and Okuda, T. (1991a). *Intervirology* **32,** 361–375.

Yamamoto, J. K., Okuda, T., Ackley, C. D., Louie, H., Pembroke, E., Zochlinski, H., Munn, R. J., and Gardner, M. B. (1991b). *AIDS Res. Hum. Retroviruses* **7,** 911–921.

Yamamoto, J. K., Hohdatsu, T., Olmsted, R. A., Pu, R., Louie, H., Zochlinski, H. A., Acevedo, V., Johnson, H. M., Soulds, G. A., and Gardner, M. B. (1993). *J. Virol.* **67,** 601–605.

Vaccination of Cats against Emerging and Reemerging Zoonotic Pathogens

CHRISTOPHER W. OLSEN

Department of Pathobiological Sciences, School of Veterinary Medicine, University of Wisconsin–Madison, 2015 Linden Drive West, Madison, Wisconsin 53706

I. Introduction
II. *Toxoplasma gondii*
III. *Bartonella henselae*
IV. *Helicobacter pylori*
V. Other Agents
VI. Summary
    Acknowledgments
    References

I. Introduction

During the last 20 years, very significant changes have occurred in the field of infectious diseases. We've gone from a time when many people felt that antibiotics and vaccines were going to make infectious diseases a thing of the past, to a point where the world's population is faced with an array of emerging and reemerging infectious agents. We are threatened by bacteria that are resistant to some of the last lines of antibiotic defense, such as methicillin-resistant *Staphylococcus aureus* and vancomycin-resistant enterococci, as well as newly discovered bacteria, such as *Escherichia coli* O157:H7. We are faced with "new" viruses, such as the Ebola viruses, the Sin Nombre group of hantapulmonary syndrome viruses, and the equine morbillivirus in Australia, that have never been encountered before. And "old" diseases, such as tuberculosis, plague, leptospirosis, toxoplasmosis, and rabies, have taken on very new significance.

One remarkable fact is that many of these infectious agents that threaten the human population are either directly zoonotic or involve animals, rather than humans, as their primary reservoir in nature. Consequently, vaccination of animals may be an important consideration for control of some of these diseases. Vaccination of animals as a public health tool is clearly not a novel idea. Over the years, the veterinary profession has developed vaccination regimes for animals against a number of diseases of public health significance (Table I). The purpose of this paper is to review specifically the feasibility and wisdom of vaccinating cats against infectious agents of emerging or reemerging importance.

On a global scale, there are many reasons, both environmental and societal, why infectious agents "emerge" (Morse, 1995). One of the most important and challenging for the purposes of this discussion is the fact that the world today includes a substantial population of immunocompromised human beings. These include individuals who are immunosuppressed for organ transplantation [the number of organ transplants in the United States rose nearly fourfold from 1982 to 1992 (United Network of Organ Sharing Scientific Registry Data, 1995)] as well as people who are immunosuppressed in the course of cancer chemotherapy or therapy for autoimmune disorders. But the single entity that has changed the picture of infectious diseases the most dramatically in the last 15 years is human immunodeficiency virus (HIV) and the AIDS pandemic. Between 1982 and 1994, HIV went from being a virtually unknown entity to being the number one cause of death among persons from 25 to 44 years of age in the United States (National Vital Statistics). The particular form of immune compromise

TABLE I

SELECTED ZOONOTIC AGENTS FOR WHICH
ANIMAL VACCINES HAVE BEEN PRODUCED

Rabies virus
Equine encephalitis viruses
Rift valley fever virus
Japanese encephalitis virus
Bacillus anthracis
Brucella abortus
Leptospira spp.

that HIV patients suffer, with intensive compromise of the cellular immune systems, puts them at particular risk for two of the agents that are discussed here, *Toxoplasma gondii* and *Bartonella henselae*.

In addition to rabies, for which vaccines already exist, cats play or may play a role in the transmission of a number of different infectious agents to humans (Table II). These include organisms for which the role of cats is very well defined, such as *T. gondii*, as well as others where the role of cats is more limited or still undefined. This review focuses on three agents that span this spectrum: *T. gondii*, *B. henselae*, and *Helicobacter pylori*. In each case, a series of considerations should be addressed in deciding whether vaccination of cats makes sense as a public health tool (Table III). First, does the agent cause a clinically significant problem in humans that warrants a large-scale control program of any kind? Second, does the agent cause clinical disease in cats as well? This has implications not only for whether vaccination of cats can be justified for their own protection, but also for how a candidate vaccine's efficacy will be tested. Third, what other sources of infection are there for humans and will vaccination of cats really make a significant difference in the overall epidemiology of the disease? Fourth, is enough known about what constitutes protective immunity to the agent in cats so that a vaccine can be engineered to elicit the appropriate response? And finally, does a candidate vaccine or vaccination strategy pose any risks for cats?

TABLE II

SELECTED INFECTIOUS AGENTS FOR WHICH CATS
MAY PLAY A ROLE IN TRANSMISSION TO HUMANS

Rabies virus
Toxoplasma gondii
Bartonella henselae
Yersinia pestis
Francisella tularensis
Coxiella burnetti
Salmonella spp.
Giardia lamblia
Cryptosporidium parvum
Microsporum canis
Sporothrix schenkii
Helicobacter pylori

TABLE III

CONSIDERATIONS FOR VACCINATION OF CATS AGAINST ZOONOTIC AGENTS

1. How clinically significant and prevalent is the disease in humans?
2. Does the agent induce clinical disease in cats?
3. How significant is the role of cats in transmission of the agent to humans?
4. Is the basis for protective immunity in cats understood?
5. Are there vaccine-associated risks for the cat?

II. *Toxoplasma gondii*

Toxoplasmosis is an example of a zoonotic disease in which cats play a very clear role in the maintenance of the organism in nature. Cats are, in fact, the ultimate source of the organism since they are the only host in which this coccidian parasite can undergo sexual replication (see Dubey, 1994, for a review of the life cycle). Sexual replication of *T. gondii* occurs in the intestinal tract of cats, leading to the production of oocysts. These oocysts are passed in their feces and are the source of infection for intermediate hosts such as rodents, birds, sheep, and pigs and for humans. The oocysts require 1–5 days of incubation in the soil for sporulation to occur and to become infectious. In the intermediate hosts and humans, *T. gondii* initially undergoes a period of rapid asexual replication as tachyzoites and then encysts in tissues as bradyzoites. These tissue cysts in the intermediate hosts are the major source for infection of cats, thus completing the life cycle. Cats can also be infected by ingestion of oocysts, as occurs in the intermediate hosts, but this route of infection is much less efficient (Dubey, 1996).

In humans, infection with *T. gondii* is relatively common, with seroprevalence rates of 3–70% in the United States and up to 90% in other parts of the world (Beaman *et al.*, 1995). Acute infections in immunocompetent individuals are largely self-limiting, but toxoplasmosis can be a devastating disease when a woman is infected during pregnancy. The tachyzoites cross the placenta to infect the fetus and can lead to fetal death and spontaneous abortion or congenital defects such as hydrocephalus, mental retardation, seizure disorders, and chorioretinitis and blindness (Beaman *et al.*, 1995). The other area of real concern is toxoplasmosis in immunocompromised hosts. Disease in these individuals most commonly occurs because of recrudescence of a previously latent infection. In particular, toxoplasma encephalitis is a common and serious complication in AIDS patients (Beaman *et al.*, 1995). In cats, *T. gondii* can infect virtually any organ in the body, leading to a diverse array of clinical presentations, including fevers of unknown origin, uveitis or chorioretinitis, myositis, encephalitis, and pulmonary disease (Taboada and Merchant, 1995). Thus, *T. gondii* is an example of a zoonotic agent that can cause serious disease in both humans and cats.

It is clear from a large number of studies in mice that resistance to and resolution of *T. gondii* infections are dependent on strong cell-mediated immune responses (Araujo, 1994; Sher *et al.*, 1995). Interferon gamma from NK cells and CD4 and CD8 T cells is an important factor in the early response to infection (Khan *et al.*, 1994; Sher *et al.*, 1995). Cell-mediated immunity to *T. gondii* can be enhanced by exogenous treatment with interleukin 7 (IL-7) (Kaspar *et al.*, 1995), IL-12 (Gazzinelli *et al.*, 1993), and IL-15 (Khan and Kaspar, 1996). In mice, a variety of different forms of *Toxoplasma* vaccines and adjuvants have been tested, including irradiated oocysts, *T. gondii* lysates + cholera toxin, ISCOMs, and mutant strains of the organism (Araujo, 1994). In domestic animals, the goal of vaccination against *T. gondii* varies for different species. In pigs, for instance, a major public health concern is to eliminate the organisms from meat. Hence, mutant strains of *T. gondii* (RH, ts-4) that do not form tissue cysts themselves and block formation of tissue cysts after challenge infection have been investigated as candidate vaccines (Lindsay *et al.*, 1993; Pinckney *et al.*, 1994; Dubey *et al.*, 1991, 1994). In sheep, both mutant strains and an ISCOM vaccine (marketed as Toxovax® in New Zealand) have been used in an attempt to reduce transplacental transmission of *T. gondii* and abortion, an economically important concern to the sheep industry (Buxton *et al.*, 1989, 1991).

In cats, the goal of vaccination from a public health point of view is to prevent the shedding of oocysts following infection. This should be feasible because cats that recover from infection are immune to the shedding of oocysts on subsequent challenge (Davis and Dubey, 1995; Dubey, 1995; Frenkel and Smith, 1982). This immunity is dependent on enteroepithelial replication of the organism (Frenkel and Smith, 1982). Consequently, an early approach to vaccination involved intentional infection of cats with *T. gondii,* followed by treatment with monensin or sulfadiazine/pyrimethamine to eliminate oocyst shedding during this immunizing infection (Frenkel and Smith, 1982). More recently, a mutant, "incomplete" strain designated T-263 has been developed and tested extensively (Pfefferkorn and Pfefferkorn, 1976; Frenkel *et al.*, 1991; Choromanski *et al.*, 1994; Freyre *et al.*, 1993). This vaccine has been shown to induce 84–100% protection from oocyst shedding after challenge infection (Frenkel *et al.*, 1991; Freyre *et al.*, 1993). Following oral administration of either tissue cysts or bradyzoites, T-263 organisms undergo partial enteroepithelial replication in the gut of the cat, but are generally blocked from being able to complete the sexual replication cycle, possibly because gametes of only one sex are produced (Frenkel et al., 1991). Thus, there should be no public health risk of oocyst shedding with this strain of the organism. The T-263 vaccine appears to be safe for most cats, but deaths have been reported in cats that were infected with FeLV 11 months prior to vaccination (Choromanski et al., 1994). Consequently, further studies in FeLV, FIV, and corticosteroid-immunosuppressed cats are warranted. In addition, because this vaccine is produced by infection of mice, there is also the potential, albeit limited, for transmission of murine viruses in the preparation, and large-scale production may be problematic. A future goal should be to further elucidate specific antigens and immune responses associated with the enteric stages of *T. gondii* (Kaspar, 1989). This information may allow development of subunit or DNA-based vaccines that will eliminate the concerns associated with a live vaccine. However, it is important to realize that any vaccination strategy against *T. gondii* is likely to have difficulty targeting the one group of cats that, from a public health perspective, may be most important to vaccinate, feral cats. Finally, although cats are the ultimate source of the organism in the overall ecology of *T. gondii*, people are also commonly infected through the consumption of tissue cysts in undercooked meats (Dubey, 1994; Beaman et al., 1995). Thus, it may be prudent to consider a multifaceted strategy in which not only cats, but also people themselves and pigs are vaccinated (Fishback and Frenkel, 1990).

III. *Bartonella henselae*

Cat scratch disease (CSD) has been recognized clinically since the early 1900s, but it was only in 1983 that a microbial organism was first identified in tissues by silver staining (Wear et al., 1983). The disease had initially been associated with a diverse group of agents, including herpesviruses, *Chlamydia*, and *Pasteurella* (Emmons et al., 1976), then more specifically *Afipia felis* (Birkness et al., 1992). However, the etiologic agent of CSD has now been defined as *Bartonella henselae*, a curved gram-negative rod previously called *Rochalimaea henselae* (as reviewed in Regnery and Tappero, 1995). Infection with *B. henselae* is most common in children. A survey in Connecticut in 1992–1993 found a prevalence rate in children <10 years of age = 9.3 cases/100,000 population, compared to 3.7/100,000 overall (Hamilton et al., 1995).

The classic form of CSD consists of an uncomplicated course of fever and regional lymphadenopathy proximal to a cat scratch or, less commonly, a bite. The scratch may persist as a nonhealing skin lesion with associated papules and/or pustules (Regnery and Tappero, 1995; Fischer, 1995). In this form, CSD is a fairly benign disease, aside from the emotional stress that an enlarged lymph node may represent a lymphoma. Unfortunately, infection can be more serious in children, including progression to Parinaud's syndrome (extensive preauricular swelling and ocular granulomas) and the development of encephalitis several weeks after the initial lymphadenopathy (Fischer, 1995). *Bartonella henselae* infection is also a serious concern in immunocompromised hosts, because these patients can develop systemic infections. Clinical presentations in these individuals include bacillary angiomatosis (a painful, nodular proliferation of blood vessels in the skin or in internal organs), peliosis hepatis (an unusual form of liver disease), endocarditis, pulmonary nodules, hepatitis, splenitis, and encephalitis (Regnery and Tappero, 1995; Adal et al., 1994; Caniza et al., 1995; Holmes et al., 1995; Liston and Koehler, 1996; Drancourt et al., 1996).

In cats, serologic data suggest that infection with *B. henselae* is quite common, with seroprevalence rates of 14–54% in the United States (Childs et al., 1994, 1995; Jameson et al., 1995). Seroprevalence rates are highest in warmer areas of the United States and areas with higher annual precipitation (Jameson et al., 1995). Mild signs, including transient fever, central nervous system (CNS) disturbance, lymphadenopathy, and anemia, developed in cats after experimental infection by intramuscular and intravenous inoculation (Kordick and Breitschwerdt, 1997). Interestingly, intradermal and subcutaneous inoculation led to raised skin lesions and lymphadenopathy reminiscent of CSD in humans (Greene et al., 1996). However, there have been no clinical signs associated with natural infection in cats, suggesting that these may not be normal routes of infection from cat to cat.

There are strong associations between the development of CSD in humans and contact with cats. Ninety percent of people with CSD have a history of contact with cats, in particular with kittens, seropositive cats (Zangwill et al., 1993), or bacteremic cats (Kordick et al., 1995). There has also been a great deal of interest in the role that fleas may play in transmission of *B. henselae* between cats and from cats to people. *Bartonella henselae* DNA has been detected by polymerase chain reaction (PCR) in fleas for up to 9 days after artificial feeding on blood containing the organism, and it is shed in flea feces in an infectious form (Higgins et al., 1996). Experimentally, *B. henselae* can be spread from cat to cat by infected fleas (Chomel et al., 1996), but not horizontally in their absence (Chomel et al., 1996; Abbott et al., 1997). Furthermore, the risk for CSD in people is increased if they own at least one kitten with fleas (Zangwill et al., 1993). However, the role of fleas in the overall epidemiology of B. henselae infection in humans is yet to be defined.

Information concerning protective immunity to B. henselae is limited. Cellular immune mechanisms are likely to be of importance since the most severe forms of infection occur in individuals with compromised cellular immunity and PBMC from AIDS patients have reduced abilities to phagocytose the organism and to initiate oxygen radical bursts (Rodriguez-Barradas et al., 1995). However, antibody-mediated mechanisms may also play a role. Immune sera enhanced phagocytosis and oxygen radical production in one study (Rodriguez-Barradas et al., 1995), and humans make strong antibody responses against specific Bartonella proteins (Anderson et al., 1995). However, the role of specific antigens in protection from infection remains unclear.

One of the most intriguing aspects of B. henselae infection in cats is the fact that they mount a strong antibody response, yet this response is insufficient to clear the organism. Bartonella henselae persists in cats in a prolonged (weeks to months) bacteremia phase in the presence of specific antibodies (Kordick et al., 1995; Abbott et al., 1997; Kordick and Breitschwerdt, 1997; Regnery et al., 1996; Breitschwerdt and Kordick, 1995). This may be due to either periodic release of organisms from an intracellular source or progressive antigenic variation (Kordick and Breitschwerdt, 1997). Cats can, however, eventually clear this bacteremia. When they do, they appear, from a number of experimental infection and antibiotic treatment studies, to be solidly immune to reinfection (Greene et al., 1996; Regnery et al., 1996; Abbott et al., 1997). So clearly some form of protective immunity eventually develops in cats, but what this is and whether it can be simulated through vaccination remains to be determined. However, in light of the strong epidemiologic links between cats and CSD development in humans, continued efforts to develop a B. henselae vaccine for cats as a public health measure is warranted. Antibiotic treatment of cats belonging to high-risk, immunocompromised persons may be a useful adjunct tool to reduce potential exposure in individual cases. However, this approach would only be useful after a person knew they were immunocompromised and sought testing of their cat, and would do nothing to reduce exposure of children and healthy adults. Incorporation of B. henselae into the routine kittenhood vaccination regimes would provide more widespread protection for all individuals and has the potential to reduce the role of cats as a reservoir for infection.

IV. Helicobacter pylori

Whereas a great deal is already known about the roles of cats in the epidemiology of *T. gondii* and *B. henselae*, the zoonotic potential of *H. pylori* is only beginning to be addressed. *Helicobacter pylori* was first isolated from humans in 1982, but since then, its discovery has revolutionized the way that gastric ulcer disease is treated. *Helicobacter pylori* can be isolated from virtually all people with peptic ulcer disease and use of appropriate antibiotics in ulcer therapy has dramatically reduced ulcer recurrence rates (Blaser, 1995). In addition, *H. pylori* infection in humans is also associated with duodenal ulcers, chronic gastritis, gastric carcinoma, and lymphoma (Blaser, 1995).

Helicobacter pylori is a gram-negative, microaerophilic curved rod that lives in the mucous layer of the stomach and duodenum and is uniquely able to survive at low pH (in part, through expression of a urease enzyme; Blaser, 1995). The rate of colonization with this organism increases with age. It is estimated that up to 50% of people in the United States are infected by 60 years of age, with even higher rates of infection in underdeveloped countries (Blaser, 1995). But in addition to humans, *Helicobacter* spp. have been isolated from a variety of animals as well: *H. pylori* from pigs with gastric ulcers (Krakowka *et al.*, 1995), *H. mustelae* from ferrets with hypergastrinemia and ulcers (Perkins *et al.*, 1996; Lee, 1994), and *H. acinomyx* from cheetahs with gastritis (Eaton *et al.*, 1993). Of most importance for this discussion, *H. pylori* has been isolated from domestic cats with gastritis (Handt *et al.*, 1995). Fox and colleagues have demonstrated that cats can be experimentally infected with *H. pylori* (Fox *et al.*, 1995), that infection is associated with gastritis (Handt *et al.*, 1995; Fox *et al.*, 1995), that infected cats shed *H. pylori* in saliva (50% of 12 infected cats were culture positive) and feces (80% of 5 infected cats) (Fox *et al.*, 1996), and that *H. pylori* from cats shares 99.7% sequence identity (16S rRNA) with isolates from humans (Fox *et al.*, 1995). Consequently, there is growing interest in the potential for *H. pylori* to be a zoonosis associated with cats. However, there is, as of yet, no proof of any role for cats in transmission of this organism to people. So until a great deal more is known, consideration of vaccination of cats is premature. However, the ability to experimentally infect cats and induce gastritis has generated interest in using cats as an experimental model for vaccine development in humans (Lee, 1996).

V. Other Agents

This review has focused on the feasibility and practicality of vaccinating cats against three specific zoonotic agents. However, cats can be involved in the transmission of a number of other infectious organisms to people (Table II). Among these, at least one biotechnology/vaccine company is currently developing a *Yersinia pestis* vaccine for cats. *Yersinia pestis* is the causative agent of plague. Zoonotic transmission of *Y. pestis* from cats has been documented, particularly in the western United States (Doll *et al.*, 1994, and as reviewed in Eidson *et al.*, 1991), but the majority of human cases of plague are acquired from contact with infected rodents and their fleas (Eidson *et al.*, 1991). Thus, it is unlikely that vaccination of cats would substantially affect the epidemiology and ecology of this disease on a national or international scale. Likewise, cats play only minor roles in the overall epidemiology of human infection with *Francisella tularensis* (tularemia), *Coxiella burnetti* (Q-fever), *Sporothrix schenckii*, and the enteric pathogens *Salmonella* spp., *Giardia lamblia*, and *Cryptosporidium parvum*. In addition, the prevalence of *Microsporum canis* infection in people is low compared to infection with dermatophytes of human origin. Thus, vaccination of cats as a public health measure is unwarranted for these agents.

VI. Summary

Many of the emerging infectious agents that threaten the human population are either directly zoonotic or involve animals, rather than humans, as their primary reservoir in nature. Vaccination of animals may be an important consideration for control of some of these diseases, and this review has specifically focused on the concept of vaccinating cats in the prevention of infection with *T. gondii*, *B. henselae*, and *H. pylori*. If we return to the considerations that were presented in Table III, *T. gondii* is really the only one of these three agents for which each of these "criteria" for vaccination is fulfilled at the present time. However, cats clearly play an important role in the epidemiology of infection with *B. henselae* and this is an organism for which we probably will and should see a vaccine for widespread and routine use in cats.

Acknowledgments

The author thanks Dr. Rhonda Pinckney for helpful discussions during the preparation of this manuscript.

REFERENCES

Abbott, R. C., Chomel, B. B., Kasten, R. W., Floyd-Hawkins, K. A., Kikuchi, Y., Koehler, J. E., and Pedersen, N. C. (1997). Experimental and natural infection with *Bartonella henselae* in domestic cats. *Comp. Immunol. Microbiol. Infect. Dis.* **20,** 41–51.

Adal, K. A., Cockerell, C. J., and Petri, W. A. (1994). Cat scratch disease, bacillary angiomatosis, and other infections due to *Rochalimaea*. *N. Engl. J. Med.* **330,** 1509–1515.

Anderson, B., Lu, E., Jones, D., and Regnery, R. (1995). Characterization of a 17-kilodalton antigen of *Bartonella henselae* reactive with sera from patients with cat scratch disease. *J. Clin. Microbiol.* **33,** 2358–2365.

Araujo, F. G. (1994). Immunization against *Toxoplasma gondii*. *Parasitol. Today* **10,** 358–360.

Beaman, M. H., McCabe, R. E., Wong, S.-Y., and Remington, J. S. (1995). *Toxoplasma gondii*. In "Mandell, Douglas and Bennett's Principles and Practice of Infectious Diseases" (G. L. Mandell, J. E. Bennett, and R. Dolin, eds.), pp. 2455–2475. Churchill-Livingstone, New York.

Birkness, K. A., George, V. G., White, E. H., Stephens, D. S., and Quinn, F. D. (1992). Intracellular growth of *Afipia felis*, a putative etiologic agent of cat scratch disease. *Infect. Immun.* **60,** 2281–2287.

Blaser, M. J. (1995). *Helicobacter pylori* and related organisms. In "Mandell, Douglas and Bennett's Principles and Practice of Infectious Diseases" (G. L. Mandell, J. E. Bennett, and R. Dolin, eds.), pp. 1956–1964. Churchill-Livingstone, New York.

Breitschwerdt, E. B., and Kordick, D. L. (1995). Bartonellosis. *J. Am. Vet. Med. Assoc.* **206,** 1928–1931.

Buxton, D., Uggula, A., Lovgren, K., Thomson, K., Lunden, A., Morien, B., and Blewett, D. A. (1989). Trial of a novel experimental *Toxoplasma* ISCOM vaccine in pregnant sheep. *Br. Vet. J.* **145,** 451–457.

Buxton, D., Thomson, K., Maley, S., Wright, S., and Bos, H. J. (1991). Vaccination of sheep with a live incomplete strain (S48) of *Toxoplasma gondii* and their immunity to challenge when pregnant. *Vet. Rec.* **129,** 89–93.

Caniza, M. A., Granger, D. L., Wilson, K. H., Washington, M. K., Kordick, D. L., Frush, D. P., and Blitchington, R. B. (1995). *Bartonella henselae:* Etiology of pulmonary nodules in a patient with depressed cell-mediated immunity. *Clin. Infect. Dis.* **20,** 1505–1511.

Childs, J. E., Rooney, J. A., Cooper, J. L., Olson, J. G., and Regnery, R. L. (1994). Epidemiologic observations on infection with *Rochalimaea* species among cats living in Baltimore, Md. *J. Am. Vet. Med. Assoc.* **204,** 1775–1778.

Childs, J. E., Olson, J. G., Wolf, A., Cohen, N., Fakile, Y., Rooney, J. A., Bacellar, F., and Regnery, R. L. (1995). Prevalence of antibodies to *Rochalimaea* species (cat-scratch disease agent) in cats. *Vet. Rec.* **136,** 519–520.

Chomel, B. B., Kasten, R. W., Floyd-Hawkins, K., Chi, B. H., Yamamoto, K., Roberts-Wilson, J., Gurfield, A. N., Abbott, R. C., Pedersen, N. C., and Koehler, J. E. (1996). Experimental transmission of *Bartonella henselae* by the cat flea. *J. Clin. Microbiol.* **34,** 1952–1956.

Choromanski, L., Freyre, A., Brown, K., Popiel, I., and Shibley, G. (1994). Safety aspects of a vaccine for cats containing a *Toxoplasma gondii* mutant strain. *J. Eukaryotic Microbiol.* **41,** 8S.

Davis, S. W., and Dubey, J. P. (1995). Mediation of immunity to *Toxoplasma gondii* oocyst shedding in cats. *J. Parasitol.* **81,** 882–886.

Doll, J. M., Zeitz, P. S., Ettestad, P., Bucholtz, A. L., Davis, T., and Gage, K. (1994). Cat-

transmitted fatal pneumonic plague in a person who traveled from Colorado to Arizona. *Am. J. Trop. Med. Hyg.* **51,** 109–114.
Drancourt, M., Birtles, R., Chaumentin, G., Vandenesch, F., Etienne, J., and Raoult, D. (1996). New serotype of *Bartonella henselae* in endocarditis and cat-scratch disease. *Lancet* **347,** 441–443.
Dubey, J. P. (1994). Toxoplasmosis. *J. Am. Vet. Med. Assoc.* **205,** 1593–1598.
Dubey, J. P. (1995). Duration of immunity to shedding of *Toxoplasma gondii* oocysts by cats. *J. Parasitol.* **81,** 410–415.
Dubey, J. P. (1996). Infectivity and pathogenicity of *Toxoplasma gondii* oocysts for cats. *J. Parasitol.* **82,** 957–961.
Dubey, J. P., Urban, J. F., Jr., and Davis, S. W. (1991). Protective immunity to toxoplasmosis in pigs vaccinated with a nonpersistent strain of *Toxoplasma gondii*. *Am. J. Vet. Res.* **52,** 1316–1319.
Dubey, J. P., Baker, D. G., Davis, S. W., Urban, J. F., and Shen, S. K. (1994). Persistence of immunity to toxoplasmosis in pigs vaccinated with a nonpersistent strain of *Toxoplasma gondii*. *Am. J. Vet. Res.* **55,** 982–987.
Eaton, K. A., Radin, M. J., Kramer, L., Wack, R., Sherding, R., Krakowka, S., and Morgan, D. R. (1993). Epizootic gastritis in cheetahs associated with gastric spiral bacilli. *Vet. Pathol.* **30,** 55–63.
Eidson, M., Thilsted, J. P., and Rollag, O. J. (1991). Clinical, clinicopathologic, and pathological features of plague in cats: 119 cases (1977–1988). *J. Am. Vet. Med. Assoc.* **199,** 1191–1197.
Emmons, R. W., Riggs, J. L., and Schacter, J. (1976). Continuing the search for the etiology of cat scratch disease. *J. Clin. Microbiol.* **4,** 112–114.
Fischer, G. W. (1995). Cat scratch disease. *In* "Mandell, Douglas and Bennett's Principles and Practice of Infectious Diseases" (G. L. Mandell, J. E. Bennett, and R. Dolin, eds.), pp. 1310–1312. Churchill-Livingstone, New York.
Fishback, J. L., and Frenkel, J. K. (1990). Prospective vaccines to prevent feline shedding of *Toxoplasma* oocysts. *Compend. Contin. Educ. Pract. Vet.* **12,** 643–648.
Fox, J. G., Batchelder, M., Marini, R., Yan, L., Handt, L., Li, X., Shames, B., Hayward, A., Campbell, J., and Murphy, J. C. (1995). *Helicobacter pylori*-induced gastritis in the domestic cat. *Infect. Immun.* **63,** 2674–2681.
Fox, J. G., Perkins, S., Yan, L., Shen, Z., Attardo, L., and Pappo, J. (1996). Local immune response in *Helicobacter pylori*-infected cats and identification of *H. pylori* in saliva, gastric fluid and faeces. *Immunology* **88,** 400–406.
Frenkel, J. K., and Smith, D. D. (1982). Immunization of cats against shedding of *Toxoplasma* oocysts. *J. Parasitol.* **68,** 744–748.
Frenkel, J. K., Pfefferkorn, E. R., Smith, D. D., and Fishback, J. L. (1991). Prospective vaccine prepared from a new mutant of *Toxoplasma gondii* for use in cats. *Am. J. Vet. Res.* **52,** 759–763.
Freyre, A., Choromanski, L., Fishback, J. L., and Popiel, I. (1993). Immunization of cats with tissue cysts, bradyzoites, and tachyzoites of the T-263 strain of *Toxoplasma gondii*. *J. Parasitol.* **79,** 716–719.
Gazzinelli, R. T., Hieny, S., Wynn, T. A., Wolf, S., and Sher, A. (1993). Interleukin-12 is required for the T-lymphocyte-independent induction of interferon gamma by an intracellular parasite and induces resistance in T-cell-deficient hosts. *Proc. Natl. Acad. Sci. USA* **90,** 6115–6119.
Greene, C. E., McDermott, M., Jameson, P. H., Atkins, C. L., and Marks, A. M. (1996). *Bartonella henselae* infection in cats: Evaluation during primary infection, treatment, and rechallenge infection. *J. Clin. Microbiol.* **34,** 1682–1685.

Hamilton, D. H., Zangwill, K. M., Hadler, J. L., and Cartter, M. L. (1995). Cat-scratch disease—Connecticut, 1992–1993. *J. Infect. Dis.* **172,** 570–573.

Handt, L. K., Fox, J. G., Stalis, I. H., Rufo, R., Lee, G., Linn, J., Li, X., and Kleanthous, H. (1995). Characterization of feline *Helicobacter pylori* strains and associated gastritis in a colony of domestic cats. *J. Clin. Microbiol.* **33,** 2280–2289.

Higgins, J. A., Radulovic, S., Jaworski, D. C., and Azad, A. F. (1996). Acquisition of the cat scratch disease agent *Bartonella henselae* by cat fleas (*Siphonaptera, Pulicidae*). *J. Med. Entomol.* **33,** 490–495.

Holmes, A. H., Greenough, T. C., Balady, G. J., Regnery, R. L., Anderson, B. E., O'Keane, J. C., Fonger, J. D., and McCrone, E. L. (1995). *Bartonella henselae* endocarditis in an immunocompetent adult. *Clin. Infect. Dis.* **21,** 1004–1007.

Jameson, P., Greene, C., Regnery, R., Dryden, M., Marks, A., Brown, J., Cooper, J., Glaus, B., and Greene, R. (1995). Prevalence of *Bartonella henselae* antibodies in pet cats throughout regions of North America. *J. Infect. Dis.* **172,** 1145–1149.

Kaspar, L. H. (1989). Identification of stage-specific antigens of *Toxoplasma gondii*. *Infect. Immun.* **57,** 668–672.

Kaspar, L. H., Matsuura, T., and Khan, I. A. (1995). IL-7 stimulates protective immunity in mice against the intracellular pathogen, *Toxoplasma gondii*. *J. Immunol.* **155,** 4798–4804.

Khan, I. A., and Kaspar, L. H. (1996). IL-15 augments CD8+ T cell-mediated immunity against *Toxoplasma gondii* infection in mice. *J. Immunol.* **157,** 2103–2108.

Khan, I. A., Ely, K. H., and Kaspar, L. H. (1994). Antigen-specific CD8+ T cell clone protects against acute *Toxoplasma gondii* infection in mice. *J. Immunol.* **152,** 1856–1860.

Kordick, D. L., and Breitschwerdt, E. B. (1997). Relapsing bacteremia after blood transmission of *Bartonella henselae* to cats. *Am. J. Vet. Res.* **58,** 492–497.

Kordick, D. L., Wilson, K. H., Sexton, D. J., Hadfield, T. L., Berkhoff, H. A., and Breitschwerdt, E. B. (1995). Prolonged *Bartonella* bacteremia in cats associated with cat-scratch disease patients. *J. Clin. Microbiol.* **33,** 3245–3251.

Krakowka, S., Eaton, K. A., and Rings, D. M. (1995). Occurrence of gastric ulcers in gnotobiotic piglets colonized by *Helicobacter pylori*. *Infect. Immun.* **63,** 2352–2355.

Lee, A. (1994). The microbiology and epidemiology of *Helicobacter pylori* infection. *Scand. J. Gastroenterol.* **201,** 2–6.

Lee, A. (1996). Vaccination against *Helicobacter pylori*. *J. Gastroenterol.* **31**(Suppl. IX), 69–74.

Lindsay, D. S., Blagburn, B. L., and Dubey, J. P. (1993). Safety and results of challenge of weaned pigs given a temperature-sensitive mutant of *Toxoplasma gondii*. *J. Parasitol.* **79,** 71–76.

Liston, T. E., and Koehler, J. E. (1996). Granulomatous hepatitis and necrotizing splenitis due to *Bartonella henselae* in a patient with cancer: Case report and review of hepatosplenic manifestations of Bartonella infection. *Clin. Infect. Dis.* **22,** 951–957.

Morse, S. S. (1995). Factors in the emergence of infectious diseases. *Emerg. Infect. Dis.* **1,** 7–15.

Perkins, S. E., Fox, J. G., and Walsh, J. H. (1996). *Helicobacter mustelae*-associated hypergastrinemia in ferrets (*Mustela putorius furo*). *Am. J. Vet. Res.* **57,** 147–150.

Pfefferkorn, E. R., and Pfefferkorn, L. C. (1976). Arabinosyl nucleosides inhibit *Toxoplasma gondii* and allow the selection of resistant mutants. *J. Parasitol.* **62,** 993–999.

Pinckney, R. D., Lindsay, D. S., Blagburn, B. L., Boosinger, T. R., McLaughlin, S. A., and

Dubey, J. P. (1994). Evaluation of the safety and efficacy of vaccination of nursing pigs with living tachyzoites of two strains of *Toxoplasma gondii*. *J. Parasitol.* **80,** 438–448.
Regnery, R., and Tappero, J. (1995). Unraveling mysteries associated with cat-scratch disease, bacillary angiomatosis, and related syndromes. *Emerg. Infect. Dis.* **1,** 16–21.
Regnery, R. L., Rooney, J. A., Johnson, A. M., Nesby, S. L., Manzewitsch, P., Beaver, K., and Olson, J. G. (1996). Experimentally induced *Bartonella henselae* infections followed by challenge exposure and antimicrobial therapy in cats. *Am. J. Vet. Res.* **57,** 1714–1719.
Rodriguez-Barradas, M. C., Bandres, J. C., Hamill, R. J., Trial, J., Clarridge, J. E., III, Baughn, R. E., and Rossen, R. D. (1995). In vitro evaluation of the role of humoral immunity against *Bartonella henselae*. *Infect. Immun.* **63,** 2367–2370.
Sher, A., Denkers, E. Y., and Gazzinelli, R. T. (1995). Induction and regulation of host cell-mediated immunity by *Toxoplasma gondii*. *Ciba Found. Symp.* **195,** 95–109.
Taboada, J., and Merchant, S. R. (1995). Protozoal and miscellaneous infections. In "Textbook of Veterinary Internal Medicine" (S. J. Ettinger and E. C. Feldman, eds.), pp. 384–397. Saunders, Philadelphia.
Wear, D. J., Margileth, A. M., Hadfield, T. L., Fisher, G. W., Schlagel, C. J., and King, F. M. (1983). Cat scratch disease: A bacterial infection. *Science* **221,** 1403–1405.
Zangwill, K. M., Hamilton, D. H., Perkins, B. A., Regnery, R. L., Plikaytis, B. D., Hadler, J. L., Cartter, M. L., and Wenger, J. D. (1993). Cat scratch disease in Connecticut. *N. Engl. J. Med.* **329,** 8–13.

Evaluation of Risks and Benefits Associated with Vaccination against Coronavirus Infections in Cats

FRED W. SCOTT

Cornell Feline Health Center and the Department of Microbiology and Immunology, College of Veterinary Medicine, Cornell University, Ithaca, New York 14853

I. Historical Perspectives of FIP
II. Current Status of FIP
III. Causative Agent of FIP
IV. Pathogenesis of Feline Coronavirus Infections
V. Immunology of Feline Coronavirus Infections
VI. Antibody-Dependent Enhancement
VII. FIP Vaccine
VIII. Risks of FIP Vaccination
IX. Benefits of FIP Vaccination
References

I. Historical Perspectives of FIP

Historically, feline infectious peritonitis (FIP) was first described in the early 1960s by Dr. Jean Holzworth (1963). At least four earlier references to clinical cases in cats that may have been FIP included (1) a conspicuous abdominal distention due to ascites, with a retrospective diagnosis of FIP, from the State Veterinary School in Utrecht, seen in 1912–1913 and reported in 1914 (de Groot and Horzinek, 1995; Jakob, 1914); (2) an infectious pleuritis in 1942 (Bonaduce, 1942); (3) a severe exudative peritonitis of unknown etiology reported in England in 1960 (Joshua, 1960); and (4) a case of chronic organizing peritonitis in 1961 (Smith and Jones, 1961). The first detailed description of a clinical case of "FIP" was reported by Feldmann and Jortner (1964). The name *feline infectious peritonitis* was coined by Wolfe and Griesemer in 1966, and it was early recognized to have an immunopathologic component (Pedersen and Boyle, 1980; Jacobse-Geels et al., 1980). Antibody-dependent enhancement (ADE) was shown to play a part in the pathogenesis of the disease (Pedersen, 1983; Weiss and Scott, 1981; Olsen *et al.*, 1993). The first vaccine for FIP, an intranasal modified live virus (MLV) vaccine, was licensed in 1991 (Gerber *et al.*, 1990; Gerber, 1995).

II. Current Status of FIP

The current status of FIP varies between types of cat populations. It is *the* most feared disease today in breeding catteries, but is less common and of less concern in the general pet population (Wolf, 1995; Kass and Dent, 1995; Addie and Jarrett, 1995). There is no effective treatment for FIP, and once classical disease occurs, mortality is nearly 100%. The only available commercial vaccine is less than 100% effective. Available laboratory tests detect feline coronavirus antibodies and therefore are not specific for FIP. Until recently, there was no test to detect FIP antigen, virus, or to identify virus carrier cats. The reliability of new antigen detection tests is still being evaluated. A cattery with enzootic FIP is difficult to manage. The great variability in incubation period of FIP (weeks, months, or even years) presents a serious challenge to prevention and control. A comprehensive review of FIP virus was published in 1993 by Olsen. This review covers the molecular biology of the virus, immunopathogenesis of infection, clinical aspects of the disease, and a discussion of vaccination for FIP. Two reviews of FIPV were published in 1995 (de Groot and Horzinek, 1995; Pedersen, 1995b). A series of manuscripts was published in *Feline Practice* as the Proceedings of the 1994 International Workshop on FIPV and FECV (Vol. 23 [3], 1995).

III. Causative Agent of FIP

The causative agent of FIP, feline infectious peritonitis virus (FIPV), is a pleomorphic, enveloped virus classified as a coronavirus. This single-stranded, positive-sense RNA genome virus contains three major structural proteins: (1) the "N" or nuclear protein (core protein), (2) the "M" or matrix glycoprotein (formerly called E1), and (3) the "S" or spike glycoprotein (formerly called E2) (de Groot and Horzinek, 1995; Olsen, 1993; Spaan *et al.*, 1988).

There are at least two cell receptors for FIPV infection of cells (de Groot and Horzinek, 1995; Holmes and Compton, 1995; Olsen et al., 1992, 1993; Olsen, 1993). One is a virus receptor on the cell membrane that is specific for epitopes on the S protein. The second is the Fc receptor on macrophages for the Fc portion of IgG antibodies.

The replication of FIPV occurs at the endoplasmic membranes of infected cells (Holmes, 1985; Olsen, 1993). The virus buds into vacuoles within the cell cytoplasm, and hence the virus remains cell-associated initially. Virus is released from infected cells after the cell is destroyed (cytolysis). Replication of virus is rapid, with the replicative cycle completed in less than 24 hours.

FIPV survives in the environment much longer than was originally thought. Infectious FIPV can be recovered from contaminated dry surfaces for 3–7 weeks at room temperature, with the amount of infectious virus present gradually decreasing with time (Scott, 1991). It is *not* a highly labile virus as is usually reported.

FIPV is one of four viruses that make up an antigenic cluster of viruses with similar genomes. The second virus, feline enteric coronavirus (FECV), generally produces a mild enteritis, but not FIP. However, there is some indication that FECV can produce FIP-like disease. Positive antibody titers against FIPV result from FECV infection. It is preferable to refer to both FIPV and FECV as feline coronavirus (FCoV). The third virus in the group is canine coronavirus (CCV), which can infect cats, but usually with a subclinical infection. CCV antibody-positive cats can experience ADE of FIPV infection (McArdle *et al.*, 1992). One U.K. isolate of CCV can produce clinical FIP in cats. Positive antibody titers against FIPV are produced with experimental CCV infection of cats. The last virus in the group is transmissible gastroenteritis virus (TGEV) of swine. TGEV can infect cats, usually with subclinical infection, but with positive antibody titers against FIPV (Woods and Pedersen, 1979).

All feline coronaviruses belong to a single serotype, but there are two subtypes of virus, FCoV-1 and FCoV-2, which can be differentiated by monoclonal antibody (mAbs) (Corapi *et al.*, 1992; Fiscus and Teramoto, 1986, 1987; Hohdatsu *et al.*, 1991; Olsen, 1993).

IV. Pathogenesis of Feline Coronavirus Infections

The pathogenesis of FCoV infection is complex and unique. The incubation period can be weeks, months, or even years, but is generally 2–3 weeks up to 3 months. Local infection or primary infection occurs in the pharyngeal and lung epithelium, and possibly the intestinal epithelium (de Groot and Horzinek, 1995; Olsen, 1993). There is minimal clinical disease during the primary infection, often just a transient fever for one to a few days. Antibodies against FIPV first appear in serum by day 7 to day 10 after infection, and then infection of macrophages occurs. Fc receptors on macrophages enable uptake of virus–antibody complexes, and infected macrophages transport the virus throughout the body. Secondary infection then occurs in many tissues, with macrophages attaching to and migrating through the walls of veins. A perivascular reaction occurs, leading to development of a pyogranuloma, the basic lesion of FIP within tissues.

Two forms of FIP are recognized (Montali and Strandberg, 1972). Early reports described FIP primarily as wet or exudative FIP. Currently, dry or granulomatous FIP is more common than the wet form of disease. The wet and dry forms of FIP are merely variations of the same disease process. In wet FIP there is an exudative reaction at the vessel walls, with exudative fluid accumulating in the peritoneal and/or the thoracic cavities.

V. Immunology of Feline Coronavirus Infections

Immunology of FCoV infections is complicated and not fully understood. Undoubtedly, all three major components of the host's immune response come into play in a fully immune cat. However, many cats develop aspects of an immune response without developing protection. In some cases, this host response makes the cat more susceptible to exposure to FIPV rather than providing protection. Humoral immunity results in serum virus neutralizing (VN) antibodies which first appear 7–10 days after infection. There is a gradual increase in VN titers until 5–6 weeks after infection, with a hypergammaglobulinemia occurring in most cats that develop clinical FIP. The VN antibodies against epitopes on the S or spike protein usually are enhancing antibodies if the right concentration of virus and antibodies exists (Olsen et al., 1992, 1993). The subclass of IgG produced may be important in determining if true immunity occurs (Corapi et al., 1992).

Cell-mediated immunity (CMI) is believed be to play an essential role in an effective immune response against FCoV (Pedersen, 1987, 1995b). Details of the CMI response have not been determined. Local immunity appears to play a significant role in preventing infection of a previously infected or vaccinated cat via anti-FCoV IgA on mucosal surfaces (Gerber, 1995; Gerber et al., 1990).

VI. Antibody-Dependent Enhancement

Immune enhancement (antibody-dependent enhancement, ADE) has been clearly shown to occur in experimental laboratory infections of cats previously infected by natural or experimental infection, and of cats previously vaccinated with Primucell FIP vaccine, experimental MLV vaccines, experimental inactivated vaccines, and experimental recombinant vaccines containing the S gene (McArdle et al., 1992, 1995; Ngichabe, 1992; Scott et al., 1992, 1995a,b; Weiss and Scott, 1981). Antibodies to the S protein produced by the host result in enhanced infection of macrophages via Fc receptors, and the infected macrophages then transport the virus throughout the body. In the enhanced infection there is a decrease in incubation time—as short as 1–2 days—after exposure to virulent FIPV. The relative amount of virus and antibodies is important in order for ADE to occur. Higher concentrations of antibody neutralize the virus, but as the concentration of antibody decreases a concentration occurs where enhanced infection results. Other related coronaviruses can cause enhanced FCoV infection in the cat, including CCV.

Infectivity of macrophages appears to be a key factor in the ability of FCoV to become a systemic infection (Pedersen, 1976; Stoddart and Scott, 1986). The infected macrophages travel in the bloodstream to various parts of the body where they attach to the walls of veins. The local infection with inflammation results in characteristic perivascular lesions identified as pyogranulomas.

VII. FIP Vaccine

A single commercial vaccine, Primucell FIP from Pfizer, is available to aid in protecting cats against FIP and FCoV infections (Christianson et al., 1989; Gerber, 1995; Gerber et al., 1990). It is a MLV, temperature-sensitive (ts) mutant produced by attenuation of the original virulent FIPV-DF2 isolate by serial passage in cell cultures at low temperature. The FIPV-DF2 isolate is a type 2 virus. Primucell FIP is licensed for intranasal administration, with two doses given 3–4 weeks apart in cats at least 16 weeks of age. Annual revaccination is recommended by the manufacturer. This vaccine stimulates local IgA and VN antibody titers in serum (Gerber, 1995; Gerber et al., 1990).

Evaluation of the risks and benefits associated with the use of this vaccine is a complicated issue. First, the severe, usually fatal nature of FIP mandates the need for a safe and effective vaccine, especially since there is no effective treatment for this disease. If a highly effective vaccine against FIP was available, it would be used routinely in feline practice.

VIII. Risks of FIP Vaccination

The risks of vaccination with Primucell FIP appear to be minimal in most situations. The vaccine has been used for the past 7 years with no increase in the incidence of FIP reported. Field safety tests and controlled field studies have documented no increase in FIP or related disease in cats vaccinated with this vaccine (Fehr et al., 1995; Hoskins et al., 1995a; Reeves, 1995; Scott et al., 1992). In experimental studies in several laboratories, however, ADE of infection has been documented in cats vaccinated with this vaccine and a variety of other experimental FIP vaccines (McArdle et al., 1995; Ngichabe, 1992; Scott et al., 1995a; Vennema et al., 1990). Under these situations, the enhanced state results in a shorter incubation period after exposure to virulent virus, a shorter course of disease with more severe clinical signs, and a greater mortality compared to unvaccinated control cats. The major factor that determines whether enhanced disease occurs is the relative concentration of virulent virus and anti-FCoV antibodies (Corapi et al., 1992; Olsen, 1993; Olsen et al., 1992, 1993; Scott et al., 1995b).

Some investigators have stated that ADE of infection does not occur under natural or field conditions, that it only occurs under laboratory conditions (Addie et al., 1995; Fehr et al., 1995; Reeves, 1995). Although ADE does not appear to be a major problem under field conditions, apparently due to the relatively small amount of virus shed from infected and carrier cats, it is virtually impossible to determine if an individual case of FIP is a result of ADE infection or nonenhanced infection. The incubation period is the only differentiating difference, with ADE infection resulting in severe clinical FIP within 12 days after exposure to virus (Scott et al., 1995b; Weiss and Scott, 1981). Nonenhanced infection does not result in severe disease until after 12 days from exposure. Since the time of exposure to virus is almost always impossible to determine in field infections, it is likewise virtually impossible to determine with certainty whether or not a particular infection is enhanced or not. In one study of the safety and efficacy of the vaccine in two high-risk cat populations in Switzerland (Fehr et al., 1995), the authors state that the vaccine did not result in enhanced disease. Yet evaluation of the reported results indicates that three cats developed FIP within the first month after vaccination, while none of the placebo vaccinated cats developed FIP within this time. Were these cases of enhanced disease? It is impossible to ascertain, but it is also impossible to rule out enhanced disease based on these data.

IX. Benefits of FIP Vaccination

The benefits of vaccination, unfortunately, are rather low. In FIP endemic catteries, two controlled studies have failed to show any decrease in the incidence of FIP in vaccinated cats compared to placebo vaccinated controls. In the original field study by the manufacturer of the vaccine (Fanton, 1991), 12 endemic catteries were evaluated in a controlled study, with 349 cats vaccinated twice with Primucell FIP, and 352 cats vaccinated with a placebo vaccine. During the observation period of 6 months, three cases of FIP occurred in the vaccinated group (0.86%) compared to four cases in the control group (1.1%). This difference was not statistically significant.

In a double-blind placebo controlled field study in Switzerland involving 138 purebred cats from 15 catteries (Fehr *et al.*, 1995), the investigators were unable to show a difference between Primucell FIP vaccinated cats and placebo vaccinated controls within these FIP endemic catteries when the vaccine was used as recommended by the manufacturer (two doses of vaccine given 3–4 weeks apart starting at 16 weeks of age). There were seven FIP deaths in the vaccinated group and five FIP deaths in the placebo group during the 15- to 21-month study period.

In endemic catteries, many kittens are infected from their carrier queens at 6–7 weeks of age, long before the vaccine can be used as licensed at 16 weeks of age (Stoddart *et al.*, 1984; Addie and Jarrett, 1995). Once a cat is infected with FCoV the vaccine will have no beneficial effect. Controlled studies on the efficacy of the vaccine in kittens younger than 16 weeks of age have not been published. Some cattery owners apparently are using the vaccine off label in kittens as young as 3 weeks of age.

In high-risk pet cat populations there is limited or no efficacy of the vaccine. In the Switzerland study mentioned earlier (Fehr *et al.*, 1995), 609 domestic and purebred household pet cats were studied in the double-blind placebo controlled field study. During the 12 months after vaccination, FIP was confirmed in 13/31 deaths in the vaccinated group compared to 17/34 deaths in the control group. The authors state that "Death losses in placebo and vaccinate groups were equal up to day 150 following immunization, while significantly more placebo-immunized animals died of FIP after that period." In this author's review of the data in this study (Fehr et al., 1995), there were two more deaths in the vaccinated group ($n = 12$) compared to the control group ($n = 10$) through 150 days, and an equal number of deaths from FIP ($n = 13$ for both groups) through 250 days after vaccination. After this 250-day period there were four deaths in the control group but no deaths in the vaccinated group. A disturbing finding is that two of the deaths in the vaccinated group occurred within the first few days after vaccination, while none of the deaths in the control group occurred within this time period. Were these cats merely incubating the disease at the time of vaccination and would have died anyway, or were these early deaths after vaccination a result of enhanced disease from the vaccine in cats that were already antibody positive? It is impossible in this review to ascertain the exact situation in these two cats. In any case, Primucell FIP did not reduce the incidence of FIP within this high-risk pet population of cats for 8 months after vaccination, or until the cats were at least 1 year of age.

The topical or intranasal vaccination with Primucell FIP stimulates VN antibodies within the sera of vaccinated cats (Fehr et al., 1995; Hoskins et al., 1995a; Scott et al., 1992, 1995a). These VN antibodies, directed against epitopes on the S protein of the virus, are also enhancing antibodies when tested in vitro against infection of peritoneal macrophages (Olsen et al., 1992; Scott et al., 1995a; Stoddart and Scott, 1988). In our studies, 48/49 cats vaccinated with Primucell FIP had enhancing antibodies within their sera after vaccination, while none of the 23 unvaccinated controls had enhancing antibodies. Enhanced infection occurred with both FIPV-1146 and FIPV-DF2.

Efficacy of Primucell FIP depends on the dose of FIPV to which vaccinated cats are exposed (McArdle et al., 1995; Scott et al., 1995a,b). Under experimental conditions where vaccinated cats are exposed to a low dose of virulent challenge virus (<10 cat infectious doses), the vaccine provides protection for some cats. If vaccinated cats are exposed to >10 cat infectious doses of FIPV (10^4 TCID$_{50}$), the vaccine provides no protection. With this higher challenge dose, many of the vaccinated cats are more susceptible to infection than the unvaccinated controls, resulting in an enhanced and more acute disease.

Some efficacy of the vaccine can be demonstrated when FCoV antibody-negative cats are vaccinated according to the manufacturer's recommendation prior to natural exposure to FCoV-infected cats. Reeves (1995) reported a significant reduction in clinical FIP in FCoV antibody-negative cats that were vaccinated twice when at least 16 weeks of age, then introduced into a large cat shelter where FIP was endemic. Of 254 vaccinates, the mortality from FIP over a 16-month period after vaccination was 0.8% compared to a FIP mortality of 3.25% in 246 placebo-vaccinated cats. The calculated efficacy of vaccination, based on preventable fractions, was 75%.

Hoskins et al., (1995a,b) evaluated the efficacy of Primucell FIP in kittens vaccinated at 16 and 19 weeks of age in two studies. In one study, the number of kittens with histopathologic indications of FIP after a low-dose FIPV-DF2 challenge was reduced from 60 to 30% in vaccinated cats compared to unvaccinated controls, a 50% efficacy based on preventable fractions. In the other study, vaccinated kittens exposed to FECV-1163 had less intestinal clinical disease, less virus in the intestine, and less histopathologic damage to the small intestine compared to unvaccinated controls.

The amount of virus to which cats are exposed under normal field conditions is unknown. However, based on studies reported to date under both natural and experimental conditions, one can only conclude that the amount of virus exposure must be low. If the exposure dose of virus was high, enhanced disease would frequently be encountered. Because enhanced disease does not occur under natural conditions, or at least it is an uncommon occurrence if at all, the exposure dose of virus must be at a low level. This is consistent with what is known about the amount of virus shed from experimentally infected cats.

There is no published information on the duration of immunity produced by Primucell FIP. As with most veterinary biologics, the manufacturer of the vaccine arbitrarily recommends annual revaccination.

In summary, Primucell FIP vaccine has efficacy in preventing some clinical FIP when FCoV antibody-negative kittens at least 16 weeks of age are vaccinated twice intranasally 3 weeks apart. The vaccine has not been shown to reduce the incidence of clinical FIP when used in endemic catteries when the vaccine is routinely given to kittens at least 16 weeks of age. The use of the vaccine appears to be limited to high-risk populations, such as breeding catteries and multicat facilities, where FCoV antibody-negative cats are vaccinated.

The American Association of Feline Practitioner's *Feline Vaccination Guidelines* (Elston et al., 1998) makes the following recommendations concerning FIP vaccination. "The panel considers this to be a non-core vaccine because of the low prevalence of disease in confined populations of cats. As a result, vaccination is recommended only for cats at risk of exposure to the causative organism. However, the panel was split as to what constituted risk of exposure to FIP-inducing coronaviruses. A minority of the panel members recommended vaccination of kittens and cats with lifestyles that resulted in substantial risk of exposure to coronaviruses. Most panel members recommended that vaccination be limited to cats in specific risk situations, such as households in which FIP had been diagnosed. Those cats for which vaccination is deemed appropriate should receive a foundation series of vaccinations as kittens, according to recognized protocols. An annual booster vaccination is recommended by vaccine manufacturers; however, to our knowledge, duration of immunity studies have not been performed."

References

Addie, D. D., and Jarrett, O. (1995). Control of feline coronavirus infections in breeding catteries by serotesting, isolation, and early weaning. *Feline Pract.* **23**(3), 92–95.

Addie, D. D., Toth, S., Murray, G. D., and Jarrett, O. (1995). The risk of typical and antibody enhanced feline infectious peritonitis among cats from feline coronavirus endemic households. *Feline Pract.* **23**(3), 24–26.

Bonaduce, A. (1942). Infektiöse Pleuritis der Katzen. *Nuova Vet.* **21**, 32.

Christianson, K. K., Ingersoll, J. D., Landon, R. M., Pfeiffer, N. E., and Gerber, J. D. (1989). Characterization of a temperature sensitive feline infectious peritonitis coronavirus. *Arch. Virol.* **109**, 185–196.

Corapi, W. V., Olsen, C. W., and Scott, F. W. (1992). Monoclonal antibody analysis of neutralization and antibody-dependent enhancement of feline infectious peritonitis virus. *J. Virol.* **66**, 6695–6705.

de Groot, R. J., and Horzinek, M. C. (1995). Feline infectious peritonitis. *In* "The Coronaviridae" (S. G. Siddell, ed), Chapter 14, pp. 293–315. Plenum, New York.

Elston, T., Rodan, I., Flemming, D., Ford, R. B., Hustead, D. R., Richards, J. R., Rosen, D. K., Scherk-Nixon, M. A., and Scott, F. W. (1998). 1998 Report of the American Association of Feline Practitioners and Academy of Feline Medicine Advisory Panel on feline vaccines. *J. Am. Vet. Med. Assoc.* **212**, 227–241.

Fanton, R. W. (1991). Field safety studies of an intranasal FIPV vaccine. *In* "New Perspectives on Prevention of Feline Infectious Peritonitis" (R. V. H. Pollock, ed.), pp. 47–50. SmithKline Beecham Animal Health, Lincoln, NB.

Fehr, D., Holznagel, E., Bolla, S., Hauser, B., Herrewegh, A. A. P. M., Horzinek, M. C., and Lutz, H. (1995). Evaluation of the safety and efficacy of a modified live FIPV vaccine under field conditions. *Feline Pract.* **23**(3), 83–88.

Feldmann, B. M., and Jortner, B. S. (1964). Clinico-pathologic conference. *J. Am. Vet. Med. Assoc.* **144**, 1409–1420.

Fiscus, S. A., and Teramoto, Y. A. (1986). Functional differences between the peplomers of two antigenically distinct feline infectious peritonitis virus isolates. *J. Cell. Biochem.* **10D**, 293 (abstr.).

Fiscus, S. A., and Teramoto, Y. A. (1987). Antigenic comparison of feline coronavirus isolates: evidence for markedly different peplomer glycoproteins. *J. Virol.* **61**, 2607–2613.

Gerber, J. D. (1995). Overview of the development of a modified live temperature-sensitive FIP virus vaccine. *Feline Pract.* **23**(3), 62–66.

Gerber, J. D., Ingersoll, J. D., Gast, A. M., Christianson, K. K., Selzer, N. L., Landon, R. M., Pfeiffer, N. E., Sharpee, R. L., and Beckenhauer, W. H. (1990). Protection against feline infectious peritonitis by intranasal inoculation of a temperature-sensitive FIPV vaccine. *Vaccine* **8**, 536–542.

Hohdatsu, T., Okada, S., and Koyama, H. (1991). Characterization of monoclonal antibodies against feline infectious peritonitis virus type II and antigenic relationship between feline, porcine, and canine coronaviruses. *Arch. Virol.* **117**, 85–95.

Holmes, K. (1985). Coronavirus replication. *In* "Virology" (B. N. Fields *et al.*, eds.), pp. 1331–1343. Raven Press, New York.

Holmes, K. V., and Compton, S. R. (1995). Coronavirus receptors. *In* "The Coronaviridae" (S. G. Siddell, ed.), Chapter 4, pp. 55–71. Plenum, New York.

Holzworth, J. (1963). Some important disorders of cats. *Cornell Vet.* **53**, 157–160.

Hoskins, J. D., Taylor, H. W., and Lomax, T. L. (1995a). Independent evaluation of a modified live feline infectious peritonitis virus vaccine under experimental conditions (Louisiana experience). *Feline Pract.* **23**(3), 72–73.

Hoskins, J. D., Henk, W. G., Storz, J., and Kearney, M. T. (1995b). The potential use of a modified live FIPV vaccine to prevent experimental FECV infection. *Feline Pract.* **23**(3), 89–90.

Jacobse-Geels, H. E., Daha, M. R., and Horzinek, M. C. (1980). Isolation and characterization of feline C3 and evidence for the immune complex pathogenesis of feline infectious peritonitis. *J. Immunol.* **125**(No. 4), 1606–1610.

Jakob, H. (1914). Therapeutische, kasuistische und statistische Mitteilungen aus der Klinik für kleine Haustiere an der Reichstierarzneischule in Utrecht (Holland). *Jahrgang 1912/13. Z. Tiermed.* **18**, 193.

Joshua, J. O. (1960). Vomiting in the cat. *Mod. Vet. Pract.* **41**(22), 36–42.

Kass, P. H., and Dent, T. H. (1995). The epidemiology of feline infectioius peritonitis in catteries. *Feline Pract.* **23**(3), 27–32.

McArdle, F., Bennett, M., Gaskell, R. M., Tennant, B., Kelly, D. F., and Gaskell, C. J. (1992). Induction and enhancement of feline infectious peritonitis by canine coronavirus. *Am. J. Vet. Res.* **53**, 1500–1506.

McArdle, F., Tennant, B., Kelly, D. F., Gaskell, C. J., and Gaskell, R. M. (1995). Independent evaluation of a modified live FIPV vaccine under experimental conditions (University of Liverpool experience). *Feline Pract.* **23**(3), 67–71.

Montali, R. J., and Strandberg, J. D. (1972). Extraperitoneal lesions in feline infectious peritonitis. *Vet. Pathol.* **9**, 109–121.

Ngichabe, C. K. (1992). Recombinant raccoon poxvirus-vectored feline vaccines. Ph.D. Thesis. Cornell University, Ithaca, NY.

Olsen, C. W. (1993). A review of feline infectious peritonitis virus: molecular biology, immunopathogenesis, clinical aspects, and vaccination. *Vet. Microbiol.* **36**, 1–37.

Olsen, C. W., Corapi, W. V., Ngichabe, C. K., Baines, J. D., and Scott, F. W. (1992). Monoclonal antibodies to the spike protein of feline infectious peritonitis virus mediate antibody-dependent enhancement of infection of feline macrophages. *J. Virol.* **66**, 956–965.

Olsen, C. W., Corapi, W. V., Jacobson, R. H., Simkins, R. A., Saif, L. J., and Scott, F. W. (1993). Identification of antigenic sites mediating antibody-dependent enhancement of feline infectious peritonitis virus infectivity. *J. Gen. Virol.* **74**, 745–749.

Pedersen, N. C. (1976). Morphologic and physical characteristics of feline infectious peritonitis and its growth in autochthonous peritoneal cell cultures. *Am. J. Vet. Res.* **37**, 567–572.

Pedersen, N. C. (1983). Feline infectious peritonitis and feline enteric coronavirus infections. Part 2: Feline infectious peritonitis. *Feline Pract.* **13**(5), 5–20.

Pedersen, N. C. (1987). Virologic and immunologic aspects of feline infectious peritonitis virus infection. *Adv. Exp. Med. Biol.* **218**, 529–550.

Pedersen, N. C., ed. (1995a). "Report from the International FIP/FECV Workshop." University of California, Davis. *Feline Pract.* **23**(3), 2–111.

Pedersen, N. C. (1995b). An overview of feline enteric coronavirus and infectious peritonitis virus infections. *Feline Pract.* **23**(3), 7–20.

Pedersen, N. C., and Boyle, J. F. (1980). Immunologic phenomena in the effusive form of feline infectious peritonitis. *Am. J. Vet. Res.* **41**, 868–876.

Reeves, N. P. (1995). Vaccination against naturally occurring FIP in a single large cat shelter. *Feline Pract.* **23**(3), 81–82.

Scott, F. W. (1991). Feline infectious peritonitis: Transmission and epidemiology. *In* "New Perspectives on Prevention of Feline Infectious Peritonitis" (R. V. H. Pollock, ed.), pp. 8–13. SmithKline Beecham Animal Health, Lincoln, NB.

Scott, F. W., Corapi, W. V., and Olsen, C. W. (1992). Evaluation of the safety and efficacy of Primucell-FIP vaccine. *Feline Health Top.* **7**(3), 6–8.

Scott, F. W., Corapi, W. V., and Olsen, C. W. (1995a). Independent evaluation of a modified live FIPV vaccine under experimental conditions (Cornell experience). *Feline Pract.* **23**(3), 74–76.

Scott, F. W., Olsen, C. W., and Corapi, W. V. (1995b). Antibody-dependent enhancement of feline infectious peritonitis virus infection. *Feline Pract.* **23**(3), 77–80.

Smith, H. A., and Jones, T. C. (1961). "Veterinary Pathology," 2nd ed., Lea & Febiger, Philadelphia.

Spaan, W., Cavanagh, D., and Horzinek, M. C. (1988). Coronaviruses: Structure and genome expression. *J. Gen. Virol.* **69**, 2939–2952.

Stoddart, C. A., and Scott, F. W. (1986). Correlation between replication of feline coronaviruses in feline macrophages in vitro and the pathogenesis of feline infectious peritonitis. *J. Cell. Biochem.* **10D**, 299 (abstr.).

Stoddart, C. A., and Scott, F. W. (1988). Isolation and identification of feline peritoneal macrophages for in vitro studies of coronavirus-macrophage interactions. *J. Leukocyte Biol.* **44**, 319–328.

Stoddart, C. A., Barlough, J. E., and Scott, F. W. (1984). Experimental studies of a coronavirus and coronavirus-like agent in a barrier-maintained feline breeding colony. *Arch. Virol.* **79**, 85–94.

Vennema, H., de Groot, R. J., Harbour, D. A., Dalderup, M., Gruffydd-Jones, T., Horzinek, M. C., and Spaan, W. J. M. (1990). Early death after feline infectious peritonitis virus challenge due to recombinant vaccinia virus immunization. *J. Virol.* **64**, 1407–1409.

Weiss, R. C., and Scott, F. W. (1981). Antibody-mediated enhancement of disease in feline infectious peritonitis: comparisons with dengue hemorrhagic fever. *Comp. Immunol. Microbiol. Infect. Dis.* **4**, 175–189.

Wolf, J. (1995). The impact of feline infectious peritonitis on catteries. *Feline Pract.* **23**(3), 21–23.

Wolfe, L. G., and Griesemer, R. A. (1966). Feline infectious peritonitis. *Vet. Pathol.* **3**, 255–270.

Woods, R. D., and Pedersen, N. C. (1979). Cross-protection studies between feline infectious peritonitis and porcine transmissible gastroenteritis viruses. *Vet. Microbiol.* **4**, 11–16.

V
EQUINE VACCINES AND DIAGNOSTICS

Diagnosis and Prevention of Equine Infectious Diseases: Present Status, Potential, and Challenges for the Future

PHILIPPE DESMETTRE

Merial, 254 rue Marcel Merieux, Lyon, France

I. Introduction
II. Viral Diseases
    A. Equine Influenza
    B. Equine Rhinopneumonitis and Equine Abortion
    C. Equine Arteritis Virus
    D. Japanese, Western, Eastern, and Venezuelan Encephalitis
    E. African Horse Sickness
    F. Equine Infectious Anemia
    G. Equine Rotavirus Infections
    H. Equine Rabies
III. Bacterial Diseases
    A. Potomac Horse Fever
    B. Lyme Disease (*Lyme borreliosis*)
    C. *Strepococcus equi* Infections (Strangles)
    D. Contagious Equine Metritis
IV. Other Viral and Bacterial Diseases
    A. Viral Diseases
    B. Bacterial Diseases
V. Summary
   References

I. Introduction

The increasing movement of horses all over the world for competition, sale, and breeding is a source of concern in terms of the risk of transmission of infectious diseases. As a counterpart, the requirement of horse breeding registries for verification of identification and passports allows individual horses to be monitored throughout their life, which in turn helps control all movements better than for any other domesticated animal species.

To prevent the spread of any equine infectious disease, movement restriction, management procedures, hygiene, and disinfection are essential. When exporting horses, quarantine either in the exporting or in the importing country is highly recommended.

With a view to limiting the risk and containing the dissemination of any infectious diseases, rapid and accurate diagnosis remains essential. For most diseases, delocalized, on-the-spot diagnostics are preferable to centralized ones. With respect to emerging diseases (e.g., equine morbillivirus pneumonia) since time is of the utmost importance, it is essential to benefit from the expertise of a network of highly dedicated laboratories capable of using the most relevant and most recent techniques and reagents.

For the control of equine diseases, vaccination has become the primary method. Existing vaccines are based on conventional ancient technologies. Such vaccines clearly need to be improved on terms of their safety and/or efficacy. There is also an increasing need for marker vaccines to differentiate vaccinated from infected animals.

This paper describes the most important viral and bacterial diseases and reviews the diagnostics and vaccines available or under development, their strengths, and their weaknesses.

II. Viral Diseases

Among the main equine infectious diseases, the viral respiratory diseases—equine influenza, equine herpesvirus type 1 or 4 infections, and equine viral arteritis^{n}doubtedly rank first in importance.

A. Equine Influenza

The equine influenza virus (EIV) is a member of the Orthomyxoviridae family and of the genus influenza virus type A. Two subtypes are recognized: H7N7 (subtype 1) and H3N8 (subtype 2).

Equine influenza has occurred in all parts of the world except Australasia. It is endemic in Europe and North and South America. All reported outbreaks in the last decade have been caused by equine 2 influenza.

Within subtype 2 (H3N8) antigenic variants appear to have arisen by genetic recombination from avian viruses (Webster and Guo, 1991). In addition, conventional H3N8 viruses have evolved by antigenic drift (Hinshaw et al., 1983).

1. Diagnosis

a. Direct The most accurate method for virus isolation remains inoculation of embryonated eggs from nasopharyngeal secretions collected during the acute phase of the disease (Ilobi et al., 1994). The identity of the isolated virus is then confirmed by hemagglutination inhibition (HI) tests using specific antisera.

Capture ELISA based on monoclonal antibody to the virus nucleoprotein (NP) is a new way of doing direct diagnosis which could be completed by other ELISA-based systems (Cook et al., 1988; Livesay et al., 1993; Chambers et al., 1994). Reverse transcriptase–polymerase chain reaction (RT–PCR) using specific primers for the detection of the viral genome is providing another new opportunity for direct diagnosis.

b. Indirect While antibodies to equine influenza virus are well detected using an HI test based on Tween-ether dissociated virus as an antigen (Burrows and Denyer, 1982), an improvement in their analysis and quantitation is provided by the use of single radial hemolysis (SRH) (Mumford et al., 1988) which is still preferred to immunoenzymatic assays (Livesay et al., 1993).

2. Vaccines

The prevention and control of equine influenza rely on vaccination and management regimes to reduce the exposure of susceptible horses. Inactivated vaccines containing representatives of the two virus subtypes are commercially available. The short-lived nature of the antibody response produced in horses by such vaccines and the continued antigenic drift of field virus are thought to be related to the limited protection provided (Burrows et al., 1977; Mumford et al., 1988; Mumford, 1992).

Improvements have been introduced through the use of single radial immunodiffusion techniques for the assay of antigens (Wood et al., 1983), of new adjuvants (oil adjuvants, polymers) (Mackowiak et al., 1990; Mumford et al., 1994a), new antigen presentation systems (IS-COMS) (Mumford et al., 1994b), and by incorporation of recent field isolates (Mumford and Wood, 1993a).

Viral vectored vaccines and, more recently, naked DNA vaccines are being evaluated and could represent new alternatives for improved vaccines.

B. EQUINE RHINOPNEUMONITIS AND EQUINE ABORTION

Five distinct herpesviruses are known to infect the horse. Two of them, EHV4 (equine rhinopneumonitis virus) and EHV1 (equine abortion virus) related to the subfamily Alphaherpesvirinae, are major causes of respiratory disease and abortion. Another member of the subfamily Alphaherpesvirinae is EHV3 (equine coital exanthema). The other two, EHV2 and EHV5, are members of the Gammaherpesvirinae subfamily.

1. Diagnosis

a. Direct Diagnosis of EHV4 or EHV1 infection requires isolation of the virus. An ELISA based on type-specific monoclonal antibodies then permits determination of the identity of the virus (Yeargan et al., 1985).

PCR using primers in the gC gene could be applied for direct identification in either foal fetal tissues or nasopharyngeal swabs. The method proved to be as sensitive as virus isolation and highly specific (Ballagi et al., 1990; O'Keefe et al., 1991; Sharma et al., 1992; Kirisawa et al., 1993).

b. Indirect Until recently the specific serodiagnosis of EHV4 and EHV1 was not possible due to strong antigenic cross-reactivity between the two viruses.

A type-specific serodiagnostic test has been developed based on type-specific epitopes located on the glycoprotein G and allows specific distinction in dually infected horses (Crabb and Studdert, 1993; Crabb et al., 1995).

2. Vaccines

a. Inactivated Inactivated EHV1 and EHV4 vaccines have been developed. Their poor initial immunogenicity has been improved with the use of relevant inactivating agents (β-propiolactone or ethyleneimine) (Bryans, 1978; Mayr et al., 1978) and adjuvants (aluminium hydroxide, mineral oil or polymers) (Thomson et al., 1979). Because the envelope plays an important role in the protective immunity, subunit vaccines based on envelope glycoproteins have been successfully developed.

Immune stimulating complex (ISCOM) vaccines prepared from detergent-treated purified virions and containing all glycoproteins are inducing full protection in the hamster model (Cook et al., 1990).

EHV1 gB and gC expressed by poxvirus vectors were shown to better protect hamsters when coexpressed than when expressed alone (Guo et al., 1989, 1990). A third major glycoprotein (gD) genetically expressed, using the baculovirus system, is inducing anti-EHV1 and EHV4 neutralizing antibodies in hamsters (Love et al., 1993).

b. Live Live attenuated marker vaccines based on gene(s) deleted viruses having a reduced virulence are proposed. The safety of such live attenuated vaccines remains an issue. Their ability to induce latency and the risk of subsequent recombination are a matter of concern.

C. Equine Arteritis Virus

The equine arteritis virus (EAV) is the type species of the recently established family of Arteriviridae. The most important consequences of infection are abortion in the mare and establishment of a carrier state in the stallion (Timoney and McCollum, 1996).

1. Diagnosis

a. Direct Due to the variability of the manifestations of EAV infections, a differential diagnosis is needed (Timoney and McCollum, 1993). The unambiguous diagnosis relies on virus isolation followed by identification using neutralization tests or monoclonal antibody-based immunoassays.

An alternative to virus isolation is to detect viral genomic RNA by RT–PCR. Such a test has been successfully applied on semen samples to detect carrier stallions (Chirnside and Spaan, 1990; St. Laurent *et al.*, 1994).

b. Indirect Virus neutralization is usually used for measuring EAV-specific antibodies (Senne *et al.*, 1985). Attempts to use ELISA tests have been made. Specificity improvement is obtained using bacterially expressed fusion G_L or N proteins (Lang and Mitchell, 1984; Cook *et al.*, 1989; Chirnside *et al.*, 1995a,b). However, such tests cannot differentiate between vaccinated and infected horses.

2. Vaccines

Live and inactivated vaccines are available. For live vaccines, attenuation of the virus has been performed by serial passages on equine and rabbit kidney cells (McCollum *et al.*, 1961, 1962). The vaccine induces mild adverse reaction. It is recommended to vaccinate horses more than 3 weeks before breeding and the vaccination of pregnant mares is discouraged (Doll *et al.*, 1968; Timoney and McCollum, 1993). The vaccine does not protect horses against infection but provides clinical protection.

The formalin inactivated vaccine was successful in protecting pregnant mares against abortion and in preventing stallions from becoming persistently infected (Fukunaga *et al.*, 1984, 1990, 1992).

Such vaccines do not allow discrimination between vaccinated and infected horses. To develop a marker vaccine, bacterially expressed G_L

protein or G_L specific synthetic peptides have been used and shown to elicit neutralizing antibodies (Chirnside et al., 1994, 1995a).

D. Japanese, Western, Eastern, and Venezuelan Encephalitis

Western (WEE), Eastern (EEE), and Venezuelan equine encephalitis (VEE) viruses are Alphaviruses belonging to the Togaviridae family. Japanese encephalitis (JE) virus is a Flavivirus belonging to the Flaviviridae family. They are all mosquito-transmitted viruses (Burke and Leake, 1989; Morris, 1989; Reisen and Monath, 1989; Walton and Grayson, 1989).

Japanese encephalitis occurs in Asia while the others occur only in the Americas.

1. Diagnosis

a. Direct Virus isolation is the main diagnostic method. But viremia is only transient and may have passed by the time neurologic signs appear. In the absence of virus isolation, viral antigens can be detected in clinical specimens by direct or indirect immunofluorescence (IF) (Monath et al., 1981).

Antigen capture ELISA can also be used but the specimens may contain sufficient quantities of viral antigens due to the limited sensitivity of the test (Monath et al., 1984).

Once more, advances are represented by the use of RT–PCR.

b. Indirect In equine encephalitis, antibodies appear at the end of the viremic period. ELISAs have been developed permitting the detection of type-specific antibodies. Common antigenic determinants within alphaviruses and flaviviruses provide a practical advantage for serodiagnosis but could also be a source of confusion (Calisher et al., 1986). Type-specific tests are consequently required to clarify the diagnosis.

2. Vaccines

The equine encephalitis viruses have the potential to spread rapidly and epidemically. As a consequence, the rapid identification of any case is crucial for developing prevention and control strategies. Inactivated monovalent or associated vaccines are available for WEE, EEE, and VEE in America. Inactivated vaccines including two antigenic variants of the JE virus are used in Asia.

Live vectored vaccines based on poxvirus vectors have recently been developed for JE, WEE, EEE, and VEE. They represent new ways of preventing the diseases, offering a decisive advantage regarding the biosafety of vaccine production processes.

E. AFRICAN HORSE SICKNESS

African horse sickness (AHS) is another arthropod-borne disease which represents a serious threat to horses in Africa, the Near and Middle East, and Southern Europe. Increasing international movement of horses involves the risk of outbreaks in other areas. The African horse sickness virus (AHSV) is an Orbivirus belonging to the Reoviridae family.

1. Diagnosis

a. Direct Presumptive diagnosis may be done on the basis of typical clinical signs and lesions associated with the seasonal prevalence of competent vectors (*Culicoides* and perhaps mosquitoes).

Virus isolation is the usual way of making a direct diagnosis of the disease (Anonymous, 1992).

Capture ELISAs based on polyclonal or monoclonal antibodies have been developed (Du Plessis et al., 1990; Hamblin et al., 1991; Laviada et al., 1992).

RT–PCR has been proposed (Stone-Marschat et al., 1994; Zientara et al., 1994). It fulfills the requirements for serogroup specificity and is at least as sensitive as virus isolation and much more rapid (results are obtained within 24 hours as opposed to 5–7 days required for virus isolation).

b. Indirect Outside endemic areas most horses die before having developed antibodies. In endemic areas and in regulating international movement of horses, identification of horses exposed to AHSV is essential.

ELISA methods have been developed to replace the complement fixation test (CF) currently used (Hamblin et al., 1991; Anonymous, 1992). A competition ELISA using recombinant VP7 protein and monoclonal antibodies to VP7 has been proposed. It avoids the need to grow live virus to manufacture it and can be safely used in virus-free areas (Chuma et al., 1992; Wade-Evans et al., 1993).

2. Vaccines

In nonendemic areas, the control of AHS involves quarantine, vector control, and vaccination.

Live attenuated vaccines were first developed using "neurotropic" virus obtained from intracerebral passages through animals (Alexander and Dutoit, 1934). Associated with significant side effects, such vaccines were then been replaced by tissue culture attenuated strains (Mirchamsy and Taslimi, 1964; Erasmus, 1978; Anonymous, 1992).

Inactivated vaccines have more recently been developed to overcome safety problems potentially linked with the use of live attenuated viruses in nonendemic areas (Mirchamsy and Taslimi, 1968; Dubourget et al., 1992; House et al., 1992; J. A. House et al., 1992).

Offering an opportunity for marker vaccines, recombinant subunit vaccines are representing a potentially safer and more effective way of inducing protective immunity. Recent evaluation of DNA-based immunization using VP2 gene is showing both humoral and cellular immune response in horses, but the immune level is not sufficient for a full protection, confirming thus the need for other antigens (Roy et al., 1992).

F. Equine Infectious Anemia

Equine infectious anemia virus (EIAV) is a member of the family Retroviridae and belongs to the genera Lentivirinae.

1. Diagnosis

Usually diagnosis of EIAV is based on clinical signs, on the evidence of sideroleukocytes, on detection of antibodies, and on detection of virus (Campbell, 1971). Antibodies to EIAV are detected by several methods including VN, HI, CF, and AGID test, but the AGID test (COGGINS test) is the most reliable one (Tanaka and Sakaki, 1962; Kono and Kobayashi, 1966; Coggins and Norcross, 1970; Pearson and Coggins, 1979; Sentsui and Kono, 1981).

ELISAs have been developed, which all use the core protein p26 or synthetic peptides (Shane et al., 1984; Shen et al., 1984; Archambault et al., 1989). The use of anti-p26 monoclonal antibodies in a competitive ELISA is a means to improve the specificity (Winston et al., 1987). Nevertheless false-positive reactions are still occurring with the ELISA-based tests and the results have to be confirmed in AGID.

Real improvements are expected from the use of RT–PCR techniques to detect viral RNA in blood (Langemeier et al., 1994).

2. Vaccines

With the exception of a live vaccine prepared from an EIAV strain attenuated by passage on donkey leukocyte and which is reported to be widely used in China, there is no other EIAV vaccine available (Shen, 1983). This reflects the global difficulty in the lentivirus vaccine development. Furthermore, if not specifically marked, the use of vaccines could confound control programs based on serodiagnostic tests alone.

G. EQUINE ROTAVIRUS INFECTIONS

Rotaviruses are recognized as the major identifiable cause of infectious diarrhea in foals around the world (Conner and Darlington, 1980). Equine rotaviruses belong to group A rotaviruses and are members of the Reoviridae family.

1. Diagnosis

a. Direct Based on the antigenic communities of group A rotaviruses, several tests developed for human diagnostics are used for the direct diagnosis of equine rotaviruses. They include ELISA and latex agglutination tests (Conner *et al.*, 1983; Dwyer *et al.*, 1988). However, little is known about the sensitivity and specificity of those different methods.

2. Vaccines

Parenteral vaccination of pregnant mare with inactivated bovine vaccine has been shown to produce an antibody response in the milk and is used for passive protection of suckling foals (Browning *et al.*, 1991).

H. EQUINE RABIES

Rabies occurs sporadically in horses and is usually transmitted by the bite of an infected wild animal (Owen, 1978). The rabies virus is a member of the Lyssavirus genus, which belongs to the Rhabdoviridae family.

1. Diagnosis

Rabies is not easy to diagnose in horses by any technique. This could be due to low concentrations of viral antigens in the brain of horses or to the production of inhibitory factors in brain and salivary gland tissues (Tabel and Charlton, 1974; Marler *et al.*, 1979).

The examination for Negri bodies is no longer used. It is advantageously replaced by immunofluorescence test (IFT) and mouse inoculation test done simultaneously (Aubert, 1982; Green *et al.*, 1992; Green, 1993).

2. Vaccines

Vaccination of horses is recommended in endemic areas. The safest and most efficient vaccines contain inactivated virus and an adjuvant. Live attenuated vaccines have to be rejected for safety reasons.

III. Bacterial Diseases

A. POTOMAC HORSE FEVER

Potomac horse fever (equine monocytic ehrlichiosis) is an acute equine diarrheal syndrome caused by a rickettsial organism, *Ehrlichia risticii*. The disease has a sporadic distribution and a seasonal incidence, suggesting that an arthropod vector is involved in the transmission of the agent (Huntingdon, 1990; Phipps, 1994).

Cases of equine ehrlichiosis not related to *E. risticii* but caused by *E. equi* have also been described in Europe.

1. Diagnosis

a. Direct Definitive diagnosis of Potomac horse fever is based on clinical signs and by demonstration of *E. risticii* within the cytoplasm of parasited monocytes and macrophages as well as by cultural evidence (Holland *et al.*, 1985).

PCR is offering a new opportunity for direct diagnosis (Biswas *et al.*, 1991).

b. Indirect Serologic tests include IFAT (indirect fluorescent antibody test) (Ristic *et al.*, 1986) and ELISA (Pretzman *et al.*, 1987; Shankarappa *et al.*, 1988).

There is a rapid raise in antibody titers a few days after infection and high antibody titers may persist up to a year after infection. Paired serum samples are therefore required to perform a diagnosis.

2. Vaccines

Several adjuvanted vaccines based on cell cultured inactivated organisms are available. Vaccination is reducing clinical signs rather than providing full protection.

B. LYME DISEASE (*Lyme borreliosis*)

Lyme disease does occur as a clinical entity in domestic animals including dogs, horses, and cattle as well as in humans. It is a vector-based infection transmitted by *Ixodes* ticks and caused by a spirochete (*Borrelia burgdorferi*) (Van Heerden and Reyers, 1984; Lindenmayer *et al.*, 1989; Parker and White, 1992).

1. Diagnosis

a. Direct Diagnosis of clinical Lyme disease is difficult and depends on the recognition of clinical signs, a history of possible exposure to the organism, and the identification of the spirochete in the affected animal. Isolation of the organism is made difficult because *B. burgdorferi* is found in extremely low number in blood and tissues and cultures take weeks to grow. If a sufficient number of organisms is present, direct observation of specimen is more rapid and practical (Barbour, 1984).

b. Indirect Serologic testing is the only practical means for confirming *B. burgdorferi* infection. IFAs as well as polyvalent or class-specific antibody ELISAs using the 41-kDa antigen are proposed. But these assays are not standardized (Magnarelli and Anderson, 1989; Golightly *et al.*, 1990). Horses are developing lower antibody titers than dogs. It is furthermore important that a correct interpretation of the results of any serologic assay be made.

Several different strains of *B. burgdorferi* have been identified. Current serologic assays have been found to detect antibodies to all of these strains but false-positive results may be due to cross-reactivity with antibodies to related species organisms in horses such as *B. theileri* or *B. hermsii* (Magnarelli *et al.*, 1987). [*Editor's note:* One needs to distinguish clearly between infection with spirochetes, including those immunologically cross-reactive with *Borrelia* species, and diseases caused by *B. burgdorferi*. Infection with spirochetes, including those with antigens shared with *B. burgdorferi*, are common in certain species like horses, dogs, and cattle. However, disease caused by *B. burgdorferi* is uncommon or nonexistent in these species (e.g., cattle).]

2. Vaccines

Based on previous development of vaccines for dogs, research is ongoing to develop vaccines for horses. A subunit vaccine, based on the outer membrane lipoprotein OspA, is produced in recombinant *Escherichia coli*. The strong immunogenicity of OspA makes it possible to induce a high level of antibodies and full protection in the absence of any adjuvants (which improve the safety of such a vaccine). The level of protection provided by such a vaccine in horses remains to be established.

C. *Streptococcus equi* Infections (Strangles)

Streptococcus equi is causing strangles, an acute contagious respiratory disease, which affects predominantly young horses (Sweeney *et al.*, 1987a,b; David Wilson, 1988).

1. Diagnosis

a. Direct Early clinical cases of strangles have to be distinguished from viral respiratory infections. Definitive diagnosis is usually based on progression of clinical signs and bacterial culture results (Schultz, 1988).

b. Indirect Serological tests to *S. equi* are not available yet and are of limited interest.

2. Vaccines

Vaccines have proven to be of some benefit in herds where the disease is endemic. Vaccination is unlikely to eliminate infection but can reduce the number and severity of cases and may slow down the spread.

Adjuvanted bacterins are inducing both local and general adverse reactions partly linked to the induction of hypersensitivity in case of repeated administration (Smith, 1994; Sezun, 1995). Purified M protein vaccines developed to overcome these reactions do not eliminate them although they reduce them (Srivastava and Barnum, 1983; Timoney and Trachman, 1985; Timoney and Mukhtar, 1993).

A vaccine based on adjuvanted enzymatic extracts of *S. equi* has also been developed (Bryant *et al.*, 1985).

None of these vaccines is completely effective in preventing strangles. This may reflect the fact that none of them induces a specific local secretory response, which may be necessary to block the infection of the upper respiratory tract mucosa (Galan and Timoney, 1985; Wallace *et al.*, 1995).

Recent research has led to the development of a modified avirulent strain of *S. equi,* which, when introduced intranasally or orally, stimulates a specific secretory response and a solid resistance to challenge. A genetically engineered *Salmonella typhimurium* expressing *S. equi* M protein antigens at its surface has also been generated for that purpose.

D. Contagious Equine Metritis

Contagious equine metritis (CEM) is a highly contagious venereal infection of equids caused by *Taylorella equigenitalis* (*Haemophilus*), a bacterium with fastidious growth requirements, causing short-term infertility. Stallions do not develop clinical signs nor become really infected. They harbor the organism at the surface of their external genitalia and contribute to the spread of CEM by a venereal route (Bowen *et al.*, 1979; Hugues *et al.*, 1978; Powell, 1981; Timoney, 1996).

1. Diagnosis

a. Direct Isolation of *T. equigenitalis* is the only actual means for establishing the diagnosis (McIntosh, 1981, 1990; Tainturier *et al.*, 1981a).

PCR has recently been developed and may provide an equally sensitive and more rapid means of confirming the infection (Bleumink-Pluym *et al.*, 1994).

b. Indirect Serologic testing for CEM is of limited interest since it can only be used for detecting infection in the acutely infected mare (Tainturier *et al.*, 1981b; Sahu *et al.*, 1983).

2. Vaccines

There is no vaccine against CEM. National control programs are in force in many countries, making the disease notifiable or reportable. They usually involve the bacteriologic screening of all stallions and mares imported (Franck *et al.*, 1979; Timoney and Powell, 1988).

IV. Other Viral and Bacterial Diseases

Other equine viral or bacterial diseases of minor importance are described.

A. Viral diseases

Among the viral diseases, vesicular stomatitis is caused by a lyssavirus, which, like the rabies virus, is a member of the Rhabdoviridae family. The disease is common to livestock and the three main serotypes of vesicular stomatitis virus (VSV) can infect horses. Although vesicular stomatitis does not have a major impact on the equine industry, it is clinically identical to the other, economically more devastating, diseases of cattle and swine.

Specific diagnosis is best made by virus isolation and there are no vaccines commercially available for horses.

B. Bacterial diseases

Salmonellosis is the most common infectious cause of diarrhea or colitis in adult horses. It can also produce epizootics, especially in foals under 8 days of age. Diagnosis requires isolation of the organism from feces. Direct identification in feces using PCR is now proposed.

In the absence of potent vaccines, the only way to prevent the disease or its spread is herd management, hygiene, and disinfection.

Leptospirosis is a zoonotic bacterial disease causing sporadic cases in horses. Among the multiple serovars, *L. bratislava* is more frequently identified in abortions and clinical diseases of foals. Periodic uveitis is also frequently recognized and could be more associated with the serovar *L. pomona*.

The diagnosis is dependent on culture, the organism being fastidious to grow. As a result of the difficulties in direct identification, diagnosis is often based on serologic tests. For serology, ELISA is replacing the standard microscopic agglutination test.

Specific vaccines are not available for horses and the use of multivalent vaccines developed for other species can cause anaphylactic reaction.

Rhodococcus (Corynebacterium) equi has the potential to cause considerable losses by inducing pneumonia in foals. Bacteriologic culture of tracheobronchial exudate or bronchial or broncheoalveolar lavage is the only way to achieve a definite diagnosis.

Despite a lot of work on the protection antigens of *R. equi* and on the development of appropriate immunization procedures, no vaccines are available to control the disease in foals.

V. Summary

The frequent transfers of horses, whether on a permanent or temporary basis, make strict control of infectious diseases essential. Such control needs a reliable and rapid means to accurately diagnose the relevant diseases.

Indirect diagnosis based on antibody detection remains certainly the best method to secure the epidemiologic surveillance of the diseases at regional, national, or even world level, while direct diagnosis is the only way to diagnose a new outbreak. New diagnostic methods resulting from advances in biochemistry, molecular biology, and immunology are now available. As far as antibody detection is concerned, the new methods are mainly based on immunoassays, especially ELISAs.

Regarding the identification of the pathogens, while isolation is still of importance, much progress has been made with immunocapture tests including capture ELISA based on monoclonal antibodies. DNA probes and amplification tests such as PCR or RT–PCR are representing a real breakthrough.

Factors common to all of these tests are specificity, sensitivity, rapid implementation, and quick results. Such tests are, however, often still at the development stage. They absolutely need to be validated under multicentric evaluations prior to being used on a larger scale. At the same time there is an obvious need for the standardization of the reagents used. The technical and economic impact of a false (either positive or negative) diagnosis justifies such an harmonization which could effectively be achieved worldwide under the aegis of the Office International des Epizooties (OIE), which is itself the primary source of disease information.

Vaccines are also essential for the control of equine infectious diseases. Most vaccines used in the prevention of viral or bacterial diseases are inactivated adjuvanted vaccines, which may cause unacceptable side effects. Also, their efficacy can sometimes be questioned. Subunit vaccines, when available, represent significant advances especially with regards to safety.

Greater progress is expected from the use of new technologies taking advantage of recent developments in molecular biology (recombinant DNA technology) and in immunology (immunomodulators). Significant results have been obtained with subunit vaccines or with live vectored vaccines using recombinant DNA technology. Good results are on the way to be achieved with genetic (or naked-DNA) vaccines.

It is therefore possible to expect the availability of a new generation of vaccines in the rather short term. Such vaccines will not only be safer and more efficacious, but they will also make it possible to differentiate vaccinated from infected animals, which will contribute to better control of the infection.

Whatever the quality of the vaccines of the future may be, vaccination alone will never be sufficient to control infectious diseases. It is therefore essential to keep on making the animal owners and their veterinarians aware of the importance of the management and the hygiene in the diseases control and to organize them under "Common Codes of Practice."

References

Alexander, R. A., and Dutoit, P. J. (1934). *Onderstepoort J. Vet. Res.* **2**, 375–391.
Anonymous, (1992) "OIE Manual of Standards for Diagnostic Tests and Vaccines," 2nd ed., pp. 91–100. Office International des Epizooties, Paris.
Archambault, D., Wang, Z. M., Lacal, J. C., Gazit, A., Yaniv, A., Dahlberg, J. E., and Tronick, S. R. (1989). *J. Clin. Microbiol.* **27**, 1167–1173.
Aubert, M. (1982). *Comp. Immunol. Microbiol. Infect. Dis.* **5**, 369–376.
Ballagi, P. A., Klingeborn, B., Flensburg, J., and Belàk, S. (1990). *Vet. Microbiol.* **22**, 373–381.
Barbour, A. G. (1984). *Yale J. Biol. Med.* **57**, 521–525.
Biswas, B., Mukherjee, D., Mattingly-Napier, B. L., et al. (1991). *J. Clin. Microbiol.* **29**, 2228.
Bleumink-Pluym, N. M. C., Werdler, M. E. B., Houwers, D. J., Parlevliet, J. M., Colenbrander, B., and Van der Zeijst, B. A. M. (1994). *J. Clin. Microbiol.* **32**, 893–896.
Bowen, J. M., Cosgrove, J. S., and Cosgrove, F. (1979). *Vet. Rec.* **104**, 441.
Browning, G. F., Chalmers, R. M., Sale, C. H. S., Fitzgerald, T. A., and Snodgrass, D. R. (1991). *Vet. Microbiol.* **27**, 231–244.
Bryans, J. T. (1978). *In* "Equine Infectious Diseases IV" (J. T. Bryans and H. Gerber, eds.), pp. 83–92. Lyon Vet. Publ., Princeton, NJ.
Bryant, S., Brown, K. K., et al. (1985). *Exact. Vet. Med.* **80**, 58–70.
Burke, D. S., and Leake, C. J. (1989). *In* "The Arboviruses: Epidemiology and Ecology" (T. P. Monath, ed.), pp. 1–20. CRC Press, Boca Raton, FL.

Burrows, R., and Denyer, M. (1982). *Arch. Virol.* **73,** 15–24.
Burrows, R., Spooner, P. R., and Goodridge, D. (1977). *Dev. Biol. Stand.* **39,** 341–346.
Calisher, C. H., Mahmud, M. I. Al-D., El-Kafrawi, A. O., Emerson, J. K., and Muth, D. J. (1986). *Am. J. Vet. Res.* **47,** 1296–1299.
Campbell, C. L. (1971). *Proc. U.S. Anim. Health Assoc.* **75,** 249–261.
Chambers, T. M., Shortridge, K. T., Il, R. H., Powell, D. G., and Watkins, K. L. (1994). *Vet. Rec.* **135,** 275–279.
Chirnside, E. D., and Spaan, W. J. M. (1990). *J. Virol. Methods* **30,** 133–140.
Chirnside, E. D., De Vries, A. A. F., Rottier, P. J. M., and Mumford, J. A. (1994). In "Equine Infectious Diseases VII" (H. Nakajima and W. Plowright, eds.), p. 298. R&W Publications, Newmarket, UK.
Chirnside, E. D., Francis, P. M., De Vires, A. A. F., Sinclair, R., and Mumford, J. A. (1995a). *J. Virol. Methods* **54,** 1–13.
Chirnside, E. D., De Vries, A. A. F., Mumford, J. A., and Rottier, P. J. M. (1995b). *J. Gen. Virol.* **76,** 1989–1998.
Chuma, T., LeBlois, H., Sanchez-Vizcaino, J. M., Diaz-Laviada, M., and Roy, P. (1992). *J. Gen. Virol.* **73,** 925–931.
Coggins, L., and Norcross, N. L. (1970). *Cornell Vet.* **60,** 330–335.
Conner, M. E., and Darlington, R. W. (1980). *Am. J. Vet. Res.* **41,** 1699–1703.
Conner, M. E., Gillespie, J. H., Schiff, E. I., and Frey, M. S. (1983). *Cornell Vet.* **73,** 280–287.
Cook, R. F., Mumford, J. A., Douglas, A., and Wood, J. M. (1988). In "Equine Infectious Diseases V" (D. G. Powell, ed.), pp. 60–65. Kentucky University Press, Lexington.
Cook, R. F., Gann, S. J., and Mumford, J. A. (1989). *Vet. Microbiol.* **20,** 181–189.
Cook, R. F., O'Neil, T., Strachan, E., Sundquist, B., and Mumford, J. A. (1990). *Vaccine* **8,** 491–496.
Crabb, B. S., and Studdert, M. J. (1993). *J. Virol.* **67,** 6332–6338.
Crabb, B. S., Drummer, H. E., Reubel, G. H., Macpherson, C. M., Browning, G. F., and Studdert, M. J. (1995). *Arch. Virol.* **140,** 245–258.
David Wilson, W. (1988). *Equine Pract.* **10,** 12–25.
Doll, E. R., Bryans, J. T., Wilson, J. C., and McCollum, W. H. (1968). *Cornell Vet.* **58,** 497–524.
Dubourget, P., Preaud, J. M., Detraz, F., Lacoste, A. C., Erasmus, B. J., and Lombard, M. (1992). In "Bluetongue, African Horsesickness and Related Orbiviruses" (T. W. Walton and B. I. Osburn, eds.), pp. 874–886. CRC Press, Boca Raton, FL.
Du Plessis, D. H., Van Wyngaardt, W., and Bremer, C. W. (1990). *J. Virol. Methods* **29,** 279–289.
Dwyer, R. M., Powel, D., Osborne, M., and Roberts, A. W. (1988). *Equine Vet. J., Suppl.* **5,** 60.
Erasmus, B. J. (1978). In "Equine Infectious Diseases IV" (J. T. Bryans and H. Gerber, eds.), pp. 401–403. Vet. Publ., Princeton, NJ.
Franck, C. J., David, J. S. E., and Smith, H. (1979). *Vet. Rec.* **105,** 395–397.
Fukunaga, Y., Wada, R., Kamada, M., Kumanomido, T., and Akiyama, Y. (1984). *Bull. Equine Res. Inst.* **21,** 56–64.
Fukunaga, Y., Wada, R., Matsumura, T., Sugiura, T., and Imagawa, H. (1990). *J. Vet. Med. Ser. B* **37,** 135–141.
Fukunaga, Y., Wada, R., Matsumara, T., Anzai, T., Imagawa, H., Sugiura, T., Kumanomido, T., Kanemaru, T., and Kamada, M. (1992). In "Equine Infectious Diseases VI" (W. Plowright, P. D. Rossdale, and J. F. Wade, eds.), pp. 239–244. R&W Publications, Newmarket, UK.

Galan, J. E., and Timoney, J. F. (1985). *Infect. Immun.* **47,** 623–628.
Golightly, M. G., Thomas, J. A., and Viciana, A. L. (1990). *Lab. Med.* **21,** 299–304.
Green, S. L. (1993). *Vet. Clin. North Am.* **9,** 337–347.
Green, S. L., Smith, L. L., Vernau, W., and Beacock, S. M. (1992). *J. Am. Vet. Med. Assoc.* **200,** 1133–1137.
Guo, P. X., Goebel, S., David, S., Perkus, M. E., Languet, B., Desmettre, Ph., Allen, G., and Paoletti, E. (1989). *J. Virol.* **63,** 4189–4298.
Guo, P. X., Goebel, S., Perkus, M. E., Taylor, J., Norton, E., Allen, G., Langet, B., Desmettre, P., and Paleotti, E. (1990). *J. Virol.* **64,** 2399–2406.
Hamblin, C., Mertens, P. P., Mellor, P. S., Burroughs, J. N., and Crowther, J. R. (1991). *J. Virol. Methods* **31,** 285–292.
Higgins, W. P., Gillespie, J. H., Schiff, E. I., Pennow, N. N., and Tanneberger, M. J. (1988). In "Equine Infectious Diseases V" (D. G. Powel, ed.). Kentucky University Press, Lexington.
Hinshaw, V. S., Naeve, C. W., Webster, R. G., Douglas, A., Skehel, J. J., and Bryans, J. (1983). *Bull. W.H.O.* **61,** 153–158.
Holland, C. J., Ristic, M., Cole, A. I., Johnson, P., Barker, G., and Goetz, T. (1985). *Science* **227,** 522–524.
House, C., House, J. A., and Mebus, C. A. (1992). *Ann. N.Y. Acad. Sci.* **653,** 228–232.
House, J. A., Lombard, M., House, C., Dubourget, P., and Mebus, C. A. (1992). In "Bluetongue, African Horsesickness, and Related Orbiviruses" (T. E. Walton and B. I. Osburn, eds.), pp. 891–895. CRC Press, Boca Raton, FL.
Hughes, K. L., Bryden, J. D., and MacDonnald, F. (1978). *Aust. Vet. J.* **54,** 101.
Huntingdon, P. J. (1990). *Aust. Equine Vet.* **8,** 112–115.
Ilobi, C. P., Henfrey, R., Robertson, J. S., Mumford, J. A., Erasmus, B. J., and Wood, J. M. (1994). *J. Gen. Virol.* **75,** 669–673.
Kirisawa, R., Endo, A., Iwai, H., and Kawakami, Y. (1993). *Vet. Microbiol.* **36,** 57–67.
Kono, Y., and Kobayashi, K. (1966). *Natl. Inst. Anim. Health Q.* **6,** 194–203.
Lang, G., and Mitchell, W. R. (1984). *J. Equine Vet. Sci.* **4,** 153–157.
Langemeier, J. L., Cook, R. F., Cook, S. J., Rushlow, K. E., Montelaro, R. C., and Issel, C. J. (1994). In "Equine Infectious Diseases VII" (H. Nakajima and W. Plowright, eds.), p. 299. R&W Publications, Newmarket, UK.
Laviada, M. D., Babin, M., Dominguez, J., and Sanchez-Vizcaino, J. M. (1992). *J. Virol. Methods* **38,** 229–242.
Lindenmayer, J., Weber, M., and Onderdonk, A. (1989). *J. Am. Vet. Med. Assoc.* **194,** 1384.
Livesay, G. J., O'Neil, T., Hannant, D., Yadav, M. P., and Mumford, J. A. (1993). *Vet. Rec.* **133,** 515–519.
Love, D. N., Bell, C. W., Pye, D., Edward, S., Hayden, M., Lawrence, G. L., Boyle, D., Pye, T., and Whalley, J. M. (1993). *J. Virol.* **67,** 6820–6823.
Mackowiak, M., Duret, C., Devaux, B., Fargeaud, D., Brun, A., and Saint-Gerand, A. L. (1990). *Bull. Mens. Soc. Vet. Prat. Fr.* **74,** 193–208.
Magnarelli, L. A., and Anderson, J. F. (1989). *J. Am. Vet. Med. Assoc.* **195,** 1365–1368.
Magnarelli, L. A., Anderson, J. F., and Johnson, R. C. (1987). *J. Infect. Dis.* **156,** 183–187.
Marler, R. J., Howard, D. R., Morris, P. G., and Johson, J. L. (1979). *J. Am. Vet. Med. Assoc.* **175,** 293–294.
Mayr, A., Thein, P., and Scheid, R. (1978). In "Equine Infectious Diseases IV" (J. T. Bryans and H. Gerber, eds.), pp. 57–67. Lyon. Vet. Publ., Princeton, NJ.
McCollum, W. H., Doll, E. R., Wilson, J. C., and Johnson, C. B. (1961). *Am. J. Vet. Res.* **22,** 731–735.

McCollum, W. H., Doll, E. R., Wilson, J. C., and Cheatham, J. (1962). *Cornell Vet.* **52,** 452–458.
McIntosh, M. E. (1981). *Vet. Rec.* **108,** 52–55.
McIntosh, M. E. (1990). In "OIE Manual of Recommended Diagnostic Techniques and Requirements for Biological Products for List A and B Diseases." Vol. 2, p. 1–5. Office International des Epizooties, Paris.
Mirchamsy, H., and Taslimi, H. (1964). *Br. Vet. J.* **120,** 481–486.
Mirchamsy, H., and Taslimi, H. (1968). *Immunology* **14,** 81–88.
Monath, T. P., McLean, R. G., Cropp, C. B., Parham, G. L., Lazuick, J. S., and Calisher, C. H. (1981). *Am. J. Vet. Res.* **42,** 1418–1421.
Monath, T. P., Nystrom, R. R., Bailey, R. E., Calisher, C. H., and Muth, D. J. (1984). *J. Clin. Microbiol.* **20,** 784–790.
Morris, C. D. (1989). In "The Arboviruses: Epidemiology and Ecology" (T. P. Monath, ed.), pp. 1–20. CRC Press, Boca Raton, FL.
Mumford, J. A. (1992). In "Equine Infectious Diseases VI" (W. Plowright, P. D. Rossdale, and J. F. Wade, eds.), pp. 207–218. R&W Publications, Newmarket, UK.
Mumford, J. A., and Wood, J. (1993). *Vaccine* **11,** 1172–1175.
Mumford, J. A., Wood, J. M., Folkers, C., and Schild, G. C. (1988). *Epidemiol. Infect.* **100,** 501–510.
Mumford, J. A., Wilson, H., Hannant, D., and Jessett, D. M. (1994a). *Epidemiol. Infect.* **112,** 421–437.
Mumford, J. A., Jessett, D., Dunleavy, U., Wood, J., Hannant, D., Sundquist, B., and Cook, R. F. (1994b). *Vaccine* **12,** 857–863.
O'Keefe, J. S., Murray, A., Wilks, C. R., and Moriarty, K. M. (1991). *Res. Vet. Sci.* **50,** 349–351.
Owen, R. (1978). *Vet. Rec.* **103,** 69.
Parker, J. L., and White, K. K. (1992). *Cornell Vet.* **82,** 253–274.
Pearson, J. E., and Coggins, L. (1979). Proc. 22nd Annu. Am. Assoc. Vet. Lab. Diag., pp. 449–462.
Phipps, L. P. (1994). *Equine Vet. Educ.* **6,** 321–322.
Powell, D. G. (1981). *Adv. Vet. Sci. Comp. Med.* **25,** 161–164.
Pretzman, C. I., Rikihiza, Y., Ralph, D., Gordon, J. C., and Bech-Nielsen, S. (1987). *J. Clin. Microbiol.* **25,** 31–36.
Reisen, W. K., and Monath, T. P. (1989). In "The Arboviruses: Epidemiology and Ecology" (T. P. Monath, ed.), pp. 89–137. CRC Press, Boca Raton, FL.
Ristic, M., Holland, C. J., Dawson, J. E., Sessions, J., and Palmer, J. (1986). *J. Am. Vet. Med. Assoc.* **189,** 39–46.
Roy, P., French, T., and Erasmus, B. J. (1992). *Vaccine* **10,** 28–32.
Sahu, S. P., Rommel, F. A., Fales, W. H., Hamdy, F. M., Swerczek, T. W., Youngquist, R. S., and Bryans, J. T. (1983). *Am. J. Vet. Res.* **44,** 1404–1409.
Schultz, J. W. (1988). *J. Comp. Pathol. Ther.* **1,** 191–208.
Senne, D. A., Pearson, J. E., and Carbrey, E. A. (1985). *Proc. 89th Annu. Meet. U.S. Anim. Health Assoc.,* Milwaukee, WI, pp. 29–34.
Sentsui, H., and Kono, Y. (1981). *Am. J. Vet. Res.* **42,** 1949–1952.
Sezun, G. S. (1995). *Aust. Vet. J.* **72,** 480.
Shane, B. S., Issel, C. J., and Montelaro, R. C. (1984). *J. Clin. Microbiol.* **19,** 351–355.
Shankarappa, B., Dutta, S. K., Sanusi, J., and Mattingley, B. L. (1988). *J. Clin. Microbiol.* **27,** 24–28.
Sharma, P. C., Cullinane, A. A., Onions, D. E., and Nicolson, L. (1992). *Equine Vet. J.* **24,** 20–25.

Shen, D. T., Gorham, J. R., and McGuire, T. C. (1984). *Am. J. Vet. Res.* **45,** 1542–1543.
Shen, R. (1983). In "International Symposium on Immunity to Equine Infectious Anemia," pp. 34–53. Harbin Veterinary Research Institute, Chinese Academy of Agricultural Sciences.
Smith, H. (1994). *Aust. Vet. J.* **71,** 257–258.
Srivastava, S. K., and Barnum, D. A. (1983). *Am. J. Vet. Res.* **44,** 41–45.
St. Laurent, G., Morin, G., and Archambault, D. (1994). *J. Clin. Microbiol.* **32,** 658–665.
Stone-Marschat, M., Carville, A., Skowronek, A., and Laegreid, W. W. (1994). *J. Clin. Microbiol.* **32,** 697–700.
Sweeney, C. R., Benson, C. E. et al. (1987a). *Compend. Contin. Educ. Pract. Vet.* **9** (Part 1), 689–695.
Sweeney, C. R., Benson, C. E. et al. (1987b). *Compend. Contin. Educ. Pract. Vet.* **9** (Part 2), 845–852.
Tabel, H., and Charlton, K. M. (1974). *Can. J. Comp. Med.* **38,** 344–346.
Tainturier, D., Delmas, C. F., and Dabernat, H. J. (1981a). *J. Clin. Microbiol.* **14,** 355–360.
Tainturier, D., Picavet, D. P., Badin de Montjoye, T., Guaguere, J., Tailliar, S., Dabernat, H. J., and Ferney, J. (1981b). *Ann. Rech. Vet.* **12,** 265–275.
Tanaka, K., and Sakaki, K. (1962). *Natl. Inst. Anim. Health Q.* **2,** 128–139.
Thomson, G. R., Mumford, J. A., and Smith, I. M. (1979). *Vet. Microbiol.* **4,** 209–222.
Timoney, P. J. (1996). *Comp. Immun. Microbiol. Infect. Dis.* **19,** 199–204.
Timoney, P. J., and McCollum, W. H. (1993). *Vet. Clin. North Am. Equine Pract.* **9,** 295–309.
Timoney, J. F., and Mukhtar, M. M. (1993). *Vet. Microbiol.* **37,** 389–395.
Timoney, P. J., and McCollum, W. H. (1996). *Equine Vet. Educ.* **8,** 97–100.
Timoney, P. J., and Powell, D. G. (1988). *J. Equine Vet. Sci.* **8,** 42–46.
Timoney, J. F., and Trachman, J. (1985). *Infect. Immun.* **48,** 29–34.
Van Heerden, J., and Reyers, F. (1984). *J. S. Afr. Vet. Assoc.* **55,** 41–43.
Wade-Evans, A. M., Woolhouse, T., O'Hara, R., and Hamblin, C. (1993). *J. Virol. Methods* **45,** 179–188.
Wallace, F. J., Emery, J. D., Cripps, A. W., and Husband, A. J. (1995). *Vet. Immunol. Immunopatho.* **48,** 139–154.
Walton, T. E., and Grayson, M. A. (1989). In "The Arboviruses: Epidemiology and Ecology" (T. P. Monath, ed.), pp. 204–231. CRC Press, Boca Raton, FL.
Webster, R., and Guo, Y. J. (1991). *Nature* **351,** 527.
Winston, S., Fiscus, S., Hesterberg, L., Matsushita, T., Mildbrand, M., Porter, J., and Teramoto, Y. (1987). *Vet. Immunol. Immunopathol.* **17,** 453–464.
Wood, J. M., Schild, G. C., Folkers, C., Mumford, J. A., and Newman, R. W. (1983). *J. Biol. Stand.* **11,** 133–136.
Yeargan, M. R., Allen, G. P., and Bryans, J. T. (1985). *J. Clin. Microbiol.* **21,** 694–697.
Zientara, S., Sailleau, C., Moulay, S., and Cruciere, C. (1994). *J. Virol. Methods.* **46,** 179–188.

The Equine Influenza Surveillance Program

J. A. MUMFORD

*Centre for Preventive Medicine, Animal Health Trust,
Newmarket, Suffolk, England*

I. Introduction
II. 1983: WHO Informal Workshop on Vaccination against Equine Influenza
III. 1992: WHO/OIE Consultation on Newly Emerging Strains of Equine Influenza
IV. 1995: Consultation of OIE and WHO Experts on Progress in Surveillance of Equine Influenza and Application to Strain Selection
V. 1996: Actions Taken by the OIE
VI. Findings of the Expert Surveillance Panel
VII. Action Taken by the European Pharmacopoeia
VIII. Actions Taken by the Committee for Veterinary Medicinal Products
IX. Actions Taken by the National Institute of Biological Standarisation and Control
X. Actions Taken by the USDA
References

I. Introduction

Equine influenza is endemic in many equine populations throughout the world with the exception of Iceland and Australasia and periodically causes explosive epizootics often associated with the introduction of subclinically affected animals into areas where the indigenous equidae have little or no immunity. In recent years most epizootics have been associated with the A/equine 2 (H3N8) subtype, however serologic evidence of A/equine 1 (H7N7) infections continues to be reported. International movement of horses for competition and breeding purposes on a worldwide basis increases the likelihood of equine influenza being transmitted long distances, and control measures require a global approach.

Inactivated vaccines against equine influenza have been available since the 1960s and generally contain representatives of both subtypes as whole virus or subunits with or without an adjuvant. Current vaccines are of variable potency with some products capable of providing immunity for at least 1 year after the initial course of three doses and others failing to stimulate detectable antibody to the hemagglutinin after a similar immunization regime. Nevertheless, mandatory vaccination policies are enforced for many groups of competition horses. The lack of standardization and poor efficacy of some products is highly relevant to the control of disease at an international level because horses travel on temporary import permits which are based on a system of health certification and compulsory vaccination schemes agreed between importing and exporting countries. For health certification to be effective, particularly in vaccinated populations, efficient laboratory diagnostic support and epidemiology surveillance is required to alert veterinarians to the presence of influenza in their locality and the necessity to exclude subclinical infections in vaccinated horses. Recognizing the variable efficacy of vaccines and laboratory diagnostic support in some countries, the Code Commission of the Office International des Epizooties (OIE) recommends that an importing country which is free of influenza should require that all horses traveling from endemic areas be fully vaccinated and have received their last booster dose within 2–8 weeks prior to shipment. In spite of this, influenza is still transmitted by subclinically affected vaccinated animals and evidence is accumulating that vaccines including out-of-date strains are ineffective in controlling virus excretion. For these reasons international surveillance and characterization of virus strains involved in outbreaks has become critical and as a result, a formal Equine Influenza Surveillance Programme has been put in place under the auspices of the OIE. This initiative was prompted by the conclusions and recommendations arising from a series of three meetings on equine influenza held under the auspices of the OIE and World Health Organisation (WHO).

II. 1983: WHO Informal Workshop on Vaccination against Equine Influenza

The poor efficacy of equine influenza vaccines was the main subject of discussion at the first meeting held in 1983, at which a number of conclusions and recommendations were made. Field and experimental studies demonstrated that vaccines were failing to protect against influenza and it was concluded that the lack of reliable *in vitro* assays to measure the HA content of vaccines and antibody responses stimulated by vaccines hampered their development. As a result two international collaborative studies were conducted to assess the reliability of the single radial diffusion (SRD) test to measure HA (Wood *et al.*, 1983) and the single radial hemolysis test for measurement of antibody to HA (Mumford, 1992). It was further concluded that more information was required on the immune response of the horse and further experimental studies in the target species were encouraged. It was recognized that many vaccines contained old prototype strains that had little relevance to the prevailing epidemiologic situation. It was concluded that there was insufficient information about circulating viruses and there was an urgent need to increase surveillance and characterization of viruses in order to select the most relevant virus for inclusion in vaccines.

III. 1992: WHO/OIE Consultation on Newly Emerging Strains of Equine Influenza

The second meeting was held following a series of major epidemics caused by H3N8 viruses in South Africa in 1986–1987, India in 1987, China in 1989, and Europe in 1989 (Mumford and Wood, 1993). The main conclusion from this meeting was that antigenic drift in A/equine 2 viruses had compromised vaccine efficacy and current vaccine strains needed updating. This was based on field and experimental observations that recently vaccinated animals with high levels of vaccinal antibody were not protected against infection and shed copious amounts of virus in nasal secretions. Disappointingly, in spite of the recommendations of the previous meeting that efforts should be made to increase surveillance and virus characterization, there was still inadequate information on which to base a reliable strain selection system for vaccines. It was agreed that the OIE and WHO reference laboratories should initiate a surveillance and virus characterization program that would be coordinated through the Animal Health Trust, an OIE Influenza Reference Laboratory.

On the basis of the results of the collaborative studies on the usefulness of SRD for vaccine potency testing and SRH for antibody measurement it was agreed that standard reagents should be prepared to enable these *in vitro* tests to be applied to the standardization of vaccine potency. It was further concluded that challenge studies in horses were a useful method of evaluating the efficacy of vaccines and establishing the relevance of antigenic and genetic drift. In spite of the technical advances reported at this meeting, it was noted that current licensing requirements precluded the rapid introduction of vaccines containing new strains of virus when drift occurred and it was recommended that licensing authorities review their procedures in the light of the information given at the meeting.

IV. 1995: Consultation of OIE and WHO Experts on Progress in Surveillance of Equine Influenza and Application to Strain Selection

The aims of the third meeting were to consider the level of surveillance operating for equine influenza, to examine recent antigenic drift, and to develop a formal review system in order to identify the need for updating vaccines since the previous meeting (Mumford et al., 1997). Procedures for informing registration authorities and vaccine manufacturers of the need to update vaccine strains were also addressed. Current requirements for registration of vaccines containing new virus strains were appraised in the light of these proposals.

From the data presented it was clear that vaccines continued to fail to protect against A/equine 2 infection and many vaccines still contained strains isolated more than a decade earlier. It was agreed that there was now sufficient information on which to base vaccine strain selection. On the basis of data presented, it was agreed that vaccines should contain an A/equine 1 virus and representatives of European and American-like A/equine 2 viruses as a result of a divergence in the genetic lineage of A/equine 2 viruses (Daly et al., 1996). The need for a formal reporting mechanism of antigenic and genetic drift and shift to inform regulatory authorities and manufacturers of the need to update strains was identified and a panel of WHO and OIE experts (Table I) agreed to meet annually to review data and make recommendations. It was recommended that enhanced dialogue should be initiated between manufacturers, experts, and regulatory authorities to identify information required to minimize vaccine licensing requirements. It was agreed that the European Pharmacopoeia (Eu Pharm), National Institute for Biological Standards and Control (NIBSC), and the OIE should develop standards for *in vitro* potency assays to process rapid licensing of updated vaccines.

Following the meeting a number of actions have been taken to fulfill

TABLE I
EXPERT SURVEILLANCE PANEL

Expert	Designation	Address
Dr. J. A. Mumford	OIE Reference Laboratory for Influenza	Animal Health Trust, Lanwades Park, Kennett, Newmarket, Suffolk, CB8 7UU, UK
Dr. T. Chambers	OIE Reference Laboratory for Influenza	Maxwell H. Gluck Equine Research Centre, University of Kentucky, Lexington, Kentucky 40546-0099, USA
Dr. W. Eichhorn	OIE Reference Laboratory for Influenza	Institute for Medical Microbiology Infections and Epidemic Disease, University of Munich, Veterinarstrasse 13, 800 Munich 22, Germany
Dr. A. Hay	WHO Influenza Research Laboratory	National Institute for Medical Research, The Ridgeway, Mill Hill, London, NW7 1AA
Dr. J. Wood	WHO International Laboratory for Biological Standards and Control	National Institute for Biological Standards and Control, Blanche Lane, Potters Bar, Hertfordshire, EN6 3QG UK
Dr. R. Webster	WHO Reference Laboratory for Animal Influenza	St. Jude Children's Research Hospital, 332 North Lauderdale, Memphis, TN 38101, USA
Dr. B. Klingeborn		Department of Virology. P.O. Box 585 BMC, S751 25, Uppsala, Sweden

the recommendations and establish a formal strain selection system and a fast track vaccine licensing system to allow vaccine manufacturers to provide vaccines of adequate potency containing epidemiologically relevant viruses.

V. 1996: Actions Taken by the OIE

Following a report of the Conclusions and Recommendations of the 1995 meeting to the Standards Commissions of the OIE, the commission agreed to support the initiative to establish a formal vaccine strain review system and to provide reference reagents for standardizing serologic assays for influenza and vaccine potency testing. They further supported the publication of an annual notice in the *OIE Bulle-*

tin on the current epidemiologic status of equine influenza and the recommendations of the Expert Surveillance Panel on the need to update vaccine strains (OIE Bulletin, 1996, 1997). The commission's recommendation was ratified at the OIE's 64th General Session in May 1996 and subsequently the *Bulletin* has published, on an annual basis, the findings of the Expert Surveillance Panel.

VI. Findings of the Expert Surveillance Panel

At the meeting held in 1995, extensive data were presented on antigenic genetic variation in A/equine 2 viruses, which had been isolated over the previous decade (Daly *et al.*, 1996). Sequence information relating to the hemagglutinin genes revealed that the genetic drift which had hitherto developed along a single lineage had diverged into two separate lineages in about 1987. Two families of viruses were identified, represented by American and European isolates. However, consistent with the international movement of horses, some American-like viruses have been isolated in Europe and at least one European-like virus was isolated in North America. Based on genetic and antigenic differences between these two families and preliminary data from experimental challenge studies in horses it was concluded that, for horses to be adequately protected, vaccines should contain viruses from both families. At that time some manufacturers had recently updated their products to contain European-like viruses such as Suffolk/89 or Borlange/91 and it was agreed that both these viruses were typical of the prevailing European isolates. For those products which contained viruses from 1983 or before it was recommended that they should be updated to contain a Modern European H3N8 virus antigenically similar to Newmarket/2/93 and it was further recommended that all products should contain a recent American H3N8 isolate antigenically similar to Kentucky/94. It was agreed that ancient viruses such as Miami/63 should be removed from current vaccines.

In 1996, the Expert Surveillance Panel met in Cairns, Australia, and reviewed new information on influenza epidemics and virus characterization. It was agreed that there had been little influenza activity and no major changes in circulating viruses to justify a change in the recommendation made in 1995. The same conclusion was reached in 1997, although it was noted that viruses from relatively isolated populations such as in Eastern Europe showed some antigenic diversity from the most prevalent types.

VII. Action Taken by the European Pharmacopoeia

Following the 1995 meeting the Eu Pharm agreed to assist the OIE with the provision of reference reagents to improve the standardization of the potency of equine influenza vaccines based on serologic assays and agreed to freeze dry equine antisera prepared against representatives of American- and European-like viruses. These reagents have now been prepared and are the subject of an international collaborative study to assign the materials agreed values.

The Eu Pharm has also recognized the need to provide vaccines with epidemiologically relevant viruses and has included this requirement in the latest monograph for equine influenza vaccines together with the requirement for standardized *in vitro* assays for *in-process* testing of antigenic content by, for example, SRD and standardized serologic assays for potency testing vaccines. The new monograph also includes a requirement for challenge studies to demonstrate efficacy of a product.

VIII. Actions Taken by the Committee for Veterinary Medicinal Products

Following the conclusions and recommendations of the 1995 meeting the Immunologicals Working Group of the Committee of Veterinary Medicinal Products established an Ad Hoc Group of Experts to develop guidelines on the "Harmonisation of Requirements for Influenza Vaccines: Specific Requirements for Substitution of an Equine Influenza Strain" in order to develop a fast-track licensing system designed to allow manufacturers to provide products containing epidemiologically relevant strains. The guidelines are based on the premise that equine influenza vaccines are well known and that substitution of one strain by another is unlikely to materially alter the composition of the product. Therefore, manufacturers can submit an application cross-referencing the original dossier for information on aspects that have remained unchanged and where necessary provide analytical, safety, and efficacy data pertaining to the new component only. The guidelines are based on manufacturers selecting a virus antigenically similar to that recommended by the OIE/WHO Expert Surveillance Panel to substitute an old strain. Control tests required during production and on the final product include in-process testing of antigenic content by SRD or a similarly validated test and potency testing in guinea pigs and horses using SRH or a similarly validated serologic assay. Because the correlation between SRH antibody and protection has been so well established (Mumford and Wood, 1992) challenge studies to demonstrate efficacy are not required. To license applications as speedily as possible, duration data are not required as a premarketing requirement, but must be provided as a postauthorization commitment to substantiate claims made.

These guidelines have now been accepted by the CVMP following a consultation period with vaccine manufacturers and have been published by the European Agency for the Evaluation of Medicinal Products.

IX. Actions Taken by the National Institute of Biological Standardization and Control

The standardization of the potency of human influenza vaccines is based on the SRD test and it has been shown to be equally valid for in-process testing of equine influenza vaccines prior to addition of adjuvant (Wood et al., 1983a,b). The monograph in the European Pharmacopoeia, the *CVMP Guidelines,* and the OIE's *Manual of Standards for Diagnostic Tests and Vaccines* all refer to the use of this test for measurement of antigenic content. This test requires the provision of a specific antiserum against the virus hemagglutinin to be assayed and a reference vaccine against which to compare test products. When vaccine strains need updating these reagents also need replacing. The NIBSC provides the reagents for human vaccines and has undertaken to provide similar products which are required for the implementation of the new monograph and CVMP *Guidelines.* Reagents are available for Prague/56 (H7N7) A/equine 1 virus and for American-like and European-like A/equine 2 (H3N8) viruses.

X. Action Taken by the USDA

At the 1995 meeting the USDA representative undertook to explore the possibility of introducing some harmonization into the *Standard Regulations for Equine Influenza Vaccines,* which were currently being drafted by the USDA.

This harmonization will be of great value in ensuring that products developed are suitable and efficacious on an international basis and that vaccinated horses traveling internationally can be expected to be provided with maximal protection induced by inactivated vaccines.

REFERENCES

Anon. (1996). Conclusions and recommendations from the consultation meeting of OIE and WHO experts on equine influenza, Newmarket, United Kingdom, 18–19 September 1995. OIE Bulletin, 108(6), 482–484.

Daly, J., Lai, A., Binns, M., Chambers, T., Barrandeguy, M., and Mumford, J. (1996). Recent worldwide antigenic and genetic evolution of equine H3N8 influenza A viruses. *J. Gen. Virol.* **77,** 661–671.

Mumford, J. A. (1992). Progress in the control of equine influenza. In "Equine Infectious Diseases VI" (W. Plowright, P. D. Rossdale, and J. F. Wade, eds.), pp. 207–218. R&W Publications, Newmarket, UK.

Mumford, J. A. (1997). Report of the expert surveillance panel on equine influenza vaccine strain selection. OIE Bulletin, 109(3), 265.

Mumford, J. A., and Wood, J. (1992). Establishing an acceptable threshold for equine influenza vaccines. *Dev. Biol. Stand.* **79,** 137–146.

Mumford, J., and Wood, J. (1993). WHO/OIE meeting: Consultation on newly emerging strains of equine influenza. 18–19 May, Animal Health Trust, Newmarket, Suffolk, UK. *Vaccine* **11,** 1172–1175.

Mumford, J. A., Chambers, T. M., and Wood, J. (1997). Equine influenza: Progress in surveillance and application to vaccine strain selection. *Vaccine.* (Submitted for publication)

Office International des Epizooties (1996). Manual of Standards for Diagnostic Tests and Vaccines, 3rd ed., Chapter 3.4.5, pp. 409–419. Office International des Epizooties, Paris.

Wood, J. M., Mumford, J. A., Folkers, C., Scott, A., and Schild, G. C. (1983a). Studies with inactivated equine influenza vaccine. 1. Serological responses of ponies to graded doses of vaccine. *J. Hyg.* **90,** 371–384.

Wood, J. M., Schild, G. C., Folkers, C., Mumford, J., and Newman, R. W. (1983b). The standardisation of inactivated equine influenza vaccines by single-radial immunodiffusion. *J. Biol. Stand.* **11,** 133–136.

Wood, J. M., Mumford, J. A., Dunleavy, U., Seagroatt, V., Newman, R. W., Thornton, D., and Schild, G. C. (1988). Single radial immunodiffusion potency tests for inactivated equine influenza vaccines. In "Equine Infectious Diseases V" (D. G. Powell, ed.), pp. 74–79. Kentucky University Press, Lexington.

Vaccination against *Strongylus vulgaris* in Ponies: Comparison of the Humoral and Cytokine Responses of Vaccinates and Nonvaccinates

C. E. SWIDERSKI,* T. R. KLEI,*‡ R. W. FOLSOM,*
S. S. POURCIAU,* A. CHAPMAN,* M. R. CHAPMAN,‡
R. M. MOORE,† J. R. McCLURE,† H. W. TAYLOR,**
AND D. W. HOROHOV*

Departments of *Veterinary Microbiology and Parasitology, †Veterinary Clinical Sciences, and **Veterinary Pathology, School of Veterinary Medicine, Louisiana State University, Baton Rouge, Louisiana 70803, ‡Department of Veterinary Science, Louisiana Agricultural Experiment Station, Louisiana State University Agricultural Center, Baton Rouge, Louisiana 70803

I. Introduction
II. Methods
    A. Experimental Design
    B. Hematology
    C. *Strongylus vulgaris* Larvae and Soluble Antigen
    D. Enzyme-Linked Immunosorbent Assay
    E. Preparation of Cells
    F. Cytokine Quantitation
    G. Statistical Analysis
III. Results
    A. Clinical Signs
    B. Larval Recoveries
    C. Eosinophil Counts
    D. ELISA
    E. Cytokine Quantitation
IV. Discussion
    Acknowledgments
    References

I. Introduction

Strongylus vulgaris is considered the most pathogenic nematode parasite of equidae due to the severe arterial lesions it causes in the mesenteric arteries during larval migration. Infective third-state larvae (L_3) ingested from contaminated pasture penetrate the mucosa of the large intestine, molt to fourth-stage larvae (L_4) in the submucosa, and then proceed along arterioles and arteries that supply the intestine to the root of the cranial mesenteric artery (Ogbourne and Duncan, 1985). Once there the larvae molt to immature adults (L_5) causing severe arteritis before returning, again via the vasculature, to the large intestine to complete their life cycle. Arterial lesions include tortuous subintimal tracts, thrombi, and, in severe cases, verminous aneurysms that can compromise perfusion of intestinal vascular beds. This syndrome known as verminous arteritis or thromboembolic disease is characterized by ischemic infarctions of the bowel, which result in toxemia, abdominal pain, and death in severe cases (White, 1985).

Although *S. vulgaris* is extremely pathogenic, its importance as an equine pathogen has diminished during the past 10 years due to the development and regular use of ivermectin (DiPietro *et al.*, 1990). This anthelimintic at normal dose levels is highly efficacious against migrating *Strongylus* larvae and its usage can eliminate the parasite from closed herds of horses (Dunsmore, 1985). Nevertheless there is some concern for the environmental and toxic effects of the anthelmintics on free living arthropod fauna (Waller, 1993). Alternate management programs such as the regular removal of feces and alternate grazing schemes have been shown to be useful (DiPietro *et al.*, 1990). These methods, however, are expensive and not adaptable to all environments. Vaccination against any infection is clearly the most economic and environmentally sound approach for the control of disease. Although progress toward this goal in the control of helminths has been slow, recent results have been encouraging (Lightowers, 1994).

Previous studies have demonstrated that oral vaccination with radiation-attenuated *S. vulgaris* L_3 induces resistance to challenge infection and prevents classical lesions of verminous arteritis (Klei, 1992). When compared to parenteral vaccination with adult or larval somatic antigen homogenates which did not confer resistance to challenge infection, the protection observed in oral radiation-attenuated L_3 vaccinated ponies correlated to both prechallenge anti-*S. vulgaris* antibody titers specific for surface antigens of L_3 stages and to induction of a postchallenge anamnestic-like eosinophilia (Monahan *et al.*, 1994). Infections by *S. vulgaris* have been shown to activate eosinophils and neutrophils *in vitro* (Dennis *et al.*, 1988) and eosinophils from *S. vulgaris* primed but not unprimed ponies kill *S. vulgaris* L_3 *in vitro* in an antibody-dependent manner (Klei *et al.*, 1992), indicating that an antibody-dependent phenomenon involving eosinophils may contribute to the resistance seen in immune ponies.

Helminth infections are characteristically associated with eosinophilia and elevated IgE production (Finkelman *et al.*, 1991), traits that also characterize the type II or Th2 response initially described in cloned mouse cells (Mosmann *et al.*, 1986; Mosmann and Coffman, 1989b). Type II responses, identified by the production of interleukin 4 (IL-4) and IL-5, which are integral to the generation of IgE (Snapper and Paul, 1987) and eosinophilia (Coffman *et al.*, 1989), respectively, have been shown to play a significant role in the protective immune response to metazoan helminths (reviewed in Mosmann and Coffman, 1989a; Cox and Liew, 1992; Urban *et al.*, 1992). A cytokine or combination of cytokines that exhibits chemotactic activity for eosinophils has been found in the supernatants of *S. vulgaris* stimulated peripheral blood mononuclear cells (PBMCs) from immune but not nonimmune ponies (Dennis *et al.*, 1993). This finding coupled with the anamnestic eosinophilia that is characteristic of immunity to *S. vulgaris* suggests that differential cytokine production, perhaps similar to that seen in classical type II responses to helminth parasites, may play a role in immunity to *S. vulgaris*. Here we compare the humoral, lymphoproliferative, and cytokine responses of vaccinated and nonvaccinated ponies challenged with *S. vulgaris*.

II. Methods

A. Experimental Design

Ten yearling ponies were raised and maintained under parasite-free conditions (Monahan *et al.*, 1997), housed in pairs on wood shavings, and fed twice daily a pelleted ration and water *ad libitum*. The experimental design consisted of three groups. Four ponies in group 1 (vaccinates) were orally vaccinated with 500 *S. vulgaris* L_3 irradiated with 90 krad ^{60}Co as described by Klei *et al.* (1982). Immunization was repeated in this group 3 weeks after the first immunization. Group 2 (nonvaccinates) consisted of four ponies and group 3 (controls) consisted of two ponies. Six weeks following the second immunization vaccinates and nonvaccinates were challenged *per os* with 1000 *S. vulgaris* L_3. Following challenge, ponies were monitored at least twice daily for fever, signs of abdominal pain, depression, and anorexia. Pyrexia was considered to be any rectal temperature above 39°C. Four days prior to challenge (D−4), on the day of challenge (D0), and on days 4 (D4) and 9 (D9) postchallenge, peripheral blood was collected for hematologic evaluation and collection of PBMCs at these time points as well as 14 days following challenge (D14). On D14 necropsy examinations were performed as described previously (Klei et al., 1982) for evaluation of lesions, and recovery of larvae from dissections of the cranial mesenteric artery (CMA) and its branches.

B. Hematology

Blood was analyzed in a routine fashion. Total white blood cell counts were analyzed using a Baker 9000 analyzer (Serono-Baker, Allentown, PA). Differential white cell counts were determined by counts of 100 cells examined at 100× under oil immersion.

C. *Strongylus vulgaris* Larvae and Soluble Antigen

Larvae for use in the irradiated immunizations were recovered as L_3 from Baermann sedimentations of fecal cultures from *S. vulgaris* monospecifically infected donor ponies (McClure et al., 1994). Larvae were washed and stored in tap water at 4°C until use. Soluble antigen preparations used in enzyme-linked immunosorbent assay (ELISA) and lymphoproliferation assays were prepared as previously described (Dennis et al., 1992) from adult *S. vulgaris* recovered from the intestines of horses.

D. Enzyme-Linked Immunosorbent Assay

Circulating antibody isotypes specific for adult *S. vulgaris* soluble antigens were monitored by ELISA using serum collected on D−4 and D14. Soluble adult *S. vulgaris* antigen (SAWA) was diluted in buffer (0.015 M Na_2CO_3, 0.035 M $NaHCO_3$, 0.003 M NaN_3) to 5 µg/ml. Antigen preparation was added to the wells of 96-well flat bottom polystyrene microtiter plates in 50-µl volumes (Dynatech Laboratories, Chantilly, VA) and incubated overnight. Unless otherwise noted, incubations were performed at 37°C in a humidified incubator. Plates were washed three times with a solution of 0.05% Tween 20 in PBS phosphate-buffered solution (PBST) and nonspecific binding sites were blocked using 1% fish gelatin (Sigma Chemicals, St. Louis, MO) in PBS (PBSG), 100 µl per well, for 1 hour at room temperature. Dilutions of serum ranging from 1:100 to 1:3200 were made in PBSG and added to triplicate wells in 50-μl aliquots. Serum from known high responders and negative responders served as standard positive and negative controls on each plate and were used to correct for interplate variability. Plates were incubated for 90 minutes then washed three times with PBST. Culture supernates from hybridoma lines specific for equine IgA, IgM, IgGa, IgGb, or IgG(T) (Lunn et al., 1996) were diluted 1:100 in PBSG, added to appropriate wells in 50-μl volumes, and incubated for 90 minutes. Following incubation, wells were washed three times with PBST. Affinity-purified, horseradish peroxidase-conjugated goat anti-mouse IgG and IgM, heavy- and light-chain specific (Jackson Immunoresearch Labs, West Grove, PA) was diluted 1:1000 in PBSG, added to wells in 50-μl aliquots, and incubated for 90 minutes. Plates were washed three times with PBST. The substrate was 3,3',5,5'-tetramethylbenzidine (TMB, Kirkeguard and Perry, Gaithersburg, MD) at 75 μl per well. Reactions were allowed to proceed for 10 minutes for optimum color development. Optical density (OD) was recorded using a Dynatech MR 700 automated microtiter plate reader (Dynatech Industries) with absorbance set at 630 nm.

E. Preparation of Cells

Equine PBMCs were isolated from venous blood by differential centrifugation over Ficoll Paque (Pharmacia LKB Biotechnology, Piscataway, NJ). After washing in calcium and magnesium free phosphate buffered saline (CMF-PBS), PBMC were suspended in media consisting of RPMI 1640 (Sigma Chemicals) supplemented with 2-mercaptoethanol (10^{-8} M), glutamine (2 mM), 100 U/ml penicillin, 100 μg/ml streptomycin, and 5% heat-inactivated fetal bovine serum (HyClone Laboratories, Logan, UT). Aliquots of PBMCs (2×10^6 cells) were frozen at $-70°C$ in RNA Stat 60 (Tel-Test, Friendswood, TX) for use in cytokine quantitation assays.

F. Cytokine Quantitation

PBMCs collected at D−4, D4, D9, and D14 and frozen in RNA Stat 60 reagent were defrosted and processed to RNA using chloroform extraction as per the reagent manufacturer's directions. RNA (0.60 μg) in a volume of 35 μl DEPC-treated water was heated to 65°C for 10 minutes to denature the RNA and then quenched in ice and pulsed in a microfuge at 4°C. Reagent master mix (45 μl) was then added, resulting in a final reaction containing 1× first strand buffer (Gibco-BRL,

Gaithersburg, MD), 0.5 mM dNTP (Perkin-Elmer, Foster City, CA), 1.0 µM oligo-dT (Promega, Madison, WI), 10 mM DTT (Gibco), 0.075 µg/µl BSA (NEB, Beverly, MA), 120 units RNasin (Promega), and 800 units Moloney Murine Leukemia Virus Reverse Transcriptase (BRL). The reaction was incubated for 1 hour at 40°C then frozen at −20°C until use.

Equine IL-2, IL-4, IL-5, IL-10, and interferon γ (IFN-γ) were quantified with the aid of the polymerase chain reaction (PCR) by interpolation against a standard curve. Standard curves were generated for each cytokine by simultaneously amplifying, from a common PCR master mix, known quantities of plasmids that contained the equine cytokine sequences of interest. Briefly, cytokines were amplified in 50-µl PCR reactions from 5 µl of each cDNA reaction using 0.5 units of Taq polymerase (Perkin-Elmer) and species-specific primers at a final concentration of 0.2 µM. Primers were commercially prepared (Genelab, Baton Rouge, LA; Baron Biotech, Milford, CT) from species-specific sequences which were determined using oligo primer analysis software (National Biosciences, Plymouth, MN). Upstream primers were biotinylated on their 5′ terminus. Cycling parameters including buffer (PCR Optimizer, Invitrogen, Carlsbad, CA) and annealing temperatures were optimized for each cytokine. All cytokines were amplified to 35 cycles, which was previously determined to maintain the reactions in their exponential amplification phase. Beta-actin, which was utilized as a housekeeping gene, was amplified from 2 µl of cDNA using 25 cycles in order to remain within its exponential amplification phase.

PCR products were quantified using the QPCR System 5000 (Perkin-Elmer) as has been described (Zhao et al., 1996; Blok et al., 1997). Five microliters of each PCR reaction was hybridized in a 50-µl reaction containing 1× PCR buffer II (Perkin-Elmer) and 10 pmol tris(2,2′-bipyridine) ruthenium II chelate labeled oligonucleotide probe whose sequence was specific to the cytokine targeted by PCR. Oligonucleotide probes were commercially prepared (Baron Biotech, Milford, CT) and purified by high-performance liquid chromatography (HPLC) to eliminate nonlabeled oligonucleotides. The reaction was heated to 95°C for 90 seconds followed by a 5-minute hold at 55°C. Streptavidin-coated iron beads (15 µl) (Dynabeads, Perkin-Elmer) were added to each reaction followed by incubation at 55°C for 30 minutes. The entire 65-µl reaction was transferred to a 175-mm polypropylene tube containing 335 µl QPCR assay buffer (Perkin-Elmer) and quantified on the QPCR System 5000 whose output is in luminosity units.

Luminosity units were converted to copy numbers using the standard curve developed for each cytokine and beta-actin. Each sample was normalized using a correction factor calculated from the beta-actin content in each sample. Results are presented as fold increases over prechallenge (D−4) levels for each cytokine mRNA.

G. Statistical Analysis

The number of days that anorexia, pyrexia, depression, and abdominal pain were exhibited by ponies in each treatment group in response to oral challenge with *S. vulgaris* larvae was subjected to one-way analysis of variance as were eosinophil counts. Two-way analysis of cytokine data was performed on both fold increases and copy numbers over time and between experimental groups. Cytokine copy number data were log transformed prior to analysis. ELISA OD data was subjected to two-way analysis of variance. Differences were considered significant when the p values were less than or equal to 0.05.

III. Results

A. Clinical Signs

Ponies in the vaccinated group exhibited significantly fewer episodes of pyrexia, depression, and abdominal pain than nonvaccinates (Table I). Pyrexia, depression, and abdominal pain exhibited by vaccinates was not significantly different from that exhibited by control ponies. There was no difference in the number of days that anorexia was exhibited by ponies in the three treatment groups.

TABLE I

Summary of the Clinical Signs Exhibited by Vaccinated, Nonvaccinated, and Control Ponies in Response to Challenge with 1000 *S. vulgaris* L_3[a]

Group	Anorexia	Pyrexia	Depression	Colic
Vaccinates	8	1	0	0
Nonvaccinates	9	5	3	3
Controls	2	0	0	0

[a]Data represent the mean number of days that ponies exhibited the respective clinical sign. Pyrexia was considered any rectal temperatures exceeding 39°C.

B. Larval Recoveries

Irradiated recipients had a 100% reduction of migration larvae compared with nonvaccinates (Table II). Larval numbers in nonvaccinates ranged from 3 to 28 per pony.

C. Eosinophil Counts

Differential white blood cell counts performed on peripheral blood smears from ponies on D−4, D0, D4, D9, and D14 following challenge demonstrated that ponies vaccinated with radiation-attenuated larvae developed an eosinophilia by day 4 following challenge (Fig. 1). In contrast, eosinophil counts from nonvaccinated ponies failed to show significant elevations within the 2 weeks following challenge.

D. ELISA

When compared to prechallenge titers of nonvaccinates, levels of IgGa, IgGb, and IgG(T) antibodies specific for SAWA were significantly greater in the sera of vaccinates (Fig. 2). Following challenge, the IgG(T) of vaccinates remained significantly greater than those of nonvaccinates but did not change significantly from prechallenge levels. Challenge with virulent *S. vulgaris* L_3 resulted in statistically significant increases in IgA in both vaccinates and nonvaccinates with nonvaccinates significantly exceeding those of vaccinates following challenge. Nonvaccinates also exhibited a statistically significant increase in IgG_a in response to challenge. There were no differences in IgM

TABLE II

Larval Recoveries from Vaccinated and Nonvaccinated Ponies Following Challenge with 1000 *S. vulgaris* L_3[a]

Immunization group	Average larval recovery	Percent protection
Vaccinates	0	100%
Nonvaccinates	16	—

[a]Larvae were recovered from scrapings of the lumen of the CMA and its branches and counted using a dissecting microscope. Larval numbers in nonvaccinates ranged from 3 to 28 per pony.

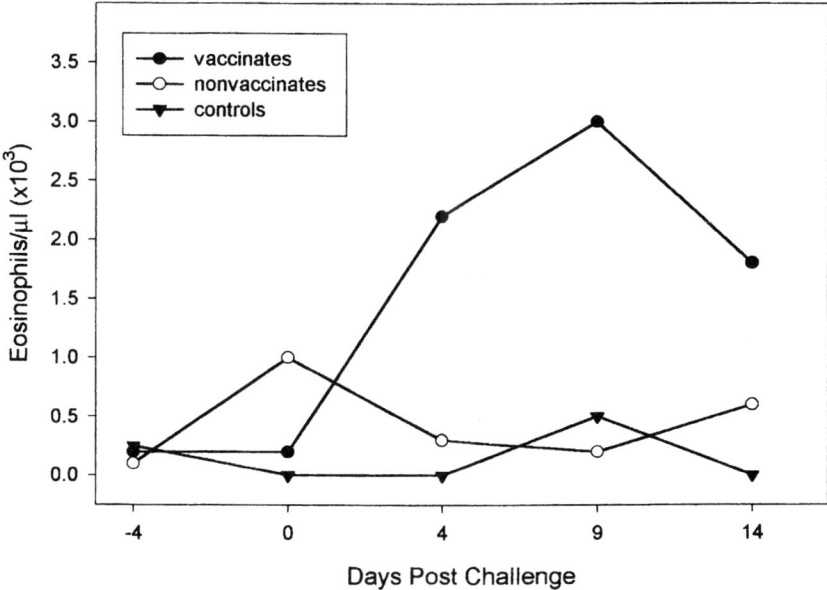

FIG. 1. Eosinophil response to *S. vulgaris* challenge in vaccinate, nonvaccinate, and control ponies. Mean eosinophil counts for each group were calculated by multiplying the WBC count/µl blood, as determined by automated counting, by the percentage of eosinophils determined by manual counting of 100 cells viewed under oil immersion.

production within or among the treatment groups prior to or following challenge.

E. Cytokine Quantitation

Fold differences in the mRNA levels of IL-2, IL-4, IL-5, IL-10, and IFN-γ in the PBMCs on D4, D9, and D14 postchallenge were compared within each treatment group (Fig. 3). The vaccinates demonstrated a significant increase in IL-4 and IL-5 mRNA levels on D14, IL-2 and IFN-γ mRNA levels significantly decreased on D9, and IL-10 on D4 and D9. PBMCs from nonvaccinates demonstrated significant increase in IL-5 by D14 and decreases in IFN-γ and IL-10 at D9.

IV. Discussion

Differences in postchallenge clinical signs and larval recoveries demonstrated that ponies immunized orally with irradiated L_3 developed a

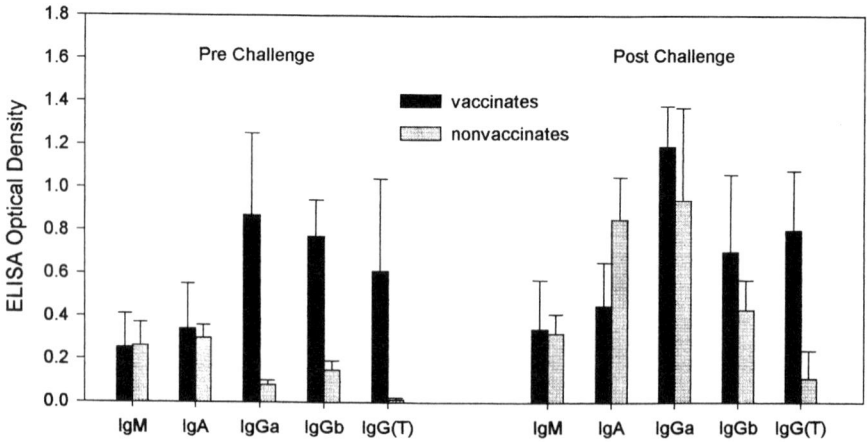

FIG. 2. Antibody isotype response of vaccinated and nonvaccinated ponies to *S. vulgaris* soluble adult worm antigen 4 days prior to and 14 days following challenge. Serum samples were analyzed by ELISA in quadruplicate. Bars represent mean OD values at 1:400 dilution of serum with standard deviation for groups of ponies in a treatment group.

protective response to challenge when compared to nonimmunized ponies. In previous studies, necropsies performed 6 weeks following challenge with virulent *S. vulgaris* L_3 demonstrated that ponies vaccinated with radiation-attenuated L_3 did not develop lesions of verminous arteritis but instead had significant periportal fibrosis, suggesting that the protective response is generated within the intestinal submucosa and that larval antigens are cleared via the portal system prior to L_4 migration into the intestinal vaculature (Monahan et al., 1994). Blocking parasite entrance to the intestinal arterioles in vaccinates should both limit peripheral exposure to parasite antigens and prevent arterial lesions, resulting in the significant reductions in fever, depression, and abdominal pain we observed in vaccinates relative to nonvaccinates. Massive killing of larvae within the wall of the intestine in vaccinates should, however, be expected to trigger the cascade of inflammatory mediators responsible for pyrexia including tumor necrosis factor α, IL-1, and IL-6. It has been postulated that clearance of killed larvae and inflammatory mediators by the portal system, which is by virtue of its network of vessels of expanding diameter mechanically less susceptible to procoagulant effects and thrombus trapping than intestinal vasculature, could also contribute to the de-

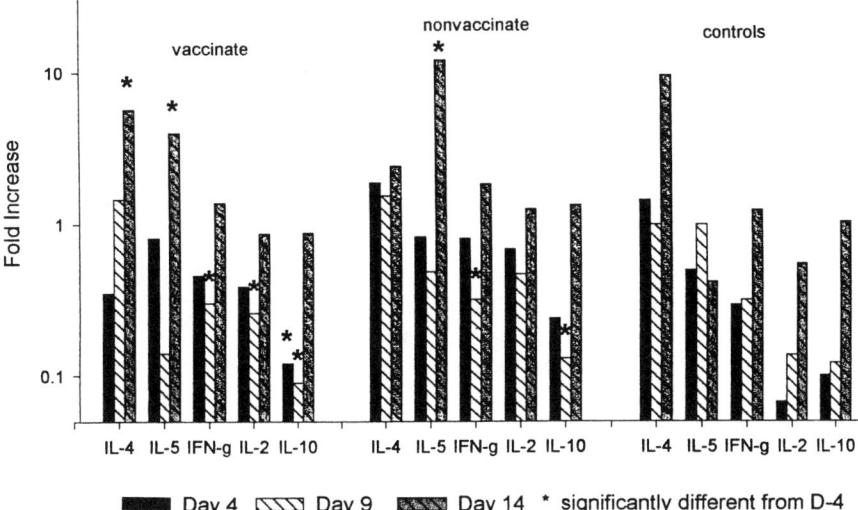

FIG. 3. Cytokine production in PBMCs from vaccinated, nonvaccinated, and control ponies after oral *S. vulgaris* L_3 challenge. Cytokine copy number was determined using a quantitative RT–PCR assay. Data are represented as the fold increase in geometric mean over prechallenge values on the indicated days following oral challenge with 1000 virulent *S. vulgaris* L_3.

crease in both pyrexia and abdominal pain observed in vaccinates as compared to nonvaccinates (Monahan et al., 1994).

Prior studies have demonstrated that following immunization with radiation-attenuated *S. vulgaris* L_3, vaccinates develop a small increase in total immunoglobulin directed against SAWA and that following challenge with virulent *S. vulgaris* L_3, this SAWA-specific antibody titer displays an anamnestic response, converging with titers of nonvaccinates by 3 weeks postchallenge (Monahan et al., 1994). We examined the differences in immunoglobulin isotypes specific for SAWA in vaccinates and nonvaccinates in response to challenge with virulent *S. vulgaris*. Our data indicate that the previously described increase in SAWA-specific immunoglobulin titers seen in response to vaccination are at least in part due to production of isotypes IgGa, IgGb, and IgG(T). Challenge resulted in significant increases in SAWA-specific IgA production in both vaccinates and nonvaccinates. By contrast, SAWA-specific IgG(T) remained significantly greater in vaccinates compared to nonvaccinates following challenge. Increases in IgG subclasses in response to *S. vulgaris* infection have been previously demonstrated with IgG(T) showing the greatest increase (Patton et al.,

1978). While the IgG(T) response of the nonvaccinates in the current study was not elevated following challenge, this could be due to the early time point in which the sera were analyzed. It is possible that the nonvaccinates would have also developed an elevated IgG(T) response later in the infection. This is consistent with the general assumption that the induction of IgG(T) antibodies requires prolonged antigenic stimulation. Though biological function and cytokine regulation of murine immunoglobulin isotypes is well characterized, neither has been done for the horse. Thus the significance of the disparate production of IgG(T) antibodies between the vaccinates and the nonvaccinates remains unknown.

Helminth infections are characteristically associated with eosinophilia (Finkelman et al., 1991) and IL-5 is required for this response (Coffman et al., 1989; Sher et al., 1990). Following challenge with virulent S. vulgaris, vaccinates developed an anamnestic eosinophilia that began to rise on D4 and appeared to be peaking at D9. The kinetics of this response preceded those previously reported (Monahan et al., 1994) and may be the result of a larval challenge that exceeded previously published doses, or perhaps improved viability of either the vaccinal or challenge larvae leading to more robust stimulation of the eosinophilic response mechanisms. The levels of mRNA for IL-5 increased significantly in the PBMC of vaccinates on D14 but could not account temporally for the eosinophilia observed in vaccinates. Levels of IL-5 in the PBMC of vaccinates did exceed those of nonvaccinates by at lease fourfold prior to challenge and on D4 and D9 following challenge. Though not statistically significant, these findings appear to be physiological significant with IL-5 production correlating to the more rapid increase in eosinophil numbers observed in vaccinates relative to nonvaccinates. Though not statistically significant, the magnitude of the IL-5 response of nonvaccinates actually exceeded that of vaccinates on D14. Analysis of the fold increase in IL-5 of nonvaccinates over prechallenge levels indicated that the IL-5 response on D14 was statistically significant. Because ponies were euthanized on D14 the significance of this finding is only speculative. However, in previous experiments eosinophil responses in nonvaccinates lagged behind that observed in vaccinates by approximately 1 week (Monahan et al., 1994). Accordingly, the slight increase in eosinophil counts observed in nonvaccinates on D14 may signal the beginning of the eosinophilic response in this group.

Like IL-5, the cytokine IL-4 plays an integral role in the protective immune response to metazoan helminths, primarily through the induction of IgE (Snapper and Paul, 1987). Elimination of the IgE response with anti-IL-4 has been shown to reverse protection in *Heligmosomoides polygyrus* immune mice (Urban et al., 1991) and primary *Trichuris muris* infections (Else et al., 1994). IL-4 production by vaccinates increased following challenge and significantly exceeded that of nonvaccinates and controls on D9. A synthesis of these findings leads to the hypothesis that anamnestic IL-4 production, possibly leading to an antigen-specific IgE response, may be one component of the immune response that differentiates. *S. vulgaris* immune from susceptible ponies. Unfortunately IgE production was not quantified in the present study because equine-specific reagents are not readily available.

We observed statistically significant decreases in IFN-γ at D9 in the nonvaccinates. Depression of IFN-γ production was also observed in the PBMCs of vaccinates though this was not statistically significant. In no cases did the decreases observed exceed the cutoff for significance that was established during standardization of this assay. Nevertheless these findings are certainly noteworthy in the light of the association between decreased IFN-γ production and the Th2 response. In the Th1/Th2 paradigm, inhibition of IFN-γ production is often attributed to IL-10. Though IL-10 was originally described as a product of mouse Th2 clones, it is now clear that IL-10 is also secreted by Th1 cells (especially in humans) and activated marcophages (Sornasse et al., 1996; Martinez et al., 1997). IL-10 is a potent inhibitor of IFN-γ production (Fiorentino et al., 1989; Urban et al., 1991); however, in our experiments a temporal relationship between increased IL-10 levels and decreased IFN-γ levels was not clear. All groups exhibited a decrease in IL-10 production following challenge that was significant on D4 and D9 in the vaccinates and D9 in nonvaccinates. Significant increases in IL-10 noted on D14 in the vaccinates and the nonvaccinates were relative to the depression on D9 and were not significantly different from prechallenge levels. It is important to remember that while IFN-γ is capable of directly inhibiting the Th2 response, IL-10 inhibition of IFN-γ production is indirect, acting on antigen-presenting cells to reduce their ability to stimulate cytokine secretion by Th1 cells. However, other cytokines including IL-3 and IL-4 (Liew et al., 1989), IL-11 (Leng and Elias, 1997), and IL-13 (Doherty et al., 1993) have also been shown to antagonize macrophage activation and may be responsible for the depression in IFN-γ production that we observed. TGF-β produced by some Th2 cells and many other cell types is also antiproliferative inhibiting leukocyte activation (Hausmann et al., 1994). Elevations in IL-4 production in the vaccinates could also account temporally for the significant decrease in IFN-γ production seen in these individuals. Similarly, the increase in IL-2 in vaccinates, though statistically significant, did not meet the predetermined cutoff for fold changes established during validation of the assay.

In summary, immunization of ponies with irradiated L_3 of *S. vulgaris* resulted in the production of the type II cytokines IL-4 and IL-5 and the reduction in the type I cytokines IL-2 and IFN-γ upon challenge. Similar, though temporally delayed, results were seen in the nonvaccinates following challenge. Anamnestic production of the type II cytokines, possibly leading to an antigen-specific IgE [or IgG(T)] response, may be one component of the immune response that provides protection from *S. vulgaris* infection in immune ponies.

ACKNOWLEDGMENTS

This work was supported in part by grants from the USDA (96-35204-3674, T.R.K.) and the Grayson Jockey Club Research Foundation, Inc. (D.W.H.). This paper was approved for publication by the director of the Louisiana Agricultural Experiment Station as manuscript No. 97-0301.

REFERENCES

Blok, H. J., Gohlke, A. M., and Akkermans, A. D. (1997). Quantitative analysis of 16S rDNA using competitive PCR and the QPCR System 5000. *BioTechniques* **22,** 700–704.
Coffman, R. L., Seymour, B. W., Hudak, S., Jackson, J., and Rennick, D. (1989). Antibody to interleukin-5 inhibits helminth-induced eosinophilia in mice. *Science* **245,** 308–310.
Cox, F. E., and Liew, F. Y. (1992). T-cell subsets and cytokines in parasitic infections. *J. Parasitol.* **8,** 371–374.
Dennis, V. A., Klei, T. R., Chapman, M. R., and Jeffers, G. W. (1988). In vivo activation of equine eosinophils and neutrophils by experimental *Strongylus vulgaris* infection. *Vet. Immunol. Immunopathol.* **20,** 61–74.
Dennis, V. A., Klei, T. R., Miller, M. M., Chapman, M. R., and McClure, J. R. (1992). Immune responses of pony foals during repeated infections of *Strongylus vulgaris* and regular ivermectin treatments. *Vet. Parasitol.* **42,** 83–99.
Dennis, V. A., Klei, T. R., and Chapman, M. R. (1993). Generation and partial characterization of an eosinophil chemotactic cytokine produced by sensitized equine mononuclear cells stimulated with *Strongylus vulgaris* antigen. *Vet. Immunol. Immunopathol.* **37,** 135–149.
DiPietro, J. A., Klei, T. R., and French, D. D. (1990). Contemporary topics in equine parasitology. *Compend. Contin. Educ. Pract. Vet.* **12,** 713–721.
Doherty, T. M., Kastelein, R., Menon, S., Andrade, S., and Coffman, R. L. (1993). Modulation of murine macrophage function by IL-13. *J. Immunol.* **151,** 7151–7160.
Dunsmore, J. P. (1985). Integrated control of *Strongylus vulgaris* infections in horses using ivermectin. *Equine Vet. J.* **17,** 191–197.
Else, K. J., Finkelman, F. D., Maliszewski, C. R., and Grencis, R. K. (1994). Cytokine-

mediated regulation of chronic intestinal helminth infection. *J. Exp. Med.* **179,** 347–351.
Finkelman, F. D., Pearce, E. J., Urban, J. F., and Sher, A. (1991). Regulation and biological function of helminth-induced cytokine responses. *Immunol. Today* **12,** A62–66.
Fiorentino, D. F., Bond, M. W., and Mosmann, T. R. (1989). Two types of mouse T helper cell. IV. Th2 clones secrete a factor that inhibits cytokine production by Th1 clones. *J. Exp. Med.* **170,** 2081–2095.
Hausmann, E. H., Hao, S. Y. Pace, J. L., and Parmely, M. J. (1994). Transforming growth factor beta 1 and gamma interferon provide opposing signals to lipopolysaccharide-activated mouse macrophages. *Infect. Immun.* **62,** 3625–3632.
Klei, T. R. (1992). Recent observations on the epidemiology, pathogenesis and immunology of equine helminth infections. *In* "Equine Infectious Diseases VI" (W. Plowright, P. D. Rossdale, and J. F. Wade, eds.), pp. 129–136. R&W Publications, Newmarket, UK.
Klei, T. R., Torbert, B. J., Chapman, M. R., and Ochoa, R. (1982). Irradiated larval vaccination of ponies against *Strongylus vulgaris. J. Parasitol.* **68,** 561–569.
Klei, T. R., Chapman, M. R., and Dennis, V. A. (1992). Role of the eosinophil in serum-mediated adherence of equine leukocytes to infective larvae of *Strongylus vulgaris. J. Parasitol.* **78,** 477–484.
Leng, S. X., and Elias, J. A. (1997). Interleukin-11 inhibits macrophage interleukin-12 production. *J. Immunol.* **159,** 2161–2168.
Liew, F. Y., Millott, S., Li, Y., Lelchuk, R., and Ziltener, H. (1989). Macrophage activation by interferon-gamma from host-protective T cells is inhibited by interleukin (IL)3 and IL4 produced by disease promoting T cells in leishmaniasis. *Eur. J. Immunol.* **19,** 1227–1232.
Lightowers, M. W. (1994). Vaccination against animal parasites. *Vet. Parasitol.* **54,** 177–204.
Lunn, D. P., Holmes, M. A., and Antczak, D. F. (1996). Summary report of the second equine leucocyte antigen workshop. *Vet. Immunol. Immunopathol.* **54,** 159–161.
Martinez, J. A., King, T. E. J., Brown, K., Jennings, C. A., Borish, L., Mortenson, R. L., Khan, T. Z., Bost, T. W., and Riches, D. W. (1997). Increased expression of the interleukin-10 gene by alveolar macrophages in interstitial lung disease. *Am. J. Physiol.* **273,** L676–L683.
McClure, J. R., Chapman, M. R., and Klei, T. R. (1994). Production and characterization of monospecific adult worm infections of *Strongylus vulgaris* and *Strongylus edentatus* in ponies. *Vet. Parasitol.* **51,** 249–254.
Monahan, C. M., Taylor, H. W., Chapman, M. R., and Klei, T. R. (1994). Experimental immunization of ponies with *Strongylus vulgaris* radiation-attenuated larvae or crude soluble somatic extracts from larval or adult stages. *J. Parasitol.* **80,** 911–923.
Monahan, C. M., Chapman, M. R., Taylor, H. W., French, D. D., and Klei, T. R. (1997). Foals raised on pasture with or without early pyrantel tartrate feed additive: comparison of parasite burdens and host responses following experimental challenge with large and small strongyle larvae. *Vet. Parasitol.* **73,** 277–289.
Mosmann, T. R., and Coffman, R. L. (1989a). Heterogeneity of cytokine secretion patterns and functions of helper T cells. *Adv. Immunol.* **46,** 111–147.
Mosmann, T. R., and Coffman, R. L. (1989b). Th1 and Th2 cells: Different patterns of lymphokine secretion lead to different functional properties. *Annu. Rev. Immunol.* **7,** 145–173.
Mosmann, T., Cherwinski, H., Bond, M. W., Giedlin, M. A., and Coffman, R. L. (1986).

Two types of murine helper T cell clone I. Definition according to profiles of lymphokine activities and secreted proteins. *J. Immunol.* **136,** 2348–2357.

Ogbourne, C. P., and Duncan, J. L. (1985). "*Stronglyus vulgaris* in the Horse: Its Biology and Veterinary Importance," Commonw. Inst. Helminthol., Misc. Publ. No. 4. Commonwealth Agricultural Bureau, Farnham Royal Bucks, England.

Patton, S., Mock, R. E., Drudge, J. H., and Morgan, D. (1978). Increase of immunoglobulin T concentration in ponies as a response to experimental infection with the nematode *S. vulgaris. Am. J. Vet. Res.* **39,** 19–23.

Sher, A., Coffman, R. L., Hieny, S., Scott, P., and Cheever, A. W. (1990). Interleukin 5 is required for the blood and tissue eosinophilia but not granuloma formation induced by infection with *Schistosoma mansoni. Proc. Natl. Acad. Sci. USA* **87,** 61–65.

Snapper, C. M., and Paul, W. E. (1987). Interferon-gamma and B cell stimulatory factor-1 reciprocally regulate Ig isotype production. *Science* **236,** 944–947.

Sornasse, T., Larenas, P. V., Davis, K. A., de Vries, J. E., and Yssel, H. (1996). Differentiation and stability of T helper 1 and 2 cells derived from naive human neonatal CD4+ T cells, analyzed at the single-cell level. *J. Exp. Med.* **184,** 473–483.

Urban, J. F., Katona, I. M., Paul, W. E., and Finkelman, F. D. (1991). Interleukin-4 is important in protective immunity to a gastrointestinal nematode infection in mice. *Proc. Natl. Acad. Sci. USA* **88,** 5513–5517.

Urban, J. F., Madden, K. B., Svetic, A., Cheever, A., Trotta, P. P., Gause, W. C., Katona, I. M., and Finkelman, F. D. (1992). The importance of Th2 cytokines in protective immunity to nematodes. *Immunol. Rev.* **127,** 205–220.

Waller, P. J. (1993). Towards sustainable nematode parasite control of livestock. *Vet. Parasitol.* **48,** 295–309.

White, N. A. (1985). Thromboembolic colic in horses. *Compend. Contin. Educ. Pract. Vet.* **7,** 156–162.

Zhao, S., Consoli, U., Arceci, R., Pfeifer, J., Dalton, W. S., and Andreeff, M. (1996). Semi-automated PCR method for quantitating MDR1 expression. *BioTechniques* **21,** 726–731.

ISCOM: A Delivery System for Neonates and for Mucosal Administration

B. MOREIN, M. VILLACRÉS-ERIKSSON, J. EKSTRÖM, K. HU, S. BEHBOUDI, AND K. LÖVGREN-BENGTSSON

Department of Veterinary Microbiology, Section of Virology, Swedish University of Agricultural Sciences, Biomedical Centre Box 585, S-751 23 Uppsala, Sweden

I. Introduction: Immune Stimulating Complex
II. The ISCOM Concept
III. Formation of ISCOM
IV. Antigen Presentation and Targeting with ISCOMs
V. Adjuvant Influence on the Transport of Antigen
VI. Adjuvants and Delivery Systems for Induction of Mucosal Immunity
VII. ISCOMs Induce a Cytokine Th1 Type Response But Also Th2
VIII. Antigens Loaded in ISCOMs Induce Immune Response in Neonates
IX. Protective Immunity
References

I. Introduction: Immune Stimulating Complex

Immune stimulating complex (ISCOM) is a delivery system for antigen and adjuvant together in the same particle. The unique component of ISCOM is a mixture of *Quillaja* saponins extracted from the bark of the tree *Quillaja saponaria* Molina. Three groups of *Quillaja* components having a triterpenoid structure named QHA, QHB, and QHC have been characterized by reverse-phase high-performance liquid chromatography (HPLC) and were functionally identified to be of particular interest for ISCOMs. Defined compositions of QHA and QHC are suitable for small companion animals while for large animals the more crude and low-priced *Quillaja* saponins are used. A commercially available ISCOM influenza vaccine for horses is available in Europe and ISCOMs are in human phase II studied in Europe and the United States.

II. The ISCOM Concept

The ISCOM was created to make antigens optimally immunogenic by (1) presenting the antigens in a physically immunogenic form, that is, several copies in a submicroscopic particle to resemble that of an infectious agent; (2) optimizing the targeting of antigen and adjuvant after both mucosal and parenteral modes of administration to lymphatic organs and cells by enclosure of adjuvant and antigen in a stable, uniform particle; and (3) optimizing the immunomodulatory capacity both for neonate and adult immune system by presenting the adjuvant and antigen components in the same particle, reducing the amount of antigen and adjuvant required for efficiently enhancing the immune response.

III. Formation of ISCOM

An ISCOM is composed of a matrix containing *Quillaja* saponins, cholesterol, and phospholipid at molar ratio of about 1:1:1. The assembly of ISCOM is mediated by hydrophobic interactions between these components and the antigen. The ISCOM antigen may be an envelope protein of a native virus, a cellular membrane protein, or peptides containing hydrophobic domains or any antigen with a hydrophobic domain. In the case of virus, the envelope is disintegrated by a detergent, preferably nonionic, to release a protein–detergent complex. On removal of the detergent (e.g., by dialysis or density gradient centrifugation) in the presence of *Quillaja* saponins, cholesterol, and additional lipid, ISCOM assembly takes place. Also, primarily nonamphipathic proteins (e.g., gp120 of HIV) may be integrated into ISCOM by refolding the protein to expose certain hydrophobic sequences to accomplish hydrophobic interactions. By efficient coupling methods, peptides may be conjugated to the external protein moiety of an ISCOM, which will then act as a carrier providing T-cell background to the peptide antigen.

By electron microscopy of ISCOMs, 40-nm cagelike structures exhibiting icosahedral symmetry with a number of morphologic subunits are demonstrated. Multiple copies of antigens are exposed on the ISCOM particle. The construct is physically remarkably stable; it can be frozen and freeze-dried and dissolved in various buffers. Lyophilization may thus be useful for storage of ISCOM vaccines.

IV. Antigen Presentation and Targeting by ISCOMs

There is a vast literature on Ag processing and presentation by antigen-presenting cells (APCs) to T cells, but there is a limited number of reports of how adjuvants influence these activities in APCs. One reason for this is that most adjuvants are toxic *in vitro* to cells, as is the case with oil adjuvants or $Al(OH)_3$. In contrast, ISCOMs are well suited for cell culture work and for immuno-electron microscopy (EM) studies where their stability makes it possible to follow the ISCOM for about 30 minutes intracellularly after their internalization in cells (Watson *et al.*, 1992). In contrast, micelles being diluted far below their critical micellar concentration disintegrate and cannot be visualized in intracellular vesicles. Using immuno-EM on biotinylated influenza virus Ag in ISCOMs, Villacres-Eriksson (1993) was able to trace these antigens in equal amounts to both cytosol and vesicles. These results were further supported by quantitative studies that determined the amount of biotinylated Ag in subcellular fractions obtained by differential centrifugation in a quantitative ELISA using a polyclonal for capture antibody, and streptavidin peroxidase was used for the detection. While macrophages take up 50%, DC 16% and B cells 13% of ISCOM-borne Ag, the corresponding values for influenza virus micelles were 25- to 50-fold lower. Activities resulting in efficient Ag uptake by APCs is probably an important function of an adjuvant. The capacity to deliver Ag to the cytosol is likely to pave the way for MHC class I restricted Ag presentation resulting in cytotoxic T-lymphocyte (CTL) response (Takahashi *et al.*, 1990). This is a feature largely confined to acid-sensitive liposomes (Harding *et al.*, 1991), ISCOMs, and synthetic lipopeptide vaccines among the nonviable Ag delivery systems (Deres *et al.*, 1989). With ISCOMs, long-lived cytotoxic memory T cells are induced in mice after one s.c. immunization (Takahashi *et al.*, 1990). Contributing to the efficiency of ISCOMs to induce CTL is their strong capacity to stimulate lymphocytes producing interferon γ (IFN-γ) and interleukin 2 (IL-2), that is, a Th1 type of response (Morein *et al.*, 1995). There are a number of reports of other adjuvants that claimed to enhance CTL, but they generally require high doses of antigen and several immunizations.

V. Adjuvant Influences on the Transport of Antigen

Following intraperitoneal administration of radiolabeled influenza ISCOMs a transient increase of neutrophils in the peritoneal fluid is recorded and subsequent higher amount of influenza virus antigen was found in the spleen up to 8 days (Watson et al., 1989).

A lipophilic fluorescent carbocyanine dye (Claassen et al., 1995) was incorporated into the lipids of liposomes, which were subsequently used to study the kinetics of the liposome uptake by macrophages. Liposomes like bacteria were found to be taken up by marginal zone macrophages (MZMs). Rabies virus (RV) glycoprotein ISCOMs labeled by this technique were followed and localized after i.v., i.p., and s.c. administrations. Two hours after i.v. and i.p. injections RV was found in MZM as expected. Conversely, RV ISCOMs were found mainly in the marginal metallophilic marcophages (MMM) located at the border of the marginal zone and, to a lesser extent, in MZM, follicular dendritic cells, and B cells. MMMs are characterized by slender processes protruding into periarteriolar lymphoid sheat (PALS) and is the main site of antibody production in the spleen and hence a strategic localization for antibody stimulation (Claassen et al., 1995).

The role of MMM in Ag handling is unclear but it has been suggested that they are involved in Ag processing (Kraal et al., 1988), while MZMs function as a means of rapid removal and elimination of Ag, thus removing it from immune responses. This difference of distribution pattern between RV and RV ISCOMs may partly explain why RV ISCOMs were 20 to 30 times more immunogenic than RV. When injected s.c. 0.08 µg of RV ISCOMs would induce a RV neutralizing antibody response while 10 µg of RV was required to obtain a similar neutralizing antibody response. Full protection to intracerebral challenge was obtained with 0.4 µg of RV ISCOMs.

Antigens are transported from the site of injection to the draining lymph nodes and subsequently specific T- and B-cell responses can be detected in various lymphatic tissues, for example, lymph nodes, spleen, and bone marrow (BM). The distributions of B- and T-cell responses after parenteral immunization with influenza virus envelope Ag incorporated into ISCOMs or as micelles adjuvanted with CFA were compared. The T-cell response, measured by proliferation and production of IL-2, IFN-γ, and IL-4, was first confined to the draining lymph node both for ISCOM-borne Ag and Ag adjuvanted with CFA. It was transient for ISCOM but comparatively long lasting for CFA adjuvanted Ag. In the spleen, however, the T-cell response was prominent for ISCOM-borne Ag but low for CFA adjuvanted Ag (Sjölander et al., 1997).

The B-cell response after immunization with ISCOMs (Sjölander et al., 1996), measured as Ab producing cells, was first detected in draining lymph nodes, and was low in the spleen with a late but prominent response in BM. The implication of a strong BM response seems to be that Ab production is retained there for a long period of time. Moreover an increasing proportion of the Ab-producing cells of all isotypes is located there with increasing age (Benner et al., 1981a,b). Possibly the BM as an organ producing antibodies is particularly important for elderly individuals and animals. The mechanisms behind distribution of B-memory response in BM and effects of various adjuvants on that distribution need to be further explored. In contrast, the use of CFA or high doses of LPS interfered with or even abolished the ongoing Ig synthesis in BM. Thus, this process can be positively or negatively influenced by adjuvants. For example, CFA can cause a delay in transfer of the antibody-producing cells from draining lymph nodes to BM due to the granulopoiesis induced in the spleen (Benner et al., 1981a).

VI. Adjuvants and Delivery Systems for Induction of Mucosal Immunity

In recent years there has been an increasing interest in adjuvants and vaccine delivery systems for induction of mucosal immune responses, especially by the oral and to a lesser extent by respiratory tract routes. There are three problems to overcome for oral vaccines: the acid pH in the stomach, the mucosal barrier, and the induction of tolerance, which is clearly observed where oral exposure precedes parenteral immunization. With regard to the respiratory tract the mucosal barrier and tolerance induction are obstacles to be overcome.

Like cholera toxin (CT) and the termolabile enterotoxin of Escherichia coli (LT) ISCOMs have also been shown to prevent induction of immunologic tolerance and to exert adjuvant activity in the digestive tract. Oral administration of ISCOMs will induce secretory IgA, CTL, and systemic immune responses (Mowat and Maloy, 1994). Using fluorochrome-labeled ISCOMs containing the G protein of rabies virus, I. J. T. M. Claassen et al. (personal communication) showed that rabies virus ISCOMs target Peyer's patches (PP) more effectively than rabies virus particles. The lymphatic system in the gut is also reached through the enterocytes, which may act as APCs (Santos et al., 1990).

Enterocytes might be used as APCs by ISCOMs but not by micelles as studies shown by Lazarova et al. (1996) indicate.

Using ovalbumin as model antigen in ISCOMs, Mowat and Maloy (1994) have demonstrated that OVA ISCOMs by the oral route induce MHC class I restricted CTL and specific IgA in the intestinal mucosa. After mucosal intranasal administration of the envelope proteins of influenza virus or from respiratory syncytial virus (RSV) in ISCOMs, a high serum antibody response is obtained that is of the same order as that following parenteral immunization. A secretory IgA response measured by ELISA is also efficiently evoked both at the local respiratory tract mucosa and distantly at the genital tract mucosa (Hu et al., 1998). In general, antigens derived from envelope viruses or cell membranes are readily incorporated into ISCOMs and they retain their biological activities and conformation provided they are isolated under mild conditions. RSV incorporated into ISCOMs also induced virus neutralizing antibodies both locally in the respiratory tract and in serum. Thus, the biological activities of influenza virus or RSV glycoproteins are conserved, facilitating the penetration of mucus, which explains their efficiency in evoking mucosal immune responses.

VII. ISCOMs Induce a Cytokine Th1 Type Response But Also Th2

ISCOMs efficiently stimulate APCs to produce IL-1 and IL-12. Also other cytokines such as IL-6, GM-CSF, and TNF-α are induced. The ISCOM induces strong T-cell responses of T-helper 1 (Th1) type characterized by delayed type of hypersensitivity (DTH) enhanced production of IFN-γ and IL-2, while IL-4 and IL-5 are not generally enhanced but often IL-10 is down-regulated. In general, ISCOMs are excellent inducers of IL-2 and IFN-γ (Morein et al., 1996).

VIII. Antigens Loaded in ISCOMs Induce Immune Response in Neonates

In several species it is shown that ISCOMs induce potent immune responses in newborns with and without maternal antibodies. Newborn seals immunized with canine distemper virus were protected against natural infection with seal morbillivirus. Puppies with maternal antibodies immunized subcutaneously at 3 and 7 weeks of age with killed canine parvovirus adjuvanted with ISCOM matrix responded with very high antibody titers when conventional, killed parvovirus vaccine was insufficiently immunogenic. Foals with maternal antibodies immunized s.c. at the age of 10 days and again 2 weeks later with Herpes Eq-2 ISCOMs responded with virus neutralizing antibodies (Nordengrahn et al., 1996) and were protected from disease (abscesses in the lungs) by secondary infection with *Rodococcus equi* bacteria.

In 2-day-old mice Sendai virus ISCOMs induced memory cells which were efficiently boosted, resulting in high antibody responses. Further mixed Th1 (IL-2 and INF-γ) and Th2 (IL-4 and IL-5) types of responses were induced by ISCOMs in newborn mice similar to the T-cell responses induced in adults. In contrast, none of these responses was induced by Sendai virus micelles when given s.c. to the 2-day-old mice. The present knowledge is that newborn mice, if they respond, do not induce a Th1 type of response but a Th2 type of response and a subsequent apoptosis of the Th1 cells (Adkins et al., 1996).

IX. Protective Immunity

Protective immunity has been induced by ISCOMs against a variety of microorganisms including viruses, bacteria, mycoplasma, and parasites. Of particular interest is that protective immunity was induced by ISCOM-borne antigens to a number of retroviruses, that is, feline leukemia virus infection in cats, to lethal infection with SIV in macaques. Of particular interest is that HIV-1 ISCOM vaccine induced protection to infection in a challenge system in rhesus macaques using a semi-heterologous challenge virus that was a chimer of HIV-1 and SIV (SHIV). This protection was correlated to virus neutralizing antibodies, Th1 type of response together with an IL-4 production and the production of the β-chemokines (MIP-α1, MIP-α2, and RANTES). These β-chemokines are produced by $CD8^+$ cells and found in HIV-1 infected long-time survivors, and they are interacting with the second HIV-1 receptor involved in the fusion and entry process of the host cell. ISCOMs also induced protection to challenge with HIV-2 in primates, and ISCOMs loaded with gp340 of a tumor-inducing herpesvirus, that is, Epstein–Barr virus (EBV), protected cotton top tamarin monkeys against challenge with EBV causing lethal tumor development. Table I lists examples of ISCOM-borne antigens having induced protective immunity.

TABLE I

PROTECTIVE IMMUNITY INDUCED BY ISCOMS AGAINST VARIOUS PATHOGENS

Antigen	Animal	Disease
Hemagglutinin measles virus	Mice	Encephalitis
Fusion protein, measles virus	Mice	Encephalitis
Hemagglutinin and fusion protein, phoid distemper virus	Seal	Lethal infection
Hemagglutinin and fusion protein, canine distemper virus	Dog	Pneumonia
gp120, Simian immunodeficiency virus	Monkey	Lethal infection
Envelope protein, bovine diarrhea virus, envelope protein	Sheep	Abortion
gp120, P24, and peptides from HIV-1, gp120 V2 and V3 regions	Monkey	Viremia (challenged with SHIV)
gp70, Feline leukemia virus	Cat	Viremia
gp360, Epstein–Barr virus	Tamarin monkey	Lethal tumor
Surface antigens, *Toxoplasma gondii*	Mice	Lethal infection
Immunoaffinity purified protein, *Trypanosoma cruzi*	Mice	Lethal infection

REFERENCES

Adkins, B. et al. (1996). Naive murine Neonatal T cells undergo apoptosis in response to primary stimulation. J. Immunol. **157,** 1343.

Benner, R., van Oudenaren, A., and Koch, G. (1981a). Chapter 14. Induction of antibody formation in the mouse bone marrow. Immunol. Methods **2,** 247–261.

Benner, R., Hijmans, W., and Haaijman, J. J. (1981b). The bone marrow: The major source of serum immunoglobulins, but still a neglected site of antibody formation. Clin. Exp. Immunol. **46,** 1–8.

Claassen, I. J. T. M., Osterhaus, A. D. M. E., and Claassen, E. (1995). Antigen detection in vivo after immunization with different presentation forms of rabies virus antigen: Involvement of marginal metallophilic macrophages in the uptake of immune-stimulating complexes. Eur. J. Immunol. **25,** 1446–1452.

Deres, K., Schild, H., Wiesmuller, K. H., Jung, G., and Rammensee, H. G. (1989). In vivo priming of virus-specific cytotoxic T lymphocytes with synthetic lipopeptide vaccine. Nature **342,** 561–564.

Harding, C. V., Collins, D. S., Slot, J. W., Geuze, H. J., and Unanue, E. R. (1991). Liposome-encapsulated antigens are processed in lysosomes, recycled, and presented to T cells. Cell **64,** 393–401.

Hu et al. (1998). Submitted for publication.

Kraal, G., Janse, M., and Claassen, E. (1988). Marginal metallophilic macrophages in the mouse spleen: Effects of neonatal injections of MOMA-1 antibody on the humoral immune response. Immunol. Lett. **17,** 139–144.

Lazarova, L., Artursson, P., Lövgren Bengtsson, K., and Sjölander, A. (1996). Influence of a particulate Quillaja saponin-containing adjuvant, ISCOM-matrix, on the handling and transport of influenza virus antigens in human intestinal epithelial cells (Caco-2). *Am. J. Physiol.* **270,** G554–G564.
Morein, B., Lövgren, K., Rönnberg, B., Sjölander A., and Villacrés-Eriksson, M. (1995). Immunostimulating complexes clinical potential in vaccine development. *Clin. Immunother.* **3,** 461–475.
Morein, B., Villacrés-Eriksson, M., Sjölander, A., and Lövgren Bengtsson, K. (1996). Novel adjuvants and vaccine delivery systems. *Vet. Immunol. Immunopathol.* **54,** 373–384.
Mowat, A. McI., and Maloy, K. (1994). Immune stimulating complexes as vectors for oral immunization. In "Novel Delivery Systems for Oral Vaccines" (D. T. O'Hagan, ed.), Chapter 3. CRC Press, Boca Raton, FL.
Nordengrahn, A., Rusvai, M., Merza, M., Ekström, J., Morein, B., and Belák, S. (1996). Equine herpesvirus type 2 (EHV-2) as a predisposing factor for *Rhodococcus equi* pneumonia in foals: Prevention of the bifactorial disease with EHV-2 immunostimulating complexes. *Vet. Microbiol.* **51,** 55–68.
Santos, L. M., Lider, O., Audette, J., Khoury, S. J., and Weiner, H. L. (1990). Characterization of immunomodulatory properties and accessory cell function of small intestinal epithelial cells. *Cell. Immunol.* **127,** 26–34.
Sjölander, A., Lövgren Bengtsson, K., Johansson, M., and Morein, B. (1996). Kinetics, localization and isotype profile of antibody responses to immune stimulating complexes (ISCOMs) containing human influenza virus envelope glycoproteins. *Scand. J. Immunol.* **43,** 164–172.
Sjölander, A., Lövgren Bengtsson, K., and Morein, B. (1997). Kinetics, localization and cytokine profile of T cell responses to immune stimulating complexes (ISCOMs) containing human influenza virus envelope glycoproteins. *Vaccine* **15,** 1030–1038.
Takahashi, H., Takeshita, T., Morein, B. Putney, S., Germain, R. N., and Berzofsky, J. (1990). Induction of CD8+ cytotoxic T cells by immunization with purified HIV-1 envelope proteins in ISCOMs. *Nature* **344,** 873–875.
Villacres-Eriksson, M. (1993). Induction of immune responses by ISCOMs. Ph.D. Thesis, Swedish University of Agricultural Sciences, Uppsala.
Watson, D. L., Lövgren, K., Watson, N. A., Fossum, C., Morein, B., and Höglund, S. (1989). The inflammatory response and antigen localization following immunization with influenza virus ISCOMs. *Inflammation* **13,** 641–649.
Watson, D. L., Watson, N. A., Fossum, C., Lövgren, K., and Morein, B. (1992). Interactions between immune-stimulating complexes (ISCOMs) and peritoneal mononuclear leucocytes. *Microbiol. Immunol.* **36,** 199–203.

An Epidemiologic Approach to Evaluating the Importance of Immunoprophylaxis

PAUL S. MORLEY

Department of Veterinary Preventive Medicine, The Ohio State University, 1900 Coffey Road, Columbus, Ohio 43210-1092

Epidemiology is the branch of medicine that focuses on interrelationships between living organisms and their environment as they relate to health. In contrast to most medical disciplines, which focus on diagnosing and treating disease in *individuals,* there is an inherent ecological nature to epidemiology because understanding, controlling and preventing disease in the *population* is emphasized. It is interesting to note that significant advances in modern epidemiology have occurred during the same period that advances in biology and medicine have led to the common use of intensive vaccination strategies by veterinarians and animal owners. Despite the concurrence of these developments, epidemiologic methods have played a relatively minor role in the development and critical appraisal of immunoprophylaxis in veterinary medicine when compared to their use in evaluating the efficacy of efforts to control diseases in humans. A critical appraisal of disease prevention strategies is paramount for these efforts to be efficient. There is no doubt that "benchtop" and clinical scientific investigation are integral to the development and evaluation of immunoprophylactic techniques. However, an epidemiologic approach can provide additional unique information that is ideally suited for objective evaluation of disease prevention strategies. Specifically, an epidemiologic approach is well suited for investigating the etiology and the relative importance of the targeted disease, identifying risk factors which influence the occurrence of disease, objectively evaluating the importance of immunoprophylaxis in preventing naturally occurring disease, and identifying prevention strategies which may be used in combination with or in place of intensive vaccination.

Three primary forces influence the occurrence of infection and infectious disease: the likelihood of exposure to an organism, the pathogenicity of the organism, and the ability of the host to counteract the pathogenic mechanisms of that agent. The health of an animal is determined by the balance of these three forces. As a result, disease prevention strategies are targeted at influencing the balance between the likelihood of exposure to infectious agents and improving specific or nonspecific immunity through immunoprophylaxis.

Because vaccines are generally intended to improve immunity to specific agents, it is critical to consider the true risk of disease associated with exposure to each specific agent targeted by vaccination. At times the association between infection and disease is apparent, but animals are commonly infected with agents whose pathogenicity is unknown or controversial. Even if infection is clearly linked to the occurrence of disease, this disease may not significantly affect production of individuals or the production unit. Disease prevention efforts will be most efficient if emphasis is placed on controlling the most important pathogens; properly conducted, epidemiologic investigations can be used to identify these agents. Epidemiologic investigations can also yield important information about the biology of the disease, which is useful in determining which stage of the life cycle of the host and the agent can be most effectively targeted in prevention strategies. The greatest benefit is likely to be realized from disease prevention strategies when prophylactic measures are targeted at animals with the greatest risk of disease.

It is equally critical to consider the efficacy of the vaccines or prophylactic therapies to be used. Vaccines are never foolproof preventive measures, and not all vaccines that are licensed and marketed commercially have equivalent efficacies. Widespread marketing, acceptance, and application of vaccines does not ensure that products are efficacious in every situation in which they will be applied. Unfortunately, it is very difficult to obtain objective information that clearly documents the efficacy of most vaccines used in veterinary medicine. Even if information is available regarding a product's ability to prevent experimentally induced disease, it is even less common to find information from properly conducted field trials which evaluate the efficacy of vaccines as they are practically applied to prevent naturally occurring disease. The success of a product in the laboratory does not ensure its success in the field. A vaccine may prove to be very immunogenic and successfully protect animals against experimental challenge, but may still fail to protect against infection and disease under conditions of natural exposure. Protection from naturally occurring disease is the most relevant outcome for measuring the value and efficacy of a vaccine. Scientists, vaccine manufacturers, veterinary practitioners, and consumers should all be greatly concerned when information is not readily available regarding objective evaluation of commercially marketed vaccines.

The occurrence of one disease in a population is usually an indication that similar agents or agents with similar routes of transmission could be a problem in this population. As such, the occurrence of infectious disease in animals is often a symptom of more than one impediment to production. Even if disease caused by one specific agent could be eliminated, production methods that enhance exposure to one pathogen will commonly enhance exposure to other pathogens. Intensive vaccination programs are not panaceas which remedy all problems created by less than optimal management conditions. Immunoprophylaxis will often be most effective when used judiciously in combination with other prevention strategies. In fact, effective and cost-efficient disease prevention programs may not even include vaccination as a control measure.

Modern vaccines have benefitted from the incredible advances realized in modern immunology and biology. There is great promise for future immunoprophylactic strategies to help overcome important disease problems which prevent animals from realizing their production potential. However, it is critical that objective information be readily available about the epidemiology of these diseases as well as the efficacy of prevention strategies in order for veterinarians and producers to make informed decisions about the appropriate role of immunoprophylaxis in disease prevention programs.

VI
SWINE VACCINES

Present Uses of and Experiences with Swine Vaccines

LENNART BÄCKSTRÖM

School of Veterinary Medicine, University of Wisconsin, Madison, Wisconsin 53706

I. Introduction
II. Material
III. Results
IV. Discussion
   Acknowledgments
   References

I. Introduction

From the first attempts by Louis Pasteur (1822–1896) to make swine erysipelas bacterins, to the first applications of hog cholera vaccines in the United States in the early 1900s, vaccines have had a place in preventive swine veterinary medicine. In the United States, hog cholera and erysipelas vaccines were commonly used as early as the 1920s and 1930s. However, not until the 1950s and 1960s would commercial vaccines against several other swine pathogens become available (A. Hogg, personal communication, 1997). Today, licensed vaccines are available for at least 6 viruses and 16 bacterial pathogens (Table I). In addition, for several pathogens vaccines are available for different serotypes of certain pathogens: six for leptospirosis, four for *Actinobacillus pleuropneumoniae* and *Escherichia coli,* and two for *Salmonella, Streptococci, Pasteurella multocida,* and *Haemophilus parasuis.* Despite general recommendations from immunologists in favor of monovalent vaccines or limited component vaccines, combina-

TABLE I

Swine Vaccines Licensed in the United States

Influenza	*Actinobacillus pleuropneumoniae* 1, 3, 5, 7
Parvovirus	*Bordetella bronchiseptica*
PRRS	*Clostridium perfringens* C
Pseudorabies	*Erysipelothrix rhusiopathiae*
Rotavirus	*Escherichia coli* K88, K99, 987P, F1
TGE	*Haemophilus parasuis* 5, 6
	Leptospira 5-way/6-way
	Mycoplasma hyopneumoniae
	Pasteurella multocida A, D
	Salmonella cholerasuis / typhimurium
	Streptococcus suis
	Serpulina hyodysenteriae
	Streptococcus equisimilis
	Pseudomonas aeuroginosa

tion vaccines with many of the above components are commonplace, most often with multiple pathogens. Vaccines containing as many as 12 components have been registered in the United States (Table II).

Unlike in many other countries, swine vaccines in the United States can be purchased over the counter or by catalog. According to the National Animal Health Monitoring System (NAHMS) only 55% of the vaccines were provided by a veterinarian (NAHMS, 1995a). However, 83% of herds using a herd veterinarian also used vaccines, compared to 48% of herds without a consulting veterinarian (NAHMS, 1995b). Thus, veterinarians appear to have a great impact on the use of vaccines.

The regulatory controls of biosafety and efficacy of vaccines, exercised by the U.S. Department of Agriculture (USDA), have increased significantly during the 1980s and 1990s. However, a conflict of interest seems to exist when efficacy assessments, challenge models, sample size, number of replications, trial allocations, and result analysis are largely sponsored and done by the biologics industry itself (Hancock, 1992; Moon and Bunn, 1993; Straw, 1994). A common critique has also been that the pharmaceutical industry generally uses experimental challenge for efficacy tests instead of natural exposure to the pathogen (Straw, 1994). Therefore, field surveys of vaccination experiences have been conducted among swine farmers and veterinarians. According to NAHMS (1992), *Leptospira interrogans* vaccines were used in 70% of the surveyed herds, parvovirus in 65%, *Erysipelothrix rhusiopathiae* in 61%, *E. coli* in 47%, *Bordetella bronchiseptica / Pasteurella multocida* (atrophic rhinitis) in 38%, TGE in 24%, and *Clostridium perfringens* C in 22% of the herds. Ratings of vaccine

TABLE II

EXAMPLES OF COMBINATION VACCINES

Bordetella bronchiseptica
Pasteurella multocida A, D

Transmissible gastroenteritis
Rotavirus A_1, A_2
Escherichia coli K88, K99, 987P, F41
Clostridium perfringens C
Bordetella bronchiseptica
Pasteurella multocida A, D

Bordetella bronchiseptica
Pasteurella multocida A, D
Actinobacillus pleuropneumoniae 1, 3, 4, 5, 7
Haemophilus parasuis
Streptococcus equisimilis
Streptococcus suis I, II
Staphylococcus aureus
Pseudomonas aeruginosa
Escherichia coli K88, K99, 987P, type 1, F41
Clostridium perfringens C
Salmonella cholerasuis
Salmonella typhimurium

efficacy varied somewhat between veterinary surveys conducted by the National Veterinary Services Laboratory (NVSL) in 1992 and by Straw in 1994. Overall, ratings were good to excellent for pseudorabies, parvovirus, influenza, erysipelas, and colibacillosis; fair to good for clostridiosis, atrophic rhinitis, and Glässer's disease (*H. parasuis*); and mixed for *Mycoplasma hyopneumoniae* and *Salmonella typhimurium* or *cholerasuis*. Efficacies were rated poor for vaccines against the porcine reproductive and respiratory syndrome (PRRS), TGE, rotavirus, *Streptococcus suis,* and *Actinobacillus pleuropneumoniae*.

This paper presents current information about the use and perceived efficacy of swine vaccines obtained in a 1997 survey of swine veterinarians in the United States and 11 other countries.

II. Material

A questionnaire was designed to include all swine pathogens for which commercial vaccines were available in the USA in the spring of

1997 (Table I). The questions included frequency of use of each vaccine, if autogenous vaccines were used, perceived efficacy, and whether the vaccine was administered monovalent or in combination with other antigens (Table III). The questionnaire was arbitrarily presented to well-known and respected swine veterinary specialists, both in private practice, in the swine industry, and in academia and extension. Most of the responses for the United States were received in conjunction with the American Association of Swine Practitioners (AASP) Convention in Quebec in March 1997. Others were sent and responded to by mail. A total of 118 questionnaires were delivered, and 105 were returned completed (89% response rate).

In addition, the same questionnaire was mailed to one well-known swine veterinary specialist in each of 11 arbitrarily chosen other countries: Canada, Mexico, Taiwan, Australia, UK, Netherlands, Belgium, France, Germany, Denmark, and Sweden (see Acknowledgements section). These specialists were asked to answer the questions in the questionnaire on behalf of their country.

III. Results

The results are compiled and summarized in Tables IV through VII. Large differences are seen between countries regarding use of swine vaccines. While licensed vaccines were available for 21 pathogens in the United States, the number declined to 17 in Canada to 5 in Australia (Table IV). Noticeably, the number of vaccines was no more than 12 in the heavily swine populated Netherlands, and 10 in Denmark.

Vaccines for colibacillosis, erysipelas, parvovirus, and *Mycoplasma* pneumonia were used in all 12 countries. *Pasteurella multocida* vac-

TABLE III

SWINE VACCINE SURVEY

Use	Efficacy	Type
All the time	Excellent	Monovalent
Frequently	Good	Combination
Sometimes	Mixed/fair	
Never	Poor	
	No opinion	

TABLE IV

SWINE VACCINES IN THE USA AND OTHER COUNTRIES

Number of swine pathogens for which vaccines are licensed in each country

United States (USA)	22	France	10
Canada	17	Belgium	10
Taiwan	16	Denmark	10
Mexico	13	United Kingdom	7
Germany	13	Sweden	7
Netherlands	12	Australia	5

Survey of vaccine use among 12 countries

	Number of countries		Number of countries
Parvovirus	12	S. cholerasuis	5
E. coli	12	S. suis	5
E. rhusiopathiae	12	Rotavirus	4
M. hyopneumoniae	11	Leptospirosis	3
P. multocida	11	H. parasuis	3
B. bronchiseptica	10	S. typhimurium	3
A. pleuropneumoniae	10	Hog cholera	2
C. perfringens C	9	Japanese encephalitis	1
Pseudorabies	7	S. hyodysenteriae	1
Influenza	6	P. aeruginosa	1
PRRS	6	S. equisimilis	1
TGE	6		

cines were used in all but 1 country, and *Bordetella* vaccines in all but 2 countries (Table V). At the bottom of the list for all countries were vaccines for TGE, rotavirus, *P. aeruginosa, S. equisimilis* and *S. suis, A. pleuropneumoniae,* and *T. hyodysenteriae.* (Dysentery vaccine was just recently licensed in the United States, so no experiences with its use are yet available.) Japanese encephalitis vaccine was only used in Taiwan, and hog cholera vaccine in Taiwan and Mexico.

Vaccines for parvovirus, *E. coli*, and erysipelas were at the top of the user lists in all countries. The 5-way *Leptospira* vaccine (*L. icterohaemorrhagiae, pomona, grippothyphosa, canicola, hardjo*) was also at the top of the list in the United States but only used in two other countries (Tables VI and VII). The 6-way *Leptospira* vaccine with *L. bratislava,* which is the most prevalent serovar in the United States, (Bolin, 1993), was used much less in this country (Table VI).

The rates of perceived vaccine efficacy were similar to the rates of

TABLE V
Use of Swine Vaccines in the USA and Other Countries

	US	C	T	D	M	NL	F	B	DK	UK	S	AUS
Reproductive diseases												
Parvovirus	Y	Y	Y	Y	Y	Y	Y	Y	Y	Y	Y	Y
Leptospirosis	Y	Y	—	—	—	—	—	—	—	—	—	Y
Japanese encephalitis	—	—	Y	—	—	—	—	—	—	—	—	—
Respiratory diseases												
Influenza	Y	Y	—	Y	—	Y	Y	Y	Y	—	—	—
M. hyopneumoniae	Y	Y	Y	Y	Y	Y	Y	Y	Y	Y	(Y)	—
A. pleuropneumoniae	Y	Y	Y	Y	Y	Y	Y	Y	Y	—	—	Y
B. bronchiseptica	Y	Y	Y	Y	Y	Y	Y	Y	—	Y	Y	—
P. multocida	Y	Y	Y	Y	Y	Y	Y	Y	Y	Y	Y	—
P. aeruginosa	Y	—	—	—	—	—	—	—	—	—	—	—
Enteric diseases												
TGE	Y	Y	Y	Y	Y	—	—	—	—	—	—	—
Rotavirus	Y	Y	Y	—	Y	—	—	—	—	—	—	—
E. coli	Y	Y	Y	Y	Y	Y	Y	Y	Y	Y	Y	Y
C. perfringens C	Y	Y	Y	Y	—	Y	Y	—	Y	Y	Y	—
S. typhimurium	Y	Y	Y	—	Y	—	—	—	—	—	—	—
S. hyodysenteriae	Y	—	—	—	—	—	—	—	—	—	—	—
Multisystemic diseases												
Pseudorabies	Y	—	Y	Y	Y	Y	Y	Y	—	—	—	—
PRRS	Y	Y	—	Y	—	Y	—	(Y)	(Y)	—	—	—
Hog cholera	—	—	Y	—	Y	—	—	—	—	—	—	—
H. parasuis	—	Y	Y	—	Y	—	—	—	Y	—	—	—
E. rhusiopathiae	Y	Y	Y	Y	Y	Y	Y	Y	Y	Y	Y	Y
S. cholerasuis	Y	Y	Y	Y	Y	—	—	—	Y	—	—	—
S. suis	Y	Y	Y	—	—	Y	—	—	Y	—	—	—
S. equisimilis	Y	—	—	—	—	—	—	—	—	—	—	—
TOTAL PATHOGENS	22	17	16	13	13	12	10	10	10	7	7	5

Key: Y, yes use; —, no use.
Country codes: US, United States; C, Canada; T, Taiwan; D, Germany; M, Mexico; NL, Netherlands; F, France; B, Belgium; DK, Denmark; UK, United Kingdom; S, Sweden; AUS, Australia

TABLE VI

Frequency of Use of Swine Vaccines

	In USA			In other countries		
Pathogen	High (%)	Low (%)	Never (%)	High (%)	Low (%)	Never (%)
Parvovirus	95	5	0	91	9	0
E. coli	91	9	0	100	0	0
E. rhusiopathiae	91	9	0	91	9	0
Lepto 5-way	90	10	0	18	0	82
M. hyopneumoniae	62	38	0	27	54	9
C. perfringens C	52	38	10	9	54	27
P. multocida	33	62	5	27	64	9
H. parasuis	33	62	5	18	0	82
B. bronchiseptica	29	67	4	18	64	18
PRRS	29	47	24	27	18	55
Pseudorabies	29	19	52	46	9	45
S. suis	10	57	33	0	36	64
Lepto 6-way (bratislava)	9	48	43	9	9	82
A. pleuropneumoniae	5	52	43	18	64	9
Influenza	5	47	48	9	27	64
S. cholerasuis	0	62	38	9	18	72
TGE	0	47	53	0	36	64
Rotavirus	0	47	53	0	27	73
S. typhimurium	0	19	81	0	18	82

use (Table VII). The efficacy rates also followed the same trend, with a high percentage of respondents reporting "no opinion" (Table VII).

Almost all veterinarians used autogenous vaccines occasionally. Also, a large majority of respondents expressed concerns about combination vaccines. Most common combinations were reported for parvovirus–erysipelas–leptospirosis, *B. bronchiseptica–P. multocida,* and *E. coli–C. perfringens*–TGE–rotavirus.

IV. Discussion

Statistical analysis was not performed since the study design of the survey differed between the United States (multiple respondents) and the other countries (one respondent for each country). Also, differentiation between killed and live vaccines, routes of administration, etc., were not covered in the survey. Nevertheless, several interesting conclusions can be made:

TABLE VII

PERCEIVED EFFICACY OF SWINE VACCINES

Pathogen	In USA			In other countries		
	High (%)	Low (%)	No opinion (%)	High (%)	Low (%)	No opinion (%)
Parvovirus	81	0	19	82	0	18
E. coli	90	0	10	91	0	9
E. rhusiopathiae	76	0	24	82	9	9
Lepto 5-way	71	0	29	9	9	82
M. hypopneumoniae	48	38	52	45	18	37
C. perfringens C	52	14	44	36	18	46
P. multocida	33	62	5	64	18	18
H. parasuis	57	10	43	18	9	73
B. bronchiseptica	38	18	44	55	18	27
PRRS	29	14	57	0	18	82
Pseudorabies	52	0	48	36	9	55
S. suis	19	19	62	0	27	73
Lepto 6-way (brataslava)	33	10	57	18	0	82
A. pleuroneumoniae	14	24	62	27	9	64
Influenza	33	10	57	9	18	73
S. cholerasuis	43	0	57	0	18	82
TGE	9	57	44	9	27	64
Rotavirus	19	48	33	9	18	73
S. typhimurium	0	0	100	0	18	82

Colibacillosis, erysipelas, and parvovirus vaccines received the highest scores for both frequency of use and perceived efficacy in all countries surveyed. For all the other pathogens, there were great variations between the different countries for use and efficacy of vaccines.

The greatest difference between the United States and other countries was found for use and efficacy of *Leptospira* vaccines (Tables V–VII). While the Lepto 5-way vaccine was used always or almost all the time by 90% and was perceived as highly efficacious by 71% of the U.S. respondents, *Leptospira* vaccines were only used in two other countries, and the perceived efficacy was low. Several commentaries indicated that such vaccines are not needed in modern confinement operations. The fact that the 6-way Lepto vaccine with *L. bratislava* (the most common serovar in the United States) was used very little according to this survey adds further questions regarding the validity of *Leptospira* vaccinations.

Different uses and opinions were also found for several other vaccines. For instance, PRRS vaccines received very low scores for efficacy in Europe compared to the United States, while *Bordetella / Pasteurella* AR vaccines received better scores than in the United States. However, at the bottom of the list for all countries were vaccines for TGE, rotavirus, *Streptococci*, and *A. pleuropneumoniae*. *Salmonella* and influenza vaccines also received quite low scores, but since they had been introduced more recently, many respondents reported "no opinion" about their use and efficacy.

Most striking was the great difference between countries regarding the numbers of licensed vaccines available against various swine pathogens (Tables IV–VII). It appears that the number of vaccines for such pathogens is much higher in the United States compared to most of the other countries. This survey was limited to veterinarians. The difference would probably be even greater if vaccine purchases and use by farmers with no veterinary involvement were added. This is commonplace in the United States (NAHMS, 1995a,b). In many of the other countries, vaccines are only available with a veterinary prescription. The difference between countries might indicate diferent prevalence and significance of the various diseases, but they might also indicate country-based differences between biologics companies regarding marketing strategies of vaccines, state regulations, and traditions and perceptions about use of vaccines.

Acknowledgments

I am grateful for the information from 105 swine veterinary practitioners in the United States about the use and efficacy of swine vaccines, as well as similar information from other countries received from Drs. N. Agger, Kruuse A/S, Marslev, and A. Holm, Odder, Denmark; G. Bergström, Swedish Animal Health Service, Skara, Sweden; W-B. Chung, Pingtung Polytech. Instit., Neipu, Pingtung, Taiwan; R. Cutler, Pig Research and Development Corp., Kingston, Australia; C. Dewey, University of Guelph, Canada; S. Done, Central Veterinary Laboratory, Surrey, UK; P. De Roose, Rhone-Merieux, Brussels, Belgium; C. Schirwel, Rhone-Merieux, Lyon, France; H-J. Selbitz, Impfstoffwerk, Dessau-Tornau, Germany; and L. Vellenga, University of Utrecht, Netherlands.

References

Bolin, C. (1993). *Leptospira interrogans* serovar *bratislava* infection in swine. *SmithKline Beecham Anim. Health Swine Consultant,* Fall, pp. 3–5, 16.

Hancock, D. (1992). Efficacy or effectiveness of vaccines. *Popul. Med. News,* September 28, Vol. 5, No. 2.

Hogg, A. Personal communication, 1997.

Moon, H., and Bunn. (1993). Vaccines against enterotoxigenic *E. coli* infections. *Vaccine* **11** (2), 213–214.

National Animal Health Monitoring System (NAHMS) (1992) "Survey of Vaccine Use." NAHMS.

National Animal Health Monitoring System (NAHMS). (1995a). "Trends in Vaccination Practices on US Swine Operations," pp. 1–2. NAHMS.

National Animal Health Monitoring System (NAHMS) (1995b). "Swine '95. Part I: Reference of 1995 Swine Management Practices," pp. 1–22. NAHMS.

National Veterinary Services Laboratory (NVSL) (1992). "Survey of Veterinarians." NVSL.

Straw, B. (1994) Vaccination of swine. *Proc. George Young Swine Conf., Annu. Nebraska SPF Swine Conf., 1994*, pp. 58–76.

Enteric Viral Infections of Pigs and Strategies for Induction of Mucosal Immunity

LINDA J. SAIF

Food Animal Health Research Program, Ohio Agricultural Research and Development Center, The Ohio State University, Wooster, Ohio 44691

I. Introduction and Background
II. Characteristics of Enteropathogenic Viruses
   A. Comparative Pathogenesis of Virulent and Attenuated Transmissible Gastroenteritis Virus, Porcine Respiratory Coronavirus, and Rotavirus
III. Mucosal Immunity to Enteropathogenic Viruses
   A. Studies of Active Immunity to Transmissible Gastroenteritis Virus and Porcine Respiratory Coronavirus
   B. New Vaccine Approaches to Induce Immunity to Transmissible Gastroenteritis Virus
   C. Studies of Active Immunity to Group A Rotavirus
   D. New Vaccine Approaches to Induce Immunity to Rotavirus
Acknowledgments
References

I. Introduction and Background

Enteropathogenic viruses, such as transmissible gastroenteritis virus (TGEV) and rotavirus, replicate and induce lesions only in the gastrointestinal tract. The susceptible target cell is the villous enterocyte (Saif, 1990). Thus active immunity against enteropathogenic viral infections depends on stimulation of local immune responses within the intestine. To date, only limited success has been achieved in the development of oral vaccines to prevent neonatal viral diarrheas, and commercial vaccines show limited efficacy in the field (Saif and Jackwood, 1990). Although use of live attenuated poliovirus is often cited as a model for an effective oral vaccine, the mechanism of viral pathogenesis and hence protective immunity differs from that needed to prevent viral diarrheas. Poliovirus undergoes primary replication in Peyer's patches or intestinal lymphoid cells (not epithelial cells), but the target cell for disease induction is the neuron (Melnick, 1990). Thus stimulation of circulating antibodies using either live oral or inactivated poliovirus vaccines can prevent the systemic spread of poliovirus to the central nervous system and the paralytic disease.

Currently only oral vaccines containing live replicating organisms have been highly effective in inducing mucosal immune responses, especially secretory, (S)IgA antibodies. The oral administration of soluble or killed antigens generally induces immunity of short duration or even systemic tolerance (reviewed in Mowat, 1994). Whether the problems encountered with oral administration of soluble protein antigens can be overcome by the use of improved mucosal adjuvants [muranyl dipeptide, immune stimulating complexes (ISCOMs), cholera or *Escherichia coli* enterotoxins, avridine, proteosomes, cytokines, etc.] or new and novel delivery systems (liposomes, live recombinant vectors, microspheres, DNA plasmids, virus-like particles) requires further investigation, and specific examples are given in this review.

Coronaviruses and rotaviruses are well-characterized enteropathogens that account for a high percentage of the viral diarrheas in many animals (Saif and Wesley, 1992; Saif et al., 1994a). In addition, rotaviruses are the leading cause of dehydrating diarrhea in young children worldwide (Kapikian and Chanock, 1990). Thus these viruses serve as important models to study mucosal immunity to enteric viruses. In this review, the impact of the site of viral replication (intestine vs respiratory tract), vaccine dose, and type (attenuated, inactivated) on the isotype, level, and distribution of virus-specific antibody-secreting cells (ASCs) and protection against viral challenge in pigs is summarized and discussed.

II. Characteristics of Enteropathogenic Viruses

Enteropathogenic viruses belonging to at least five different families have been associated with diarrhea in pigs (Table I, Saif, 1990). Each of these viruses infects mainly the villous enterocytes of pigs and, with the possible exception of astroviruses (Saif *et al.*, 1980), induces villous atrophy and a malabsorptive diarrhea (Saif, 1990). None of these viruses causes systemic infections; hence, the localized nature of these intestinal viral infections is of prime consideration for designing effective strategies to induce mucosal immunity. A potential explanation for

TABLE I

CLASSIFICATION AND CHARACTERISTICS OF PORCINE ENTEROPATHOGENIC VIRUSES

Family/virus	Size (nm)	Nucleic acid	Discovery Year	Discovery Investigator	Intestinal replication site Villous	Intestinal replication site Crypt
Enveloped						
Coronaviridae/transmissible gastroenteritis virus (TGEV)	60–220	ssRNA	1946	Doyle and Hutchings (Solf and Wesley, 1992)	+	−
Nonenveloped						
Reoviridae/rotavirus (group A)	55–70	dsRNA	1976	Woode *et al.*	+	−
Rotavirus (Group B)	55–70	dsRNA	1980	Bridger	+	−
Rotavirus (Group C)	55–70	dsRNA	1980	Saif *et al.*	+	−
Rotavirus (Group E)	55–70	dsRNA	1986	Chasey *et al.*	?	?
Caliciviridae/calicivirus	30–40	ssRNA	1980 1980	Bridger Saif *et al.*	+	−
Astroviridae/Astrovirus	28–30	ssRNA	1980 1980	Bridger Saif *et al.*	+	−
Adenoviridae/Adenovirus	70–90	DNA	1981	Coussement *et al.*	+	±

the localized nature of many enteric viral infections was highlighted in a recent study (Rossen *et al.*, 1996). The authors found that TGEV enters and exits from polarized epithelial cells *in vitro* via the apical surface; in contrast, another coronavirus, mouse hepatitis virus (MHV), enters the same cells apically but exits basolaterally. The investigators speculated that similar differences in the mode of release of coronaviruses from infected host cells could contribute to the nature of the localized intestinal infections induced by TGEV (released apically into the gut lumen) or the systemic infections associated with MHV (released basolaterally into the blood and lymph).

A. Comparative Pathogenesis of Virulent and Attenuated Transmissible Gastroenteritis, Porcine Respiratory Coronavirus, and Rotavirus

Exposure of pigs to attenuated TGEV (TGEV-A), virulent TGEV (TGEV-V) or porcine respiratory coronavirus (PRCV) results in distinct disease patterns related to differences in virulence and tissue tropisms between the viruses (Table II) (Pensaert and Cox, 1989; Saif and

TABLE II

VERTICAL AND LONGITUDINAL SITES OF REPLICATION AND VILLOUS ATROPHY IN THE INTESTINE FOR PORCINE CORONAVIRUSES AND GROUP A ROTAVIRUSES

Virus	Diarrhea	Longitudinal		Vertical small intestine (site)		Villous atrophy small intestine		Respiratory tract		Reference
		Small intestine	Colon	Villous	Crypt	Site	Extent	Upper	Lower	
Coronavirus										
Virulent TGEV	Severe	D,J,I	−	+ Entire	−	D,J,I	Severe	±	±	Frederick *et al.* (1976)
Attenuated TGEV	Mild/none	J,I	−	+ Entire	−	J,I/none	Mild	+	±	Frederick *et al.* (1976) Furuuchi *et al.* (1979)
PRCV	None	−	−	− NA	−	None	None	+	+	Pensaert and Cox (1989)
Rotavirus										
Group A	Mild-severe	D,J,I	±	+ Entire	−	J,I	Moderate–severe	−	−	Theil *et al.* (1978)

Wesley, 1992; Saif et al., 1994b). Virulent TGEV replicates in villous epithelial cells throughout the small intestine, inducing severe villous atrophy and a malabsorptive diarrhea leading to nearly 100% mortality in seronegative, neonatal pigs. Attenuated strains of TGEV replicate in scattered villous epithelial cells in the distal portion of the small intestine of neonatal pigs and induce mild or no diarrhea (Frederick et al., 1976). They also replicate more extensively in the respiratory tract compared to virulent TGEV strains (Furuuchi et al., 1979). In contrast, PRCV strains replicate in the upper and lower respiratory tract, with little or no replication in the intestine, and generally cause subclinical infections or mild respiratory disease (Pensaert and Cox, 1989). TGEV infections remain a leading cause of piglet diarrhea and mortality in swine herds in North America, and commercial vaccines, even live attenuated oral TGEV vaccines, are of limited efficacy in the field (Saif and Wesley, 1992). In previous studies, PRCV induced partial protection against experimental challenge with TGEV (Van Nieuwstadt et al., 1989; Cox et al., 1993), but the mechanisms involved were not elucidated. These three antigenically related porcine coronaviruses with distinct differences in virulence and tissue tropisms (enteric TGEV-A or TGEV-V or respiratory PRCV) provided an ideal model to study interactions between bronchus-associated lymphoid tissues (BALT) and gut-associated lymphoid tissues (GALT) in the induction of mucosal immunity and protection against virulent TGEV challenge (Brim et al., 1995; Saif, 1996; VanCott et al., 1993, 1994).

By comparison, porcine group A rotaviruses also replicate throughout the small intestine and occasionally the colon, inducing moderate to severe villous atrophy in the distal small intestine (Table II, Theil et al., 1978). Similar to enzootic infections with TGEV-V in seropositive herds, rotaviruses are a frequent cause of diarrhea in 2- to 3-week-old pigs with morbidity rates approaching 100%, but with lower mortality rates (5–20%) (Paul and Stevenson, 1990; Saif et al., 1994a).

To date, commercial and experimental candidate vaccines have not been highly effective in preventing enteric viral infections and gastroenteritis in humans or animals (reviewed in Saif and Jackwood, 1990; Kapikian and Chanock, 1990). Poor efficacy has frequently been encountered in the field using live oral or parenterally administered vaccines to prevent coronavirus and rotavirus-induced diarrhea in swine (Saif and Jackwood, 1990). Likewise, clinical trials of candidate rotavirus vaccines in infants have often failed in various aspects of safety, immunogenicity, or efficacy, especially when tested in developing countries (Kapikian and Chanock, 1990). These results suggest that more research is needed to optimize enteric vaccines to induce local mucosal immune responses that more closely mimic ones elicited after exposure to the virulent organism.

III. Mucosal Immunity to Enteropathogenic Viruses

A unique mucosal immune system separate from the systemic immune system has evolved to protect mucosal surfaces from pathogens and to exclude environmental antigens and foreign proteins, thereby preventing them from evoking systemic-type inflammatory immune responses (reviewed by Brandtzaeg, 1992; Husband, 1993; McGhee et al., 1992; Mestecky, 1987). The mucosal immune system is characterized by a preponderance of SIgA antibodies selectively secreted onto mucosal surfaces by an active transport mechanism (polyimmunoglobulin receptor, PIgR). The SIgA antibodies play a major role in preservation of mucosal integrity by down-regulation of systemic-type immune responses, preventing invasion of pathogens from the mucosa by blocking of attachment or invasion, neutralization (in the lumen or intracellularly), and "immune exclusion." These functions are in contrast to systemically induced IgG antibodies that mediate inflammatory reactions leading to the killing and elimination of pathogens, thereby maintaining systemic sterility.

Although in earlier studies SIgA was envisioned to act mainly at the luminal mucosal surfaces, recent data suggest that dimeric IgA may bind antigens on the basolateral side of intestinal epithelial cells (Kaetzel et al., 1992). These immune complexes would then be transported across the epithelial cell via the PIgR and secreted back into the intestinal lumen, thereby eliminating foreign antigens that have penetrated the epithelium, Other recent reports suggest that SIgA may function intracellularly in host defense by inhibiting viral replication or assembly in vitro (Armstrong and Dimmock, 1992; Marzanec et al., 1992) and in vivo (Burns et al., 1996). If further confirmed in vivo, such findings imply that SIgA can promote recovery from viral infections as well as initial protection.

Another unique feature of the mucosal immune system compared to the systemic immune system is the induction of antigen-specific B and T cells in IgA inductive organized lymphoid tissue (GALT, BALT, etc.) and their distribution to remote mucosal effector sites (i.e., lamina propria regions of the intestine, bronchi, genitourinary tract, and secretory glands). This cellular distribution pathway linking distant mucosal sites is referred to as the common mucosal immune system (Mestecky, 1987). Thus antigen taken up (via M cells) and processed via GALT [Peyer's patches (PP) and aggregates of lymphoid tissue in the lamina propria] induces activated T and B cells which migrate from the PP through the MLN and via the thoracic duct into the systemic circulation, subsequently repopulating distant mucosal tissues. Maturation of these B cells into IgA plasma cells occurs within the mucosal effector sites in response to antigen, T cells, and cytokines (Lebman and Coffman, 1994). Key studies in rabbits confirmed that PP are an enriched source of IgA precursor cells which repopulate the lamina propria of the intestine and distant mucosal sites (Craig and Cebra, 1971).

Among the first reports to document that antigenic stimulation at one mucosal site (intestine) leads to SIgA antibody responses at a distinct mucosal site (mammary gland) were the studies of lactogenic immunity to TGEV in swine by Bohl et al. (1972) and Saif et al. (1972). The discovery and subsequent confirmation (Weisz-Carrington et al., 1978) of the interrelationship between the SIgA system of the intestine and mammary gland was an important tenet of the common mucosal immune system, and this system was later confirmed in humans and other species (Mestecky, 1987; McGhee et al., 1992). Thus antigen-specific B and T cells induced in IgA inductive lymphoid tissues (GALT, BALT, etc.) are disseminated to remote mucosal effector sites (i.e., lamina propria of the gut, mammary gland, bronchi, genitourinary tract, etc.).

Recent studies, including further studies of immunity to porcine coronaviruses (Van Cott et al., 1993, 1994; Saif et al., 1994b; Saif, 1996), have suggested that functional compartmentalization and limited reciprocity may exist within some components of the common mucosal immune system. For example, migration of cells from BALT is more limited than from GALT (Sminia et al., 1989) and BALT exposure often leads to dissemination of non-IgA committed secondary B cells (Cebra et al., 1984). In addition, IgA precursor cells derived from GALT more readily repopulate the gut lamina propria than distant mucosal sites (Cebra et al., 1984; Brandtzaeg, 1992). Such observations have important implications for the design of effective mucosal vaccines, but information on effective and practical procedures to induce protective immunity at mucosal surfaces is lacking. In the following sections, our studies of the induction of mucosal immunity and protection using the antigenically related porcine coronaviruses, TGEV and PRCV, are reviewed as are results of studies comparing different types of rotavirus vaccines.

A. Studies of Active Immunity to Transmissible Gastroenteritis Virus and Porcine Respiratory Coronavirus

To analyze the interrelationships between BALT and GALT related to protective immunity, we used as a model the three antigenically related porcine coronaviruses. Virulent TGEV replicates primarily in the intestine and induces diarrhea; attenuated TGEV replicates in the intestine and the upper respiratory tract but induces no diarrhea; and PRCV replicates in the upper and lower respiratory tract, but induces only a subclinical infection (Table II; Frederick et al., 1976; Pensaert and Cox, 1989; Saif and Wesley, 1992). These questions were addressed: Is PRCV a more effective candidate vaccine for TGEV than attenuated TGEV? Does a high dose of attenuated TGEV administered orally induce greater ASC responses in GALT than a lower dose (comparable or higher virus titer than commercial TGEV vaccines)? What are the comparative IgA and IgG ASC responses induced in GALT and BALT and the level of protection after inoculation with PRCV, TGEV-A, or TGEV-V and challenge with TGEV-V? In pigs recovered from infection with TGEV-V and reexposed to TGEV-V, what are the correlates of protective immunity?

We first investigated the comparative immune responses to live PRCV versus TGEV-V, the degree of protection induced against TGEV-V challenge, and potential correlates of protection. Three groups of 11-day-old TGEV seronegative pigs were oronasally inoculated with virulent TGEV, PRCV, or mock-infected cell-culture fluids, respectively, and challenged 24 days later with virulent TGEV (Brim et al., 1995; Saif et al., 1994b; Saif, 1996; VanCott et al., 1993, 1994). Immune responses in intestinal (gut lamina propria and mesenteric lymph nodes (MLNs) and respiratory (bronchial lymph nodes, BLN) lymphoid tissues were assessed at challenge and postchallenge day (PCD) 4 by enumeration of IgA and IgG TGEV-specific ASC by ELISPOT and by lymphoproliferative assays (LPAs) using inactivated TGEV as antigen. The major ASC responses and percent of pigs protected are summarized in Table III. All pigs inoculated with TGEV-V developed diarrhea, shed TGEV in feces, and recovered. The presence of high numbers of IgA-ASC in the gut lamina propria (LP) and high LPA responses in the MLN at challenge (PCD 0) was associated with 100% protection against diarrhea after TGEV challenge. No significant increases were observed in numbers of ASC or LPA responses in the gut LP or MLN, respectively, after TGEV challenge (PCD 4), reflecting the lack of viral replication associated with complete protection. In contrast, pigs inoculated with PRCV had no clinical disease and shed virus in nasal secre-

TABLE III

COMPARISON OF INTESTINAL AND RESPIRATORY ASC RESPONSES AND PROTECTIVE IMMUNITY INDUCED BY TGEV AND PRCV STRAINS IN NEONATAL PIGS AT POSTCHALLENGE DAY (PCD) 0 AND 4[a]

Virus inoculum group	Mean No. ASC/5 × 10^5 MNC at PCD 0							Mean No. ASC/5 × 10^5 MNC at PCD 4							Percent protection against TGEV challenge	
	Intestinal lamina propria			Bronchial lymph node				Intestinal lamina propria			Bronchial lymph node			Diarrhea (%)	Shedding (%)
	IgG	IgA	G/A[b]	IgG	IgA	G/A		IgG	IgA	G/A	IgG	IgA	G/A		
Virulent TGEV	109	620	0.18	25	1	25		15	109	0.14	300	94	3.2	100	80
PRCV	<1	1	UD	223	1	223		150	4	38	320	7	46	58	17
Controls	<1	<1	—	<1	<1	—		<1	<1	—	<1	<1	—	10	22

[a] Data summarized from VanCott et al. (1994).
[b] G/A, ratio of IgG to IgA ASCs; UD, undetermined because numerator <1.

tions but not feces. At challenge (PCD 0), the PRCV-exposed pigs had mainly IgG ASC and high LPA responses in the BLN, but low ASC numbers and LPA responses in the intestine (gut LP or MLN, respectively). About 58% of the pigs were protected against diarrhea (compared to 10% of controls) and only 17% were protected against fecal TGEV shedding (comparable to controls). After TGEV challenge (PCD 4), the numbers of IgG-ASC and to a lesser extent IgA-ASC increased rapidly in the intestinal lamina propria of the PRCV-exposed pigs, suggesting that virus-specific IgG-ASC precursors derived in BALT or systemic lymphoid tissues of the PRCV-exposed pigs may migrate to the intestine in response to TGEV challenge and contribute to the partial protection observed. The higher numbers of IgA-ASC in BALT of TGEV-exposed pigs compared to PRCV-exposed pigs at PCD 4 probably reflects TGEV replication and restimulation in the gut resulting in trafficking of IgA precursor cells from GALT to BALT (Husband, 1994; Mestecky, 1987; McGhee *et al.*, 1992). Thus TGEV infections or vaccines that induce immunity via GALT and secondarily via BALT may prevent PRCV infections. Whether the more frequent use of live attenuated TGEV vaccines in the United States (which induce IgG ASC in BLN, Table IV) compared to Europe has had an impact on limiting the

TABLE IV

Comparison of Intestinal and Respiratory Primary and Memory ASC Responses Induced by Virulent TGEV and Low Versus High Doses of Attenuated TGEV in Neonatal Pigs[a]

		Mean No. ASC/5 × 10⁵ MNC					
Virus inoculum group/response[b]		Mesenteric lymph node			Bronchial lymph node		
		IgG	IgA	G/A[c]	IgG	IgA	G/A[c]
Virulent TGEV	Primary	48	9	5	7	1	7
	Memory	5295	1159	5	2989	327	9
Attenuated TGEV							
Low dose (10⁶ pfu)	Primary	2	<1	UD	16	1	16
	Memory	60	<10	UD	866	34	25
High dose (10⁸ pfu)	Primary	9	1	9	28	1	28
	Memory	1133	79	14	4475	159	28

[a]Data summarized from VanCott *et al.* (1993).
[b]Primary immune responses were assayed by ELISPOT directly on mononuclear cells (MNCs) obtained from pigs at PID 12 and 24 and the mean numbers of ASC per 5 × 10⁵ MNC are shown. Memory or secondary immune responses were assayed by ELISPOT after *in vitro* TGEV stimulation (5 days) of MNC obtained from pigs at PID 24 and 40 and the mean numbers of ASC per 5 × 10⁵ MNC are shown.
[c]G/A, ratio of IgG to IgA ASCs; UD, undetermined.

spread of PRCV infections in the United States is unknown, but at present PRCV infections appear to be less widespread among swine in the United States than in Europe. Thus our major conclusions were that functional compartmentalization exists in the BALT and GALT responses: immunization via BALT (PRCV infection) induced a systemic type of response (IgG-ASC) with low numbers of ASC and LPA responses in the gut and provided incomplete protection against TGEV-V. Immunization via GALT (TGEV-V infection) induced high numbers of IgA-ASC and high LPA responses in the gut and provided complete protection against TGEV-V induced diarrhea. Further studies on the induction and immune regulation of responses to TGEV and PRCV that affect the distribution of ASC and T lymphocytes should provide important insights to optimize oral vaccine regimens to elicit protective mucosal immune responses against enteric pathogens.

In a similar series of experiments, we also examined the effect of the dose (10^6 versus 10^8 pfu) of live TGEV-A administered oronasally to 11-day-old TGEV seronegative pigs, on primary and memory ASC responses in the MLN and BLN (Saif *et al.*, 1994b; Saif, 1996; VanCott *et al.*, 1993). Our findings (summarized in Table IV) revealed that the high dose of TGEV-A (10^8 pfu) induced 2–4 times more primary IgG ASC and about 5–20 times more memory IgG ASC in the MLN and BLN than the lower dose. Only the high dose of TGEV-A elicited low numbers of primary or memory IgA ASC in the MLN, but numbers were 9–15 times lower than after inoculation with TGEV-V. Of interest were the two- to fourfold higher numbers of primary and memory IgG ASC induced in BALT by the high-dose TGEV-A compared to TGEV-V consistent with reports that attenuated strains of TGEV replicate more extensively in the respiratory tract compared to virulent TGEV strains (Furuuchi *et al.*, 1979). Thus the high degree of attenuation of TGEV vaccines leading to reduced viral replication in the intestine of the sow (Saif and Jackwood, 1990; Saif and Wesley, 1992) and the use of low doses ($\leq 10^6$ pfu/ml) of live attenuated TGEV vaccines orally in piglets (Saif *et al.*, 1994b; VanCott *et al.*, 1993) were major determinants in their failure to induce SIgA antibodies in sow's milk or IgA ASC in the piglets' intestines, respectively. Such factors presumably contribute to the corresponding lack of efficacy of TGEV vaccines in the field.

B. New Vaccine Approaches to Induce Active Immunity to TGEV

Several new potential vaccine approaches to induce immunity to TGEV have recently been reported based on delivery of antigenic peptides of the TGEV S protein in orally administered live bacterial vectors (Der Vartanian et al., 1997; Smerdou et al., 1996). In studies by Der Vartanian et al., (1997), two antigenic peptides of the TGEV S protein, TGEV S_A and S_C, were tandemly inserted (25 amino acids) into the major CIpG subunit of the CS 31A fibrillae of *Escherichia coli* K-12 strain. The responses of mice to these constructs were as follows: (1) The two TGEV epitopes were immunogenic when injected intraperitoneally (IP) into mice as hybrid CIpG subunits, chimeric CS31A polymers, or recombinant bacteria; (2) the chimeric CS31A fibrillae elicited TGEV antibodies in the serum of mice reactive with TGEV peptides and native TGEV; and (3) mice inoculated orally with the recombinant bacteria produced IgA intestinal antibodies reactive against the CS31A fibrillae and TGEV S_C peptide.

In another approach, a recombinant live attenuated *Salmonella typhimurium* was used for oral delivery of a TGEV peptide vaccine in rabbits (Smerdou et al., 1996). An antigenic peptide of the TGEV S protein (S_D, amino acids 378–395) was expressed as a fusion protein with *E. coli* LT-B in the Salmonella. The rationale for fusion with LT-B was to enhance the immunogenicity of the bivalent vaccine since LT-B also functions as an oral adjuvant. Studies of immune responses of rabbits inoculated with purified LT-B/S_D fusion products expressed from Salmonella or the recombinant Salmonella revealed that neutralizing antibodies to TGEV were induced by the purified LT-B/S_D and TGEV antibodies were elicited in serum and intestinal secretions after oral inoculation with the recombinant Salmonella.Thus, if similar TGEV neutralizing IgA antibody responses can be induced in the intestines of pigs by the recombinant bacterial vaccines, such vaccines warrant further study to access their ability to induce protective immunity to TGEV in pigs.

In our laboratory, we are currently exploring optimal oral adjuvants and delivery systems for recombinant TGEV S and M protein vaccines. Preliminary data indicate S and M protein vaccines administered IP with incomplete Freund's adjuvant (IFA) induced higher numbers of memory ASCs in GALT than an inactivated TGEV vaccine administered IP (Sestak et al., 1997).

C. Studies of Active Immunity to Group A Rotavirus

The gnotobiotic piglet model of porcine and human rotavirus-induced diarrhea has been used to further evaluate the influence of vaccine type (attenuated or binary-ethyleneimine inactivated) compared to wild-type virus infection on induction of intestinal ASC responses and protective immunity (Chen et al., 1995; Saif et al., 1996; Yuan et al., 1996, 1998). Results of oral or IM inoculation of 3- to 5-day-old pigs with Wa human rotaviruses (G1, P1A) and homologous virulent rotavirus oral challenge at postinoculation day (PID) 21 are summarized in Table V. B-cell responses (ASC) were measured by ELISPOT for intestinal (lamina propria) and systemic (peripheral blood lymphocytes, PBL) lymphoid tissues at challenge (PID 21). The major findings were that the numbers of IgA ASCs in the intestinal lamina propria and PBL were significantly greater in virulent-rotavirus inoculated pigs (mimic natural infection) than in the other groups (attenuated, inactivated, controls) and were correlated ($r = 0.9$) with the high degree of protection against diarrhea (89% of piglets protected). The transient appearance of IgA ASC in the blood mirrored the IgA ASC responses in the gut and could serve as an indicator for IgA ASC intestinal responses after rotavirus infection. Piglets inoculated with attenuated rotavirus had partial protection against diarrhea (44% protected) and the second highest numbers of IgA and IgG ASC in the intestinal lamina propria. Interestingly pigs inoculated IM or perorally (PO) with inactivated rotavirus in IFA had a very high number of IgG ASCs in PBL, but few IgG or IgA ASCs in the intestinal lamina propria and, like pigs given inactivated virus PO, minimal protection (0–17%) against diarrhea. Thus, the vaccine type influenced the site, isotype,

TABLE V

Comparison of Mucosal and Systemic ASC Responses and Protective Immunity Induced by Virulent, Attenuated and Inactivated Rotavirus Vaccines in Neonatal Pigs at PCD 0 (PID 21)[a]

Virus inoculum group	Mean No. ASC/5 × 10^5 MNC						Percent protection against rotavirus challenge	
	Intestinal lamina propria			Peripheral blood lymphocytes			Diarrhea	Shedding
	IgG	IgA	G/A[b]	IgG	IgA	G/A[b]		
Live								
Virulent rotavirus (PO)	64	53	1.2	2	6	0.3	89%	100%
Attenuated rotavirus (PO)	41	6	6.8	2	1	2	44%	19%
Inactivated								
Rotavirus (PO)	0.7	5	0.14	88	1	88	0%	0%
Rotavirus (IM)	4	3	1.3	237	2	119	17%	0%
Controls	<1	<1		<1	<1		14%	0%

[a]Data summarized from Yuan *et al.,* (1996, 1998); Saif *et al.,* (1996, p. 199).
[b]G/A, ratio of IgG to IgA ASCs.

and level of the ASC response and, similar to the results of the TGEV studies, the degree of protection was correlated with the numbers of IgA ASCs induced in the intestine.

Similarly, in studies of natural rotavirus infections in children, higher fecal IgA antibody titers to rotavirus were associated with protection against infection and illness (Matson *et al.*, 1993). Mouse studies of rotavirus-induced infection revealed similar findings: induction of intestinal IgA antibody responses were positively associated with protection against rotavirus shedding (Feng *et al.*, 1994).

D. New Vaccine Approaches to Induce Immunity to Rotavirus

Although not yet evaluated in swine, a new strategy for rotavirus vaccines is the creation of recombinant virus-like particles (VLPs) produced by the coexpression of the four rotavirus capsid genes (VP2/4/6/7) in a baculovirus expression system (Crawford *et al.*, 1994). The VLP vaccines administered with IFA have been tested in rotavirus seronegative mice and rabbits (Conner *et al.*, 1996) and as a maternal vaccine to enhance passive immunity in rotavirus seropositive cows (Fernandez *et al.*, 1996). The VLP vaccines were shown to be noninfectious (no RNA), stable, antigenically authentic, and highly immunogenic in the above species. They induced protective immunity against rotavirus shedding in vaccinated mice and rabbits (Conner *et al.*, 1996) and passive immunity against rotavirus diarrhea in calves fed colostrum from the VLP-vaccinated cows (Fernandez *et al.*, 1998). Thus VLP vaccines show promise as novel vaccines designed to induce mucosal immunity against rotavirus. Further research is needed to identify the optimal delivery systems and mucosal adjuvants for use with the VLP vaccines to most effectively stimulate mucosal immunity.

Acknowledgments

Research in our laboratory on mucosal immunity to coronavirus and rotavirus infections in pigs was supported in part by grants from the National Institutes of Health (R01Al33561 and R01Al37111) and the World Health Organization (GPV/V27/181/24). Salaries and research support were provided by state and federal funds appropriated to the Ohio Agricultural Research and Development Center, The Ohio State University.

References

Armstrong, S. J., and Dimmock, N. J. (1992). Neutralization of influenza virus by low concentrations of hemagglutinin-specific polymeric immunoglobulin A inhibits viral fusion activity, but activation of the ribonucleoprotein is also inhibited. *J. Virol.* **66**, 3823–3832.

Bohl, E. H., Gupta, R. K. P., Olquin, M. V. F., and Saif, L. J. (1972). Antibody responses in serum, colostrum and milk of swine after infection or vaccination with transmissible gastroenteritis virus. *Infect. Immun.* **6,** 289–301.

Brandtzaeg, P. (1992). Humoral immune response patterns of human mucosae: Induction and relation to bacterial respiratory tract infections. *J. Infect. Dis.* **165,** S167–S176.

Bridger, J. C. (1980). Detection by electron microscopy of caliciviruses, astroviruses and rotavirus-like particles in the faece of piglets with diarrhoea. *Vet. Rec.* **107,** 532.

Brim, T. A., VanCott, J. L., Lunney, J. K., and Saif, L. J. (1995). Cellular immune responses of pigs after primary inoculation with porcine respiratory coronavirus or transmissible gastroenteritis virus and challenge with transmissible gastroenteritis virus. *Vet. Immunol. Immunopathol.* **48,** 35–54.

Burns, J. W., Siadat-Pajouh, M., Krishnaney, A. A., and Greenberg, H. B. (1996). Protective effect of rotavirus VP6-specific IgA monoclonal antibodies that lack neutralizing activity. *Science* **272,** 104–107.

Cebra, J. J., Fuhrman, J. A., Griffin, P., Rose, F. V., Schweitzer, P. A., and Zimmerman, D. (1984). Changes in specific B cells and the dissemination of the primed state in vivo following antigenic stimulation by different mucosal routes. *Ann. Allergy* **53,** 541–549.

Chasey, D., Bridger, J. C., and McCrae, M. A. (1986). A new type of atypical rotavirus in pigs. *Arch. Virol.* **89,** 235.

Chen, W. K., Campbell, T., VanCott, J. L., and Saif, L. J. (1995). Enumeration of isotype-specific antibody-secreting cells derived from gnotobiotic piglets inoculated with porcine rotaviruses. *Vet. Immunol. Immunopathol.* **45,** 265–284.

Conner, M. E., Zarley, C. D., Hu, B., Parsons, S., Drabinski, D., Greiner, S., Smith, R., Jiang, B., Corsaro, B., Barniak, V., Madore, H. P., Crawford, S. E., and Estes, M. K. (1996). Virus-like particles as a rotavirus subunit vaccine. *J. Infect. Dis.* **174,** S88–S92.

Coussement, W., Ducatelle, R., Charlier, G., and Hoorens, J. (1981). Adenovirus enteritis in pigs. *Am. J. Vet. Res.* **42,** 1905–1911.

Cox, E., Pensaert, M. B., and Callebaut, P. (1993). Intestinal protection against challenge with transmissible gastroenteritis virus of pigs immune after infection with the porcine respiratory coronavirus. *Vaccine* **11,** 267–272.

Craig, S. W., and Cebra, J. J. (1971). Peyer's patches: An enriched source of precursors for IgA-producing immunocytes in the rabbit. *J. Exp. Med.* **134,** 188–200.

Crawford, S. E., Labbe, M., Cohen, J., Burroughs, M., Zhou, Y. J., and Estes, M. K. (1994). Characterization of virus-like particles produced by the expression of rotavirus capsid proteins in insect cells. *J. Virol.* **68,** 5945–5952.

Der Vartanian, M., Girardeau, J.-P., Martin, C., Rousset, E., Chavarot, M., Laude, H., and Contrepois, M. (1997). An *E. coli* CS31A fibrillum chimera capable of inducing memory antibodies in outbred mice following booster immunization with the enteropathogenic coronavirus transmissible gastroenteritis virus. *Vaccine* **15,** 111–120.

Feng, N., Burns, J. W., Bracy, L., and Greenberg, H. B. (1994). Comparison of mucosal and systemic humoral immune responses and subsequent protection in mice orally inoculated with a homologous or a heterologous rotavirus. *J. Virol.* **68,** 7766–7773.

Fernandez, F. M., Todhunter, D., Parwani, A. V., Crawford, S. E., Conner, M. E., Smith, K. L., Estes, M. K., and Saif, L. J. (1996). Isotype-specific antibody responses to rotavirus in cows inoculated with subunit vaccines comprised of recombinant SA11 rotavirus corelike particles (CLP) or virus-like particles (VLP). *Vaccine* **14,** 1303–1312.

Fernandez, F. M., Conner, M. E., Parwani, A. V., Nielsen, P. R., Hodgins, D., Crawford, S. E., Estes, M. K., and Saif, L. J. (1998). Passive immunity to bovine rotavirus in new born calves fed colostrum supplements from cows immunized with recombinant SA11 rotavirus core-like particles (CLP) or virus-like particles (VLP) vaccines. *Vaccine* **16,** 507–516.

Frederick, G. T., Bohl, E. H., and Cross, R. F. (1976). Pathogenicity of an attenuated strain of TGE virus for newborn pigs. *Am. J. Vet. Res.* **37,** 165–169

Furuuchi, S., Shimizu, Y., and Kumagai, T. (1979). Multiplication of low and high cell culture passaged strains of TGE virus in organs of newborn piglets. *Vet. Microbiol.* **3,** 169–178.

Husband, A. J. (1993). Novel vaccination strategies for the control of mucosal infection. *Vaccine* **11,** 107–112.

Kaetzel, C. S., Robinson, J. K., Chintalacharuvir, K. R., Vaerman, J. P., and Lamm, M. E. (1992). The polymeric immunoglobulin receptor secretory component mediates transport of immune complexes across epithelial cells a local defense function for IgA. *Proc. Natl. Acad. Sci. USA* **88,** 8796–8800.

Kapikian, A. Z., and Chanock, R. M. (1990). Rotaviruses. *In* "Virology" (B. N. Fields *et al.*, eds.), pp. 1353–1404. Raven Press, New York.

Lebman, D. A., and Coffman, R. L. (1994). Cytokines in the mucosal immune system. *In* "Handbook of Mucosal Immunology" (P. L. Ogra *et al.*, eds.), pp. 243–250. Academic Press, San Diego, CA.

Marzanec, M. B., Kaetzel, C. S., Lamm, M. E., Fletcher, D., and Nedrud, J. G. (1992). Intracellular neutralization of virus by immunoglobulin A antibodies. *Proc. Natl. Acad. Sci. USA* **89,** 6901–6905.

Matson, D. O., O'Ryan, M. L., Herrera, I., Pickering, L. K., and Estes, M. K. (1993). Fecal antibody responses to symptomatic and asymptomatic rotavirus infections. *J. Infect. Dis.* **167,** 577–583.

McGhee, J. R., Mestecky, J., Dertzbaugh, M. T., Eldridge, J. H., Hirasawa, M., and Kiyono, H. (1992). The mucosal immune system: from fundamental concepts to vaccine development. *Vaccine* **10,** 75–88.

Melnick, J. L. (1990). Enteroviruses: Polioviruses, coxsackie viruses, echoviruses and new enteroviruses. *In* "Virology", (B. N. Fields *et al.*, eds.), pp. 549–606. Raven Press, New York.

Mestecky, J. (1987). The common mucosal immune system and current strategies for induction of immune responses in external secretions. *J. Clin. Immunol.* **7,** 265–276.

Mowat, A. M. (1994). Oral tolerance and regulation of immunity to dietary antigens. *In* "Handbook of Mucosal Immunology" (P. L. Ogra *et al.*, eds.), pp. 185–201. Academic Press, San Diego, CA.

Paul, P. S., and Stevenson, G. W. (1992). Rotavirus and reovirus. *In* "Diseases of Swine" (A. D. Leman *et al.*, eds.), pp. 331–348. Iowa State Univ. Press, Ames.

Pensaert, M. B., and Cox, E. (1989). Porcine respiratory coronavirus related to transmissible gastroenteritis virus. *Agri-Practice* **10,** 17–21.

Rossen, J. W. A., Bekker, C. P. J., Strous, G. J. A. M., Horzinek, M. C., Dveksler, G. S., Holmes, K. V., and Rottier, P. J. M. (1996). A murine and a porcine coronavirus are released from opposite surfaces of the same epithelial cells. *Virology* **224,** 345–351.

Saif, L. J. (1990). Comparative aspects of enteric viral infections. *In* "Viral Diarrheas of Man and Animals" (L. J. Saif and K. W. Theil, eds.), pp. 9–31. CRC Press, Boca Raton, FL.

Saif, L. J. (1996). Mucosal immunity: An overview and studies of enteric and respiratory coronavirus infections in a swine model of enteric disease. *Vet. Immunol. Immunopathol.* **54,** 163–169.

Saif, L. J., and Jackwood, D. J. (1990). Enteric virus vaccines: Theoretical consider-

ations, current status and future approaches. *In* "Viral Diarrheas of Man and Animals" (L. J. Saif and K. W. Theil, eds.), pp. 313–329. CRC Press, Boca Raton, FL.
Saif. L. J., and Wesley, R. (1992). Transmissible gastroenteritis virus. *In* "Diseases of Swine" (A. D. Leman *et al.,* eds.), 7th ed., pp. 362–386. Iowa State University Press, Ames.
Saif, L. J., Bohl, E. H., and Gupta, R. K. P. (1972). Isolation of porcine immunoglobulins and determination of the immunoglobulin classes of transmissible gastroenteritis viral antibodies. *Infect. Immun.* **6,** 600–609.
Saif, L. J., Bohl, E. H., Theil, K. W., Cross, R. F., and House, J. A. (1980). Rotavirus-like, calicivirus-like, and 23-nm virus-like particles associated with diarrhea in young pigs. *J. Clin. Microbiol.* **12,** 105–111.
Saif, L. J., Bohl, E. H., Theil, K. W., Kohler, E. M., and Cross, R. F. (1980). 30 nm viruslike particles resembling astrovirus in intestinal contents of a diarrheic pig. *Proc. Conf. Res. Workers Anim. Dis.* Chicago, 1980, Abstr. No. 149.
Saif, L. J., Rosen, B., and Parwani, A. (1994a). Animal rotaviruses. *In* "Virus Infections of the Gastrointestinal Tract" (A. Z. Kapikian, ed.), 2nd ed., pp. 279–367. Dekker, New York.
Saif, L. J., VanCott, J. L., and Brim, T. A. (1994b). Immunity to transmissible gastroenteritis virus and porcine respiratory coronavirus infections in swine. *Vet. Immunol. Immunopathol.* **43,** 89–97.
Saif, L. J., Ward, L. A., Rosen, B. I., and To, T. L. (1996). The gnotobiotic piglet as a model for studies of disease pathogenesis and immunity to human rotaviruses. *Arch. Virol.* **12,** 153–161.
Sestak, K., Lewis, P., Gillespie, A., and Saif, L. J. (1997). Immune responses in pigs intraperitoneally (IP) inoculated with baculovirus-expressed transmissible gastroenteritis virus S and M proteins. Conf. Res. Workers Anim. Dis., Chicago, 1997, Abstract.
Smerdou, C., Anton, I. M., Plana, J., Curtiss, R., III, and Enjuanes, L. (1996). A continuous epitope from transmissible gastroenteritis virus S protein fused to *E. coli* heatlabile toxin B subunit expressed by attenuated Salmonella induces serum and secretory immunity. *Virus Res.* **41,** 1–9.
Sminia, T., van der Brugge-Gamelkoom, G. J., and Jeurissen, S. H. M. (1989). Structure and function of bronchus-associated lymphoid tissue (BALT). *Crit. Rev. Immunol.* **9,** 119–150.
Theil, K. W., Bohl., E. H., Cross, R. F., Kohler, E. M., and Agnes, A. G. (1978). Pathogenesis of porcine rotaviral infection of experimentally-inoculated gnotobiotic pigs. *Am. J. Vet. Res.* **39,** 213–220.
VanCott, J. L., Brim, T. A., Simkins, R. A., and Saif, L. J. (1993). Antibody-secreting cells to transmissible gastroenteritis virus and porcine respiratory coronavirus in gut- and bronchus-associated lymphoid tissues of neonatal pigs. *J. Immunol.* **150,** 3990–4000.
VanCott, J. L., Brim, T. A., Lunney, J., and Saif, L. J. (1994). Contribution of antibody secreting cells induced in mucosal lymphoid tissues of pigs inoculated with respiratory or enteric strains of coronavirus to immunity against enteric coronavirus challenge. *J. Immunol.* **152,** 3980–3990.
Van Nieuwstadt, A. P., Zetstra, T., and Boonstra, J. (1989). Infection with porcine respiratory coronavirus does not fully protect pigs against intestinal TGE virus. *Vet. Rec.* **125,** 58–60.
Weisz-Carrington, P., Roux, M. E., McWilliams, M., Phillips-Qualiata, J. M., and Lamm, M. E. (1978). Hormonal induction of the secretary immune system in the mammary gland. *Proc. Natl. Acad. Sci. USA* **75,** 2928–2932.

Woode, G. N., Bridger, J., Hall, G. A., Jones, J. M., and Jackson, G. (1976). The isolation of reovirus-like agents (rotaviruses) from acute gastroenteritis of piglets. *J. Med. Virol.* **9,** 203–209.

Yuan, L., Ward, L. A., To, T. L., and Saif, L. J. (1996). Systemic and intestinal antibody-secreting cell responses to inactivated human rotavirus in a gnotobiotic pig model of diarrheal disease. *J. Virol.* **70,** 3075–3083.

Yuan, L., Kang, S.-Y., Ward, L. A., To, T. L., and Saif, L. J. (1998). Antibody-secreting cell responses and protective immunity assessed in gnotobiotic pigs inoculated orally or intramuscularly with inactivated human rotavirus. *J. Virol.* **72,** 330–338.

Use of Interleukin 12 to Enhance the Cellular Immune Response of Swine to an Inactivated Herpesvirus Vaccine

FEDERICO A. ZUCKERMANN,* STEPHEN MARTIN,† ROBERT J. HUSMANN,* AND JULIE BRANDT*

Department of Veterinary Pathobiology, University of Illinois, Urbana, Illinois 61801, †Pharmacia / Upjohn, Kalamazoo, Michigan

I. Introduction
II. Cytokines as Vaccine Adjuvants
III. Interleukin 12 and Its Ability to Modulate Acquired Immunity
IV. Porcine Model to Examine the Adjuvant Effect of Interleukin 12
V. Immune Mechanism(s) of Protective Immunity against Herpesviruses
VI. Interleukin 12-Mediated Enhancement of the Cell-Mediated Immune Response to an Inactivated PrV Vaccine
VII. Summary and Conclusion
References

I. Introduction

Commercially available vaccines against pseudorabies virus (PrV) are comprised of either the inactivated pathogen or modified live virus (MLV). We have examined the humoral and cellular immune responses of swine to vaccination against PrV by measuring the titer of virus-neutralizing antibodies, the frequency of PrV-specific interferon γ (IFN-γ) producing cells and the intensity of the lymphoproliferative response *in vitro*. Our data demonstrate that while both MLV and inactivated PrV vaccines are able to induce similar levels of humoral immunity, formulations designed to mimic commercially available inactivated vaccines are not as effective as MLV vaccines in stimulating either cell-mediated or protective immunity. To examine the immunoenhancing effect of interleukin 12 (IL-12), pigs were immunized with inactivated PrV with our without human recombinant IL-12 as the only source of adjuvant. Treatment with IL-12 enhanced the cellular immune response and protective immunity induced by the inactivated vaccine to levels similar to those obtained when using an inactivated vaccine mixed with an oil-in-water adjuvant. These studies quantitatively demonstrate that an inactivated herpesvirus vaccine is less efficient than a MLV vaccine at inducing cell-mediated and protective immunity in pigs. Our results provide evidence for the existence of a dichotomy in the regulation of porcine humoral and cellular immune responses. The data also demonstrate an enhancing effect of IL-12 on the porcine cellular immune response to viral antigens and show a positive correlation between the intensity of this response and protective immunity.

II. Cytokines as Vaccine Adjuvants

In vaccine design, formulations must be devised that will produce a protective immune response, and not an ineffective or deleterious one. Although a humoral immune response may be effective against some microbial invaders, it may not protect the host against others, mainly obligate intracellular pathogens like viruses. It is thought that infections by cytopathic viruses are primarily cleared by a combination of T cells that produce IFN-γ and mediate delayed type hypersensitivity (DTH) responses (Zinkernagel, 1996). A predominant humoral immune response against a cytopathic virus might not be sufficiently protective or may even be deleterious, since the uptake of virus–antibody complexes via Fc receptors on macrophages can lead to an enhanced infection of these cells by arteriviruses such as the porcine respiratory and reproductive virus (Yoon et al., 1997).

Experimentally, the quality of the immune response induced by a vaccine can be modulated by the type of adjuvant included in the vaccine. The mechanism(s) of adjuvant action are poorly understood. Adjuvants are available that help induce preferentially either a cellular or humoral immunity in response to a vaccine (Cooper, 1994). Thus, it is theoretically possible to obtain the optimal type of immune response to effectively control and eliminate a particular pathogen by simply choosing the correct adjuvant. The application of this knowledge to the development of vaccines for domestic animals is hampered by the fact that most of the research on adjuvant action has focused on laboratory animals, and the resulting data may not accurately portray their effect in other species. A large number of vaccine adjuvants have been developed (for a compendium of vaccine adjuvants, see Vogel and Powell, 1995). Cytokines are a prominent example, not only because of their great potential as adjuvants, but also for their potential to provide insight into the possible mechanisms of action of adjuvants (Pape et al., 1997).

The administration of cytokines as therapeutic and/or prophylactic agents is possible due to the availability of recombinant cytokines in sufficient quantities and purity. Early experimentation examined the effects of IL-1, IL-2, and IFN-γ in rodents (reviewed in Heath, 1995), although testing in large animals has also been performed (reviewed by Campos et al., 1994; Lofthouse et al., 1996). Recently, IL-12 has attracted attention as a possible adjuvant due to its known ability to stimulate cell-mediated immunity. This cytokine is known to direct the differentiation of T cells following their exposure to antigen into IFN-γ-producing cells. IL-12 has been shown to promote the development of protective immunity against several intracellular pathogens, including viruses (Schijns et al., 1995; Tang and Graham, 1995), and thus has potential as a vaccine adjuvant (Bliss et al., 1996a,b; Scott and Trinchieri, 1997). Although this effect has been clearly shown in mice, the ability of this cytokine to stimulate protective immunity in a large animal model has only been demonstrated recently (Zuckermann et al., 1998a).

III. Interleukin 12 and Its Ability to Modulate Acquired Immunity

The innate immune system has a critical influence on the development of adaptive immunity (Trinchieri, 1997). Cytokines produced by marcophages, dendritic cells, and other accessory cells directly affect the development of adaptive immune responses. IL-12 is the single most important factor required for the efficient differentiation of naive T cells into IFN-γ-producing memory/effector T cells (reviewed by Scott and Trinchieri, 1997). IL-12 is a cytokine with proinflammatory functions, which is produced by phagocytic cells, dendritic cells, and Langerhans' cells in response to infections by microbes, including viruses (Coutelier et al., 1995). IL-12 is a heterodimeric cytokine consisting of 35- and 40-kDa subunits. Both subunits are highly conserved across species. The predicted amino acid sequences of the two porcine IL-12 subunits are approximately 85% homologous with their human counterparts (Foss and Murtaugh, 1997). Due to this high degree of similarity, recombinant human (rHu) IL-12 is able to enhance porcine NK cell cytolytic activity (Cho et al., 1996) and IFN-γ production (Y.-B. Kim and F. Zuckermann, unpublished observations).

The production of IL-12 is strictly regulated by both positive and negative feedback mechanisms. IFN-γ, primarily a product of T and NK cells, represents the most potent up-regulator of IL-12 production (Ma et al., 1996). IL-10, a product of macrophages, lymphocytes, and other cell types, is the most physiologically relevant inhibitor of IL-12 production by accessory cells (D'Andrea et al., 1993). Because of its potential toxic effects, overproduction of IL-12 is down-regulated by IL-10 (Tripp et al., 1993). Elimination of this negative regulatory mechanism can lead to a potentially lethal and toxic syndrome resulting from an overexuberant IL-12 response, such as that seen in IL-10 knockout mice infected with *Toxoplasma gondii* (Gazzinelli et al., 1996). Recent reports have shown that the interactions between IL-12, IL-10, and IFN-γ are complex, involving reciprocal induction and suppression events. These interactions significantly affect the quality of the immune response (reviewed by Trinchieri, 1997). The interactions of these cytokines during a viral infection is not well defined and is just beginning to be investigated. During acute infection of mice with murine cytomegalovirus (MCMV), IL-12 has been shown to be essential for induction of the early NK-cell-mediated and IFN-γ-dependent mechanisms of antiviral defense. In contrast, the late and most efficient production of IFN-γ by T cells during this primary acute viral infection appears to be largely IL-12 independent, at least in this murine system (Orange and Biron, 1996). This observation is distinct from those obtained from the examination of immune responses to certain intracellular bacteria and parasites (reviewed in Trinchieri and Scott, 1994). In these infections, IL-12 is responsible for NK cell production of IFN-γ and is necessary for the development of the Th1 lymphocyte responses mediated by IFN-γ (reviewed by Romani et al., 1997). The production of IL-12 followed by IFN-γ expression, creates a microenvironment in which antigen-specific $CD4^+$ T cells are preferentially induced to differentiate into IFN-γ-producing cells, resulting in even higher levels of this cytokine (McKnight et al., 1994). Despite the disparate results observed in mice during an acute MCMV infection, IL-12 has been shown to have an adjuvant effect on the immune response to inactivated PrV, promoting the T-cell-mediated IFN-γ response in mice (Schijns et al., 1995) and pigs (Zuckermann et al., 1998a,b). Thus IL-12 appears to play a role in modulating anti-viral T-cell-mediated immunity (CMI) in mice and pigs.

IV. Porcine Model to Examine the Adjuvant Effect of Interleukin 12

Pseudorabies virus is an alpha herpesvirus, whose natural host is the pig. Infection of pigs by this virus causes Aujeszky's disease. The clinical response to virus challenge depends on the immune status and age of the animal (Kluge et al., 1992). Aujeszky's disease is characterized by a fatal encephalitis in newborn pigs and a milder syndrome in older swine, which is manifested as severe depression, anorexia, pyrexia, ataxia, and respiratory distress (Baskerville et al., 1973). Protective immunity can be induced by inoculation with both live and inactivated PrV vaccines (Donaldson et al., 1987; Wardley et al., 1991; Van Oirschot and de Leeuw, 1985). Although both vaccines prevent mortality resulting from virus challenge, the clinical outcome differs significantly for the two vaccines. Challenge of vaccinated animals results in a reduced rate of weight gain, and even weight loss, depending on the severity of the challenge and the level of protective immunity conferred by the vaccine (Vannier, 1985; Van Oirschot and de Leeuw, 1985; Wardley et al., 1991; Zuckermann et al., 1998a). Weight changes observed within 7 days after PrV challenge have been shown to be a sensitive, reproducible and statistically sound parameter, which allows the quantification of the level of protective immunity conferred by different PrV vaccines (Stellman et al., 1989). Measurement of this parameter has demonstrated that indeed inactivated vaccines are less effective than live vaccines at inducing protective immunity.

Immunization with live virus vaccines generates a robust cytotoxic T-lymphocyte response (Zuckermann et al., 1990). Several studies have also demonstrated induction of a strong lymphoproliferative response to vaccination (Van Oirschot, 1978; Kimman et al., 1995; Zuckermann and Husmann, 1996a; Zuckermann et al., 1998a). Characterization of the T cells mediating these responses has shown that the virus-specific cytotoxic T-lymphocyte response is mediated by CD8 single positive (SP) lymphocytes, while the lymphoproliferative response is mediated by both CD4 SP and CD4/CD8 double positive (DP) lymphocytes (Zuckermann et al., 1990; Pescovitz et al., 1994; Kimman et al., 1995; Zuckermann and Husmann, 1996a,b). Other investigations have examined the humoral immune response and found that it is strongly induced by both infection and vaccination (reviewed by Chinsakchai and Molitor, 1994). We have evaluated the humoral and CMI response to vaccination with commercially available modified live or inactivated PrV vaccines. We have determined that while the intensity of the lymphoproliferative response to either of these two types of vaccines may or may not differ, the MLV vaccine induces a three- to fivefold higher frequency of virus-specific IFN-γ-producing cells than does an inactivated vaccine (Fig. 1C; Zuckermann et al., 1998a). Remarkably, there is a dichotomy between the humoral and cellular immune responses to these two types of vaccines. While in some instances the inactivated vaccine is capable of inducing an equal (Fig. 1A) or even higher (Zuckermann et al., 1998b) titer of virus neutralizing antibodies than the MLV vaccine, the latter stimulates the generation of a greater number of virus-specific IFN-γ-producing cells (Fig. 1C; Zuckermann et al., 1998a,b). These results indicate that the humoral and cellular immune responses of a pig, at least to this viral vaccine, are independently regulated. Further evidence of this phenomenon is the observation that while immunization with an unadjuvanted, inactivated PrV vaccine is capable of inducing a humoral immune response which is not significantly different in titer than that induced by an adjuvanted inactivated vaccine, it only induced a weak cellular immune response (F. A. Zuckermann and R. Husmann, unpublished observations).

V. Immune Mechanism(s) of Protective Immunity against Herpesviruses

Live PrV vaccines are more effective than inactivated vaccines in generating protective immunity. Changes in body weight that occur during the 7 days after virus challenge demonstrate this fact (Stellman et al., 1989). The extent to which an individual immune mechanism contributes to the control of Aujeszky's disease has been difficult to establish. The fact that humoral immunity is capable of mediating disease protection was shown by the demonstration that passive immunization of swine with mAbs specific for PrV was able to protect pigs from lethal virus challenge (Marchioli et al., 1988). However, other evidence suggests that humoral immunity might not be the main mechanism mediating protection. There is a poor correlation between the titer of vaccine-induced virus-neutralizing antibody in pigs and the level of protection against disease (Kimman et al., 1994). Humans suffering from agammaglobulinemia are not predisposed to severe life-threatening viral infection with either measles or herpes simplex virus (HSV) infection. This contrasts with the outcome in patients suffering from T-cell immunodeficiency such as DiGeorge's syndrome (congenital absence of the thymus) or in nude athymic mice, where infections with HSV are more severe and life threatening than in normal mice. Transfer of passive neutralizing antibodies delays the disease process, but only the adoptive transfer of HSV-specific T cells that protects the nude mice by resolving the infection (reviewed by Schmid and Rouse, 1992). Direct evidence that CMI is essential in mediating protective immunity against herpesviruses has been obtained in murine experimental models of HSV and PrV infection. For HVS, adoptive transfer experiments performed by several groups of investigators have shown that both CD4 and CD8 positive lymphocytes are able to provide protective immunity (Nash and Cambouropoulos, 1993; Rouse et al., 1988; Schmid and Rouse, 1992). For PrV, the administration of IFN-γ-neutralizing antibody at the time of vaccination significantly decreased vaccine-induced protective immunity in mice (Schijns et al., 1995), suggesting that the development of Th1-type responses is necessary for the generation of protective immunity. Based on these observations a strong case for CMI as a major contributor to protective immunity against herpesviruses can be made.

A major component of the cellular immune response is the secretion by T helper cells of inflammatory cytokines such as IFN-γ. IFN-γ can have a direct antiviral effect by inhibiting virus growth and also by inducing the expression of MHC class II antigens. Both HSV and PrV are susceptible to the inhibitory effects of IFN-γ (Schijns et al., 1991; Schmid and Rouse, 1992). Administration of IFN-γ in combination with an inactivated PrV vaccine into mice, was able to modulate the immune response by increasing the production of virus-specific IgG_{2a}, and enhancing the resistance to PrV challenge (Schijns et al., 1995). Once again, CMI and in particular IFN-γ-producing cells, have been shown to be important in conferring protective immunity against herpesviruses. The effector mechanism(s) by which these cells mediate their protective effect *in vivo* is unknown. Based on available evidence it is reasonable to assume a central role for IFN-γ-producing cells (either CD4 or CD8 positive lymphocytes and NK cells) in controlling herpesvirus infections, and in the development of protective immunity (Biron, 1997; Nash and Cambouropoulos, 1993; Schmid and Rouse, 1992). The effector functions by which these cells are likely to provide antiviral protection include activation and attraction of phagocytic cells, B-cell helper activity for generating complement fixing antibodies, and direct cytolytic action. As cited earlier, our studies clearly show that there are differences in the quality and quantity of the immunity induced by a live versus inactivated PrV vaccine (Fig. 1; Zuckermann et al., 1998a). While inactivated vaccines are equally efficient as a MLV vaccine at inducing humoral immunity, they only induce a weak and transient virus-specific IFN-γ response, compared to

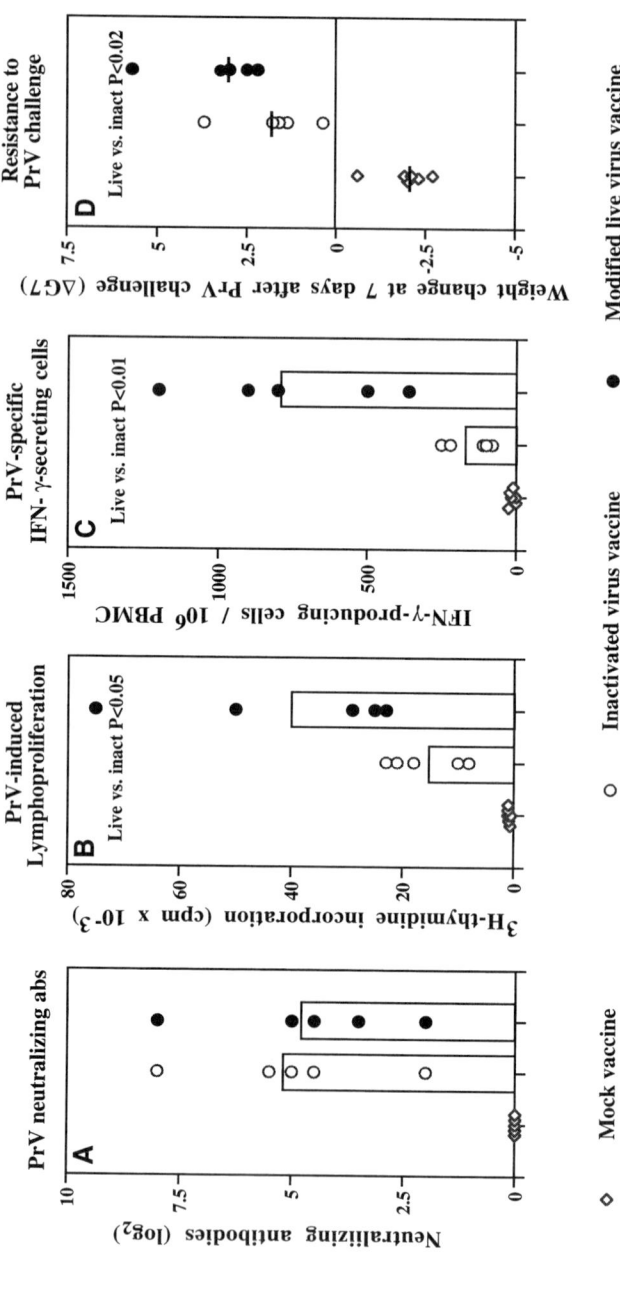

a MLV vaccine which induces a robust virus-specific IFN-γ response. It is tempting to speculate that the differential induction of CMI is responsible for the distinct levels of protective immunity conferred by these two types of vaccines (Fig. 1D). At the very least, a strong IFN-γ response to PrV vaccine is a good predictor that a pig has developed a strong protective immune response against this virus.

VI. Interleukin 12-Mediated Enhancement of the Cell-Mediated Immune Response to an Inactivated PrV Vaccine

Recent studies in mice have demonstrated that the intensity of the CMI response to an inactivated PrV vaccine can be enhanced by the administration of IL-12 (Schijns et al., 1995). The observation that an inactivated PrV vaccine is inefficient at inducing the generation of virus-specific IFN-γ-producing memory T cells, and even more inefficient in the absence of an adjuvant, provides an ideal model for studying the possible adjuvant effect of IL-12 in the induction of these subsets of T cells in pigs. We have developed an ELISPOT assay to enumerate IFN-γ-producing cells. This assay provides a sensitive and accurate method for measuring even slight changes in the frequency of antigen-specific IFN-γ-producing cells generated in response to different formulations of a vaccine (Zuckermann et al., 1998a). Utilizing this model we have demonstrated that, while the injection of IL-12 alone does not stimulate a PrV-specific immunity, the injection of IL-12 in

---

FIG. 1. Comparison of the humoral, cellular, and protective immunity generated by an inactivated and a modified live pseudorabies virus vaccines. Three groups of 8-week-old cross-bred pigs (five pigs per group) were immunized twice, four weeks apart with either saline (◇), inactivated PrV vaccine (○; Triad, Oxford Labs), or a modified live PrV vaccine (●; Tolvid, Upjohn). Fourteen days after the second vaccination, serum and peripheral blood mononuclear cells were isolated and the humoral and cellular immune responses were measured as previously described (Zuckermann and Husmann, 1996a). The frequency of virus-specific IFN-γ-producing cells was determined with an ELISPOT assay (Zuckermann et al., 1998a). While the humoral immune response between the two immunized groups does not differ (panel A), both the virus-induced lymphoproliferative response (panel B) and the frequency of PRV-specific IFN-γ-producing cells (panel C) are significantly higher in animals receiving the MLV vaccine. All of the animals were challenged at this time with 10 LD_{50} of wild-type PrV (strain Rice). The animals were weighed at the time of challenge and 7 days later. The $\Delta G7$ value (i.e., the weight change during this period) was calculated (panel D) as described by Stellman et al. (1989). This value was used as a measurement of the potency of the vaccine, and was significantly different between the two immunized groups ($P<0.02$).

combination with an inactivated PrV vaccine enhances the strength of the cellular immune response over the response induced by the inactivated vaccine alone. The increased CMI response was manifested as an increase in the frequency of virus-specific IFN-γ-secreting cells (Fig. 2). Furthermore, this effect was associated with an increased level of protective immunity (Zuckermann et al., 1998b).

VII. Summary and Conclusion

Vaccination is the single most successful medical measure against infectious disease. However, the major barrier for achieving the full protective effect or immunization is how to render attenuated, killed,

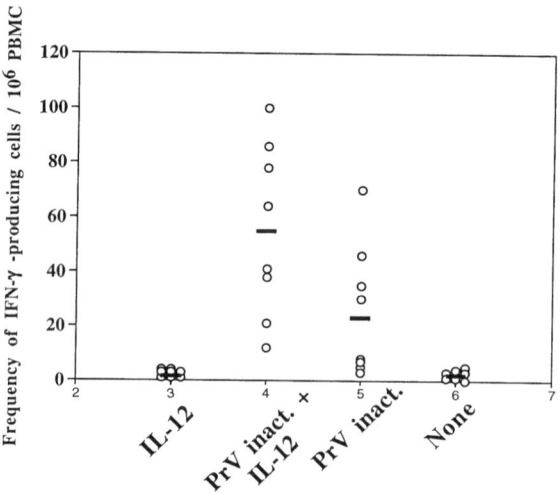

FIG. 2. IL-12 enhances the generation of virus-specific IFN-γ-producing cells in response to vaccination with an inactivated PrV vaccine. Pigs at 6 weeks of age were allocated to four treatment groups (eight pigs per group). One group was left untreated. Pigs in two other groups were injected intramusculary in the gluteal region with one dose per pig (2 cc) of an inactivated PrV vaccine (PR-Vac®-Killed, SmithKline Beecham). Pigs in one of these two groups were also injected with 1 μg of human recombinant IL-12 (R&D Systems, Minneapolis, MN) in a 1 cc volume at a site adjacent to the vaccine site. The IL-12 treatment was repeated 20 hours after the vaccination. The fourth group of pigs received only the IL-12 injection. The vaccination was repeated 2 weeks later. Peripheral blood monuclear cells were isolated 2 weeks after the second vaccination and the frequency of PrV-specific IFN-γ-producing cells enumerated in an ELISPOT assay as previously described (Zuckermann et al., 1998a). Animals inoculated with the inactivated PrV vaccine in combination with IL-12 had a significantly higher frequency of PrV-specific IFN-γ-secreting cells ($P<0.05$).

or subunit vaccines as immunogenic as the fully infectious versions of these microbes (Hughes and Babiuk, 1995; Rabinovich et al., 1994). In the case of PrV, infection with wild-type virus induces an immune response superior to vaccination with a live modified vaccine. After primary intranasal infection with wild-type PrV, the replication of a homologous secondary virus challenge is completely inhibited, and the much sought "sterile immunity" is generated (Kimman et al., 1994). In contrast, the immune response of pigs similarly exposed to PrV mutants, which have been attenuated by removal of the thymidine kinase (TK) and the envelope glycoprotein gE gene (McGregor et al., 1985; Zuckermann et al., 1988), is insufficient for preventing the replication of a homologous wild-type virus challenge (Kimman et al., 1994). Furthermore, inactivated PrV vaccines are even less effective at inducing protective immunity than are live modified PrV vaccines (de Leeuw and Van Orischot, 1985; Stellman et al., 1989; Vannier, 1985). The importance of inactivated and subunit vaccines resides in their stability and safety, since no infectious microbe is being introduced into the animal. However, because of the recognized lower effectiveness of inactivated vaccine types, they usually fall in disfavor when a modified live vaccine alternative is available. There is a critical need to develop strategies to enhance the immunogenicity of live, inactivated, and subunit vaccines for human and veterinary use (Hughes and Babiuk, 1995; Rabinovich et al., 1994). Although the inoculation of an animal with a virulent microbe is obviously not the desired method to produce sterile immunity, the immune response generated to infection with wild-type PrV clearly demonstrates that this type of immunity is possible. Research directed at devising strategies to increase the immunogenicity of different types of vaccines is necessary. Because of the wealth of information available on PrV immunity (reviewed by Chinsakchai and Molitor, 1994; Nauwynck, 1997), on PrV vaccines (Kimman et al., 1992, 1994; Mettenleiter, 1991; Scherba and Zuckermann, 1996) and increasingly on the porcine immune system (Lunney, 1993; Lunney et al., 1996; Saalmüller, 1995), the swine herpesvirus model is ideal for investigating the development of vaccine formulations with enhanced immunogenicity.

Among the strategies currently being examined for the enhancement of the immunogenicity of inactivated and subunit vaccines is the use of recombinant cytokines administered together with antigen (Hughes and Babiuk, 1995; Rabinovich et al., 1994). The ability to regulate the development of an immune response by cytokines such as IL-12 provides the theoretical basis to use these cytokines as adjuvants to immunopotentiate the response to an inactivated vaccine. More importantly, it provides a model to investigate the mechanisms behind the induction of protective immunity and the components of a vaccine necessary for stimulating such a response. By providing cytokines such as IL-12 or IFN-γ in combination with the vaccine inoculum, it is reasonable to expect that they will be able to direct the differentiation of T cells during the primary immune response. Modulation, in a predictable and desired manner of the quality and quantity of the induced protective immunity, should be achievable. The ability to manipulate a vaccine-induced immune response in the direction of a predominantly cellular response (Th1-like) instead of a predominantly humoral one (Th2-like) is perhaps best illustrated by the need to develop an effective vaccine against the porcine reproductive and respiratory syndrome (PRRS) virus, whose infectivity can be significantly enhanced *in vitro* and *in vivo* by antibody induced by vaccination against this virus (Yoon *et al.*, 1997). Examination of the ability of IL-12 to enhance the CMI response of swine to an inactivated or MLV vaccine should have application not only to controlling PrV but also other pathogens such as PRRS virus.

References

Baskerville, A., McFerran, J. B., and Dow, C. (1973). Aujeszky's disease in pigs. *Vet. Bull.* **43,** 465–480.

Biron, C. A. (1997). Activation and function of natural killer cell responses during viral infections. *Curr. Opin. Immunol.* **9,** 24–34.

Bliss, J., Maylor, R., Stokes, K., Murray, K. S., Ketchum, M. A., and Wolf, S. F. (1996a). Interleukin-12 as adjuvant. Characteristics of primary, recall, and long-term responses. *Ann. N. Y. Acad. Sci.* **795,** 26–35.

Bliss, J., Van Cleave, V., Murray, K., Wiencis, A., Ketchum, M., Maylor, R., Haire, T., Resmini, C., Abbas, A. K., and Wolf, S. F. (1996b). IL-12, as an adjuvant, promotes a T helper 1 cell, but does not suppress a T helper 2 cell recall response. *J. Immunol.* **156,** 887–894.

Campos, M., Godson, D. L., Hughes, H. P. A., and Babiuk, L. A. (1994). Cytokine applications in infectious diseases. *In* "Cell-mediated Immunity in Ruminants" (B. M. L. Goddeeris and W. I. Morrison, eds.), pp. 229–240. CRC Press, Boca Raton, FL.

Chinsakchai, S., and Molitor, T. W. (1994). Immunobiology of pseudorabies virus infection in swine. *Vet. Immunol. Immunopathol.* **43,** 107.

Cho, D., Lee, W. J., Halloran, P. J., Trinchieri, G., and Kim, Y. B. (1996). Enhancement of porcine natural killer cell activity by recombinant human and murine IL-12. *Cell. Immunol.* **172,** 29–34.

Cooper, P. D. (1994). The selective induction of different immune responses by vaccine adjuvants. *In* "Strategies for Vaccine Design" (G. L. Ada, ed.), pp. 125–158. R. G. Landes, Austin, TX.

Coutelier, J. P., Van Broeck, J., and Wolf, S. F. (1995). Interleukin-12 gene expression after viral infection in the mouse. *J. Virol.* **69,** 1955–1958.

D'Andrea, A., Aste-Amezaga, M., Valiante, N. M., Ma, X., Kubin, M., and Trinchieri, G. (1993). Interleukin-10 inhibits human lymphocyte IFN-g production by suppressing

natural killer cell stimulatory factor/interleukin-12 synthesis in accessory cells. *J. Exp. Med.* **178,** 1041–1048.
de Leeuw, P. W., and Van Oirschot, J. T. (1985). Vaccines against Aujeszky's disease: Evaluation of their efficacy under standardized laboratory conditions. *Vet. Q.* **7**(3), 191.
Donaldson, J. T., Wardly, R. C., Martin, S., and Harkness, J. W. (1987). Influence of vaccination of Aujeszky's disease virus and disease transmission. *Vet. Rec.* **115,** 121.
Foss, D. L., and Murtaugh, M. P. (1997). Molecular cloning and regulation of expression of porcine interleukin-12. *Vet. Immunol. Immunopathol.* **57,** 121–134.
Gazzinelli, R. T., Wysocka, M., Hieny, S., Scharton-Kersten, T., Cheever, A., Kuhn, R., Muller, W., Trinchieri, G., and Sher, A. (1996). In the absence of endogenous IL-10, mice acutely infected with *Toxoplasma gondii* succumb to a lethal immune response dependent on CD4$^+$ T cells and accompanied by overproduction of IL-12, IFN-γ and TNF-α. *J. Immunol.* **157,** 798–805.
Heath, A. W. (1995). Cytokines as immunological adjuvants. *Pharm. Biotechnol.* **6,** 645–658.
Hughes, H. P. A., and Babiuk, L. A. (1995). Potentiation of the immune response by cytokines. *In* "Cytokines in Animal Health and Disease" (M. J. Myers, and M. P. Murtaugh, eds.), pp. 183–202. Dekker, New York.
Kimman, T. G., deWind, N., Pol, J. M. A., Berns, A. J. M., and Gielkens, A. L. J. (1992). Contribution of single genes within the unique short region of Aujeszky's disease virus (suid herpesvirus type 1) to virulence, pathogenesis and immunogenicity. *J. Gen. Virol.* **73,** 243.
Kimman, T. G., deWind, N., deBruin, T., deVisser, Y., and Voermans, J. (1994). Inactivation of glycoprotein gE and thymidine kinase or the US3-encoded protein kinase synergistically decreases in vivo replication of pseudorabies virus and the induction of protective immunity. *Virology* **205,** 511.
Kimman, T. G., deBruin, T. M. G., Voermans, J. J. M., Peeters, B. P. H., and Bianchi, A. T. J. (1995). Development and antigen specificity of the lymphorproliferation response of pigs to pseudorabies virus: dichotomy between secondary B- and T-cell responses. *Immunology* **86,** 372.
Kluge, J. P., Beran, G. W., Hill, H. T., and Platt, K. B. (1992). Pseudorabies (Aujeszky's disease). *In* "Disease of Swine" (A. D. Leman, B. E. Straw, W. L. Mengeling, S. D'Allaire, and D. J. Taylor, eds.), pp. 312–323. Iowa State University Press, Ames.
Lofthouse, S. A., Andrews, A. E., Elhay, M. J., Bowles, V. M., Meeusen, E. N., and Nash, A. D. (1996). Cytokines as adjuvants for ruminant vaccines. *Int. J. Parasitol.* **26,** 835–842.
Lunney, J. K. (1993). Characterization of swine leukocyte differentiation antigen. *Immunol. Today* **14,** 147–148.
Lunney, J. K., Saalmüller, A., Pauly, T., Boyd, P., Hyatt, S., Strom, D., Martin, S., and Zuckermann, F. A. (1996). Cellular immune responses controlling infectious diseases. *In* "Swine in Biomedical Research" (L. B. Schook and M. E. Tumbleson, eds.), pp. 307–318. Plenum, New York.
Ma, X., Chow, J. M., Gri, G., Carra, G., Gerosa, F., Wolf, S. F., Dzialo, R., and Trinchieri, G. (1996). The interleukin-12 p40 gene promoter is primed by interferon-g in monocytic cells. *J. Exp. Med.* **183,** 147–157.
Marchioli, C., Yancey, R. J., Timmins, J. G., Post, L. E., Young, B. R., and Povendo, D. A. (1988). Protection of mice and swine from pseudorabies virus induced mortality by administration of pseudorabies virus specific mouse monoclonal antibodies. *Am. J. Vet. Res.* **49,** 860–869.

McGregor, S., Easterday, B. C., Kaplan, A. S., and Ben-Porat, T. (1985). Vaccination of swine with thymidine kinase-deficient mutants of pseudorabies virus. *Am. J. Vet. Res.* **46,** 1494–1497.

McKnight, A. J., Zimmer, G. J., Fogelman, I., Wolf, S. F., and Abbas, A. K. (1994). Effects of IL-12 on helper T cell-dependent immune responses *in vivo. J. Immunol.* **152,** 2172–2181.

Mettenleiter, T. C. (1991). Molecular biology of pseudorabies (Aujeszky's disease) virus. *Comp. Immunol. Microbiol. Infect. Dis.* **14,** 151–158.

Nash, A. A., and Cambouropoulos, P. (1993). The immune response to herpes simplex virus. *Semin. Virol.* **4,** 181–189.

Nauwynck, H. J. (1997). Functional aspects of Aujeszky's disease (pseudorabies) viral proteins with relation to invasion, virulence and immunogenicity. *Vet. Microbiol.* **55,** 3–11.

Orange, J. S., and Biron, C. A. (1996). An absolute and restricted requirement for IL-12 in natural killer cell IFN-γ production and antiviral defense. *J. Immunol.* **156,** 1138–1142.

Pape, K. A., Khoruts, A., Mondino, A., and Jenkins, M. K. (1997). Inflammatory cytokines enhance the in vivo clonal expansion and differentiation of antigen-activated CD4+ T cells. *J. Immunol.* **159,** 591–598.

Pescovitz, M. D., Sakopoulos, A. G., Gaddy, J. A., Husmann, R., and Zuckermann, F. A. (1994). Porcine peripheral blood CD4+/CD8+ dual expressing T-cells. *Vet. Immunol. Immunopathol.* **43,** 53.

Rabinovich, N. R., McInnes, P., Klein, D. L., and Hall, B. F. (1994). Vaccine technologies: View to the future. *Science* **265,** 1401.

Romani, L., Puccetti, P., and Bistoni, F. (1997). Interleukin-12 in infectious diseases. *Clin. Microbiol. Rev.* **10,** 611–636.

Rouse, B. T., Norley, S., and Martin, S. (1988). Antiviral cytotoxic T lymphocyte induction and vaccination. *Rev. Infect. Dis.* **10,** 16–25.

Saalmüller, A. (1995). Characterization of swine leukocyte differentiation antigens. *Immunol. Today* **17,** 352.

Scherba, G., and Zuckermann, F. A. (1996). Aujeszky's disease (pseudorabies). A natural host-virus model of human herpesviral disease. *In* "Advances in Swine in Biomedical Research" (M. E. Tumbleson and L. B. Schook, eds.), pp. 385–394. Plenum, New York.

Schijns, V. E., Van der Neut, R., Haagmans, B. L., Bar, D. R., Schellekens, H., and Horzinek, M. C. (1991). Tumour necrosis factor-α, interferon-γ, and interferon-β exert antiviral activity in nervous tissue cells. *J. Gen. Virol.* **72,** 809.

Schijns, V. E., Haagmans, B. L., and Horzinek, M. C. (1995). IL-12 stimulates an antiviral type-1 cytokine response but lacks adjuvant activity in IFN-γ-receptor-deficient mice. *J. Immunol.* **155,** 2525–2532.

Schmid, D. E., and Rouse, B. T. (1992). The role of T cell immunity in control of herpes simplex virus. *In* "Herpes Simplex Virus" (B. T. Rouse, ed.), pp. 57–74. Springer-Verlag, Berlin.

Scott, P., and Trinchieri, G. (1997). IL-12 as an adjuvant for cell-mediated immunity. *Semin. Immunol.* **9,** 285–291.

Stellman, C., Vannier, P., Chappuis, G., Brunn, A., Dauvergne, M., Fargeaud, D., Bugand, M., and Colson, X. (1989). The potency testing of pseudorabies vaccines in pigs: A proposal for a quantitative criterion and a minimum requirement. *J. Biol. Stand.* **17,** 17.

Tang, Y. W., and Graham, B. S. (1995). Interleukin-12 treatment during immunization

elicits a T helper cell type 1-like immune response in mice challenged with respiratory syncytial virus and improves vaccine immunogenicity. *J. Infect. Dis.* **172,** 734–738.
Trinchieri, G. (1997). Cytokines acting on or secreted by marcophages during intracellular infection (IL-10, IL-12, IFN-γ). *Curr. Opin. Immunol.* **9,** 17–23.
Trinchieri, G., and Scott, P. (1994). The role of interleukin-12 in the immune response, disease and therapy. *Immunol. Today* **15,** 460.
Tripp, C. S., Wolf, S. F., and Unanue, E. R. (1993). Interleukin-12 and tumor necrosis factor-α are co-stimulators of interferon-γ production by natural killer cells in severe combined immunodeficiency mice with listeriosis, and interleukin-10 is a physiologic antagonist. *Proc. Natl. Acad. Sci. USA* **90,** 3725.
Vannier, P. (1985). Experimental infection of fattening pigs with pseudorabies (Aujeszky's disease) virus: Efficacy of attenuated live- and inactivated-virus vaccines in pigs with or without passive immunity. *Am. J. Vet. Res.* **46,** 1498.
Van Oirschot, J. T. (1978). In vitro stimulation of pig lymphocytes after infection and vaccination with Aujeszky's disease virus. *Vet. Microbiol.* **3,** 255.
Van Oirschot, J. T., and de Leeuw, P. W. (1985). Intranasal vaccination of pigs against Aujeszky's disease. 4. Comparison with one or two doses of an inactivated vaccine in pigs with moderate maternal antibody titers. *Vet. Microbiol.* **10,** 401.
Vogel, F. R., and Powell, F. (1995). A compendium of vaccine adjuvants and excipients. In "Vaccine Design" (M. F. Powell and M. J. Newman, eds.), pp. 141–228. Plenum, New York.
Wardley, R. C., Thomsen, D. R., Berlinski, P. J., Post, L. E., Meyer, A. L., Petrovskis, E. A., and Chester, S. T. (1991). Immune response in pigs to Aujeszky's disease viruses defective in glycoprotein gI or gX. *Res. Vet. Sci.* **50,** 178.
Yoon, K. J., Wu, L. L., Zimmerman, J. J., and Platt, K. B. (1997). Field isolates of porcine reproductive and respiratory syndrome virus (PRRSV) vary in their susceptibility to antibody dependent enhancement (ADE) of infection. *Vet. Microbiol.* **55,** 277–287.
Zinkernagel, R. (1996). Immunology taught by viruses. *Science* **271,** 173–176.
Zuckermann, F. A., and Husmann, R. (1996a). Functional and phenotypical analysis of swine peripheral blood CD4$^+$CD8$^+$ double positive T cells. *Immunology* **87,** 500–512.
Zuckermann, F. A., and Husmann, R. J. (1996b). The human and Swine CD4/CD8 double positive T-lymphocyte. In "Advances in Swine in Biomedical Research" (M. E. Tumbleson and L. B. Schook, eds.), pp. 331–343. Plenum, New York.
Zuckermann, F. A., Mettenleiter, T. C., and Ben-Porat, T. (1988). Role of pseudorabies virus glycoproteins in immune response. In "Vaccination and Control of Aujeszky's Disease" (J. T. Van Oirschot, ed.), pp. 107–117. Kluwer Academic Publishers, Dordrecht, The Netherlands.
Zuckermann, F. A., Zsak L., Mettenleiter, T. C., and Ben-Porat, T. (1990). Pseudorabies virus glycoprotein gIII is a major target anitgen for murine and swine virus-specific cytotoxic T. lymphocytes. *J Virol.* **64,** 802.
Zuckermann, F. A., Husmann, R., Schwartz, R., Brandt, J., Mateu de Antonio, E., and Martin, S. (1998a). Interleukin-12 enhances the virus specific interferon gamma response of pigs to an inactivated pseudorabies virus vaccine. *Vet. Immunol. Immunopathol.* **56,** 57.
Zuckermann, F. A., Husmann, R., Schwartz, R., Brandt, J., and Martin, S. (1998b). Interleukin-12 enhances the cellular immune response and protective immunity induced in pigs by an inactivated herpesvirus vaccine. Submitted for publication.

Swinepox Virus as a Vaccine Vector for Swine Pathogens

DEOKI N. TRIPATHY

Department of Veterinary Pathobiology, University of Illinois, Urbana, Illinois 61802

I. Introduction
II. Conventional Vaccines
III. Recombinant Virus Vectored Vaccines
   A. Vaccinia and Avianpox Viruses
   B. Swinepox Virus as Vaccine Vector
   C. Procedure for Creation of Recombinant Swinepox Virus
   D. Recombinant Swinepox Virus Vaccines
   E. Future of Swinepox Virus Vectored Vaccines
IV. Summary
   Acknowledgments
   References

I. Introduction

Pigs are efficient converters of feed grains into valuable animal protein. As a result the swine industry provides about 25% of the energy and 9% of the protein that human beings obtain from animal sources (Pond, 1983). The industry annually provides about 30–35 kg of high-quality protein for each person in the United States. To increase production efficiency, innovative management practices have been instituted and biologicals are used to reduce disease-related losses. Moreover, swine production has changed from a large number of small farms to a relatively small number of large operations. Because of the intensive nature of these production units, losses due to contagious disease have been magnified, especially those manifested in the respiratory, reproductive, and enteric systems. To prevent diseases, swine are routinely vaccinated against common pathogens that are responsible for significant mortality, morbidity, and reduced weight gain. Some of the live vaccines, used, for example, against the viral diseases, are rotavirus, transmissible gastroenteritis virus, pseudorabies virus, and parvovirus. In spite of the availability of effective vaccines for some diseases, novel pathogens (e.g., the porcine respiratory and reproductive syndrome virus) continue to emerge and some of the attenuated virus (e.g., pseudorabies virus) vaccines can become latent.

Although the economic impact of swine diseases varies, significant losses due to infectious agents still occur. While current information on disease-related losses is not available, according to a 1986 report of the Committee on CSRS Animal Health Research Programs, major disease-related losses attributed to respiratory infections were $400 million, to reproductive disorders were $200 million, and to enteric infections were $214 million annually in 1976. In this regard, three distinct swine diseases that are viral in origin are briefly described next.

Aujeszky's disease is caused by the herpesvirus (pseudorabies virus) and is responsible for significant economic losses to the swine industry. This disease is contagious and is characterized by encephalomyelitis and inflammation of the upper respiratory tract. Mortality can reach 100% in piglets under 2 weeks of age. The respiratory form of the disease is common in growing and adult pigs. In pregnant sows, abortion, mummification of fetuses, or stillbirths can occur depending on the stage of pregnancy. Recovered or subclinically infected pigs continue to shed virus leading to persistent herd infection. The annual cost of pseudorabies for swine producers was more than $21 million in the mid-1980s (Miller *et al.*, 1996). Most vaccines do provide clinical protection against disease but do not prevent shedding or multiplication of the virus. Thus, some animals remain carriers for variable periods and become a source of infection for susceptible animals. Therefore, it is important for vaccination to prevent or reduce virus shedding to the extent that transmission to other susceptible animals is reduced. To attenuate the virus further, several genetically engineered deletion mutants have been developed and evaluated for their ability to reduce virus shedding. The impact of such vaccines is not yet fully known. Several countries are attempting to eradicate pseudorabies virus infection in their swine populations with or without the use of vaccines. However, in certain regions feral swine harboring latent virus can still be a potential source of infection for the domestic swine. Eradication of pseudorabies virus from such a population is practically impossible.

Diarrheal disease is a common and significant problem among neonatal pigs. Economic losses in the United States due to neonatal diarrhea are estimated in excess of $200 million annually. Similar problems are encountered in other countries, such as Australia. Here, Mullan *et al.* (1994) estimated a loss of $260 to $330 per breeding sow in the ensuing 12 months after infection with transmissible gastroenteritis virus (TGEV). According to the National Animal Monitoring System of USDA, TGE cost the pork industry located in Iowa alone $10 million annually in 1987 and 1988 (Hill, 1989). TGEV causes a highly contagious enteric disease affecting pigs of all ages. In case of neonatal pigs, TGE is characterized by severe diarrhea, vomiting, and mortality approaching 100%. This disease is caused by a coronavirus that is shed in feces and nasal secretions. Current vaccines consisting of attenuated or inactivated TGEV are inadequate (Saif and Jackwood, 1989; Saif and Wesley, 1992).

In recent years an economically important emerging pathogen, porcine respiratory and reproductive syndrome (PRRS) virus, has been responsible for significant losses to the swine industry. Clinical signs of the disease range from inapparent infection to severe losses of more than 20% pig production and can occur in all types of swine production systems (Becker and Schwartz, 1996). PRRS virus strains of variable virulence cause reproductive and respiratory tract disease. Modified live virus vaccines against PRRS are available although a considerable amount of evidence suggests that vaccines are clearly not the entire solution to the PRRS problem. A severe form of PRRS recently emerged in Iowa, despite vaccination (Halbur and Bush, 1997).

II. Conventional Vaccines

The greatest triumph in the history of disease prevention and eradication came toward the end of the eighteenth century when Edward Jenner introduced inoculation against smallpox. Jenner's work has led to the development of many human and animal vaccines and the rapid advancement of the sciences of immunology, virology, and vaccinology. Vaccines represent an important tool for the prevention or eradication of diseases. In this regard, the greatest achievement in this century has been the eradication of smallpox.

Prevention of diseases through vaccination has been shown to be extremely beneficial, not only in reducing mortality and morbidity but also in reducing the cost of animal production. In addition to providing protection to vaccinated animals, vaccines also reduce the spread of infection. Due to their relatively low cost, vaccines are popular instruments of disease prevention that offer an important form of investment for the long-range success of the animal industry. The impact of vaccines for the prevention and control of diseases is becoming increasingly recognized and research on all aspects of vaccination has increased considerably in recent years.

The traditional approach to vaccine development includes both modified live (attenuated) and inactivated vaccines. In many instances attenuation has been attempted by serial passage of the virulent virus in an *in vitro* cell culture system. Live attenuated vaccines are developed by trying to establish the balance between maximum immunogenicity and minimum virulence for the host. Because the genetic makeup of such vaccine strains is not known, in many cases there is the risk of reversion to virulence under field conditions. Efficacy of vaccines is usually determined by the protective immune response in the host. Because immunity can be broken by an overwhelming challenge, no vaccine can be claimed as "perfect." Recovery from a natural infection usually leads to a strong and long-lasting protection against reinfection by the same pathogen. Although a vaccine that is completely safe normally does not induce as strong an immunity as the natural infection, to be efficacious it should induce an appreciable degree of protective immunity in the host when exposed to a reasonable natural challenge.

Attenuated modified live vaccines are generally preferred since they provide strong, long-lasting immunity and are more easily produced than inactivated vaccines. However, as mentioned earlier, they often pose the threat of reversion to virulence and transmissibility to other species and often are required to be maintained at a correct passage level. Moreover, the same vaccine produced by different biological manufacturers may have a varied potential for reversion to virulence. Furthermore, it is probable that *in vivo* recombination either between different vaccine strains and/or field strains, for example, pseudorabies virus, will result in the generation of a strain with greater virulence than the original vaccine strain(s). These concerns are unwarranted when using inactivated or subunit vaccines; however, frequent administration is required and, therefore, application of killed vaccines becomes more expensive than the use of live vaccines. Although in designing a vaccine the main emphasis is to protect the host against the disease, it is also important to consider the duration of immunity, lack of any adverse effects, ease of administration, and low cost. The increasing number of vaccines against common and emerging pathogens has made their individual administration impractical. In this regard, the biological companies have realized for a long time the advantages of combined vaccines that will protect against several diseases. Because multi-antigenic vaccines require a relatively lower number of injections, the cost of packaging, storage, delivery equipment, and labor are reduced.

The problems associated with the use of current vaccines can be overcome by the development of a new generation of live vaccines in which only the protective antigen(s) of a pathogen is presented to the immune system of the host and chances of reversion to virulence are eliminated. In such a vaccine the beneficial properties of both live and killed vaccines can be retained.

III. Recombinant Virus Vectored Vaccines

The availability of molecular methods and knowledge enable us to overcome the limitations of traditional approaches in vaccine development. Using modern genetic engineering techniques, it is possible to isolate, identify, and sequence important genes of pathogenic organisms and place them into new vectors in which they can be faithfully expressed. Both bacterial and viral vectors can be used for the expression of foreign genes. Bacteria are easy to manipulate and can provide a high level of expression, but in bacteria glycosylation, proteolytic processing, and subunit assembly of eukaryotic proteins may not occur properly. Consequently, the use of such vectors may not result in the production of an authentic protein(s). Therefore, the genomes of both small and large viruses, for example, baculoviruses, adenoviruses, herpesviruses, and poxviruses, have been manipulated for expression of foreign proteins. The genomes of the large-sized viruses (e.g., poxviruses and herpesviruses) are difficult to alter but have the capacity to express a significant amount of foreign genetic material (i.e., several foreign genes). When considering herpesviruses as vaccine vectors, however, it is important to consider their potential for delayed persistence and oncogenesis. On the other hand the small-sized viruses (e.g., adenoviruses) have the limitation that their genomes can accommodate only a small amount of foreign genetic material without the virus becoming defective in replication.

A. Vaccinia and Avianpox Viruses

Although poxviruses have been of concern for many years by virtue of their impact on human and animal health, the recent increasing interest in these viruses stems from their usefulness as viral vectors

(Moss and Flexner, 1987). Extensive experience obtained with the use of vaccinia virus as a live vaccine, as well as its wide host range, large size genome capable of accommodating a substantial amount of foreign DNA and inability to induce oncogenic transformation have been some of the important features favoring its potential use as a vector for immunization against important pathogens. Additionally, genetically altered vaccinia viruses maintain their infectivity after insertion of foreign genes, and induce both humoral and cellular immunity. Vaccinia vectored vaccines are relatively inexpensive to produce and easy to administer.

Since the first demonstration in 1982 of the ability of vaccinia virus to express an inserted herpes simplex virus thymidine kinase (TK) gene (Panicali and Paoletti, 1982), a large variety of foreign genes have been expressed by recombinant vaccinia viruses. When those genes encode for antigens responsible for eliciting an immune response, the resultant recombinant viruses have been shown to elicit protective immunity in animals against the respective pathogens. Thus, when foreign genes are expressed under vaccinia virus regulation, the proteins are produced in a native state. Because the genome of vaccinia virus has the ability to accept up to 25,000 bp of inserted foreign DNA, more than 10 foreign genes could conceivably be expressed by a single live recombinant virus (Flexner and Moss, 1997). Thus the demonstrated and potential success with vaccinia virus as a gene expression vector has kept its popularity unchallenged for so many years. However, because of its wide host range, there has been reluctance in accepting vaccinia-vectored vaccines due to postvaccination complications in humans (Gurvich, 1992), which were observed during the use of vaccinia for the eradication of smallpox. Consequently, this technology has been applied to host-specific avianpox viruses (e.g., fowlpox and canarypox viruses). Use of such vectors with their greatly restricted replicative ability provides a safety advantage for the host as well as those who come in contact with them.

In spite of limited basic information about the genome of these viruses, remarkable progress has been made in the development of avianpox virus vectored vaccines. Some key events that led to these early successes were (1) the continuous use of live fowlpox virus vaccines by the poultry industry for more than 50 years to prevent fowlpox in chicken and turkeys, (2) the effective substitution of heterologous vaccinia virus promoters in lieu of homologous avianpox virus promoters (Tripathy and Wittek, 1990; Schnitzlein and Tripathy, 1990) in the development of recombinant fowlpox virus vectors, (3) the ability of primary as well as a permanent cell line of avian origin to support the growth of fowlpox virus, and (4) the availability of several genes encoding for specific proteins from poultry pathogens. Consequently, several recombinant fowlpox viruses expressing specific proteins from a variety of avian pathogens were created. Immunization of susceptible birds with such recombinants resulted in the development of specific antibodies and enabled protection to subsequent challenge with the respective virulent pathogen (Tripathy, 1996). These developments have resulted in the licensing of a recombinant fowlpox virus vaccine expressing the hemagglutinin of Newcastle disease virus. Additionally, it has been shown that avianpox viruses (e.g., fowlpox and canarypox viruses) expressing foreign antigens can induce an immune response in mammalian hosts (Taylor *et al.*, 1988, 1991) without causing a productive infection.

B. SWINEPOX VIRUS AS VACCINE VECTOR

Before considering swinepox virus as an expression vector for genes from swine pathogens, it is necessary to mention that several other viruses have been engineered to express genes encoding for protective antigens of swine pathogens. For example, Tuboly *et al.* (1994) showed that baculovirus-expressed spike protein of the transmissible gastroenteritis virus was capable of inducing TGEV-specific antibodies of the IgG class in pigs. A recombinant pseudorabies virus expressing the enveloped glycoprotein E1 of hog cholera virus provided protection against both pseudorabies and hog cholera (classical swine fever) viruses (van Zijl *et al.*, 1991). A highly attenuated strain of vaccinia virus was developed by Tartaglia *et al.* (1992) and used as a vector for the expression of pseudorabies virus proteins, which were evaluated for protection (Brockmeier *et al.*, 1993, 1997; Mengeling *et al.*, 1994).

Successes with vaccinia and avianpox viruses as expression vectors provided further impetus to consider using swinepox virus for the expression of antigens from swine pathogens. Like vaccinia virus, swinepox virus is brick-shaped and its genome consists of a double-stranded DNA of approximately 175 kb (Massung and Moyer, 1991a). It is the only distinct member of the genus *Suipoxvirus* that has worldwide distribution. As with vaccinia and fowlpox virus, modification of swinepox virus into an expression vector is practical due to certain biological attributes. These include its host specificity, mild pathogenicity, thermo stability, and low transmissibility (Tripathy *et al.*, 1981; House and House, 1992; Tripathy, 1993). The restricted replicative ability of swinepox virus with its attenuated nature provides a safety advantage not only to the recipient but also to nonvaccinated contacts.

In the case of modified viral vectors, the main safety concern has been whether the vector is virulent and capable of producing clinical disease in the host. The development, production, and application of a conventional vaccine is based on several factors: its safety, efficacy, and cost as well as the seriousness of the disease. Because swinepox virus infection is so mild and occurs so rarely, a need to develop a vaccine against it was never realized. Only isolated cases of swinepox have been reported in recent years (Olffumi et al., 1981; Borst et al., 1990) and genetic and antigenic differences of those viruses are not known. Unlike vaccinia virus, very limited basic or applied research had been done on this virus in recent years. However, due to the recent interest in poxviruses as vaccine vectors, some studies on the molecular biology of this virus have been conducted (Massung and Moyer, 1991a,b).

Because of the large size of its genome, like vaccinia virus, swinepox virus can accommodate a significant amount of foreign genetic material with the possibility of developing polyvalent vaccines. A cell line of porcine origin that will support the growth of the swinepox virus has been available for many years. As with other poxvirus vectored vaccines, the antigens expressed by swinepox virus should be properly processed and glycosylated. Finally, the recombinant swinepox virus vectored vaccines would be safer than conventional live vaccines since they will only contain a minor portion of the genome of the foreign pathogen.

So far the only nonessential locus that has been identified in the swinepox virus genome is the TK gene (Schnitzlein and Tripathy, 1991). Since studies with other poxviruses including fowlpox virus (Tripathy and Schnitzlein, 1991; Beard et al., 1991) have shown that following insertional inactivation of TK gene, the recombinant viruses become less pathogenic than the parent virus, the resulting recombinant swinepox virus should become less virulent than the parent virus. Thus, the swine industry would benefit from a new generation of recombinant swinepox virus vectored vaccines because of some desirable features of this virus.

C. Procedure for Creation of Recombinant Swinepox Virus

The basic recombinant DNA techniques used to construct vaccinia virus and fowlpox virus recombinants which have been modified and extended to swinepox virus are briefly described below.

The first step in the production of recombinant viruses is to create plasmids that can direct the insertion of the foreign transcriptional unit(s) into the virus genome. Insertion occurs by homologous recombination and thus requires that the foreign DNA be flanked by a contiguous virus genomic region. In the case of vaccinia and fowlpox virus recombinants, the most commonly used insertion site for foreign genes is the TK gene of these viruses. In this regard, the TK gene of swinepox virus has been identified and sequenced (Schnitzlein and Tripathy, 1991; Feller et al., 1991). This gene (TK) is nonessential for virus replication. The development of swinepox virus as a recombinant vector capable of inducing immunity against various swine pathogens requires that the inserted foreign gene(s) be expressed. Because the transcriptional machinery of the virus will not recognize host—cell promoters, foreign genes much be linked to poxvirus promoters.

Because of the unique and conserved nature of poxvirus transcriptional regulatory elements, two defined vaccinia virus promoters (P11 and P7.5) have been predominantly used in the creation of recombinant vaccinia and avianpox viruses. As with the distantly related fowlpox virus (Tripathy and Wittek, 1990; Schnitzlein and Tripathy, 1990), both promoters are also recognized by swinepox virus. Generally the vaccinia virus late P11 promoter is used to regulate transcription of the marker gene, whereas expression of the other gene (encoding the protective antigen) is controlled by the early–late promoter P7.5. These transcriptional units are positioned immediately adjacent to each other and are flanked by virus DNA sequences to ensure insertion into the virus genome.

Once the plasmid has been generated, it is transfected into pig kidney (PK-15) cells that have been previously infected with swinepox virus. Recombinants are generated by homologous recombination between the replicating swinepox virus genomes and the transfected plasmid as shown schematically in Fig. 1. Because more than 99% of the progeny from such an infection retain the parental genome, a procedure for the screening/selection of the recombinant progeny is required. One method of identification utilizes the *Escherichia coli lacZ* gene as a marker. In this case, a chromogenic substrate, 5-bromo--4chloro-3 indolylβ-D-galactosidase (X-gal), which is converted to a blue compound by the action of the expressed enzyme (β-galactosidase) is then used to identify the virus plaques produced by the recombinant virus in the progeny against a background of colorless plaques generated by nonrecombinant viruses. Alternatively, recombinants carrying the *E. coli* xanthine-guanine phosphoribosyl transferase gene (*gpt*) as a marker can be selected due to their resistance to mycophenolic acid

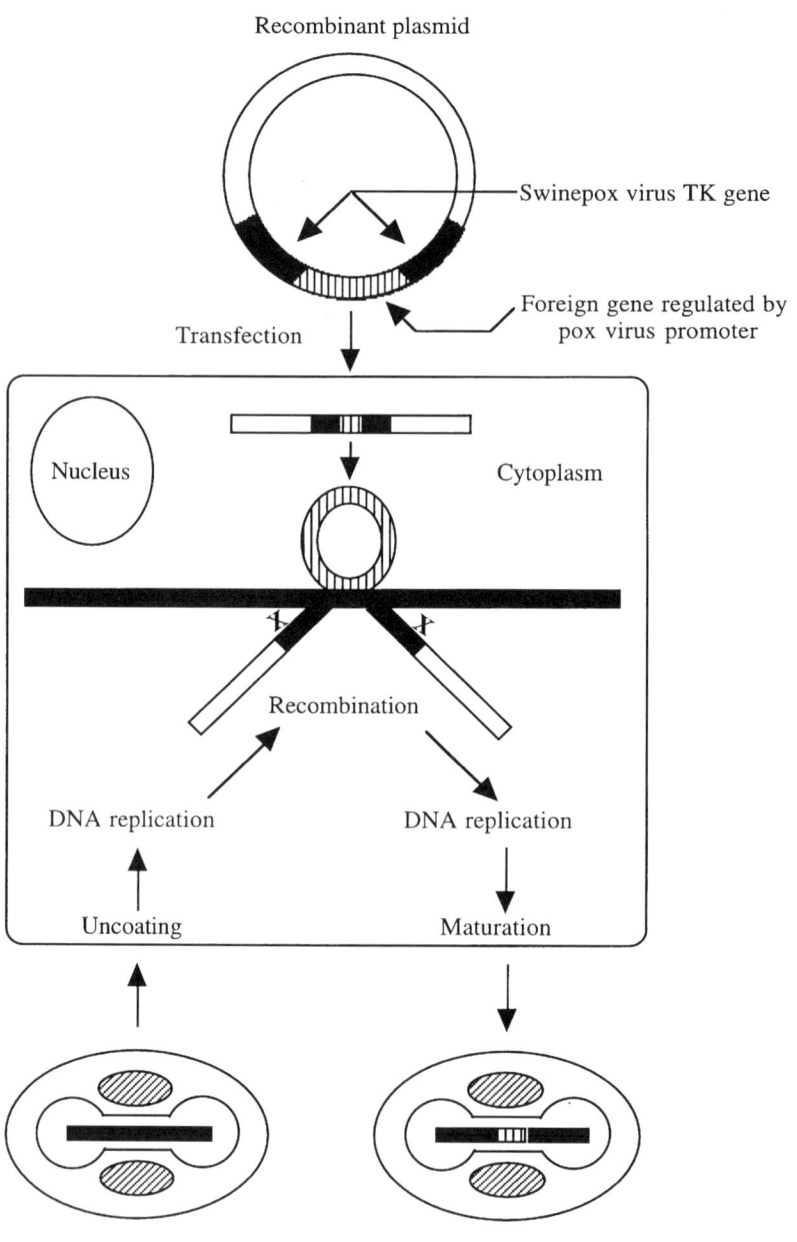

Fig. 1. Schematic representation of generation of recombinant swinepox virus.

(Boyle and Coupar, 1988). In addition, recombinant viruses can be identified by plaque hybridization using a DNA probe specific for the inserted foreign gene. Thymidine kinase-deficient (TK$^-$) cells have been used effectively in the selection of recombinant viruses in which foreign genes were introduced within the TK gene. Since a TK$^-$ cell line of swine origin which will allow selective growth of the recombinant TK$^-$ swinepox virus is not available, inclusion of the marker gene becomes important to facilitate selection/screening of the recombinant virus.

Regardless, of the identification procedure, virus stocks are prepared from plaques after no wild-type virus can be detected in two consecutive rounds of infection. The production of authentic antigen by the recombinant swinepox virus is verified by using specific antibodies against the respective protein(s) in an immunoprecipitation assay or by detecting immunofluorescence on the surface of cells infected with the recombinant viruses (Fig. 2).

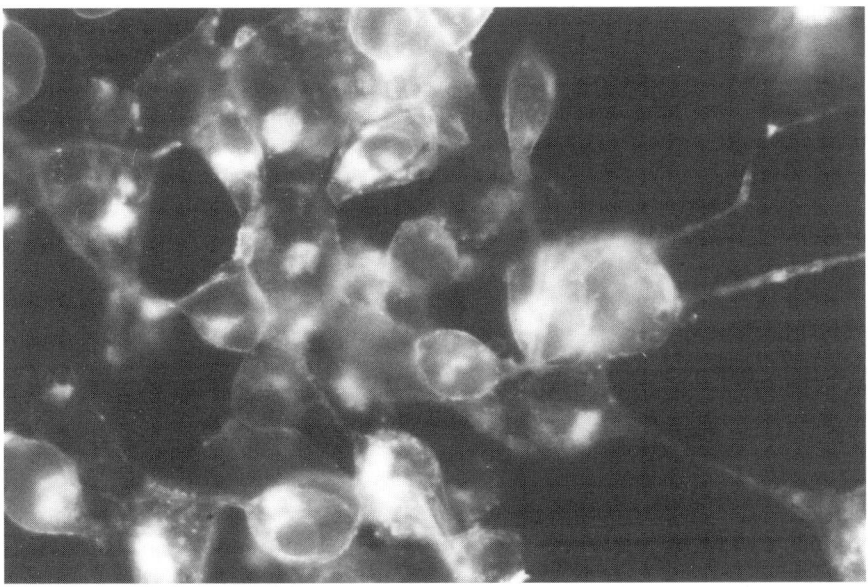

FIG. 2. PK-15 cells infected with recombinant swinepox virus expressing gIII glycoprotein of pseudorabies virus. Specific immunofluorescence is observed following reaction with antibody against gIII glycoprotein of pseudorabies virus.

D. Recombinant Swinepox Virus Vaccines

Using the above-mentioned protocols, recombinant swinepox viruses expressing the gp50 or gIII glycoprotein of pseudorabies virus have been generated (Tripathy et al., 1994). These enveloped glycoproteins were chosen due to their importance in the induction of protective immunity. Of the several pseudorabies virus glycoproteins that have been identified, gIII and gp50 have been shown to induce antibodies that neutralize the virus (Marchioli et al., 1987; Zuckermann et al., 1990) and also provide protection from subsequent lethal challenge (Marchioli et al., 1987). Genes encoding these glycoproteins were separately inserted into the swinepox virus genome.

To evaluate the protective ability of recombinant swinepox viruses expressing pseudorabies virus genes, susceptible pigs were vaccinated with either of the two recombinants or a mixture of both. All animals vaccinated with the recombinant viruses were protected when challenged with virulent pseudorbies virus. In contrast, either mortality occurred or severe clinical disease developed in those swine mock vaccinated or inoculated with unaltered swinepox virus (Tripathy et al., 1994). All animals injected with either the unaltered swinepox virus or recombinant swinepox virus developed local lesions at the site of inoculation which regressed within 7–10 days. No secondary lesions were observed in any of the animals. Moreover, the primary lesions resulting from recombinant swinepox virus infection regressed faster than those produced by the parental unmodified swinepox virus. This phenomenon has been observed with other poxviruses in which their TK gene has been insertionally inactivated. Further evidence of protection by recombinant swinepox virus expressing glycoproteins gp50 and gp63 of pseudorabies virus has been reported by van der Leek et al. (1994).

These studies indicate that the swinepox virus vectored recombinant vaccines are capable of expressing the foreign genes to a level that will induce protective immunity. The host specificity of such recombinants should favor their use as ideal immunizing agents for swine.

E. Future of Swinepox Virus Vectored Vaccines

Vaccination against diseases is carried out to limit the economic losses caused by mortality and morbidity and growth retardation in infected animals. The ultimate aim, however, if possible, is to eradicate the disease by regular use of attenuated or inactivated vaccines. In spite of the regular use of conventional vaccines, significant losses do still occur. Therefore, new or improved vaccines are needed for many current and emerging infectious swine diseases which contribute to unnecessary mortality and morbidity and significant economic loss.

A new approach to vaccine development is the expression of genes of foreign pathogens using live attenuated viral vaccines as vectors. In this regard, large size viruses such as pox and herpesviruses have been very promising. However, as mentioned earlier, herpesviruses have the potential of reactivation from latency resulting in recurrent infections and possible shedding of the virus. It is possible that in certain instances *in vivo* recombination of different vaccine strains may result in the generation of recombinants with virulence greater than that of involved vaccine strains. Vaccinia virus vectored vaccines, on the other hand, have proved to be highly successful immunizing agents, but regular application of such live recombinant vaccinia viruses has been controversial because of minor complications associated with the vaccination program against smallpox. Therefore, the development of live vaccines that are self-restricted, which cannot be transmitted from vaccinated animals to contact animals, and which are not released into the environment would be ideal. Limited experimental studies with TK-inactivated swinepox virus indicate that such live swinepox virus monovalent or polyvalent vaccine vectors can be created.

Lack of adequate local immunity by parenterally administered vaccines against those pathogens whose portal of entry is respiratory or gastrointestinal tract has been realized for a long time. Indeed vaccinia vectored vaccines administered parenterally stimulate protective levels of serum IgG antibody and cellular immunity but do not induce mucosal immunity associated with IgA antibody. Because oral immunization with recombinant vaccinia virus containing the rabies glycoprotein gene provided protection against rabies (Rupprecht *et al.,* 1986), field trials are being conducted with an oral bait for vaccinating wild animals for the control of rabies. In this case, immunization probably occurs due to viral replication in the tonsils. Safety and efficacy of oral vaccination of raccoons by feeding raccoonpox virus containing rabies virus glycoprotein in sponge-baits has been described by Esposito *et al.* (1988). Enteric immunization of mice with recombinant vaccinia virus containing the influenza virus hemagglutinin gene induced mucosal IgA antibody, serum IgG antibody, and cell-mediated immunity (Meitin *et al.,* 1994). Similarly, mice immunized intragastrically with recombinant vaccinia virus containing the hemagglutinin and nucleoprotein genes of influenza virus were protected against influenza (Bender *et al.,* 1996).

In natural cases of swinepox in neonatal piglets, lesions in the mouth and respiratory tract have been observed indicating that the virus can replicate in sites other than skin. Additionally, antibodies in the sera and antibodies of IgA class in the intestinal contents of swine infected orally with swinepox virus have been observed providing further evidence of virus multiplication after oral administration (Tuboly *et al.*, 1993). This study suggests that swinepox virus can be a potential vector for the expression of antigens from enteric and respiratory viral agents of swine. Use of an oral vaccine for enteric pathogens would not only be convenient for mass administration through the drinking water but less expensive than conventional vaccines. Additionally, swinepox virus could be used in the construction of a multivalent recombinant vaccine because of its ability to accommodate a large amount of foreign genetic material. Alternatively, combined recombinant vaccines expressing different antigens could be used to protect against multiple pathogens. Whether oral or intranasal administration of recombinant swinepox virus vaccine expressing antigens of enteric/respiratory pathogens would induce a desirable protective immune response needs to be determined. If such an effective recombinant vaccine is ever developed, the swine industry would benefit tremendously.

Like avianpox viruses, swinepox virus has never been isolated from any other host than the pig. Interestingly, host-restricted recombinant avianpox viruses have been shown to express foreign antigens in mammalian hosts without producing a productive infection. In a similar manner, if recombinant swinepox viruses containing foreign antigens were able to express them optimally in hosts other than swine without a productive infection, this virus could become a potential candidate for vectored vaccines for a wide range of species including man.

IV. Summary

Several small and large viruses (e.g., adenovirus, poxvirus, and herpesviruses) have been investigated as vaccine vectors. Each viral system has its advantages and disadvantages. One major advantage for viral vector vaccines is their ability to elicit a protective cell-mediated immunity as well as a humoral response to the antigen delivered by the vector. One major problem to using recombinant viruses as vaccines is the pathogenic potential of the parent virus. Therefore, it is important that along with the optimal expression of the foreign genes and ability to provide protection, the pathogenicity of the vector virus must be reduced during genetic manipulation without affecting its multiplication.

The requirements to develop a viral vector, for example, swinepox virus, are a cell culture system that will support the growth of the virus, a suitable nonessential region(s) in the virus genome for insertion of foreign DNA so that virus replication is not affected, a foreign gene(s) that encodes for an immunogenic protein of a swine pathogen, strong transcriptional regulatory elements (promoters) necessary for optimal expression of the foreign genes, a procedure for delivering the foreign gene(s) into the nonessential locus, and a convenient method of distinguishing the recombinant viruses from the parent wild-type virus.

Using this methodology, recombinant swinepox virus vaccines expressing pseudorabies virus antigens have been developed and shown to provide protection against challenge. These studies and evidence of local infection of the oral tract by swinepox virus indicate its potential as a recombinant vector for providing immunity against various swine pathogens including those that infect the respiratory and gastrointestinal tracts.

ACKNOWLEDGMENTS

The author would like to thank D. William M. Schnitzlein for helpful suggestions and for Fig. 1.

REFERENCES

Beard, C. W., Schnitzlein, W. M., and Tripathy, D. N. (1991). Protection of chickens against highly pathogenic avian influenza (H_5N_2) virus by recombinant fowlpox viruses. *Avian Dis.* **35,** 356–359.

Becker, B. J., and Schwartz, K. J. (1996). "Survey of Porcine Reproductive and Respiratory Syndrome," Winter, pp. 69–78. Iowa State University Veterinarian, Ames.

Bender, B. S., Rowe, C. A., Taylor, S. F., Wyatt, L. S., Moss, B., and Small, P. A., Jr. (1996). Oral immunization with a replication-deficient recombinant vaccinia virus protects mice against influenza. *J. Virol.* **70,** 6418–6424.

Borst, G. H. A., Kimman, T. G., Gielkens, A. L. J., and van der Kamp, J. S. (1990). Four sporadic cases of congenital swinepox. *Vet. Rec.* **127,** 61–63.

Boyle, D. B., and Coupar, B. E. H. (1988). A dominant selectable marker for the construction of recombinant poviruses. *Gene* **65,** 123–128.

Brockmeier, S. L., Larger, K. M., Tartaglia, J., Riviere, M., Paoleti, E., and Mengeling, W. L. (1993). Vaccination of pigs against pseudorabies with highly attenuated vaccinia (NYVAC) recombinant viruses. *Vet. Microbiol.* **38,** 41–58.

Brockmeier, S. L., Larger, K. M., and Mengeling, W. L. (1997). Successful pseudorabies vaccination in maternally immune piglets using recombinant vaccinia virus vaccines. *Res. Vet. Sci.* **62,** 281–285.

Committee on CSRS Animal Health Research Programs, Board of Agriculture, National Research Council (1986). "Animal Health Research Programs of the Cooperative State

Research Service, Strengths, Weaknesses, and Opportunities," p. 26. National Academy Press, Washington, DC.
Esposito, J., Knight, J. C., Shaddock, J. H., Novembre, F. J., and Baer, G. M. (1988). Successful oral rabies vaccination of raccoons with raccoon poxvirus recombinants expressing rabies virus glycoprotein. *Virology* **165,** 313–316.
Feller, J. A., Massung, R. F., Turner, P. C., Gibbs, E. P. J., Bockamp, E. O., Beloso, A., Talavera, A., Vinuela, E., and Moyer, R. W. (1991). Isolation and molecular characterization of the swinepox virus thymidine kinase gene. *Virology* **183,** 578–585.
Flexner, C., and Moss, B. (1997). Vaccinia virus as a live vector for expression of immunogens. *In* "New Generation of Vaccines" (M. M. Levine, G. C. Woodrow, J. B. Kaper, and G. S. Cobon, eds.), (2nd rev. expanded ed.), pp. 297–314. Dekker, New York.
Gurvich, E. B. (1992). The age-dependent risk of postvaccination complications in vaccines with smallpox vaccine. *Vaccine* **10,** 96–97.
Halbur, P. G., and Bush, E. (1997). Update on abortion storms and sow mortality. *Rep.-Swine Health Prod.* **5,** 73.
Hill, H. T. (1989). Preventing epizootic TGE from becoming enzootic TGE. *Vet. Med.* **84,** 432–436.
House, J. A., and House, C. A. (1992). *In* "Diseases of Swine" (A. D. Leman, B. E. Straw, W. L. Mengeling, S. D'Allaire, and D. J. Taylor, eds.), 7th ed., pp. 358–361. Iowa State Univ. Press, Ames.
Marchioli, C. C., Yancey, R. J., Petrovskis, E. A., Timmins, J. G., and Post, L. E. (1987). Evaluation of pseudorabies virus glycoprotein gp50 as a vaccine for Aujeszky's disease in mice and swine: Expression by vaccinia virus and Chinese hamster ovary cells. *J. Virol.* **61,** 3977–3982.
Massung, R. F., and Moyer, R. W. (1991a). The molecular biology of swinepox virus. I. A characterization of the viral DNA. *Virology* **180,** 347–354.
Massung, R. F., and Moyer, R. W. (1991b). The molecular biology of swinepox virus. II. The infectious cycle. *Virology* **180,** 355–364.
Meitin, C. A., Bender, B. S., and Small, P. A., Jr. (1994). Enteric immunization of mice against influenza with recombinant vaccinia. *Proc. Natl. Acad. Sci. USA* **91,** 11187–11191.
Mengeling, W. L., Brockmeier, S. L., and Larger, K. M. (1994). Evaluation of a recombinant vaccinia virus containing pseudorabies (PR) virus glycoprotein genes gp50, gII and gIII as a PR vaccine for pigs. *Arch. Virol.* **134,** 259–269.
Miller, G. Y., Tsai, J. S., and Foster, D. L. (1996). Benefit-cost analysis of the national pseudorabies virus eradication program. *J. Am. Vet. Med. Assoc.* **208,** 208–213.
Moss, B., and Flexner, C. (1987). Vaccinia virus expression vectors. *Annu. Rev. Immunol.* **5,** 305–324.
Mullan, B. P., Davies, G. T., and Cutler, R. S. (1994). Simulation of the economic impact of transmissible gastroenteritis on commercial pig production in Australia. *Aust. Vet. J.* **71,** 151–154.
Olffumi, B. E., Ayoade, G. O., Ikede, B. O., and Akpavie, S. O. (1981). Swinepox in Nigeria. *Vet. Rec.* **109,** 278–280.
Panicali, D., and Paoletti, E. (1982). Construction of poxviruses as cloning vectors: Insertion of the thymidine kinase gene from herpes simplex virus into the DNA of infectious vaccinia virus. *Proc. Natl. Acad. Sci. USA* **79,** 4927–4931.
Pond, W. G. (1983). Modern pork production. *Sci. Am.* **248,** 96–103.
Rupprecht, C. E., Wiktor, T. J., Johnston, D. N., Hamir, A. N., Dietzchold, B., Wunner,

W. H., Glickman, L. T., and Koprowski, H. (1986). Oral immunization and protection of raccoons (Procyon lotor) with a vacciniarabies glycoprotein recombinant virus vaccine. *Proc. Natl. Acad. Sci. USA* **83,** 7947–7950.

Saif, L. J., and Jackwood, D. J. (1989). Enteric virus vaccines: Theoretical considerations, current status, and future approaches. *In* "Viral Diarrheas of Man and Animals" (L. J. Saif and K. Theil, eds.), pp. 313–329. CRC Press, Boca Raton, FL.

Saif, L. J., and Wesley, (1992). Transmissible gastroenteritis. *In* "Diseases of Swine" (A. D. Leman, B. E. Straw, W. L. Mengeling, S. D'Allaire, and D. J. Taylor, eds.), 7th ed., pp. 362–386. Iowa State University Press, Ames.

Schnitzlein, W. M., and Tripathy, D. N. (1990). Utilization of vaccinia virus promoters by fowlpox virus recombinants. *Anim. Biotechnol.* **2,** 164–174.

Schnitzlein, W. M., and Tripathy, D. N. (1991). Identification and nucleotide sequences of the thymidine kinase gene of swinepox virus. *Virology* **181,** 727–732.

Tartaglia, J., Perkus, M. E., Taylor, J., Norton, E. K., Audonnet, J.-C., Cox, W. I., Davis, S. W., Van der Hoeven, J., Meigner, B., Riviere, M., Languet, B. and Paoletti, E. (1992). NYVAC: A highly attenuated strain of vaccinia virus. *Virology* **188,** 217–232.

Taylor, J., Weinberg, R., Languet, B., Desmetire, P., and Paoletti, E. (1988). Recombinant fowlpox virus inducing protective immunity in non-avian species. *Vaccine* **6,** 497–503.

Taylor, J., Trimarchi, C., Weinberg, R., Languet, B., Guillemin, F., Dsmettre, P., and Paoletti, E. (1991). Efficacy studies on a canarypox-rabies recombinant virus. *Vaccine* **9,** 190–193.

Tripathy, D. N. (1993). Swinepox. *Curr. Vet. Ther.* **3,** 482–483.

Tripathy, D. N. (1996). Fowlpox virus vectored vaccines for control of poultry diseases. *Proc. World's Poult. Cong., 20th,* New Delhi, India, Vol. II, pp. 497–503.

Tripathy, D. N., and Schnitzlein, W. M. (1991). Expression of avian influenza hemagglutinin gene by recombinant fowlpox virus. *Avian Dis.* **35,** 186–191.

Tripathy, D. N., and Wittek, R. (1990). Regulation of foreign gene in fowlpox virus by a vaccinia virus promoter. *Avian Dis.* **34,** 218–220.

Tripathy, D. N., Hanson, L. E., and Crandell, R. A. (1981). Poxviruses of veterinary importance, diagnosis of infections. *In* "Comparative Diagnosis of Viral Diseases" (E. Krustak, ed.), Chapter 6, pp. 267–346. Academic Press, New York.

Tripathy, D. N., Schnitzlein, W. M., and Zuckermann, F. A. (1994). Recombinant swinepox viruses expressing pseudorabies virus genes protect swine against virulent pseudorabies virus. *Abstr. Int. Conf. Proxviruses and Iridoviruses, 10th,* Banff, Alberta, Canada, p. 191, April 30–May 5, 1994.

Tuboly, T., Nagy, E., and Derbyshire, J. B. (1993). Potential viral vectors for the stimulation of mucosal antibody responses against enteric viral antigens in pigs. *Res. Vet. Sci.* **54,** 345–350.

Tuboly, T., Nagy, E., Dennis, J. R., and Derbyshire, J. B. (1994). Immunogenicity of the S protein of transmissible gastroenteritis virus expressed in baculovirus. *Arch. Virol.* **137,** 55–67.

van der Leek, M. L., Feller, J. A., Sorensen, G., Isaacson, W., Adams, C. L., Borde, D. J., Pfeiffer, N., Tran, T., Moyer, R. W., and Gibbs, E. P. J. (1994). Evaluation of swinepox virus as a vaccine vector in pigs using an Aujeszky's disease (pseudorabies) virus gene insert coding for glyoproteins gp50 and gp63. *Vet. Rec.* **134,** 13–18.

van Zijl, M., Wensvoort, G., deKluyver, E., Hulst, M., van der Gulden, H., Gaielkens, A., Berns, A., and Moorman, R. (1991). Live attenuated pseudorabies virus expressing envelope glycoprotein E1 of hog cholera virus protects swine both against pseudorabies and hog cholera. *J. Virol.* **65,** 2761–2765.

Zuckermann, F. A., Zsak, L., Mettenleiter, T. C., and Ben-Porat, T. (1990). Pseudorabies virus glycoprotein III is a major target antigen for murine and swine virus-specific cytotoxic T lymphocytes. *J. Virol.* **64,** 802–812.

VII
POULTRY VACCINES

Introduction to Poultry Vaccines and Immunity

J. M. SHARMA

Veterinary PathoBiology, College of Veterinary Medicine, University of Minnesota, St. Paul, Minnesota 55108

I. Introduction
II. Disease Prevention by Vaccination
III. Vaccination Programs
IV. Vaccine Delivery
V. New Developments in Poultry Vaccines
VI. General Features of the Avian Immune System
VII. Humoral Immunity
VIII. Cell-Mediated Immunity
IX. Summary
References

I. Introduction

Poultry constitutes a vital segment of world agriculture. Yearly global production exceeds 30 billion birds. This number is expected to increase steadily in the foreseeable future. In the United States, poultry is a $20 billion industry, with a yearly production of more than 8 billion birds. About 63% of the total value of U.S. poultry is attributed to broiler production. Other significant segments of the industry include egg laying chickens and turkeys (Table I). More recently, ratites, principally ostriches, have entered the commercial market, and although their numbers are quite small, they may increase in due course.

Yearly production of large populations of birds creates a number of management challenges. A vast majority of commercial birds are raised under the system of intensive rearing. In this system, large numbers of birds, often flocks containing tens of thousands of birds, are

TABLE I

COMMERCIAL POULTRY IN THE UNITED STATES[a]

Bird	Number produced per year (millions)	Feed conversion	Market age or productive age	Value ($ billions)
Broiler chickens	7300	1.8	45 days	13
Layer chickens	294	2.15 (=1 lb eggs)	52 weeks	4
Turkeys	293	2.4	14–21 weeks	3

[a]Data obtained from 1995–1996 summaries of the U.S. Department of Agriculture, National Agriculture Statistics Service.

placed under one roof in closed houses. This type of housing places birds in proximity to each other and increases the risk of rapid spread of infectious disease. Thus, poultry producers must constantly vaccinate birds to minimize the threat of disease outbreaks. Proper management of flock health by biosecurity and vaccination is one of the critical factors in profitable poultry production.

In this article, I give a brief overview of the commonly used poultry vaccines. In addition, I share some of the general features of the avian immune system that are relevant to the host response to vaccines.

II. Disease Prevention by Vaccination

Birds are susceptible to numerous viral, bacterial, fungal, and parasitic diseases (Calnek et al., 1997). Exposure of susceptible birds to infectious agents may result in clinical disease and death. Many agents tend to cause subclinical infections associated with immunosuppression and poor flock performance. Disease-associated stress and reduced flock performance are a common cause of economic loss to the industry.

Certain viral and bacterial agents are endemic in poultry-producing areas and tend to cause recurring infections in commercial flocks. Of particular interest are certain respiratory viruses such as Newcastle disease virus (NDV), infectious bronchitis virus (IBV), and infectious laryngotracheitis virus (ILTV); neoplastic disease of chickens such as Marek's disease virus (MDV) and avian leukosis viruses; and immunosuppressive viruses such as infectious bursal disease virus (IBDV), infectious anemia virus, and hemorrhagic enteritis virus (HEV). Flocks must be routinely protected against these common pathogens. The most common method of protection is vaccination.

Both active and passive immunization are practiced to control diseases in poultry. Active immunization using live vaccines is the current industry standard although passive immunization is also used in certain special circumstances. For example, passive immunization is widely used to protect young chicks against IBD (Lukert and Saif, 1997). IBDV is an economically important, immunosuppressive virus that is endemic in most poultry-producing areas in the world. Newly hatched chickens are highly susceptible to the immunosuppressive effects of IBDV (Allan et al., 1972; Faragher et al., 1974; Pertile and Sharma, 1998). Thus early post-hatch protection against the virus is critical. The protection is achieved by hyperimmunizing hens so that protective levels of maternal antibody are transmitted to progeny chickens. The disadvantage of passive immunization is that it is difficult to determine the optimum time for actively immunizing a flock that has chickens with varying levels of residual maternal antibody.

Table II shows some of the commonly used vaccines in poultry. Most live vaccines are either mild isolates that induce a protective immune response against their pathogenic counterparts or pathogenic agents that have been attenuated.

TABLE II

Commonly Used Poultry Vaccines

Vaccine	Species vaccinated
Marek's disease	Chickens
Newcastle disease	Chickens, turkeys
Infectious bronchitis	Chickens
Infectious bursal disease	Chickens
Infectious laryngotracheitis	Chickens
Fowl pox	Chickens
Avian encephalomyelitis	Chickens
Reovirus	Chickens
Hemorrhagic enteritis	Turkeys
Cholera	Chickens, turkeys
Bordetellosis	Turkeys

III. Vaccination Programs

The selection of appropriate vaccines and the regimen of their use vary widely among commercial flocks. Many factors must be considered in designing a vaccination program. These factors include flock history, endemic infectious agents, proximity to other birds, age and genetic background of the birds, health status of the parent flock, the level of biosecurity being practiced, and other management and environmental influences.

Table III shows the vaccination program being used in three selected flocks of broiler chickens, layer chickens, and turkeys. The data in Table III are based on specific flocks. The vaccination program shown should not be considered optimum for all flocks.

As can be seen from the data in Table III, chickens and turkeys may need to be vaccinated repeatedly against certain agents. The most commonly used vaccines elicit a strong primary immune response in unexposed birds. This response eventually wanes, leaving behind immunologic memory which can be boosted by subsequent vaccinations or environmental exposure to the same agent. Certain vaccines elicit a

TABLE III

VACCINATION PROGRAMS BEING PRACTICED IN THREE SPECIFIC FLOCKS OF POULTRY

Birds being vaccinated	Vaccine	Age when vaccine is administered
Broiler chickens	Marek's disease	*In ovo*
	Infectious bronchitis and Newcastle disease	Day 1, day 14
	Infectious bursal disease	Day 21
Layer chickens	Marek's disease	Day 1
	Infectious bursal disease	Week 2, week 6, week 12
	Newcastle disease and infectious bronchitis	Week 2, week 6, week 12
	Infectious laryngotracheitis and fowlpox	Week 12
	Mycoplasma gallisepticum	Week 15
Turkeys	Newcastle disease	Day 1 (recombinant), week 3, week 8, week 12, week 16
	Hemorrhagic enteritis	Week 4

lifelong immunity following a single administration. For example, a single exposure to the Marek's disease vaccine virus initiates a persistent infection and a lifelong immunity.

IV. Vaccine Delivery

Vaccines are administered to eggs, that is, *in ovo* vaccination, or to chicks after they have hatched. Ease of administration of a vaccine is an important requisite for considering the vaccine desirable for poultry. Because large populations of animals must be vaccinated, the most beneficial vaccines are those that can be delivered simultaneously to large numbers of birds with minimum amount of labor. Post-hatch delivery systems include aerosol, spray, drinking water, eye drop, and injection.

In ovo vaccination is a relatively new concept in vaccine administration. This method of vaccine delivery has replaced post-hatch injection of Marek's disease vaccine in broiler chickens. In *in ovo* vaccination, the vaccine is injected in eggs during later stages of embryonation, usually at 17–18 days of incubation (Sharma and Burmester, 1982). There are several advantages of *in ovo* vaccine delivery over parenteral application of vaccines in hatched birds. Because the developing embryo is exposed to an immunizing dose of the vaccine virus while still in the egg, the vaccinal protection is well established by the time the chick hatches and is first exposed to environmental pathogens. *In ovo* vaccination also substantially reduces the labor cost associated with individual handling of chicks in the conventional vaccination procedure. The development of multiple-head egg injection machines has facilitated simultaneous inoculation of large numbers of eggs, up to 50,000 eggs per hour (Gildersleeve *et al.*, 1993). The *in ovo* injection machines can deliver precise quantities of vaccines into each individual egg and the injection needles are automatically cleaned and disinfected between injections.

The initial work that established the concept of safely introducing live viral vaccines into embryonated eggs was done with the Marek's disease vaccine (Sharma and Burmester, 1982). This vaccine continues to be the principal vaccine currently delivered via the *in ovo* technology in commercial hatcheries. Successful *in ovo* use of vaccines against a number of other disease agents or mixtures of agents has also been demonstrated under laboratory conditions (Sharma and Witter, 1983; Sharma, 1985; Wakenell and Sharma, 1986; Fadly and Nazerian,

1989; Ahmad and Sharma, 1992, 1993; Stone, 1994; Sarma *et al.*, 1995; Reddy *et al.*, 1996; Karaca *et al.*, 1998).

V. New Developments in Poultry Vaccines

Vaccine development is an area of active research. The advent of molecular technology has resulted in the generation of a number of recombinant poultry vaccines. Live fowlpox virus (FPV) and turkey herpesvirus (HVT) have been used most extensively as vectors with inserts of genes of immunogenic proteins from a number of avian agents including NDV, MDV, IBDV, ILTV, avian influenza, reticuloendotheliosis virus, turkey rhinotracheitis, and hemorrhagic enteritis. Most of these recombinants express the immunogenic genes and are protective under laboratory conditions. One recombinant, FPV containing HN and F gene inserts of NDV has been licensed by the U.S. Department of Agriculture and is available commercially (Boyle and Heine, 1993; Nazerian *et al.*, 1992). The recombinant is not being extensively used in chickens because anti-NDV antibodies present at the time of vaccination tend to interfere with the recombinant's protective efficacy against NDV. However, certain commercial turkey flocks have used this recombinant with good success. Commercial availability of additional recombinant vaccines appears imminent. Vectors other than FPV and HVT are also being tested (Vakharia *et al.*, 1993; Jenkins *et al.*, 1991; Curtiss, 1990) as are chemical and deletion mutants, immunogenic subunits, naked DNA, and antigen–antibody complexes (Curtiss and Kelly, 1987; Lillehoj and Trout, 1993; Whitfill *et al.*, 1995; Robinson *et al.*, 1993; Jackwood *et al.*, 1995; Ahmad and Sharma, 1993). "Designer" vaccines that have matching immunogenic peptides with major histocompatibility complex (MHC) haplotypes for optimum antigen presentation to immune cells are also being considered (Witter and Hunt, 1994; Thacker *et al.*, 1995).

There is much current interest in using adjuvants to enhance the immune response against vaccines and thus enhance vaccine efficacy. Genetic cloning and expression of several avian cytokines in recent years have made cytokines attractive candidates as adjuvants. Cytokines are soluble proteins produced by a variety of cells. These proteins play a critical regulatory role in normal immunologic and certain physiologic functions. Therapeutic administration of extraneous cytokines has been shown to enhance vaccinal immunity (Anderson *et al.*, 1987; Weinberg and Merigan, 1988; Nunberg *et al.*, 1989; McCullough *et al.*, 1992; Heath and Playfair, 1992; Balkwill, 1993; Reddy *et al.*, 1993;

Blecha *et al.*, 1995; Gao *et al.*, 1995; Karaca *et al.*, 1998) and to qualitatively improve the nature of protective immunity (Afonso *et al.*, 1994). Recently, several FPV constructs coexpressing NDV and type I interferon (IFN) genes have been developed and we have examined one such construct *in vivo* (Karaca *et al.*, 1998). We studied the response of chickens hatching from eggs inoculated at embryonation day 17 with rFPV-NDV-IFN or rFPV-NDV. Several interesting observations were made. *In ovo* administration of the virus/cytokine recombinant induced protective immunity against NDV. Although the presence of IFN reduced anti-NDV antibody levels in chickens, there were no apparent quantitative differences in protection between the cytokine-containing and cytokine-lacking constructs. Most notable was the observation that chickens hatching from eggs injected with FPV-NDV had significantly lower body weight at 2 weeks of age than unvaccinated controls. This loss in body weight was not detected in chickens that hatched from eggs inoculated with rFPV-NDV-IFN. This result indicated that the cytokine may have reduced the stress associated with replication of a live FPV vector while maintaining protective efficacy.

VI. General Features of the Avian Immune System

Among the avian species, the immune system of the chicken has been studied most extensively and will be briefly discussed here. The reader is referred to other reviews for additional information (Toivanen and Toivanen, 1987; Sharma, 1991, 1997; Davison *et al.*, 1996; Vainio and Imhof, 1996; Pastoret *et al.*, 1998). Available data indicate that the basic mechanisms of immunity are shared by all birds. The overall organization and functions of the immune system are similar in chickens and mammals.

The primary lymphoid organs in the chicken include the bursa of Fabricius and the thymus. The bursa is of particular interest because this organ is unique to birds. An equivalent organ does not exist in mammals. Bursa is a saclike structure located in the region of the hind gut. B-cell differentiation takes place in the bursa. The thymus gland, which is a bilateral, multilobular structure that spans the cervical and thoracic areas, is the primary lymphoid organ for T-cell differentiation. Although birds lack lymph nodes, the secondary lymphoid system is well developed and lymphoid tissue is scattered through the body (Fletcher and Barnes, 1998). The important secondary lymphoid structures include the Harderian gland, bone marrow, and conjunctival-associated, gut-associated, head-associated, and bronchial-associated lymphoid tissues. In addition, a number of visceral organs have diffusely scattered lymphoid cells that tend to proliferate during an immune reaction.

The development of the immune system begins early during embryogenesis. Bursal precursor cells can be detected in the embryo at around 7 days of embryonation. Cells expressing surface IgM, IgG, and IgA can be detected at 10, 14, and 16 days of embryonation respectively. The T-cell precursors enter the thymus in three closely regulated waves, one wave each at 6.5, 12, and 18 days (Coltey et al., 1989). Each wave lasts for about 2 days and cells of each wave are able to differentiate into $\alpha\beta$ or $\gamma\delta$ cells. T cells with surface CD3 molecules appear in the embryo at 9 days and those with T-cell receptor (TCR) at 12 days of embryonation. Because the late-stage embryos are able to respond immunologically to antigens, it is possible to immunize chickens or turkeys by *in ovo* vaccination.

VII. Humoral Immunity

Surface immunoglobulins serve as receptors for antigen recognition by B cells (Kincade and Cooper, 1971). Because birds have an extremely limited number of Ig genes, immunoglobulin diversity is attained by the process of gene conversion. In this process, segments of pseudo genes are inserted into the V_L and V_H regions of the genome (McCormack and Thompson, 1990).

Upon stimulation with an antigen, responsive B cells differentiate into plasma cells that secrete antibodies specific to the antigen. The successful "take" of a vaccine is often monitored by demonstrating a rise in antibody titer within a few days of vaccination. Birds produce three classes of antibodies: IgM, IgG, and IgA. Primary antibody response is initiated by the development of IgM antibody. Subsequently, IgG and IgA are produced. The mechanism of antibody class switching is not known although, as in mammals, the switch is probably mediated by cytokines. Although avian and mammalian IgG have similar biological functions, the avian IgG molecule is longer than its mammalian counterpart and lacks a genetically encoded hinge. Because of these differences, avian IgG is sometimes referred to as IgY. IgA is considered critical for local immunity in the respiratory and intestinal tracts. In birds, IgA is transported to the liver and stored in the bile.

A number of serologic procedures have been developed to assay antibodies in birds. These assays include enzyme-linked immunosorbent assay (ELISA), gel defusion, agglutination, hemagglutination, serum neutralization, immunofluorescence, and Western blotting. The development of ELISA has greatly facilitated serologic monitoring of commercial flocks.

VIII. Cell-Mediated Immunity

T-lymphocytes are the principal cells of cell-mediated immunity. The avian TCR is a multichain complex and consists of the TCR chains that recognize the antigen and CD3 complex that is important for signal transduction. The antigen binding sites of the TCR complex are formed by glycoprotein chains designated as TCRα and TCRβ ($\alpha\beta$ T cells) and TCRγ and TCRδ ($\gamma\delta$ T cells). In the adult chicken, $\gamma\delta$ T cells, detectable by TCR 1 monoclonal antibodies (Chen et al., 1991), constitute about 20–50% of the circulating T cells. These cells are also present in the spleen and the intestinal epithelium.

The constant region of the TCR molecule is encoded by a single nonpolymorphic gene whereas three gene fragments encode the variable region, which is composed of V, J, and D segments. As in mammals, the avian T-cell diversity is generated by somatic recombination, imprecise joining of the V, J, and D elements, and by combining different TCRα and TCRβ or TCRγ and TCRδ chains (Tjoelker et al., 1990; Gobel et al., 1994).

There are two functional subsets of T cells: CD4 helper cells and CD8 cytotoxic/suppressor cells. The functions of $CD4^+$ and $CD8^+$ cells are MHC restricted. The mechanisms of $CD4^+$ effector cell functions in chickens are not well elucidated. Isolation and biological characterization of avian cytokines will facilitate the study of helper T-cell functions.

Although T-cell-mediated immune responses are important in vaccine-induced protective immunity in birds, assay procedures to quantitate these responses are cumbersome and not routinely used. Live viral vaccines have been shown to elicit detectable cytotoxic T-cell responses (Maccubin and Scheirman, 1986; Omar and Schat, 1996; Seo and Collisson, 1997). Cultured immune T helper cells when exposed to specific antigens may proliferate and secrete cytokines (Karaca et al., 1996). Mitogen proliferation assays are commonly used to assess non-antigen-specific immune responsiveness in birds. Mitogen proliferation can be conducted in isolated lymphoid cells or whole blood (Hovi et al., 1978; Lee, 1978; Sharma and Belzer, 1992).

Natural killer cells and the cells of antibody-dependent cellular cytotoxicity constitute important cells of the innate immunity in chickens. The NK-cell activity that can be quantitated by ^{51}Cr-release cytotoxicity assays using tumor cell targets has been studied in some detail

(Sharma and Schat, 1991). Vaccination with Marek's disease vaccines up-regulates the NK-cell activity in chickens (Sharma, 1981; Heller and Schat, 1987).

IX. Summary

The poultry industry constitutes a significant sector of world agriculture. In the United States, more than 8 billion birds are produced yearly with a value exceeding $20 billion. Broiler chickens are the largest segment of the industry. Birds raised under commercial conditions are vulnerable to environmental exposure to a number of pathogens. Therefore, disease prevention by vaccination is an integral part of flock health management protocols. Active immunization using live vaccines is the current industry standard. Routinely used vaccines in chickens include MDV, NDV, IBV, and IBDV, and in turkeys NDV and HEV. Newer vaccines, including molecular recombinants in which genes of immunogenic proteins from infectious agents are inserted into a live viral vector, are also being examined for commercial use. Efforts are under way to enhance vaccine efficacy by the use of adjuvants, particularly cytokines. The vaccine delivery systems include *in ovo* injection, aerosol, spray, drinking water, eye drop, and wing web injection. The *in ovo* vaccination procedure is relatively new and at the present time it is used primarily to vaccinate broiler chickens against MDV. Birds respond to vaccines by developing humoral and cellular immune responses. Bursa of Fabricius and the thymus serve as the primary lymphoid organs of the immune system. B cells use surface immunoglobulins as antigen receptors and differentiate into plasma cells to secrete antibodies. Three classes of antibodies are produced: IgM, IgG (also called IgY), and IgA. Successful vaccinal response in a flock is often monitored by demonstrating a rise in antibody titer within a few days of vaccination. ELISA is used most commonly for serologic monitoring. T cells are the principal effector cells of specific cellular immunity. T cells differentiate into $\alpha\beta$ and $\gamma\delta$ cells. In adult birds, $\gamma\delta$ cells may constitute up to 50% of the circulating T cells. Functionally, $CD4^+$ cells serve as helper cells and $CD8^+$ cells as cytotoxic/suppressor cells.

References

Afonso, L. C. C., Scharton, T. M., Vieira, L. Q., Wysocka, M., Trinchieri, G., and Scott, P. (1994). The adjuvant effect of interleukin-12 in a vaccine against *Leishmania major*. Science **263**, 235–237.

Ahmad, J., and Sharma, J. M. (1992). Evaluation of a modified-live virus vaccine administered *in ovo* to protect chickens against Newcastle disease. *Am. J. Vet. Res.* **53**, 1999–2004.

Ahmad, J., an Sharma, J. M. (1993). Protection against hemorrhagic enteritis and Newcastle disease in turkeys by embryo vaccination with monovalent and bivalent vaccines. *Avian Dis.* **37**, 485–491.

Allan, W. H., Faragher, J. T., and Cullen, G. A. (1972). Immunosuppression by the infectious bursal agent in chickens immunized against Newcastle disease. *Vet. Rec.* **90**, 511–512.

Anderson, G., Urbano, O., Fedorka-Cray, P., Nunberg, J., and Doyle, M. (1987). Interleukin 2 and protective immunity in *Haemophilus pleuropneumonia*. Preliminary studies. *In* "Vaccines, 87," pp. 22–25. Cold Spring Harbor Press, Cold Spring Harbor, NY.

Balkwill, F. (1993). Cytokines in health and disease. *Immunol. Today* **14**, 149–150.

Blecha, F., Reddy, D. N., Chitkomekown, C. G., Movey, D. S., Chengappa, M. M., Goodband, R. D., and Nelssen, J. L. (1995). Influence of recombinant bovine interleukin-1 beta and interleukin-2 in pigs vaccinated and challenged with *Streptococcus suis*. *Vet. Immunol. Immunopathol.* **44**, 329–346.

Boyle, D. B., and Heine, H. G. (1993). Recombinant fowlpox virus vaccine for poultry. *Immunol. Cell Biol.* **71**, 391–397.

Calnek, B. W., Barnes, H. J., Beard, C. W., McDougald, L. R., and Saif, Y. M., eds. (1997). "Disease of Poultry," 10th ed. Iowa State University Press, Ames.

Chen, C. H., Pickel, J. M., Lahti, J. M., and Cooper, M. D. (1991). Surface markers on avian immune cells. *In* "Avian Cellular Immunology" (J. M. Sharma, ed.), pp. 1–22. CRC Press, Boca Raton, FL.

Coltey, M., Bucy, R. P., Chen, C. H., Cihak, J., Losch, U., Char, D., Le Douarin, N. H., and Cooper, M. D. (1989). Analysis of the first two waves of thymus homing stem cells and their T cell progeny in chick-quail chimeras. *J. Exp. Med.* **170**, 543–557.

Curtiss, R. (1990). Attenuated *Salmonella* strains as live vectors for the expression of foreign antigens. *In* "New Generation Vaccines" (G. C. Woodrow and M. M. Levine, eds.), pp. 161–188. Dekker, New York.

Curtiss, R., and Kelly, S. M. (1987). *Salmonella typhimurium* deletion mutants lacking adenylate cyclase and cyclic AMP receptor proteins are avirulent and immunogenic. *Infect. Immun.* **55**, 3035–3043.

Davison, T. F., Morris, T. R., and Payne, L. N. (1996). "Poultry Immunology." Carfax Publishing Company, Abingdon, Oxfordshire.

Fadly, A. N., and Nazerian, K. (1989). Hemorrhagic enteritis of turkeys: Influence of maternal antibodies and age at exposure. *Avian Dis.* **33**, 778–786.

Faragher, J. T., Allan, W. T., and Wyeth, C. J. (1974). Immunosuppressive effect of infectious bursal agent on vaccination against Newcastle disease. *Vet. Rec.* **95**, 385–388.

Fletcher, O. J., and Barnes, H. J. (1998). Lymphoid organs and their anatomical distribution. *In* "Handbook of Vertebrate Immunology" (P. P. Pasoret, H. Bazin, A. Govererts, and P. J. Griebel, eds.). Academic Press, San Diego, CA (in press).

Gao, Y., Daley, M., and Splitter, G. A. (1995). BHV-1 glycoprotein 1 and recombinant interleukin 1 beta efficiently elicit mucosal IgA response. *Vaccine* **13**, 871–877.

Gildersleeve, R. P., Hoyle, C. M., Miles, A. M., Murray, D. L., Ricks, C. A., Secrest, M. N., Williams, C. J., and Womack, C. L. (1993). Developmental performance of an egg injection machine for administration of Marek's disease vaccine. *J. Appl. Poult. Res.* **2**, 337–346.

Gobel, T. W. F., Chen, C. H., Lahti, J., Kubota, T., Kuo, C., Aebersold, R., and Cooper,

M. D. (1994). Identification of T cell receptorα-chain genes in the chicken. *Proc. Natl. Acad. Sci. USA* **91,** 1094–1098.

Heath, A. W., and Playfair, J. H. L. (1992). Cytokines as immunological adjuvants. *Vaccine* **10,** 427–434.

Heller, E. D., and Schat, K. A. (1987). Enhancement of natural killer cell activity by Marek's disease vaccines. *Avian Pathol.* **16,** 51–60.

Hovi, T., Suni, J., Hortling, L., and Vaheri, A. (1978). Stimulation of chicken lymphoytes by T- and B-cell mitogens. *Cell. Immunol.* **39,** 70–78.

Jackwood, M. L., Hilt, D. A., and Webster, R. G. (1995). Vaccination of SPF leghorn chickens with a plasmid vector containing the S1 glycoprotein gene of infectious bronchitis virus. *Proc. 132nd Annu. Meet. Am. Vet. Med. Assoc.,* Pittsburgh, PA, pp. 143.

Jenkins, M. C., Castle, M. D., and Danforth, H. D. (1991). Protective immunization against the intestinal parasite *Eimeria acervulina* with recombinant coccidia antigen. *Poult. Sci.* **70,** 539–547.

Karaca, K., Kim, I. J., Reddy, S. K., and Sharma, J. M. (1996). Nitric oxide inducing factor as a measure of antigen, and mitogen specific T cell responses in chickens. *J. Immunol. Methods* **192,** 97–103.

Karaca, K., Sharma, J. M., Winslow, B. J., Junker, D. E., Reddy, S., Cichran, M., and McMillen, J. (1998). Recombinant fowlpox viruses coexpressing chicken type 1 IFN and Newcastle disease virus HN and F genes: Influence of IFN on protective efficacy and humoral responses of chickens following *in ovo* or post-hatch administration of recombinant viruses. *Vaccine* (in press).

Kincade, P. W., and Cooper, M. D. (1971). Development and distribution of immunoglobulin containing cells in the chicken: An immunofluorescent analysis using purified antibodies μγ and light chains. *J. Immunol.* **106,** 371–382.

Lee, L. F. (1978). Chicken lymphocyte stimulation by mitogens A microassay with whole-blood cultures. *Avian Dis.* **22,** 296–307.

Lillehoj, H. S., and Trout, J. M. (1993). Coccidia: A review of recent advances on immunity and vaccine development. *Avian Pathol.* **22,** 331.

Lukert, P. H., and Saif, Y. M. (1997). Infectious bursal disease. *In* "Diseases of Poultry" (B. W. Calnek, H. J. Barnes, C. W. Beard, L. R. McDougald, and Y. M. Saif, eds.), 10th ed., pp. 721–738. Iowa State University Press, Ames.

Maccubin, D. L., and Scheirman, L. W. (1986). MHC restricted cytotoxic response of chicken T cells: Expression, augmentation, and clonal characteristics. *J. Immunol.* **136,** 12–16.

McCormack, W. T., and Thompson, C. B. (1990). Somatic diversification of the chicken immunoglobulin light chain genes. *Adv. Immunol.* **48,** 41–61.

McCullough, K. C., Pullen, L., and Parkinson, D. (1992). The immune response against foot-and-mouth disease virus: Influence of the T lymphocyte growth factors IL-1 and IL-2 on the murine humoral response *in vivo*. *Immunol. Lett.* **31,** 41–46.

Nazerian, K., Lee, L. F., Yanagida, N., and Ogawa, R. (1992). Protection against Marek's disease by a fowlpox virus recombinant expressing the glycoprotein B of Marek's disease virus. *J. Virol.* **66,** 1402–1408.

Nunberg, J., Doyle, M. V., York, S. M., and York, C. J. (1989). Interleukin 2 acts as an adjuvant to enhance the potency of inactivated rabies virus vaccine. *Proc. Natl. Acad. Sci. USA* **86,** 4240–4243.

Omar, A. R., and Schat, K. A. (1996). Syngeneic Marek's disease virus (MDV)-specific cell mediated immune responses against immediate, early, late, and unique MDV proteins. *Virology* **222,** 87–99.

Pastoret, P. P., Bazin, H. Govererts, A., and Griebel, P. J. eds. (1998). "Handbook of Vertebrate Immunology." Academic Press, San Diego, CA.

Pertile, T. L., and Sharma, J. M. (1998). Effect of viruses on immune functions. In "Handbook of Vertebrate Immunology" (P. M. Pastoret, H. Bazin, A. Govererts, and P. J. Griebel, eds.), pp. 73–136. Academic Press, San Diego, CA.

Reddy, D. N., Reddy, P. G., Xue, W., Minocha, H. C., Daley, M. J., and Blecha, F. (1993). Immunopotentiation of bovine respiratory disease virus vaccines by interleukin-1 beta and interleukin-2. Vet. Immunol. Immunopathol. **37,** 25–38.

Reddy, S. K., Sharma, J. M., Ahmad, J., Reddy, D. N., McMillen, J., Cook, S., Wild, M., and Schwartz, R. (1996). Protective efficacy of a recombinant herpesvirus of turkeys as an in ovo vaccine against Newcastle and Mareks disease in specific pathogen free chickens. Vaccine **14,** 469–477.

Robinson, H. L., Hunt, L. A., and Webster, R. G. (1993). Protection against a lethal influenza virus challenge by immunization with hemagglutination-expressing plasmid DNA. Vaccine **11,** 957–960.

Sarma, G., Greer, W., Gildersleeve, R. P., Murray, D. L., and Miles, A. M. (1995). Field safety and efficacy of in ovo administration of HVT+SB-1 bivalent Mareks disease vaccine in commercial broilers. Avian Dis. **39,** 211–217.

Seo, S. H., and Collisson, E. W. (1997). Specific cytotoxic T lymphoytes are involved in in vivo clearance of infectious bronchitis virus. J. Virol. **71,** 5173–5177.

Sharma, J. M. (1981). Natural killer cell activity in chickens exposed to Marek's disease virus: Inhibition of activity in susceptible chickens and enhancement of activity in resistant and vaccinated chickens. Avian Dis. **25,** 882–893.

Sharma, J. M. (1985). Embryo vaccination with infectious bursal virus alone or in combination with Marek's disease vaccine. Avian Dis. **29,** 1155–1169.

Sharma, J. M., ed. (1991). "Avian Cellular Immunology." CRC Press, Boca Raton, FL.

Sharma, J. M. (1997). The structure and function of the avian immune system. Acta. Vet. Hung. **45,** 229–238.

Sharma, J. M., and Belzer, S. W. (1992). Blatogenic response of whole blood cells of turkeys to a T-cell mitogen. Dev. Comp. Immunol. **16,** 77–84.

Sharma, J. M., and Burmester, B. R. (1982). Resistance to Marek's disease at hatching in chickens vaccinates as embryos with the turkey herpes virus. Avian Dis. **26,** 134–149.

Sharma, J. M., and Schat, K. A. (1991). Natural immune functions. In "Avian Cellular Immunology" (J. M. Sharma, ed.), pp. 51–70. CRC Press, Boca Raton, FL.

Sharma, J. M., and Witter, R. L. (1983). Embryo vaccination against Marek's disease with serotypes 1, 2, and 3 vaccines administered singly or in combination. Avian Dis. **27,** 453–463.

Stone, H. D. (1994). In ovo vaccination of chicken embryos with experimental Newcastle disease virus oil emulsion vaccine. Proc. 131st Ann. Meet. Am. Vet. Med. Assoc., San Francisco.

Thacker, E. L., Fulton, J. E., and Hunt, H. D. (1995). In vitro analysis of a primary, major histocompatibility complex (MHC)-restricted, cytotoxic T-lymphocyte response to avian MHC class I cDNA inserted into a recombinant ALV vector. J. Virol. **69,** 6439–6444.

Tjoelker, L. W., Carlson, M., Lee, K., Lahti, J., McCormack, W. T., Leiden, J. M., Chen, C. H., Cooper, M. D., and Thompson, C. B. (1990). Evolutionary conversation of antigen recognition: the chicken T cell receptor β-chain. Proc. Natl. Acad. Sci. USA **87,** 7856–7860.

Toivanen, A., and Toivanen, P. eds. (1987). "Avian Immunology: Basis and Practice," Vols. I and II. CRC Press, Boca Raton, FL.

Vainio, O., and Imhof, B. A., eds. (1996). "Immunology and Developmental Biology of the Chicken," Springer-Verlag, Berlin.
Vakharia, V. N., Snyder, D. B., He, J., Edwards, G. H., Savage, P. K., and Mengel-Whereat, S. A. (1993). Infectious bursal disease virus structural proteins expressed in a baculovirus recombinant confer protection in chickens. *J. Gen. Virol.* **74,** 1–6.
Wakenell, P. S., and Sharma, J. M. (1986). Chicken embryonal vaccination with avian infectious bronchitis virus. *Am. J. Vet. Res.* **47,** 933–938.
Weinberg, A., and Merigan, T. C. (1988). Recombinant interleukin 2 as an adjuvant for vaccine-induced protection. Immunisation of guinea pigs with herpes simplex virus subunit vaccines. *J. Immunol.* **140,** 294–298.
Whitfill, C. E., Haddad, E. E., Ricks, C. A., Skeeles, J. K., Newberry, L. A., Beasley, J. N., Andrews, P. D., Thoma, J. A., and Wakenell, P. S. (1995). Determination of optimun formation of a novel infectious bursal disease antibody with IBDV. *Avian Dis.* **39,** 687–699.
Witter, R. L., and Hunt, H. D. (1994). Poultry vaccines of the future. *Poult. Sci.* **73,** 1087–1093.

In Ovo Vaccination Technology

C. A. RICKS, A. AVAKIAN, T. BRYAN, R. GILDERSLEEVE,
E. HADDAD, R. ILICH, S. KING, L. MURRAY, P. PHELPS,
R. POSTON, C. WHITFILL, AND C. WILLIAMS

*EMBREX, Inc., P.O. Box 13989, Research Triangle Park,
North Carolina 27709*

I. Introduction
II. Technology Discovery
III. Commercialization of *In Ovo* Marek's (HVT/SB1) Vaccination
    A. Automated Egg Injection Equipment
    B. Field Efficacy
    C. Process Optimization
IV. Commercialization of Other *In Ovo* Live Viral Vaccines
    A. Infectious Bronchitis Virus
    B. Newcastle Disease Virus
    C. Infectious Bursal Disease (IBDV, Gumboro Disease)
    D. Fowlpox
V. Maternal Antibody Effects on Viral Vaccine Efficacy
VI. Safe and Effective Vaccination in Presence of Maternal Antibodies
    A. Bursaplex™, Infectious Bursal Disease Virus Antibody Complex Vaccine
    B. Other Virus–Antibody Complex Vaccines
    C. Mechanism of Action
VII. Bacterial Vaccines
VIII. Coccidial Vaccines
IX. Summary
    References

I. Introduction

In 1992, the first automated egg injection system, the Inovoject® system, was introduced into the United States. This provided the poultry industry with the ability to vaccinate birds *in ovo,* that is, in the

---

The authors are employees and consultants of Embrex, Inc., which disclaims responsibility for any private publication of its personnel. The views expressed herein do not necessarily reflect the views of Embrex, Inc. or its management.

egg. Today, more than 80% of the U.S. broiler industry vaccinate their birds *in ovo* against Marek's disease, thereby replacing the conventional method of vaccinating newly hatched birds by subcutaneous injection. Benefits of *in ovo* vaccination compared to post-hatch vaccination include earlier immunity, reduction in bird stress, precise and uniform injection, reduced labor costs, and reduced contamination. *In ovo* vaccination technology is now being introduced into the European, Asian, and Latin American markets. This paper reviews (1) the discovery and commercialization of the *in ovo* vaccination process for prevention of Marek's disease; (2) the utility of this technology for use with other viral vaccines typically used in the industry such as Newcastle disease, infectious bronchitis, and infectious bursal disease (IBDV); (3) issues relating to maternal antibody inactivation of viral vaccines in young birds or embryos; (4) the use of viral neutralizing factor technology to facilitate safe and effective vaccination *in ovo*; and (5) the use of bacterial and protozoal vaccines *in ovo*. The results cited in this paper are based on the limited trials described. They are not determinative of future results or the results that would be experienced in any specific hatchery.

II. Technology Discovery

Marek's disease is an ubiquitous neoplastic disease caused by a highly contagious herpesvirus. In 1982 Sharma and Burmester demonstrated that chickens vaccinated with Marek's herpesvirus of turkey (HVT serotype 3) as embryos were better protected to a 3-day neonatal Marek's (MDV) challenge than those vaccinated at hatch. When the challenge was administered to birds at 7 days of age, protection was equivalent irrespective of route of vaccination. This was the first demonstration that *in ovo* vaccination might provide earlier protection to a Marek's field challenge. Sharma and Witter (1983) extended these observations to show that Marek's disease virus vaccines of serotypes 1 and 2 administered to 18-day-old embryonated eggs also induced better protection against post-hatch challenge at 3 days with virulent MDV than vaccines given at hatch. Sharma et al. (1984) further demonstrated that HVT administered at day 18 of incubation rapidly infected the embryos and replicated to a high titer in the embryonic lung. This early viral replication presumably provided the means for the embryo to establish an appreciable degree of immunity by the time of hatch.

III. Commercialization of *in ovo* Marek's (HVT/SB1) Vaccination

Prior to 1992, Marek's vaccination was typically accomplished by manual vaccination of each newly hatched bird by the subcutaneous route employing a bivalent HVT/SB1 (serotypes 3 and 2, respectively) vaccine. In some countries an attenuated serotype 1 virus was also employed. Following the demonstration by Sharma and Burmester in 1982 that embryonal vaccination was possible, scientists and engineers at Embrex undertook the development of an automated egg injection and transfer system, the Inovoject®, capable of high-speed inoculation of eggs and eliminating the need for manual post-hatch injection. It was expected that the benefits of using such a system would be reduced labor costs and healthier chicks due to the earlier development of immunity and the reduced stress associated with the elimination of manual handling and injection.

A. Automated Egg Injection Equipment

The Inovoject is an automated vaccinating system that requires two operators, one loading incubated eggs held within their respective flats onto a load conveyor for injection and a second to remove injected eggs that have been transferred into the hatching basket. The process is accomplished during normal transfer from incubators to hatcheries and typically conducted between 17.5 and 18.5 days of incubation. The machine is designed in a modular fashion to accommodate a variety of egg flat configurations ranging from 36 to 150 eggs per injection and to withstand the harsh environment of the hatchery.

Operation of the equipment begins with manually loading flats of eggs onto the main conveyor system. The eggs are moved by the conveyor to a controlled position under the injection head where the eggs are sensed and the injection cycle begins. At that time the injection head, carrying anywhere from 36 to 150 tooling injectors, is lowered onto the flat of eggs. Egg penetration is accomplished by a series of steps which include punching a hole in the top of the egg followed by introduction of a needle through the punch mechanism. The injectable solution is then delivered to the egg through the needle by a peristaltic pump in predetermined volumes of 50–200 μl. Following completion of the injection process, the egg flat is carried by conveyor to the transfer operation while the next flat requiring injection is brought into position under the injection head. Simultaneously with the conveying of the flat of eggs to the transfer operation and for the next injection cycle, a sanitation cycle is initiated. This process step consists of flushing a buffered sanitizing solution through the injection tooling assembly to sanitize the injection needle and punch prior to the next injection.

Eggs which have been injected and moved by conveyor onto the transfer table are positioned and accumulated under the vacuum transfer head. The transfer head is designed to lift and separate the eggs from their flats and then gently deposit the eggs into a hatching basket. After the eggs are lifted up and separated from the flats, the empty hatching basket is put into position by the second Inovoject system operator. The basket is then sensed, as in position and ready to receive eggs, by the basket sensor. As the eggs are being unloaded into the hatching basket the empty egg flats are removed by the operator or automatically ejected from the end of the transfer system depending on the needs of the hatchery. The loaded hatching basket is then removed and replaced by a second empty hatching basket to repeat the cycle.

B. Field Efficacy

Based on the work of Sharma and Burmester (1982) it was expected that, as mentioned earlier, automated egg vaccination would result in reduced labor costs and healthier chicks due to earlier protection to a Marek's field challenge and the reduced stress associated with the elimination of manual handling and injection. In a series of eight field trials conducted between 1989 and 1992 at two integrated broiler companies in North Carolina using 1.3 million eggs per treatment group, consistent improvements in feed conversion and reduced settlement costs were observed in the Inovoject system compared to the conventionally (at hatch) Marek's vaccinated group (Gildersleeve et al., 1993; Gildersleeve and Fluke, 1995). Hatchability, 2-week mortality, livability to time of processing, and live body weight were similar in the two treatment groups. Leukosis and total condemnations were similar in the two treatment groups. Sarma and colleagues (1995) reported on the field safety and efficacy of a bivalent (HVT and SB1) vaccine administered *in ovo* compared to conventional (at hatch) vaccination. Overall livability and feed efficiency tended to be improved in the *in ovo* treatment groups and total condemnations and condemnations due to air sacculitis and septicemia/toxemia were lower. No treatment differences in leukosis condemnation rates were observed. Birds challenged with a very virulent strain of MD-V (RB1/B) were protected irrespective of vaccination route. In both studies leukosis rates were low indicating a low level of Marek's challenge. Under these conditions *in ovo* vaccination is equally as effective as the conventional post-hatch method of vaccine administration. Additional controlled studies under more severe exposures are needed to determine whether *in ovo* vaccination is more efficacious than conventional vaccination.

C. Process Optimization

Process control is very important when introducing the *in ovo* vaccination process into a hatchery. Particular attention must be paid to controlling mold levels in the hatchery environment, using strict aseptic techniques for vaccine preparation, and ensuring that the vaccine is introduced into eggs at the correct stage of incubation.

1. Aspergillus and Egg Injection

Aspergillus fumigatus is the most common etiologic agent of aspergillosis, however *Aspergillus flavus* and *Aspergillus niger* have been isolated. Aspergillosis is the most common fungal disease of poultry and most often occurs in a pulmonary form. Disease in broiler chickens is characterized by an acute onset, usually within 10 days of placement. The hatchery or moldy litter or grain at the growout farm is most often the source of extensive exposure to the *Aspergillus* organisms that cause disease. "Brooder pneumonia," as it is often called, is characterized generally with high mortality and morbidity. Clinical signs include gasping, cyanosis, dyspnea, and accelerated breathing. Improperly maintained hatchery conditions have been shown to affect a hatchery adversely after *in ovo* injection.

For optimal utilization of *in ovo* technology, the hatchery must address biosecure air flow patterns within its building (positive and negative pressures, clean to dirty direction), routine air handling equipment maintenance and sanitation, the elimination of biofilms on equipment, moisture management (floors, humidifiers, drains, etc.), and effective sanitation and disinfection of all eggs handling equipment in order to control *Aspergillus* levels. The hatchery has the additional task of enforcing the quality of product allowed in the system. This can only be accomplished through a coordinated managerial effort including microbiological assessment from breeder farm through the hatchery, to the broiler farm.

2. Vaccine Preparation for the Inovoject Egg Injection System

Correct preparation of vaccine for vaccination by the Inovoject egg injection system is mandatory. No single set of procedures is more important to the process of egg injection. Preparation must ensure sterility of the vaccine through aseptic techniques and minimal trauma to cell-associated vaccine (Marek's), thereby maintaining the highest possible titer or plaque forming units (pfu). Vaccine used in the Inovoject system must be sterile.

Cell-associated or "live" Marek's vaccine ampules are frozen and stored in liquid nitrogen. The ampules must be thawed and broken open for usage. Diluent should be stored at room temperature. Any diluent that is discolored or cloudy should be discarded. Marek's vaccine is available in 1000 and 2000 dose ampules, with different types including serotype 3 (HVT), serotype 2 (SB-1), serotype 1 (CVI-988, Rispens), and combinations of serotypes 2 and 3, (HVT and SB-1). Marek's vaccine should be completely used within 1 hour of hanging on the machine. Prepared vaccine should be used within 1.5 hours from the time the first ampule is thawed. Following the Inovoject system setup cleaning cycle, sterile saline is aseptically attached to the vaccine delivery system, and forced through the system. Vaccine can then be introduced to the Inovoject vaccine delivery system.

3. Day of Injection

The 18th day of incubation is the recommended time for Inovoject system vaccination and transfer of eggs from the setter to the hatcher. However, commercial broiler practices sometimes result in the need to transfer and inject on either day 17 or 19 of incubation. Unpublished observations (P. Phelps) described later suggest hatchability is depressed if eggs are transferred on day 17 instead of day 18 of incubation, regardless of whether eggs are vaccinated or not. Future studies are needed to determine the effect of transfer and injection on day 19 of incubation.

Trials conducted at three different commercial hatcheries suggested hatchability of eggs transferred but not injected on day 18 of incubation averaged 1.4% better than eggs transferred on day 17. The number of eggs evaluated was 632,880 and 772,686 for days 17 and 18, respectively. If eggs were injected at the time of transfer, the hatchability depression seen on day 17 versus day 18 was 1.73%. The number of injected eggs evaluated was 600,480 and 772,880 for days 17 and 18, respectively. Thus the hatchability depression due to transfer on day 17 versus day 18 was increased by only 0.33% if eggs were injected. Therefore, the majority of the hatchability depression seen when eggs are vaccinated by the Inovoject system on day 17 instead of day 18 is due to earlier transfer not egg injection. These studies were not controlled for flock, incubator, or season but based on multiple hatches of large numbers of eggs.

Hatchability differences between eggs injected on days 17 and 18 for 14 separate flocks at a commercial hatchery were evaluated over a 2-month period. Hatchability was increased by 1.18% if eggs were injected on day 18 of incubation instead of day 17.

If eggs must be vaccinated by the Inovoject system on day 17 instead of day 18 due to hatchery schedules, injection as late on day 17 as possible is preferred. Data collected at a commercial hatchery suggested hatchability decreases of 2.20% if eggs were injected on day 17 instead of day 18 but decreases of only 0.97% if eggs were injected on day 17.5 instead of day 18.5. Studies were not controlled for season, breeder flock, or incubator but based on multiple observations of large numbers of eggs.

4. Upside Down Eggs

Prior to transfer, hatching eggs are incubated vertically with the large or air cell end up. This orientation ensures that the embryo develops with its head toward the large end of the egg, enabling it to emerge through the air cell and hatch successfully. Sometimes eggs are mistakenly placed in the setter trays upside down or with the pointed end up. Bauer and coworkers (1990) found the incidence of eggs set upside down in a commercial hatchery varied between 0.3 and 3.4%. Setting eggs upside down results in increased embryonic malpositions and reduced hatchability (Byerly and Olsen, 1931). Hatchability decreases between 10 and 32% have been reported for eggs set upside down (Bauer *et al.*, 1990; Cain and Abbott, 1971; Talmadge, 1977; El-Ibiary *et al.*, 1966). Muller and Williams (1975) also reported that setting eggs upside down decreased chick quality.

The impact of vaccination by the Inovoject system on hatchability and chick quality of eggs set upside down was not known. Moreover the advantage or disadvantage of turning eggs which were set upside down, right side up before injection is unknown. Unpublished studies of P. Phelps and S. Bryan were conducted to address these two issues.

Hatchability depressions of eggs set upside down ranged between 10 and 18.5%. These data agree with the literature previously cited. Vaccination by the Inovoject system resulted in additional hatchability decreases of 2.6% in trial 1 and 6.0% in trial 2, however, these differences were not significant. If one assumes only 0.5% of all eggs set are set upside down then the impact of vaccination by the Inovoject system would only decrease total hatchability by 0.215%. Vaccination by the Inovoject system had no increased effect on the 1-week mortality of chicks that hatched from eggs set upside down but did have a slight numerical decrease on 1-week body weight. The decreased body weights were due to an increase in the number of cull or unthrifty chicks.

Experiments were also conducted to determine if eggs set upside down should be turned right side up before Inovoject system vaccination. Embryonated eggs that have been set upside down often bleed when injected, however experimental results indicated eggs should be left upside down because righting the eggs before injection further decreased hatchability by 8%.

5. Moisture Loss

Optimum hatchability is achieved when hatching eggs lose between 11 and 14% of their preincubation weight as moisture (Mauldin and Wilson, 1988). Egg weight or moisture loss is an indirect measure of embryo oxygen uptake. Weight loss is not constant during incubation. The final stages of embryo development require increased oxygen. Other factors that affect egg weight loss during incubation include egg size, shell porosity, relative humidity, incubation temperature, and egg storage time and conditions (Wineland, 1996). The impact of Inovoject system vaccination, specifically the hole left in the egg following injection, on moisture loss during incubation is minimal. Inovoject system vaccination increases total egg moisture loss by 0.30% if eggs are injected on day 17 of incubation and by 0.15% if eggs are injected on day 18 of incubation (P. Phelps, unpublished data). This slight change in moisture loss is not thought to impact hatchability or chick quality.

6. Rispens

CVI 988/Rispens-type vaccines are typically administered *in ovo* when field MD challenges exceed the level of protection afforded by HVT alone or in combination with serotype 2 vaccines such as SB1 or 301B. HVT+Rispens is used extensively in broilers in the Delmarva region of the United States, in Italy, and in Japan where field MD challenges are often greater than other regions (Embrex, Inc., unpublished data; Rosenberger, 1996). *In ovo* HVT MD vaccination has been shown to maximize the interval between vaccination and early field challenge and causes an earlier viremia *in ovo* vaccinates than in subcutaneous vaccinates because the *in ovo* administration of the vaccine precedes subcutaneous administration by 3 days and HVT replicates rapidly *in ovo* (Sharma and Burmester, 1982; Fabris *et al.*, 1994). Fabris *et al.* (1994) reported similar viremia results for HVT+Rispens MD vaccine when administered *in ovo*, though the difference between the two routes and times of vaccination were not as robust for Rispens viremia when compared to HVT viremia. Sharma (1987) reported that when he inoculated serotype 1 and 2 MD viruses *in ovo,* the viruses did not replicate rapidly like HVT but instead showed a replication that was delayed until just after hatch. Fabris *et al.* (1994) agreed with Sharma (1987) but reported that Rispens vaccination *in ovo* resulted in an earlier viremia than subcutaneous vaccination with Rispens after hatch. This observation could be explained by viral replication near the time of hatch in the *in ovo* vaccinates, while viral replication after subcutaneous vaccination would likely occur several hours to one full day after hatch. It is not known if the delayed replication of serotype 1 and 2 MD vaccine viruses, relative to the rapid replication of HVT, affects the field efficacy of *in ovo* serotype 1 and 2 MD vaccines by reducing the effective interval between vaccination and field challenge. The field efficacy of *in ovo* HVT+Rispens vaccination must be comparable if not superior to that of subcutaneous vaccination. In the regions around the world where MD field challenges are increasing and where *in ovo* MD vaccination with Rispens type vaccines is practiced, a high degree of protection against MD exist (Embrex, Inc., unpublished data).

IV. Commercialization of Other *in ovo* Live Viral Vaccines

In addition to Marek's vaccination, a typical vaccine program used in a commercial broiler operation in the United States includes the use of Newcastle and bronchitis vaccines administered via spray cabinet to the hatchling. In some cases the newly hatched bird may be vaccinated with a mild bursal vaccine. An additional vaccine administered in the drinking water or by course spray at 14–21 days of age is often required during growout, since the presence of maternal antibodies in the hatchling will inactivate most viral vaccines administered immediately after hatch. The injection of many of these posthatch vaccines *in ovo* is unsafe for the embryo. In the United States only those vaccines that are approved by APHIS for use *in ovo* should be used.

A. Infectious Bronchitis Virus

Infectious bronchitis virus (IBV) causes a highly contagious respiratory disease and occasionally a nephrosis/nephritis syndrome in chickens. Wakenell and Sharma (1986) demonstrated that a commercial IBV vaccine injected *in ovo* was pathogenic to avian embryos. These investigators suggested that this was not surprising since the vaccine employed was an egg propagated vaccine. Using a tissue culture attenuation system, pathogenicity to the embryo could be eliminated. Following passage in tissue culture (40 times) the vaccine could be safely administered into SPF embryos. Birds were protected following challenge at 4 weeks of age to a virulent strain of Massachusetts 41. The pathologic and immunologic effects of vaccination with the attenuated strain were comparable to those induced by conventional vaccination of chicks (Wakenell et al., 1995). Based on personal observations made by Dr. Wakenell, embryos may not be vaccinated with bronchitis vaccines typically employed in the hatchling. These vaccines must be attenuated and the type of attenuation is critical, that is, tissue culture attenuation may be employed, but embryo attenuation is contraindicated. To our knowledge there are no APHIS approved bronchitis vaccines available for in ovo administration in the United States.

B. Newcastle Disease Virus

Newcastle disease is a highly contagious respiratory disease of chickens and turkeys that continues to have a substantial economic impact on the poultry industry throughout the world. Neither the B1 or La Sota vaccine strains currently used in chickens can be employed in ovo in chickens or turkeys due to embryo toxicity (Ahmad and Sharma, 1992, 1993). Birds that do hatch from in ovo vaccinated eggs exhibit high early mortality. Several groups are investigating alternative approaches for providing an in ovo-compatible Newcastle vaccine. Treatment of the B1 strain of Newcastle disease virus with an alkylating agent, ethylmethane sulfonate, markedly reduces the virulence of the virus for the 18-day chick embryo (Ahmad and Sharma, 1992). Hatched chicks developed antibody to Newcastle and are protected against challenge at 4 weeks of age with a highly virulent GB-Texas strain of Newcastle. Stone (1993) is evaluating the use of a killed Newcastle vaccine for embryonic use. Whitfill and coworkers are exploring the use of viral neutralizing factor technology (see later) as a means to develop a safe and effective Newcastle vaccine for in ovo use. There are no Newcastle vaccines yet registered for in ovo use in the United States.

C. Infectious Bursal Disease (IBDV, Gumboro Disease)

Infectious bursal disease is an acute, highly contagious disease of chickens that occurs worldwide. Infections before 3 weeks of age are normally subclinical but cause immunosuppression due to widespread destruction of lymphocytes. Very virulent strains of IBD have, since the late 1980s, caused high mortalities, particularly after day 30 of growout, in Europe, Asia, and Africa. As early as 1985, Sharma demonstrated that if viruses of low virulence were employed, embryos from SPF chickens could be inoculated on the 18th day of incubation. The resulting hatched chicks were resistant to challenge with virulent IBD at 3 weeks of age or older. Viruses of moderate virulence such as 2512 were not safe to administer to embryos lacking maternal antibody. Studies by Sharma (1985) and more recently by Soleno et al. (1997) have demonstrated that Marek's and mild infectious bursal vaccines may be administered *in ovo* with no indication of interference between them. Combination vaccines registered for this use are available in the United States.

D. Fowlpox

Vaccination *in ovo* against fowlpox has been practiced in the United States, Mexico, Italy, and Japan. The most thorough evaluation of pox vaccination *in ovo* was conducted during the development of an *in ovo* fowlpox vaccine during the past several years in Japan (Embrex, Inc. and Nisseiken Co., Ltd., unpublished data). The studies conducted in Japan will be submitted for publication this year by the vaccine manufacturer (Nisseiken Co., Ltd., Ome, Tokyo, Japan). The data from the studies show that the efficacy and safety of chicken fowlpox vaccine for *in ovo* use can be greatly enhanced by injecting a highly diluted, mild tissue culture adapted vaccine into the embryo proper on day 19 of incubation. The vaccine was both safe and efficacious in small-scale field tests. It is likely that extensive field data will become available from the manufacturer in 1998 after publication of their studies and government approval of their vaccine.

V. Maternal Antibody Effects on Viral Vaccine Efficacy

Maternal antibodies have been shown to interfere with the effectiveness of Marek's embryonal vaccination (Sharma and Graham, 1982). The interference was greater when cell-free Marek's vaccine was used compared to that elicited when cell-associated Marek's vaccine was employed. The authors suggest that maternal antibodies likely neutralized the vaccine virus more readily when it was in cell-free form.

Several studies have demonstrated maternal antibody interference when mild bursal vaccines are employed in embryonal vaccination (Sharma, 1985).

Chickens all over the world are commonly vaccinated for Newcastle disease and infectious bronchitis using attenuated live virus. These vaccinations commonly take the form of a spray administration at day of hatch followed by drinking water administration at 1–2 weeks of age. Day of hatch vaccination by spray administration for both Newcastle disease and infectious bronchitis appears to produce some local immunity in the respiratory tract despite the possible interference of maternal immunity. This local immunity is usually short lived. Maternal immunity has been shown to reduce the degree and duration of immunity stimulated by day of age vaccinations for Newcastle disease (Giambrone and Closser, 1990; Sharma et al., 1989; Holmes, 1979) and infectious bronchitis (Klieve and Cumming, 1988).

VI. Safe and Effective Vaccination in Presence of Maternal Antibodies

Typically, in poultry a mild vaccine (effective in low maternal antibody birds) is given at day of age and a more virulent (intermediate) vaccine is give at 1–2 weeks of age, a time when maternal antibody levels have declined. It would be advantageous to have vaccines that effectively immunize commercial broiler chickens with one administration given either *in ovo* or on the day of hatch, irrespective of the maternal antibody status of the bird.

Embrex has developed a novel proprietary technology, viral neutralizing factor (VNF®), which improves the safety and efficacy of live viral vaccines. The VNF technology utilizes specific viral neutralizing antibody in hyperimmune antiserum and a vaccine virus. The specific antiserum is mixed with a vaccine virus to form a virus–antibody complex vaccine. The amount of antiserum used is not sufficient to neutralize the vaccine virus, but is sufficient to delay the pathologic effects caused by the vaccine virus when compared to the use of the vaccine virus without antiserum (naked virus). This delaying of pathologic effects allows young chickens to be vaccinated more effectively and without the adverse consequences sometimes associated with live viral vaccines. It also allows for administration of vaccine viruses that would otherwise be too virulent to use *in ovo* or at day of hatch.

A. Bursaplex™, Infectious Bursal Disease Virus Antibody Complex Vaccine

Infectious bursal disease virus (IBDV) causes a highly contagious acute infection of chickens. The primary lymphoid organ, the bursa of Fabricius (bursa), is where B-cell maturation occurs and this organ is severely affected by IBD (Lukert and Saif, 1997). Susceptible birds infected under 2 weeks of age with virulent IBDV have few clinical signs, but usually become immunosuppressed, predisposing these birds to various other diseases (Saif, 1991). Despite extensive vaccination of breeders and progeny, outbreaks causing significant economic losses still occur. Maternal antibodies protect birds from IBD for the first few weeks of life, but also prevent effective immunization with live IBD vaccines during this time. Maternal antibody levels vary from flock to flock and within individuals of the same flock. Thus, it is often difficult to predict the best time to vaccinate chickens in the field with a live IBD vaccine (Lukert and Saif, 1997). It would be advantageous to have an IBD vaccine that could be given in the hatchery and have it provide protection for life without a booster, regardless of the maternal antibody level of the chicks. Bursaplex™ is a vaccine developed specifically for that purpose.

Extensive studies have been conducted utilizing Bursaplex *in ovo* and at day of age (Avakian *et al.*, 1993; Haddad *et al.*, 1993; Whitfill *et al.*, 1992, 1995a,b). The vaccine is licensed for sale in the United States and is presently being used to vaccinate millions of embryos each week.

Experimental data from Whitfill *et al.* (1995a) have shown that *in ovo* vaccination of specific pathogen-free (SPF) chickens with Bursaplex results in a normal hatch, no clinical signs in the hatchlings, and delayed gross lesions in bursa until day 7 of age. When the same naked vaccine virus is administered *in ovo* at the same dose it causes reduced percent hatch, bursal lesions by day 1 of age, and 15% post-hatch mortality. Thus, the addition of antibody delays pathologic changes caused by the vaccine virus, but does not prevent infection and replication. SPF birds vaccinated *in ovo* with Bursaplex have been shown to have an elevated antibody titer and be protected from standard very virulent IBDV challenge. The use of the virus–antibody complex vaccine Bursaplex is designed to result in a safer immunization, which is particularly important in birds with low maternal antibody levels. Furthermore, studies have shown that the vaccine was efficacious.

In one study broiler chickens with high levels of maternal antibody to IBDV were vaccinated *in ovo* with either Bursaplex or the same dose of naked vaccine virus. Birds vaccinated with Bursaplex exhibited IBDV antibody mean titer values greater than birds vaccinated with naked vaccine virus when measured at day 42 of age (2379 versus 1297). A separate group of vaccinated birds was challenged at day 21 of age with vvIBDV strain DV86. The nonvaccinated control birds had 20% mortality and 30% clinical illness, while both vaccinated groups exhibited 100% protection from the challenge. Hence, the presence of maternally derived antibody against IBDV did not interfere with the ability of Bursaplex to stimulate a protective immune response, even before an active humoral immune response was measurable.

In another study, Bursaplex was given *in ovo* to commercial broilers. The mean maternal antibody titer in these day old broilers was 3689. On day 34 of age, the Bursaplex vaccinates had antibody to IBDV and 19/20 were protected from a standard USDA IBDV challenge. In a similar but different study, commercial broilers with a day 1 of age mean maternal antibody titer to IBDV of 6492 were vaccinated *in ovo* with Bursaplex. On day 35 of age, birds were challenged with the USDA standard IBDV challenge strain and 21/21 vaccinates were protected from the challenge.

These studies have shown that the antibody complex vaccine Bursaplex is both safe and effective in birds with no maternal antibody, moderate maternal antibody, and high maternal antibody. Even when the starting maternal antibody titer was over 6000, Bursaplex was able to break through and elicit immunity in the broilers. The time at which the vaccine virus breaks through maternal antibody depends on the starting maternal antibody level in each bird, with a higher titer resulting in a later release of the Bursaplex vaccine virus. Thus, test results indicate that one administration of Bursaplex at the hatchery protects broilers with varying levels of maternal antibody and eliminates the need to guess when to give a field vaccination.

A number of commercial field trials have been conducted using Bursaplex. The tests (Table I) were conducted using a time period on Bursaplex (usually a week of production from one hatchery) and a time period on the conventional IBD program (controls). In these studies the farm was the unit of treatment separation. Bursaplex vaccinated birds received no other IBD vaccination. Production data were supplied by the producer.

In the eight trials (Table I) approximately 22.5 million birds were given Bursaplex and were compared to the same number of controls. Bursaplex use resulted in an average cost reduction of 0.24 cents per pound produced. This was primarily due to the 0.02 points lower feed

TABLE I

PRODUCTION VALUES FROM LARGE-SCALE TRIALS[a] USING THE IBD VACCINE BURSAPLEX IN COMMERCIAL BROILER CHICKENS

Trial number	Vaccine[b] treatment	Route[c]	No. of birds (millions)	Percent live	Average weight	FCR[d]	Average age	Percent cond	SC[e]
1	Bursaplex	IO	5.6	93.7	5.36	2.185	53	0.81	19.41
	Control	IO	2.8	93.3	5.29	2.212	53	1.04	19.66
2	Bursaplex	IO	2.1	95.6	4.84	1.974	47	2.22	20.57
	Control	IO	2.4	95.9	4.83	2.001	47	2.33	20.84
3	Bursaplex	IO	3.9	95.1	4.01	1.865	41	1.51	22.61
	Control	IO	2.1	95.0	3.97	1.870	41	1.51	22.74
4	Bursaplex	AH	1.2	94.4	5.10	2.050	49	0.88	18.58
	Control	AH	4.4	94.1	5.22	2.085	50	0.82	18.79
5	½Bursaplex[f]	IO	2.5	93.7	5.51	2.124	53	1.05	18.93
	Control	IO	3.1	93.7	5.46	2.128	53	1.06	18.98
6	Bursaplex	IO	1.0	96.4	6.26	2.130	55	0.88	22.96
	Control	IO	1.1	96.2	6.14	2.132	55	1.04	23.01
7	Bursaplex	IO	5.6	94.3	4.45	1.979	46	0.92	18.82
	Control	IO	6.0	92.8	4.20	2.019	45	1.11	19.56
8	Bursaplex	IO	0.5	96.5	4.02	1.913	43	0.76	21.71
	Control	IO	0.7	95.7	3.96	1.930	43	0.69	21.94

[a]Eight trials consisting of approx. 22.5 million birds. Hatchery conditions were monitored to ensure proper conditions. Trial results are not necessarily determinative of future results.

[b]Birds were given Bursaplex or a conventional IBD vaccination (control) that consisted of IBD at the hatchery and/or IBD via drinking water/spray at the farm. No Bursaplex vaccinated birds were given a field boost. In trials 6 and 7 the controls did not receive an IBD vaccination.

[c]IO, *in ovo;* AH, at hatch.

[d]FCR, feed conversion ratio.

[e]SC, standard settlement cost in cents per pound as supplied by the poultry producer.

[f] ½Bursaplex, a half dose of Bursaplex; in trial 5 a half dose of Bursaplex was compared to a conventional vaccine program consisting of an *in ovo* IBD vaccination followed by a field boost.

conversion ratio on average. Furthermore, Bursaplex treated birds were on average 0.06 pounds heavier, had 0.07% lower field condemnations and 0.4% better livability than controls in these trials. These eight trials were conducted under a variety of conditions and the data indicate that the use of Bursaplex is both safe and efficacious and cost effective in commercial broiler production as assessed in these trials.

B. OTHER VIRUS–ANTIBODY COMPLEX VACCINES

The infectious virus–antibody complex concept of vaccine development has been tested with three IBD vaccine viruses, Newcastle disease virus (NDV), and three avian reoviruses. Development of experimental vaccines using NDV and avian reovirus is in the early stages, and vaccine formulations have not been optimized. Data using these experimental complex vaccines are similar to that presented for Bursaplex with respect to delay and improved safety for administration *in ovo* and day of hatch. In addition to Bursaplex, two other IBDV antibody complex vaccines have been developed for selected international markets. The technology also appears to work with viruses that infect mammals. Human reovirus (stain TD3) antibody complexes were used in mice and resulted in a delay of pathologic changes and improved safety. A canine parvovirus complex vaccine demonstrated delay and increased safety in SPF pups.

In addition to improved safety, there has been a trend toward enhanced immunity in each of these model systems. This enhanced immunity was evidenced by higher antibody titers and/or higher levels of protection following a challenge, when compared to individuals that were vaccinated with the naked vaccine virus. This enhanced immunity may be related to the way in which immune complexes are processed by the immune system. There is a body of research using antigen–antibody complexes (immune complexes) that may provide some clues. Previous work *in vitro* and in mammals using inactivated antigen–antibody complexes have shown that B-lymphocyte memory (Klaus, 1978; Taylor *et al.*, 1979; Terres and Wolins, 1961), antigen-specific T-lymphocyte proliferation (Celis and Chang, 1984; Marusic-Galesic *et al.*, 1991; Schalke *et al.*, 1985), and antibody response (Heyman *et al.*, 1982) can be enhanced when compared to administration of the antigen alone. It has been suggested that enhanced immunity resulting from administration of immune complexes occurs because a primed state is achieved within hours of an antigen–antibody complex injection (Terres *et al.*, 1972). In a typical immune response, immune complexes are formed upon second antigen exposure and are considered an important step in the generation of the secondary immune response (Szakal *et al.*, 1991). Thus, in the case of infectious virus–antibody complex vaccines, the initial presentation as an immune complex followed by the subsequent replication of the vaccine virus in the target tissue may trigger an immune response with elements of a secondary response along with the primary response.

C. Mechanism of Action

Research into the mechanism of action of IBDV antibody complex vaccines is ongoing and is not yet completely understood. However, some aspects have been determined. In one study, various tissues and cell types from SPF birds were identified with monoclonal antibodies (Jeurissen et al., 1988a,b) for the presence of vaccine virus at numerous times postvaccination. It was shown that replication of the IBDV antibody complex vaccine was delayed about 5 days when compared to the same dose of vaccine virus without antibody. The mechanism behind this delay is still unknown. The addition of antibody did not seem to cause a different initial localization of replication, since both antibody complexed and naked virus began replicating in the same lymphoid organs. IBDV was detected in/on B lymphocytes, macrophages, and follicular dendritic cells. Chicks given antibody complex vaccine *in ovo* always had small clusters of lymphocytes in some bursal follicles. Furthermore, repopulation of bursal lymphocytes occurred much faster in complex vaccinates that in birds given naked virus. The most striking difference between the complex vaccine and the naked virus was that antibody complex vaccinates had much larger numbers of follicular dendritic cells in their bursae and spleen with surface localized IBDV and a much higher number of germinal centers in the spleen.

In conclusion, the antibody complex vaccine Bursaplex has been shown in studies to be both safe and effective when given *in ovo* or at day of age in experimental trials. Commercial field data have shown that Bursaplex is safe and suggests that it may be more effective than other commonly used IBD prevention programs. The use of the antibody complex vaccine technology has been shown to work with two other IBD vaccine viruses, Newcastle disease vaccine viruses, three avian reoviruses, a human reovirus in a murine model, and a canine parovirus. Thus, the technology appears to be broadly applicable and may be used to increase the safety and efficacy of a number of live viral vaccines in both avian and mammalian subjects. Additional research is needed to understand the mechanism(s) of action of these complex vaccines and their possible immune enhancing capacities.

VII. Bacterial Vaccines

Very little published work is available on the use of bacterial vaccines for *in ovo* use. Chick embryos inoculated on day 16 of incubation with heat-killed *Campylobacter jejuni* organisms into the amniotic fluid developed an early antibody response associated primarily with the IgA isotype, with high titers in both systemic and mucosal compartments of immunized chicks (Noor et al., 1995). *Campylobacter* and salmonella contamination of chicken carcasses poses a public health risk to individuals eating improperly cooked chicken. Preliminary studies using a genetically engineered (gene deletion) salmonella vaccine have demonstrated that *in ovo* administration appears to be a safe and effective route to administer this vaccine (Coloe et al., 1994).

VIII. Coccidial Vaccines

Coccidiosis is a protozoan parasite of poultry that causes massive economic loss to the global poultry industry. Partial protection of birds immunized as embryos with *Eimeria tenella* extracts or recombinant *E. tenella* antigens (Ruff et al., 1988; Fredericksen et al., 1989) suggests that embryos are capable of mounting an immune response to the protozoal coccidial parasite *Eimeria*. However, when live intact oocysts or sporocysts of *E. maxima* were administered *in ovo* at day 15, 17, or 18 of incubation there was no evidence of protection to subsequent coccidial challenge (Watkins et al., 1995). In contrast, inoculation of chicks post-hatch provided significant protection against subsequent challenge. Additional studies are needed in order to develop appropriate methodologies for effective use of live coccidial vaccines *in ovo*.

IX. Summary

More than 80% of the U.S. broiler industry has converted to the *in ovo* vaccination process for control of Marek's disease. Providing certain criteria are met, including timing and site of vaccine placement, vaccine mixing, machine sanitization, and hatchery management specifications, this has proven to be an efficacious and convenient method of vaccination. Efforts to extend the technology for other viral vaccines including Newcastle, bronchitis and bursal disease, and bacterial and parasitic vaccines are in progress. Collectively, these studies demonstrate that *in ovo* vaccination technology using approved vaccine is a safe, efficacious, and convenient method for vaccination of poultry.

References

Ahmad, J., and Sharma, J. M. (1992). Evaluation of a modified-live virus vaccine administered *in ovo* to protect chickens against Newcastle disease. *Am. J. Vet. Res.* **53,** 1999–2004.

Ahmad, J., and Sharma, J. M. (1993). Protection against hemorrhagic enteritis and Newcastle disease in turkeys by embryo vaccination with monovalent and bivalent vaccines. *Avian Dis.* **37,** 485–491.

Avakian, A. P., Whitfill, C. E., Haddad, E. E., Ricks, C. A., Skeeles, K., Andrews, P., and Thoma, J. (1993). Efficacy of a novel infectious bursal disease (IBD) vaccine administered *in ovo* to broiler chickens. *Poult. Sci.* **72**(Suppl. 1), 49.

Bauer, F., Tullet, S. G., and Wilson, H. R. (1990). Effects of setting eggs small end up on hatchability and posthatching performance of broilers. *Br. Poult. Sci.* **31,** 715–724.

Byerly, T. C., and Olsen, M. W. (1931). The influence of gravity and air-hunger on hatchability. *Poul. Sci.* **10,** 281–287.

Cain, J. R., and Abbott, U. K. (1971). Incubation of avian eggs in an inverted position. *Poult. Sci.* **50,** 1223–1226.

Celis, E., and Chang, T. W. (1984). Antibodies to Hepatitis B surface antigen potentiate the response of human T lymphocyte clones to the same antigen. *Science* **224,** 297.

Coloe, P. J., Gerraty, N. L., Christopher, W., Alderton, M. R., and Smith, S. C. (1994). Vaccination of poultry with Salmonella Typhimurium STM-1. Laboratory and field evaluations. *Proc. West Poult. Dis. Conf.,* Sacramento, CA, *1994,* pp. 37–39.

El-Ibiary, H. M., Shaffner, C. S., and Godfrey, E. F. (1966). Hatchability of eggs set small end up. *Poult. Sci.* **45,** 419–420.

Fabris, G., Pattarello, I., and Figarolli, V. (1994). *In ovo* Marek's vaccination. Unpublished presentation at the *In ovo* Vaccination Workshop, 9th European Poultry Conference, Glasgow, Scotland, 1994.

Fredericksen, T. L., Thaxton, J. P., Gildersleeve, R. P., Rowe, D. G., Ruff, M. D., Strohlein, D. A., and Danforth H. D. (1989). *In ovo* administration of a potential recombinant coccidial antigen vaccine in poultry. *Coccidia* and *Intestinal Coccidiomorphs, Int. Coccidiosis Conf., 5th,* Tours, France, pp. 655–660.

Giambrone, J. J., and Closser, J. (1990). Effect of breeder vaccination on immunization of progeny against Newcastle disease. *Avian Dis.* **34,** 114–119.

Gildersleeve, R. P., and Fluke, D. R. (1995). *In ovo* technology for vaccine delivery. *North Cent. Avian Dis. Conf. Symp. New Vaccines Delivery Syst., 46th,* Minneapolis/St. Paul, MN, p. 35.

Gildersleeve, R. P., Hoyle, C. M., Miles, A. M., Murraym, D. L., Ricks, C. A., Secrest, M. N., Williams, C. J., and Womack, C. L. (1993). Developmental performance of an egg injection machine for administration of Marek's disease vaccine. *J. Appl. Poult. Res.* **2,** 337–346.

Haddad, E. E., Whitfill, C. E., Ricks, C. A., Avakian A. P., Skeeles, K., Andrews, P., and Thoma, J. (1993). Efficacy of a novel infectious bursal disease (IBD) vaccine administered *in ovo* to SPF chickens. *Poult. Sci.* **72**(Suppl. 1), 49.

Heyman, B., Andrighetto, G., and Wigzell, H. (1982). Antigen-dependent-IgM-mediated enhancement of the sheep erythrocyte response in mice. *J. Exp. Med.* **155,** 994.

Holmes, H. C. (1979). Resistance of the respiratory tract of the chicken to Newcastle disease virus infection following vaccination: The effect of passively acquired antibody on its development. *J. Comp. Pathol.* **89,** 11–19.

Jeurissen, S. H. M., Janse, E. M., Ekino, S., Nieuwenhuis, P., Koch, G., and de Boer, G F.

(1988a). Monoclonal antibodies as probes for defining cellular subsets in the bone marrow, thymis, bursa of Fabricius, and spleen of the chicken. *Vet. Immunol. Immunopath.* **19,** 225–238.

Jeurissen, S. H. M., Janse, E. M., Koch, G., and de Boer, G. F. (1988b). The monoclonal antibody CVI-ChNL-68.1 recognizes cells of the monocyte-macrophage lineage in chickens. *Dev. Comp. Immunol.* **12,** 855–864.

Klaus, G. G. B. (1978). Generation of memory cells II. Generation of B-memory cells with preformed antigen-antibody complexes. *Immunology* **34,** 643.

Klieve, A. V., and Cumming R. B. (1988). Infectious bronchitis: Safety and protection in chickens with maternal antibody. *Aust. Vet. J.* **65,** 396–397.

Lukert, P. D., and Saif, Y. M. (1997). Infectious bursal disease. *In* "Diseases of Poultry" (B. W. Calnek, H. J. Barnes, C. W. Beard, W. M. Reid, and Y. M. Saif, eds.), 10th ed., p. 721. Iowa State University Press, Ames.

Marusic-Galesic, S., Pavelic, K., and Pokric, B. (1991). Cellular immune responses to the antigen administered as an immune complex. *Immunology* **72,** 526.

Mauldin, J. M., and Wilson, J. L. (1988). Watch egg during incubation. *Poult. Dig.* **47**(557); 242–345.

Muller, H. D., and Williams, M. (1975). The fate of hatching eggs set upside down. *Poult. Sci.* **54,** 1344–1345.

Noor, S. M., Husband, A. J., and Widders, P. R. (1995). In ovo oral vaccination with *Campylobacter jejuni* establishes early development of intestinal immunity in chickens. *Br. Poult. Sci.* **36,** 563–573.

Ruff, M. D., Fredericksen, T. L., Thaxton, J. P., Strohlein, D. A., Danforth, H. D., and Gildersleeve, R. G. (1988). *In ovo* vaccination against *Eimeria tenella*. *Poult. Sci.* **67,** 147 (abstr.).

Saif, Y. M. (1991). Immunosuppression induced by infectious bursal disease virus. *Vet. Immunol. Immunopathol.* **30,** 45.

Sarma, G., Greer, W., Gildersleeve, R. P., Murray, D. L., and Miles, A. M. (1995). Field safety and efficacy of *in ovo* administration of HVT and BS-1 bivalent Marek's disease vaccine in commercial broilers. *Avian Dis.* **39,** 211–217.

Schalke, B. C. G., Klinkert, W. E. F., Wekerle, H., and Dwyer, D. S. (1985). Enhanced activation of a T cell line specific for acetylcholine receptor (AChR) by using anti-AChR monoclonal antibodies plus receptors. *J. Immunol.* **134,** 3643.

Sharma, J. M. (1985). Embryo vaccination with infectious bursal disease virus alone or in combination with Marek's disease vaccine. *Avian Dis.* **29,** 1155–1169.

Sharma, J. M. (1987). Delayed replication of Marek's disease virus following *in ovo* inoculation during late stages of embryonal development. *Avian Dis.* **31,** 570–576.

Sharma, J. M., and Witter, R. L. (1982). Resistance to Marek's Disease at hatching in chickens vaccinated as embryos with the turkey herpesvirus. *Avian Dis.* **26,** 134–149.

Sharma, J. M., and Graham, C. K. (1982). Influence of maternal antibody on efficacy of embryo vaccination with cell-associated and cell-free Marek's disease vaccine. *Avian Dis.* **26,** 860–870.

Sharma, J. M., and Burmester, B. R. (1983). Embryo vaccination against Marek's disease with serotypes 1, 2 and 3 vaccines administered singly or in combination. *Avian Dis.* **27,** 453–463.

Sharma, J. M., Lee, L. F., and Wakenell, P. S. (1984). Comparative viral, immunologic, and pathologic responses of chickens inoculated with herpesvirus of turkeys as embryos or at hatch. *Am. J. Vet. Res.* **45,** 1619–1623.

Sharma, P. N., Muneer, M. A., and Cho, Y. (1989). Role of maternal antibodies in immunization of chicks against Newcastle disease virus. *Vet. Sci. Zootec. Int.,* pp. 51–55.

Soleno, W., Sarma, G., Winans, R., and Wasmoen, T. (1997). The *in ovo* host animal interference testing of live infectious bursal disease vaccine vs live chicken (SB1) and turkey herpesvirus (HVT) Marek's disease vaccine (abstr.) *Int. Vet. Vaccines Diagn. Conf., 1st,* Madison, WI, *1997.*

Stone, H. D. (1993). Efficacy of experimental animal and vegetable oil-emulsion vaccines for Newcastle disease and avian influenza. *Avian Dis.* **37,** 399–405.

Szakal, A. K., Burton, G. F., Smith, J. P., and Tew, J. G. (1991). Antigen processing and presentation *in vivo.* In "Topics in Vaccine Adjuvant Research" (D. R. Spriggs and W. C. Koff, eds.), pp. 11–23. CRC Press, Boca Raton, FL.

Talmadge, D. W. (1977). The effect of incubating eggs narrow end up on malposition II and hatchability. *Poult. Sci.* **56,** 1046–1048.

Taylor, R. B., Tite, J. P., and Manzo, C. (1979). Immunoregulatory effects of a covalent antigen-antibody complex. *Nature* **281,** 488.

Terres, G., and Wolins, W. (1961). Enhanced immunological sensitization of mice by the simultaneous injection of antigen and specific antiserum I. Effect of varying the amount of antigen used relative to the antiserum.*J. Immunol.* **86,** 361.

Terres, G., Morrison, S. L., Habicht, G. S., and Stoner, R. D. (1972). Appearance of an early "primed state" in ice following the concomitant injections of antigen and specific antiserum. *J. Immunol.* **108,** 1473.

Wakenell, P. S., and Sharma, J. M. (1986). Chicken embryonal vaccination with avian infectious bronchitis virus. *Am. J. Vet. Res.* **47,** 933–938.

Wakenell, P. S., Sharma J. M., and Slocombe, R. F. (1995). Embryo vaccination of chickens with infectious bronchitis virus: Histologic and ultrastructural lesion response and immunologic response to vaccination. *Avian Dis.* **39,** 752–765.

Watkins, K. L., Brooks, M. A., Jeffers, T. J., Phelps, P. V., and Ricks, C. A. (1995). The effect of *in ovo* oocyst or sporocyst inoculation on response to subsequent coccidial challenge. *Poult. Sci.* **74,** 1597–1602.

Whitfill, C. E., Ricks, C. A., Haddad, E. E., Andrews, P. A., and Skeeles, J. K. (1992). Infectious bursal disease (IBD) vaccine for day of age administration in broiler chickens. *Poult. Sci.* **71**(Suppl. 1), 59.

Whitfill, C. E., Avakian, A. P., Haddad, E. E., and Ricks, C. A. (1995a). Virus-antibody complex vaccines. *Proc. North Cent. Avian Dis. Conf., 46th, 1995,* pp. 42–51.

Whitfill, C. E., Haddad, E. E., Ricks, C. A., Skeeles, J. K., Newberry, L. A., Beasley, J. N., Andrews, R., Thoma, J. A., and Wakenell, P. S. (1995b). Determination of optimum formulation of a novel infectious bursal disease virus (IBDV) vaccine constructed by mixing bursal disease antibody with IBDV. *Avian Dis.* **39,** 687–699.

Wineland, M. (1996). Factors influencing embryo respiration. *Poult. Dig.,* September, pp. 16–20.

Current and Future Recombinant Viral Vaccines for Poultry

MARK W. JACKWOOD

Department of Avian Medicine, College of Veterinary Medicine, University of Georgia, Athens, Georgia 30602

I. Introduction and Background
II. Virus Vectors
   A. Fowlpox Virus
   B. Herpesvirus of Turkeys
III. Future Recombinant Vaccines for Poultry
   A. Subunit Vaccines and Synthetic Peptides
   B. Nucleic Acid Vaccines
IV. Summary
   References

I. Introduction and Background

Biotechnology has changed the way scientists approach the development of new vaccines. The ability to manipulate genes directly (biotechnology) has allowed scientists to create nonpathogenic vaccines capable of inducing a protective immune response. The poultry industry leads the way in the development and use of these recombinant vaccines. The first commercially available recombinant viral vector vaccine was produced for poultry.

II. Virus Vectors

Virus vectors are nonpathogenic viruses carrying a foreign gene inserted into a region of the viral genome that is not required for viral replication. When the virus vector infects the host, the foreign protein is expressed and the host immune system responds to the virus vector as well as the foreign protein. There are several advantages to using a virus vector. Virus vectors are live viruses that replicate in the host and induce an immune response usually with only a single vaccination. In addition, the amount and the presentation of the foreign protein is such that a strong protective response is induced in the vaccinated animal. Another advantage to viral vectors as vaccines is that more than one foreign gene can be inserted into the vector, allowing for the construction of a multivalent vaccine. The main disadvantage of virus vector vaccines are that they are expensive to develop.

A. Fowlpox Virus

The first commercially available virus vector vaccine was a fowlpox virus. Fowlpox Virus is a double-stranded DNA virus in the *Poxviridae* family, genus *Avipoxvirus*. Avian pox is the disease caused by fowlpox virus, and numerous vaccine strains have been shown to be safe and effective over many years of use. Scientists have found that inactivation of the thymidine kinase gene in fowlpox virus does not affect the replication of that virus. Thus, foreign genes inserted into that region of the genome have no affect on the ability of the vaccine viruses to infect, replicate, or induce an immune response in the host.

The first commercially available fowlpox virus vaccine vector contains the hemagglutinin neuraminidase (HN) and fusion (F) genes from Newcastle disease virus (NDV). Newcastle disease virus is an orthomixovirus that has a single-stranded negative sense RNA genome. That virus causes an upper respiratory tract disease in poultry. The HA and F proteins, located on the surface of the virus, have been shown to induce neutralizing antibodies that protect the host from disease.

To construct the fowlpox virus vaccine vector, copy DNA was prepared to the NDV genes coding for the HA and F proteins and that cDNA was inserted into a transfer plasmid. The transfer plasmid contained fowlpox virus nucleic acid sequences flanking the inserted NDV genes, and was used to insert the HA and F genes into a vaccine strain of fowlpox virus using a technique called homologous recombination. When the fowlpox virus vaccine vector containing the NDV genes coding for the HA and F proteins is used to vaccinate poultry, the birds respond immunologically to the poxvirus as well as the NDV proteins.

The major advantage of the fowlpox virus vector for NDV is that there is absolutely no chance of an upper respiratory reaction. That is because the pox virus replicates in the skin not in the upper respiratory tract where live NDV vaccines replicate. Furthermore, since the pox virus vector was originally a vaccine strain, it is a safe and effective vaccine for pox. The major disadvantage of the fowlpox virus vector for NDV is that maternal antibodies to NDV interfere with the immune response. The vaccine works best in maternal antibody negative birds.

Currently there are two commercially available fowlpox virus vectors for NDV. Both contain the HN and F genes of NDV. VectorVax FP-N (Hoechst-Roussel Vet, Somerville, NJ) is a lyophilized product that was licensed in July 1994. Trovac-NDV (SELECT Laboratories, Gainesville, GA) is a liquid N_2 frozen product that was licensed in October 1995. Other fowlpox virus vectored vaccines likely to be licensed in the near future will contain genes from avian influenza virus, infectious laryngotracheitis virus (ILTV), and possibly avian immune modulator genes (cytokines).

B. HERPESVIRUS OF TURKEYS

Another virus vector being developed for poultry is based on herpesvirus of turkeys (HVT), an alpha herpesvirus used as a vaccine against Marek's disease (Morgan *et al.,* 1993). Several nonessential regions have been identified in the unique short and unique long region of the viral genome of HVT. Genes from NDV, avian influenza virus, Marek's disease virus, and ILTV are being inserted into HVT to develop a virus vector against those viruses. Because HVT causes a persistent systemic infection and stimulates both humoral as well as cell-mediated immunity it is hoped that a single vaccination with that vector will induce long-lasting immunity.

III. Future Recombinant Vaccines for Poultry

A. SUBUNIT VACCINES AND SYNTHETIC PEPTIDES

Several other approaches to recombinant vaccines for poultry are currently being pursued. Subunit vaccines, which are the immunogenic proteins of disease agents, are being pursued using baculovirus or the T7 transient expression systems.

Synthetic peptides are short amino acid chains containing only the neutralizing epitope of an immunogenic protein from a disease agent. They can be synthesized in the laboratory and must be linked to a carrier molecule, such as bovine serum albumin of keyhole limpet hemocyanin.

B. Nucleic Acid Vaccines

Perhaps the most promising new recombinant vaccine technology is the development of nucleic acid vaccines. Nucleic acid vaccines, also called gene vaccines or DNA vaccines, are usually a bacterial plasmid containing a cloned gene from a disease agent. Eukaryotic promoters allow the gene to be expressed when the DNA is injected directly into the animal. Generally nucleic acid vaccines are injected intramuscularly, taken up by the skeletal muscle fibers, and the immu-

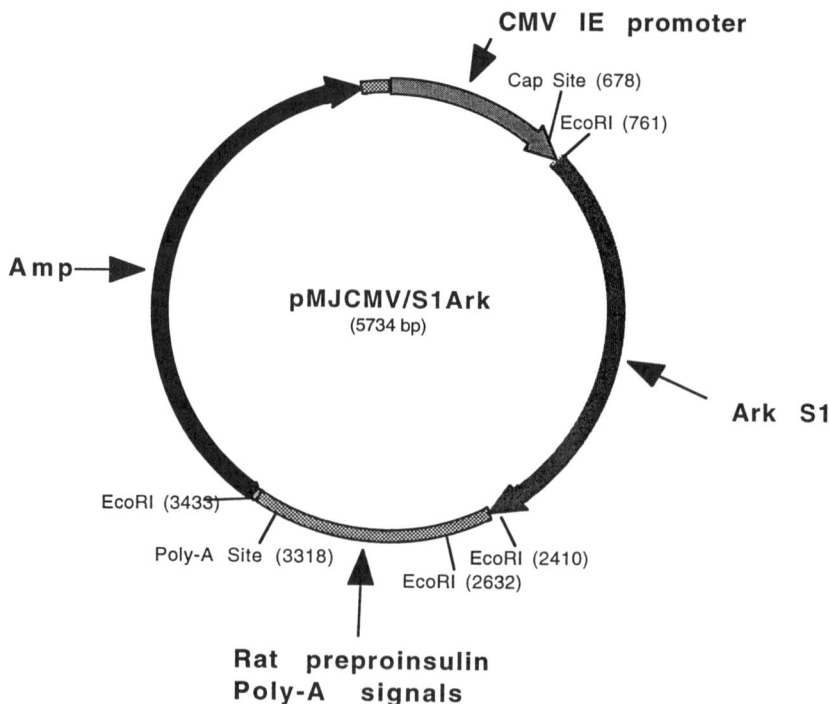

FIG. 1. Infectious bronchitis virus nucleic acid vaccine containing the S1 glycoprotein gene (Ark S1) from the Arkansas serotype of the virus. The plasmid contains the ampicilin resistance gene (Amp), the cytomegalovirus immediate early promoter (CMV IE promoter) including the cap site at position 678bp, and the rat preproinsulin polyadenylation and processing signals (Rat preproinsulin Poly-A signals) including the poly-A site at 3318 bp. The numbers following restriction enzyme sites are in base pairs.

TABLE I

DNA Vaccine Titration and Protection of Chickens against Ark IBV Challenge[a]

Group	Vaccine given at 14 days of age	Vaccine given at 35 days of age	Clinical signs at 7 days postchallenge[b]	Average tracheal lesion scores[c]
1	50 µg pMJAS1	100 µg pMJAS1	6/6	2.71[A]
2	100 µg pMJAS1	200 µg pMJAS1	2/5	2.26[B]
3	150 µg pMJAS1	300 µg pMJAS1	0/5	2.16[B]
4	150 µg pBC12[d]	300 µg pBC12	6/6	2.53[A]
5	Live Ark	Live Ark	0/6	1.16[B]
6	TE buffer	TE buffer	6/6	2.03[B]

[a]See text for details; data for nonchallenged birds are not presented.
[b]Number of birds with clinical signs/number of birds examined.
[c]Numbers within the column with different superscripts are statistically different ($p<0.1$).
[d]pBC12, the pBC12/CMV/IL-2 plasmid.

nogenic protein encoded by the cloned gene is expressed. When the bird mounts an immune response to the expressed protein, that immature response has been shown to be protective and persistent.

One of the first nucleic acid vaccines was described and developed by Fynan et al. (1993) to avian influenza virus. Those scientists cloned the H7 (hemagglutinin) gene into an expression plasmid containing the cytomegalovirus immediate early promoter and showed that birds were protected from a lethal challenge with H7 influenza virus. Nucleic acid vaccines have also been developed to a number of other avian diseases including one developed in our laboratory (Jackwood et al., 1995) to infectious bronchitis virus (IBV). That vaccine contained the base plasmid described by Cullen (1986) and the IBV S1 gene subunit of the immunogenic spike glycoprotein (see Fig. 1). In vitro transfection studies showed that the S1 glycoprotein could be expressed in COS cells. Following vaccination of specific pathogen-free chickens twice at 14 and 35 days of age with 150 and 300 µg of DNA, respectively, the birds were shown to be protected from the disease following challenge with the homologous serotype of the virus (Table I).

IV. Summary

The use of biotechnology to create recombinant viral vaccines holds many promises for the future. But, to be practical, new vaccines must have a selective advantage over traditional vaccines. A vaccine that is novel because it is a recombinant vaccine is not enough. Recombinant vaccines must be safer, or more efficacious, or less expensive to produce in order for them to gain a niche in the marketplace.

REFERENCES

Cullen, B. R. (1986). Trans-activation of human immunodeficiency virus occurs via a bimodal mechanism. *Cell* **46,** 973–982.

Fynan, E. F., Robinson, H. L., and Webster, R. G. (1993). Use of DNA encoding influenza hemagglutinin as an avian influenza vaccine. *DNA Cell Biol.* **12,** 785–789.

Jackwood, M. W., Hilt, D. A., and Moore, K. M. (1995). Vaccination of SPF leghorn chickens with a plasmid vector containing the S1 glycoprotein gene of infectious bronchitis virus. *Proc. 132nd Annu. Meet. Am. Vet. Med. Associ.,* Pittsburg, PA, p. 143.

Morgan, R. W., Gelb, J., Jr., Pope, C. R., and Sondermeijer, P. J. A. (1993). Efficacy in chickens of a herpesvirus of turkeys recombinant vaccine containing the fusion gene of Newcastle disease virus: Onset of protection and effect of maternal antibodies. *Avian Dis.* **37,** 1032–1040.

VIII
FISH, EXOTIC, AND WILDLIFE VACCINES

Development and Use of Modified Live *Edwardsiella ictaluri* Vaccine against Enteric Septicemia of Catfish

PHILLIP H. KLESIUS AND CRAIG A. SHOEMAKER

United States Department of Agriculture–Agricultural Research Service, Fish Diseases and Parasites Research Laboratory, Auburn, Alabama 36830

I. Introduction
II. Materials and Methods
    A. Production of an Attenuated *E. ictaluri* RE-33 Vaccine
    B. RE-33 Safety and Experimental Animals
    C. RE-33 Vaccination
    D. *E. ictaluri* Isolates and Experimental Challenge
III. Results
    A. Characteristics of *E. ictaluri* RE-33
    B. RE-33 Safety and Clearance
    C. Protective Immunity Induced by *E. ictaluri* RE-33 Against Enteric Septicemia of Catfish
IV. Discussion
V. Summary
    Acknowledgments
    References

I. Introduction

Enteric septicemia of catfish (ESC) is responsible for about $20–$30 million annual losses to catfish farmers in the southeastern United States (Plumb and Vinitnantharat, 1993). The causative agent, a gram-negative rod-shaped bacteria, *Edwardsiella ictaluri*, was first isolated and described in 1976 (Hawke, 1979; Hawke *et al.*, 1981). Control of *E. ictaluri* has been by feeding antibiotic medicated feed.

This practice is expensive and ineffective because sick fish do not eat and antibiotic resistance to oxytetracycline and ormethoprim–sulfamethoxine (the approved drugs for use on food fish) has been observed (Waltman and Shotts, 1986; Plumb and Vinitnantharat, 1990). In salmonid culture and other animal husbandry industries, effective vaccines have been developed and marketed. Vaccines have significantly reduced antibiotic use and economic loss to poultry, swine, beef, and salmon producers. Vaccines developed for use in salmon/trout culture are typically killed products injected intraperitoneally in an oil adjuvant. An effective vaccine is desirable to the channel catfish industry to prevent ESC. However, the vaccine must be low cost and not injected (i.e., immersion or feed vaccines).

Vaccination of channel catfish (*Ictalurus punctatus*) with *E. ictaluri* bacterins has not resulted in acquired immunity of long duration against *E. ictaluri* infection (Thune *et al.*, 1994; Shoemaker and Klesius, 1997). Nusbaum and Morrison (1996) also demonstrated that killed *E. ictaluri* was not entering the fish and suggested this affected vaccine efficacy. *Edwardsiella ictaluri* has been described as a facultative intracellular pathogen of channel catfish (Miyazaki and Plumb, 1985; Shotts *et al.*, 1986; Morrison and Plumb, 1994). Because of the nature of *E. ictaluri* as an intracellular pathogen it is not unusual that killed vaccines have not been successful. Nonliving vaccines [killed *Brucella abortus* or antigen preparations, i.e., lipopolysaccharides (LPS) or outer-membrane proteins (OMPs)] were not capable of inducing protective immunity of long duration against *B. abortus* (an intracellular pathogen) in cattle or mice presumably because only humoral immunity was induced (Montaraz and Winter, 1986). Other researchers have suggested that cell-mediated immunity is needed for development of protective immune responses to intracellular pathogens (North, 1974, 1975; Eisenstein *et al.*, 1984; Montaraz and Winter, 1986; Antonio and Hedrick, 1994; Shoemaker *et al.*, 1997). Klesius and Sealey (1995) and Shoemaker and Klesius (1997) demonstrated that antibody alone was not responsible for protective immunity to enteric septicemia of catfish. Antonio and Hedrick (1994), Shoemaker *et al.* (1997), and Shoemaker and Klesius (1997) suggested acquired immunity to ESC is dependent on the cellular immune response.

Protective immunity to *B. abortus* in cattle was demonstrated using modified live *Brucella* species vaccines (Montaraz and Winter, 1986; Schurig *et al.*, 1991; Winter *et al.*, 1996). These vaccines typically rely on an attenuated mutant of *Brucella* species. One vaccine that was developed lacks the O-side chain of LPS, which is believed to be one of the virulence factors involved in *B. abortus* pathogenesis. Thornton *et al.* (1994) utilized A-layer and O-antigen-deficient strains or mutants of *Aeromonas salmonicida* as vaccine strains and demonstrated protection to furunculosis in rainbow trout (*Oncorhynchus mykiss*). Recent work in our laboratory demonstrated that a short immersion exposure to a low number of living *E. ictaluri* could induce strong acquired immunity to ESC for at least 6 months (Klesius and Shoemaker, 1997). Klesius and Shoemaker (1997) demonstrated that protective immunity against ESC was dependent on the isolate of *E. ictaluri* used to immunize the fish, suggesting antigenic heterogeneity. A multivalent vaccine may be needed to protect channel catfish against all *E. ictaluri* isolates present in production ponds. We hypothesized an O-antigen-deficient *E. ictaluri* would be an effective vaccine against enteric septicemia of catfish.

The objective of this study was to develop an attenuated *E. ictaluri* vaccine that was safe and provided long-lasting acquired immunity in channel catfish to enteric septicemia of catfish.

II. Materials and Methods

A. Production of an Attenuated *E. ictaluri* RE-33 Vaccine

The parental microorganism is *E. ictaluri* EILO isolate originally isolated from the walking catfish *Clarius batrachus* from Thailand. EILO isolate is weakly virulent for channel catfish and requires at least 2×10^7 cfu/ml for 60-minute immersion exposure to cause about 20% ESC mortality. Plumb and Klesius (1988) showed that this isolate was different from 13 U.S. isolates by polyacrylamide gel analysis of reduced proteins. Kasornchandra *et al.* (1987) indicated serologic similarities between EILO isolate and other U.S. isolates of *E. ictaluri*. Biochemically, the isolate is the same as U.S. isolates (Kasornchandra *et al.*, 1987; Bader *et al.*, 1998).

The procedure used to produce the RE-33 vaccine was modified from that described by Schurig *et al.* (1991) to produce the rough *B. abortus* vaccine strain RB-51. Briefly, *E. ictaluri* isolate EILO was passaged on increasing concentrations of rifampicin (3-[4-methylpiperazinyl-iminomethyl] rifamycin SV; (Sigma Chemical Company, St. Louis, MO) supplemented brain heart infusion (BHI) agar to a final concentration of 320 µg/ml rifampicin for 33 passages.

LPS was extracted as described by Schurig *et al.* (1991). Bacterial growth from a 250-ml BHI broth culture of both the parent EILO and RE-33 was harvested by centrifugation. The cell pellet was then suspended in 25 ml of 10 mM Tris, pH 8.0, and bacteria were killed by addition of an equal volume of acetone and stirred overnight. Cells were centrifuged at 13,000g for 5 minutes at 4°C and then washed with 100 ml sterile distilled water. The bacteria were then resuspended in 45 ml sterile distilled water and then 55 ml phenol was added (incubated at 68°C for 40 minutes). The mixture was then centrifuged at 17,000g at 4°C for 10 minutes. The phenol layer was removed and the process repeated three times. The phenol phases were pooled and washed with hot distilled water (about 66°C), centrifuged at 17,000g at 4°C for 10 minutes (five washes). LPS was precipitated by addition of 5 volumes cold methanol (99 ml): sodium acetate saturated methanol (1 ml) with stirring at 4°C for 1 hour. This solution was then centrifuged again at 17,000g at 4°C for 10 minutes and the resulting pellet harvested in 10 ml sterile distilled water. This was then frozen at −70°C and subsequently lyophilized.

Samples of phenol extracted LPS from *E. ictaluri* EILO and RE-33 were treated as described by Schurig *et al.* (1991) prior to electrophoresis. SDS–PAGE was then carried out using the Phast system (Pharmacia-Biotech, Uppsala, Sweden) with 10–15% gradient gels. The gels were then silver stained to examine presence or absence of the O-side chain of LPS.

B. RE-33 Safety and Experimental Animals

All catfish utilized in this study were raised at the USDA ARS facility in Auburn, Alabama, in tanks or aquaria supplied with well water. Catfish were free of *E. ictaluri* as determined by ELISA (Klesius, 1993) or culture (Klesius, 1992a) prior to use. A total of 16,460 channel catfish, blue catfish (*Ictalurus furcatus*), and the blue catfish × channel catfish were vaccinated by immersion with about 1×10^5 cfu/ml RE-33 vaccine for 2 minutes at a density of about 50 fish/liter in 24–26°C water to monitor for any signs of ESC. Fish utilized in the experiments ranged in size from 10 to 50 g and were about 3–9 months of age. Channel catfish (125 USDA and 60 Marion strain) were vaccinated at 200 times the normal vaccine dose (2×10^7/ml) for a 30 times longer exposure (1 hour) to determine the safety of the vaccine. Twenty catfish (blue × channel catfish) were injected intraperitoneally (IP) with 50 µl of a 24-hour culture of RE-33 that was isolated from a channel catfish after vaccination to determine if reversion to virulence occurred. Five serial passages of *E. ictaluri* RE-33 were also conducted in five groups of 10 channel catfish each. Briefly, *E. ictaluri* RE-33 was obtained from a vaccinated channel catfish and grown in BHI broth for 24 hours before use. The first group of 10 fish was vaccinated at the normal dose and time. Two to 3 days following vaccination, two to three fish were euthanized and cultured for *E. ictaluri* RE-33. After a pure culture was obtained, this process was repeated five times. Fifty channel catfish were immersed in 2 ml/L RE-33 culture that had been streaked on 320 µg/ml rifampicin-supplemented BHI agar plates to non-rifampicin-supplemented plates for 16 passages to examine reversion to virulence.

C. RE-33 Vaccination

Catfish were immersed for 2 minutes in water (24–26°C) containing 2 ml RE-33 vaccine/liter of water (about 1×10^5 cfu/ml of water). The RE-33 vaccine was cultured in BHI broth at 25–27°C in a water bath shaking at 50 rpm for 24 hours before vaccination. Prior to vaccination, the *E. ictaluri* RE-33 vaccine was tested for resistance to 320 µg/ml rifampicin and purity. Survival and clearance of RE-33 vaccine within the host was determined by an enrichment technique described by Klesius (1992b).

D. *Edwardsiella ictaluri* Isolates and Experimental Challenge

The isolates used to infect the channel catfish were obtained from cases of enteric septicemia of catfish in the southeastern United States, except for the isolate from Thailand (Table I). All isolates were passed in fish and not grown in or on culture media for more than three passages before use. Experimental challenge was carried out by immersion with *E. ictaluri* isolates (24-hour BHI broth cultures) at a concentration of 1 to 2×10^7 cfu/ml for 1 hour as described by Klesius and Sealey (1995), Klesius and Shoemaker (1997), and Shoemaker and Klesius (1997). An equal number of nonvaccinated (naive) fish were challenged with each group of vaccinated fish challenged. Results of experimental challenge are presented as relative percent survival (RPS) as described by Amend (1981). RPS is calculated according to the following formula:

$$\text{RPS} = 1 - \frac{\% \text{ vaccinate mortality}}{\% \text{ control mortality}} \times 100$$

III. Results

A. Characteristics of *E. ictaluri* RE-33

Edwardsiella ictaluri RE-33 mutant is differentiated from the parent microorganism because it is resistant to 320 µg/ml rifampicin.

TABLE I

Edwardsiella ictaluri Isolates Used in Experiments

Isolate	Source	Location
AL-93-75	Channel catfish with ESC	Alabama
AL-95-58	Channel catfish with ESC	Alabama
AL-96-25	Channel catfish with ESC	Alabama
ATCC-33202	American Type Culture Collection	Georgia[a]
S94-629	Channel catfish with ESC	Mississippi
S94-649	Channel catfish with ESC	Mississippi
S94-707	Channel catfish with ESC	Mississippi
S94-827	Channel catfish with ESC	Mississippi
S94-873	Channel catfish with ESC	Mississippi
S94-1017	Channel catfish with ESC	Mississippi
S94-1051	Channel catfish with ESC	Mississippi
S94-1034	Channel catfish with ESC	Mississippi
EILO	Walking catfish[b]	Thailand

[a]Original isolate described by Hawke (1979) which was isolated from channel catfish.
[b]Reisolated from a channel catfish at our laboratory.

Biochemical characteristics of the *E. ictaluri* RE-33 are identical to *E. ictaluri* as described in *Bergey's Manual of Determinative Bacteriology* (Holt *et al.*, 1994). Silver-stained SDS–PAGE profiles of LPS extracted from *E. ictaluri* RE-33 indicated that the O-chain was not present as compared to the parent *E. ictaluri* EILO lipopolysaccharide. A lower antibody response (data not shown) in channel catfish to *E. ictaluri* RE-33 was observed as determined by agglutination (Chen and Light, 1994) and ELISA (Klesius, 1993).

B. RE-33 Safety and Clearance

Safety of the vaccine administered by immersion was determined in 16,460 channel catfish, blue catfish, and blue catfish × channel catfish (Tables II and III). All vaccinates were alive and free of signs of ESC or mortality resulting from vaccination. Vaccinates have been held at least 4 months without any mortality or signs of ESC at the current vaccine dose of about 10^5 *E. ictaluri* RE-33/ml for a 2-minute exposure. Safety of the vaccine was evaluated at 200 times the dose and 30 times the exposure time (Tables IV and V). No mortality or signs of ESC occurred in the 185 fish tested at this dose and exposure time. Rever-

TABLE II

Safety of Vaccine Dose of *Edwardsiella ictaluri* RE-33 in Channel Catfish, Blue Catfish, and Blue Catfish × Channel Catfish Vaccinates[a,b,c]

Experiment	Date	Species or strains	Numbers vaccinated	Number alive/day postvaccination (DPV)	Signs of ESC or mortality
1	7/96	Marion	60	60/14	None
2	10/96	USDA	125	125/14	None
3	10/96	USDA	1050	1050/60	None
4	11/96	USDA	525	525/30	None
5	12/96	USDA	1100	1100/120	None
6	1/97	USDA	1100	1100/90	None
7	5/97	USDA	3500	3500/90	None
8	5/97	Norris × USDA	3000	3000/90	None
9	5/97	Blue × USDA	3000	3000/90	None
10	5/97	Blue	3000	3000/90	None
		Total	16460	16460	None

[a] Immersion vaccination with about 1×10^5 *E. ictaluri* RE-33/ml for 2 minutes at a density of about 50 fish per liter water at 24–26°C. Immunized fish were kept in 1500-liter fiberglass tanks supplied with recirculating or well water at 18 or 26°C with a flow rate of 0.5 liter/minute. Fish were daily fed commercial catfish ration at 4% of their body weight. Fish were daily observed for mortalities and abnormal behavior.
[b] Minus the fish used for experimental purpose (i.e, challenge experiments).
[c] Fingerlings weighing from 10 to 50 g depending on experiment.

sion to virulence did not occur in the 20 fish tested after *E. ictaluri* RE-33 was passaged in channel catfish. No mortality or signs of ESC were observed in the fish for 35 days after injection. Reversion to virulence did not occur in channel catfish in which the *E. ictaluri* RE-33 was serially passaged five times through channel catfish. The fish used in this experiment were held in the laboratory without signs of ESC or adverse behavior for at least 20 days. *Edwardsiella ictaluri* RE-33 was not isolated from 5 of the fish sampled 12 days after injection. No mortality or signs of ESC were observed for 35 days in 50 fish immersed in *E. ictaluri* RE-33 passed on rifampicin-supplemented and non-rifampicin-supplemented BHI agar plates for 16 passages.

Survival and clearance of *E. ictaluri* RE-33 was evaluated in some vaccinated groups. *Edwardsiella ictaluri* RE-33 was isolated from 6 of 10 channel catfish sampled 14 days after vaccination at 200 times the vaccine dose for 30 times longer exposure (Table V). *Edwardsiella ic-*

TABLE III

Safety of Vaccine Dose of *Edwardsiella ictaluri* RE-33 in Channel Catfish, Blue Catfish, and Blue Catfish × Channel Catfish

Days postvaccination[a]	Total number of fish	Number of fish cultured/culture positive[b]	Number of fish alive minus sampled	Signs of ESC/mortality	Relative percent survival 2 months postvaccination[c]
5	12,500	35/11	12,500	None	ND[d]
11	12,500	35/6	12,500	None	ND
28	12,500	ND	12,500	None	100

[a]Immersion vaccination with 1×10^5 *E. ictaluri* RE-33/ml for 2 minutes at a density of about 50 fish per liter water. Immunized fish were kept in 1500-liter fiberglass tanks supplied with recirculating or well water at 18 or 26°C with a flow rate of 0.5 liter/minute.
[b]Culture technique as described by Klesius (1992b).
[c]Relative percent survival determined on a subsample of 60 vaccinates as compared to 60 nonvaccinated controls (mortality in controls was 67%). Only channel catfish were challenged at 28 days postvaccination.
[d]Not done.

TABLE IV

Two Hundred Times Vaccine Dose of EDWARDSIELLA ICTALURI RE-33 in Channel Catfish at 14 Days Postvaccination (DPV) and Relative Percent Survival[a,b]

Treatment	Number of fish	Signs of ESC/mortality	Alive 12 DPV	Relative percent survival[c]
Vaccinates	125	None	125	94.7
Controls	125	None	125	

[a]Fish immersion vaccinated with 2×10^7 E. ictaluri RE-33/ml for 60 minutes. All conditions as previously described.

[b]Challenge with E. ictaluri AL-93-75 at 2×10^7/ml for 60 minutes as described by Klesius and Sealey (1995) and Shoemaker and Klesius (1997).

[c]Relative percent survival (RPS) as described by Amend (1981); RPS $\geq$ 50% is protection by vaccination. Mortality in the controls was 92%.

taluri RE-33 was also isolated from 11 of 35 catfish sampled 5 days after immersion vaccination at the normal dose and exposure time. However, 11 days after vaccination E. ictaluri RE-33 was only isolated from 6 of 35 fish sampled.

C. Protective Immunity Induced by E. ICTALURI RE-33 Against Enteric Septicemia of Catfish

Vaccination of channel catfish with E. ictaluri RE-33 at 200 times the vaccine dose and for 30 times longer resulted in relative percent

TABLE V

Two Hundred Times Vaccine Dose of EDWARDSIELLA ICTALURI RE-33 on the Survival of 60 Channel Catfish at 12 Days Postvaccination (DPV)

Treatment	Number of fish	Signs of ESC/ mortality	Alive 12 DPV	Culture positive for E. ictaluri[a]
Vaccinates[b]	60	None	60	6 of 10 sampled
Controls[c]	60	None	60	0

[a]E. ictaluri isolated from trunk kidney by enrichment technique of Klesius (1992b). One E. ictaluri colony or more per fish considered culture positive. Mean weight of fingerlings was 15 g.

[b]Vaccinated at 200 times (2×10^7/ml) E. ictaluri RE-33/ml for 30 times longer (60 minutes).

[c]Controls immersed in brain heart infusion broth only.

survival (RPS) of 94.7% after challenge with virulent *E. ictaluri* (Table IV) at 14 days postvaccination (DPV). No fish vaccinated at 200 times the vaccine dose died after vaccination. A positive effect by vaccination is a RPS greater than 50% (Amend, 1981). Of the 12,500 fish vaccinated, the 60 channel catfish challenged showed a relative percent survival of 100% at 28 DPV at the normal dose (about 1×10^5 *E. ictaluri* RE-33/ml) for the 2-minute immersion vaccination (Table III). We also examined the ability of *E. ictaluri* RE-33 to induce protective immunity to 12 different *E. ictaluri* isolates obtained from diseased channel catfish throughout the Southeast and the parent isolate from Thailand (Table I). Relative percent survivals were greater than or equal to 50% at 14 DPV for 8 of 13 isolates tested (RPSs ranged from 50 to 98.3%; Table VI). Five of the 13 isolates had RPS values less than 50% (Table VI). However, by increasing the time after vaccination, protection to isolate S94-694 and S94-707 was demonstrated with RPSs of 96 and 93%, respectively (Table VI). We have demonstrated protective immunity by a single vaccination for as long as 105 days following vaccination (Table VI).

IV. Discussion

The data presented indicate that *E. ictaluri* RE-33 appears to have many properties of the attenuated *B. abortus* RB-51. *Brucella abortus* RB-51 is an approved vaccine strain for brucellosis in cattle. Biochemically, *E. ictaluri* RE-33 is the same as the parent EILO isolate of *E. ictaluri*. *Edwardsiella ictaluri* RE-33 lacks the O-polysaccharide as does *B. abortus* RB-51 (Schurig et al., 1991) and both are resistant to rifampicin. Vaccination with *E. ictaluri* RE-33 resulted in a lower antibody response to *E. ictaluri* presumably because of the lack of the O-polysaccharide. The lack of antibody response to O-polysaccharide has been demonstrated in cattle vaccinated with RB-51 (Schurig et al., 1991). Specific antibody response to intracellular pathogens may have little or no protective effect (Olivier et al., 1985; Portnoy, 1992) and in the case of *B. abortus* and *E. ictaluri* infection it may have a negative effect (Schurig et al., 1991; Klesius and Sealey, 1995).

Survival of *E. ictaluri* RE-33 in vaccinates was demonstrated. Smith (1990) and Thornton et al. (1994) suggest live vaccine strains must gain entry and persist in the host to be effective vaccines. However, *E. ictaluri* RE-33 was cleared at about 14 days postvaccination. Thornton et al. (1991) also demonstrated clearance of live vaccine strains. They felt the clearance was a result of the increased susceptibility of the live

TABLE VI

PROTECTION AGAINST ENTERIC SEPTICEMIA OF CATFISH AFTER IMMERSION VACCINATION[a] OF CHANNEL CATFISH WITH EDWARDSIELLA ICTALURI RE-33 VACCINE

Experiment	Days postvaccination (DPV)	Number of fish vaccinated	E. ictaluri challenge isolate[b]	Relative percent survival[c]	Percent mortality in controls
1	14	60	EILO[d]	51.7	100.0
2	14	60	AL-93-75	98.3	100.0
3	14	125	AL-93-75	96.6	94.6
4	14	60	AL-93-75	96.8	94.4
5	84	60	AL-93-75	93.9	26.4[e]
6	105	60	AL-93-75	78.9	79.3
7	14	60	S94-873	54.0	68.3
8	14	60	S94-1017	71.7	100.0
9	14	60	S94-1051	50.0	100.0
10	14	60	AL-96-25	50.4	96.7
11	14	60	S94-827	47.0	96.7
12	14	60	S94-1034	27.0	98.3
13	14	60	ATCC-33202	6.7	100.0
14	14	125	AL-95-58	78.4	40.8[e]
15	14	125	S94-629	53.3	24.0[e]
16	14	60	S94-694	18.3	100.0
17	42	60	S94-694	96.0	55.0[e]
18	63	60	S94-694	100.0	55.0[e]
19	14	60	S64-707	29.0	91.7
20	42	60	S94-707	93.0	70.0
Total		1395	13 isolates	Mean Protection = 64.6%	

[a] Immersion vaccination with 1×10^5 E. ictaluri RE-33/ml for 2 minutes at a density of about 50 fish per liter of water.
[b] Challenge with E. ictaluri isolates at 1 to 2×10^7/ml for 1 hour described by Klesius and Sealey (1995) and Shoemaker and Klesius (1997).
[c] Relative percent survival (RPS) as determined by Amend (1981). RPSs ≥50% are considered protection by vaccination (Amend, 1981).
[d] Parent of RE-33 vaccine strain.
[e] Data do not fit Amend's criteria of 60% mortality in controls.

vaccine to host lytic factors. Wise et al. (1997) showed clearance of virulent *E. ictaluri* from the blood at 216 hours after infection. *Edwardsiella ictaluri* RE-33 was more susceptible to serumcidal activity (complement) than were virulent *E. ictaluri* isolates (P. H. Klesius, unpublished observation).

Edwardsiella ictaluri RE-33 is safe and no mortality has resulted from vaccination of 16,460 channel catfish, blue catfish, and blue × channel catfish hybrids. Even at 200 times the vaccine dose and for 30 times longer exposure, no ESC or mortality resulted in 185 channel catfish tested at this dose and time. No ESC or mortality was observed in fish injected with RE-33 after passage in channel catfish and/or after passage on media lacking rifampicin. Five serial *in vivo* passages of *E. ictaluri* RE-33 were completed in channel catfish with no reversion to virulence.

In two previous studies, Shoemaker et al. (1997) and Shoemaker and Klesius (1997) demonstrated that acquired immunity was dependent on the ability of channel catfish macrophages to kill *E. ictaluri*. Mean percent killing of *E. ictaluri* by macrophages (1:6 ratio) at 2.5 hours was significantly greater for macrophages obtained from channel catfish vaccinated with live *E. ictaluri* when *E. ictaluri* used in the assay was either opsonized (93.8%) or nonopsonized (75.9%) with anti-*E. ictaluri* antibody. Bactericidal activity of macrophages from nonvaccinated channel catfish for *E. ictaluri* was 46.2 and 55.9% for opsonized and nonopsonized *E. ictaluri*, respectively.

Acquired immunity of long duration to enteric septicemia of catfish was proven possible by live *E. ictaluri* vaccination of channel catfish (Klesius and Shoemaker, 1997; Shoemaker and Klesius, 1997). Protective immunity as demonstrated by RPS indicated that *E. ictaluri* RE-33 vaccine provided protection to 8 of 13 isolates tested. Protection to all isolates was not expected as Klesius and Shoemaker (1997) recently showed that protective immunity to enteric septicemia of catfish was dependent on the isolate used to immunize catfish. However, this study demonstrated that acquired immunity against two of the five isolates, in which protection was not demonstrated at 14 days postvaccination, was produced by holding the fish for longer periods (Table VI).

V. Summary

The present study showed that *E. ictaluri* RE-33 vaccine does not cause ESC but does stimulate protective immunity. The RE-33 vaccinates were protected against ESC for at least 4 months following a single bath immersion in a low number of *E. ictaluri* RE-33 without booster vaccination. Antibody responses are weak after RE-33 vaccination. *Edwardsiella ictaluri* RE-33 vaccine presents no risk or hazard to catfish. RE-33 vaccine will prevent ESC caused by most isolates of *E. ictaluri* in catfish. We recently obtained from USDA, Animal Plant Health Inspection Service (APHIS), and the state veterinarians of Alabama and Mississippi, approval to field test the RE-33 vaccine in young catfish. About 2–3 million 10- to 30-day-old channel catfish in Alabama and Mississippi have been vaccinated since June 1997 with no adverse effects of vaccination.

ACKNOWLEDGMENT

The authors wish to thank Drs. John Plumb and John Grizzle (professors of the Department of Fisheries and Allied Aquacultures, Auburn University, AL) and Andrew Goodwin (assistant professor of the Department of Aquaculture/Fisheries, University of Arkansas at Pine Bluff, AR) for critical review of the manuscript.

REFERENCES

Amend, D. (1981). Potency testing of fish vaccines. *Dev. Biol. Stand.* **49,** 447–454.

Antonio, D. B., and Hedrick, R. P. (1994). Effects of the corticosteroid Kenalog on carrier state of juvenile channel catfish exposed to *Edwardsiella ictaluri. J. Aquat. Anim. Health* **6,** 44–52.

Bader, J. A., Shoemaker, C. A., Klesius, P. H., Connolly, M. A., and Barbaree, J. M. (1998). Genomic subtyping of *Edwardsiella ictaluri* isolated from diseased channel catfish by arbitrarily primed polymerase chain reaction. *J. Aquat. Anim. Health* **10,** 22–27.

Chen, M. F., and Light, T. S. (1994). Specificity of the channel catfish antibody to *Edwardsiella ictaluri. J. Aquat. Anim. Health* **6,** 266–270.

Eisenstein, T. K., Killar, L. M., and Sultzer, B. M. (1984). Immunity to infection with *Salmonella typhimurium:* Mouse-strain differences in vaccine-and serum-mediated protection. *J. Infect. Dis.* **150,** 425–435.

Hawke, J. P. (1979). A bacterium associated with disease of pond cultured channel catfish, *Ictalurus punctatus. J. Fish. Res. Board Can.* **36,** 1508–1512.

Hawke, J. P., McWhorter, A. C., Steigerwalt, A. G., and Brenner, D. J. (1981). *Edwardsiella ictaluri* sp. nov., the causative agent of enteric septicemia of catfish. *Int. J. Syst. Bacteriol.* **31,** 396–400.

Holt, J. G., Krieg, N. R., Sneath, P. H. A., Stanley, J. T., and Williams, S. T. (1994). "Bergey's Manual of Determinative Bacteriology," 9th ed. Williams & Wilkins, Baltimore, MD.

Kasornchandra, J., Rogers, W. A., and Plumb, J. A. (1987). *Edwardsiella ictaluri* from walking catfish *Clarias batrachus* L., in Thailand. *J. Fish Dis.* **10,** 137–138.

Klesius, P. H. (1992a). Immune system of channel catfish: An overture on immunity to *Edwardsiella ictaluri. Annu. Rev. Fish Dis.* **2,** 235–338.

Klesius, P. H. (1992b). Carrier state of channel catfish infected with *Edwardsiella ictaluri. J. Aquat. Anim. Health* **4,** 227–230.

Klesius, P. H. (1993). Rapid enzyme-linked immunosorbent tests for detecting antibodies to *Edwardsiella ictaluri* in channel catfish *Ictalurus punctatus* using exoantigen. *Vet. Immunol. Immunopathol.* **36,** 359–368.
Klesius, P. H., and Sealey, W. M. (1995). Characteristics of serum antibody in enteric septicemia of catfish. *J. Aquat. Anim. Health* **7,** 205–210.
Klesius, P. H., and Shoemaker, C. A. (1997). Heterologous isolates challenge of channel catfish, *Ictalurus punctatus*, immune to *Edwardsiella ictaluri*. *Aquaculture* **157,** 147–155.
Miyazaki, T., and Plumb, J. A. (1985). Histology of *Edwardsiella ictaluri* in channel catfish *Ictalurus punctatus*, Rafinesque. *J. Fish Dis.* **8,** 389–392.
Montaraz, J. A., and Winter, A. J. (1986). Comparison of living and nonliving vaccines for *Brucella abortus* in BALB/c mice. *Infect. Immun.* **53,** 245–251.
Morrison, E. E., and Plumb, J. A. (1994). Olfactory organ of channel catfish as a site of experimental *Edwardsiella ictaluri* infection. *J. Aquat. Anim. Health* **6,** 101–109.
North, R. J. (1974). Cell-mediated immunity and the response to infection. *In* "Mechanisms of Cell-mediated Immunity" (R. T. McCluskey and S. Cohen, eds.), pp. 185–200. Wiley, New York.
North, R. J. (1975). Nature of "memory" in T-cell-mediated antibacterial immunity: Anamnestic production of mediator T-cells. *Infect. Immun.* **12,** 754–760.
Nusbaum, K. E., and Morrison, E. E. (1996). Entry of ^{35}S-labelled *Edwardsiella ictaluri* into channel catfish (*Ictalurus punctatus*). *J. Aquat. Anim. Health* **8,** 146–149.
Olivier, G., Evelyn, T. P. T., and Lallier, R. (1985). Immunogenicity of vaccines from a virulent and an avirulent strain of *Aeromonas salmonicida*. *J. Fish Dis.* **8,** 43–55.
Plumb, J. A. and Klesius, P. (1988). An assessment of the antigenic homogeneity of *Edwardsiella ictaluri* using monoclonal antibody. *J. Fish Dis.* **11,** 499–510.
Plumb, J. A., and Vinitnantharat, S. (1990). Dose titration of sarafloxicin (A-56620) against *Edwardsiella ictaluri* infection in channel catfish. *J. Aquat. Anim. Health* **2,** 194–197.
Plumb, J. A., and Vinitnantharat, S. (1993). Vaccination of channel catfish, *Ictalurus punctatus*, (Rafinesque), by immersion and oral booster against *Edwardsiella ictaluri*. *J. Fish Dis.* **16,** 65–71.
Portnoy, D. A. (1992). Innate immunity to facultative intracellular bacterial pathogens. *Curr. Opin. Immunol.* **4,** 20–24.
Schurig, G. G., Roop, R. M., II, Bagchi T., Boyle, S., Buhrman, D., and Sriranganathan, N. (1991). Biological properties of RB51; a stable rough strain of *Brucella abortus*. *Vet. Microbiol.* **28,** 171–188.
Shoemaker, C. A., and Klesius, P. H. (1997). Protective immunity against enteric septicemia in channel catfish, *Ictalurus punctatus* (Rafinesque), following controlled exposure to *Edwardsiella ictaluri*. *J. Fish Dis.* **20,** 101–108.
Shoemaker, C. A., Klesius, P. H., and Plumb, J. A. (1997). Killing of *Edwardsiella ictaluri* by macrophages from channel catfish immune and susceptible to enteric septicemia of catfish. *Vet. Immunol. Immunopathol.* **58,** 181–190.
Shotts, E. B., Blazer, V. S., and Waltman, W. D. (1986). Pathogenesis of experimental *Edwardsiella ictaluri* infections in channel catfish (*Ictalurus punctatus*). *Can. J. Aquat. Sci.* **43,** 36–42.
Smith, H. (1990). Pathogenicity and the microbe in vivo. *J. Gen. Microbiol.* **136,** 377–383.
Thornton, J. C., Garduno, R. A., Newman, S. G., and Kay, W. W. (1991). Surface-disorganized, attenuated mutants of *Aeromonas salmonicida* as live furunculosis vaccines. *Microb. Pathog.* **11,** 85–99.
Thornton, J. C., Garduno, R. A., and Kay, W. W. (1994). The development of live vaccines

for furunculosis lacking the A-layer and O-antigen of *Aeromonas salmonicida. J. Fish Dis.* **17,** 195–204.

Thune, R. L., Hawke, J. P., and Johnson, M. C. (1994). Studies on vaccination of channel catfish, *Ictalurus punctatus,* against *Edwardsiella ictaluri. J. Appl. Aquacul.* **3,** 11–23.

Waltman, W. D., and Shotts, E. B. (1986). Antimicrobial susceptibility of *Edwardsiella ictaluri. J. Wildl. Dis.* **22,** 173–177.

Winter, A. J., Schurig, G. G., Boyle, S. M., Sriranganathan, N., Bevins, J. S., Enright, F. M., Elzer, P. H., and Kopec, J. D. (1996). Protection of BALB/c mice against homologous and heterologous species of *Brucella* by rough strain vaccines derived from *Brucella melitensis* and *Brucella suis* biovar 4. *Am. J. Vet. Res.* **57,** 677–683.

Wise, D. J., Schwedler, T. E., and Terhune, J. S. (1997). Uptake and clearance of *Edwardsiella ictaluri* in the peripheral blood of channel catfish *Ictalurus punctatus* fingerlings during immersion challenge. *J. World Aquacult. Soc.* **28,** 45–51.

Fish Vaccines

SOMSAK VINITNANTHARAT,* KJERSTI GRAVNINGEN,† AND EVAN GREGER*

Alpharma NW, Aquatic Animal Health Division, Bellevue, Washington 98005, †Alpharma AS, Aquatic Animal Health Division, Oslo, Norway

I. Introduction
II. Immunization Methods in Fish
    A. Immersion Vaccination
    B. Injection Vaccination
    C. Oral Vaccination
III. Benefit of Using Vaccine
IV. Vaccines against Some Specific Diseases
    A. Vaccine against *Yersinia ruckeri*
    B. Vaccine against *Vibrio viscosus*
    C. Polyvalent Vaccine against *Vibrio anguillarum* Serotypes I and II, *V. Salmonicida,* and *A. Salmonicida*
V. Summary
    References

I. Introduction

During the past 30 years, aquaculture has grown into a very significant industry in many parts of the world. Salmonid production in Norway increased from 50,000 metric tons in 1986 to 241,000 metric tons in 1995. Japan produced 150,000 metric tons of yellowtail in 1996. Production of channel catfish in the United States increased from 21,000 metric tons in 1980 to 214,000 metric tons in 1996 (Fig. 1). With increased aquaculture, production husbandry practices have evolved from extensive or semi-intensive to intensive or superintensive. Under conditions of high population densities, infectious diseases pose a constant and costly threat to successful animal husbandry. Even when environmental conditions are good and fish are healthy, certain infec-

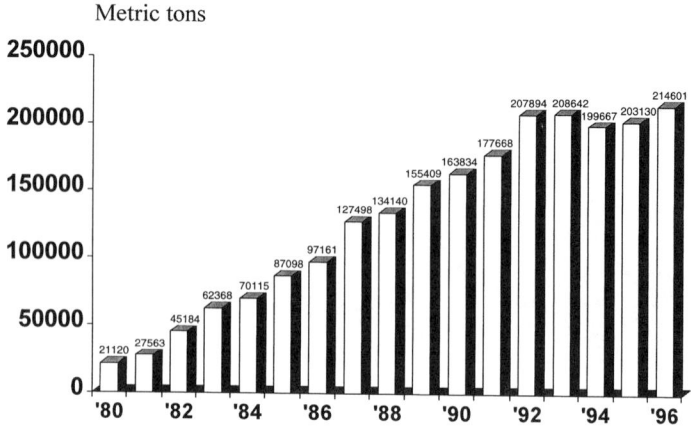

FIG. 1. Channel catfish production in United States from 1980 to 1996.

tious agents, if introduced into the farm, are so virulent that mass mortalities can and do occur. Antibiotics provide a useful means to control many bacterial diseases but often do not provide a satisfactory solution. Many problems are associated with antibiotic use including the development of antibiotic resistance. Recurrent outbreaks due to bacterial resistant strains necessitate further, costly treatments. Antibiotics combined into the feed is the only mass means of administration. Chemotherapy is not as successful as desired, because highly infected fish do not eat and cannot be medicated. The general lack of registered and approved therapeutic chemical and antibiotics, only oxytetracycline and Romet, for aquaculture necessitates other means of disease control. The successful use of vaccines in warm-blooded animals and the decreasing effectiveness of antibiotics in controlling bacterial fish diseases have led to the development of vaccines for fish.

In 1976 a vaccine against enteric redmouth (ERM) disease, caused by gram-negative bacteria *Yersinia ruckeri,* was the first commercial product licensed by the U.S. Department of Agriculture for fish. Since then there have been many more fish vaccines commercialized including vaccines against *Vibrio anguillarium, Vibrio ordalii, Vibrio salmonicida, Vibrio viscosus, Aeromonas salmonicida,* and *Photobacterium damsela* subsp. *piscicida.* A variety of vaccine compositions are available. Methods of administration, efficacy, and benefits of using vaccine in fish are discussed.

II. Immunization Methods in Fish

In general, there are three different methods for vaccination in fish (Ellis, 1988): immersion, injection, and oral. Each method has advantages and disadvantages (Table I).

A. Immersion Vaccination

Immersion vaccination is an easy and effective method for immunizing fish. Fish are immersed in a dilute vaccine solution for a short period of time and released into the culture unit, typically ponds, raceways, or net pens. Use of this method is somewhat limited to certain operations. For instance, for fish that will not be moved after they are stocked into the culture unit, the immersion vaccination can be used only at stocking time. The duration of protection, between 3 and 12 months, is not long enough for the culture cycle of some fish species. Immersion vaccination is also cost prohibitive for larger size fish.

B. Injection Vaccination

Intraperitoneal injection (IP) vaccination is the most effective way of immunizing fish. The injection method also allows the use of adjuvants, especially oil adjuvant, that will prolong the protection compared to immersion vaccination. Fish are anesthetized and injected IP with vaccine then returned to clean water to recover. Commercial operations use repeating injection guns, both manually and automatically operated, that allow each operator to inject 1000–2000 fish per hour. The system typically consists of an anesthetization tank, an injection table with injecting gun connected to a vaccine bottle or bag, and a

TABLE I

Comparison of Three Different Methods of Fish Vaccination

	Immersion	Injection	Oral
Application	Easy	Moderate	Very easy
Stress	Light	Moderate	None
Labor	Inexpensive	Intensive	None
Effectiveness	Good	Excellent	Fair
Duration	3–12 Months	12–24 Months	2–4 Months

recovery tank. Injection vaccination is very labor intensive. The size of the fish limits the use of the injection method. In general, for a fish weighing less than 5 g the injection method is not practical. Some other disadvantages of injection vaccination are adhesion formation, temporary reduced feeding, inadvertent puncture of the intestine, and creation of a wound that could provide a portal of entry for infectious agents.

C. Oral Vaccination

Oral vaccination is the most convenient way to immunize fish because the vaccine can be administered at any time on any size of fish during the culture cycle and in all types of culture systems. The vaccine is either incorporated in or adhered to feed and then fed to fish. It is the least stressful method because handling is not required. As with immersion, oral vaccination is not cost effective when attempting to immunize larger fish. Oral vaccination gives the least efficacy compared to the immersion and injection methods. The major problem seems to be the destruction and absorption of antigens by the fish digestive system. Further research is needed to develop methods to protect antigens from getting destroyed by the digestive system and improve absorption in order to improve efficacy of oral vaccination.

III. Benefit of Using Vaccine

Efficacious vaccines protect immunized fish against specific diseases and, as a result, the reduction of losses due to diseases. Once the fish acquires protection against certain diseases, production becomes more predictable. Efficacy testing of fish vaccine is accomplished by actual challenge with live organism at a specific time interval postvaccination. Relative percent survival (RPS) is used to evaluate vaccine efficacy. The higher the RPS value the better the protection:

$$\text{RPS} = 1 - \frac{(\% \text{ mortality in vaccinated fish})}{\% \text{ mortality in control fish}} \times 100$$

Vaccines are used as a preventive measure, that is, before disease is anticipated, as opposed to treatment with antibiotics after disease has occurred. In the past before vaccines were introduced on the fish farm, antibiotics were used as a means for controlling diseases. Antibiotics proved useful by helping to control many bacterial diseases but they are not the final answer. Chemotherapy is not as successful as desired, because highly infected fish do not eat and cannot be medicated. Many problems are associated with antibiotics including the development of antibiotic resistance and recurrent outbreaks necessitating further, costly treatments.

Vaccines for aquaculture are very successful in reducing the amount of antibiotics used such as is done for aquaculture in Norway. In 1990 the amount of antibiotic used was 231g for each kilogram of salmonids produced when furunculosis disease occurred. A vaccine against furunculosis was introduced in late 1990, and 4 years later the amount of antibiotics used per kilogram of salmonids produced in Norway dropped to only 6.4 g.

Since the first commercial vaccine, vaccine against ERM, was introduced to aquaculture, many more vaccines have become commercially available to fish farmers, including vaccines against the viral disease infectious pancreatic necrosis (IPN), vaccine against bacterial diseases caused by *P. damsela* subsp. *piscicida, V. anguillarum, V. ordalii, V. salmonicida, V. viscosus,* and *A. salmonicida.*

IV. Vaccines against Some Specific Diseases

A. Vaccine against *Yersinia ruckeri*

Yersinia ruckeri is the causative agent of enteric redmouth disease. It is a gram-negative rod bacteria of the family enterobacteriaceae. It was first isolated from rainbow trout (*Onchorhyncus mykiss*) in the 1950s by Rucker (Ross et al., 1966). Enteric redmouth is a subacute to acute systemic infection. Clinical signs include reddening of the mouth and opercula, inflammation and erosion of the jaws and palate, hemorrhaging at the base of fins, and exophthalmia. Internally, hemorrhages may occur in muscle and intestine that may also contain a yellow fluid.

Vaccine against ERM disease was the first commercial product licensed by the U.S. Department of Agriculture for fish in 1976. In the past vaccine against ERM was commonly delivered by immersion. In this experiment we have compared the efficacy of immersion vaccine to oral vaccine. Eight groups of 250 rainbow trout were vaccinated by different vaccination regimes. Seven groups were vaccinated by immersion or orally at 6 g, and two groups were given an oral booster 3 months later; fish size was 30 g. The immersion vaccine contained 3–6 $\times 10^9$ colony forming units (cfu)/ml. All groups, including the nonvaccinated controls, were marked, mixed in two parallel tanks, and challenged with *Y. ruckeri* serotype -O1 three and six months after the second vaccination. The challenge was introduced by intraperitoneal injection with 4.6×10^5 and 1.5×10^5 cfu/fish at 3 and 6 months, respectively, after second vaccination. Cumulative mortality in the nonvaccinated group was 90% in the first and 56% in the second challenge. The RPS against *Y. ruckeri* in the group vaccinated once by immersion was low, 6–53%, 6 months after vaccination, and nearly zero 9 months after vaccination. Both of the groups that were given an oral booster using either immersion or oral as the primary dose of vaccination showed good protection against *Y. ruckeri* in challenges 3 and 6 months after second vaccination.

All vaccinated groups showed significantly better survival than the nonvaccinated controls after the first challenge. Only groups given an oral booster (groups 5 and 7, Table II) were significantly better than the nonvaccinated control fish in the second challenge. No significant difference between groups 5 and 7 was observed in any challenge, nor were differences observed between the groups vaccinated once by immersion (groups 2, 3 and 4, Table II). Fish vaccinated once orally at 6 g survived significantly better than fish immersed at 6 g after the first challenge, but in the second challenge neither group vaccinated only once showed significant protection. Vaccination of rainbow trout is normally performed once by immersion in the hatchery. The fish are transferred to the growout farm at average weight of less than 6 g. This means that vaccination will be performed on smaller fish during transfer from hatchery to the growout farm. The results obtained in this experiment show that the longevity of vaccination by immersion at this stage will protect the fish for less than 9 months, and will therefore not protect the fish throughout the production cycle. To make fish farming profitable, it is critical to protect the high value fish of 100g and more. To maintain the protection against yersiniosis throughout the production cycle, a booster vaccination is needed. Oral vaccination would be preferable and the only realistic alternative for revaccination of rainbow trout. A vaccination regime based on two oral vaccinations should give sufficient protection and is easy for farmers to administer. This immunization method causes minimum stress on the fish.

B. Vaccine against *Vibrio viscosus*

Vintersår or "winter ulcer" was first identified in Norwegian Atlantic salmon operations in the early 1980s, and was registered at 50 fish farms by the National Veterinary Institute in 1990 (Lunder, 1992; Anonymous, 1991). The disease typically occurs at sites in Iceland and Norway from February through April when water temperatures are

TABLE II
EFFICACY OF VACCINE AGAINST *YERSINIA RUCKERI*[a]

Group	Regime — First vaccination	Regime — Second vaccination	Challenge 1 (3 month) Mortality/total	Challenge 1 (3 month) RPS	Challenge 1 (3 month) RPS average	Challenge 2 (6 month) Mortality/total	Challenge 2 (6 month) RPS	Challenge 2 (6 month) RPS average
1a	PBS	No	45/50			24/50		
1b		No	45/50			32/50		
2a	Fatro	No	24/50	47[b]		29/50	-21	
2b		No	21/50	53[b]	50[b]	30/50	6	-7
3a	Al1	No	38/50	16[b]		18/50	25	
3b		No	30/50	33[b]	25[b]	29/50	9	17
4a	Al2	No	31/50	31[b]		21/50	13	
4b		No	32/50	29[b]	30[b]	28/50	13	13
5a	Al2	AO	15/50	67[b]		8/50	67	
5b		AO	8/50	82[b]	75[b]	9/50	72	69[b]
6a	AO	No	14/50	69[b]		20/50	17	
6b		No	17/50	62[b]	65[b]	34/50	-6	5
7a	AO	AO	13/50	71[b]		9/50	63	
7b		AO	11/50	76[b]	74[b]	7/50	78	70[b]
8a	No	AO	30/50	33[b]		26/50	-8	
8b		AO	22/50	51[b]	42[b]	39/50	-22	-15

[a]Number of dead fish relative to total number of fish within each group during challenge 3 and 6 months after last vaccination. Eight groups are designated 1–8 with parallel tanks a and b. RPS is calculated based on the mortality of vaccinated groups compared to nonvaccinated fish group 1.
[b]Significant difference from group 1 ($p < 0.05$).

below 8°C. Survivors of infection typically recover in the spring when water temperatures rise above 8°C. The disease infects juvenile and adult salmon and trout raised in saltwater, and also occurs at freshwater hatcheries when seawater is added for smolt acclimation (Lunder, 1992). The disease typically involves shallow, superficial lesions on scale-covered tissue which develop into penetrating ulcers. Early investigations suggested that winter ulcers resulted from mechanical disruption of vesicles formed after vascular thrombosis of dermal vessels, and were influenced by high levels of dietary iron (Salte et al., 1994). However, an infectious etiology was strongly suggested when two Vibrio species, V. viscosus and V. wodanis, were frequently isolated from kidneys of Atlantic salmon during winter ulcer outbreaks at eight different farms, and identified in situ in degenerative muscle tissue using immunohistochemistry (Lunder, 1992; Lunder et al., 1995). Subcutaneous injection of Atlantic salmon in saltwater with V. viscosus produced a disease condition that resembled winter ulcer, providing additional evidence supporting a bacterial etiology. Healthy Atlantic salmon held in tanks with fish infected during a natural outbreak also developed ulcers, indicating that the disease could be transmitted horizontally. In addition, mechanical disruption of the skin was shown to be a predisposing factor (Lunder et al., 1995). Though mortality during winter ulcer outbreaks is typically 0–10%, up to 20% mortality has occurred in a 1-month period at certain farms. The disease has been reported in up to 50% of adult fish prior to slaughter (Lunder, 1992). Lesion formation necessitates downgrading from "superior" to "ordinary" or "production" quality fish, and a reduction in market value. Infections of adult fish and losses due to mortality and downgrading represent a significant economic loss in affected Atlantic salmon farms in Norway and Iceland.

Though the scientific name is not currently accepted nomenclature, V. viscosus is used here for practical purposes, referring to Vibrio phenon 1 in Tor Lunder's Ph.D. thesis (1992). An IP challenge model for V. viscosus was developed in rainbow trout. Injection of live bacteria produced signs typical of a septicemia, concentration-dependent mortality, and the ability to reisolate the bacteria from liver tissue of mortalities. The challenge model was used to evaluate potency of a multivalent, formalin-killed, mineral oil adjuvanted vaccine. Rainbow trout (9.5 g) were vaccinated with 0.2 cc of a vaccine containing V. viscosus, V. anguillarum (serotypes I and II), V. salmonicida, and A. salmonicida. Vaccinated fish were held in 32-gal freshwater tanks (10°C) at a 25 gallon per hour (gph) flow rate and fed 1/16-in. pelleted feed at 2% body weight.

At 21 and 43 days postvaccination, 10.1 and 19.1 g fish were injected intraperitoneally with *V. viscosus* cultured in a specialized media to an optical density of 2.2–2.4 at 560 nm. Injections were performed with a 1.0-cc syringe using a 26-guage, 3/8-in. needle. Four replicate vaccinate and control groups of 24 fish each were placed in flow-through 16.6-liter tanks supplied with 10°C freshwater at a 5 gph flow rate, and mortality was recorded daily for 21 days. Challenge 21 days postvaccination produced a cumulative percent mortality of 2% in vaccinates and 52% in control fish, with an RPS of 96%. At 43 days, a challenge that killed 83% of control fish produced a 1% mortality in vaccinated fish, and an RPS of 98%. The high RPS in vaccinated fish challenged with *V. viscosus* demonstrates that vintersår can be successfully prevented by vaccination.

C. POLYVALENT VACCINE AGAINST *VIBRIO ANGUILLARUM* SEROTYPES I AND II, *V. SALMONICIDA,* AND *A. SALMONICIDA*

At the beginning of the fish vaccine era, most of the commercial vaccines were introduced to fish by immersion of monovalent vaccine but by the early 1990s the oil adjuvant technology was incorporated into fish vaccines and the fish were immunized by IP with polyvalent vaccines instead of monovalent vaccine. This chapter briefly discusses the etiologic agents and the efficacy of the polyvalent vaccine against *V. anguillarum, A. salmonicida,* and *V. salmonicida.*

Vibrio salmonicida is the causative agent of cold-water vibriosis or Hitra disease (Egidius *et al.,* 1981). It is classified in the family Vibrioaceae. It is a short, 0.3–1.0 μm by 1.5–1.8 μm, curved rod, peritrichous flagellated. The bacterium is halophilic and requires at least 0.5% NaCl in the medium for growth. It is sensitive to vibriostat 0/129 (2,4-diamino 6,7-diisopropyl pteridine phosphate) and novobiocin. Hitra was one of the most serious diseases in Norwegian fish farming in the 1980s. The disease derives its nickname from heavy outbreaks in the island of Hitra in 1979 and 1980. Outbreaks of cold-water vibriosis have also been reported from Shetland, Faroe Islands, and Canada. The typical external signs include hemorrhaging of the skin, the area around the gills, and the vent. Internally hemorrhaging may be evident in all the organs and also muscle. The liver may be pale. Histologically, necrosis can be observed in the kidney, muscle, gastrointestinal tract, spleen, and gills. These clinical signs were different from the clinical signs of disease caused by *A. salmonicida.*

Aeromonas salmonicida is gram negative, nonmotile, and cytochrome oxidase positive. The majority of the strains produce a brown diffusing pigment. *Aeromonas salmonicida* is the causative agent for furunculosis disease (Bullock and Stuckey, 1975). The disease is named after the raised liquefactive muscle lesions (furuncles) that sometimes occur in chronically infected fish although these lesions are rarely seen in acute infection (McCarthy and Roberts, 1980). The clinical signs include darkening and going off feed. Internally, the viscera are hemorrhagic, the kidney tissue is very soft, the spleen is enlarged and the liver is very pale or mottled with petechiae. In the subacute form skin lesions are present. Internally, there is intestinal inflammation and hemorrhaging in various organs.

Vibrio anguillarum is gram negative and of the family Vibrioaceae. It is a short curved rod, is motile, and is cytochrome oxidase positive. It is sensitive to vibriostat 0/129 and novobiocin. *Vibrio anguillarum* is the causative agent of vibriosis disease (Bullock, 1977). This disease is distributed worldwide in cultured fish, principally in marine environments, but sporadic outbreaks have occurred in freshwater. The disease is normally a generalized septicemia with clinical signs indistinguishable from other bacterial septicemias. The disease ranges from peracute (mortalities without gross lesion) and acute (hemorrhaging of the eyes, gills, vent, skin and internal organs, ascites fluid in body cavity) to subacute and chronic (hemorrhagic ulcerations of the skin and underlying muscle).

Vaccine against vibriosis, cold-water vibriosis, and furunculosis has been available to fish farmers for some time but as monovalent and immersion vaccine. We discuss the efficacy of the polyvalent oil adjuvant injectible vaccine against *V. salmonicida, A. salmonicida,* and *V. anguillarum* (Fig. 2). Four hundred rainbow trout, average size of 5.36 g, were immunized by IP with the polyvalent oil adjuvant commercial product, Bioject 1900. They were kept at 15°C in a flow-through circular tank with the flow rate of 24 gallons per hour (gph).The other 400 fish were kept as nonimmunized controls. Three weeks postvaccination they were challenged with *V. anguillarum* serotypes I and II (causative agent of vibriosis), and *V. salmonicida* (causative agent of Hitra disease) by IP. Fish challenged with *V. anguillarum* were kept in a flow through tank at 15°C with a flow rate of 5 gph. Fish challenged with *V. salmonicida* were kept in the flow-through tank at 10°C with a flow rate of 5 gph. The mortality was recorded for a period of 14 days. The dead fish were confirmed for specific mortality with antiserum. At the end of 14 days the mortality in the control fish challenged with *V. anguillarum* serotype I was 87% and there was no mortality in vaccinated fish. The RPS was 100. The results of the challenge with *V. anguillarum* serotype II also gave an RPS of 100 (80% mortality in

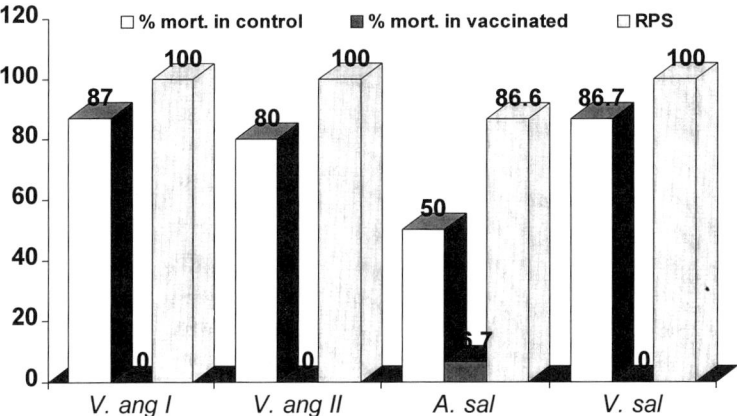

FIG. 2. Potency of polyvalent vaccine in rainbow trout challenged with four bacterial pathegens.

control fish and no mortality in vaccinated fish). The RPS from *V. salmonicida* challenged was 100% (86.7% mortality in control and no mortality in vaccinated fish). The fish were challenged with *A. salmonicida* (causative agent of furunculosis) 5 weeks postimmunization by intramuscular injection. They were kept at 15°C in a flow-through circular tank with a flow rate of 5 gph. The mortality was recorded for a period of 14 days. The specific mortality was confirmed with antiserum. At the end of 14 days the mortality in control fish due to *A. salmonicida* was 50% and the mortality in vaccinated fish was 6.7%, resulting in an RPS of 86.6.

Data indicated that this oil adjuvant polyvalent vaccine protects the fish from each disease very well. The effectiveness of vaccine from different commercial companies varies somewhat depending on how they grow the organism. Media composition, growth conditions, and the downstream processing will all affect the efficacy of final product.

V. Summary

Fish vaccines can be delivered the same way we immunize warmblooded animals. Fish can be immunized by immersion in vaccine for a short period of time—30 seconds to 2 minutes. They can be immunized by injection, intramuscularly or intraperitoneally, and orally by mixing vaccines with feed either by top dressing or by incorporating into feed as an ingredient. Fish also respond to vaccine the same way as other animals do, but since fish are cold-blooded animals, the response to vaccine depends largely on the water temperature. In general, the higher the water temperature, the faster the immune response of fish to the vaccine.

During the past 20 years fish vccines have become an established, proven, and cost-effective method of controlling certain infectious diseases in aquaculture worldwide. Fish vaccines can significantly reduce specific disease-related losses resulting in a reduction of antibiotics use. The final result is the decrease of overall unit costs and more predictable production. Fish vaccines are advantageous over antibiotics because they are natural biological materials that leave no residue in the product or environment, and therefore will not induce a resistant strain of the disease organism. Fish vaccines are licensed by the federal government and closely regulated in the same manner as all other veterinary vaccines to ensure safety, potency, and efficacy.

Even though commercial vaccines for aquaculture work really well in terms of protecting the fish against certain diseases, they should be used only as part of the overall fish health management program, because fish vaccines are not a cure-all. Animal husbandry is still the key to success in aquaculture.

References

Anonymous (1991). "RAPP." The Royal Norwegian Ministry of Agriculture, Department of Veterinary Services, Oslo.
Bullock, G. L. (1977). Vibriosis in fish. *U.S. Fish Wild. Serv., Fish Dis. Leaf.* **50,** 1–11.
Bullock, G. L., and Stuckey, H. M. (1975). *Aeromonas salmonicida*: Detection of asymptomatially infected trout. *Prog. Fish-Cult.* **37,** 237–239.
Egidius, E., Andersen, K., Clausen, E., and Raa, J. (1981). Cold Water vibriosis or "Hitra disease" in Norwegian salmon farming. *J. Fish Dis.* **4,** 353–354.
Ellis, A. E. (1988). General principle of fish vaccination. In "Fish Vaccination" (A. E. Ellis, ed.), pp. 1–19. Academic Press, Orlando, FL.
Lunder, T. (1992). "Winter ulcer" in Atlantic salmon: A study of pathological changes, transmissibility, and bacterial isolates. Ph.D. Thesis, National Veterinary Institute.
Lunder, T., Evensen, O., Holstad, G., and Hastein, T. (1995). 'Winter ulcer' in the Atlantic salmon *Salmo salar*. Pathological and bacteriological investigations and transmission experiments. *Dis. Aqua. Org.* **23,** 39–40.
McCarthy, D. H., and Roberts, R. J. (1980). Furunculosis of fish-the present state of our knowledge. In *"Advances in Aquatic Microbiology"* (M. R. Droop and H. W. Jannasch, eds.), pp. 293–341. Academic Press, London.
Ross, A., Rucker, R., and Ewing, W. (1966). Description of a bacterium associated with redmouth disease of rainbow trout (*Salmo giardneri*). *Can. J. Microbiol.* **12,** 763–770.
Salte, R., Rorvik, K. A., and Reed, E. (1994). Winter ulcers of the skin in Atlantic salmon, *Salmo salar* L: Pathogenesis and possible aetiology. *J. Fish Dis.* **17,** 661–665.

Cross-Species Vaccination in Wild and Exotic Animals

JOSEPH F. CURLEE, JR.

United Vaccines, Inc., Madison, Wisconsin 53713

I. Introduction
II. Canine Distemper Virus
III. *Clostridium botulinum* Type C
IV. Summary
References

I. Introduction

Organisms exist that may produce morbidity and mortality in a wide variety of animal species. In some cases, vaccines intended for immunization against a particular disease agent in one species may also confer protection to unrelated species. This article discusses the disease caused by canine distemper virus and *Clostridium botulinum* type C in different types of animals. Historical laboratory and field data are cited concerning the safety and efficacy of off-lable use of these two particular vaccines.

II. Canine Distemper Virus

Canine distemper virus is a morbillivirus classified in the family Paramyxoviridae. The disease has a worldwide distribution with varying forms of morbidity and mortality, depending on the species affected. It is known to affect 8 of the 11 families of carnivores. These include Ailuridae, Ailuropodidae, Canidae, Hyaenidae, Mustelidae,

Procyomidae, Viverridae, and Felidae (Montali et al., 1987). It has also been reported in Ursidae (bears), but is usually subclinical (Poston and England, 1992). Canine distemper virus has been associated with marine mamals such as seals, dolphins, and porpoises. In 1988 in the North and Baltic Sea area of Northern Europe there was a massive outbreak of distemper in Phocine species. Particularly affected were the harbor seal (Phoca vitulina). The same year in the Lake Baikal region of Siberia there was significant mortality in Siberian seals (Phoca sibircca). There was no clear epidemiologic connection between the outbreak in marine seals in Europe and the outbreak in freshwater Lake Baikal seals, which were several thousand kilometers away. Further work established that the European and Siberian isolates were very similar to canine distemper virus (Mamaev, et al., 1996). Further studies (Blixenkrone-Møller et al., 1989) in Denmark showed evidence that the Phocine distemper virus would infect naive mink and that vaccination with modified live canine distemper virus vaccine licensed for mink would provide protection against Phocine distemper virus (Blixenkrone-Møller et al., 1989). It was also speculated that several distemper outbreaks in mink were caused by contact with infective Phocine distemper virus tissues. That same year, a study found that harbor seals were protected from Phocine distemper virus challenge using certain inactivated canine distemper virus vaccines (Visser et al., 1989). A study in the United Kingdom showed that gray seals (Halichoerus grypus) produced high antibody levels to attenuated canine distemper virus following intramuscular injection (Hughes et al., 1992).

There have been numerous accounts of canine distemper virus vaccination in terrestrial exotic species. Some articles have described canine distemper virus infection following inoculation of modified live canine distemper virus vaccines, usually of canine origin. The affected species were red pandas (Ailurus fulgens), black-footed ferrets (Mustela nicripes), kinkajous (Potos flavus), gray foxes (Urocyon cineieoargenteus), possibly African cape hunting dogs (Lycaon pictus), and fennec foxes (Montali et al., 1987). Currently, only killed canine distemper virus vaccine is used in the black-footed ferret. The new generation of genetically altered vaccines may prove to be of benefit, but much data must be collected, especially with regard to efficacy.

Killed canine distemper virus vaccines have been used, but follow-up studies in many species have shown little or no serologic evidence of protection. Avian origin modified live canine distemper virus vaccine was shown safe and/or immunogenic in gray foxes, bush dogs, maned wolves, and fennec foxes (Montali et al., 1987). Safety was evident in a study conducted in North American badgers (Taxidea taxus) (Goodrich et al., 1994). A field study was conducted in raccoons (Procyon lator) by

United Vaccines and the Willowbrook Wildlife Haven to access the safety of a combination avian origin modified live canine distemper virus and killed mink enteritis vaccine (Hulsebos, 1993). No adverse reactions were noted in any of the 262 animals vaccinated up to three times. Because animals were not experimentally challenged, the efficacy of vaccination could not be assessed; however, canine distemper virus and canine parvovirus were endemic in the area and were manifested clinically in numerous unvaccinated raccoons delivered to the shelter. United Vaccines and Dr. Christine Miller of the Miami Metro Zoo measured serum antibody response to vaccination of fennec fox pups with avian origin modified live canine distemper virus. All pups showed an SN titer response of >1:64. Challenge studies were not performed to determine if the titer was protective (Hulsebos, 1994).

Canine distemper virus in large felids was first reported in 1981 by Cook and Wilcox. In the past few years there have been canine distemper virus outbreaks in large felids in zoos in the United States. Exposure is thought to be from local wildlife. A study by Harder *et al.*, (1996) found that CDV isolates from recent outbreaks in the United States and Europe were distinct from vaccine strains and that isolates clustered according to geographical locations instead of host species origin. In 1994 an epidemic of canine distemper virus in Serengeti lions claimed approximately one-third of the lion population. In addition, large numbers of bat-eared foxes, spotted hyenas, and possibly leopards have died (DVM Newsmagazine, 1995). A vaccination program has been initiated in the wild dog population in an attempt to prevent future outbreaks.

III. *Clostridium botulinum* Type C

Toxins of *C. botulinum* are some of the most potent known. Eight different types of botulism have been identifed (A, B, C_1, C_2, D, E, F, and G) based on antigenically distinct neurotoxins. Identification is performed by inoculating mice with type-specific antitoxin. Animals are then inoculated with sera or intestinal contents of the dead animal. Types A, B, and E are usually confined to human disease, whereas types C and D are usually associated with animals (Barsanti, 1990). Botulism is usually caused by ingestion of preformed toxin. The toxin is abosrbed in the stomach where it is distributed to the neuromuscular junctions of cholinergic nerves and prevents presynaptic release of acetylcholine. General clinical signs are an ascending paralysis. Death is usually caused by respiratory paralysis.

A variety of animals have been reported to be susceptible to type C botulism. Among those are the domestic dog, mink cattle, horse, green sea turtles (*Chelonia mydas*), sheep, and numerous waterfowl, poultry, pheasants, gulls, and cranes. It has also been reported in lions (Barsanti, 1990).

In cattle (types C and D) and sheep (type C) the disease has generally been seen in areas of the world deficient in phosphorus (South Africa, Australia, southwest United States and South America). Animals suffering from pica may ingest the toxin from animal remains. Other reports have been from the consumption of silage containing dead carcasses (Gray and Bulgin, 1982) or contaminated broiler litter (Pugh *et al.*, 1994; McIlroy *et al.*, 1987). In the latter case, animals destined to graze an area covered in contaminated litter, where several deaths were seen, were vaccinated with 10 ml of United Vaccines type C toxoid 14 days apart. Clinical signs of botulism were not seen in the vaccinated animals.

In green sea turtles, type C botulism causes paralysis of the pectoral muscles and is referred to as "floppy flipper disease." United Vaccines (personal communication between Roger Brady, United Vaccines, and Fern Wood, Cayman Turtle Farm Ltd., February 1993) has supplied type C toxoid for more than 20 years for immunization of sea turtles in the Carribean. There is little, if any, published data available concerning safety and efficacy.

Mink are extremely susceptible to type C botulism. Once mink ingest preformed toxin, paralysis of the rear limbs can occur as early as 6 hours with death in 12 hours (personal experience/observation). In a recent outbreak of botulism in Norway, approximately 150,000 unvaccinated mink died; whereas a few farms that vaccinated were protected. Type C specific botulism vaccine has been available to the mink producers for approximately 50 years. Efficacy and saftety are well documented.

Outbreaks of type C botulism have been reported in poultry (Harrigan, 1980; Wilcox and Pass, 1980; Haagsma, 1974). In each case the cause was the lack of removal of carcasses that were then cannibalized by others of the flock. Good sanitation practices eliminated the problem. However, one study showed that chickens vaccinated at 2 weeks of age with a type C botulism mink vaccine were provided with acceptable protection following toxin challenge at 6 and 8 weeks of age, but not at 3 weeks of age (Dohms *et al.*, 1981).

In pheasants, type C botulism is usually ingested from toxin-laden maggots found in decomposing carcasses. The larval stages of blowflies (Calliphoridae) and flesh flies (Sarcophagidae) usually carry the highest toxin load (Forey and Abinanti, 1980). It was shown that one decomposing bird carcass may contain up to 5000 maggots, which would be sufficient to kill 40,000 pheasants (Graham, 1978).

Type C botulism epidemics in waterfowl (ducks, geese, swans) have been dramatic. Avian botulism has been reported in every continent except Antarctica. Outbreaks in the United States have been recorded since 1910, most occurring west of the Mississippi River. In 1952 it was estimated that 4 to 5 million waterfowl died in the western United States (Friend et al., 1985). The disease is characterized by a flacial paralysis, which ascends to the neck (limberneck). Most animals die due to respiratory paralysis or drowning. Waterfowl ingest the toxin from maggots found in dead carcasses in contaminated brackish water. It has been shown that a duck can be intoxicated by ingesting only two to four maggots (Olsen, 1994).

Many zoos, which maintain susceptible birds, routinely vaccinate against type C botulism. In 1992 Dr. Richard Cambre of the Denver Zoological Gardens conducted a vaccine trial using United Vaccines mink botulsim vaccine (Cambre and Kenny, 1993). The Denver Zoological Gardens and surrounding areas had 7 consecutive years of summer botulism outbreaks. Two hundred seventy-three birds of 42 species were inoculated with 1 ml of vaccine SQ on the craniodorsal throat between the wings. A booster vaccination was given approximately 1 month later. None of the vaccinated birds became sick or died due to botulism. During that same time, five unvaccinated nesting ducks and numerous wild waterfowl were found dead.

IV. Summary

These are two examples of organisms which may cause morbidity and/or mortality among numerous unrelated species. Since it is cost prohibitive in most instances to have a biological licensed for wild or exotic species, it remains a challenge to the zoo or wildlife veterinarian to determine if a licensed vaccine for other species is safe and efficacious for a particular exotic species.

References

Barsanti, J. A. (1990). "Infectious Diseases of the Dog and Cat". Saunders, Philadelphia.
Blixenkrone-Møller, M., Svansson, V., Have, P., Botner, A., and Nielsen, J. (1989). Infection studies in mink with seal-derived morbillivirus. *Arch. Virol.* **106,** 165–170.
Cambre, R. C., and Kenny D. (1993). Vaccination of zoo birds against avian botulism with mink botulism vaccine. *Proc. Am. Assoc. Zoo Vet. 1993,* pp. 383–385.
Cook, R. D., and Wilcox, G. E. (1981). A paramyxovirus-like agent associated with demyelinating lesions in the CNS of cats. *J. Neuropathol. Exper. Neurol.* **40,** 328.

Dohms, J. E., Allen, P. H., and Cloud, S. S. (1981). The immunization of broiler chickens against Type C botulism. *Avian Dis.* **26**(2), 340–345.

DVM Newsmagazine (1995). "Distemper Epidemic Strikes Serengeti Lions," September, p.22. DVM Newsmag. Cleveland Ohio.

Foreyt, W. J., and Abinanti, F. R. (1980). Maggot associated Type C botulism in game farm pheasants. *J. Am. Vet. Med. Assoc.* **177**(9), 827–828.

Friend, M., Locke, L. N., and Kennelly, J. J. (1985). "Avian Botulism." Fish and Wildlife Service, U.S. Department of Interior, U.S. Government Printing Office, Washington, DC.

Goodrich, J. M., Williams, E. S., and Burkirk, S. W. (1994). Effects of a modified-live virus canine distemper vaccine on captive badgers *(Taxidea taxus)*. *J. Wild. Dis.* **30**(4), 492–496.

Graham, J. M. (1978). Clostridium botuinum Type C and its toxin in fly larvae. *Vet. Res.* **102**, 242–243.

Gray, T. C., and Bulgin, M. S. (1982). Botulism in an Oregon dairy cow herd. *J. Am. Vet. Med. Assoc.* **180**(2), 160–162.

Haagsma, J. (1974). An outbreak of botulism in broiler chickens. *Tijdsch. Diergeneeskd.* **99**(20), 1069–1070.

Harder, T. C., Kenta, M., Vas, H., Siebelink, K., Huisman, W., van Amerongen, G., Örvell, C., Barrett, D., Appel, M. J. G., and Osterhaus, A. D. M. E. (1996). Canine distemper virus from diseased large felids: Biological properties and phylogenetic relationships. *J. Gen. Virol.* **77**, 397–405.

Harrigan, K. E. (1980). Botulism in broiler chickens. *Aust. Vet. J.* **56**, 603–605.

Hughes, D. E., Carter, S. D. Robinson, I., Clarke, D. D., and Clarke, C. J. (1992). Anticanine distemper virus antibodies in common and grey seals. *Vet. Rec.* May 16, pp. 449–450.

Hulsebos, J. K. (1993). "Raccoon Field Safety Studies of Commercial Distemper and Parvovirus Vaccines." Proprietary data of United Vaccines, Inc., Madison, WI.

Hulsebos, J. K. (1994). "Fennec Fox Serum Neutralization Study." Proprietary data of United Vaccines, Inc., Madison, WI.

Mamaev, L. V., Visser, I. K. G., Beikov, S. I., Denikiva, N. N., Harder, T., Goatley, L., Rima, B., Edginton, B., Österhaus, A. D. M. E., and Barrett, T. (1996). Canine distemper virus in Lake Baikal seals *(Phoca sibirica)*. *Vet. Rec.* May 4, pp. 437–439.

McIlroy, S. G., McCracken, R. M. and Huey, J. A. (1987). Botulism in cattle grazing pasture dressed with poultry litter. *Ir. Vet. J.* **41**(3), 245–248.

Montali, R. J., Bartz, C. R., and Bush, M. (1987). "Virus Infections of Carnivores, Canine Distemper Virus," pp. 437–443. Elsevier, Amsterdam.

Olsen, J. H. (1994). "Avian Medicine: Principles and Applications" Wingers Publishing, Inc., Lake Worth, FL.

Poston, R. P., and England, J. J. (1992). "Veterinary Diagnostic Virology" pp. q135–138. Mosbey Year Book, St. Louis, MO.

Pugh, D. G., Rankins, D. L., and Powe, T. (1994). Feeding broiler litter to beef cattle. *Vet. Med.* July, pp. 661–667.

Visser, I. K. G., van de Bildt, M. W. G., Brugge, H. M., Reijnders, P. J. H., Vedder, E. J., Kuiper, J., deVries, P., Groen, J., Walvoost, H. C., Uytde Haag, F. G. C. M., and Österhaus, A. D. M. E. (1989). Vaccination of harbour seals *(Phoca vitulina)* against phocid distemper with two different inactivated canine distemper virus (CDV) vaccines. *Vaccine* **7**, 521–526.

Wilcox, G. E., and Pass, D. A. (1980). An outbreak of botulism in broiler chickens. *Aust. Vet. J.* **56**, 612–613.

Some Experiments and Field Observations of Distemper in Mink and Ferrets

JOHN R. GORHAM

Agriculture Research Service, USDA, and the Department of Veterinary Microbiology and Pathology, Washington State University, Pullman, Washington 99164

I. Introduction
II. Vaccines
   A. Formalized Tissue Vaccines
   B. Live Chicken Embryo Propagated Vaccines
   C. Estimated Minimal Number of Egg Infectious Doses Required for Immunization of Ferrets and Mink
   D. Duration of CDV Immunity in Isolated Animals Following CEP Virus Vaccination
   E. Inference between Chicken Embryo Propagated and Virulent CDV
   F. Production of CDV Inclusion Bodies by CEP Vaccine
III. Routes of Vaccination
IV. Maternal Antibody and Vaccination
V. Vaccination of Pregnant Female Mink
VI. Transplacental and Neonatal Attempts to Immunize Ferrets against CDV
VII. Time Interval Required to Infect Ferrets by Direct Contact
VIII. Experimental Epidemiology
   A. Infectious Period
   B. Induced Outbreaks
IX. Future Research
   Acknowledgment
   References

I. Introduction

The purpose of this report is to describe some observations on the occurrence of canine distemper (CD) on mink farms and experimental trials using mink and ferrets. Ferrets, because of their high susceptibility to canine distemper virus (CDV), show an invariably fatal course, which makes them the most satisfactory animal for conducting many distemper experiments.

II. Vaccines

Vaccination is the only suitable means for the control of mink and ferret distemper. Attenuated live CDV vaccines produce a more dependable, longer lasting immunity and have replaced inactivated vaccines for ranch-raised mink and pet ferrets.

A. Formalized Tissue Vaccines

Formalized aqueous or oil emulsion vaccines evoke an uncertain immunity in ferrets (Laidlaw and Dunkin, 1928, 1931; Ott et al., 1959). No one can argue with Laidlaw who said "the injection of live virus was indispensable for the production of a strong and durable immunity."

In our experience and others in the United States, Canada, and Russia, the use of formalized CDV tissue vaccines was disastrous in attempts to control outbreaks of mink distemper. The onset of immunity was delayed and the duration of protection was uncertain. Prophylactic vaccination failed to prevent outbreaks and ongoing outbreaks were almost impossible to control. Because virulent CDV replicates in ferrets and mink vaccinated with inactivated vaccines, susceptible contacts are at risk for infection. Furthermore, the formalized CDV tissue vaccines prepared from mink spleens and livers frequently contained the Aleutian disease parvovirus that survived formalization. Massive outbreaks of Aleutian disease, particularly in Washington and Oregon, resulted in losses of millions of dollars in mink pelts (Gorham et al. 1965).

B. Live Chicken Embryo Propagated Vaccines

Adaption of CDV to embryonated eggs by Haig in South Africa (1948, 1949), Cabasso and Cox (1949), and West et al. (1956) marked the end of distemper's dark ages. Here was a convenient, inexpensive method to produce vaccines. Moreover, the chorioallantois of the developing embryo could be used for the titration of CDV antibody in neutralization tests. If the reports of attenuation in embryonated eggs are compiled, the passage level at which the chorioallantoic membranes first became thickened and showed whitish plaques was from the 3rd to 10th passage. At passage level 20–33, the virulence was reduced for the ferret. The passage level when CDV virulence was lost for the ferret varied from 22 to 110 (Gorham, 1960).

Neither ourselves nor G. R. Hartsough (personal communication, 1991) have observed or have had reports of the chicken embryo propagated (CEP) vaccine reverting to virulence in vaccinated farm-raised mink. Moreover, several trials have shown that commercial mink CEP vaccines are not shed by the vaccinate. Contact mink and ferrets have not developed distemper antibody or exhibited CD signs. On the other hand, there is a single report in which a commercial mink vaccine induced fatal distemper in black-footed ferrets (Carpenter et al., 1976). In all probability the genotype of this ferret species was the determining factor that resulted in its susceptibility to the mink vaccine. Commercial mink distemper vaccines have been used since the early 1950s for immunizing European ferrets (*Mustela putorius*) against distemper and to our knowledge have not caused distemper. In one trial, it required 14 serial ferret back passages to revert a mink vaccine strain to virulence (Goto et al., 1976; Gorham and Goto, 1997).

C. Estimated Minimal Number of Egg Infectious Doses Required for Immunization of Ferrets and Mink

A divergence in the minimal immunizing egg infectious dose (EID_{50}) needed to confer resistance in dogs has been reported. In our trials, ferrets were immunized with about 2 and mink were immunized with 32 EID_{50} (Svehag and Gorham, 1962). An interval of 15 days was selected between vaccination and challenge to allow sufficient viral replication for a limited number of EID_{50} values of attenuated vaccine virus to evoke immunity. Low levels of vaccine virus replication were not overwhelmed by challenges as high as 100,000 ferret lethal doses (FLD_{50}).

D. Duration of CDV Immunity in Isolated Animals Following CEP Virus Vaccination

In our trials (Burger and Gorham, 1964) mink and ferrets were found to be immune to virulent CDV challenge 6 years following vaccination. Cabasso et al. (1953) also reported that ferrets resisted challenge at 5.5 years after CEP vaccination. For comparative purposes dogs had high levels of CDV antibody at 6.5 years following vaccination with the Rockborn (1958) strain of vaccine (Carmichael, 1997).

For prophylaxis almost all young mink in the United States and Canada receive a single vaccination in the summer after they reach 10 weeks of age to preclude the neutralization of the CEP vaccine virus by maternal antibody. If the mink have been successfully immunized by primary vaccination at 10 weeks of age, we feel that revaccination is not required. Because the majority of mink on farms are pelted at ~3 years of age, a level of herd immunity is established that greatly reduces the risk of an outbreak.

Present-day distemper vaccines for mink are inexpensive and effectively immunize susceptible mink, and they can be combined with other vaccines to protect against mink virus enteritis, botulism, and *Pseudomonas* infections. The CEP vaccine protects against all field CDV isolates that have been tested including the harbor seal (Blixenkrone-Møller, 1993) and the Serengeti and California lion isolates reported by Appel and Montali (1994) in which CEP vaccine protected ferrets against Serengeti and lion isolate challenge trials (Evermann *et al.*, 1997).

E. Interference between Chicken Embryo Propagated and Virulent CDV

When graded doses (~3–300,000 EID_{50}) of CEP distemper virus were given to ferrets at intervals prior to, simultaneously with, or after challenge with a low dosage of ferret lethal doses (~5 FLD_{50}) of virulent virus, a correlation between dosage of CEP and onset of resistance was observed (Burger and Gorham, 1964). If ~3 EID_{50} were given, three of four ferrets were immunized when the interval prior to challenge was 13 days, whereas 30 EID_{50} evoked protection after a 5-day interval. Interference of the CEP with the virulent virus challenge was observed when dosages of ~300 EID_{50} or higher were injected one day prior to, or simultaneously with, the challenge virus. When an estimated 30,000 EID_{50} were given 1 or 2 days after the challenge virus, interference with the virulent virus challenge was recorded (Fig. 1).

If the level of virulent virus challenge was increased to ~1000 FLD_{50}, the course of the natural disease was not altered and interference was not observed. Interference of CEP vaccine with virulent ferret distemper virus was expressed as full protection, prolonged incubation, and/or death time and occasional recovery. Full protection was the most common observation.

The results of mink experiments were consistent with those of ferrets in that there was an inverse relationship between the dose of virulent challenge virus and interference by CEP vaccine. In these

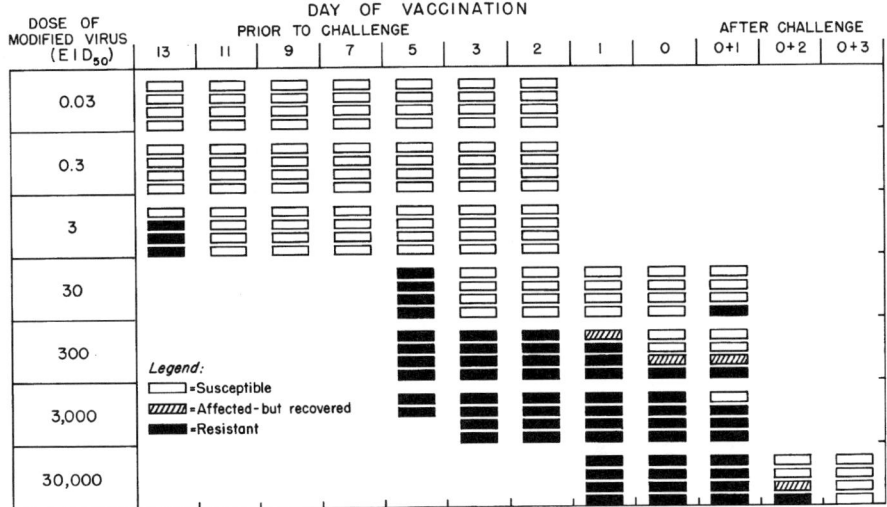

FIG. 1. Correlation between vaccine dose and time of vaccination prior to, or simultaneously with, virulent virus challenge (day 0), and after challenge.

experiments, interference was used in the broad sense, since the mechanism involved in early protection was not clear.

F. PRODUCTION OF CDV INCLUSION BODIES BY CEP VACCINES

The cornerstone of CD diagnoses has been the demonstration of CDV inclusion bodies by a variety of histologic staining methods including H and E and trichrome tissue stains (Page and Green, 1942; Gorham, 1948). Currently, immunofluorescence (IF) techniques (Coffin and Liu, 1957; Moulton, 1956) for the detection of CDV antigen have supplemented or replaced older conventional procedures in many diagnostic laboratories. Coupled with a clinical impression of CD in ferrets and mink, these laboratory procedures are an excellent means to diagnose CD. But there is a delay in the conduct of the laboratory procedures and the notification of the results to the attending veterinarian. I would hazard the speculation that in a few years there will be a simple "pen side" test for the detection of CDV antigen.

When investigating disease outbreaks, it is important to know whether commercial mink vaccines produce inclusion bodies using conventional stains and specific IF. When mink and ferrets were vaccinated with CEP, CDV inclusion bodies were not observed but we detected CDV antigen by IF in the spleen 5–7 days after vaccination and in the respiratory epithelium on days 9–11 postvaccination.

Blixenkrone-Møller (1989) found viral antigen by IF in lymphoid tissue 6–12 days after vaccination but not in epithelium, respiratory, or urinary tracts or in the nervous tissue of vaccinated mink. A study of the distribution of viral antigen in mink infected with virulent CDV revealed IF after 6–12 days following exposure in a wide variety of tissues including lymphoid tissues, epithelial cells of the skin, mucous membranes, lung, kidney, and cells of the central nervous system.

III. Routes of Vaccination

Resistance to virulent CDV can be stimulated by aerosol (airborne) exposure to CEP vaccines. The onset of resistance following aerosolization in ferrets occurred at 5 days (Gorham *et al.*, 1954a). Similarly, mink were protected by aerosol exposure against virulent CDV challenge (Gorham *et al.*, 1954b). Using commercially available CEP vaccines parenterally, we anticipate that the onset of protection in mink occurs within 2–3 days after vaccination. This is a reasonable assumption if the vaccine dose contains 1000 or more EID_{50}. If distemper occurs on unvaccinated mink farms when the kittens are less than 6 weeks of age, multiple aerosol vaccinations offer the best control of the transmission of CDV. Because the mink are not handled during vaccination, the opportunity of viral spread by direct and indirect contact is markedly reduced.

Chicken embryo vaccine can also be administered by the airborne route as a dust. Following exposure to a chicken embryo propagated virus in a dust, 7 of 10 mink developed CDV neutralizing antibodies, and 12 of 14 ferrets were immune to challenge with virulent CDV. The experimental dust vaccine was never used in the field because the dust particles were hygroscopic (Gorham *et al.*, 1995).

IV. Maternal Antibody and Vaccination

In the case of dogs, foxes, mink, and ferrets and perhaps all other *Mustelidae, Canidae,* and *Procyonidae,* maternal antibody is a double-edged sword protecting young animals from distemper and conversely blocking early CEP vaccination.

Ott and Gorham (1955) observed the response of neonatal and young ferrets from distemper susceptible and immune dams to intranasal inoculation of CEP vaccine. Ferret kits from CD susceptible dams were immunized after 8 days of age; young from immune dams were not immunized until 36–47 days after birth.

In a later study, Farrell and his coworkers (1971) showed that all young ferrets were immunized by aerosol CEP vaccine and by subcutaneous injection of CEP vaccine at 10 weeks or more of age. At younger age groups, there was a greater likelihood of the vaccine being neutralized by maternal antibody.

The level of maternal antibody that mink and ferret kittens receive from the female varies between kittens in a single litter and between kittens in other litters. Although the decline in maternal immunity and its ability to block distemper vaccination varies between kittens, it is safe to say that almost all kittens can be effectively immunized after 10 weeks of age. If the female mink have not been vaccinated against CD, mink kittens as young as 4 weeks can be protected by CEP vaccine (Hansen and Lund, 1972).

But there is a window of vulnerability. Although we have no solid research to verify our field observations, we feel that during the 8- to 10-week period following birth, some mink kittens from vaccinated females are susceptible to highly virulent invasive strains of CDV. At the same level of declining antibody, immunity with an attenuated vaccine strain can be blocked.

Also, it is highly likely that mink infected with Aleutian disease parvovirus are immunosuppressed and are unlikely to survive CD. Also CDV immunization is uncertain. Aleutian disease has a worldwide distribution and has infected millions of mink. It is characterized by an immune disorder (Bloom et al., 1994).

V. Vaccination of Pregnant Female Mink

In CD disease outbreaks that occur on farms during late February, March, April, and May, pregnant females must be vaccinated to control the spread of the virus. Tennison (1954) reported that egg-propagated distemper virus did not inferfere with mink reproduction if given either immediately before or 2–55 days after breeding. We conducted a trial in which pregnant female mink were inoculated subcutaneously with CEP vaccine (Hagen et al., 1970). Both susceptible and immune females were vaccinated during the 21st to 30th days of pregnancy. We found that there was no difference in the average number of kittens in the litters of the CDV susceptible and immune groups of females when their kittens reached 21 days of age. Female adult mink without litters were equally distributed in both the susceptible and immune groups. Fetal abnormalities were not found in either susceptible or immune adult females that were vaccinated during pregnancy.

VI. Transplacental and Neonatal Attempts to Immunize Ferrets against CDV

The possibility of transplacental immunization seemed reasonable since a viremia occurs following vaccination with CEP vaccine (Gorham, 1957; Gorham et al., 1957).

Susceptible pregnant female ferrets were vaccinated with CEP vaccine at intervals from 2 to 16 days prior to parturition. All of the kittens were susceptible to virulent CDV when they reached 10 weeks of age. Either sufficient vaccine did not transverse the placental barrier to immunize the unborn kittens, or perhaps the vaccine virus was qualitatively deficient. In another trial young ferret kittens were vaccinated within 48 hours after birth. When these ferret kittens were challenged with virulent CDV at 10 weeks of age, only 10 of 22 survived, which is probably a reflection of immunologic immaturity (Gorham and Ott, 1964).

VII. Time Interval Required to Infect Ferrets by Direct Contact

The experiment was designed to favor transmission (Shen and Gorham, 1978). In a series of contact exposure trials, two distemper-infected ferrets, which we designated as donor ferrets, were placed in a single wire mesh pen measuring $2 \times 3 \times 5$ feet with 10 distemper susceptible recipient ferrets. The number of days the donor ferrets were infected after exposure varied from 6 to 11 days. The length of time the infected donors were in direct contact with the susceptible recipient ferrets varied from 15 minutes to 8 hours.

Infected distemper donor ferrets in the later stages of the disease, that is, 9, 10, and 11 days after exposure, transmitted distemper more effectively, especially when longer contact times of 4–8 hours were employed. A ferret infected for 11 days and placed with 10 susceptible ferrets for 8 hours of exposure transmitted distemper to all 10 ferrets. However, only 1 of 10 ferrets was infected if the contact time was reduced to 15 minutes. Because there is a high level of virus in the lungs of ferrets late in the incubation period, more virus was probably shed into the air to infect susceptible contact ferrets.

VIII. Experimental Epidemiology

While there are many variables in the transmission trials and the induced CDV outbreaks discussed next, the limited results may provide some insight on the spread of CDV in naturally occurring outbreaks.

A. Infectious Period

It is relevant to know when CDV first appears in the secretions of infected mink and ferrets. Similarly, the time interval of virus shedding following the disappearance of clinical signs is also important. Experiments have shown that CDV was first demonstrable in the nasal exudate of mink on the fifth day following inoculation and persisted from 46 to 51 days (Gorham and Brandly, 1953). In these trials CD signs were observed in mink until the 36th day (Fig. 2). Thus, there is a time when mink appear healthy but are capable of transmitting the disease.

Virus was first detected in the nasal exudate of ferrets on the fifth day postinoculation and persisted until death of the ferret. The conjunctival exudate of ferrets was shown to contain CDV on the 16th day following inoculation. Virus was present in the conjunctival exudate collected from mink on days 21 and 30 postinoculation. When nasal exudates are infectious, conjunctival exudates also contain virus.

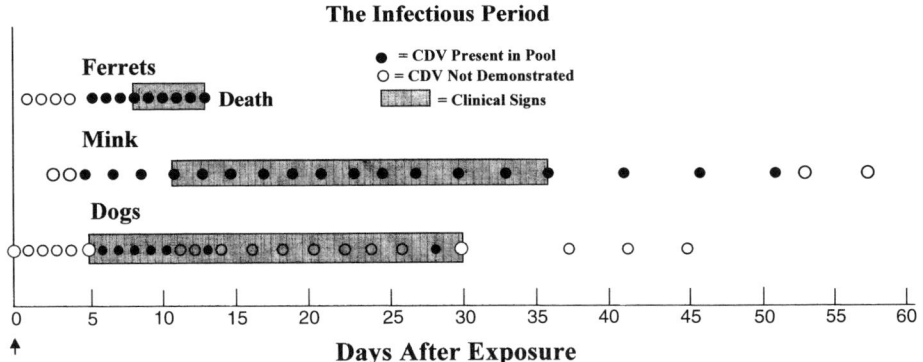

Fig. 2. Results of tests for distemper virus in pooled samples of nasal exudate from CDV susceptible ferrets (10) mink (12) and 8 dogs. Canine distemper virus was demonstrated by inoculation of the nasal exudate into susceptible ferrets.

We have demonstrated CDV in dog urine on days 10, 11, and 12 postexposure to virulent CDV. Since the urinary bladders of ferrets and mink contain IF antigen, it is likely that the urine of infected mink and ferrets also contains CDV.

In a small trial, eight CD susceptible dogs were exposed to a virulent CDV aerosol and their nasal exudates were pooled and injected into susceptible ferrets. Virus was demonstrated on days 6, 7, 8, 9, 10, 13, and 28 (Fig. 2). It is likely that there may have been CDV in the pooled inocula on days 14–27 but it was diluted below an infectious dose in their nasal exudate by dogs that were not shedding CDV.

B. Induced Outbreaks

1. The influence of immune ferrets on the spread of CDV in small susceptible CDV ferret populations was investigated by observing experimentally induced outbreaks (Kelker, 1980).

The pen arrangements for 100 ferrets along with the feeding and observation route are shown as a broken line in Fig. 3. The outbreak

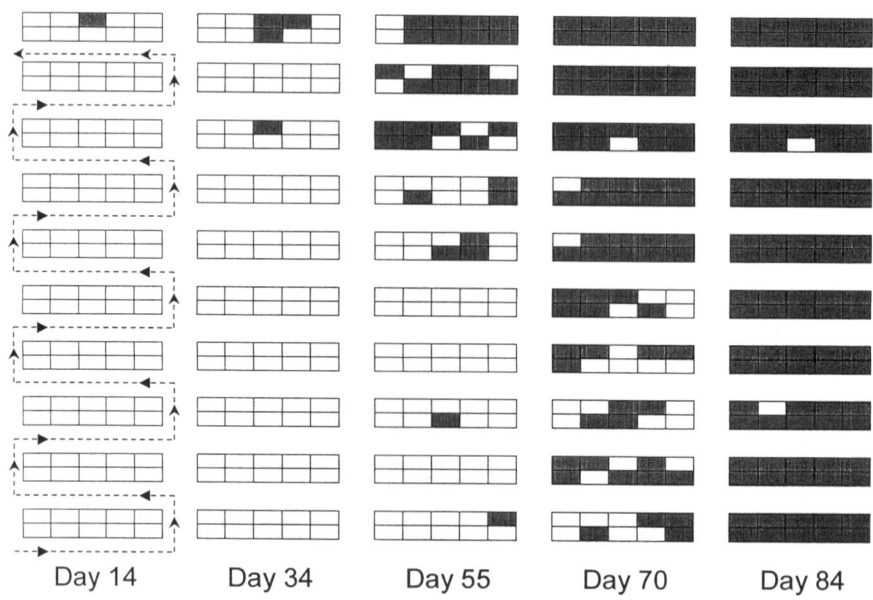

FIG. 3. Pen arrangement for experimental ferret distemper outbreaks, showing the feeding and observation route and the position in the pens of ferrets dead of distemper (black squares).

was started by inoculating one ferret (day 0) in row 1. Ferrets that died of CD in the pens are designated as black squares. The number of virus hits probably reached a threshold by day 34, and the disease spread rapidly until the experiment was terminated on day 84.

2. The second induced outbreak was conducted using the same pen arrangement. Fifty susceptible ferrets were placed in cages alternating with empty cages in a checkerboard configuration. The outbreak was started by infecting a single ferret in the center of the first row as in the above trial. CDV spread through the population and all ferrets were dead of CD by day 94.

3. A third trial was set up in the same configuration of pens as the previous trial except that CDV immune ferrets were placed in empty pens and the outbreak was started in the same manner as in items 1 and 2. The last ferret in this induced outbreak died of CD on day 71. The progression of the outbreak was somewhat more rapid in this induced outbreak that contained immune ferrets. While it is possible, we feel rather doubtful that the immunized ferrets played a significant role in CDV transmission. Obviously, these preliminary trials should be repeated.

4. In many instances the first cases of a distemper are overlooked or misdiagnosed during a mink farm outbreak. To simulate this situation, an outbreak was started by infecting 1 ferret in a population of 60 susceptible ferrets. We waited until 3 additional ferrets were showing signs of CD, then we vaccinated 60, 70, and 80 percent of the remaining ferrets in separate trials. In these three trials, almost all of the unvaccinated ferrets succumbed to CDV. Clearly, it is difficult to control an outbreak if virulent CD has a major advantage over the establishment of immunity by vaccination. On the other hand, we found it was difficult to start an outbreak when 70% or more of the population was immune. Although there many variables in these ferret induced CDV outbreaks, the limited results may provide some background on the spread of CDV in naturally occurring outbreaks.

IX. Future Research

Mink offer a unique opportunity for CD study. It is not unusual for farms to have 25,000 or more mink in which 7000–8000 mink are raised on an acre of ground. When a CD outbreak occurs, "shoe leather" epidemiology allows reasonable impressions of CD activity.

Most veterinarians familiar with mink farming and mink farmers in North America and Scandinavia have observed that distemper mortality in disease outbreaks is almost invariably greater in the pastel (bb or gg) color phase mutation mink than in natural dark mink (Hansen, 1971, 1985; and Hunter and Lemieux, 1996). While the Aleutian (aa) Chediak Higashi gene (Padgett et al., 1964) has increased the susceptibility of this color phase mutation to bacterial diseases, we have not noted an increase of CDV virulence in this genotype. While the increased CDV disease mortality of the pastel mink is a solid clinical worldwide observation, molecular tools are available to determine if the pastel genes for color are linked to genes for CD susceptibility.

There seems to be a gradual change in the clinical course of CD in mink during the past 50 years even though the case/fatality rate is about the same. The catarrhal signs (nasal, ocular exudates, hyperkeratotic foot pads) are less severe. Interestingly, the occurrence of neurotropic episodes has apparently not decreased.

If the observation of the reduction in the severity of the catarrhal signs is valid, the genotypic properties of CDV strains circulating on mink farms should also be considered. A study of the interaction between CDV and the mink would provide some insight on the reduced severity of the catarrhal forms of the disease. One wonders if widescale CDV vaccination of millions of mink each year influenced the disease picture.

Large populations of mink on farms where the breeding history of each mink is recorded allow critical observation of the circumstances under which clinical disease occurs on a farm and the factors that influence the frequency, spread, and distribution on a single farm or from farm to farm. The environmental effects of temperature and humidity and the influence of nutrition on the disease can be observed.

Acknowledgment

Research in our laboratory on mink diseases has been supported in part by the Mink Farmers Research Foundation.

References

Appel, M. J. G., and Montali, R. J. (1994). Canine distemper and emerging morbillivirus diseases in exotic species. *Proc. Assoc. Reptilian Amphibian Vet., Am. Assoc. Zoo Vet.,* Pittsburgh, PA, 1994, pp. 336–339.
Blixenkrone-Møller, M. (1989). Detection of intracellular canine distemper virus antigen in mink inoculated with an attenuated or a virulent strain of canine distemper virus. *Am. J. Vet. Res.* **50,** 1616–1620.
Blixenkrone-Møller, M. (1993). Biological properties of phocine distemper virus and canine distemper virus. *APMIS R. Vet. Agric. Univ. Frederiksberg, Den.* **101**(Suppl. 36), 5–51.

Bloom, M. E., Kanno, H., Mori, S., and Wolfinbarger, J. B. (1994). Aleutian mink disease: Puzzles and paradigms. *Infect. Agents Dis.* **3,** 279–301.

Burger, D., and Gorham, J. R. (1964). Response of ferrets and mink to vaccination with chicken embryo-adapted distemper virus. *Arch. Gesamte Virusforsch.* **15,** 449–461.

Cabasso, V. J., and Cox, H. R. (1949). Propagation of canine distemper virus on the chorio-allantoic membrane of embryonated hen eggs. *Proc. Soc. Exp. Biol. Med.* **71,** 246–250.

Cabasso, V. J., Stebbins, M. R., and Cox, H. R. (1953). Onset of resistance and duration of immunity to distemper in ferrets following a single injection of avianized distemper vaccine. *Vet. Med.* **48,** 147–150.

Carmichael, L. (1997). Personal communication.

Carpenter, J. W., Appel, M. J. G., Erickson, R. C., and Norilla, M. N. (1976). Fatal vaccine induced canine distemper virus infection in black-footed ferrets. *J. Am. Vet. Med. Assoc.* **169,** 961–964.

Coffin, D. L., and Liu, C. (1957). Studies on canine distemper infection by means of fluorescein-labeled antibody. II. The pathology and diagnosis of the naturally occurring disease in dogs and the antigenic nature of the inclusion body. *Virology* **3,** 132–145.

Evermann, J. F., Gorham, J. R., Leathers, C., McKeirnan, A., and Appel, M. J. G. (1998). Felid morbillivirus: Pathogenesis in experimentally infected ferrets and protection by CDV vaccination. *In preparation.*

Farrell, R. K., Skinner, S. F., Gorham, J. R., and Laurerman, L. H. (1971). The aerosol and subcutaneous administration of attenuated egg-adapted distemper vaccine to ferret kits from distemper immune females. *Res. Vet. Sci.* **12,** 392–393.

Gorham, J. R. (1948). Pollak's trichrome stain for demonstrating distemper inclusion bodies in tissue sections. *Science* **107,** 175.

Gorham, J. R. (1957). Egg-propagated distemper virus. *Vet. Med.* **52,** 339–346.

Gorham, J. R. (1960). Canine distemper (*La Maladie de Carré*). *Adv. Vet. Sci.* **6,** 287–351.

Gorham, J. R., and Brandly, C. A. (1953). The transmission of distemper among ferrets and mink. *Proc. Book, Am. Vet. Med. Assoc. 90th Annu. Meet.,* pp. 129–141.

Gorham, J. R., and Goto, H. (1997). Reversion to virulence of an attenuated distemper virus vaccine strain induced by rapid serial passage in ferrets. *Proc. Soc. Trop. Vet. Med.,* 2386.

Gorham, J. R., and Ott, R. L. (1964). Transplacental and neonatal attempts to immunize ferrets against distemper. *Vet. Rec.* **76,** 434.

Gorham, J. R., Leader, R. W., and Gutierrez, J. C. (1954a). Distemper immunization of ferrets by nebulization with egg-adapted virus. *Science* **119,** 125–126.

Gorham, J. R., Leader, R. W., and Gutierrez, J. C. (1954b). Distemper immunization of mink by air-borne infection with egg-adapted virus. *J. Am. Vet. Med. Assoc.* **125,** 134–136.

Gorham, J. R., Farrell, R. K., and Ott, R. L. (1955). Preliminary studies on living virus dust vaccine for distemper. *Cornell Vet.* **45,** 326–330.

Gorham, J. R., Farrell, R. K., Ott, R. L., and Parisot, T. (1957). Multiplication of attenuated egg-adapted distemper virus in the vaccinated host. *Vet. Med.* **55,** 289–292.

Gorham, J. R., Leader, R. W., Padgett, G. A., and Burger, D. (1965). Some observations on the natural occurrence of aleutian disease. *U.S., Public Health Serv. Publ.* **1378,** 279–285.

Goto, H., Shen, D. T., and Gorham, J. R. (1976). Reversion to virulence of an attenuated distemper virus vaccine strain induced by rapid serial passage in ferrets. *Fed. Proc.* **35,** 1021.

Hagen, K. W., Goto, H., and Gorham, J. R. (1970). Distemper vaccine in pregnant ferrets and mink. *Res. Vet. Sci.* **11**, 458–460.

Haig, D. A. (1948). Preliminary note on the cultivation of Green's Distemperoid virus in fertile hen eggs. *Onderstepoort J. Vet. Sci. Anim. Ind.* **23**, 149–155.

Haig, D. A. (1949). Further observations on the growth of Green's Distemperoid virus in developing hen eggs. *J. S. Afr. Vet. Med. Assoc.* **19**, 73–80.

Hansen, M. (1971). The antibody response and the protection obtained by using different types of vaccines in two genotypes of mink. *Nord. Veterinaer.* **21**, 374–382.

Hansen, H. (1985). Virus and virus diseases. In "Mink Production" (G. Joergensen, ed.), pp. 219–340. Danish Fur Breeders Assoc., Tjele, Denmark.

Hansen, M., and Lund, E. (1972). The protective capacity of neutralizing antibodies by distemper virus infections in mink. *Acta Pathol. Microbiol Scand.* **80**, 795–800.

Hunter, D. B., and Lemieux, N. (1996). "Mink-biology, Health and Disease." Graphic and Print Services, University of Guelph, Guelph, Ontario, Canada.

Kelker, D. (1980). The effect of immunes on the spread of distemper in small ferret populations. *Comput. Biol. Med.* **10**, 53–60.

Laidlaw, P. P., and Dunkin, G. W. (1928). Studies in dog distemper. IV. The immunization of ferrets against dog distemper. *J. Comp. Pathol. Ther.* **41**, 1–17.

Laidlaw, P. P., and Dunkin, G. W. (1931). Studies in dog distemper. VI. Dog distemper antiserum. *J. Comp. Pathol. Ther.* **44**, 1–25.

Moulton, J. E. (1956). Fluorescent antibody studies of demyelination in canine distemper. *Proc. Soc. Exp. Biol. Med.* **91**, 460–464.

Ott, R. L., and Gorham, J. R. (1955). The response of newborn and young ferrets to intranasal administration with egg-adapted distemper virus. *Am. J. Vet. Res.* **16**, 571–572.

Ott, R. L., Svehag, S. E., and Burger, D. (1959). Resistance to experimental distemper in ferrets following the use of killed tissue vaccine. *West. Vet.* **6**, 107–111.

Padgett, G. A., Leader, R. W., Gorham, J. R., and O'Mary, C. C. (1964). The familiae occurrence of the Chediak-Higashi syndrome in mink and cattle. *Genetics* **49**, 505–512.

Page, W. G., and Green, R. G. (1942). An improved diagnostic stain for distemper inclusions. *Cornell Vet.* **32**, 265–268.

Rockborn, G. (1958). Canine distemper virus in tissue culture. *Arch. Gesamte Virusforsch.* **8**, 1–8.

Shen, D. T., and Gorham, J. R. (1978). Contact transmission of distemper virus in ferrets. *Res. Vet. Sci.* **24**, 118–119.

Svehag, S. E., and Gorham, J. R. (1962). An attempt to demonstate the minimal number (estimated by probit analysis) of biological units of chicken embryo-propagated distemper virus required to immunize ferrets and mink. *Arch. Gesamte Virusforsch.* **12**, 250–258.

Tennison, L. B. (1954). Vaccination of mink with distemper vaccine, modified live virus, during gestation. *Vet. Med.* **49**, 294–295.

West, J. L., McNutt, S. H., and Brandly, C. A. (1956). The adaptation of fox distemper virus to embryonating chicken eggs. *Cornell Vet.* **46**, 39–50.

Vaccination of Wildlife against Rabies: Successful Use of a Vectored Vaccine Obtained by Recombinant Technology

M. MACKOWIAK,* J. MAKI,* L. MOTES-KREIMEYER,* T. HARBIN,* AND K. VAN KAMPEN†

*Merial, Limited, 115 Transtech Drive, Athens, Georgia 30601,
†The Van Kampen Group, Ogden, Utah

I. Introduction and Background
II. Raboral V-RG: A Rabies Vaccine Created by Recombinant Technology
III. USDA Testing of Recombinant Vaccines
IV. Controlling Raccoon Rabies
V. Rabies in Texas: Coyotes and Gray Foxes
VI. Summary
References

I. Introduction and Background

Rabies is a deadly disease that has plagued mankind since early antiquity (Artois et al., 1990). Although safe and effective conventional vaccines are available to protect domestic animals and people (i.e., veterinarians and other at-risk public health personnel), rabies is still a serious threat to human health. In the United States there have been 15 confirmed human deaths attributed to rabies infection since 1987. Several cases were due to exposures occurring outside the United States, whereas others were due to rabies-infected bats or contact with nonvaccinated kittens or puppies that were harboring the virus.

These case numbers pale in comparison to the approximate 60,000 human deaths occurring each year in underdeveloped countries where prophylactic postexposure rabies treatment is rarely available. Rabies remains a disease of public concern in countries without sufficient public health services and with roaming populations of unvaccinated feral dogs or other vector species. In the United States and Europe, pets and domestic species such as sheep, cattle, and horses are routinely vaccinated against rabies. However, rabies continues to remain endemic in these countries due to wildlife species that serve as reservoirs. In Europe, the red fox and raccoon dog are the primary vector species for rabies; in the United States the raccoon, skunk, fox, and coyote harbor the virus in their specific ecological niches (Krebs et al., 1996). Rabies is considered to be endemic in bats both in Europe and the United States (Anonymous, 1992, 1993, 1994).

II. Raboral V-RG: A Rabies Vaccine Created by Recombinant Technology

During the late 1980s, a new era of biological science began as recombinant technology emerged from the laboratory and into practical application in human and veterinary medicine (Maki and Mackowiak, 1994). Prior to this time, rabies control in wild species consisted of depopulation or individual trapping and vaccinating of raccoons and foxes using conventional rabies vaccines. However, these methods are labor intensive and have only been applied in small-scale situations. To vaccinate wildlife species on a grander scale, a rabies vaccine was needed that could be safely and easily distributed into the environment as well as administered with minimal human interference.

Raboral V-RG was developed as an alternative rabies vaccine that proved to have the unique and novel attribute of being effective by the oral route. The vaccine consists of a modified live vaccinia virus containing the rabies surface glycoprotein gene inserted in its genome. The first experimental use of the recombinant vaccine in wildlife was initiated in Europe. Baiting campaigns were conducted in Luxembourg, Belgium, and France from 1989 to 1991 (Brochier et al., 1996; Brochier and Pastoret, 1993). The vaccine was contained within a plastic sachet surrounded by an edible fishmeal bait and deployed into areas known to contain rabies-infected red fox populations. These campaigns resulted in a dramatic reduction in rabies cases in red foxes and the use of Raboral V-RG was considered a success. Baiting campaigns were continued biannually for several years and rabies in red foxes continued to decline. Based on these results Raboral V-RG was licensed in France in 1995 and in Belgium and Luxembourg in 1996, which marked the first licensure of a recombinant rabies vaccine in the world.

The first USDA-approved experimental use of the vaccine in the United States was conducted in 1990 and 1991 on a barrier island off the coast of Virginia (Parramore Island). The vaccine was tested primarily for bait preference, raccoon characteristics, and safety in the environment but it was also demonstrated to be immunogenic in the selected target species, the raccoon (Hanlon *et al.*, 1989). Following this small but landmark use of the vaccine, USDA approved its use in two additional trials, which took place on protected game lands in Westmoreland County, Pennsylvania, in 1992. Through these small-scale trials, vaccine safety was demonstrated by its repeated use in the environment (Anonymous, 1992, 1993, 1994) without adverse events nor deleterious impact on the flora or fauna. Additionally, the initial apprehension and concern of government officials and environmentalists toward using a recombinant vaccine began to decrease. Thus, the feasibility of vaccinating wild populations of raccoons against rabies was becoming a reality. Raboral V-RG was conditionally licensed in 1995 and granted full licensure for use in raccoons in 1997. Today, Raboral V-RG remains the only licensed rabies vaccine for wildlife in the United States and the only rabies vaccine effective by the oral route.

III. USDA Testing of Recombinant Vaccines

USDA classifies recombinant vaccines into three categories based on their complexity. Type I vaccines consist of subunit proteins produced by recombinant bacteria or other organisms which often require purification prior to use. These vaccines frequently require adjuvants which are necessary to increase the effectiveness of the purified protein. Type I vaccines offer the benefit of excellent safety over whole organism vaccines and are commonly used in veterinary medicine against diseases such as Lyme disease and feline leukemia.

Type II recombinant vaccines consist of the actual pathogenic organism which has been specifically modified by recombinant technology to be less virulent. An example of a type II vaccine is the gene-deleted modified live pseudorabies vaccine for pigs. Deletion of specific gene sequences from the wild-type virus results in a safer vaccine, which induces a limited repertoire of antibodies in the recipient pig. The use of this vaccine allows discrimination of vaccinated pigs from infected pigs and has played a major role in success of the pseudorabies eradication program in the United States.

Type III vaccines are unique in that they consist of a modified live virus or bacteria that contain additional genetic information, often a gene or genes encoding an immunogenic protein from a different organism. The modified live organism that contains the gene is called a vector and serves as a genetic delivery mechanism in the recipient animal. For example, the Raboral V-RG vaccine is a type III, live vaccinia virus vector containing the rabies glycoprotein gene. Following administration, this recombinant virus expresses the rabies virus gene product, which is a membrane protein recognized by the host's immune system.

To construct Raboral V-RG, a genetic recombinant element called a plasmid was used in a process known as homologous recombination. The rabies glycoprotein gene was initially removed from the rabies virus as a sequence of ribonucleic acids (RNA). This gene was transcribed into a more stable form consisting of the same information encoded by deoxyribonucleic acids (cDNA), which was further modified and inserted into a circular DNA molecule called a plasmid. The plasmid was specifically designed to transfer the rabies gene into the vaccinia virus and allow the incorporation of the foreign gene into the virus's genetic makeup, or genome. Using a process called transfection, the plasmid was introduced into vaccinia virus and the recombinant virus was amplified in cell culture. The recombinant virus, now containing the rabies gene in its genome, produced the rabies glycoprotein during it replication cycle and thus expressed the foreign gene product in vaccinia infected cells (Desmettre et al., 1990; Rupprecht and Kieny, 1988; Wiktor et al., 1988).

Following administration of the recombinant virus to an animal, cells take up the virus and it begins a replication cycle; however, the replication cycle is self-limiting and few progeny viruses are actually made. During this replication process the rabies glycoprotein is expressed and subsequently recognized by the host's immune system and antibodies are made against it. Subsequently, the host's immune response is "primed" against the rabies glycoprotein and will respond to future exposure to wild-type rabies virus by producing specific antibodies against it. In this way the animal becomes vaccinated against infection by the virulent rabies virus.

Due to genetic manipulations that occurred during the construction of the vaccine, the recombinant vaccinia virus cannot replicate at the same rates and with the same degree of virulence as the native form of the virus. Raboral V-RG contains only the rabies glycoprotein gene; thus, it cannot cause rabies. The recombinant vaccine contains only one gene from the rabies virus and is therefore lacking enough genetic information to make a complete rabies virus. Because Raboral V-RG is such a novel vaccine against a deadly zoonotic agent, the regulatory process for approval has been long and rigorous. Because the vaccine was developed for use in wildlife, extensive safety trials were performed to ensure that it would not cause lesions in the recipient animals. The vaccine has been tested by various routes of administration in more than 60 species (Brochier et al., 1989, 1996; Desmettre et al., 1990) including rodents, carnivores, ruminants, omnivores, birds of prey, and primates (Table I).

While product safety was being proven in the field, the vaccine was also being developed for industrial production and USDA approval. Efficacy and safety trials were conducted using raccoons housed in controlled environmental conditions to satisfy USDA regulations. Such trials are a necessary part of the licensing procedure for rabies vaccines and require the monitoring of serum antibodies in the recipient animals as well as daily observations for safety. To prove the efficacy of Raboral V-RG, a single dose of vaccine was enclosed in a fishmeal bait and fed to individually caged raccoons. Nine months later, the vacci-

TABLE I

Species Involved in Raboral V-RG Safety Testing

Flying squirrel	Red fox	Domestic cat
Cotton rat	Gray fox	Seagull
Marsh rice rat	Artic fox	Opossum
Syrian hamster	Domestic dog	Short-tailed shrew
Groundhog	Coyote	Squirrel monkey
Gray squirrel	Raccoon dog	Chimpanzee
European field mouse	Bobcat	Rabbit
Yellow-necked mouse	Woodchuck	Porcupine
Wood mouse	Black bear	Field mole
Woodland jumping mouse	Raccoon	Meadow mole
Carrion crow	Cattle	Common mole
Ring-billed gull	Sheep	Bank mole
Great horned owl	White-tailed deer	Red-backed mole
Deer mouse	River otter	Laboratory mouse
Ferret	European badger	Hamster
Mink	Javelina	Daubenton bat
Polecat	Magpie	Nude mouse
Vampire bat	Jay	Domestic pig and wild boar
Red-tailed hawk	Water vole	Laboratory guinea pig
Common buzzard	Striped skunk	Eurasian badger
Kestrel	Horse	

nates and a group of unvaccinated controls were injected with virulent rabies virus. The animals were observed for 90 days and during this time period 10/12 controls (83%) died and 21/27 (78%) of vaccinated raccoon survived. This rigorous test demonstrated that the vaccine protected the majority of vaccinated raccoons against rabies. Reports describing test results from controlled studies as well as additional field experiments were submitted to USDA to allow larger studies to take place in areas of the eastern United States that were experiencing an epizootic of rabies in indigenous wild raccoon populations.

IV. Controlling Raccoon Rabies

Prior to 1983 raccoon rabies was considered a sporadic enzootic disease in the southeastern United States. However, the epidemiology of the disease abruptly changed following the transportation of rabies-infected raccoons to southeastern West Virginia for hunting purposes. Once released into native populations of raccoons, an epizootic of rabies occurred in several states along the Atlantic coast. From the initial focus, the infection spread rapidly northward following the natural river valleys of the Appalachian Mountains to New Jersey, New York, and other northeastern states (Brochier et al. 1996).

As a state affected by the epizootic, New Jersey was the first state to use the yet unlicensed rabies vaccine in raccoons. By 1992, rabies virus had infected the majority of raccoon populations in the western part of the state. To test the vaccine in an uninfected area, the geographically isolated peninsula of Cape May County was selected as a testing site. Biannual campaigns were carried out during the spring and fall of each year from 1992 to 1994 under the direction of the New Jersey Department of Health. Vaccine-filled baits were distributed at a density of 1 bait/acre over an area of 213 square miles (552 square kilometers). Post-bait distribution safety data, bone samples for biomarker testing, and serology samples were collected from raccoons from within the vaccination zone. Over the 3 years that baits were distributed, no cases of raccoon rabies occurred below the vaccination zone (Table II). A few rabid raccoons were detected at the periphery of the barrier. Raboral V-RG demonstrated protective efficacy within the vaccination zone and halted the expansion of the rabies epizootic for 2.5 years.

Due to funding costs and apparent reduction in the number of rabid raccoons detected from within the vaccination zone, the number of baits distributed in Cape May, New Jersey, were decreased during the fall 1994 campaign and baits were not distributed during the spring of

TABLE II

New Jersey Rabies Surveillance (1994–1995)

	Epizootic		
	Positive	Negative	Total
Treated area	5	56	61
Nontreated area	36	11	47

	Enzootic		
	Positive	Negative	Total
Treated area	5	56	61
Nontreated area	23	20	43

1995. Shortly thereafter, in July 1995, a rabid raccoon was detected beyond the vaccine zone and within Cape May county. Clearly, the threat of rabies infection had not yet subsided and the epizootic quickly spread into the remainder of the peninsula. In the fall of 1995 baits were distributed throughout the entire Cape May area using varying densities as determined by location of suitable raccoon habitat. Successive yearly single baiting campaigns were continued in 1996 and 1997. This second round of vaccine application took place after rabies had entered the area and thus tested the efficacy of the vaccine during an enzootic rabies infection as compared to the 1992–1994 campaigns which were against an advancing epizootic. The vaccine proved to be efficacious in both applications as determined by a variety of methods. As of mid-1998, no new case of rabies have been confirmed in Cape May County since January 1997.

During 1992–1997, rabies serology data collected from nonvaccinated areas were compared to serology data collected from raccoons within the vaccinated zones. Rabies incidence data collected after the epizootic had entered New Jersey (in 1994) demonstrated 53% (23/43) of raccoons from nonvaccinated areas were infected with rabies. During the same period, only 8% (5/61) of raccoons from within the vaccination zone were infected with rabies. These comparative data indicate that Raboral V-RG had established a level of population immunity that resulted in greater than a six-fold decrease in rabies infection rate. In areas in which rabies was enzootic, rabies incidence in raccoons was demonstrated to have infection rates of 8% (5/61) in treated areas as compared to 76.5% (36/47) raccoons in nontreated areas (Table II).

As the rabies epizootic continued to spread northward, New York was the next state to use Raboral V-RG to immunize raccoons against rabies. The first experiments in New York compared rabies incidence rates in two geographically distinct areas separated by the Hudson River. Albany and Rennselaer counties received vaccine-filled baits at 75–100 baits per square kilometer in two separate test sites of approximately 3000 square kilometers per country. Baits were distributed from 1994 to 1996 in biannual campaigns under the direction of the New York State Department of Health. As in New Jersey, post-bait distribution safety data, bone samples for biomarker testing, and serology samples were collected from raccoons from within the vaccination zone. Rabies incidence data were collected from vaccinated and nonvaccinated areas of the counties. In treated areas of Albany county, 7% (3/41) of raccoons were infected with rabies. In nontreated areas of the same county 35% (22/62) were infected. Also, in Rennselaer county treated areas were less infected 5% (1/19) compared to nontreated areas demonstrating 56% (31/55) (Table III). These two separate trial areas confirmed the New Jersey findings of efficacy and safety of Raboral V-RG in raccoons and efficacy in an enzootic rabies area.

A third state, Massachusetts, began baiting against raccoon rabies in 1994 with the purpose of protecting the Cape Cod isthmus from

TABLE III

RACCOONS EXAMINED FOR RABIES (1993–1995)

	Albany County, New York		
	Positive	Negative	Total
Treated area	3	38	41
Nontreated area	22	40	62
	Rennselaer County, New York		
	Positive	Negative	Total
Treated area	1	18	19
Nontreated area	31	24	55

infection. The geographical formation of Cape Cod provided adequate natural boundaries to set up an isolated test site similar in theory to New Jersey's Cape May peninsula. In a study conducted by Tufts University, vaccine-filled baits were applied in the spring and fall from 1994 to 1997 using densities of 50–200 baits per square kilometer based on raccoon habitat. Although the baited area is only 8 miles across, the additional natural barrier of the Cape Cod canal has contributed to the prevention of rabies beyond the vaccine-treated area. During the summer of 1996, the rabies epizootic entered Massachusetts and challenged the vaccine barrier. To date, rabid raccoons have not been detected beyond the treated area and surveillance methods, similar to techniques described above, continue to monitor the safety and efficacy of the vaccine. Rabies incidence data from Massachusetts are equally impressive in that 1% (1/100) of raccoons collected from 1994 to 1996 were positive for rabies within the vaccination zone. This one animal was detected at the western border of the zone and was most likely an intrusion from the infected area. In nonvaccinated areas of western Cape Cod county rabies incidence rates of 24.5% (14/57) infected raccoons were reported during the same time period (1994–1996) (Table IV). Thus, these data confirm the previously mentioned findings of product efficacy in raccoons. Additionally, no safety concerns have been reported in nontarget species from any of the described test zones.

In 1995, Florida began using Raboral V-RG to combat a rabies epizootic in raccoons which involved the heavily populated Pinellas County. Approximately 100,000 vaccine-filled baits were distributed using mosquito control helicopters under the direction of the Pinellas County Animal Control Division. Baits were distributed by air throughout the county including drainage ditches supplemented by hand placement of

TABLE IV

Raccoons Examined for Rabies (1994–1996)

	Cape Cod, Massachusetts		
	Positive	Negative	Total
Treated area	1	99	100
Nontreated area	14	43	57

baits in heavily populated suburbs. Rabies incidence began to decrease later that year and a subsequent campaign in 1996 has contributed to a continued decline in rabid raccoons. As of mid-1998, no new cases of rabies have been detected since August 1997. This campaign continues to date and has demonstrated the successful use of Raboral V-RG into a subtropical environment in which raccoons remain active throughout the year.

Another Raboral V-RG trial currently under way in raccoons involves the distribution of vaccine in two different baiting formats in upstate New York. In this trial, orchestrated by Cornell University, six different counties are receiving baits in specific designated vaccination zones. In this study, the vaccine has been applied using two different bait formats and is distributed only once a year. The goal of this study is to broaden the methods of vaccine application as well retard the expansion of the raccoon rabies epidemic into southern Canada. Since these campaigns are relatively new, data have not yet been analyzed. Two additional campaigns have recently been initiated in Ohio and Vermont due to the entrance of the raccoon epizootic into these new areas. These campaigns were initiated in 1997 and vaccine was distributed based on knowledge gained from previous trials. Currently, Raboral V-RG is being used in six states to combat raccoon rabies.

The safety of the vaccine has been proven and the efficacy of the product has been demonstrated by stopping the spread of a rabies outbreak as well as decreasing the incidence of infection in rabies endemic areas.

V. Rabies in Texas: Coyotes and Gray Foxes

In 1988, an epizootic of canine rabies developed along the south Texas border near the Rio Grande River (Meehan, 1995). These cases were of great concern to the Texas Department of Health since previously rabies in coyotes had been sporadic and infrequent (Clark et al., 1994). Additionally, due to decreased hunting pressure on gray fox populations, numbers of this vector species were increasing in the southwestern part of the state and a gray fox strain of rabies was emerging as a public health threat. Thus, the occurrence of two major rabies epizootics was imminent and in July 1994 the governor of Texas declared a state of emergency releasing state funds to combat rabies in Texas and protect the threatened city of San Antonio. Due to the dire nature of the rabies problem, permission was granted by USDA to use Raboral V-RG experimentally in coyotes (APHIS-ADC, 1995; Meehan,

1995) and gray foxes once the vaccine had passed preliminary immunogenicity and safety testing. Texas Department of Health officials, in conjunction with USDA personnel and Merial Limited, the manufacturer of the vaccine, conducted preliminary yet critical experiments in caged wild-caught coyotes and gray foxes to determine bait acceptance (Meehan, 1995) and vaccine efficacy.

In 1995, 830,000 vaccine-filled baits were distributed across south Texas in a 15,000-square-mile band covering the epizootic wave of coyote rabies that had expanded northward across the south central portion of the state. Surveillance efforts and rabies case monitoring was reported via a county-based system coordinated by the Texas Department of Health. Data collected post-baiting indicated that by the fall of 1995 rabies cases were declining and the epizootic had not spread beyond the northern vaccination border. Additional campaigns conducted in 1996 and 1997 distributed more than 1 million vaccine-filled baits per year over 21,000 square miles and rabies in coyotes continues to decrease.

The use of Raboral V-RG in gray foxes lagged slightly behind the coyote program with the first vaccine-filled baits being dropped in southwest Texas in 1996 and again in 1997. Approximately 1.2 million baits were distributed each year in a 30-mile-wide band surrounding the perimeter of a > 25,000-square-mile area of known gray fox rabies cases. The bait container was modified to be acceptable to the smaller gray fox, and post-baiting acceptance data based on biomarker and rabies serology indicate excellent vaccine uptake and immunogenicity. Cases of rabies in gray foxes continue to decrease and surveillance of rabies cases continues, as with the coyote program, under the direction of the Texas Department of Health.

The described rabies vaccination programs using Raboral V-RG exemplify federal, state and private companies working together for the common good. The potential deleterious impact that these epizootics may have had on public health will never be known due in part to the successful use of Raboral V-RG. Data collected by several different groups of scientists during the past 7 years of field research indicate that Raboral V-RG has earned its place in rabies control programs.

VI. Summary

The impact of recombinant technology in veterinary and human medicine can only be hypothesized at this time. The development of vaccines and other biological products that go beyond the abilities of conventional products demonstrates the benefits of this new technology. Raboral V-RG was developed as an alternative rabies vaccine with the novel attribute of being effective by the oral route. Within 10 years after its first application as an experimental vaccine in European, red foxes it developed into a useful tool and is being used to curtail rabies epizootics in three wildlife species in the United States. The use of this vaccine can be considered as monumental in contributing to the control of rabies in species that were at one time considered to be incapable of vaccination in large-scale campaigns.

REFERENCES

Anonymous (1992). Availability of environmental assessment and finding of no significant impact. *Fed. Regist.* **57**(56), 10002–10003.
Anonymous (1993). *Fed. Regist.* **58**(232), 64287.
Anonymous (1994). *Fed. Regist.* **59**(73), 18097.
APHIS-ADC (1995). Environmental assessment and finding of no significant impact (South Texas).
Artois, M., Charlton, K. M. et al. (1990). Vaccinia recombinant virus expressing the rabies virus glycoprotein: Safety and efficacy trials in Canadian wildlife. *Can. J. Vet. Res.* **54**, 504–507.
Brochier, B., and Pastoret, P.-P. (1993). Rabies eradication in Belgium by fox vaccination using vaccinia-rabies recombinant virus. *Onderstepoort J. Vet. Res.* **601**, 469–475.
Brochier, B., Blancou, J., Thomas, E. et al. (1989). Use of recombinant vaccinia-rabies glycoprotein virus for oral vaccination of wildlife against rabies: Innocuity to several non-target bait consuming species. *J. Wildl. Dis.* **25**(4), 540–547.
Brochier, B., Aubert, M. A., Pastoret, P. P. et al. (1996). Field use of a vaccinia-rabies recombinant vaccine for the control of sylvatic rabies in Europe and North America. *Rev. Sci. Tech. Off. Epizoot.* **15**(3), 947–970.
Clark, K. A., Neill, S. U., Smith, J. S. et al. (1994). Epizootic canine rabies transmitted by coyotes in South Texas. *J. Am. Vet. Med. Assoc.* **204**(4), 536–540.
Desmettre, P., Languet, B., Chappuis, G. et al. (1990). Use of vaccinia rabies recombinant for oral vaccination of wildlife. *Vet. Microbiol.* **23**, 227–236.
Fishbein, D. B., and Robinson, L. E. (1993). Current concepts: Rabies. *N. Engl. J. Med.* 1632–1638.
Hanlon, C. L., Hayes, D. E., Hamir, A. N., et al. (1989). Proposed field evaluation of a rabies recombinant vaccine for raccoons (*Procyon Lotor*): Site selection, target species characteristics, and placebo baiting trails. *J. Wildl. Dis.* **25**(4), 555–567.
Krebs, J. W., Strine, T. W., Smith, J. S. et al. (1996). Rabies surveillance in the United States during 1995. *J. Am. Vet. Med. Assoc.* **209**(12), 2031–2044.
Maki, J., and Mackowiak, M., (1994). Recombinant technology: Vaccine development. *Proc. Emerg. Technol. Symp. North Am. Vet. Conf. 1994*.
Meehan, S. (1995). Rabies epizootic in coyotes combated with oral vaccination program. *J. Am. Vet. Med. Assoc.* **206**(8), 1097–1099.
Pastoret, P. P., Brochier, B., Chappuis, G., and Desmettre, P. (1993). Vaccines against rabies. *In* "Veterinary Vaccines," pp. 139–155. Springer-Verlag, Berlin.

Rupprecht, C. E., and Kieny, M. P. (1988). Development of a vaccinia-rabies recombinant virus vaccine. *In* "Rabies: Development in Veterinary Virology," pp. 335–364. Kluwer Academic Publishers, Boston.

Wiktor, T. J., Kieny, M. P., and Lathe, R. (1988). New generation of rabies vaccine, vaccinia-rabies glycoprotein recombinant virus. *Appl. Virol. Res.* **1,** 69–90.

IX
REGULATION, LICENSING, AND STANDARDIZATION OF VACCINES AND DIAGNOSTICS

Authorities and Procedures for Licensing Veterinary Biological Products in the United States

DAVID A. ESPESETH AND THOMAS J. MYERS

U.S. Department of Agriculture, Animal and Plant Health Inspection Service, Veterinary Services, Center for Veterinary Biologics, Riverdale, Maryland 20782

I. Introduction
II. Organization
III. Licensing Procedures for Conventional Vaccines and Bacterins
IV. Licensing Procedures for Nonconventional Products
   A. Recombinant Products
   B. Immunomodulators
V. Conditional Licenses
VI. Licenses for Further Manufacture
VII. Sublicensing
VIII. Exemptions to Licensure
IX. Autogenous Products
X. Summary
   Reference

I. Introduction

In the United States, veterinary biological products are regulated in accordance with the Virus Serum Toxin (VST) Act of 1913, as amended in 1985 (21 United States Code 151–159). The VST Act makes it unlawful to:

1. Prepare, sell, or ship any worthless, contaminated, dangerous, or harmful veterinary biological product in or from the United States.
2. Prepare, sell, or ship any veterinary biological product in or from

the United States, unless it is prepared in a licensed establishment in compliance with U.S. Department of Agriculture regulations.

The regulations developed to administer this Act are published in Title 9, Code of Federal Regulations (9 CFR), Parts 101–118 (January 1, 1996). The regulations define veterinary biological products to be:

> all viruses, serums, toxins, and analogous products of natural or synthetic origin, such as diagnostics, antitoxins, vaccines, live microorganisms, killed microorganisms, and the antigenic or immunizing components of microorganisms intended for use in the diagnosis, treatment, or prevention of diseases of animals (9 CFR 101.2).

For a biologics producer to demonstrate that a product is not worthless, contaminated, dangerous, or harmful, the producer must demonstrate that the product is pure, safe, potent, and efficacious [9 CFR 102.3(b)].

II. Organization

Within the USDA, the VST Act is administered by the Animal and Plant Health Inspection Service (APHIS), Veterinary Services, Center for Veterinary Biologics (CVB). Three primary units within CVB are responsible for the licensing, inspection, and testing functions of the Veterinary Biologics program:

1. *Center for Veterinary Biologics—Licensing and Policy Development* (CVB-LPD) in Riverdale, Maryland, is responsible for prelicensing activities, such as the review of license applications and the development and publication of licensing requirements and program policies.
2. *Center for Veterinary Biologics–Inspection and Compliance* (CVB-IC) in Ames, Iowa, is responsible for postlicensing activities, such as inspection of establishments, release of product serials (lots) for marketing, receiving consumer complaints, and coordinating investigations of suspected violations of the VST Act.
3. *Center for Veterinary Biologics–Laboratory* (CVB-L) in Ames, Iowa, is responsible for conducting confirmatory tests on Master Seeds, Master Cells, and product serials (both prelicense and postlicense), and for developing new test methods, references, and reagents used in such tests.

Further information concerning CVB and the Veterinary Biologics Program (including regulations, memorandums, and notices) can be obtained from the CVB home page on the Internet at http://www.aphis.usda.gov/vs/cvb/index.html.

III. Licensing Procedures for Conventional Vaccines and Bacterins

Two types of licenses are required in order to produce and market a veterinary biological product in the United States (9 CFR 102.2):

1. A United States Veterinary Biologics Establishment License.
2. A United States Veterinary Biological Product License.

Procedures for the issuance of establishment and product licenses in the United States are designed to define and document what is being licensed and who will be responsible for the production and distribution of the product that is authorized. These procedures are also intended to ensure the purity, safety, potency, and efficacy of each product and the accuracy of labeling.

To obtain a United States Veterinary Biologics Establishment License, an application (APHIS form 2001) must be filed with APHIS identifying the name and address of the applicant and any subsidiaries or divisions that will be doing business under the license. The application also identifies the person responsible for the license, including corporate officers for corporations.

The following information must be submitted in support of an application for an establishment license (9 CFR 102.3):

1. An application for a U.S. Veterinary Biological Product License.
2. A copy of the articles of incorporation if the applicant or subsidiaries are incorporated.
3. Plot plans and blueprints of the facilities to be licensed and a legend that provides a brief description of the activities performed and equipment located in each room, including decontamination procedures and other precautions against cross-contamination.
4. A certificate from the appropriate water pollution control agency indicating that the establishment is in compliance with applicable water quality control standards.
5. A short resume describing the training and experience of those employees at the establishment who will be responsible for essential steps in production, testing, and initial distribution of product.

Prior to issuing an establishment license, CVB-IC personnel conduct a prelicensing inspection of the facilities. Inspectors review the adequacy of record-keeping systems designed to document each step in production. Inspectors also review the construction and operation of the establishment to validate that they are as represented on the blueprints and legends and to ensure that conditions are acceptable for production of the veterinary biological product intended to be licensed. Laboratory practices are observed to ensure that facilities are being operated at an acceptable standard. CVB-IC personnel train a person at the establishment to collect and submit valid samples from each serial (lot) of product for testing at CVB-L. Quality control testing and other compliance requirements are also reviewed at this time. The report of the prelicensing inspection is forwarded to CVB-LPD for consideration in the licensing process. If the establishment is found satisfactory in all regards, the establishment license is issued with the first product license [9 CFR 102.4, (a), (b)].

An application for a United States Veterinary Biological Product License (APHIS Form 2003) and the following information and data must be submitted during the licensing process for a new veterinary biological product [9CFR 102.3, (b) & 102.5]:

1. *An Outline of Production:* The Outline of Production is the detailed protocol for manufacturing and testing the product. The outline includes information on each microorganism (Master Seed) found in the product; the methods for culturing and harvesting the microorganisms; a stepwise description of the preparation of the product; and a description of the purity, safety, and potency tests conducted on the product (9 CFR 114.8 & 114.9).

2. *Purity data:* The production of a pure and uniform product is based on the Master Seed concept (9 CFR 113.8), in which a stock of a specifically identified microorganism is the source of all seed materials for production. In most cases, the final product must not be more than five serial passages from the Master Seed. The Master Seed, Master Cell Stock, primary cells, ingredients of animal origin, and final product must be tested for viral, bacterial, mycoplasmal, and fungal contamination. Eggs used in the production of biological products must be derived from specific-pathogen-free flocks. The purity and identity of Master Seeds and Master Cells are confirmed by testing conducted at CVB-L (9 CFR 113.25–113.32, 113.34, 113.36, 113.37, 113.42, 113.43, 113.46, 113.47, 113.51–113.53, 113.55, & Veterinary Biologics Memorandum No. 800.65 and 800.88).

3. *Laboratory safety data:* Safety testing may include a combination of various studies. Live virus vaccines are typically evaluated in the host animal using the product at a $10\times$ dose, while killed products are usually evaluated at a $1\times$ dose. The vaccinated animals should be

evaluated both for systemic and local reactions. Live products must be characterized to determine if they have the ability to shed from the host and be transmitted to contact animals. Reversion-to-virulence (backpassage) studies are required to provide information on the genetic stability of live attenuated vaccines. Adjuvants for products used in food-producing animals must be approved by the USDA Food Safety Inspection Service (9 CFR 113.33, 113.38–113.41, 113.44, 113.45 & Veterinary Biologics [General Licensing Considerations] No. 800.201).

4. *Efficacy data:* Product efficacy must be demonstrated by statistically valid host animal vaccination and challenge studies. The vaccination must be conducted using an experimental product containing the minimum level of antigen prepared from the highest allowable passage level from the Master Seed, as defined in the Outline of Production. The precise challenge method and the criteria for determining protection vary with the immunizing agent. Sufficient data must be collected to validate each label claim; that is, each recommended route or method of administration, each recommended species or age of animal to be vaccinated, and any claims for degree or duration of protection must be supported (9 CFR 113.64–113.455 & Veterinary Biologics [General Licensing Considerations] No. 800.200).

5. *Potency data:* Potency tests are designed to measure the relative strength of a product. Data must be submitted to demonstrate that a proposed potency test is correlated with the host animal efficacy study. Potency tests are then conducted on each serial (lot) of product prior to release for marketing. For release of live vaccines, virus titrations or bacterial counts are used. For killed viral or bacterial products, potency tests may be conducted in laboratory or host animals (e.g., challenge or serology), or with quantitative *in vitro* methods (9 CFR 113.64–113.455).

6. *Field safety data:* Field safety studies are designed to detect unexpected reactions (local lesions, morbidity, mortality, etc.) that may not have been observed during the development of the product. The tests are done on the host animal, at a variety of geographic locations, using large numbers of susceptible animals that the vaccine producer does not own. The test animals should represent all the ages and husbandry practices for which the product is indicated. A request to conduct a field safety trial must be reviewed and approved both by APHIS and by authorities in the cooperating States (9 CFR 103.3).

7. *Three prelicensing serials:* Licensees are required to produce and test three consecutive serials (lots) of final product in their licensed establishment in accordance with the approved Outline of Production. Samples of these serials are tested at CVB-L to confirm the purity,

safety, and potency test results submitted by the producer. These serials are also used by the producer to conduct the field safety studies.

Upon satisfactory completion of all requirements, including review and acceptance of labels and circulars, a U.S. Veterinary Biological Product License may be issued.

IV. Licensing Procedures for Nonconventional Products

A. Recombinant Products

Master Seed viruses or bacteria developed through genetic recombination techniques present unique licensing considerations. These products must meet the same standards of purity, safety, potency, and efficacy required of conventional products. Additionally, the licensee must conduct studies to evaluate any potential effects on the human environment that could result from release of a live recombinant microorganism. This would include studies to:

1. Biochemically characterize the recombinant microorganism.
2. Evaluate its genetic stability (both *in vitro* and *in vivo*).
3. Examine for any changes in the tissue tropism or virulence of the microorganism in the host.
4. Assess its potential to shed from the host and spread to target and nontarget host species.
5. Evaluate its ability to persist in the environment.
6. Examine its potential to undergo recombination with similar field strains of the microorganism.

The data from these studies are used by APHIS to conduct a risk analysis and to prepare an Environmental Assessment, in accordance with the National Environmental Policy Act (NEPA), prior to release of the product for field testing or licensure. The NEPA procedures also require public notification, through the *Federal Register,* of any recombinant microorganism release action to be taken by APHIS.

B. Immunomodulators

Products designed to enhance or suppress host immune responses are also licensed by APHIS. To be licensed, however, the product must meet the definition of a veterinary biological product and carry a claim for the prevention or treatment of a specific veterinary disease. Because of the unique nature of these products, specific standard requirements are not in place for licensing this product category. Nevertheless, the licensee must demonstrate the purity, safety, potency, and efficacy (as defined by the product claim) of the product prior to licensure. Clinical efficacy trials may be employed in place of typical vaccination-challenge trials to evaluate such products if an acceptable laboratory challenge model cannot be developed.

V. Conditional Licenses

APHIS may issue a conditional license in order to meet an emergency condition, limited market, local situation, or other special circumstance (9 CFR 102.6). Such a license may be approved under an expedited procedure, provided the product is shown to be pure and safe, and to have a reasonable expectation of efficacy. This process allows APHIS to:

1. Respond to emergency disease outbreaks.
2. License products for minor species and other limited market situations where the cost of establishing full efficacy before marketing would prohibit the development of needed products.
3. License needed products when host animal efficacy has been established, but difficulty in the development of a fully satisfactory potency test would result in undue delay in the issuance of a regular license.

Conditional licenses are issued for a period of 1 year. Before reissuance, the licenses must demonstrate acceptable progress toward completion of host animal efficacy and/or potency tests. Labels for conditionally licensed products must bear a statement that the product is under conditional license and that potency and efficacy studies are in progress. Conditional licenses are not issued for any product already marketed under a regular license by another manufacturer.

VI. Licenses for Further Manufacture

Licensing products for further manufacture has permitted split manufacturing procedures, where two or more licensed establishments work together to produce a product. These are regular licenses for products that are only permitted to be shipped from one licensed establishment to another licensed establishment or for export. This procedure permits one company to obtain a license for further manufacture to prepare a product to a certain stage of production and ship it to a second company. The second company finishes the product and releases it under a regular license. Licensing in this manner has permitted the industry to take the best advantage of its production capacity and to expand company product lines without extensive development costs [9 CFR 114.3 (d) & Veterinary Services Memorandum No. 800.61].

VII. Sublicensing

Sublicensing of a licensed product from one company to another is also permitted. In this process the company that has a license for the product contracts to transfer to a second company the data, technology, and materials necessary to produce the product. The Outline of Production must be transferred along with Master Seed and Master Cell Stock. The receiving company must repeat purity testing of the Master Seed and Master Cell Stock and do an immunogenicity test in a reduced number of animals to confirm previous data. Additional field safety studies are not required. This process has been useful in the transfer of products and technology from one company to another (Veterinary Services Memorandum 800.58).

VIII. Exemptions to Licensure

Biological products may be produced and used in the absence of federal establishment and product licenses only under the following three circumstances:

1. Products used by USDA, or under USDA supervision, in a USDA disease control program (9 CFR 106).
2. Products prepared under state license, issued by a state program considered by APHIS to be equivalent to the federal program (9 CFR 107.2). Such products are limited to distribution only within the licensing state. (Currently, only California has such a program.)
3. Products prepared by a person for use in animals owned by that person or by a veterinarian for use under a veterinarian–client–patient relationship (9 CFR 107.1).

IX. Autogenous Products

Microorganisms isolated from diseased animals by a veterinarian or appropriate specialist may be submitted to a licensed establishment for preparation of an autogenous vaccine or bacterin (9 CFR 113.113). Such products carry the following restrictions:

1. They must carry a warning statement what the potency and efficacy of autogenous biologics have not been established.
2. They may only be used by or under the direction of a veterinarian or approved specialist.
3. Unless specifically exempted by APHIS, they may only be used in the herd or flock of origin. Groups of animals under the same ownership but at different locations are considered separate herds or flocks.
4. Unless specifically exempted by APHIS, production seed viruses or bacteria for autogenous products may not be older than 15 months from the date of isolation, or 12 months from the date of harvest of the first serial (lot) of product prepared from the seed, whichever comes first.
5. The expiration date for a serial of an autogenous product may not exceed 18 months from the date of harvest.

X. Summary

The licensing procedures reviewed above provide a framework for the production of pure, safe, potent, and efficacious veterinary biological products. The licensing, inspection, and testing activities of the Veterinary Biologics program provide the oversight necessary to ensure the continued availability of high-quality veterinary biological products in the United States.

Reference

Anonymous (1993). Autogenous biologics guidelines issued by AVMA. *J. Am. Vet. Med. Assoc.* **203,** 175–176.

Licensing Procedures for Immunological Veterinary Medicinal Products in the European Union

P. P. PASTORET* AND F. FALIZE[†]

Department of Immunology-Vaccinology, Faculty of Veterinary Medicine, University of Liège, Sart Tilman, B-4000 Liège, Belgium, †Pharmaceutical Inspectorate, Ministry of Public Health, Belgium

I. Introduction
II. Role of the European Agency for the Evaluation of Medicinal Products
III. Available European Procedures
IV. New Definitions of Veterinary Biologicals
   A. Definition of a New Active Substance
V. Revision-Validation of Vaccines already on the Market
VI. Manufacturing Authorization
VII. Batch Control/Release
VIII. Special Case of Equine Influenza Vaccines
IX. Role of the European Pharmacopoeia
X. Summary
   References

I. Introduction

The pharmaceutical legislation of the European Community, which has evolved over a 30-year period, covers both medicinal products for human and veterinary use. Harmonization of requirements in the area of veterinary medicines began in 1981 with the adoption of Directives 81/851/EEC and 81/852/EEC, laying down common requirements for manufacturing and marketing authorization, based on the evaluation of the quality, safety, and efficacy of the product. Many additional measures were taken to harmonize further the procedures and the criteria for the evaluation of veterinary medicinal products, such as framework requirements and interpretive guidelines for their testing, principles, and guidelines of Good Manufacturing Practice, and a community procedure for the evaluation of high-technology products. However, granting of authorizations remained national. As a consequence, although applications were evaluated on the basis of these harmonized criteria and procedures, and in some cases in common by the authorities of the Member States, there were differences in the decisions reached by the Member States on individual products. This was why the commission made proposals in 1990 for a new system for marketing authorization of medicinal products, which was adopted by the Council of Ministers in 1993 and entered into force on January 1, 1995.

One of the first consequence was the creation of the European Agency for the Evaluation of Medicinal Products (EMEA) in London.

II. Role of the European Agency for the Evaluation of Medicinal Products

In 1995 a new European system for the authorization of medicinal products came into force. After 10 years of cooperation between national registration authorities at the European Union (EU) level and 4 years of negotiations, the Council of the EU adopted in June 1993 three directives and one regulation, which together form the legal basis of the system (Brunko, 1997).

The EMEA was established by Council Regulation (EEC) 2309/93 of July 22, 1993. (OJL 214, 24.8.1993), and London was chosen as its seat by decision of the heads of state and government on October 29, 1993.

This agency formulates opinions and, apart from the administrative staff and the management board, is composed of two scientific committees, the CPMP in charge of medicinal products for humans and the CVMP in charge of the animal health sector.

The CVMP is responsible for the evaluation of applications for marketing authorization for products derived from biotechnology, for productivity enhancers, new chemical entities intended for use in food-producing animals, and other innovative new products. In addition, the CVMP makes recommendations on MRLs (maximum residue limits) for substance used in food-producing animals. Nevertheless, due to the fact that it is impossible to discriminate between the results of a vaccination versus a natural infection, there is no need to determine MRLs for immunologic products. To support its activities, the CVMP relies on a pool of 400 experts put at the disposal of the agency by the EU Member States. These experts may participate in any of the CVMP working parties. Among the working parties, the Immunological Veterinary Medicinal Working Party (IVMP/WP) has a double mandate: to examine any request for scientific advice made by a company during the development of a new vaccine, and to advise the CVMP on more general policy issues such as the elaboration and revision of guidelines on immunologic products. The guidelines for the testing of veterinary medicinal products are contained within Volume VII of the rules governing medicinal products in the EU, published by the European commission in 1994.

Among the working program of the IVMP/WP for 1998, appear the following:

- Potency testing of biologicals
- Reduction in the number of animals in safety testing
- Use of adjuvants in veterinary biologicals
- Production and quality control of veterinary medicinal products derived by recombinant DNA technology, including DNA vaccines
- Revision of guidelines on duration of protection and vaccination schemes

The role of the European Pharmacopoeia will be discussed separately.

III. Available European Procedures

Since 1995, two new registration procedures for human and veterinary medicinal products have become available through the European Union: the centralized and the decentralized procedures.

The centralized procedure is compulsory for medicinal products derived from biotechnology (Part A), and available at the request of applicants for other innovative new products (Part B). Applications are submitted directly to the agency in London. At the conclusion of the scientific evaluation undertaken in 210 days within the agency, the opinion of the Scientific Committee is transmitted to the commission to be transformed in a further 90 days into a single market authorization applying to the whole European Union.

Applications may be submitted under the rules of Part A of the annex to the regulation or under the rule of Part B of the same annex.

Article 3 of Reg. 2309/93 states that no medicinal product referred to in Part A of the annex to the regulation may be placed on the market without being submitted to the centralized procedure; on the other hand, the person responsible for placing on the market a medicinal product referred to in Part B of the annex may request to pass through the centralized procedure. Part B of the annex states the following criteria:

- Medicinal products developed by other biotechnological processes which, in the opinion of the agency, constitute a significant innovation
- Medicinal products administered by means of new delivery systems which, in the opinion of the agency, constitute a significant innovation
- Medicinal products presented for an entirely new indication which, in the opinion of the agency, is of significant therapeutic interest
- Medicinal products based on radioisotopes which, in the opinion of the agency, are of significant therapeutic interest
- New medicinal products derived from human blood or human plasma

The decentralized procedure, applying to a majority of conventional medicinal products, is based on the principle of mutual recognition of national authorizations. It provides for the extension of a marketing authorization granted by one Member State to one or more other Member States identified by the applicant.

Should the original national authorization not be recognized by other Member States, the points in dispute are to be submitted to the agency's scientific committees for arbitration. In this case, the final decision is adopted by the European Commission, with the assistance of the regulatory committee or, in the event of serious disagreement between the Member States, by the Council of the European Union. National procedures remained available to applicants during the transition period, that is, up to January 1, 1998, after which they may be used exclusively for nationally marketed medicinal products, the decentralized procedure being mandatory.

The Member States recently created a mutual recognition facilitation group in order to facilitate admission of medicinal products under the decentralized procedure. Meetings are being held monthly at the EMEA.

IV. New Definitions of Veterinary Biologicals

To anticipate forthcoming products on the market, the IVMP/WP has produced new definitions of immunologic products eligible for Part

B of the annex. According to Council Directive 90/677/EEC extending the scope of Directive 81/851/EEC on the approximation of the laws of the Member States relating to veterinary medicinal products and laying down additional provisions for immunologic veterinary medicinal products, the term *immunologic veterinary medicinal product* means a veterinary medicinal product administered to animals in order to produce active or passive immunity or to diagnose the state of immunity. To fulfill the requirements for admission of Part B of the centralized procedure a new active substance had to meet the following criteria.

A. DEFINITION OF A NEW ACTIVE SUBSTANCE

A new chemical, biological or radiopharmaceutical active substance includes:

1. A chemical, biological or radiopharmaceutical substance not previously contained in a veterinary medicinal product authorized in the European Community
2. An isomer, mixture of isomers, a complex or derivative or salt of a chemical substance previously authorized as a veterinary medicinal product in the European Community but differing substantially in properties with regard to safety and efficacy from that chemical substance previously authorized
3. A biological substance previously authorized as a veterinary medicinal product in the European Community, but differing substantially in molecular structure, nature of the source material or manufacturing process
4. A radiopharmaceutical substance which is a radionucleide, or a ligand not previously authorized as a veterinary medicinal product in the European Community, or the coupling mechanism to link the molecule and the radionuclide has not been previously authorized in the European Community

The proposed definition of a new biological substance that constitutes a new active ingredient in the context of Part B of the annex to Council Regulation 2309/93 is expanded as follows:
A biological active substance is considered new if:

- The antigen is contained within a product indicated against a newly emerging species of pathogen, or
- A new antigen is contained within a product indicated against a disease where existing products are proven and recognized not to alleviate suffering from diseases in the target species;

- An antiserum is contained within a product indicates against a newly emerging species of pathogen, or
- The antiserum is contained within a product indicated against a disease where existing products are proven and recognized not to alleviate suffering from disease in the target species;
- The product consists of a substance which through immunologic mechanisms affects the physiologic function of an animal (e.g., one which causes immunocastration), either if the substance has a new efficacy claim or if a new active substance has a different mode of action;
- The product consists of a substance, which modulates the function of the immune system either if the substance has a new efficacy claim or if a new active substance has a different mode of action.

V. Revision-Validation of Vaccines Already on the Market

Immunologic veterinary medicinal products were initially excluded from the scope of the legislation and continued to be regulated nationally until 1993. With the objective of the single market, it became necessary to include these products into the scope of the harmonized legislation. Upon proposal from the commission, the council adopted Directive 90/677/EEC extending the scope of the pharmaceutical legislation to immunologics, which entered into force on April 1, 1993, for new products. For already existing products, a transitional period of 5 years was granted, during which these products are being reviewed for compliance with the requirements of the directives, according to an agreed-on harmonized chronology with regard to the different species-specific vaccines.

Because these time constraints could not be implemented it was decided that the nondefended products would still be withdrawn from the market on April 1, 1998, whereas the defended ones could be examined within a mutual recognition procedure, at the request of the applicant.

Note, however, that, since the entry into force of Directive 87/22/EEC, veterinary vaccines derived from biotechnology were already covered by the Community requirements and had access to the Community "concertation" procedure by virtue of their biotechnological nature. On these matters, the CVMP was assisted by a specialized group of experts from the Member States.

Draft legislation is generally proposed by the European Commission and adopted by the Council of Ministers. This legislation (i.e., regulations and directives) is binding on all Member States. On some specific matters, however, the commission has been empowered by the council to adopt legislation itself, by a regulatory process involving a committee of governmental experts from the Member States. This legislation has the same value and status as council legislation.

In the veterinary medicines sector, the commission has been charged with the updating of testing requirements, whenever the need may arise. This was done in 1991 to take into account the technical progress achieved since the adoption of the original testing Directive 81/852/EEC and to cover immunologics newly introduced into legislation. This resulted in directive 92/18/EEC, describing the testing requirements to be followed by manufacturers intending to file an application for marketing authorization. Special requirements for immunologics were agreed to in the directive, relating to the demonstration of quality, safety, and efficacy of the product. Given the framework aspect of the directive, it was felt necessary to supplement it by a series of guidelines representing the detailed and harmonized interpretation of these requirements. These guidelines are intended to assist manufacturers in complying with the framework provisions of the directive. Compliance with these guidelines provides assurance to the industry that the research and development work undertaken will be considered valid by the Member States. To avoid placing too many constraints on scientific and technical developments, other approaches to those described in a guideline can be followed if it can be shown that this is justified. The current guidelines address both live and inactivated vaccines in general, with specific provisions for a series of species-specific vaccines. Besides guidelines specifically addressing testing of immunologics, some guidelines of a general, horizontal nature apply, such as the guidelines of good clinical practice.

VI. Manufacturing Authorization

In accordance with Directive 81/851/EEC, authorization is also required for manufacture of veterinary medicinal products, including immunologics. This directive provides for regular inspections and that manufacture must be supervised by a "qualified person," who certifies that each batch is in conformity with the approved specifications for the product. For the implementation of these requirements, the commission has adopted Directive 91/412/EEC relating to the principle and guidelines of Good Manufacturing Practice (GMP), and published a detailed guide on GMP developed by a group of pharmaceutical inspectors from the Member States.

Unlike conventional medicinal products, which are produced using chemical and physical techniques capable of a high degree of consistency, the production of biological medicinal products involves processes and materials that are subject to variability and, by virtue of their biological nature, provide good substrates for the growth of microbiological contaminants. Therefore, in-process controls take on a major importance. This line of thought led to the adoption of supplementary provisions addressing the manufacture of immunologic veterinary medicinal products.

VII. Batch Control/Release

As stated earlier, manufacturers are required to have the services of a qualified person at their disposal to certify that each batch of product has been manufactured and checked in accordance with the conditions for marketing authorization. This is a basic requirement of the pharmaceutical legislation. In the case of batches imported from third countries, each batch has to undergo a full qualitative analysis and a quantitative analysis of at least the active ingredients in the first Member State of import into the European Union, under the supervision of a qualified person. Not until this control by the qualified person has been carried can a batch circulate within the European Union without further control.

In the special case of immunologic veterinary medicinal products, an additional step may be introduced. Directive 90/677/EEC allows those Member States which consider it necessary to ask for the submission of samples of each production batch of the bulk and/or finished product for examination by a control laboratory before that batch is placed on the market. This official batch release is not meant to waive the requirement of batch control by the qualified person.

Except in specially justified circumstances, batch release carried out by one national control laboratory must normally be recognized without repetition by the other Member States. To ensure the smooth operation of this provision, an administrative information exchange procedure has been agreed between the competent authorities. Although all Member States do not require official batch release for veterinary immunologics, it was felt by all that they had to be involved in this information exchange scheme.

VIII. Special Case of Equine Influenza Vaccines

Equine influenza has remained among the main acute contagious respiratory diseases of horses worldwide. Equine influenza is represented by two subtypes: influenza A/equine 2 virus (H_3N_8), which is the most important cause of respiratory diseases in the horse, and influenza A/equine 1 virus (H_7N_7), which is still circulating subclinically but is almost considered to be extinct.

However, a divergence in the evolution of A/equine 2 (H_3N_8) viruses has occurred since 1987 and two families of viruses are now circulating. These were designated European-like and American-like, although representatives of both families have been isolated in both continents. There is increasing evidence from field studies that antigenic drift in the gene coding for the hemagglutinin (HA), which is the major surface protein of these influenza A strains, eventually renders vaccine strains obsolete and is likely to compromise vaccine efficacy.

A formal reporting mechanism on antigenic/genetic drift or shift of equine influenza viruses and a vaccine strain selection system has been set up, so that vaccine manufacturers and regulatory authorities are informed of the potential need to update vaccine virus strains.

An Expert Surveillance Panel, including representatives from three WHO reference laboratories and from three OIE reference laboratories, reviews every year the epidemiologic and virologic information and makes recommendations about suitable vaccine strains. These recommendations are published annually by the OIE in its *Bulletin*. Because antigenic drift in equine influenza occurs at a slower rate than in human influenza, it is considered that a regular update of the strains could be necessary every 3–5 years.

The development of effective vaccines can now be facilitated by the availability of reliable *in vitro* assays such as single radial diffusion (SRD) to measure vaccine bulk antigen content in terms of HA content, or single radial hemolysis (SRH) to measure serologic responses.

For in-process controls, SRD provides a reliable method of measuring hemagglutinin content of equine influenza bulk antigens, although it cannot be used on final adjuvanted products. Use of SRD tests is therefore limited to the in-process control of adjuvanted vaccines. SRD tests can provide a great improvement on the chick cell agglutination (CCA) test because it is not susceptible to wide test variation and measures immunologically active HA. SRH is a sensitive and reproducible method for measuring antibody to hemagglutinin.

A new outbreak associated with a breakdown of existing vaccines may require a change in the formulation of equine influenza vaccines. It is expected that manufacturers will wish to make such changes in response to evidence of an antigenic drift and on the need for such a change from the report and recommendation from the Expert Surveillance Panel.

Equine influenza vaccines are well known, and it is unlikely that the replacement of one strain by another would lead to such substantial changes so as to justify a new full set of safety and efficacy tests to be carried out. In addition, there is a need to consider reduction of the number of animals used in the testing of medicinal products whenever possible.

Therefore, provided there have been no or few adverse reactions with the previous formulation, a twofold approach is proposed for the testing of the new formulation:

1. Cross-references to the original dossier would be accepted for those parts that remain unchanged.
2. Where necessary, the analytical, safety, and efficacy sections of the original dossier would need to be amended and new additional data generated.

IX. Role of the European Pharmacopoeia

The last 30 years have seen profound changes in the organization of the European people and the regulation of medicinal products (Artiges, 1997). Thirty years ago, each country had its own regulations, and between them the European countries had two-thirds of the world's pharmacopoeias with all the possible variations. The European Pharmacopoeia Convention has now been signed by 24 parties: twenty-three countries,[1] and just recently by the Commission of the European Communities; moreover 10 European and non-European countries,[2] and the World Health Organization (WHO) have observer status. Close relations are maintained with the licensing authorities of the European Economic Area, where integration is developing via the implementation of common directives and guidelines of medicines for human and veterinary use. The European Pharmacopoeia cofounded, with the Japanese Pharmacopoeia and the United States Pharmacopoeia, the Pharmacopoeial Discussion Group (PDG) in 1990; this group is working assiduously for harmonization at the world level, and it partici-

---

[1]Austria, Belgium, Cyprus, Croatia, Denmark, Finland, France, Germany, Greece, Iceland, Ireland, Italy, Luxembourg, Netherlands, Norway, Portugal, Slovenia, Spain, Sweden, Switzerland, Turkey, United Kingdom, the Former Yugoslav Republic of Macedonia; Member States must apply the standards of the European Pharmacopoeia.

[2]Albania, Australia, Bulgaria, Canada, Hungary, Czech Republic, People's Republic of China, Lithuania, Slovakia, Poland; Observer States do not have to apply the European Pharmacopoeia standards. Some of them apply the standards on a voluntary basis.

pates in the International Conference on Harmonisation of Technical Requirements for Registration of Pharmaceuticals for Human Use (ICH) program. Unfortunately this will have no consequences in the field of veterinary vaccines as there are no monographs on these preparations in the U.S. and Japanese pharmacopoeias.

In Europe, during the 1960s, mainly within the framework of the two major international organizations, the European Union and the Council of Europe, it was agreed to pool technical and scientific expertise. This led to the formation of a coherent body of regulations covering marketing and quality control of medicines for human and veterinary use, manufactured locally or imported. Its main components are marketing authorization, granted case by case for medicines manufactured industrially, and the Pharmacopoeia, a tool for standardization. The regulations surrounding marketing authorization were elaborated by the European Economic Community after extensive public consultations with professional pharmaceutical associations as well as the European Free Trade Association countries and the Nordic countries. The European Pharmacopoeia was developed under the aegis of the Council of Europe by means of specific international convention, which from the start allowed a larger number of European countries to participate.

The convention on the Elaboration of a European Pharmacopoeia is based on a dual commitment by its signatory states:

- A commitment to elaborate a common pharmacopoeia, by contributing financially to its budget and by sending experts
- A commitment to make official on their territories, the specifications of the European Pharmacopoeia replacing, where applicable, the existing national requirements

This commitment has been made official and integrated into the regulations for the registration of medicines manufactured industrially since the adoption in 1975 of the first directive (75/318/EEC) on the standards and protocols for analytical, pharmacotoxicological, and clinical studies on medicines for human use; this principle has also been applied in the area of drugs for veterinary use according to Directive 81/852/EEC and also when these requirements were extended to immunologic products (human vaccines, immunosera, and allergens; Directive 89/342/EEC), to veterinary vaccines (Directive 90/677/EEC), and to homeopathic medicines for human use (Directive 92/73/EEC) and for veterinary use (Directive 92/74/EEC)).

The harmonization envisaged involves real integration and the creation of supranational European specifications. This has been the decision taken by the European Pharmacopoeia Commission since its first meeting. To this end, it created groups of experts in the various pharmaceutical areas; the chairman of each group is usually a delegate at the commission, and the experts are proposed by the national delegations and appointed by the commission.

In the field of veterinary vaccines, a specific group of experts has been set up.

As far as vaccines are concerned the functions of the European Pharmacopoeia have been affected in recent years by the increasing degree of integration of the internal market of the European Union.

For the first 25 years of its life, the European Pharmacopoeia took a "classical" view of vaccine monographs, concentrating on the finished product available as a commercial article that an independent analyst could obtain and test according to the specifications; as far as vaccines were concerned many of the essential aspects of quality were virtually ignored and had then to be dealt with by each country in parallel systems in compendia of regulations and guidelines applicable to biologicals. Inevitably the decisions differed and this provided the spark necessary to initiate a change of course. It was decided that monographs would in future treat all aspects of quality throughout the manufacturing process. A new section, Production, has been added to each monograph, which sets out the essential features and control requirements along the manufacturing process.

The production of new harmonized monographs and the revision of existing ones has accelerated with the concomitant harmonization of European registration legislation. New or revised monographs or general chapters are constantly added. In 1997, around 50 monographs on veterinary immunologics were described in the edition of the European Pharmacopoeia and around 25 new monographs are in preparation. Before their adoption, all monographs are published for inquiry and comments in the Forum of the European Pharmacopoeia, *PHARMEUROPA*, issued four times a year by the Council of Europe. In addition, the European Pharmacopoeia endorsed the aims of the European Convention for the Protection of Vertebrate Animals used for Experimental and Other Scientific Purposes and encouraged its groups of experts to take into account the three R's rule and review all existing animal tests.

For the time being no monographs on combined veterinary vaccines have been prepared because many combinations are available and the resources to prepare monographs on all these combinations have not so far been available. Nevertheless, in the near future such monographs will be prepared in order to harmonize the approach of the different countries.

X. Summary

With the adoption of the new authorization system, all regulatory needs for veterinary medicinal products have been fulfilled with the European Union. This system, indeed, provides access to a continent-wide market to innovative products, in particular vaccines, and facilitates access to the markets of the Member States for other products. This should have a clearly favorable impact on the veterinary vaccines industry.

References

Artiges, A. (1997). The role of the pharmacopoeias. *In* "Veterinary Vaccinology" (P. P. Pastoret, J. Blancou, P. Vannier, and C. Verschueren, eds.), pp. 705–717. Elsevier, Amsterdam.

Brunko, P. (1997). Procedures and technical requirements in the European Union. *In* "Veterinary Vaccinology," (P. P. Pastoret, J. Blancou, P. Vannier, and C. Verschueren, eds.), pp. 674–679. Elsevier, Amsterdam.

International Association of Biological Standardization and International Harmonization

DANIEL GAUDRY

Merial, Gainesville, Georgia 30501

I. Introduction
II. Introducing the IABS
   A. The Mission
   B. The Organization
   C. The Tools of the IABS
III. A Case Study: Report of the Avian Products Standardization Committee (March 1979)
   A. Marek's Subcommittee
   B. SPF Poultry Flocks Subcommittee

I. Introduction

During the past several years, the regulatory authorities of the United States, Japan, and the European Union (EU) met on several occasions in an attempt to harmonize the regulations dealing with human and veterinary biologics. In veterinary biologics, the discussions were held between U.S. and EU authorities. The obstacles being so numerous, the current trend is to aim for *mutual recognition* rather than a change of the existing standards within a given geographic area. Even with this "smoother" approach, several years will be needed to achieve this objective. A proposed way to speed up this process is to use the tools offered by the International Association of Biological Standardization (IABS).

In human biologicals, several of the existing requirements took their roots in meetings organized by the IABS and especially in the meeting cosponsored by the IABS, the World Health Organization (WHO) and/or the Federal Drug Administration (FDA). The latest ones have been dealing with the purity of plasma derivatives (Cannes, 1992; Washington, DC, 1996) and the safety of biologicals (Bethesda, 1995; Annecy, 1996).

Surprisingly, despite the number of meetings of interest for veterinary medicine, the IABS is not well recognized by veterinary scientists who perceive the IABS as an association dealing with human products only: Less than 5% of the current membership are veterinarians.

II. Introducing the IABS

The IABS was founded in 1955 by a group of independent experts who identified an urgent need for an improvement in the quality and comparability of data being exchanged between scientists working in research, development, production, regulation, and standardization of biological products.

Today the IABS is a nonprofit association administered by the Swiss Civil Code and specific statutes with a main office located in Geneva. The IABS is a branch of the International Union of Microbiological Societies and has close links with the WHO. There are about 460 members in 45 countries.

The IABS may be considered as a platform for discussion and consensus of temporary issues related to biologicals. This organization is unique in its ability to bring together for informal discussion state controllers, manufacturers, and research workers from academic institutions and public health organizations.

A. The Missions

According to its statutes, the main missions of the IABS are:

- To encourage the appropriate control and standardization of biological products
- To promote uniform methods for the international control of biological products
- To organize congresses, symposia, and other scientific meetings and to publish reports of such meetings

B. The Organization

The supreme authority of the association is vested in the general assembly, which is constituted from the membership. The general assembly meets every 2 years on the occasion of a scientific meeting organized by the association. The general assembly elects the council, which is the executive board of the association. The council elects the following officers:

- President
- Two vice-presidents
- Secretary general
- Chairman of the scientific committee
- Chairman of the editorial committee
- Treasurer

The council meets at least twice a year to discuss the items proposed on an agenda sent by the secretary general. Experience shows that most of the time spent by the council is dedicated to matters dealing with the organization of scientific meetings and with the publication of the proceedings of these meetings and of the journal.

C. Tools of the IABS

1. Meetings and Proceedings

In 1965, S. Karger was selected as the official publisher of the association. Since that time, the IABS has convened more than 90 international meetings of contemporary interest, which are then published as proceedings of those meetings.

2. The Journal

Biologicals is an international journal published quarterly, devoted to the timely publication of broad ranging reports relevant to the development, preparation, and quality control of biologicals used in human and veterinary medicine. Reports on biologicals derived from new technologies are especially encouraged.

Three types of papers are acceptable: original research reports, short papers, and review articles dealing with topics of current interest.

3. The Newsletter

The *Newsletter* is a quarterly publication which reports the current activities of the IABS regarding the membership, scientific activities, announcements and programs of future symposia, and the reports of the general assembly.

III. A Case Study: Report of the Avian Products Standardization Committee (March 1979)

As already reported, in human biologicals, several of the existing requirements took their roots in meetings organized by the IABS. In the veterinary field, we would like to report a "case study" with the hope that it will be used as an example for further progress in the international harmonization process.

In 1973, the IABS organized in Lyon, France, an international symposium on the Requirements for Poultry Virus Vaccines. The proceedings of this symposium were published in 1974 as *Developments in Biological Standardization,* Vol. 25. At the conclusion of the symposium, it was resolved that an Avian Products Standardization Committee, composed of four subcommittees, should be established. Each subcommittee, under the direction of an experienced chairman, considered the following topics:

- Marek's disease (Dr. B. R. Burmester)
- SPF poultry flocks (Dr. R. Luginbuhl, succeeded by Dr. J. B. McFerran)
- Infectious bronchitis (Dr. R. Winterfield)
- Newcastle disease (Dr. R. P. Hanson, succeeded by Mr. W. H. Allan)

Each chairman was responsible for forming his subcommittee by the selection of four to six recognized experts in the appropriate field. Representatives of the United States and European Union were present in each subcommittee. All of these experts were to advise on the problems that had been discussed at the symposium and outlined in the conclusions of the proceedings and to make recommendations. Where appropriate, they would also arrange for scientific studies and assays to be conducted to establish parameters for vaccine production and control.

The recommendations of three of the subcommittees were submitted and published in a report in March 1979. Through unforeseen circumstances, it has not been possible to finalize the report on Newcastle disease.

A summary follows of the work carried out by the SPF and Marek's groups, underlining the most interesting steps:

A. Marek's Subcommittee

A plan was immediately developed with actual succession of events as given here. A questionnaire designed to obtain information of all aspects of the manufacture and testing of Marek's disease vaccine was sent to 59 licensing authorities and manufacturers. Usable replies were obtained from 24. Information was collated, studied, and a tentative draft standard prepared. After thorough study by the working group, a second draft was prepared and sent for critical review to all those who responded to the questionnaire. Subsequently, a third draft was prepared with due consideration of all respondents and others.

All of these consultations allowed the subcommittee not to recommend the PD50 test, as initially planned, for the evaluation of the efficacy of the vaccines, due largely to the variations observed in test results. At that time, a lot of discrepancies were observed in the evaluation of the PFUs in Marek's assays. This group made available samples of a reference batch and recommended that a reference sample should be included in any titration process.

These recommendations, made from numerous international practical observations were, at that time, of great help for the Marek's disease scientists in comparing their results.

B. SPF Poultry Flocks Subcommittee

Two hundred questionnaires were distributed and 32 replies received from virtually all the important egg producers, national control authorities, and various experts.

The variety of replies indicated that there were many ways to produce eggs free from microorganisms and that there is not one "golden" way to success.

Detailed recommendations were made regarding the management and maintenance of SPF poultry flocks as well as their microbiological surveillance.

The proposals of the SPF subcommittee were used a few years later as a reference document for the setting up the new EU regulations regarding the poultry SPF flocks.

More recently, at Ploufragan, France, in 1992, another meeting co-sponsored by the IABS and dealing with veterinary biologics was very well received. The proceedings of this meeting, published in Vol. 72 of the IABS series and titled The First Steps Towards an International Harmonisation: 1993 and the Free Circulation of Vaccines within the E.E.C., were used by the experts working on the international harmonization of veterinary biologics. In a certain way, we may say that this meeting was a good presentation of the existing problems to be solved for international harmonization. Five years later, the need for another follow-up of this meeting is obvious. In this regard, we would propose another meeting to be organized by the same team and dealing with a limited topic. The proceedings of a meeting on *Quality Control of Seeds and Raw Materials of Biological Origin* would certainly help the competent authorities to set up the rules allowing exchange of seeds between the United States and Europe. That would represent a first step toward the reality of international harmonization.

In conclusion, I want to stress that the IABS does not attempt to intrude on the responsibilities of regulatory agencies. On the contrary, the strength of this association resides in the meetings organized under its umbrella, allowing *nonofficial* discussions between the interested parties. The tools offered by the IABS have been largely and successfully used by the scientists dealing with human biologics. We hope that this example will be followed in veterinary biologics, especially at a time when international harmonization is such a crucial topic for the progress of veterinary medicine.

Technical Requirements for the Licensing of Pseudorabies (Aujeszky's Disease) Vaccines in the European Union

P. VANNIER

Cneva Zoopôle Les Croix, BP 53 22400 Ploufragan, France

I. Introduction
II. Safety Testing
    A. Principles
    B. Laboratory Testing
    C. Field Testing
III. Efficacy Testing
    A. Laboratory Trials
    B. Field Trials
IV. Batch Release Controls
V. Summary
    References

I. Introduction

In general, vaccine properties cannot be evaluated by a few simple tests. Many series of experiments have to be done in order to provide adequate information on the biological properties of a vaccine. These properties cannot be established without a review of all tests done on the vaccine and have to be determined by objective and quantifiable criteria. This approach is totally applicable to Aujeszky's disease (AD) vaccines.

So, testing of AD vaccines is founded on several trials to define the properties of this kind of vaccine. But, in a second step and particularly in the case of a marketing authorization, it is necessary to define an acceptability threshold for safety as well as for efficacy. This problem is not easy to solve. How do we determine this threshold? What criteria can be taken into consideration? Different options have been choosen in the existing regulatory and technical texts which are mainly the 92/18 E. U. directive and the European Pharmacopoeia (EP) monographs about live and inactivated AD vaccines used in pigs. In that text, because the quality part of the file is not specific to AD vaccines, it is not particularly developed, contrary to the safety and efficacy ones.

II. Safety Testing

For a vaccine, local and general reactions have to be determined. When a live vaccine is used, it is necessary to differentiate the exact safety properties of the vaccinal strain from those of the finished product including the adjuvant.

A. Principles

In general, safety is tested initially under experimental conditions. When the results of these preliminary tests are known, it is necessary to enlarge the number of animals vaccinated in order to evaluate, under practical conditions, the safety of the vaccine. In all the cases, safety testing must be done on the pigs under the usual conditions of use or under abnormal conditions in order to reveal any reactions not discovered previously. In a first step, laboratory trials have to be performed and in a second step only, field trials can be done. These trials supplement, on a greater scale, the laboratory tests.

B. Laboratory Testing

General and local reactions must be examined.

1. General Effects Assessment

a. Live Vaccines. The definition of the properties of an AD viral strain depends on the characteristics of this virus; so specific tests need to be performed to better understand the behavior of the vaccinal strain.

Because the AD virus is neurotropic and is particularly pathogenic for young piglets, intracerebral tests and vaccination of 3-day-old piglets are very useful for determining the degree of safety of a strain. So in the EP monograph, five piglets, 3–5 days old each, received $10^{4.5}$ $TCID_{50}$ of the vaccine virus intracerebrally. None of the piglets should die or show signs of neurologic disorders.

All of the information about intrinsic properties of AD strains cannot be provided because they are part of confidential marketing authorization applications files. But, as examples, the following data will illustrate this text.

Some strains such as Alfort 26, which is thermosensitive, and ADV Omnimark have no effect on 2-day-old piglets inoculated by the intracerebral route (Toma et al., 1979; Kit, 1989). These strains as the 783 (TK^-, gI^-) and the Begonia ones do not provoke nervous signs when they are injected by the intramuscular route to very young piglets (2–4 days old) even if the 783 strain can induce slight depression and fever (Van Oirschot et al., 1990; Visser and Lütticken, 1989). But, nervous signs and mortality depend on a dose-effect law: When the Begonia strain is injected by the intracerebral route to 2-day-old piglets (titer: $10^{6.3}$ $TCID_{50}$/0.1 ml), five out of six piglets die, whereas no mortality is observed with a $10^{5.3}$ $TCID_{50}$ dose. This means that even TK^- strains remain pathogenic particularly for the central nervous system when high doses of virus are used (Visser and Lütticken, 1989).

It is essential, too, to assess the properties of a vaccine and specifically of live ones in the target animals (e.g., in normal conditions of use for fattening pigs), generally vaccinated when they are between 9 and 12 weeks old and for pregnant sows when it is claimed by the manufacturer and authorized. Assays have been performed in pigs (4–10 weeks old) with Alfort 26, Bartha, and Begonia strains (Toma, 1979; McFerran and Dow, 1975; Visser and Lütticken, 1989). No clinical signs including thermic reactions were observed after vaccination.

So, in the EP monograph the animals used in the test for immunogenicity are also used to evaluate safety. The rectal temperature of the vaccinated animal is measured at the time of vaccination and 6, 24, and 48 hours later. No animal shows a temperature rise greater than 1.5°C and the number of animal showing a temperature greater than 41°C does not exceed 10% of the group. At slaughter, the injection site is examined for local reactions. No abnormal local reactions attributable to the vaccine are produced. The animals used for field trials are also used to evaluate safety. A test is carried out in each category of animals for which the vaccine is intended (sows, fattening pigs). Not fewer than three groups each of not fewer than 20 animals are used with corresponding groups of not fewer than 10 controls. The rectal temperature of each animal is measured at the time of vaccination and 6, 24, and 48 hours later. No animal shows a temperature rise greater than 1.5°C and the number of animals showing a temperature greater than 41°C does not exceed 25% of the group. Again, at slaughter, the injection site is examined for local reactions. No abnormal local reactions attributable to the vaccine are produced.

In addition, 10 piglets, 3–4 weeks old that do not have antibodies against AD virus or against a fraction of the virus, each receive by a recommended route a quantity of virus corresponding to 10 doses of vaccines. Ten piglets of the same origin and age are kept as controls. The animals are observed for 21 days. The piglets have to remain in good health. The weight curve of the vaccinated piglets does not differ significantly from that of the controls.

Ten piglets, 3–5 days old, which do not have antibodies against AD virus or against a fraction of the virus, each receive by the intranasal route a quantity of virus corresponding to 10 doses of vaccine. The animals are observed for 21 days. None of the piglets dies or shows signs of neurologic disorder attributable to the vaccine virus.

Ten piglets, 3–4 weeks old, and which do not have antibodies against AD or against a fraction of the virus, each receive a daily injection of 2 mg of prednisolone per kilogram of body mass for 5 consecutive days. On the third day, each piglet receives a quantity of virus corresponding to one dose of vaccine by a recommended route. The animals are observed for 21 days following administration of the virus. The piglets must remain in good health.

To complement these studies in the target species, the real degree of attenuation of an AD viral strain can be evaluated by inoculation into other species such as chickens, dogs, cats, mice, etc.

Reversion to virulence following serial passage has to be examined. Primary vaccination is done by the recommended route of administration, which is most likely to be followed by reversion to increased virulence. A series of at least five passages in piglets are made.

The objective of these assays is to test the genetic stability of live vaccinal strains. They seem to be less necessary or unuseful when a genetically modified live strain is involved, especially when it was developed by gene deletion.

However, it may be necessary to examine the possibility of recombination or genomic rearrangement with strains existing in the field or with other strains (Henderson et al., 1991).

The virus excretion of a vaccinal strain by a vaccinated pig is interesting. Obviously the more the vaccine strain is disseminated throughout the body of a vaccinated animal, the greater the risk of spreading and shedding. Vaccinal strains such as Alfort 26 and Bartha are, in most cases, only recovered from the site of inoculation and the satellite lymph nodes (Toma et al., 1979; McFerran and Dow, 1975). In the EP monograph, 14 pigs, 3–4 weeks old and which do not have antibodies against AD virus or against a fraction of the virus, are vaccinated with one dose of vaccine by the recommended route and at the recommended site. Four pigs are kept as contact controls. Nasal and oral swabs are collected daily from the day before vaccination until 10 days after vaccination. The vaccine is acceptable if the virus is not isolated from the secretions collected.

The ability of the AD vaccine strain to spread from a vaccinated pig to unvaccinated ones (transmissibility) must be tested by using the recommended route of administration. In the EP monograph, the same test is carried out on four separate occasions. Each time, four piglets of the same age (without any AD antibody) are kept together with them. Antibodies are not detected in any group of contact controls (5 weeks later).

Until recently, it appeared interesting to know more about the possibility that attenuated AD vaccinal strains can also become latent with the initiation of eradication programs in different countries (Mengeling, 1991). Studies demonstrated that an attenuated TK negative strain of AD virus can establish a reactivatable latent infection in pigs (Mengeling, 1991). No reactivation was observed after vaccination of pigs with vaccinal strains that had a naturally occurring gene deletion for viral glycoprotein I (gI) (Mengeling, 1991; Van Oirschot and Gielkens, 1984).

If live vaccines are used on pregnant sows, the effects on the progeny have to be studied. The born piglets should not become infected by the vaccinal strain.

b. Inactivated Vaccines. As for live ones, it is essential to test the inactivated vaccines in the target animals in normal conditions of use for fattening pigs and for sows when it is claimed by the manufacturer and authorized in the different countries. As described previously, it is fundamental to use objective and quantifiable criteria to detect and to measure adverse reactions such as temperature before and after the vaccinations on vaccinated and control groups, weight performances, litter size, and reproductive performance. So the tests have to be performed by administering the vaccine in the recommended dose and at each recommended route of administration to the pigs for which it is intended.

The pigs or sows are usually kept under observation and submitted to examinations until any reaction has disappeared and the period of observation must not be less than 14 days after administration. This period has to be extended when, for example, the vaccine is used in pregnant sows and it is necessary to assess the putative effects of the vaccine on the reproductive performance, which means the period of observation is, for those conditions, the duration of pregnancy and lasts until the farrowing.

Moreover, it is generally requested to vaccinate with a double dose to have a better opportunity to detect adverse reactions which could be at the limit of a detectable level when a single dose is administered.

2. Local Reactions

Local reactions are often associated with the use of inactivated vaccines as these side effects can be induced by adjuvants and particularly oil adjuvants. But some AD live vaccines are mixed with different adjuvants which modify what was observed up to now.

The local reactions are of two types: allergic or inflammatory. In the case of AD vaccines, local reactions are mainly inflammatory and can be more or less complicated (necrotic or suppurative) depending on the nature of the adjuvants used and the aseptic conditions of the vaccination. Oil adjuvants can induce a great variety of effects ranging from muscular degeneration to granuloma, fibrosis, and abscessation. In addition to the nature of the oil used (the intensity of the reaction is reduced when metabolizable oils are used in the vaccine), it is the type of the emulsion (water/oil, oil/water, water/oil/water) which induces these reactions to a greater or lesser extent (Hall *et al.*, 1989; Vannier, 1986).

In consequence, it is necessary to observe not only from the outside the site of injection, but also by dissection when slaughtering the pigs and particularly the finishing ones.

C. Field Testing

Field trials are necessary to assess the safety of an AD vaccine in a high number of pigs or sows. In Europe, tests are carried out in each category of animals for which the vaccine is intended (sows, fattening pigs). Not fewer than three groups each of not fewer than 20 animals are used with corresponding groups of not fewer than 10 controls. The rectal temperature of each animal is measured at the time of vaccination and 6, 24, and 48 hours later. At slaughter, the injection site has to be examined for local reactions.

If the vaccine is intended to be used in sows, reproductive performances have to be recorded. [*Editor's note:* The extensive battery of safety tests described here for vaccines licensed by the European Union is not required by the USDA for licensing of similar vaccines in the U.S.]

III. Efficacy Testing

A. Laboratory Trials

The biological properties of vaccines are generally based on the clinical protection they confer through passive immunity by vaccinating the dams or by actively immunizing the growing pigs.

1. Assessment of Passive Immunity

To test the efficacy of vaccines, it is important to mimic the natural infection conditions. AD infection provokes important losses in young piglets from nonimmune sows. So, when vaccinating sows, the main goal is to protect the young piglets through passive immunity confered by the colostrum ingested immediately after birth.

To measure this passive immunity and the protection induced by vaccinating the sows, experimental models were carried out. Eight sows are vaccinated according to the vaccinal scheme and by the recommended route during pregnancy and, when the piglets are between 6 and 10 days old they are given an intranasal challenge exposure with a virulent AD strain (Andries et al., 1978; Vannier et al., 1976). Different values of virulent virus titers were used in such assays: 2×10^5 $TCID_{50}/2$ ml or 10^3 to 10^4 UFP/ml. It is better to use a strain titrated in lethal dose 50. It is recommended to inoculate by the nasal route, 10^2 pig LD 50 per pig in 1 ml. The efficacy of the vaccine is assessed by comparing clinical signs, mainly mortality on piglets from unvaccinated dams (minimum of four in EP monographs) with the ones observed on piglets from vaccinated sows.

The vaccine is satisfactory if not less than 80% protection against mortality is found in the piglets from the vaccinated sows compared to those from the control sows. The test is not valid if the average number of piglets per litter for each group is less than six.

2. Assessment of Active Immunity

a. Clinical protection. Several criteria can be taken under consideration to measure the active immunity induced by vaccinating the pigs. Generally, the pigs are vaccinated at the beginning of the growing period, which means when they are between 9 and 12 weeks old. The laboratory trials are performed by challenging the pigs at the end of the finishing period when they weigh between 80 and 90 kg.

In general, at least three criteria such as rectal temperature, weight loss, and clinical signs with mortality are used to measure the clinical protection of pigs after vaccination and challenge. The antibody titers have little predictive value for the efficacy of the vaccines (de Leeuw and Van Oirschot, 1985). Weight loss compared between the vaccinated and control groups is certainly the parameter that is the most reproducible and quantifiable when the challenge conditions are well standardized. The measure of the difference of weight gain or loss between the two groups of pigs and in the interval of time between challenge (D0) and D7 (7 days later) has a very good predictive value for the efficacy of the vaccines (de Leeuw and Van Oirschot, 1985). In the EP monograph, it is indicated that each animal (10 vaccinated pigs, 5 controls) is weighed and then challenged by the intranasal route with a suitable quantity of a virulent strain of AD virus (at least 10^6 $TCID_{50}$ of a virulent strain having undergone not more than three passages and administered in not less than 4 ml of diluent). Each animal is weighed 7 days after challenge or at the time of death if this occurs earlier and the average daily gain is calculated as a percentage.

The vaccine complies with the test if:

- All the vaccinated pigs survive and the difference between the averages of the daily gains for the two groups is not less than 1.5
- The geometrical mean titers and the duration of excretion of the challenge virus are significantly lower in vaccinates than in controls.

The test is not valid unless all the control pigs display signs of AD and the average of their daily gains is less than -0.5.

Mean titers and the duration of excretion of the challenge virus are determined in swabs taken from the nasal cavity of each animal daily from the day before challenge until virus is no longer detected.

This method to evaluate the efficacy of AD vaccine is now well tested, which allows us to lay down an objective index that provides the opportunity to determine the level of efficacy of a vaccine (Stellmann et al., 1989). This index, which compares the relative weight losses between vaccinated pigs and control ones, can also be used in releasing batch controls (potency testing) as in efficacy testing. But the acceptable value of the index is different in the two tests (1.5 for the efficacy testing, 1 for potency testing).

Figure 1 shows the mean virus titers excreted by pigs from different vaccinated and control groups. Different synthetic index can be used to express the quantity of virulent virus excreted by pigs taking into consideration the duration and the level of viral excretion, as the number of pigs excreting virulent virus. Differences between vaccines can be observed using similar protocols (Pensaert et al., 1990; Vannier et al., 1991).

The effects of vaccine with regard to viral shedding were compared when the vaccines where used in the presence or absence of passive immunity. The comparison suggested that the clinical protection provided by the vaccines was relatively lower when the pigs were first vaccinated while posessing passive antibodies, which is a well-known phenomenon. Likewise, viral excretion appeared to be elevated when passive antibodies were present, but the relative position of the vaccines was the same when they were compared in the presence or absence of maternal immunity. Thus, it may be prudent to compare vaccines without passive antibodies to better standardize the assay.

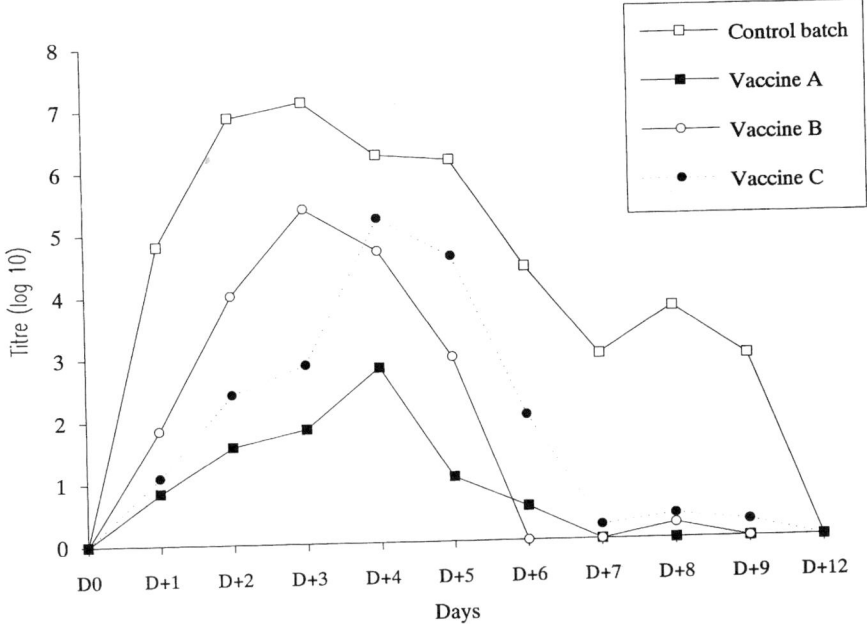

FIG. 1. Mean titers of virus excreted by pigs vaccinated with three different vaccines compared with the level of virus shedding by control pigs.

B. Field Trials

In general terms, it is extremely difficult to assess vaccine efficacy in animal populations. To do this, it would be necessary to vaccinate the animals in the absence of the pathogen that the vaccines protects against, then to await the moment of infection and to compare the effects of infection in vaccinated animals (or the offspring of vaccinated dams) with the effects in the unvaccinated animals of the same age, in the same building, and in the same batch as the vaccinated animals (or those protected passively). All of these conditions are difficult to realize in the field. That is why field trials are certainly more appropriate to safety testing than to efficacy testing. In Directive 92/18 for efficacy, field trials are not absolutely necessary if good experimental data are provided.

IV. Batch Release Controls

It is essential to differentiate the tests that are carried out on a routine basis to release the produced batches from those performed to define the biological properties of a vaccine. The trials carried out for batch releasing are not the same as the ones carried out once to determine safety and efficacy of a vaccine. The batch release controls are always short-term trials, as cheap as possible, and not systematically carried out in the pigs. Their purpose is mainly to attest to the consistency of the quality of the finished product, which has to be in conformity with the quality initially defined in the marketing authorization application.

For safety, these batch release trials can be performed on guinea pigs, pigs, rabbits, or other species depending on whether the vaccine is inactivated or live.

For potency, if good correlation was established between trials carried out on a routine basis and the efficacy ones done once, it can be used *in vitro* tests (titration, etc.) *in vivo* trials by challenging pigs or other susceptible species (mice, rabbits, etc.), by measuring the antibody response after vaccination.

In that kind of control, the most difficult point is to determine an acceptability threshold to accept or to reject the batch according to the results obtained. [*Editor's note:* Batch release controls involving animals are not required by USDA, but many of my colleagues feel they should be required with vaccines used in food animals and companion animals. A small number of animals should be used to assess the continued immunogenicity of the vaccine.]

V. Summary

Under the light of current scientific knowledge, particularly with the progress of molecular biology and of the definition of assays to be performed, it is possible to know, as accurately as possible, the biological properties of a vaccine. Most requirements of EP monographs and Directive 92/18 are founded on that concept. It is clear that there is a balance between safety and efficacy in the case of a live attenuated viral strain that means the more efficient a strain, the less safe it can be. Nevertheless, the problem is more complex; considerable progress has been done to set up new finished products and particularly with the adjuvants which are used now even in combination with live attenuated AD strains. The efficacy of a vaccine can be greatly enhanced, maintaining good local and general safety.

But a debate always occurs when it is necessary to determine the acceptability threshold of a vaccine with regard to its safety and efficacy. The points of view are often very divergent. But, in any case, this threshold depends on the local conditions in the different countries. It is clear that objectives of a vaccination program and the requirements about a vaccine cannot be the same in heavily infected countries with a compulsory vaccination program as in countries or regions with a low prevalence of AD infection or with an absence of any infection. Moreover, it must also be considered that vaccines constitute only one element of a control or eradication program targeted against Aujeszky's disease virus.

REFERENCES

Andries, K., Pensaert, M. B., and Vandeputte, J. (1978). Effect of experimental infection with pseudorabies (Aujeszky's Disease) virus on pigs with maternal immunity from vaccinated sows. Am. J. Vet. Res. **39**(8), 1282–1285.

de Leeuw, P. W., and Van Oirschot, J. T. (1985). Vaccines against Aujeszky's disease: Evaluation of their efficacy under standardized laboratory conditions. Vet. Q. **7**, 780–786.

Hall, W., Molitor, T. W., Joo, H. S., and Pijoan, C. (1989). Comparison of protective immunity and inflammatory responses of pigs following immunization with different Actinobacillus pleuropneumoniae preparations with and without adjuvants. Vet. Immunol. Immunopathol. **22**, 175–186.

Henderson, L. M., Randall, L. L., David, A. J., and Stutz, D. R. (1991). Recombination of pseudorabies virus vaccine strains in swine. Am. J. Vet. Res. **52**(6), 820–825.

Kit, S. (1989). Safety and efficacy of genetically engineered Aujeszky's disease vaccines. In "Vaccination and Control of Aujeszky's Disease," (J. T. Van Oirschot, ed.), pp. 45–55.

McFerran, J. B., and Dow, C. (1975). Studies on immunisation of pigs with the Bartha strain of Aujeszky's disease virus. Res. Vet. Sci. **19**, 17–22.

Mengeling, W. L. (1991). Virus reactivation in pigs latently infected with a thymidine kinase negative vaccine strain of pseudorabies virus. Arch. Virol. **120**, 57–70.

Pensaert, M. B., de Smet, K., and de Waele, K. (1990). Extent and duration of virulent virus excretion upon challenge of pigs vaccinated with different Glycoprotein-Deleted Aujeszky's disease vaccines. Vet. Microbiol. **22**, 107–117.

Stellmann, C., Vannier, P., Chappuis, G., Brun, A., Dauvergne, M., Fargeaud, D., Bugaud, M., and Colson, X. (1989). The potency testing of pseudorabies vaccines in pigs. A proposal for a quantitative criterion and a minimum requirement. J. Biol. Stand. **17**, 17–27.

Toma, B. (1979). Obtention et caracterisation d'une souche thermosensible de virus de la maladie d'Aujeszky (souche ALFORT 26). Recl. Med. Vet. **155**(2), 131–137.

Toma, B., Brun, A., Chappuis, G., and Terre, J. (1979). Propriétés biologiques d'une souche thermosensible (ALFORT 26) de virus de la maladie d'Aujeszky. Recl. Med. Vet. **155**(3), 245–252.

Vannier, P. (1986). Immunisation de porcs charcutiers contre la maladie d'Aujeszky avec deux vaccins à adjuvants huileux. Etude des réactions locales. Recl. Med. Vet. **162**(1), 37–44.

Vannier, P., Tillon, J. P., Toma, B., Delagneau, J. F., Loquerie, R., and Prunet, P. (1976). Protection conférée au porc par un nouveau vaccin huileux à virus inactivé contre la maladie d'Aujeszky. Conséquences pratiques. *J. Rech. Porc Fr.* pp. 281–290.

Vannier, P., Hutet, E., Bourgueil, E., and Cariolet, R. (1991). Level of virulent virus excreted by infected pigs previously vaccinated with different glycoprotein deleted Aujeszky's diseases vaccines. *Vet. Microbiol.* **29,** 213–223.

Van Oirschot, J. T., and Gielkens, A. L. J. (1984). Intranasal vaccination of pigs against pseudorabies: Absence of vaccinal virus latency and failure to prevent latency of virulent virus. *Am. J. Vet. Res.* **45**(10), 2099–2103.

Van Oirschot, J. T., Terpstra, C., Moormann, R. J. M., Berns, A. J. M., and Gielkens, A. L. J. (1990). Safety of an Aujeszky's disease vaccine based on deletion mutant strain 783 which does not express thymidine kinase and glycoprotein I. *Vet. Rec.* **127,** 443–446.

Visser, N., and Lütticken, D. (1989). Experiences with a gI-/TK- modified live pseudorabies virus vaccine: Strain Begonia. *In* "Vaccination and Control of Aujeszky's Disease" (J. T. Van Oirschot, ed.), 37–44.

Specific Licensing Considerations for Modified Live Pseudorabies Vaccines in the United States

DIANE L. SUTTON

U.S. Department of Agriculture, Animal Plant Health Inspection Service, Veterinary Services, Center for Veterinary Biologics, Licensing and Policy Development, Riverdale, Maryland 20737

I. Introduction
II. Licensing Considerations
   A. Master Seed and Cell Testing
   B. Testing of Prelicensing Serials
   C. Field Safety Studies
   D. Stability
   E. Labels
   Reference

I. Introduction

All veterinary biologics licensed in the United States, including pseudorabies vaccines, must be in compliance with the Virus Serum Toxin Act (VSTA) of 1913, as amended by the Food Security Act of 1985. Thus, they must be in compliance with the regulations in Title 9, Chapter 1, Subchapter E of the Code of Federal Regulations (9 CFR), written to implement these acts. The VSTA prohibits the importation and interstate movement of worthless, contaminated, dangerous, or harmful veterinary biological products. The Food Security Act extends this authority to all products shipped into, within, or from the United States. In the United States, licenses for veterinary biological products are issued by the United States Department of Agriculture (USDA), Animal Plant Health Inspection Service (APHIS), Veterinary Services (VS), Center for Veterinary Biologics (CVB).

The regulations that specifically apply to the licensing of live pseudorabies vaccines are found in 9 CFR 113.300, General Requirements for Live Virus Vaccines, and 9 CFR 113.318, Pseudorabies Vaccine, and the sections referenced therein.

Guidelines to clarify the requirements for licensing these products are found in Veterinary Services Memorandums, Notices, and General Licensing Considerations. The most notable of these are General Licensing Considerations 800.200, Efficacy Studies; General Licensing Considerations 800.201, Back Passage Studies; and Veterinary Services Memorandum 800.84, Guidelines for Submission of Materials in Support of Licensure.

The Center for Veterinary Biologics—Laboratory (CVB-L) provides Supplemental Assay Methods (SAM) which are recommended protocols for conducting some of the tests required by the regulations. These include SAM 117, Titration of Pseudorabies Antibody (Constant Virus-Varying Serum Method); SAM 118, Supplemental Assay Method for Titration of Pseudorabies Virus; SAM 119, Method for Titration of Pseudorabies Virus Neutralizing Antibody (Constant Virus-Varying Serum Method); and SAM 123, Supplemental Assay Method for Pseudorabies Virus Challenge Test in Swine.

The National Environmental Policy Act (NEPA) of 1969 must also be considered in the licensing of any live product, particularly products produced through biotechnology. NEPA governs the release of organisms into the environment as a result of field testing and/or licensing.

II. Licensing Considerations

The establishment where the vaccine will be produced must either have an establishment license or be able to qualify for an establishment license at the time the pseudorabies product license is issued (9 CFR 102.2). The general requirements for a product license are described in 9 CFR 102.3(b). They include submission of the following: a product license application, APHIS Form 2003; an Outline of Production in accordance with 9 CFR 114.8 and 114.9; study reports; and evidence that ingredients of animal origin used to produce the product were sterilized or tested for purity (9 CFR 113.53).

A. Master Seed and Cell Testing

Master Seeds must be tested for purity, safety, identity, and immunogenicity (9 CFR 113.300 and 113.318). Purity testing includes freedom from bacteria and fungi (9 CFR 113.27), mycoplasma (9 CFR 113.28), and extraneous agents (9 CFR 113.55). The safety tests include host animal safety at 10 times the normal dose in minimum age animals (9 CFR 113.44), mouse safety [9 CFR 113.33(a)], and a back-passage study (Veterinary Biologics General Licensing Consideration 800.201). Identity testing may be done by fluorescent antibody, serum neutralization, or an APHIS-approved outline test [9 CFR 113.300(c)].

The immunogenicity test is done in the host animal [9 CFR 113.-318(b) and SAM 123], using 25 test swine (20 vaccinates and 5 controls). The test swine must each have a serum neutralization titer of less than 1:2, and the swine must be of the minimum age that will be recommended on the label. The serial of product used to conduct the immunogenicity study should be produced at the highest passage of the virus and cell stock. The titer must be established both prior to and after administration of the serial to the test animals. The product must be administered as recommended on the label.

The challenge virus must be supplied or approved by APHIS. Pigs are challenged 14–28 days postvaccination and are observed for 14 days postchallenge. Four of five controls must develop severe central nervous system signs or die, and at least 19 of 20 vaccinates must remain free of signs of pseudorabies for a satisfactory test. If four of five controls do not develop severe central nervous system signs, the test is inconclusive and may be repeated. Supplemental Assay Method 123 details the clinical signs that are considered significant in evaluating the study.

The master seed must be retested for immunogenicity in three years [9 CFR 113.318(b)(4)]. Only five vaccinates and five controls need to be used. The serum neutralization titer of each animal is determined 14–28 days postvaccination. If the five controls have not remained negative at 1:2, the test is inconclusive and may be repeated. At least four of five vaccinates in a valid test must have titers of 1:8 final serum dilution or higher, and the remaining vaccinate a titer of 1:4 or higher for the test to be satisfactory, unless the master seed is shown to be effective by challenge of the controls and vaccinates [9 CFR 113.318(b)(4)(v)].

Cell stock testing for primary cells or cell lines used to produce the product is done in accordance with 9 CFR 113.51 or 9 CFR 113.52, respectively. Cells are tested for identity and purity.

The CVB-L will conduct confirmatory tests on Master Seeds and Cells. This is usually limited to purity and identity, but may include any of the required tests.

B. Testing of Prelicensing Serials

Prior to licensing, three consecutive serials of product produced in accordance with a filed outline of production must be tested and found satisfactory. The testing is the same as that required for serial release (9 CFR 113.300 and 113.318) and is done on product taken from final containers. The product is tested for purity, identity, safety, and virus titer. The virus titer must be sufficiently greater than the titer of vaccine used in the immunogenicity test to ensure that when tested at any time within the expiration period, each serial and subserial must have a virus titer at least $10^{0.7}$ TCID$_{50}$ per dose greater than that used in the immunogenicity test, but not less than $10^{2.5}$ TCID$_{50}$ per dose. CVB-L will conduct confirmatory tests on the three prelicensing serials.

C. Field Safety Studies

Field safety studies are done to assess the safety of a biological product when used according to label directions in animals kept under conditions typical for that species. An acceptable safety study should include animals in at least three geographically distinct locations and, when applicable, different breeds or conditions of husbandry. The study population should include sufficient numbers of all classes of animals that are included in the label recommendations, such as pregnant animals (representing each trimester), lactating animals, and minimum age animals. A sufficient number is the number of animals needed to provide a 95% assurance that the reaction rate is not greater than an acceptable level for a particular reaction. Such levels are established on a case-by-case basis depending on the type of product and the population for which it is recommended. The product should be administered as recommended on the label, including administration of multiple doses. At least two prelicensing serials should be evaluated. Disposition of study animals should be in accordance with 9 CFR 103.2.

Field safety studies must meet the requirements of 9 CFR 103.3 and the National Environmental Policy Act. To conduct a field safety test, the manufacturer must submit for approval an application to ship an experimental biological product in accordance with 9 CFR 103.3 and Veterinary Services Memorandum No. 800.67. The application should include the following: a copy of a permit or letter of permission from the proper state or foreign animal health authorities of each state or foreign country involved, a tentative list of the names of the proposed recipients and quantity of experimental product that is shipped to each individual, a description of the product, recommendations for use, and results of preliminary work including the method of preparation and testing presented in the appropriate outline format described in 9 CFR 114.9, label or label sketches, and an experimental protocol. The application should also include the identity of the serials to be shipped and the test results for these serials.

NEPA compliance must be adequately addressed prior to the initiation of a field study for products that are not exempted by categorical exclusion from the preparation of an environmental assessment or an environmental impact statement by 7 CFR 372.5(c). NEPA compliance is documented by submission of a completed risk analysis, which should include a completed Summary Information Format, as applicable, for either Veterinary Biologics (conventional live products which have potential for a significant environmental effect), Category II Veterinary Biologics (gene deleted or marker gene inserted), or Category III Veterinary Biologics (live vectored products); a Summary Information Format for Environmental Releases; and a Hazard Identification for Veterinary Biologics as described in "Risk Analysis for Veterinary Biologics" by Gay and Orr (1994). The current edition of this publication can be obtained from APHIS-VS-CVB-LPD.

The experimental protocol for field safety studies should include the following:

1. Specific objectives of the study as related to the recommended use of the product
2. A description of the animals to be used including the number of subjects, species, age, sex, breed, and any other distinguishing features
3. The source, housing, and management of the subjects during the study
4. The procedures for administering the product
5. The observations that will be made, at what intervals, and by whom
6. How long the animals will be observed and what follow-up work will be done if reactions occur (Reactions should be observed until they have resolved. For live products the observation period should take into account the incubation period of the disease, and in-contact animals should be observed for any adverse reactions that might result from lateral transmission of the vaccine virus.)
7. An explanation of how the data will be evaluated, including the statistical analysis that will be done

8. Slaughter withdrawal time if the product is for use in food-producing animals
9. Data acceptable to APHIS demonstrating that the use of the product in food-producing animals is unlikely to result in an unwholesome condition in the edible parts of animals subsequently slaughtered
10. A statement from the researcher or sponsor agreeing to furnish on request, prior to movement of the animals, the information needed to locate and identify the animals at slaughter
11. Methods and procedures for maintaining records of the quantities of the experimental product prepared, shipped, and used.

D. Stability

Stability data must be provided to support the expiration date of the product (9 CFR 114.13). If real-time stability data are not available, accelerated stability data should be provided. Stability dating must be confirmed prelicensure or postlicensure by testing serials at release and at or after the dating requested.

E. Labels

Labels must be submitted for approval in accordance with 9 CFR, part 112, prior to licensure.

Reference

Gay, C. G., and Orr, R. L. (1994). "Risk Analysis for Veterinary Biologics." U.S. Department of Agriculture, Animal and Plant Health Inspection Service, Washington, DC.

Why Do Vaccine Labels Say the Funny Things They Do?

DAVID R. HUSTEAD

Fort Dodge Animal Health, Overland Park, Kansas 66225

I. Introduction
II. Vaccine Label Expectations
III. Vaccine Label Reality
IV. Recommendations for Standardization Improvements

I. Introduction

Every day, veterinary practitioners pick up bottles of vaccine, read the label, and still wonder how they should use this product in their patients. In the United States these curious practitioners then call manufacturers for advice. As a veterinarian that gets to answer those questions, I use this experience as the impetus for this paper. Why is it that intelligent veterinary school educated practitioners have so many questions about how to use vaccines? Obviously, some of the blame must lie with the labels they are reading. If the label contained the information that they needed, their questions would be resolved. While some readers would claim that this problem is due to greedy manufacturers working in cahoots with inept regulators, this is really not the case. This is not to say that the regulations that govern vaccine labels do not need improvement or that manufacturers are not pressured by economic factors to write vaccine labels toward the lowest common denominator. It is fair to say that the sum total of the system, as it exists today, does create some very interesting vaccine labeling issues.

II. Vaccine Label Expectations

Users of biological agents have very high expectations for a product label. They expect a label to discuss practical immunology, preventive medicine theory and application, applied clinical medicine, product liability, owner compliance, and the economics of veterinary medicine. That's a pretty tall order for document that might be as small as a postage stamp. If that were not enough, users would also like to see contingency plans in case they need to use products in less than conventional manners. The sellers of a vaccine believe the vaccine label should be able to offer unique advantages, so that buyers will understand why their product, and no other, should be purchased. Regulators want labels only to make those claims that are beyond scientific scrutiny. Lay users want all of the above along with explanations about what a vaccine is, what a vaccine can or cannot do, all potential side effects, why they should use this product, and when to use it. Clearly vaccine labels cannot fulfill all of these demands. Unfortunately we must accept compromise, and with each compromise we find disappointment.

III. Vaccine Label Reality

It is a fact of life in the 1990s that the information that appears on a vaccine label is greatly influenced by the approval process. It is not my intent to conduct an in-depth review of the approval process for vaccines, but without an understanding of the process, labeling issues lack a frame of reference.

The USDA was granted authority to regulate veterinary vaccine by Congress through the Virus Serum Toxin Act of 1913. This means that the USDA has been regulating labels for a long time. Many of its procedures have developed slowly over time. Like most bureaucratic institutions, changes in procedures occur slowly, usually following the discovery of inadequacies in the process that are newly defined under the light of new knowledge or understanding. In addition, changes to the system are often implemented to ease the burden of the regulator and not the regulated industry or the customer. The USDA requires that any biologic agent that is intended to enter interstate commerce be approved by the agency prior to its sale. The approval process covers four main areas: purity, potency, efficacy, and safety. Because purity and potency do not to any reasonable degree affect the text that is found on the label, I will concentrate on efficacy and safety.

Efficacy testing to support approval is normally an *in vivo* assessment of the product's ability to stimulate the immune system. It is the goal of this immune response to provide clinically important protection to the animal if subsequent challenge occurs. Efficacy studies are normally either serologic quantification of antibody formed in response to vaccine administration or they are challenge studies comparing the clinical disease observed in vaccinates compared to nonvaccinates after challenge. Sometimes both serologic and challenge testing are performed. Efficacy studies are normally performed in the target species, but there are diseases where this *in vivo* efficacy testing is performed in nontarget (i.e., laboratory animals) species. The major reason for nontarget species testing would be where challenges in the target species are not well perfected, for example, equine influenza and where laboratory animal testing has been shown to have correlation to target animal efficacy.

From a regulatory perspective, efficacy studies that are performed to obtain approval can be either codified or noncodified. A study protocol is codified when the USDA determines the protocol parameters that must be utilized to demonstrate efficacy for approval and then publishes that protocol in the Code of Federal Regulations (9 CFR 113). Some of the parameters that will be defined in the code include the species of animal, the number of vaccinates and controls, the dose of vaccine administered, the dose and source of challenge organism, the clinical scoring system used to evaluated disease signs observed, or the serologic tests that must be done. The code will also list the level to which the study must demonstrate effectiveness to be considered eligible for approval. These efficacy levels are commonly 80% of controls must show illness while 80% of the vaccinates should not. If titer is the effectiveness measurement then 75% of the vaccinates must have a titer greater than a set level. If a codified study exists for an antigen, then that is the study that must be performed to support product approval. The situation is very different if no codified study exists for the antigen you are concerned about.

If the antigen in question does not have a codified efficacy test, then the manufacturer is free to develop its own test to demonstrate that its product has efficacy. The efficacy test procedure and the results of that testing will have to be approved by the USDA. At first glance this may appear overly manufacturer friendly, but I would argue it is reasonable given the level of knowledge and the rate of change of knowledge in this area. If noncodified effectiveness studies could not be performed, there would be a incredible delay in getting novel products into the marketplace. An example: At the time of this writing there are no approved products in the United States to prevent feline immunodeficiency virus (FIV) disease in cats. If a manufacturer wanted to bring such a vaccine to the marketplace they would have to go through the effort that it takes to develop a satisfactory efficacy test protocol. This efficacy test will most likely include postvaccination serology and challenge information. If the manufacturer had to wait until an FIV challenge protocol and vaccine efficacy test were developed by academics or the USDA, such a product might never make it to the marketplace. A manufacturer would have to hope that academics or the USDA was interested enough in this disease to develop this type of testing. Without a financial reward (selling the vaccine) to generate study in this area, maybe such testing protocols would never be developed. Even if there is interest by the USDA and academics, the time it takes for the USDA to codify a test protocol can be measured in decades. At the time of this writing there is still no codified protocol for feline leukemia virus efficacy testing, even thought the first commercial vaccine has been available since the mid-1980s.

While the use of noncodified efficacy testing has allowed the vaccine industry to provide new and novel vaccines to be brought to the marketplace, their use also causes problems. The first challenge protocols are most likely developed by the vaccine company. It is in the best interest of that company not to divulge too many details of the challenge protocol. This is to prevent competitors from using the same challenge system to get another product on the market. This less than complete disclosure creates confusion in the minds of users and interferes with the amount of technical data on the label and its interpretation. As additional products enter the marketplace each manufacturer will have developed its own efficacy testing protocols. Each efficacy model may measure different parameters so that different label claims can be supported. This can create lots of confusion when users read label claims on products. An example: Early feline leukemia vaccine efficacy testing did not include mechanisms to measure FeLV infection in bone marrow. Therefore, these early approved vaccines did not have prevention of latent infection claim. Years later when techniques for determining bone marrow infection and latency were more readily available, some product approvals utilized these techniques. Therefore, these subsequently approved products obtained label indications for the prevention of latency. A user reading a vaccine label from an early approved product and comparing it to a later approved product would conclude that the vaccine with the prevention of latency claim was able to do something that the product without the label claim was not able to do. While this conclusion might be correct, is it based on science or reading habits? Is it more plausible that the difference in the label text is a reflection of the measuring system used and not in the immunologic responses associated with one vaccine and not the other? Either way, the user cannot tell from reading the label.

The results that are obtained during the efficacy testing for a vaccine then determine the wording that is used on the label to describe the indications for this product. If this testing results in evidence that the immune response of the vaccinates prevents all tested indications of infection, then a product may claim that it *prevents infection.* I am not aware of any USDA-approved products that can make this label claim. If this efficacy testing indicates that the immune response clearly prevents the expression of the majority of the signs of disease in vaccinates then the claim will be *prevention of disease* due to infection with the agent in the test. This same claim would be obtained if the efficacy test was a serology limit and the vaccinates were all above that limit. Notice that there is no way for the vaccine reader to know from reading the label which standard (serology vs. challenge) was used to determine efficacy. If the immune response in the vaccinates was only able to demonstrate that the clinical scores in the vaccinates were better than the controls and this difference was statistically significant then the label will claim *as an aid in the prevention of disease* associated with infection agent in the test. For the aid in prevention claim, there is no requirement for this difference in the disease observed in the vaccinates vs. the controls to be clinically apparent or relevant. Without a careful reading of the label and an understanding of the differences in the similar phrases used in the indications sections of these labels, the reader could easily not know how the vaccine was tested. In addition, many users do not understand that the phrases in the indication section of the label tell them a lot about what to expect from the product in terms of clinical effectiveness.

In addition to efficacy, vaccine labels have information about the safety behavior of the product. The standard text on vaccine labels is an announcement that vaccine use can be associated with anaphylaxis along with a treatment recommendation to use epinephrine if anaphylaxis is observed. Until very recently, no other safety information was provided. It is interesting to compare the single sentence found on most veterinary vaccines with the safety information that is provided with vaccines intended for use in humans. One of the common DTP vaccines for children has 23 column inches on its package insert to discuss issues of vaccine safety. While one can argue that this is excessive and it is due to the concerns of litigation that surround human medicine, one could also argue that the average medical practitioner has ready access to considerably more information concerning vaccine safety than does the average veterinary practitioner.

Recently the amount of safety information on labels has increased. The USDA is requiring manufacturers to add to labels those vaccine mediated events that are seen in safety studies performed for approval. Therefore, you will see more comments about fever, lethargy, or injection site swelling after vaccination appearing on labels. While this is a step in the right direction, are the events that are seen in a safety study on a few hundred animals a good reflection of what a veterinarian in private practice should expect? These additions to labels will only occur for newly approved products or for old products where a manufacturer submits changes to the USDA, and then the USDA decided that changes to the safety information should be made. The end result of this new attitude by USDA is that nearly identical products can have very different safety information on their labels. This will be true when at the same time, the user has every reason to expect that these differing products would have identical, or nearly identical, biological behavior. In addition, some manufacturers have added safety comments describing events that are commonly reported to them as part of their postapproval adverse event reporting system. Some manufacturers have added safety comments about reported events that are not commonly reported, but are serious. An example: Some feline vaccines now have statements that fibrosarcomas have been reported after use of a product. While I am in favor of more information to users, what is a user supposed to think when these statements are on some vaccines and not others? Isn't it reasonable to conclude that a product without a fibrosarcoma warning would have no such reports associated its use while a product with this warning has these reports? How is a user to know?

In addition to information about efficacy and safety, vaccine users need labels to tell them how to use the product in animals. The directions that are found on the label are for the most part the way the vaccine was used during its efficacy and safety testing. This use pattern may or may not be clinically relevant or convenient. This use pattern may or may not be the best way to use the product. Normally little work is done by manufacturers to determine how products should be used when confronted with clinically common problems like owners that do not return animals to the clinic in the 2- to 4-week window that the label says is the interval during which boosters should be given. The end result is a vaccine label that tells users about one effective way to use a product, but is says very little, or nothing, about how to use the product within the realities of clinical practice or common animal husbandry practices.

This logically brings up the question of appropriate long-term revaccination internals. Vaccines approved by the USDA have contained statements that annual revaccination is recommended. The original source of this label recommendation is not known to me. Those older than I tell me that this recommendation was first put on rabies vaccine based on experimental evidence and then the practice spread to other products. With the exception of rabies vaccines, manufacturers have not been under any obligation to prove that this revaccination interval was effective. This is again an area where the USDA has recently changed its standards. New novel (antigens that currently are not approved) products that are approved must provide data to support the revaccination interval that they recommend. So if a vaccine were to be approved tomorrow for FIV and the revaccination interval on the label was annual, the manufacturer would have provided data to show that animals kept in isolation for 1 year demonstrate the same efficacy performance as those tested just a few weeks postvaccination. Another confusing point exists for established products. Some manufacturers have conducted studies to determine that the annual revaccination interval is supported by objective data, while others have not. The product labels would read identically. So readers see the revaccination claim and do not know if it is objectively supported or historically applied. Finally, readers need to understand that the revaccination interval that is claimed is not the most efficient or the best revaccination interval. Studies to determine the best revaccination interval are very complex and are, at present, outside of the scope of manufacturer's resources.

Vaccine labels have historically not had much information about the age of animals that should be vaccinated. Recently the USDA has determined that labels need to have a minimum age information. This standard is being applied to all newly approved novel products. This standard is also being applied to currently approved products if the manufacturer submits any changes for approval to the USDA. At that time the USDA can require them to add minimum age information to the label. This minimum age will be determined by the ages of the animals in the efficacy and safety studies. The minimum age on the label should be the age of the animals in the efficacy study and at least one-third of the animals in the safety study must be at this age, too. This has created many interesting product labels. The company I work for recently added an updated A2 influenza strain to its products. Our original product label indicated it should be used in healthy horses.

Because of the ages of the animals in our safety study, our new and improved product now carries a label that reads *for vaccination of healthy horses nine months of age or older.* The reason for this label change was not newly discovered biological behavior of the product; it was due to changes in regulatory attitude. It is very unlikely that the reader of the label will know of this difference. I can assure you that there were many confused vaccine users after they read this new vaccine label.

There is another vaccine label issue that will bring about changes in what users read on labels. Historically the USDA has said that for most diseases efficacy demonstrated against the accepted challenge would provide support for a label claim that was not restricted if multiple clinical syndromes existed. An example: Clinical BVD creates several syndromes including mucosal disease, persistent infection, acute disease, and reproductive disorders. The challenge study used for vaccine efficacy testing looks only at the acute syndrome. Vaccine labels have ignored the multiple syndromes and claimed efficacy against disease cause by infection with BVD virus. The USDA now wants vaccine labels only to claim those syndromes that are demonstrated in the efficacy testing. If they restrict this attitude to new novel products then there won't be too much confusion, but if they follow the lead they have created on age claims it could get very ugly in the marketplace as some vaccine get labels restricted to the syndromes they demonstrate, but others continue to ignore the issue. Just how this new position will impact vaccine labels is still unclear.

IV. Recommendations for Standardization Improvements

So how do we get our vaccine labels to stop saying the funny things they do? First, we need improvements in the way we test vaccines for efficacy and safety. The challenge protocols that are used need to be improved so that they more closely resemble the reality of the infectious disease. Safety protocols need to be expanded to better gather safety data so that more precise safety information can make its way to the product labels. The safety data that are accumulated from adverse event reports need to get better access to vaccine labels, but this access needs to be standardized and not haphazard. The USDA needs to improve its handling of codified protocols. Similar products need to be tested by similar test methods. The USDA needs to make users more aware of its standards and the impact of those standards. The USDA needs to understand the impact that its changes in attitudes have on the marketplace. At present, the USDA gives advantages to companies that do not change products. If they like their label they know that if they don't ask for other changes their label probably won't get reviewed. The company that wants to improve its product, gets its label reviewed when the changes are submitted for approval. To fix this the USDA needs to alter its present regulatory style. Vaccines made from similar techniques (i.e., killed vs. MLV) and for the same antigens share more characteristics than not. These similar vaccines will have similar immunologic activity and similar uses in animal health programs. Therefore, they should have identical testing procedures. This requires an increase and improvement in the codified efficacy protocols. If the product is approved, then all products within the same category should have similar, or even, identical labels. The safety data on each product within the group should be similar based on the collective evidence of the product group. There should be flexibility in the process, so that if a company can demonstrate compelling evidence that the product has some unique characteristic, that activity could be presented on the label. In the absence of this compelling evidence, the labels should remain the same. If the USDA wants to change the approval process or label recommendations, these changes should occur simultaneously in all products within the same category, so that the impact of these changes is seen at once in the marketplace. There is no question that this increase in standardization will require more regulatory activity and personnel. With more standardization we would have less user confusion and better use of these products in animals. Better use of vaccines would give us healthier animals, and having healthier animals is a goal that we all can share.

Standardization of Diagnostic Assays for Animal Acute Phase Proteins

P. DAVID ECKERSALL,[1] SUSAN DUTHIE,[1]
MATHILDA J. M. TOUSSAINT,[2] ERIK GRUYS,[2] PETER
HEEGAARD,[3] MARIA ALAVA,[4] CORNELIA LIPPERHEIDE,[5]
AND FRANCOIS MADEC[6]

[1]*University of Glasgow Veterinary School, Bearsden Road, Glasgow, G61 1QH, UK,* [2]*Faculty of Veterinary Medicine, Utrecht University, Yalelaan 1, 3508 TD Utrecht, The Netherlands,* [3]*Danish Veterinary Laboratory, 27 Bulowsvej, 1790 Frederiksburg, Denmark,* [4]*Universidad de Zaragoza 12 Pedro Cerbuno, 50009 Zaragoza, Spain,* [5]*University of Bonn, Regina-Pacis-Weg 3, Bonn D5311, Germany,* [6]*CNEVA Ploufragan, BP 53, Ploufragan 22440, France*

I. Introduction and Background
II. Acute Phase Proteins in Farm Animals
   A. Production
   B. Acute Phase Proteins in Cattle
   C. Acute Phase Proteins in Pigs
III. Methods of Acute Phase Protein Assay
   A. Biochemical Assays
   B. Immunochemical Assays
IV. Standardization of Acute Phase Protein Assays
   A. Need for International Standardization
   B. International Reference Preparation for Animal Plasma Proteins
   Acknowledgment
   References

I. Introduction and Background

Monitoring the plasma concentration of acute phase proteins in animals is now established as providing valuable diagnostic information in conditions involving inflammation, infection, or trauma (Eckersall and Conner, 1988; Gruys et al., 1993, 1994; Kent, 1992). The investigation of these proteins is becoming even more widespread as commercial diagnostic kit producers provide veterinary laboratories with assay systems that have been validated for particular species. In addition, it has been suggested that the determination of the plasma concentration of these proteins is of particular value in farm animal production as an indicator of clinical and subclinical diseases and as an aid to meat inspection when carcasses of potential risk to public health can be identified (Saini and Webert, 1991; Eckersall, 1992; Gruys et al., 1993).

Development of assay methods to quantify acute phase proteins in animals has been pioneered by a number of laboratories on at least three continents. Thus methods for the analysis of acute phase proteins such as haptoglobin, serum amyloid A (SAA), and α_1-acid glycoprotein (AGP) have been developed in the United States and Canada (Eurell et al., 1992; Young et al., 1995; Godson et al., 1996), Japan (Tamura et al., 1989; Itoh et al., 1992, 1993; Ohwada and Tamura, 1995), and Europe (Skinner et al., 1991; Conner et al., 1988a,b; Horadagoda et al., 1993; Alsemgeest, 1993; Alsemgeest et al., 1993). In these laboratories calibration of the assays has been achieved by isolation of the protein and determination of its concentration prior to its use as a primary standard. Unlike assays for biochemical analytes where the molecular structure is identical in all species, protein assays should not share standards between species especially when immunoreactivity forms the basis of the methodology. Lack of cross-reaction or only partial cross-reaction between the antiserum and target antigen in the new species has the potential to lead to gross inaccuracies.

It is important for veterinary medicine that assays for each species should be as accurate as possible. International harmonization of calibration by use of a common reference preparation would be an important step toward this goal. Indeed, if the measurement of acute phase protein is to fulfill its potential as a marker for infection or inflammation in animals at slaughter or as a marker for the identification of illness in animals prior to travel across international boundaries, then agreement on standards will be vital. For example, this will allow the concentration of haptoglobin in porcine serum measured in Denmark to agree with the result of the same test carried out on the same sample in Japan. Only if assays are calibrated with the same international reference material will it be possible to agree on acceptable reference limits of acute phase protein concentration above which it would indicate that an animal is unfit for transport or that a carcass should be marked for special inspection before being passed for human consumption.

Before describing an approach to be taken to rectify the current absence of international harmonization, the biochemistry of selected acute phase proteins in animals is briefly reviewed to emphasize their importance in aiding veterinary medicine and public health. The excellent reviews on this topic in the literature during the last few years can be consulted for further details of the acute phase reaction (Sehgal et al., 1989; Thompson et al., 1992; Gruys et al., 1994; Baumann and Gauldie, 1994; Eckersall, 1995).

II. Acute Phase Protein in Animals

A. Production

The acute phase proteins are a group of plasma proteins, produced in the hepatocyte, the concentration of which varies during the host's response to infection, inflammation or trauma (Sehgal et al., 1989; Baumann and Gauldie, 1994). The change in acute phase protein concentration in plasma can show either a major acute phase response by increasing from very low levels by 100- to 500-fold (e.g., bovine SAA), a moderate response by increasing three to four times the normal concentration (e.g., porcine haptoglobin), or a minor response when the maximum increase is only twice the normal concentration (e.g., bovine ceruloplasmin). The concentration of an acute phase protein may decrease instead of increase, thus albumin which reduces in concentration during the acute phase response is a negative acute phase protein. It has been established that there is significant variation between species in the acute phase protein profile as not all of the proteins respond in the same way in all species.

The production of the acute phase proteins is stimulated by pro-inflammatory cytokines released into the circulation from the site of infection or inflammation. Interleukin 6 (IL-6) has been identified as the predominant cytokine capable of stimulating the response, but other cytokines can also lead to the production of the proteins, notably IL-1 and tumor necrosis factor α (TNF-α), IL-11, and macrophage inflammatory protein either directly or indirectly via stimulation of IL-6 (Heinrich et al., 1990; Baumann and Schendel, 1991; Richards et al., 1992). Indeed, the assay of the acute phase proteins can be regarded as providing a robust alternative to cytokine assay as a means of quantifying the outcome of the overall cytokine induced systemic response to infection or inflammation.

B. Acute Phase Proteins in Cattle

In cattle and other ruminants, haptoglobin has been the acute phase protein most commonly monitored as a marker of inflammation (Skinner et al., 1991; Wittum et al., 1996; Gray et al., 1996). Haptoglobin concentration rises up to 300-fold during the acute phase response in cattle, increasing from a concentration of less than 0.01 g/liter to reach 2–3 g/liter within 48 hours of infection.

Serum amyloid A, which is an apolipoprotein associated with high-density lipoprotein, is a major acute phase protein in cattle. This has been shown to be more sensitive than haptoglobin because it can be induced earlier following infection, and in studies of clinical cases, SAA was shown to be the most efficient diagnostic test for the determination of inflammatory lesions (Alsemgeest, 1994). SAA is the precursor of the amyloid A protein that is associated with amyloidosis.

α_1-Acid glycoprotein is a highly glycosylated protein which has a moderate acute phase reaction in cattle and is raised in inflammatory conditions and in the presence of natural or experimentally induced liver abscesses (Motoi et al., 1992).

Negative acute phase proteins in addition to albumin have been identified in cattle. α_2-Macroglobulin has been shown to decrease in the blood during an acute phase response, while trace elements Fe and Zn are also reduced. Combination of the results from positive and negative acute phase proteins have been used to calculate an acute phase index for maximizing the diagnostic efficiency of these markers (Toussaint et al., 1995).

C. Acute Phase Proteins in Pigs

C-Reactive protein (CRP) is the most requested acute phase protein assayed in human medicine and is also a major reactant in pigs. In pigs, it has been shown to respond to turpentine injection with a 100- to 200-fold increase being observed (Lampreave et al., 1994; Eckersall et al., 1996). In the same studies it has been shown that porcine haptoglobin is a moderate acute phase protein, increasing from a normal level of 1–2 mg/ml and rising to over 5 mg/ml at the peak of the response after turpentine injection.

The acute phase in the pig has unusual features in that the response of AGP to turpentine injection did not show the type of response found in other species (Eckersall *et al.,* 1996; Lampreave *et al.,* 1994), although raised AGP levels have been recognized in pigs with a range of infectious or inflammatory conditions (Itoh *et al.,* 1992). A further unusual aspect of the porcine acute phase reaction is that a protein, called major acute phase protein (MAP) has been identified and shown to be the most sensitive protein to use as a marker of disease in this species. Porcine MAP was related by amino acid sequence analysis to a trypsin inhibitory protein in man (Gonzalez-Ramon *et al.,* 1995).

Pathophysiologic reactions of the acute phase proteins in the presence of disease in cattle and pigs have now been established. There is little doubt that the diagnostic value that comes from determination of their concentration would be greatly enhanced if there were to be international harmonization of calibration and standards. It is imperative that the worldwide veterinary diagnostic community undertake the task of harmonizing calibration of assays. A problem exists in that assay methods and calibration procedures have evolved independently in different laboratories, but this should not be an insurmountable obstacle. A similar situation faced clinical biochemists in human reference laboratories in the 1970s and the procedures they established to produce a reference preparation for human serum proteins (Johnson, 1993) should be a blueprint for present-day veterinary clinical biochemists.

III. Methods of Acute Phase Protein Assay

A. Biochemical Assays

While most assays for acute phase proteins in animals are based on immunochemical methodology, some assays rely on biochemical properties of the proteins and may be of continuing value. The advantages that biochemical assays have over immunochemical systems is that they can be applied to all species, after appropriate validation, and that they can be a better measure of biologically active protein.

Haptoglobin can be measured on the basis of its ability to bind hemoglobin, which it does with a high affinity, and this characteristic has been utilized as the basis for routine biochemical assays (Eurell *et al.,* 1992; Harvey and Gaskin, 1978). There are variations in the way in which this innate activity is used for quantification. The simplest system is to monitor the effect of haptoglobin on the characteristic hemoglobin (or methemaglobin) absorbance in a spectrophotometer. There is a shift in the maximum absorbance when haptoglobin binds to hemoglobin and this can be used to quantify the concentration of haptoglobin. A disadvantage is that each sample has to be accompanied by reagent and sample controls. In a variation of this approach, use is made of the property of the hemoglobin–haptoglobin complex to conserve the peroxidase activity of the hemoglobin at low pH, whereas free hemoglobin is inactivated (Makimura and Suzuki, 1982; Conner et al., 1988a; Skinner et al., 1991). Peroxidase substrates such as guaiacol or tetra methyl benzidine have been used for production of chromogen, although these have disadvantages of either being noxious (guaiacol) or so sensitive to peroxidase that samples have to be diluted several hundred times before assay (tetra methyl benzidine).

There is a difference in the ways in which results of assays based on hemoglobin binding are calibrated and reported. Because the assays are based on the binding of hemoglobin and because a controlled amount of hemoglobin is added, some assays use the proportion of hemoglobin bound as an indicator of the quantity of haptoglobin present in the serum sample. The results of these assays should be reported as weight of hemoglobin binding capacity per volume of serum, e.g., mg HbBC/100 ml rather than weight per volume (mg/ml). To avoid confusion, in biochemical haptoglobin assays abbreviations of weight per volume should only be used with assays which are calibrated with purified haptoglobin or a secondary standard serum. A further factor to consider is that hemolysis in samples can lead to erroneous results with assays based on hemoglobin–haptoglobin binding.

Determination of the acid soluble glycoprotein content is another assay that does not rely on immunoreactions. This assay involves the precipitation of the majority of serum protein by perchloric acid with the concentration of the remaining, soluble protein being determined by Lowry or bicinchoninic acid methods. This has been developed as a diagnostic test and could potentially be used as a routine clinical biochemical test. Other acute phase proteins that have been identified in animals using biochemical methodologies are α_1-antiprotease, α_1-antichymotrypsin, α_2-macroglobulin, and ceruloplasmin.

B. Immunochemical Assays

The predominant methodology used for quantification of acute phase proteins has been based on immunoassay. Many types of immunoassay have been described for analysis of these proteins. A number of assays, particularly for AGP, are based on immunodiffusion in agarose (Itoh et al., 1992; Motoi et al., 1992), which have the disadvantage of requiring

24 or 48 hours for diffusion to be complete. The precipitation of the antibody–antigen complex is also the basis for immunoturbidimetry, which has been used for the measurement of canine CRP (Eckersall *et al.*, 1991), equine and canine haptoglobin (Weidmeyer *et al.*, 1994), and has been used for quantification of porcine CRP. This has the advantages of being rapid and adaptable to biochemical analyzers. Solid phase immunoassays have also been described, either as enzyme-linked immunosorbent assays in microtiter plates (Eckersall *et al.*, 1989) or as latex agglutination inhibition systems (Yamamoto *et al.*, 1993).

Assays using antiserum to the analagous human protein are justified if there is sufficient cross-reaction with the animal protein and if the assay is validated for use in a particular species. However, there is little doubt that species-specific antiserum and standards are essential if consistent and comparable assays are to be developed to allow the widespread use of these assays.

IV. Standardization of Acute Phase Protein Assays

A. Need for International Standardization

In animal clinical biochemistry, the majority of analytes are biochemical compounds that have the same chemical structure in all species or are enzymes with a common activity. Glucose is the same in cattle, dogs, and pigs; aspartate transaminase has the same activity in all species. However, a new era is dawning when there will soon be a demand to quantify specific proteins in animal plasma by the use of antibody-based tests. It is important that the assays used to measure such proteins be calibrated against the protein from the relevant species. No doubt the commercial diagnostic companies, who will increasingly market these immunoassays in the future, will produce the necessary reagents and take care in preparation of their own calibrants. However, quantification of the concentration of a purified protein is not straightforward. Factors such as the true purity of the sample, the choice of total protein assay, whether based on the Lowry reaction, Coomassie blue binding, or the bicinchoninic acid reaction, and the choice of albumin or immunoglobulin to calibrate the total protein assay will affect the derived concentration for the acute phase primary standard.

There is already evidence that assays developed in different laboratories can give substantially different results when the same samples are analyzed. The results for bovine haptoglobin assayed by a peroxidase activity method when compared to an ELISA method varied by 30% (Fig. 1) (Young et al., 1995). Furthermore, in a collaborative study undertaken in Europe it was clear that there were wide differences in the results obtained when the same samples were assayed for porcine haptoglobin and CRP (Tables I and II). This survey revealed that for haptoglobin there was up to a sixfold difference in the concentration of haptoglobin as measured in different laboratories in the same porcine serum samples. A 40-fold difference was found in the concentrations of porcine CRP in the same samples as determined by two of the laboratories.

Such variation between laboratories leads to confusion in results. To allow data from all veterinary laboratories to converge toward a consensus in calibration, an international effort should be undertaken to provide laboratories and diagnostic kit manufacturers with reference materials for international harmonization of acute phase protein assays. This is essential if the acute phase protein assays are to be fully utilized in providing accurate diagnostic information for animal health and welfare.

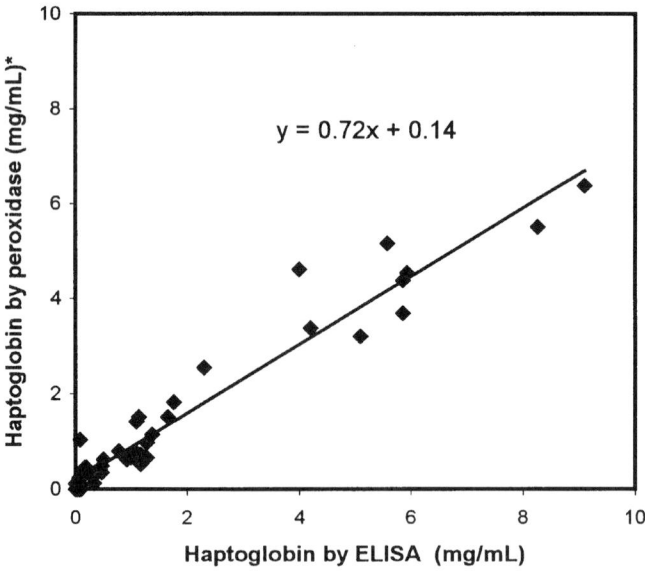

FIG. 1. Bovine serum samples analyzed for the concentration of haptoglobin by an ELISA assay (ordinate) and by a hemoglobin binding assay method (abscissa), which was calibrated with a secondary standard of haptoglobin in acute phase serum. Data reanalyzed from Young et al. (1995). Haptoglobin in peroxidase assay was quantified by calibration with a secondary standard.

TABLE I

CRP (mg/liter) Measured in Control Serum Samples by Four Different Laboratories

	Control A	Control B	Control C
Lab 1	3.8	1.2	5.6
Lab 2	69	33	125
Lab 3	79	35	99
Lab 4	101	16	235

B. International Reference Preparations for Animal Plasma Proteins

In human medicine, protein assays have been harmonized across the world for at least two decades and recently a third international reference preparation for proteins in human serum was prepared, calibrated, and released to reference laboratories and diagnostic kit manufacturers (Johnson, 1993). Assays for specific protein in human serum in clinical biochemistry laboratories across the world are now calibrated indirectly with this preparation so that an assay for human CRP carried out in a hospital in Wisconsin ought to give the same result as an assay carried out in Moscow, Hong Kong, or Santiago.

The approach taken in human clinical laboratories was to pool and aliquot a large volume of serum. Aliquots were then distributed to reference laboratories, which analyzed the serum for a full range of specific proteins using their own calibration material for the assays. Results were collated and a consensus value derived for each protein, which became the accepted value for the concentration of the serum protein. The reference preparation was then used to develop secondary and tertiary standards for assay calibration. This approach may not

TABLE II

Haptoglobin (mg/ml) Measured in Control Serum Samples by Three Different Laboratories

	Control A	Control B	Control C
Lab 1	1.3	0.66	1.75
Lab 2	0.38	0.43	6.0
Lab 3	2.56	1.58	4.64

lead to the most accurate estimate of the concentration of each protein and to a certain extent the agreed values are arbitrary, but experience shows that once the assigned values for the reference material were used the precision of protein assays between laboratories improved significantly. The human reference preparation is not only being used for calibration of acute phase protein assays but also for other specific protein assays, such as transferrin, complement C3, C4, and the immunoglobulins IgA, IgG, and IgM.

In veterinary medicine there are no equivalent reference preparations for immunoglobulins or complement for any species. Although commercial kits are available for the analysis of the immunoglobulins classes in cattle, pig, horses, dogs, and cats, there are no procedures available to develop interlaboratory agreement on results obtained.

International collaboration is urgently required to remedy this state of affairs and our group from interested European laboratories invites collaboration from colleagues around the world in order to initiate the preparation and distribution of International Reference Preparations for Proteins in Animal Serum. A summary of the process required to develop reference preparations is given in Table III. While the initial emphasis should be on animals of agricultural importance, the same process should be applied to establishing reference preparations for serum protein of all species of domestic animals. Individual laboratory scientists may contribute to this process, but it will be important for relevant learned societies and national and international organizations to become aware of this problem and the proposals whereby it can be overcome. Relevant organizations that may wish to support this essential process are listed in Table IV.

TABLE III

A Proposal to Establish a Reference Preparation for Animal Serum Protein

1. Establish international network of reference libraries performing specific serum protein assays.
2. Prepare pooled acute phase serum from species of interest.
3. Aliquot and distribute to reference laboratories.
4. Combine results for each protein as an "all-laboratory mean."
5. Agree that "all-laboratory mean" values be assigned to the reference preparation.
6. Utilize the reference preparations as primary standards for calibration of assays.

TABLE IV

ORGANIZATIONS POSSIBLY CONCERNED WITH A REFERENCE
PREPARATION FOR ANIMAL SERUM PROTEIN

European Union
World Veterinary Association
Office International des Epizooties
Food and Agricultural Organisation
International Standards Organisation
International Federation of Clinical Chemistry
National Departments or Ministries of Agriculture
International Association of Biological Standardization
International Society for Animal Clinical Biochemistry

ACKNOWLEDGMENT

Collaboration between the authors was made possible by a European Union Concerted Action AIR3-CT994-2255 on "The use of blood analysis in the integrated pig production chain."

REFERENCES

Alsemgeest, S. P. M. (1993). Peripartum acute-phase protein serum amyloid A concentration in plasma of cows and fetuses. *Am. J. Vet. Res.* **54,** 164–167.
Alsemgeest, S. P. M. (1994). Concentrations of serum amyloid-A (SAA) and haptoglobin (Hp) as parameters of inflammatory diseases in cattle. *Vet. Q.* **16,** 21–23.
Alsemgeest, S. P. M., Taverne, M. A. M., Boosman, R., Vanderweyden, B. C., and Gruys, E. (1993). Peripartum acute-phase protein serum amyloid-A concentration in plasma of cows and fetuses. *Am. J. Vet. Res.* **54,** 164–167.
Baumann, H., and Gauldie, J. (1994). The acute phase response. *Immunol. Today* **15,** 74–80.
Baumann, H., and Schendel, P. (1991). Interleukin-11 regulates the hepatic expression of the same plasma protein genes as interleukin-6. *J. Biol. Chem.* **266,** 20424–20427.
Conner, J. G., Eckersall, P. D., Wiseman, A., and Douglas, T. A. (1988a). Bovine acute phase response following turpentine injection. *Res. Vet. Sci.* **44,** 82–88.
Conner, J. G., Eckersall, P. D., and Douglas, T. A. (1988b). Inhibition of elastase by canine serum: demonstration of an acute phase response. *Res. Vet. Sci.* **44,** 391–393.
Eckersall, P. D. (1992). Meat inspection: The potential of acute phase protein assay. *Meat Foc. Int.* **1,** 279–283.
Eckersall, P. D. (1995). Acute phase proteins as markers of inflammatory lesions. *Comp. Haematol. Int.* **5,** 93–97.
Eckersall, P. D., and Conner, J. G. (1988). Bovine and canine acute phase proteins. *Vet. Res. Commun.* **12,** 169–178.
Eckersall, P. D., Conner, J. G., and Parton, H. (1989). An enzyme-linked immunosorbent assay for canine C-reactive protein. *Vet. Rec.* **124,** 490–491.

Eckersall, P. D., Conner, J. G., and Harvie, J. (1991). An immunoturbidimetric assay for canine C-reactive protein. *Vet. Res. Commun.* **15,** 17–24.

Eckersall, P. D., Saini, P. K., and McComb, C. (1996). The acute phase response of acid soluble glycoprotein, alpha-1 acid glycoprotein, ceruloplasmin, haptoglobin and C-reactive protein in the pig. *Vet. Immunol. Immunopathol.* **51,** 377–385.

Eurell, T. E., Bane, D. P., Hall, W. F., and Schaeffer, D. J. (1992). Serum haptoglobin concentration as an indicator of weight gain in pigs. *Can. J. Vet. Res.* **56,** 6–9.

Godson, D. L., Campos, M., Attah-Poku, S. K., Redmond, M. J., Cordeiro, D. M., Sethi, M. S., Harland, R. J., and Babiuk, L. A. (1996). Serum haptoglobin as an indicator of the acute phase response in bovine respiratory disease. *Vet. Immunol. Immunopathol.* **51,** 277–292.

Gonzalez-Ramon, N., Alava, M., Sarsa, A., Pineiro, M., Esko, J. D., Garcia-Gil, A., Lampreave, F., and Pineiro, A. (1995). The major acute phase serum protein in pigs is homologous to human plasma kallikrein sensitive PK-120. *FEBS Lett.* **371,** 227–230.

Gray, M. L., Young, C. R., Stanker, L. H., and Bounous, D. I. (1996). Measurement of serum haptoglobin in neonatal farm-raised and bob veal calves using two immunoassay methods. *Vet. Clin. Pathol.* **25,** 38–42.

Gruys, E., Vanederen, A. M., Alsemgeest, S. P. M., Kalsbeek, H. C., and Wensing, T. (1993). Acute-phase protein values in blood of cattle as indicator of animals with pathological processes. *Arch. Lebensm. Hyg.* **44,** 107–107.

Gruys, E., Obwolo, M. J., and Toussaint, M. J. M. (1994). Diagnostic significance of the major acute phase proteins in veterinary clinical chemistry: A review. *Vet. Bull.* **64,** 1009–1018.

Harvey, J. W., and Gaskin, J. M. (1978). Feline haptoglobin. *Am. J. Vet. Res.* **39,** 549–553.

Heinrich, P. C., Castell, J. C., and Andus, T. (1990). Interleukin-6 and the acute phase response. *Biochem. J.* **265,** 621–636.

Horadagoda, A., Eckersall, P. D., Alsemgeest, S. P. M., and Gibbs, H. A. (1993). Purification and quantitative measurement of bovine serum amyloid-A. *Res. Vet. Sci.* **55,** 317–325.

Itoh, H., Tamura, K., Izumi, M., Motoi, Y., Kidoguchi, K., and Funayama, Y. (1992). The influence of age and health-status on the serum alpha-1-acid glycoprotein level of conventional and specific pathogen-free pigs. *Can. J. Vet. Res.* **57,** 74–78.

Itoh, H., Tamura, K., Izumi, M., Motoi, Y., and Funayama, Y. (1993). Characterization of serum alpha-1-acid glycoprotein in fetal and newborn calves during development. *Am. J. Vet. Res.* **54,** 591–595.

Johnson, A. M. (1993). A new international reference preparation for proteins in human serum. *Arch. Pathol. Lab. Med.* **117,** 29–31.

Kent, J. (1992). Acute phase proteins: Their use in veterinary diagnosis. *Br. Vet. J.* **148,** 279–282.

Lampreave, F., Gonzalez-Ramon, N., Martinez-Ayensa, S., Hernandez, M.-A., Lorenzo, H.-K., Garcia-Gil, A., and Pineiro, A. (1994). Characterisation of the acute phase serum protein response in pigs. *Electrophoresis* **15,** 672–676.

Makimura, S., and Suzuki, N. (1982). Quantitative determination of bovine serum haptoglobin and its elevation in some inflammatory diseases. *Jpn. J. Vet. Sci.* **44,** 15–21.

Motoi, Y., Itoh, H., Tamura, K., Miyamoto, T., Oohashi, T., and Nagasawa, S. (1992). Correlation of serum concentration of alpha-1-acid glycoprotein with lymphocyte blastogenesis and development of experimentally induced or naturally acquired hepatic-abscesses in cattle. *Am. J. Vet. Res.* **53,** 574–579.

Ohwada, K., and Tamura, K. (1995). Usefulness of alpha 1 acid glycoprotein (α1-AG) values in screening pound dogs acquired from animal shelters for experimental use. *Exp. Anim.* **42,** 627–630.

Richards, C. D., Brown, T. J., Shoyab, M., Baumann, H., and Gauldie, J. (1992). Recombinant oncostatin-M stimulates the production of acute phase proteins in HepG2 cells and rat primary hepatocytes *in vitro. J. Immunol.* **148,** 1731–1736.

Saini, P. K., and Webert, D. (1991). Application of acute phase reactants during antemortem and postmortem meat inspection. *J. Am. Vet. Med. Assoc.* **198,** 1898–1901.

Sehgal, P. B., Greininger, G., and Tosato, G. (1989). Regulation of the acute phase and immune responses. *Ann. N.Y. Acad. Sci.* **557,** 1–583.

Skinner, J. G., Brown, R. A. L., and Roberts, L. (1991). Bovine haptoglobin response in clinically defined field conditions. *Vet. Rec.* **128,** 147–149.

Tamura, K., Yatsu, T., Itoh, H., and Motoi, Y. (1989). Isolation, characterization and quantitative measurement of serum a1-acid glycoprotein in cattle. *Jpn. J. Vet. Sci.* **51,** 987–994.

Thompson, D., Milford-Ward, A., and Whicher, J. (1992). The value of acute phase measurements in clinical practice. *Ann. Clin. Biochem.* **29,** 123–131.

Toussaint, M. J. M., Van Ederen, A. M., and Gruys, E. (1995). Implication of clinical pathology in assessment of animal health and in animal production and meat inspection. *Comp. Haematol. Int.* **5,** 149–157.

Weidmeyer, C. E., Solter, P. F., and Hoffmann, W. E. (1994). Validation of a human Immunoturbidimetric assay for detection of haptoglobin in equine and canine serum. *In* "Proceedings of the VIth Congress of International Society for Animal Clinical Biochemistry" (J. H. Lumsden, ed.), University of Guelph, p. 98. Guelph, Ontario, Canada.

Wittum, T. E., Young, C. R., Stanker, L. H., Griffin, D. D., Perino, L. J., and Littledike, E. T. (1996). Haptoglobin response to clinical respiratory tract disease in feedlot cattle. *Am. J. Vet. Res.* **57,** 646–649.

Yamamoto, S., Shida, T., Miyaji, S., Santsuka, H., Fujise, H., Mukawa, K., Fukukawa, E., Nagae, T., and Naiki, M. (1993). Changes in serum C-reactive protein levels in dogs with various disorders and surgical traumas. *Vet. Res. Commun.* **17,** 85–93.

Young, C. R., Eckersall, P. D., Saini, P. K., and Stanker, L. H. (1995). Validation of immunoassays for bovine haptoglobin. *Vet. Immunol. Immunopathol.* **49,** 1–13.

Vaccination Practices in Veterinary Medicine: Standardization versus Tailored to Needs?

SERGE MARTINOD

Pfizer Central Research, 1 Eastern Point Road, Groton, Connecticut 06340

I. Introduction
II. Can We Standardize Vaccination Practices?
    A. Standardization of Veterinary Practices
    B. Limitations
III. Are We Vaccinating Too Much?
    A. Risks
    B. Benefits
IV. Consequences for the Animal Health Industry
    A. Future Challenges
    B. Partnership for Better Vaccines
V. Summary

I. Introduction

Vaccines have already achieved great success in controlling many diseases of importance to farm or companion animals or which threaten human health. Vaccines have brought eight major human diseases under various degrees of control: smallpox (complete eradication), diphtheria, tetanus, yellow fever, whooping cough, polio, measles, and rabies. More than 80% of the world's children are now being immunized against the polio virus and the annual number of cases has been cut from 400,000 in 1980 to 90,000 in the mid-1990s. Measles is another possible candidate for eradication. Veterinary vaccines also have had a profound influence in the world. They can control devastating diseases such as foot-and-mouth disease in cattle, canine distemper, feline and canine parvovirus, pseudorabies in swine, and rabies in all species, as well as the economic losses due to many respiratory, reproductive, or enteric pathogens.

Prevention is always better than cure, to minimize the suffering of animals and losses among livestock. Vaccines, coupled with precautions such as quarantine, movement control, and sound management, can achieve this objective. However, immunization should never be seen as routine. It is important to recognize that biologicals can cause adverse reactions or that inappropriate use could promote the production of escape mutants. The animal owners and the public have a right to expect that the preparation and use of animal biologicals is reliably based on the highest standards of quality, safety, and efficacy. In this paper we review some of the issues surrounding standardization of veterinary vaccine practices.

II. Can We Standardize Vaccination Practices?

Immunization is one of the most important tools in veterinary medicine and probably the most cost effective. All animals around the world are at risk from a multitude of infectious diseases. Furthermore, many different products are available in countries around the world. Therefore, an effort to define vaccination schemes and protocols, while needed, must take into account numerous issues.

A. STANDARDIZATION OF VETERINARY PRACTICES

1. Label Claims

Before being administered to a patient, vaccines have to be researched, developed, field tested, licensed, produced, quality controlled, regulated, purchased, transported, stored, and delivered. Most of these activities are subject to national or international legislation. The data generated during the development phase and reviewed during the licensing process is the basis for the labeling and product information literature. Full instructions on how to use the product, details of the claim for it, dosing regimes, and contraindications are given in the label. It is the primary source of information provided by the manufacturers and authorized by government licensing agencies and should always be taken into consideration by the users.

2. Guidelines

For many species professional organizations such as the American Veterinary Medical Association (AVMA) regularly publish updated guidelines to help the veterinarians in the design of the best vaccination practices. These recommendations are based on information from the label claims, the published scientific literature, the experience from the field, and the epizootiology of each disease.

3. International Organizations

Many international organizations are concerned with animal health worldwide: the Office International des Epizooties (OIE), the Food and Agriculture Organization (FAO), and the World Health Organization (WHO) deal with the control of animal diseases. Furthermore, regional organizations such as the Commission of the European Union or the Pan-American Health Organization are also involved in decision making and recommendations on animal health policies including vaccination based on political, economic, and technical considerations. Government policies can also made defined vaccination programs mandatory or exclude particular vaccines from use.

4. "Global" Diseases

Many viruses for example are distributed around the world: Canine distemper and parvovirus, feline parvovirus, bovine herpesvirus 1 (BHV1) or bovine viral diarrhea virus (BVDV), pseudorabies virus, and Marek's disease virus infect animals worldwide. At least in theory, global recommendations for vaccination programs could be made for these diseases.

5. Lessons from Human Vaccines

The use of pediatric vaccines has been standardized successfully by the WHO and local medical pediatric associations. These vaccines have brought major disease such as poliomyelitis under control in many countries. This could set up an example for veterinary vaccines.

With well-defined label claims, government or international legislation and guidelines, and professional association recommendations, vaccination programs could be easily standardized. However, such policies have many limitations where the biologicals are put into action to protect diverse populations against infectious agents.

B. LIMITATIONS

Generally, routine vaccination is undertaken is young animals with booster vaccinations at various intervals depending on manufacturers' or government instructions. However, in practice, which vaccines are given and precisely when they are given varies according to many factors.

1. Multiplicity of Target Species and Husbandry Conditions

The market for veterinary vaccines is very fragmented. Very few individual vaccines can be administered to more than one species. Within a given species different disease conditions, distribution channels, or husbandry conditions in different countries contribute to further fragmentation. Customers require different products, use patterns, and administration routes (e.g., oral, intradermal, intranasal, intramuscular, in ovo, etc.) depending on the management system, housing, age of the animals, traditions, or opinions of leading vaccinologists in each country.

2. Difference in Disease Conditions

The prevalence of diseases within a given species varies from one region to the other. The epidemiology of different serotypes (e.g., *Actinobacillus pleuropneumoniae*) restricts the use of some products to certain geographic areas. Moreover, the disease incidence is often of a transient nature (e.g., transmissible gastroenteritis in pigs) and therefore the need for a particular vaccine may change rapidly.

3. Strain Variations and Antigenic Diversity

A certain degree of antigenic diversity is always identified among isolates of the same pathogen. This variation may concern only a few epitopes which are differentiated by a panel of monoclonal antibodies and not relevant for a vaccine formulation or it may require the development of a different vaccine strain. Foot-and-mouth disease virus (FMDV) occurs as seven serotypes, within each of which there is the potential for wide antigenic diversity. In the face of FMDV outbreak, it is critical to demonstrate that cross-protection is induced by the vaccines. Pestiviruses also exhibit wide differences. Two biotypes (cytopathic and noncytopathic), two genotypes (types I and II), and a continuous spectrum of antigenic variations can be found in BVDV isolates.

4. Lifestyle

Multiple-cat households are defined as homes with three or more cats. By nature cats are solitary creatures, comingling only at times of mating or territorial disputes. Many health problems seen in multiple-cat households result from artificial environments in which the cat must live and do not exist in single-cat households. Therefore, the recommendations for vaccines should be different for cats living in multiple- or single-cat households.

5. Age at First Vaccination

Puppies become susceptible to viral infections as the maternal antibody titer declines to nonprotective levels. Currently, parvovirus in dogs is more commonly encountered in puppies from 6 weeks to 6 months of age. High levels of maternal antibodies can prevent some vaccines from being effective. Conversely, most adult dogs have become immune through vaccination or natural infection and are refractive to boosting with modified live vaccines.

6. Breed Susceptibility

Certain breeds are reported to be more susceptible to the development of parvovirus disease despite antibody titers considered protective in other breeds. Doberman pinschers, Labrador retrievers, and Rottweilers are reported to be more severely affected by parvoviral infections. For those breeds at high risk, it is sometimes recommended that the vaccination protocol be continued to 20 weeks instead of 12 weeks of age.

7. Economical Issues

BHV-1 is responsible for infections of the respiratory and reproductive tract. The infections are widespread. The virus causes severe clinical signs and economic losses in the United States, but the disease is mostly subclinical in most countries of the European Union. Therefore, the control of the clinical disease is critical in North America but not in Europe. However, European Union Directive 64/32/EEC allows countries to restrict trade in live animals on the basis of their health status. As a consequence, a marker vaccine strategy is being developed in Europe not only to control the disease, but to limit the virus circulation and eradicate the virus from the cattle population. The vaccine strategies are totally different between the two continents. Zoosanitary legislation may also limit the types of vaccines used in control programs to a particular strain or marker, thereby distorting competition in the marketplace.

8. Difference in Regulatory Requirements

National or regional regulatory requirements further contribute to the market complexity of veterinary vaccines. For example, the use of vaccines against clinical swine fever (CSF) is forbidden in the European Union despite a severe outbreak of the disease in 1997 in the

Netherlands. Vaccines against CSF are allowed in other parts of the world as a tool to control the disease outbreaks.

9. Influence of the Climate or Ecosystems

The geographic distribution of tick-borne diseases is highly dependent on the existence of the vectors. Pathogens, vectors, and reservoir hosts exist in ecologic assemblages and a combination of events have led to increasing human and companion animals interaction with these assemblages. For example, *Amblyomma americanum,* which transmits the agent of human ehrlichiosis, is found throughout the southeastern United States while *Ixodes dammini,* which transmits the agent of Lyme disease, is found in the northeastern and north central (Wisconsin, Minnesota) part of the country. The recommendation for the prevention of each disease will vary dramatically depending on the geographic location of the population being vaccinated.

The choice of a vaccine and a vaccination regimen should always be selected with the best interest of the animals and their owners in mind. Appropriate immunization should be given in accordance with manufacturers' guidelines, government regulations, and veterinarian recommendations. The local veterinarian is usually the best qualified to make the recommendation. Vaccination programs need to be tailored to each individual animal or livestock operation.

III. Are We Vaccinating Too Much?

A. RISKS

Recently, concerns about vaccine reactions have raised questions and doubts about the best immunization practices.

1. Safety

Vaccination safety is of utmost importance. The use of biological products has become so routine that vaccination is sometimes believed to be innocuous. However, varying levels of immunity, adverse reactions, and other unexpected events are a reality with biological products. The major safety problems reported are injection site reactions, systemic reactions, allergic reactions, immunosuppression, inadequate inactivation, residual pathogenicity, genetic recombinations, and contaminations. An increased incidence of fibrosarcomas in cats has been linked to vaccination since many of the lesions occur in common sites of immunization. The yearly prevalence has been estimated at 2 cases per 10,000 vaccinated cats. Although there are still many unknowns, the potential risk is to be taken into consideration before making the decision to vaccinate.

2. Duration of Immunity

Following primary immunization, natural exposure or virus persistence will result in a boosting of immunity in a population. However, in most situations, such natural boosting is unreliable. Booster vaccinations are often given at intervals to ensure immunologic memory for a rapid immune response to pathogens. The timing may vary depending on the targets and the vaccines. Currently, the conventional recommendation is for an annual booster for most vaccines with exceptions for, for instance, some rabies vaccines when a 3-year duration of immunity has been demonstrated.

It is known that some vaccines such as leptospirosis vaccines provide short-lived immunity. Furthermore, because mucosal memory is usually short lived, vaccines designed to replicate on mucosal surface may need to be repeated frequently. However, there may be a lack of scientific documentation to back up annual boosters of vaccines such as feline and canine parvovirus or canine distemper. Concerns about side effects should be incentive to conduct the research necessary to determine the optimal intervals between vaccinations. Animal serologic examination could also help define the proper immunization intervals for individual animals or herds, in diseases where antibody titers correlate with protection.

3. Vaccine Components

Effective vaccination of young puppies against canine parvovirus, distemper, and adenovirus 1 and 2 is critical. The benefits from canine coronavirus vaccines or from Lyme disease vaccines outside endemic areas is more controversial.

4. Cost

Cost is another disadvantage to overvaccination. The owner of the animal could be paying for something the animal does not really need. Therefore, the selection, strategic need, route, and program of administration of any vaccine, as well as care in handling and inoculating, should receive much more attention.

B. BENEFITS

As mentioned earlier, vaccines have played a very important role in bringing many diseases under control. Before decisions are made to reduce the number of vaccinations, some factors should be carefully considered.

1. Herd Immunity

Vaccines are effective in preventing diseases not only in individuals but also in populations. This type of protection is called herd immunity. When a virus spreads from one animal to another, it requires both an infected host to spread it and a susceptible host to catch it. Herd immunity works by decreasing the number of susceptible animals. When the number of susceptible hosts drops low enough, the disease will disappear from the population because there are not enough animals to continue the infection cycle. It is therefore critical to maintain a high percentage of the host population immunized to achieve herd immunity and hence disease control.

2. Wild Reservoirs

Many species of wild carnivores, such as mustelids, are susceptible to canine distemper virus and therefore represent a potential source of infection for dogs. There are many examples of outbreaks of diseases occurring in dog populations without sufficient protection.

3. Protection of Other Species

In addition to protecting the target species, vaccination is also used to arrest the cycle within a species or interrupt transmission from one susceptible species to another or from animal to man. Human public health is dependent on the control of zoonotic diseases in animals. In some instances such as rabies, it is the only possible option for control.

4. Drug Resistance and Residues

The development of safe and efficacious antibiotics for the control of bacterial infections has probably limited the development of bacterial vaccines. However, with the increased frequency of bacterial resistance to drugs and an increase of the public awareness to antimicrobial residues in animal products, there is a need for greater use of antibacterial vaccines. The economic advantage for producers to offer residue free animal products without compromising the animal welfare and public health should not be underestimated.

5. Animal Welfare

Veterinarians have a duty to protect animal health and welfare. Successful vaccination involves much more than the administration of the product. They should also provide the customer with information, recommend preventive health programs, understand the proper handling of products, report adverse reactions, and increase the public's awareness of food safety and zoonotic issues.

Balancing risks and benefits as well as tailoring vaccination programs to the individual patient or population should be taken into consideration by the veterinarians before administering any biological products.

IV. Consequences for the Animal Health Industry

A. FUTURE CHALLENGES

1. Technologies

The search for the ideal vaccine must continue. Many technologies such as biotechnology, delivery technology (slow or intermittent pulse release of antigens at appropriate sites), or immunomodulation hold great promises for the future. The mastering of these new technologies should result in the following improvements:

- Safer and more efficacious vaccines
- Vaccines with longer duration of immunity
- Vaccines that can bypass inhibitory maternal antibodies
- Vaccines for major diseases that cannot be currently prevented
- Vaccines for emerging diseases
- More convenient vaccines that are well adapted to husbandry managements or animal owner lifestyles
- Less dependency on refrigeration (cold chain)

2. Design of New Vaccines

Recent discoveries in vaccinology and immunology will allow a better design for new vaccines. Factors that are irrelevant or detrimental to the protective immune response can be removed from the antigenic fractions of biologicals by recombinant DNA technology. It is likely that many microorganisms contain molecules capable of biasing the immune response in a fashion advantageous to the parasite.

Optimal antigen presentation is of key importance for successful immunization. Researchers involved in the design of new products should pay attention to the cells involved in the immune response, the surface molecules expressed by antigen-presenting cells and T cells, the cytokines and chemokines produced by the cells, and the peptides binding to MHC class I and II molecules.

Vaccines also need to induce long-lasting memory and should include multiple helper and cytotoxic T-cell epitopes. The persistence of antigens may also be important.

Infectious agents have developed multiple mechanisms for escaping the immune response, including the production of escape mutants, the interference with MHC antigen presentation, the exhaustion of cytotoxic T cells, or the mimicking of effects of cytokines. Decreasing the likelihood of escape mechanisms coming into play will be critical in the design of vaccines against viruses such as porcine reproductive and respiratory syndrome virus.

3. Epidemiological Implication of Vaccinations

The choice of a vaccine strain should be justified on the basis of epidemiologic data. Genomic and antigenic diversity among bovine viral diarrhea viruses may contribute to vaccine failure. Furthermore, the situation may be changing all the time: new strains of feline calicivirus different from the F9 strains are now isolated from the field. Likewise, serovars of leptospira not contained in current vaccines are being reported from diagnostic laboratories. An epidemiologic surveillance program must be developed rapidly. Strategies for the use of and measurement of the success of vaccination in the control of animal disease have to be considered in the context of the epidemiology of that disease.

4. Strategies for Vaccine Use

Prevention of disease can reach one of the three following goals: diminution of incidence, of prevalence, or of disease consequences. For example, a vaccine that only protects against clinical disease, but not against infection, allows propagation of the pathogen in the host population. Secondarily, reducing the severity of the symptoms could also contribute to the diminution of incidence by reducing shedding of the pathogens (for example, nasal discharge of BHV1). A clear definition of the goals of the vaccination and a good knowledge of the property of the biological product are needed before the decision to immunize an animal is taken. Vaccination should usually be used and judged in a population context. The development of national or regional policies and the development in the trade of animals and animal products will play a role in the development of vaccine prophylaxis at all levels. The strategic thinking is moving from "think globally, act locally" to "think locally, act globally."

5. Cost

Fragmented markets, pressures on prices, development of a monopolistic competition, increasing cost for discovering, developing, registering, producing, and marketing biological products will impact the relatively small veterinary vaccine industry at a time when large investments are required.

B. Partnership for Better Vaccines

The availability of better biological products is dependent on a partnership, summarized in Fig. 1, in which there are many roles and responsibilities. Consumers, animal owners, veterinarians, the scientific community, national and international agencies, government regulatory offices, and the vaccine industry must communicate and collaborate. Because the control of animal diseases is becoming more and more complex, a formal structure to promote collaboration between parties might be needed.

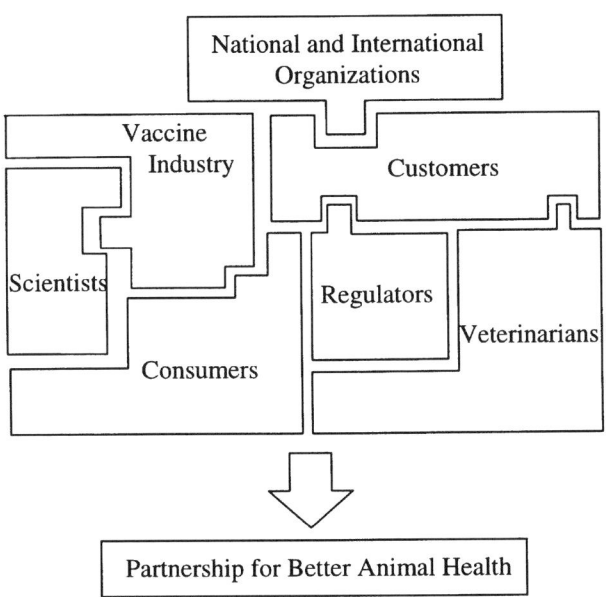

FIG. 1. Partnership for better animal health.

V. Summary

Significant achievements have been made during the nineteenth century to improve animal health and welfare through vaccination. Vaccination standards and practices will change significantly in the next century. Vaccination programs need to be tailored to each livestock operation or individual animal in accordance with manufacturers' instructions, government regulations, scientific standards, professional organization guidelines, and veterinarian recommendations. Vaccination should always be considered in a population context. A better use of vaccine will require a different approach and a significant investment in research. Finally, all the partners involved in animal health should consider a close collaboration to meet future animal and public health challenges.

International Harmonization of Standards for Diagnostic Tests and Vaccines: Role of the Office International des Epizooties (OIE)

PETER F. WRIGHT

Canadian Food Inspection Agency, National Centre for Foreign Animal Disease, 1015 Arlington Street, Suite T2300, Winnipeg, Manitoba, Canada, R3E 3M4

I. Introduction
II. Organization and Structure
III. Specialist Commissions
IV. Standards Commission
   A. *Manual of Standards for Diagnostic Tests and Vaccines*
   B. Prescribed Tests for International Trade
   C. OIE Reference Laboratories and Collaborating Centers
   D. Guidelines for Veterinary Laboratories
   E. International Reference Reagents
   F. Provision of Scientific and Technical Advice
   G. World Trade Organization
V. Summary
   Acknowledgments
   References

I. Introduction

In 1920, a consignment of zebu cattle in transit from Africa to South America arrived at the port of Antwerp in Belgium. While in port, the cattle were off-loaded and held at a quarantine station before their journey was to resume. It was soon discovered that these cattle were afflicted with rinderpest; however, it was too late to prevent its spread beyond the quarantine station. Fortunately, the strain of rinderpest virus was not especially virulent and the epizootic was controlled but not before reaching the French border. This event served as a grim reminder of the scourge of rinderpest that swept Europe in the latter part of the 1800s. The following year, France hosted an international conference on epizootics which laid the groundwork for what was to become the Office International des Epizooties (OIE). Three years later in 1924, the OIE was established under an international agreement originally signed by 28 countries. As of 1997, membership in the OIE has grown to include 145 countries.

The primary objectives of the OIE are as valid today as when they were first drafted 73 years ago. They are:

1. To inform governments of the occurrence and course of animal diseases throughout the world and of ways to control these diseases
2. To coordinate, at the international level, studies devoted to the surveillance and control of animal diseases
3. To harmonize regulations for trade in animals and animal products among its member countries

In working toward these objectives, the OIE has undertaken a major role in the international harmonization of standards for diagnostic tests and vaccines. At the center of this role is one of the OIE's specialist commissions, the Standards Commission.

II. Organization and Structure

To appreciate the role of the Standards Commission in harmonization, it is pertinent to understand the structure of the OIE (Fig. 1). The OIE operates under the authority and control of the International Committee, which is comprised of the permanent delegates, usually the chief veterinary officer, from each of the member countries. The International Committee is presided over by a president and a vice-president elected from among the delegates.

The activities of the OIE are conducted by the Central Bureau. Headed by a director general, the bureau is responsible for the daily operations of the office and is also responsible for implementing the resolutions approved by the International Committee.

Resolutions presented to the International Committee for ratification are prepared by the Central Bureau with the support from several elected commissions, which include:

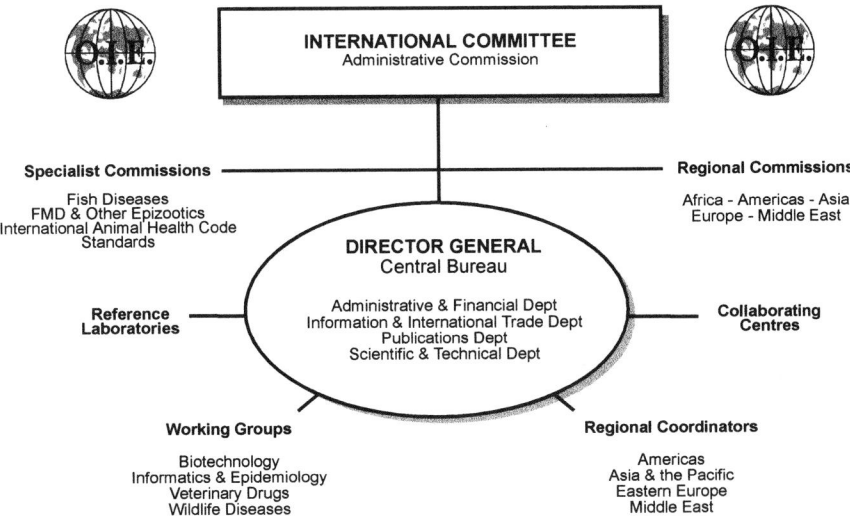

FIG. 1. Organization and structure of the Office International des Epizooties (OIE). The OIE operates under the authority and control of the International Committee, which is comprised of the permanent delegates from each of 145 member countries. The activities of the OIE are conducted by the Central Bureau, headed by a director general. The Central Bureau is supported by several elected commissions, permanent regional offices, and designated working groups, reference laboratories, and collaborating centers.

1. The administrative commission—executive of the International Committee plus additional members from among permanent delegates
2. Four specialist commissions—members from among world experts
3. Five regional commissions—members from within designated regions

To better serve member countries, the Central Bureau has also established and coordinates networks of specialized groups which include:

1. Reference laboratories—providing scientific and technical expertise on specific diseases (e.g., rinderpest, brucellosis, etc.)
2. Collaborating centers—providing scientific and technical expertise in certain fields (e.g., diagnostic techniques, vaccinology, etc.)
3. Working groups—monitoring and analyzing recent developments in specific areas (e.g., biotechnology, wildlife diseases, etc.)

4. Regional coordinators—providing expertise and services out of regional offices to meet specific needs (e.g., disease surveillance, control and eradication programs, etc.)

The financial resources of the OIE are derived primarily from regular annual contributions and exceptional contributions from member countries.

III. Specialist Commissions

The roles of the specialist commissions are to study problems related to the epidemiology and control of animal diseases, and issues related to the harmonization of international regulations applicable to trade in animals and animal products. Four specialist commissions have been established and include:

1. Foot and Mouth Disease and other Epizootics Commission—Originally established in 1946 and expanded to include other epizootics in 1988, this commission assists in identifying the most appropriate strategies and measures for disease prevention and control.
2. Standards Commission—Established in 1949, this commission defines standards for diagnostic tests and vaccines as applied to mammals, birds, and bees.
3. International Animal Health Code Commission—Established in 1960, this commission defines the regulatory standards that apply to international trade in animal and animal products.
4. Fish Diseases Commission—Established in 1960, this commission specializes in prevention, diagnosis, and control of diseases as well as regulatory standards that pertain to international trade in fish, crustaceans, and mollusks.

IV. Standards Commission

The Standards Commission acts as the primary scientific and technical reference point for the OIE. In close collaboration with the Scientific and Technical Department of the Central Bureau, the commission is active on several, interdependent fronts.

A. *Manual of Standards for Diagnostic Tests and Vaccines*

The purpose of the *Manual* is to define standards for diagnostic tests and for production of biological products, principally vaccines. By doing so, the *Manual* contributes to the harmonization of methods of surveillance and control of important animal diseases. The *Manual* also acts as the principal reference for the *International Animal Health Code,* prepared by the Code Commission. Wherever there is a reference in the *Code* to a diagnostic test or vaccine, it is fully described in the *Manual.*

The *Manual* was first published in three volumes in a looseleaf format between 1989 and 1991. The second edition was compiled and published as a bound book in 1992. The latest edition, which contains more than 700 pages, was published in 1996. Subsequent additions will appear approximately every 4 years.

The *Manual* targets scientific and technical staff working at the bench and the latest edition begins with seven introductory chapters covering a broad range of topics:

1. Sampling methods
2. Good laboratory practice, quality control, and quality assurance
3. Principles of validation of diagnostic tests for infectious diseases
4. Tests for sterility and freedom from contamination of biological materials
5. Human safety in the veterinary microbiology laboratory
6. Principles of vaccine production
7. Biotechnology in the diagnosis of infectious diseases and vaccine development

The 93 chapters that follow the introductory chapters deal with individual diseases. All diseases appearing in the *Code* are covered, in addition to a number of others which are not in the *Code*. Although the diseases are too numerous to list, these chapters are organized in the following manner, with the number of diseases in parentheses:

1. List A Diseases (15)
2. List B Diseases
    a. Multiple Species (8)
    b. Cattle (13)
    c. Sheep and Goats (6)
    d. Horses (14)
    e. Pigs (5)
    f. Birds (11)
    g. Lagomorphs (3)
    h. Bees (5)
3. Diseases not in the Code (13)

List A includes transmissible diseases that have the potential to spread rapidly within and beyond national borders and result in very serious socioeconomic and/or public health consequences. They include diseases such as foot-and-mouth disease and Newcastle disease and are of major importance to international trade. List B includes transmissible diseases usually contained within national borders and are considered to be of socioeconomic and/or public health concern. They include diseases such as brucellosis and rabies and are of importance to international trade. Some diseases, such as Q fever and porcine reproductive and respiratory syndrome, which are not covered in the *Code* are included because they also have international trade implications in many instances.

Individual chapters are written and reviewed by internationally recognized experts. In many cases, a chapter is authored by an expert in the field of diagnostics and an expert in the field of vaccinology. As part of the process, each member country is also given the opportunity to review and make comments on all chapters before publication.

Each chapter includes a general overview of the disease, agent, diagnostic tests, and available vaccines. Following are detailed descriptions of diagnostic techniques and, where applicable, requirements for vaccines or other biological products. The descriptions and relevant references represent the internationally accepted standards for diagnostic tests and vaccines.

B. Prescribed Tests for International Trade

The Standards Commission is also responsible for evaluating diagnostic tests with respect to their potential application in international trade. Based on this evaluation, new tests may be recommended for incorporation into the *Code* and, in some cases, older tests may be recommended for retirement. The commission ensures that all new tests proposed by developing laboratories are properly standardized and internationally validated before being considered. New and existing tests are grouped into two categories:

1. Prescribed tests—those required by the Code for the testing of animals before they are moved internationally
2. Alternative tests—those suitable for the diagnosis of disease within a local setting, and can be used in the import/export of animals after bilateral agreement

Prescribed and alternative tests are listed in the *Manual*. However, this list is updated as required and circulated to member countries between editions of the *Manual*.

C. OIE REFERENCE LABORATORIES AND COLLABORATING CENTERS

The Standards Commission, and indeed the whole of the OIE, depends very much on international experts for scientific and technical advice. To this end, a voluntary network of reference laboratories and collaborating centers has been established. The commission reviews and recommends the designation of OIE reference laboratory or collaborating center status based on a defined set of criteria. To qualify, institutions must meet the majority of the following:

1. Be internationally recognized as a center of expertise
2. Be willing to prepare and/or distribute reference strains and diagnostic standards, antisera, antigens, or other reagents to member countries
3. Be actively involved in diagnostic testing and the analyses of epizootiological data, both nationally and internationally
4. Be actively involved in the development of new methods
5. Be actively involved in the coordination of collaborative studies at the international level
6. Be actively involved in the publication and dissemination of scientific and technical information.
7. Be willing to provide consultative assistance to the OIE
8. Be willing to provide international training in specific areas of expertise
9. Be willing to organize scientific meetings on behalf of the OIE

OIE reference laboratories are recognized for their expertise on specific diseases. Wherever possible, multiple laboratories for any given disease have been designated on a regional basis to ensure that their particular expertise is available to most member countries. To date, more than 110 laboratories have requested and received OIE reference laboratory status. This includes designations for all 15 List A diseases and approximately 25 List B and other diseases.

OIE collaborating centers, on the other hand, are recognized for their expertise in technologies such as ELISA and other molecular techniques for diagnosis, in disciplines such as vaccinology or in programs such as surveillance and control of diseases prevalent in certain regions of the world. Six collaborating centers have now been established and the list is expected to expand.

By creating this network, international harmonization is promoted through the sharing of knowledge and the establishment of collaborative projects related to methods development, standardization, quality assurance, and validation.

OIE reference laboratories and collaborating centers are also listed in the *Manual*. Again, the list is updated as required and circulated to member countries between editions of the *Manual*.

D. Guidelines for Veterinary Laboratories

The Standards Commission has drafted a series of guidelines to promote harmonization among member country laboratories and to assist OIE-designated laboratories in the delivery of their services. Guidelines have been prepared for:

1. International reference standards for antibody assays (1993)
2. Data sheets for reference standards (1994a)
3. Validation of diagnostic tests for infectious diseases (1994b)
4. Laboratory quality evaluation (1995a)
5. Laboratory proficiency testing (1996)

These guidelines have been distributed to member countries through the biannual reports of the Standards Commission and pertinent guidelines have been sent directly to designated laboratories. Additional guidelines are currently under development. New and existing OIE guidelines, which fulfill the requirements of the ISO 9000 series of standards and relevant ISO/IEC guides, will be compiled and made available worldwide through a special issue of the *OIE Scientific and Technical Review* slated for publication in late 1998.

E. International Reference Reagents

The Standards Commission has been actively encouraging OIE reference laboratories to prepare and distribute reference reagents for diagnostic tests. Priority has been given to a number of important diseases for which there are no available international standard sera for particular diagnostic techniques. The intent is to have available to member country laboratories fully characterized and well-documented reference reagents for the calibration of diagnostic tests used for international trade and for the preparation of national standards. The major emphasis to date has been on the need for reference sera for antibody assays. Standard sera have either been prepared or are under development for the following diseases tests: brucellosis, rinderpest, contagious bovine pleuropneumonia, infectious bovine rhinotracheitis, enzootic bovine leukosis, pest des petits ruminants, Aujeszky's disease, classical swine fever, rabies, African horse sickness, equine influenza, equine viral arteritis, and equine infectious anaemia.

The selection, characterization, standardization, and distribution of reference reagents by an OIE reference laboratory is a very large but necessary commitment to international harmonization. Those laboratories taking on this commitment, without subsidy from the OIE, deserve much credit for their initiative.

F. Provision of Scientific and Technical Advice

Over the years, the OIE has established permanent working relationships with more than 20 international organizations including the Food and Agricultural Organization of the United Nations (FAO), the World Health Organization (WHO), the Inter-American Institute for Cooperation on Agriculture (IICA), and the Pan-American Health Organization (PAHO). In matters of a scientific and technical nature concerning diagnostic tests and vaccines, the Standards Commission provides information and advice through the Scientific and Technical Department to these organizations. In many instances, members of the Standards Commission represent the OIE at international and regional consultative meetings. By maintaining cooperative working relationships, the OIE is able to promote harmonization at the international program level.

G. World Trade Organization

The Sanitary and Phytosanitary Agreement of the World Trade Organization, as included in the Marrakech agreement (1994), specifically recommends the use of standards, guidelines, and recommendations developed under the auspices of the OIE in order to promote harmonization of regulations for international trade in animals and animal products.

The *Manual of Standards for Diagnostic Tests and Vaccines* (1996b), prepared by the Standards Commission, represents one of the four normative works published by the OIE that promote such international harmonization. The others include the *International Animal Health Code* (1992), the *International Aquatic Animal Health Code* (1995c), and the *Diagnostic Manual for Aquatic Animal Diseases* (1995b).

V. Summary

The OIE is recognized as the world organization for animal health. Serving 145 member countries, the OIE provides current information on disease occurrence, coordinates studies on disease surveillance and control, and harmonizes regulations for trade in animals and animal products. This paper focuses on the role of one the OIE's specialist commissions, the Standards Commission.

The Standards Commission works in close collaboration with the Scientific and Technical Department of the OIE's Central Bureau on the international harmonization of standards for diagnostic tests and vaccines. The *Manual of Standards for Diagnostic Tests and Vaccines*, approved by the International Committee, defines the international standards for diagnostic tests and for the production of biological products as applied to mammals, birds, and bees. The *Manual* lists and details those tests which are prescribed for international trade and others which are suitable for bilateral trade agreements. The *Manual* represents one of the key scientific and technical references for harmonization of regulations for trade in animals and animal products.

The commission coordinates the activities of a network of some 110 OIE reference laboratories and six collaborating centers. By creating and nurturing this network, international harmonization is promoted through the sharing of knowledge and the establishment of collaborative projects related to methods development and standardization, production and distribution of international reference standards, quality assurance, and assay validation. Through a series of guidelines provided to participants, the commission ensures the quality and focus of these projects.

In matters of a scientific and technical nature concerning diagnostic tests and vaccines, the Standards Commission collaborates with other international organizations such as the FAO, WHO, IICA, and PAHO, thus promoting harmonization at the international program level. Underscoring the important role of the OIE at this level, the Sanitary and Phytosanitary Agreement of the World Trade Organization, as included in the Marrakech agreement (1994), specifically recommends the use of standards, guidelines, and recommendations developed under the auspices of the OIE in order to promote harmonization of regulations for trade in animals and animal products.

ACKNOWLEDGMENTS

On behalf of the OIE, the author wishes to thank the organizers of the First International Veterinary Vaccine and Diagnostics Conference for the opportunity to present an overview of the OIE's activities in the area of international harmonization. The author also wishes to thank the OIE for the privilege and opportunity to represent the Office and the work of the Standards Commission.

REFERENCES

Office International des Epizooties (OIE) (1992 with 1993–1996 update). "OIE International Animal Health Code," 6th ed. OIE, Paris (ISBN 92-9044-315-4).

Office International des Epizooties (OIE) (1993). "OIE Guidelines for International Reference Standards for Antibody Assays," Report of the Meeting of the Standards Commission. OIE, Paris.

Office International des Epizooties (OIE) (1994a). "OIE Guidelines for Data Sheets for Reference Standards," Report of the Meeting of the Standards Commission. OIE, Paris.

Office International des Epizooties (OIE) (1994b). "OIE Guidelines for Validation of Diagnostic Tests for Infectious Diseases," Report of the Meeting of the Standards Commission. OIE, Paris.

Office International des Epizooties (OIE) (1995a). "OIE Guidelines for Laboratory Quality Validation," Report of the Meeting of the Standards Commission. OIE, Paris.

Office International des Epizooties (OIE) (1995b). "OIE Diagnostic Manual for Aquatic Animal Diseases." 1st ed. OIE, Paris, (ISBN 92-9044-383-9).

Office International des Epizooties (OIE) (1995c). "OIE International Aquatic Animal Health Code," 1st ed. OIE, Paris (ISBN 92-9044-384-7).

Office International des Epizooties (OIE) (1996a). "OIE Guidelines for Laboratory Proficiency Testing," Report of the Meeting of the Standards Commission. OIE, Paris.

Office International des Epizooties (OIE) (1996b). "OIE Manual of Standards for Diagnostic Tests and Vaccines," 3rd ed. OIE, Paris (ISBN 92-9044-423-1).

X
ADVERSE VACCINE REACTIONS, FAILURES, AND POSTMARKETING SURVEILLANCE

Mechanistic Bases for Adverse Vaccine Reactions and Vaccine Failures

JAMES A. ROTH

Institute for International Cooperation in Animal Biologics, College of Veterinary Medicine, Iowa State University, Ames, Iowa 50011

I. Introduction
II. Adverse Vaccine Reactions
    A. Contamination of Vaccines with Extraneous Agents
    B. Failure to Inactivate the Vaccine Organism in a Killed Vaccine
    C. Adverse Vaccine Reactions Due to Residual Virulence of Vaccine Organisms
    D. Adverse Vaccine Reactions Due to Vaccine-Induced Immune Suppression
    E. Adverse Vaccine Reactions Due to Excessive Induction of Cytokine Release
    F. Hypersensitivity Responses to Vaccine Antigens
    G. Vaccine-Induced Triggering or Exacerbation of Hypersensitivity Disease to Nonvaccine Antigens
    H. Vaccine-Induced Neoplastic Disease
    I. MLV BVD Vaccine Triggering Mucosal Disease in Persistently Infected Cattle
    J. Adverse Reactions Due to Multiple Vaccines Administered Concurrently
    K. Injection Site Lesions
III. Vaccine Failure
    A. Insufficient Time to Develop Immunity
    B. Vaccine Failure Due to Alterations in the Vaccine
    C. Host Factors Responsible for Vaccine Failure
    D. Vaccine Failure Due to Exposure to an Overwhelming Challenge Dose
    E. Vaccine Failure Due to Inadequate Duration of Immunity
    F. Vaccine Failure Due to Antigenic Differences between Vaccine and Field Strains
    G. Vaccine Failure Due to Interference When Multiple Vaccines Are Administered Concurrently
IV. Summary
References

I. Introduction

Vaccines have proven to be very beneficial for controlling diseases in domestic animals. Their widespread use has dramatically reduced the incidence of severe and fatal diseases in companion animals (canine distemper, canine parvovirus, infectious canine hepatitis, and feline panleukopenia). They have also enabled the intensification of livestock production, thus enabling great increases in efficiency in animal origin food and fiber production. In addition, animal vaccines have improved human health through control of zoonotic diseases such as rabies, brucellosis, and leptospirosis. Indeed, it can be argued that animal vaccines have had a profound impact on modern society. Without effective rabies vaccines many people would not opt to keep companion animals in their homes, and without effective vaccines for controlling major diseases in food-producing animals the availability of animal proteins for human consumption would be greatly reduced. However, in spite of the success of animal vaccines, vaccines sometimes induce adverse reactions in animals and sometimes they fail to protect animals. When making decisions regarding vaccination programs for animals, veterinarians and animal owners must weigh the risks of vaccinating vs. the risks of not vaccinating. They must also use vaccines in a manner that induces optimal protection. This article provides an overview of some of the reasons why vaccines occasionally produce adverse reactions (Table I) and reasons why vaccines sometimes fail to protect animals from disease (Table II).

To produce protective immunity, a vaccine must stimulate a reaction in the animal. There usually must be a reaction both at the site of injection and systemically in order to produce an effective immune response. This reaction involves extensive activity by antigen-presenting cells, production of a variety of cytokines, and alterations in the trafficking of lymphocytes within the body. In addition, if the vaccine contains live organisms, they probably need to replicate to induce effective immunity. Live viruses must infect and replicate within cells. These essential reactions to a vaccine may induce observable clinical signs. Hopefully, the reaction to the vaccine will be mild and either unnoticeable or acceptable to the animal owner.

To understand vaccine safety and efficacy, it is important to understand the process by which vaccines are developed and tested by vaccine producers, and licensed by the United States Department of Agriculture (USDA), Animal and Plant Health Inspection Service (APHIS), Center for Veterinary Biologics (CVB). The federal government regulations for the United States of America regarding veterinary vaccines

TABLE I

POTENTIAL MECHANISMS RESPONSIBLE FOR ADVERSE VACCINE REACTIONS

- Contamination with extraneous agents
- Failure to inactivate agent in killed vaccine
- Residual virulence of vaccine organisms
- Vaccination of immunosuppressed animal
- Immune suppression induced by the vaccine
- Excessive induction of cytokine release
- Multiple vaccines administered concurrently
- Hypersensitivity to vaccine antigens
    Type I—immediate type
    Type II—cytotoxic type
    Type III—immune complex type
    Type IV—delayed type
- Triggering or exacerbation of hypersensitivity to nonvaccine antigens
    Allergies
    Autoimmune disease
- Induction of neoplastic changes
- MLV BVD vaccine triggering mucosal disease in persistently infected cattle

are found in the Virus Serum Toxin Act (VSTA) in Title 9 of the Code of Federal Regulations (9 CFR). The VSTA gives the USDA the authority to regulate veterinary vaccines in the United States. According to the 9 CFR a USDA licensed biological must be "pure, safe, potent, and efficacious, and not be worthless, contaminated, dangerous, or harmful." To understand this statement, it is important to understand what is meant by *safe* and *efficacious*. The definition found in the 9 CFR for safe or safety regarding veterinary biologics is "freedom from proper ties causing undue local or systemic reactions when used as recom-

TABLE II

POTENTIAL REASONS FOR VACCINE FAILURE

- Insufficient time after vaccination to develop immunity
- Something happened to the vaccine to make it ineffective
- The physiologic status of the animal impaired the response to the vaccine
- The animal was immunosuppressed at some point after vaccination
- The animal was exposed to an overwhelming challenge dose of infectious agent
- The duration of immunity after vaccination was not adequate
- Important antigenic differences exist between the vaccine and field strains
- Interference when multiple vaccines are administered concurrently

mended or suggested by the manufacturer." This definition has two important qualifiers for the term *safety*. It does not state that a vaccine should produce no reaction, rather it states that a vaccine should not cause "undue local or systemic reactions." This is a recognition of the fact that stimulating a potent immune response is likely to produce at least a mild local and systemic reaction in the animal. The second important point is that according to the definition the safety of the vaccine is only ensured when it is used as recommended or suggested by the manufacturer. The recommendations and suggestions can be found on the label for the vaccine. Most vaccine label statements will indicate that a particular vaccine is only for use in healthy animals of a particular species. Healthy is defined as "apparently normal in all vital functions and free of signs of disease."

The 9 CFR definition for efficacious or efficacy is "specific ability or capacity of the biological product to effect the result for which it is offered when used under the conditions recommended by the manufacturer." The label found on the vaccine will indicate the "result for which the vaccine is offered" and will also indicate the conditions under which the vaccine is recommended for use. Therefore, it is very important to read and follow label instructions in order to achieve maximum safety and efficacy from vaccine usage.

II. Adverse Vaccine Reactions

When animals develop adverse clinical signs within a few days to weeks after vaccination it is important to determine whether those clinical signs were vaccine induced or were not due to vaccination and only coincidentally occurred after the vaccine was administered. Animals commonly experience adverse clinical signs from a wide variety of causes and animals are commonly vaccinated. Therefore, it is to be expected that occasionally adverse clinical signs will occur after animals have been vaccinated for reasons unrelated to vaccine administration. There are also many reasons why vaccines may induce adverse reactions in the animal. It is important to differentiate true adverse vaccine reactions from false adverse vaccine reactions. Some of the causes of true adverse vaccine reactions are summarized in Table I and explained next.

A. Contamination of Vaccines with Extraneous Agents

A prominent example of this occurred when it was discovered that some lots of the live oral human poliomyelitis vaccine were contaminated with live simian virus 40 (SV40) in the 1950s (Pennisi, 1997; Shah and Nathanson, 1976). Millions of people were potentially exposed to live SV40 through administration of polio vaccine. To date, there is no solid epidemiologic evidence that any adverse health affects can be attributed to exposure to this agent. The SV40 virus had not yet been discovered when the human polio vaccine was produced. This raises the question of how does one test for all potential known and unknown viruses in each production lot of modified live virus vaccines. There have been numerous examples of extraneous agents contaminating veterinary vaccines. A list of these examples with appropriate references is given in Table III.

B. Failure to Inactivate the Vaccine Organism in a Killed Vaccine

A dramatic example of this cause of adverse vaccine reactions occurred with the killed poliovirus vaccine in people. Formaldehyde, used to inactivate the poliovirus in the vaccine, failed to completely inactivate the vaccine virus (Gard and Lycke, 1957; Nathanson and Langmuir, 1963). This resulted in several cases of poliomyelitis in people that had received the vaccine. There have also been cases where formaldehyde failed to inactivate the foot-and-mouth disease virus (Beck and Strohmaier, 1987; King *et al.*, 1981) and the Venezuelan equine encephalitis virus (Kinney *et al.*, 1992) in their respective vaccines. In both of these cases the vaccine was shown to induce disease because of the lack of complete inactivation of the virus by the formaldehyde (Brown, 1993). An example of a failure to completely inactivate a bacterial pathogen in a killed bacterin occurred when thimerosol was used to inactivate *Haemophilus somnus* in an *H. somnus* vaccine. The thimerosol failed to kill the *H. somnus*. Approximately half the ani-

TABLE III

Examples of Adverse Vaccine Reactions due to Extraneous Agents in Vaccines

- Live SV40 in human polio vaccine (Pennisi, 1997; Shah and Nathanson, 1976)
- Killed hog cholera virus in pseudorabies vaccine (Jensen, 1981)
- Live *Mycoplasma* in multiple live virus veterinary vaccines (Thornton, 1986)
- Live border disease virus in Orf vaccine (Loken *et al.*, 1991)
- Live bovine leukemia virus in babesiosis and anaplasmosis vaccines (Rogers *et al.*, 1988)
- Live bovine viral diarrhea virus in hog cholera vaccine (Wensvoort and Terpstra, 1988)
- Live border disease virus in pseudorabies vaccine (Vannier *et al.*, 1988)
- Live blue tongue virus in a canine vaccine (Evermann *et al.*, 1994; Wilbur *et al.*, 1994)
- Live bovine viral diarrhea virus in bovine vaccines (Lohr *et al.*, 1983; Neaton, 1986)

mals on one farm that were injected with vaccine shortly after its production developed thromboembolic meningoencephalitis and died.

C. ADVERSE VACCINE REACTIONS DUE TO RESIDUAL VIRULENCE OF VACCINE ORGANISMS

Modified live vaccine organisms have been attenuated to have reduced virulence. The attenuation must be shown to be stable when passaged through animals; therefore, reversion to virulence is thought to be a rare event. However, the attenuated vaccine strains may be capable of producing disease in immunosuppressed animals. Induction of disease by the vaccine organism has occasionally been reported when modified live virus (MLV) vaccines have been administered to healthy animals. However, it has occurred much more frequently when MLV vaccines are administered to unhealthy animals, by a nonrecommended route of exposure, to animals younger than the intended age for use of the vaccine, or when the vaccine is used in other than the intended species. Examples of MLV vaccines occasionally causing disease in healthy animals of the recommended species without apparent predisposing causes include the induction of rabies in dogs and cats after administration of an MLV rabies vaccine (Bellinger et al., 1983; Esh et al., 1982; Erlewein, 1981; Whetstone et al., 1984; Pedersen et al., 1978) and the induction of ovarian lesions and infertility in seronegative heifers administered MLV bovine herpesvirus 1 (BHV1) vaccine during estrus (Smith et al., 1990; Chiang et al., 1990; Miller et al., 1989; Van der Maaten et al., 1985). Since most heifers already have antibody to BHV1 due to either vaccination or previous exposure, this is thought to be a rare occurrence.

An example of vaccine-induced disease resulting from administration of vaccine to unhealthy animals is the induction of encephalitis by MLV canine distemper virus vaccine in dogs infected with canine parvovirus (Krakowka et al., 1982). An example of adverse vaccine reaction after exposure of an animal to an MLV vaccine by a nonrecommended route of exposure is the induction of clinical feline viral rhinotracheitis after inadvertent exposure by the intranasal route to an MLV vaccine that was intended for intramuscular administration only (Povey and Wilson, 1978). MLV vaccines that have been shown to be safe in older animals may not be safe in neonatal animals. An MLV BHV-1 vaccine induced fatal BHV1 infection in neonatal purebred Salers calves (Bryan et al., 1994). This may have been partially due to the breed of the animals since there are other reports that MLV BHV1 vaccines are apparently safe in neonatal calves (Schuh and Walker, 1990).

There have been several examples of MLV vaccines inducing lethal disease when administered to a species other than the target species. An MLV pseudorabies virus vaccine produced fatal pseudorabies in lambs (Clark et al., 1984; Van Alstine et al., 1984). This occurred when a syringe that had been used to administer the pseudorabies vaccine to pigs was used without proper disinfection to vaccinate lambs with another vaccine 3 days later. The MLV canine distemper virus vaccine has been shown to induce canine distemper infection in gray foxes (Halbrooks et al., 1981), kinkajous (Kazacos et al., 1981), and lesser pandas (Bush et al., 1976). An MLV rabies vaccine has been shown to induce rabies in a pet skunk (Debbie, 1979). An MLV feline panleukopenia vaccine induced cerebellar hypoplasia when given experimentally to neonatal ferrets (Duenwald et al., 1971).

D. Adverse Vaccine Reactions Due to Vaccine-Induced Immune Suppression

An MLV bovine viral diarrhea (BVD) virus vaccine has been shown to suppress neutrophil function and lymphocyte blastogenesis in cattle (Roth and Kaeberle, 1983). This correlates with the observation that cattle tend to be somewhat more susceptible to bacterial pneumonia after administration of MLV BVD vaccines, especially if the animals are stressed at the time of vaccination. Several commercially available canine vaccines have been shown to be capable of inducing lymphopenia and suppressing blastogenesis of peripheral blood lymphocytes (Phillips et al., 1989; Mastro et al., 1986; Kesel and Neil, 1983). Lymphopenia and suppression of blood lymphocyte blastogenesis must be interpreted with caution, however, because it may only be an indication of changes in lymphocyte trafficking between the blood and lymphatic systems rather than an indication of depressed lymphocyte function. Vaccination with an MLV BHV1 vaccine has been shown to exacerbate the lesions of infectious bovine keratoconjunctivitis after experimental intraocular challenge with *Moraxella bovis* (George et al., 1988).

E. Adverse Vaccine Reactions Due to Excessive Induction of Cytokine Release

Interleukin 1 (IL-1), IL-6, and tumor necrosis factor α (TNF-α) are potent proinflammatory cytokines that are released by macrophages and other cells in response to infection, endotoxin and other bacterial components, and some vaccine adjuvants. These proinflammatory cytokines can induce a wide range of clinical signs. They may induce acute inflammation at the local site of production, they may induce rapid synthesis and secretion of acute phase proteins by the liver, they may act on the hypothalamus to induce fever and malaise, they may reduce rate of gain and feed efficiency, and in sufficiently high concentrations they may induce hypoglycemia, reduce cardiac output, cause hypovolemic shock, and cause disseminated intravascular coagulation. Lipopolysaccharide (or endotoxin) from gram-negative bacteria is one of the most potent inducers of the proinflammatory cytokines (Cullor, 1994; Ellis and Yong, 1997; Galanos and Freudenberg, 1993). A number of other bacterial components, listed in Table IV, have also been shown to induce proinflammatory cytokine production (Erdos *et al.*, 1975; Henderson and Wilson, 1995; Allison and Eugui, 1995). These components are generally the most active if they are released from the degraded bacterial cell. Killed bacterins that contain excessive amounts of these bacterial components can induce clinical signs due to excessive induction of cytokine release. This is more likely to occur if multiple killed bacterins are administered at the same time and if these bacterins contain adjuvants that also induce cytokine release. The production of small amounts of proinflammatory cytokines is beneficial to the induction of a protective immune response. However, overproduction of the proinflammatory cytokines can have mild to very severe adverse side affects.

F. Hypersensitivity Responses to Vaccine Antigens

Animals may develop any of the four types of immune-mediated hypersensitivity reactions to vaccine antigens. Systemic anaphylaxis

TABLE IV

Bacterial Components that Induce Proinflammatory Cytokines

- Lipopolysaccharide
- Lipid A
- Porins
- Muramyl peptides
- Peptidoglycan
- Mycoplasma lipoproteins
- Teichoic acid
- Lipoteichoic acids
- Lipoarabinomannans
- Protein A
- Superantigens

due to type I (immediate type) hypersensitivity is the most dramatic type of adverse vaccine reaction. This can occur as a result of the induction of IgE class antibody to essentially any component of a vaccine (Bonin et al., 1973; Wilson et al., 1968; Erdos et al., 1975). As with all of the hypersensitivity reactions, the animal will not react on first exposure to an antigen (unless it has received passive antibody responsible for the reaction). It will only react after there has been sufficient time to produce the sensitizing antibody or memory T cells.

A local type I hypersensitivity reaction may occur due to IgE induced against infectious agents by the vaccine. Immunization against bovine respiratory syncytial virus under experimental conditions was shown to induce IgE antibodies specific for BRSV which apparently contributed to the development of symptoms following aerosol challenge with BRSV (Stewart and Gershwin, 1989a,b).

Vaccine-induced type II (cytotoxic type) hypersensitivity reactions can occur when vaccines are used that contain normal cell antigens. For example, vaccines that contain erythrocyte antigens may induce anti-erythrocyte antibodies leading to immune-mediated hemolytic anemia.

Type III (immune complex type) hypersensitivity can occur when circulating antibody specific for vaccine antigens is present at the time of vaccination. This can lead to an Arthus reaction at the site of injection due to complement fixation and neutrophil recruitment to the site. This mechanism is commonly responsible for the local inflammatory reaction at the site of injection, especially when administering booster vaccinations with killed vaccines. Sometimes, hypersensitivity can be one component of a more complex adverse vaccine reaction. Antibody induced by the vaccine may lead to immune complex type hypersensitivity reactions after the animal becomes infected when the antibody binds to replicating infectious agents. Examples include anterior uveitis and corneal edema (blue eye) after vaccination with canine adenovirus (Carmichael et al., 1975; Wright, 1976) and the sensitization to the effusive form of feline infectious peritonitis after vaccination with experimental killed vaccines (Pedersen and Black, 1983).

Sometimes, hypersensitivity may be one component of a more complex adverse vaccine reaction. Bacterins for *Pasteurella haemolytica* which were marketed and widely sued for several years were of marginal efficacy and were even capable of increasing the severity of lesions in animals either experimentally (Wilkie et al., 1980) or naturally exposed (Bennett, 1982) to the *P. haemolytica*. There are at least two hypothesized mechanisms by which the immune response induced by the bacterin could potentiate pneumonia after *P. haemolytica* challenge. First, the high concentration of complement-fixing antibody induced by vaccination with a bacterin could rapidly activate complement if a large number of *P. haemolytica* organisms were introduced into the lung either naturally or artificially. This could cause a type III hypersensitivity response leading to acute inflammation in the lung and severe pneumonia. Second, antibody against cell surface antigens will opsonize the *P. haemolytica* in the lung and enhance phagocytosis by alveolar macrophages and neutrophils. Because there may be insufficient leukotoxin-neutralizing antibody or cell-mediated immunity to activate phagocytes, the bacteria present in the alveoli and ingested by phagocytes are not efficiently killed and may produce leukotoxin that could destroy the phagocytes. This destruction would cause the phagocytes to release their hydrolytic enzymes into the lung.

G. Vaccine-Induced Triggering or Exacerbation of Hypersensitivity Disease to Nonvaccine Antigens

In the last few years concern has been expressed that vaccination may trigger or exacerbate autoimmune disease or allergies (hypersensitivities), especially in dogs and cats (see article by Dr. Jean Dodds in this volume). Vaccination has been shown to augment production of IgE antibody to pollen in inbred atopic dogs (Frick and Brooks, 1983). Remember that animals with allergies or autoimmune diseases are not healthy animals, and that vaccines are only recommended for use in healthy animals. Dr. Harm HogenEsch addresses the topic of vaccine-induced autoimmunity in another article in this volume.

H. Vaccine-Induced Neoplastic Disease

In recent years, an increased incidence of fibrosarcoma occurring at sites commonly used for vaccination in cats has been observed (Hendrick *et al.*, 1992, 1994; Kass *et al.*, 1993). The causal relationship and mechanistic basis for vaccine-associated fibrosarcomas in cats has not been firmly established (Ellis *et al.*, 1996).

I. MLV BVD Vaccine Triggering Mucosal Disease in Persistently Infected Cattle

Shortly after MLV BVD vaccines were introduced, it was recognized that a very small percentage of cattle developed a syndrome 7–20 days after vaccination that closely resembled BVD mucosal disease (Lambert, 1973; Peter *et al.*, 1967). Based on the current understanding of the pathogenesis of mucosal disease (Bolin *et al.*, 1985; Brownlie *et al.*, 1984) this was almost certainly due to the cytopathic BVD virus in the vaccine triggering mucosal disease in calves that were immunotolerant to, and persistently infected with, a noncytopathic BVD virus. The mechanistic basis for the induction of the lesions of mucosal disease is not clearly understood. This unique syndrome is primarily due to abnormalities in the animal rather than to a defect in the vaccine.

J. ADVERSE REACTIONS DUE TO MULTIPLE VACCINES ADMINISTERED CONCURRENTLY

Vaccines are tested for safety and efficacy when administered to healthy animals in the formulation in which they are packaged to be sold. Vaccines are not required to be tested for safety and efficacy when administered concurrently with other vaccines. This would not be practical since there are too many possible vaccines that may potentially be used in combination. An example of a safety problem that occurred when two different vaccines were administered concurrently involved a newly developed MLV canine coronavirus and parvovirus vaccine given at the same time as an MLV canine distemper–hepatitis virus vaccine. The evidence indicated that the other MLV components allowed the canine coronavirus in the vaccine to induce neurologic disease in some vaccinated animals (Wilson et al., 1986).

K. INJECTION SITE LESIONS

Injection site lesions are a common occurrence and are of great concern in food-producing animals. They may lead to unacceptable blemishes in, or decreased quality of, meat intended for human consumption. There are many possible causes of injection site lesions, including organisms introduced with a contaminated needle, live contaminating organisms in the vaccine, adjuvant induced reactions, cytokine release, hypersensitivity reactions (types I, II, III, or IV), trauma, and hemorrhage (Straw et al., 1985, 1990; Droual et al., 1993; Littledike, 1993; Stokka et al., 1994; Dexter et al., 1994; Apley et al., 1994; Straw, 1986).

III. Vaccine Failure

Vaccines that are licensed by the USDA have been tested to determine that they are safe and effective. However, "effective" is a relative term. It does not mean that the vaccine must be able to induce complete immunity under all conditions which may be found in the field. This would not be realistic since the immune system is not capable of such potent protection under adverse conditions.

To be federally licensed, the vaccine must have been tested under controlled experimental conditions. The vaccinated group must have had significantly less disease than the nonvaccinated control group. This testing is typically done on healthy, nonstressed animals under good environmental conditions and with a controlled exposure to a single infectious agent. Vaccines may be much less effective when used in animals that are under stress, incubating other infectious diseases, or exposed to a high dose of infectious agents due to overcrowding or poor sanitation.

It is important to remember that for most diseases the relationship between the infectious agent and the host is sufficiently complicated that vaccination cannot be expected to provide complete protection. The vaccine can increase the animal's resistance to disease, but this resistance can be overwhelmed if good management practices are not followed. Some of the causes for vaccine failure are summarized in Table II and explained next.

A. Insufficient Time to Develop Immunity

The host requires several days after vaccination before an effective immune response will develop. If the animal encounters an infectious agent near the time of vaccination, the vaccine will not have had time to induce immunity. The animal may come down with clinical disease resulting in apparent vaccination failure. In this situation, disease symptoms will appear shortly after vaccination and may be mistakenly attributed to vaccine virus causing the disease (McKercher et al., 1968).

B. Vaccine Failure Due to Alterations in the Vaccine

Improperly handled and administered vaccines may fail to induce the expected immune response in normal, healthy animals. Modified live bacterial and viral vaccines are only effective if the agent in the vaccine is viable and able to replicate in the vaccinated animal. Observing proper storage conditions and proper methods of administration are very important for maintaining vaccine viability. Failure to store the vaccine at refrigerator temperatures, or exposure to light, may inactivate the vaccine. Even when stored under appropriate conditions, the vaccine loses viability over time. Therefore, vaccines that are past their expiration date should not be used. The use of chemical disinfectants on syringes and needles can inactivate modified live vaccines if there is any residual disinfectant.

The use of improper diluent or the mixing of vaccines in a single syringe may also inactivate modified live vaccines. Diluents for lyophilized vaccines are formulated specifically for each vaccine. A diluent that is appropriate for one vaccine may inactivate a different vaccine. Some vaccines and diluents contain preservatives that may inactivate other modified live vaccines. For these reasons, multiple vaccines should not be mixed in a single syringe unless that particular combination has been adequately tested to ensure there is no interference.

C. Host Factors Responsible for Vaccine Failure

Vaccine failures may occur because a vaccinated animal is not able to respond appropriately to the vaccine. Vaccine failure in young animals may be due to the presence of maternal antibody which prevents adequate response to vaccination. It can also be due to immunosuppression from a variety of causes.

Maternal antibodies derived from colostrum are a well-known cause of vaccine failure (Greene, 1990). These antibodies in the young animal's circulation may neutralize or remove the antigen before it can induce an immune response. Typically, virulent infectious agents are capable of breaking through maternal immunity earlier than modified live or killed vaccines. This means that even if young animals are immunized frequently, there still may be a period when they are vulnerable to infection. Vulnerability occurs between the time that young animals lose their maternal antibody and before they develop their own active immune responses. This period can be shortened by the use of less-attenuated and/or higher titered modified live vaccines or the use of killed vaccines with high antigenic mass and strong adjuvants (Smith-Carr et al., 1997; Larson and Schultz, 1996).

A high challenge dose of infectious agents will break through maternal immunity sooner than low exposure to infectious agents. Therefore, overcrowding and poor sanitation exacerbate the problem of inducing immunity in young animals before they come down with clinical disease.

Veterinarians commonly recommend that puppies and kittens be vaccinated every 3 weeks between approximately 6 and 18 weeks of age. However, for large domestic animals, a single vaccination is commonly recommended to induce immunity during the first few weeks or months of life. There is no inherent difference between large and small domestic animals in their responses to vaccination in the face of maternal immunity. The frequent vaccinations recommended in puppies and kittens minimizes the period of vulnerability to infectious diseases.

Because only one vaccination is commonly recommended for large domestic animals, the timing of vaccination is important. If the vaccine is administered too soon, it may be ineffective because of the presence of maternal antibody. If the vaccine is administered after all maternal antibodies are gone from animals in the group, there may be a prolonged period of vulnerability before they develop their own immune response. The optimal age to vaccinate young animals is highly variable. It will depend on the antibody titer of the mother and the amount of colostrum ingested. It is impossible to predict an optimal age to vaccinate a young animal, unless its antibody titers are determined. Most veterinarians and producers decide that because of time and expense considerations it is impractical to vaccinate young food-producing animals frequently to minimize their period of vulnerability to infection. However, frequent vaccination may be justified in cases of unusually high disease incidence in young animals.

Immunosuppression due to a variety of factors including stress, malnutrition, concurrent infection, or immaturity or senescence of the immune system may also lead to vaccination failure. If the immunosuppression occurs at the time of vaccination, the vaccine may fail to induce an adequate immune response. If the immunosuppression occurs sometime after vaccination, then disease may occur due to reduced immunity in spite of an adequate response to the original vaccine. Therapy with immunosuppressive drugs (e.g, glucocorticoids) may also cause this to occur.

D. Vaccine Failure Due to Exposure to an Overwhelming Challenge Dose

Most vaccines do not produce complete immunity to disease. They provide an increased ability to resist challenge by infectious agents. If a high-challenge dose of organisms is present due to overcrowding or poor sanitation, the immune system may be overwhelmed, resulting in clinical disease.

E. Vaccine Failure Due to Inadequate Duration of Immunity

The peak response to a vaccine typically occurs 2–6 weeks after vaccination. The level of immunity then begins to gradually decline. A common recommendation is to revaccinate annually. However, if the animal did not have a strong initial immune response due to stress at the time of vaccination, or if it is stressed and exposed to a high-challenge dose several months after vaccination, there may not be enough residual immunity to protect the animal. This is especially true for certain killed vaccines. Under these circumstances, it may be necessary to revaccinate more frequently than once per year.

F. Vaccine Failure Due to Antigenic Differences between Vaccine and Field Strains

For certain types of infectious agents, particularly bacteria that are vulnerable to control by the development of antibodies against surface components and viruses which use RNA as their genetic material and consequently have high mutation rates, there are often several antigenic variants of each agent. For antibody-mediated protection to be effective, the antibodies formed must bind the important strain-specific antigens on the surface of the bacteria or virus. Cell-mediated immunity is usually not as strain specific as antibody-mediated immunity. To determine if a vaccine's failure to protect is due to antigenic differences between the vaccine and field strains it is necessary to isolate the field strain and compare it to the vaccine strain. Antigenic differences between strains leading to lack of vaccine efficacy are usually more of a problem with killed vaccines than modified live vaccines.

G. Vaccine Failure Due to Interference When Multiple Vaccines Are Administered Concurrently

As mentioned earlier, vaccines are tested for safety and efficacy when administered singly to animals. However, multiple vaccines are commonly administered concurrently to animals. Very little published data are available concerning the efficacy of vaccines when used in combination. One study demonstrated that there was no detrimental effect on the antibody response to a bovine respiratory syncitial virus vaccine when administered in combination with up to 17 different immunogens (Carmel et al., 1992). In contrast, an MLV BHV1 vaccine when administered in combination with an experimental *Pasteurella haemolytica* vaccine containing outer membrane proteins and genetically attenuated leukotoxin significantly reduced the antibody response to the leukotoxin and the efficacy of the *P. haemolytica* vaccine in preventing morbidity and mortality due to bovine respiratory disease (Harland et al., 1992).

IV. Summary

Mild local and systemic reactions to vaccines are to be expected as a natural consequence of vigorously stimulating the immune system. Dramatic adverse reactions to vaccines are occasionally due to mistakes during the production or handling of vaccines. More often, they are due to not following label instructions, particularly the restriction to only use vaccines in healthy animals. It is important to publish well-documented instances of adverse vaccine reactions so that producers and users of vaccines can all learn from the experience and avoid similar problems.

Vaccine failure to protect from disease is usually due to problems with either client education or compliance with good animal management practices. It is important for clients to understand the proper timing and method of vaccine administration, what to realistically expect for vaccine efficacy, and the importance of minimizing immunosuppressive factors and exposure to high doses of infectious agents in vaccinated animals.

Veterinary vaccines have produced dramatic benefits in terms of animal health, human health, and efficiency of food production. Advances in research and the accumulating experience with vaccines are leading to safer and more effective vaccines. Proper usage of vaccines and adherence to good management practices will continue to be essential to achieve maximal vaccine safety and efficacy.

REFERENCES

Allison, A. C., and Eugui, E. M. (1995). Induction of cytokine formation by bacteria and their products. In "Virulence Mechanisms of Bacterial Pathogens," (J. A. Roth, C. A. Bolin, K. A. Brogden, F. C. Minion, and M. J. Wannemuehler, eds.) 2nd ed., pp. 303–332. American Society for Microbiology Press, Washington, D.C.

Apley, M., Wray, M., and Armstrong, D. (1994). Subcutaneous injection site comparison of two multiple valent clostridial bacterin/toxoids in feedlot cattle. *Agri-Practice* **15**, 9–12.

Beck, E., and Strohmaier, K. (1987). Subtyping of European foot-and-mouth disease virus strains by nucleotide sequence determination. *J. Virol.* **61**, 1621–1629.

Bellinger, D. A., Chang, J., Bunn, T. O., Pick, J. R., Murphy, M., and Rahija, R. (1983). Rabies induced in a cat by high-egg-passage Flury strain vaccine. *J. Am. Vet. Med. Assoc.* **183**, 997–998.

Bennett, B. W. (1982). Efficacy of *Pasteurella* bacterins for yearling feedlot cattle. *Bovine Pract.* **3**, 26–30.

Bolin, S. R., McClurkin, A. W., Cutlip, R. C., and Coria, M. F. (1985). Severe clinical disease induced in cattle persistently infected with noncytopathic bovine viral diarrhea virus by superinfection with cytopathic bovine viral diarrhea virus. *Am. J. Vet. Res.* **46**, 573–576.

Bonin, O., Schmidt, I., and Ehrengut, W. (1973). Sensitization against calf serum proteins as a possible cause of allergic reactions after vaccination. *J. Biol. Stand.* **1,** 187–193.
Brown, F. (1993). Review of accidents caused by incomplete inactivation of viruses. *Dev. Biol. Stand.* **81,** 103–107.
Brownlie, J., Clarke, M. C., and Howard, C. J. (1984). Experimental production of fatal mucosal disease in cattle. *Vet. Rec.* **114,** 535–536.
Bryan, L. A., Fenton, R. A., Misra, V., and Haines, D. M. (1994). Fatal, generalized bovine herpesvirus type-1 infection associated with a modified-live infectious bovine rhinotracheitis parainfluenza-3 vaccine administered to neonatal calves. *Can. Vet. J.* **35,** 223–228.
Bush, M., Montali, R. J., Brownstein, D., and James, A. E. J. (1976). Vaccine-Induced canine distemper in a lesser panda. *J. Am. Vet. Med. Assoc.* **169,** 959–960.
Carmel, D. K., Barao, S. M., and Douglass, L. W. (1992). Effects of vaccination against 18 immunogens in beef replacement heifers at weaning. *J. Am. Vet. Med. Assoc.* **201,** 587–590.
Carmichael, L. E., Medic, L. S., Bistner, S. I., and Aguirre, G. D. (1975). Viral-antibody complexes in canine adenovirus type 1 (Cav-1) ocular lesions: Leukocyte chemotaxis and enzyme release. *Cornell Vet.* **65,** 331–351.
Chiang, B. C., Smith, P. C., Nusbaum, K. E., and Stringfellow, D. A. (1990). The effect of infectious bovine rhinotracheitis vaccine on reproductive efficiency in cattle vaccinated during estrus. *Theriogenology* **33,** 1113–1120.
Clark, L. K., Molitor, T. W., Gunther, R., and Joo, H. S. (1984). Pathogenicity of a modified-live pseudorabies vaccine virus in lambs. *J. Am. Vet. Med. Assoc.* **185,** 1535–1537.
Cullor, J. S. (1994). Safety and efficacy of gram-negative bacterial vaccines. *Bovine Proc.* **26,** 13–26.
Debbie, J. G. (1979). Vaccine-induced rabies in a pet skunk. *J. Am. Vet Med. Assoc.* **175,** 376–377.
Dexter, D. R., Cowman, G. L., Morgan, J. B., Clayton, R. P., Tatum, J. D., Sofos, J. N., Schmidt, G. R., Glock, R. D., and Smith, G. C. (1994). Incidence of Injection-site blemishes in beef top sirloin butts. *J. Anim. Sci.* **72,** 824–827.
Droual, R., Bickford, A. A., and Cutler, G. J. (1993). Local reaction and serological responses in commercial layer chickens injected intramuscularly in the leg with oil-adjuvanted *Mycoplasma gallisepticum* bacterin. *Avian Dis.* **37,** 1001–1008.
Duenwald, J. C., Holland, J. M., Gorham, J. R., and Ott, R. L. (1971). Feline panleukopenia: Experimental cerebellar hypoplasia produced in neonatal ferrets with live virus vaccine. *Res. Vet. Sci.* **12,** 394–396.
Ellis, J. A., and Yong, C. (1997). Systemic adverse reactions in young Simmental calves following administration of a combination vaccine. *Can. Vet. J.* **38,** 45–47.
Ellis, J. A., Jackson, M. L., Bartsch, R. C., McGill, L. G., Martin, K. M., Trask, B. R., and Haines, D. M. (1996). Use of immunohistochemistry and polymerase chain reaction for detection of oncornaviruses in formalin-fixed, paraffin-embedded fibrosarcomas from cats. *J. Am. Vet. Med. Assoc.* **209,** 767–771.
Erdos, L., Lang, C., Jaszovsky, I., and Nyerges, G. (1975). The demonstration of the sensitizing effect of the residual animal serum content of vaccines. *J. Biol. Stand.* **3,** 77–82.
Erlewein, D. L. (1981). Post-vaccinal rabies in a cat. *Feline Pract.* **11,** 16–21.
Esh, J. B., Cunningham, J. G., and Wiktor, T. J. (1982). Vaccine-induced rabies in four cats. *J. Am. Vet. Med. Assoc.* **180,** 1336–1339.

Evermann, J. F., McKeirnan, A. J., Wilbur, L. A., Levings, R. L., Trueblood, E. S., Baldwin, T. J., and Hughbanks, F. G. (1994). Canine fatalities associated with the use of a modified live vaccine administered during late stages of pregnancy. *J. Vet. Diagn. Invest.* **6,** 353–357.
Frick, O. L., and Brooks, D. L. (1983). Immunoglobulin E antibodies to pollens augmented in dogs by virus vaccines. *Am. J. Vet. Res.* **44,** 440–445.
Galanos, C., and Freudenberg, M. A. (1993). Mechanisms of Endotoxin Shock and Endotoxin hypersensitivity. *Immunobiology* **187,** 346–356.
Gard, S., and Lycke, E. (1957). Inactivation of poliovirus by formaldehyde. Analysis of inactivation curves. *Arch. Gesamte Virusforsch.* **7,** 471–482.
George, L. W., Ardans, A., Mihalyi, J., and Guerra, M. R. (1988). Enhancement of infectious bovine keratoconjunctivitis by modified-live infectious bovine rhinotracheitis virus vaccine. *Am. J. Vet. Res.* **49,** 1800–1806.
Greene, C. E. (1990). Immunophophylaxis and immunotherapy. *In* "Infectious Diseases of the Dog and Cat" (C. E. Greene, ed.), pp. 21–54. Saunders, Philadelphia.
Halbrooks, R. D., Swango, L. J., Schnurrenberger, P. R., Mitchell, F. E., and Hill, E. P. (1981). Response of gray foxes to modified live-virus canine distemper vaccines. *J. Am. Vet. Med. Assoc.* **179,** 1170–1174.
Harland, R. J., Potter, A. A., van Drunen-Little-van den Hurk, S., Van Donkersgoed, J., Parker, M. D., Zamb, T. J., and Janzen, E. D. (1992). The effect of subunit or modified live bovine herpesvirus-1 vaccines on the efficacy of a recombinant *Pasteurella haemolytica* vaccine for the prevention of respiratory disease in feedlot calves. *Can. Vet. J.* **33,** 734–741.
Henderson, B., and Wilson, M. (1995). Modulins: A new class of cytokine-inducing, proinflammatory bacterial virulence factor. *Inflammation Res.* **44,** 187–197.
Hendrick, M. J., Goldschmidt, M. H., Shofer, F. S., Wang, Y. Y., and Somlyo, A. P. (1992). Postvaccinal sarcomas in the cat: Epidemiology and electron probe microanalytical identification of aluminum. *Cancer Res.* **52,** 5391–5394.
Hendrick, M. J., Shofer, F. S., Goldschmidt, M. H., Haviland, J. C., Schelling, S. H., Engler, S. J., and Glaitto, J. M. (1994). Comparison of fibrosarcomas that developed at vaccination sites and at nonvaccination sites in cats: 239 cases (1991–1992). *J. Am. Vet. Med. Assoc.* **205,** 1425–1429.
Jensen, M. H. (1981). Hog cholera antibodies in pigs vaccinated with an Aujeszky-vaccine based on antigen produced in IB-RS-2 cells. *Acta Vet. Scand.* **22,** 517–523.
Kass, P. H., Barnes, W. G., Spangler, W. L., Chomel, B. B., and Culbertson, M. R. (1993). Epidemiologic evidence for a causal relation between vaccination and fibrosarcoma tumorigenesis in cats. *J. Am. Vet. Med. Assoc.* **203,** 396–405.
Kazacos, K. R., Thacker, H. L., Shivaprasad, H. L., and Burger, P. P. (1981). Vaccination-induced distemper in kinkajous. *J. Am. Vet. Med. Assoc.* **179,** 1166–1169.
Kesel, M. L., and Neil, D. H. (1983). Combined MLV canine parvovirus vaccine: Immunosuppression with infective shedding. *VM/SAC, Vet. Med. Small Anim. Clin.* **78,** 687–691.
King, A. M. Q., Underwood, B. O., McCahon, D., Newman, J. W. I., and Brown, F. (1981). Biochemical identification of viruses causing the 1981 outbreaks of foot-and-mouth disease virus in the U.K. *Nature* **293,** 479–480.
Kinney, R. M., Tsuchiya, K. R., Sneider, J. M., and Trent, D. W. (1992). Molecular evidence for the origin of the widespread Venezuelan equine encephalitis epizootic of 1969 to 1972. *J. Gen. Virol.* **73,** 3301–3305.
Krakowka, S., Olsen, R. G., Axthelm, M., Rice, J., and Winters, K. (1982). Canine par-

vovirus infection potentiates canine distemper encephalitis attributable to modified live-virus vaccine. *J. Am. Vet. Med. Assoc.* **180,** 137–139.

Lambert, G. (1973). Bovine viral diarrhea: Prophylaxis and postvaccinal reactions. *J. Am. Vet. Med. Assoc.* **163,** 874–876.

Larson, L. J., and Schultz, R. D. (1996). High-titer canine parvovirus vaccine: Serologic response and challenge-of-immunity study. *Vet. Med.* **19,** 210–218.

Littledike, E. T. (1993). Variation of abscess formation in cattle after vaccination with a modified-live *Pasteurella haemolytica* vaccine. *Am. J. Vet. Res.* **54,** 1244–1248.

Lohr, C. H., Evermann, J. F., and Ward, A. C. (1983). Investigation of dams and their offspring inoculated with a vaccine contaminated by bovine viral diarrhea virus. *VM / SAC, Vet. Med. Small Anim. Clinic.* **78,** 1263–1266.

Loken, T., Krogsrud, J., and Bjerkas, I. (1991). Outbreaks of border disease in goats induced by a pestivirus-contaminated orf vaccine, with virus transmission to sheep and cattle. *J. Comp. Pathol.* **104,** 195–209.

Mastro, J. M., Axthelm, M., Mathes, L. E., Krakowka, S., Ladiges, W., and Olsen, R. G. (1986). Repeated suppression of lymphocyte blastogenesis following vaccination of CPV-immune dogs with modified-live CPV vaccines. *Vet. Microbiol.* **12,** 201–211.

McKercher, D. G., Saito, J. K., Crenshaw, G. L., and Bushnell, R. B. (1968). Complications in cattle following vaccination with a combined bovine viral diarrhea-infectious bovine rhinotracheitis vaccine. *J. Am. Vet. Med. Assoc.* **152,** 1621–1624.

Miller, J. M., Van der Maaten, M., and Whetstone, C. A. (1989). Infertility in heifers inoculated with modified-live bovine herpesvirus-1 vaccinal strains against infectious bovine rhinotracheitis on postbreeding day 14. *Am. J. Vet. Res.* **50,** 551–554.

Nathanson, N., and Langmuir, A. D. (1963). The Cutter incident. Poliomyelitis following formaldehyde-inactivated poliovirus vaccination in the United States during the spring of 1955. *Am. J. Hyg.* **78,** 16–28.

Neaton, H. J. (1986). Which BVD vaccine should I use? *Vet. Med.* **81,** 876–881.

Pedersen, N. C., and Black, J. W. (1983). Attempted immunization of cats against feline infectious peritonitis, using avirulent live virus or sublethal amounts of virulent virus. *Am. J. Vet. Res.* **44,** 229–234.

Pedersen, N. C., Emmons, R. W., Selcer, R., Woodie, J. D., Holliday, T. A., and Weiss, M. (1978). Rabies vaccine virus infection in three dogs. *J. Am. Vet. Med. Assoc.* **172,** 1092–1096.

Pennisi, E. (1997). Monkey virus DNA found in rare human cancers. *Science* **275,** 748–749.

Peter, C. P., Tyler, D. E., and Ramsey, F. K. (1967). Characteristics of a condition following vaccination with bovine virus diarrhea vaccine. *J. Am. Vet. Med. Assoc.* **150,** 46–52.

Phillips, T. R., Jensen, J. L., Rubino, M. J., Yang, W. C., and Schultz, R. D. (1989). Effects of vaccines on the canine immune system. *Can. J. Vet. Res.* **53,** 154–160.

Povey, R. C., and Wilson, M. R. (1978). A comparison of inactivated feline viral rhinotracheitis and feline caliciviral disease vaccines with live-modified viral vaccines. *Feline Pract.* **8,** 35–42.

Rogers, R. J., Dimmock, C. K., de Vost, A. J., and Rodwell, B. J. (1988). Bovine leucosis virus contamination of a vaccine produced *in vivo* against bovine babesiosis and anaplasmosis. *Aust. Vet. J.* **65,** 285–287.

Roth, J. A., and Kaeberle, M. L. (1983). Suppression of neutrophil and lymphocyte function induced by a vaccinal strain of bovine viral diarrhea virus with and without the concurrent administration of ACTH. *Am. J. Vet. Res.* **44,** 2366–2372.

Schuh, J., and Walker, S. (1990). Outbreaks of neonatal infectious bovine rhinotracheitis. *Can. Vet. J.* **31,** 592.

Shah, K., and Nathanson, N. (1976). Human exposure to SV40: Review and comment. *Am. J. Epidemiol.* **103,** 1–12.

Smith, P. C., Nusbaum, K. E., Kwapien, R. P., Stringfellow, D. A., and Driggers, K. (1990). Nicrotic oophoritis in heifers vaccinated intravenously with infectious bovine rhinotracheitis virus vaccine during estrus. *Am. J. Vet. Res.* **51,** 969–972.

Smith-Carr, S., Macintire, D. K., and Swango, L. J. (1997). Canine Parvovirus. Part I. Pathogenesis and vaccination. *Compend. Contin. Educ. Pract. Vet.* **19,** 125–133.

Stewart, R. S., and Gershwin, L. J. (1989a). Role of IgE in the pathogenesis of bovine respiratory syncytial virus in sequential infections in vaccinated and nonvaccinated calves. *Can. J. Vet. Res.* **50,** 349–355.

Stewart, R. S., and Gershwin, L. J. (1989b). Detection of IgE antibodies to bovine respiratory syncytial virus. *Vet. Immunol. Immunopathol.* **20,** 313–323.

Stokka, G. L., Edwards, A. J., Spire, M. F., Brandt, R. T., and Smith, J. E. (1994). Inflammatory response to clostridial vaccines in feedlot cattle. *J. Am. Vet. Med. Assoc.* **204,** 415–419.

Straw, B. (1986). Injection reactions in swine. *Anim. Health Nutr.* **41,** 10–14.

Straw, B. E., McLachlan, N. J., Corbett, W. T., Carter, P. B., and Schey, H. M. (1985). Comparison of tissue reactions produced by *Haemophilus pleuropneumoniae* vaccines made with six different adjuvants in swine. *Can. J. Comp. Med.* **49,** 149–151.

Straw, B. E., Shin, S., Callihan, D., and Petersen, M. (1990). Antibody production and tissue irritation in swine vaccinated with *Actinobacillus* bacterins containing various adjuvants. *J. Am. Vet. Med. Assoc.* **196,** 600–604.

Thornton, D. H. (1986). A survey of mycoplasma detection in veterinary vaccines. *Vaccine* **4,** 237–240.

Van Alstine, W. G., Anderson, T. D., Reed, D. E., and Wheeler, J. G. (1984). Vaccine-induced pseudorabies in lambs. *J. Am. Vet. Med. Assoc.* **185,** 409–410.

Van der Maaten, M. J., Miller, J. M., and Whetstone, C. A. (1985). Ovarian lesions induced in heifers by intravenous inoculation with modified-live infectious bovine-rhinotracheitis virus on the day after breeding. *Am. J. Vet. Res.* **46,** 1996–1999.

Vannier, P., Leforban, Y., Carnero, R., and Cariolet, R. (1988). Contamination of a live virus vaccine against pseudorabies (Aujeszky's disease) by an ovine pestivirus pathogen for the pig. *Ann. Rech. Vet.* **19,** 283–290.

Wensvoort, G., and Terpstra, C. (1988). Bovine viral diarrhoea virus infections in piglets born to sows vaccinated against swine fever with contaminated vaccine. *Res. Vet. Sci.* **45,** 143–148.

Whetstone, C. A., Bunn, T. O., Emmons, R. W., and Wiktor, T. J. (1984). Use of monoclonal antibodies to confirm vaccine-induced rabies in ten dogs, two cats, and one fox. *J. Am. Vet. Med. Assoc.* **185,** 285–288.

Wilbur, L. A., Evermann, J. F., Levings, R. L., Stoll, I. R., Starling, D. E., Spillers, C. A., Gustafson, G. A., and McKeirnan, A. J. (1994). Abortion and death in pregnant bitches associated with a canine vaccine contaminated with bluetongue virus. *J. Am. Vet. Med. Assoc.* **204,** 1762–1765.

Wilkie, B. N., Markham, R. J. F., and Shewen, P. E. (1980). Response of calves to lung challenge exposure with *Pasteurella haemolytica* after parenteral or pulmonary immunization. *Am. J. Vet. Res.* **41,** 1773–1778.

Wilson, R. B., Kord, C. E., Holladay, J. A., and Cave, J. S. (1986). A neurologic syndrome associated with use of a canine coronavirus-parvovirus vaccine in dogs. *Compend. Contin. Educ. Pract. Vet.* **8,** 117–123.

Wilson, S., Weber, J. C. W., Aprile, M. A., and Macmorine, H. G. (1968). Detection of sensitizing tissue culture components in viral vaccines. *J. Allergy* **42,** 319–329.

Wright, N. G. (1976). Canine adenovirus: Its role in renal and ocular disease: A review. *J. Small Anim. Pract.* **17,** 25–33.

Weighing the Risks and Benefits of Vaccination

LARRY T. GLICKMAN

Department of Veterinary Pathobiology, Purdue University, West Lafayette, Indiana 47907

I. Defining the Problem
II. Vaccine Risk Assessment
III. Current Sources of Data on Adverse Reactions
IV. Suggested Improvements in Postmarketing Surveillance
V. Risk Management and Risk Communication
VI. Summary
Acknowledgment
References

I. Defining the Problem

There is increasing concern about the safety of vaccines for companion animals. The *Journal of the American Veterinary Medical Association* recently addressed this topic in its section on Current Concepts in an article titled "Are we vaccinating too much?" (Smith, 1995). Two sides of the issue were discussed: (1) Are the currently used vaccination schedules excessive? (2) Are the adverse reactions to vaccines unacceptable? One point raised by several experts was that there is little scientific evidence to back up label claims for annual administration of most vaccines. While in the past most veterinarians believed that annual vaccination was beneficial and would do no harm, this attitude has begun to change. Several events appear to underlie increasing concern for the safety of veterinary vaccines, including a possible nationwide epidemic of postvaccinal sarcomas in cats (Hendrick and Goldschmidt, 1991) and an apparent increase in the frequency of autoimmune diseases in dogs (Dodds, 1985), one of which (e.g., immune-mediated hemolytic anemia) has been associated in controlled epidemiologic studies with recent vaccinations (Duval and Giger, 1996).

While few veterinarians challenge the fact that most vaccines are highly effective and have been responsible for a marked decrease in the incidence of common infectious diseases of dogs and cats, many have proposed reducing the frequency of vaccine use by either vaccinating every other year or by performing annual antibody titer screens to determine which vaccines to administer (Smith, 1995). In essence, by only vaccinating dogs when indicated by antibody titers, veterinarians would tailor vaccination schedules to individual patients. While these approaches will no doubt continue to be debated for some time to come, the need now is for accurate quantitative estimates of the nature and frequency of adverse effects associated with vaccines (i.e., the risks) as well as the best way to vaccinate individual animals according to their age, general health, and environment. This type of information can only be ascertained through a formal process of risk assessment. Another aspect of the issue that must be considered is how to communicate risk estimates to the veterinary community and companion animal owners, so they can make informed decisions and become more involved in managing the risks posed by vaccines.

II. Vaccine Risk Assessment

Risk assessment is the use of scientific evidence to define the health effects of exposures of individuals or populations to hazardous substances and situations (National Research Council, 1983). In this context the hazard in question is the vaccine and the situation is the act of vaccination. Four steps are usually undertaken to assess risk: (1) *hazard identification,* the determination of whether a particular substance or procedure is or is not causally linked to a particular health effect; (2) *dose–response assessment,* the determination of the relation between the magnitude of the exposure (e.g., frequency or time since last vaccination) and the probability of occurrence of the health effects in question; (3) *exposure assessment,* the determination of the extent of exposure; a description of the population at risk; and (4) *risk characterization,* a description of the nature and the magnitude of risk, including the uncertainty associated with it. This last step is usually performed by combining the results of exposure and dose–response assessments. A risk assessment might stop with the first step, hazard identification, if no adverse effect is found, or if it is decided to take action without further analysis.

This question arises: Where will the data come from for a risk assessment of individual veterinary vaccines? There is general agreement that safe vaccines are those which do not induce local or systemic adverse reactions, are not excreted or are excreted only at low levels for modified live organisms, do not revert to virulence, do not affect the fetus, and are completely inactivated for killed vaccines. However, there are few published studies specifically designed to estimate the frequency of adverse reactions following vaccination.

Clinical trials conducted to evaluate the efficacy of veterinary vaccines prior to their licensing often include less than 100 animals. While this may be sufficient for a manufacturer to demonstrate efficacy and obtain a license from the U.S. Department of Agriculture for their product, it is unlikely to reveal adverse effects even when the incidence is high. For example, if a disease like immune-mediated hemolytic anemia (IMHA) occurs at a rate of 2 per 10,000 dogs per year independent of vaccination in the general population, and if a new vaccine induces IMHA at a rate of 50 per 10,000 dogs per year, then a clinical trial of approximately 2500 dogs will be needed to have a 90% probability of detecting this adverse effect. If these two rates are lower, for example, 2 per 100,000 dogs and 50 per 100,000 dogs, respectively, then approximately 20,000 dogs will be required. Surely no studies of this magnitude are feasible or likely to be conducted in the future. Where then will accurate data come from to measure the safety of a vaccine so that any adverse effects can be detected before they occur in near-epidemic proportions following its widespread use?

III. Current Sources of Data on Adverse Reactions

The types of adverse reactions to animal vaccines have been well characterized in the veterinary literature (Brooks, 1991; Tizard, 1990). These reactions can generally be categorized as either systemic or local. Systemic reactions include type I hypersensitivity or anaphylaxis, type III complex-mediated hypersensitivity, diluent and contamination problems, and reactions due to endotoxins. Local site reactions include type I hypersensitivity, type IV cell-mediated (delayed-type hypersensitivity), reactions to adjuvents such as granulomas and possibly even cancer (Hendrick et al., 1992), diluent and contamination related problems, and faulty administration techniques. The failure of a vaccine to protect against the disease for which it is intended can also be considered an adverse consequence. Evaluating the risk of adverse reactions for an individual animal following vaccination is based on knowledge of the frequency of adverse reactions in the general population of similar animals and the host and environmental risk factors that are associated with these events.

Surveillance is defined as the continuing scrutiny of all aspects of occurrence of disease that are pertinent to effective control (Last, 1988). This approach is routinely used in human and veterinary medicine to monitor the use and untoward effects of vaccines. To be effective as a method of disease control and prevention, surveillance should involve four interrelated components, namely, data collection, analysis, interpretation, and timely dissemination. In the U.S. veterinarians are requested to voluntarily report all potential adverse reactions and vaccine failures to vaccine manufacturers. These data are collated by the manufacturers and compared with the number of vaccines of that type that were sold to determine the adverse reaction rate, usually by lot number, on a monthly or yearly basis. Inherent to this approach, however, is a marked degree of underreporting and a bias toward more severe adverse reactions. Also, since the actual number of animals that were vaccinated with the vaccine in question in a given time period is unknown, it is not possible to calculate a true adverse reaction incidence or risk rate. This method of data collection can be effective, however, for identifying changes in the frequency of adverse effects from vaccine lot to vaccine lot, epidemics of adverse reactions associated with new products (Martin, 1985), and geographic clusters of excess adverse reactions. Routinely collected veterinary health data such as that in the Veterinary Medical Data Base (Priester and McKay, 1980), and hospital or diagnostic laboratory records are also potential sources of information regarding unusual disease occurrences.

Surveillance is a form of descriptive epidemiology in that it is designed only to describe the frequency and distribution of adverse reactions without regard to predetermined causal or other hypotheses. Its main value is to reveal potential causal relationships that can be further examined in controlled analytic epidemiologic studies that attempt to quantify the association between health effects and specific exposure(s) (Last, 1988). For example, concern was first raised about possible vaccine site reactions and fibrosarcomas in cats by veterinary pathologists who regularly reviewed surgical biopsy specimens from their own teaching hospital and from regional veterinary practices (Hendrick and Goldschmidt, 1991). A clear trend of increasing vaccine site-associated fibrosarcoma prevalence was later demonstrated among tissue specimens evaluated histologically at a state veterinary diagnostic laboratory where the vaccine site fibrosarcomas were found to differ morphologically and biologically from those occurring at nonvaccine sites (Doddy et al., 1996). In a case-control study (Kass et al., 1993) it was shown that the vaccine-site fibrosarcomas were significantly associated with injection of several types of inactivated vaccines. Experimental studies are currently in progress to identify the mechanisms responsible for these cancers so that preventive action can be taken. Great uncertainty remains, however, regarding the actual risk of fibrosarcoma attributable to feline vaccines, with estimates ranging from as little as 1 per 10,000 vaccinated cats (Burton and Mason, 1997) to as high as 1.3 per 1000 cats (Lester et al., 1996). This illustrates the inadequacy of currently used passive surveillance systems in veterinary medicine for accurately estimating vaccine-associated risks.

Most surveillance efforts are designed to relate specific adverse reactions to recently administered vaccines, that is, to vaccination in the previous few hours, days, or even months. However, in the past few years, there appears to be a considerable increase in the number of dogs and cats recognized to have immune-mediated diseases with onset in middle and older ages. While a variety of causes or predisposing factors are known or thought to precipitate immune-mediated diseases, it has been suggested that some of these can be triggered by frequent exposure to modified live vaccines (Dodds, 1985). The longer the latency period between vaccination and disease, the less likely it is that the relationship will be detected by routine postmarketing vaccine surveillance.

Most reports of an association between vaccination and autoimmune disease in animals are anecdotal in nature such as "A number of cases (10) have come to my attention over the last 2–3 years which appear to have an autoimmune–vaccination relationship. Specifically, a number of dogs have been observed to develop conditions which appear to be immune-mediated, following annual vaccination, at an incidence which is greater than that which has been observed in previous years" (Albritton, 1996). A controlled epidemiologic study of this possible relationship found that, when compared with a randomly selected hospital control group of dogs, dogs with IMHA were more likely to have been vaccinated within the previous month ($p < 0.0001$) and the dogs with IMHA that had been vaccinated in the previous month had more severe disease than those with IMHA that had been vaccinated more than 1 month previously (Duval and Giger, 1996). All of the recently vaccinated dogs with IMHA in the study had received combination vaccines from various manufacturers against canine distemper, adenovirus type 2, leptospirosis, parainfluenza, and parvovirus.

While a single epidemiologic study does not by itself establish a cause-and-effect relationship between vaccination and autoimmune disease, it indicates a need to develop better reporting systems and for experiments to define possible underlying immunologic mechanisms. In such a recent experiment, a group of Beagle dogs vaccinated with a commercial multivalent canine vaccine at 8, 10, 12, 16, and 20 weeks of age, and with a rabies vaccine at 16 weeks of age, developed a significant rise of IgG autoantibodies against fibronectin, laminin, cardiolipin, sphingomyelin, DNA, and collagen compared with a group of unvaccinated control dogs (HogenEsch et al., 1998). Because all of the dogs were euthanatized at 22 weeks of age, it was not possible to determine the clinical significance of these postvaccinal autoantibody responses. However, many of these autoantibodies and in particular anti-fibronectin, are present in elevated concentration in humans with systemic lupus erythematosus and rheumatoid arthritis (Henane et al., 1986; Girard et al., 1995). Another interesting finding in this study was that the autoantibody response appeared to have a genetic basis, in that it was more pronounced in dogs from some litters than others.

IV. Suggested Improvements in Postmarketing Surveillance

Reports are received by vaccine manufacturers by telephone and in writing from animal owners and veterinarians concerning health problems that appear to be temporally associated with vaccination of their animal(s). These contacts may be handled by veterinarians or by trained veterinary technicians who record the facts, usually on standardized forms. Data from these forms are often coded and entered into a computerized database. The data are periodically analyzed and reports generated showing adverse reaction rates by vaccine lots over time. Often, the reaction rates are categorized as to local or systemic, allergic, lethal, lack of efficacy, etc. No industry-wide standard currently exists for characterizing adverse reactions or for determining what constitutes an excess adverse reaction rate for a particular vaccine. Also, since the number of animals that were actually vaccinated with the vaccine is unknown, adverse reaction rates are calculated based on the number of vaccines sold and distributed over a given time period, rather than by the actual population at risk.

The present system of post-marketing surveillance for veterinary vaccines can be improved considerably if industry-wide standards are adopted for characterizing and counting adverse reactions and if individual manufacturers consider the following:

1. Increase ascertainment of adverse reactions by encouraging more complete reporting by veterinarians. This can be accomplished by supplying prestamped postcards with all vaccines sold, so that veterinary practices can more easily record and report adverse reactions. The postcard should provide a place to record the species, breed, age, sex, and health of the animal; the type and lot number of the vaccine; route of administration; time from vaccination to clinical signs; clinical signs; outcome; and name, address, and telephone number of the reporting person. For food animal vaccines the postcard should also record herd information such as number of animals in the herd, number vaccinated, and number affected; and changes in productivity and feed consumption. More complete reporting will facilitate the interpretation of clusters of adverse events by time and place that may be artifacts of incomplete or sparse data. Also, a method is needed to score the reliability of each report that is received based on supporting documentation such as laboratory and pathologic findings, follow-up investigation, and qualifications of the reporting party.

2. Vaccine manufacturers should standardize their reporting systems to be consistent with each other in terms of the type and severity of adverse reactions. This standardization requires the training of veterinary paraprofessionals to collect and code adverse reaction reports.

3. Criteria should be developed for analyzing and reporting adverse reaction data on a regular basis. This should include establishing statistical methods to determine when an adverse reaction rate exceeds the expected value, which is the fundamental definition of an epidemic.

No universal criteria can necessarily be applied to determine the excess number of adverse reactions sufficient to warrant further investigation. The decision to investigate is influenced by factors such as the severity of the health consequences and the particular circumstances of the events. This analysis may exclude reports thought to be invalid, because they are either incomplete or unreliable. Also, a standard needs to be established for defining what goes into the denominator when calculating adverse reaction rates (e.g., total number of a vaccine lot sold, number of a vaccine lot sold in the previous month or the current month, etc.).

In addition to improvements in current postmarketing surveillance systems, new approaches should be developed to measure more accurately the risk of vaccine-associated adverse reactions. The following should be considered:

1. Manufacturers need to decide when surveillance data indicate the need for additional investigation. Procedures have been described to investigate outbreaks of disease in animals (Kahrs, 1978) as well as geographic clusters of health-related events (Fiore et al., 1990). Veterinary epidemiologists should lead these investigations of adverse reactions and, therefore, need to be part of the professional and technical services teams in industry. In some instances, confirmation of the occurrence of a vaccine-associated epidemic should be followed by analytic epidemiologic studies (e.g., case-control) so that multiple etiologic hypotheses can be tested concurrently using appropriate field investigations. This is particularly true where the population at risk cannot be unequivocally defined and or fully enumerated (Dwyer et al., 1994).

2. Because postmarketing surveillance is likely to reveal apparent space–time clusters of adverse reactions, increasing use should be made of statistical software such as CLUSTER in interpreting the patterns observed and in deciding whether further investigation is warranted (Aldrich and Drane, 1990).

3. Special postmarketing surveillance systems need to be established that are capable of defining the population at risk in order to determine true adverse reaction incidence rates. One possible method is to prospectively monitor over a specified time period all animals in selected veterinary practices that are vaccinated. The growth of large corporate practices such as VetSmart, which see as many as 1 million dogs and cats yearly in multiple geographic locations and which utilize a common computerized medical record system, make such a reporting system more cost effective. Each vaccine manufacturer will need to identify a group of veterinary practices that use their products and will provide accurate information on potential adverse reactions on a regular basis. The number of animals that need to be included in such a monitoring system should be based on the statistical probability of detecting some specified minimum adverse reaction rate with a given level of confidence.

4. Existing postmarketing industry surveillance programs need to utilize individuals with expertise in database management, epidemiology, and computer programming to provide more timely analysis and dissemination of adverse reaction reports. Such individuals should work with industry and practicing veterinarians to improve data collection procedures and with the U.S. Department of Agriculture to

make current surveillance programs more suitable for analytic studies, while at the same time retaining their administrative value.

V. Risk Management and Risk Communication

Risk assessment is designed to draw extensively on scientific evidence linking specific exposures and health effects. However, as indicated earlier, the available information may be incomplete and lacking in quality such that there is uncertainty in the nature and magnitude of health effects associated with vaccination. This problem has no immediate solution. Nevertheless, decisions regarding the use of certain vaccines still must be made by veterinarians, animal owners, and governmental agencies in the face of this uncertainty. The process of evaluating alternative strategies and selecting among them has been termed risk management. Risk management entails consideration of not only the scientific facts regarding vaccine safety and efficacy, but also political, social, economic, and technical concerns.

A decision to vaccinate companion animals against a particular disease may involve consideration of the efficacy of the vaccine, the likelihood of the animal being exposed to the disease-causing agent, the age and health of the animal, and the probability of side effects. However, veterinarians might also take into account the revenue that will be lost by not including this vaccine in their routine protocol for all animals, while owners are interested in the cost to them. For example, should all cats be vaccinated yearly for rabies when there are rabies vaccines licensed that provide 3 years duration of immunity? Though the exact risk of fibrosarcoma in cats following vaccination for rabies and other infectious diseases is uncertain, repeated immunizations of cats has been associated with an increased risk of fibrosarcoma (Kass et al., 1993). Does the risk of rabies and the ensuing public health threat in areas where rabies is endemic outweigh the risk of fibrosarcoma from yearly rabies vaccination? Yearly vaccination of cats against rabies is required by law in some states, despite the fact that no formal risk assessment was ever done justifying this regulation in light of rabies vaccines that provide 3 years of protection. Similar issues can be raised about vaccines for Lyme disease and for vaccines currently being developed for heart worm infection, *Toxoplasma, Giardia,* etc. Who will do the risk assessments for these vaccines and how will they be done?

Veterinarians are increasingly expected to discuss the benefits and risks of vaccination with their clients and this topic is frequently highlighted in newspapers and magazines. The animal-owning public depends on their veterinarians to interpret the results of scientific studies and to share with them their opinions. This is made more difficult when the public's perception of the risk is disproportionate to the scientific facts, that is, the client overestimates the risk to their animal. Effective risk communication is not something veterinarians are taught in school.

Several guidelines for risk communication have been developed for epidemiologists (Sandman, 1991) and most of these could benefit veterinarians. These include:

1. Tell people who are most affected what you know—and tell them first.
2. Make sure people understand what you are telling them and what you think its implications are.
3. Acknowledge uncertainty promptly and thoroughly.
4. Seek advice from experts in the field when you have questions.
5. Show respect for public concerns even when they are not "scientific."
6. Decide that risk communication is part of your job, and learn its rudiments—it is easier than dealing with disgruntled clients. Poor communication may compromise even the best trained veterinarian.

VI. Summary

The following summarizes this author's current thoughts regarding veterinary vaccines and their safety:

1. Every licensed animal vaccine is probably effective, but also produces some adverse effects.
2. Prelicensing studies of vaccines are not specifically designed to detect adverse vaccine reactions.
3. An improved system of national postmarketing surveillance is required to identify most adverse vaccine reactions that occur at low and moderate frequency.
4. Even a good postmarketing surveillance system is unlikely, however, to detect delayed adverse vaccine reactions, and the longer the delay the less likely they will be associated with vaccination.
5. Analytic epidemiologic (field) studies are the best way to link vaccination with delayed adverse reactions, but these are often hindered by incomplete vaccination histories in medical records in veterinary practice and by a lack of veterinarians in industry trained in epidemiologic methods.

6. Each licensed veterinary vaccine should be subjected to a quantitative risk assessment, and these should be updated on a regular basis as new information becomes available.
7. Risk assessment should be used to identify gaps in information regarding the safety and efficacy of vaccines, and appropriate epidemiologic studies conducted to fill these gaps that contribute to the uncertainty in risk estimates.
8. Risk assessment is an analytical process that is firmly based on scientific considerations, but it also requires judgments to be made when the available information is incomplete. These judgments inevitably draw on both scientific and policy considerations.
9. Representatives from industry, government, veterinary medicine, and the animal-owning public should be involved in risk management, that is, deciding between policy options.

The controversy regarding vaccine risks is intensifying to the point that some animal owners have stopped vaccinating their animals. They offer as justification the belief that current vaccines are "just too dangerous." Some owners report that since they completely stopped vaccinating their animals, they have been healthy. What they fail to realize is that a high percentage of animal owners are responsible and do vaccinate their animals, thus providing "herd immunity" protection to the unvaccinated animals whom they contact. The solution to the vaccine controversy is not to abandon vaccination as an effective means of disease prevention and control, but rather to encourage vaccine research to answer important questions regarding safety and to identify the biological basis for adverse reactions. Key questions to be answered include these: What components of vaccines are responsible for adverse reactions? What is the genetic basis for susceptibility to adverse health effects in animals? How can susceptible individuals be identified? Do multivalent vaccines cause a higher rate of adverse reactions than monovalent vaccines? Is administration of multiple doses of monovalent vaccines really any safer than administering a single multivalent vaccine? These and other vaccine-related questions deserve our attention as veterinarians so we can fulfill our veterinary oath to relieve animal suffering and "above all else, do no harm."

ACKNOWLEDGMENT

Research at Purdue University on vaccine-related adverse reactions in animals is supported in part by the John and Winifred Hayward Foundation.

References

Albritton, A. R. (1996). Autoimmune disease and vaccination? *Vet. Allergy Clin. Immunol.* **4,** 16–17.

Aldrich, T. E., and Drane, J. W. (1990). "CLUSTER: User's Manual for Software to Assist Investigators with Investigations of Rare Health Events." Agency for Toxic Substances and Disease Registries, Atlanta, GA.

Brooks, R. (1991). Adverse reactions to canine and feline vaccines. *Aust. Vet. J.* **68,** 342–344.

Burton, G., and Mason, K. V. (1997). Do postvaccinal sarcomas occur in Australian cats? *Aust. Vet. J.* **75,** 100–106.

Dodds, W. J. (1985). Immune-mediated blood diseases. *Dog News,* June, pp. 8–26.

Doddy, F. D., Glickman, L. T., Glickman, N. W., and Janovitz, E. B. (1996). Feline fibrosarcomas at vaccination sites and non-vaccination sites. *J. Comp. Pathol.* **114,** 165–174.

Duval, D., Giger, U. (1996). Vaccine-associated immune-mediated hemolytic anemia in the dog. *J. Vet. Intern. Med.* **10,** 290–295.

Dwyer, D. M., Strickler, H., Goodman, R. A., and Armenian, H. K. (1994). Use of case-control studies in outbreak investigations. *Am. J. Epidemiol.* **16,** 109–123.

Fiore, B. J., Hanrahan, L. P., Anderson, H. A. (1990). Public health response to reports of clusters. *Am. J. Epidemiol.* **132,** S14–S22.

Girard, D., Raymond, Y., Labbe, P., and Senecal, J.-L. (1995). Characterization of a novel human antibody xenoreactive with fibronectin. *Clin. Immunol. Immunopathol.* **77,** 149–161.

Henane, T., Tigal, D., and Monier, J. C. (1986). Anti-fibronectin autoantibodies in systemic lupus erythematosus, rheumatoid polyarthritis, and various viral or bacterial infectious diseases. *Pathol. Biol.* **34,** 165–171.

Hendrick, M. J., and Goldschmidt, M. H. (1991). Do injection site reactions induce fibrosarcomas in cats? Letter to the editor. *J. Am. Vet. Med Assoc.* **199,** 968.

Hendrick, M. J., Goldschmidt, M. H., Shofer, F. S., Wang, Y., and Somlyo, A. P. (1992). Postvaccinal sarcomas in the cat: Epidemiology and electron probe microanalytical identification of aluminum. *Cancer Res.* **52,** 5391–5394.

HogenEsch, H., Azcona-Oliveira, J., Scott-Moncrieff, C., Snyder, P. W., and Glickman, L. T. (1998). Vaccine-induced autoimmunity in the dog. *Adv. Vet. Sci. Comp. Med* (in press).

Kahrs, R. F. (1978). Techniques for investigating outbreaks of livestock disease. *J. Am. Vet. Med. Assoc.* **173,** 101–103.

Kass, P. H., Barnes, W. G., Spangler, W. L., Chomel, B. B., and Culbertson, M. R. (1993). Epidemiologic evidence for a causal relation between vaccination and fibrosarcoma tumorigenesis in cats. *J. Am. Vet. Med. Assoc.* **203,** 396–405.

Last, J. M. (1988). "A Dictionary of Epidemiology." Oxford University Press, New York.

Lester, S., Clemett, T., and Burt, A. (1996). Vaccine site-associated sarcomas in cats: Clinical experience and a laboratory review (1982–1993). *J. Am. Anim. Hosp. Assoc.* **32,** 91–95.

Martin, M. L. (1985). Canine coronavirus enteritis and a recent outbreak following modified live virus vaccination. *Compend. Contin. Educ. Comp. Anim.* **7,** 1012–1017.

National Research Council (1983). "Risk Assessment in the Federal Government: Managing the Process." National Academy Press, Washington, DC.

Priester, W. A., and McKay, F. W. (1980). "The Occurrence of Tumors in Domestic Animals," Natl. Cancer Inst. Monogr. No. 54. U.S. Dept. of Health and Human Services, Washington, DC.

Sandman, P. M. (1991). Session III: Ethical considerations and responsibilities when communicating health risk information. *J. Clin. Epidemiol.* **44**(Suppl. I), 41S–50S.

Smith, C. (1995). Are we vaccinating too much? *J. Am. Vet. Med. Assoc.* **207,** 421–425.

Tizard, I. (1990). Risks associated with the use of live vaccines. *J. Am. Vet. Med. Assoc.* **196,** 1851–1858.

More Bumps on the Vaccine Road

W. JEAN DODDS

Hemopet, 938 Stanford Street, Santa Monica, California 90403

I. Introduction and Background
II. Overview of Adverse Effects of Vaccines
III. Breed Study Examples
    A. Vaccine-Associated Disease in Old English Sheepdogs
    B. Vaccine-Associated Disease in a Family of Young Akitas
    C. Vaccine-Associated Disease in Young Weimaraners
IV. Periodicity of Booster Vaccination
V. Alternative Strategies to Conventional Vaccination
    A. Monitoring Serum Antibody Titers
    B. Reducing the Number of Vaccine Antigens Used or Given Simultaneously
    C. Avoid Vaccinating or Overvaccinating Certain Populations
    D. Alternative Methodologies
VI. Summary and Future Directions
    References

I. Introduction and Background

The challenge to produce effective and safe vaccines for the currently prevalent infectious diseases of humans and animals has become increasingly difficult (Bloom, 1994; Cohen, 1994; Stratton *et al.*, 1994). In veterinary medicine, evidence implicating vaccines in triggering immune-mediated and other chronic disorders (vaccinosis) is growing (Dodds, 1983, 1993, 1995a, 1997; Phillips and Schultz, 1992; Alderink *et al.*, 1995; Schultz, 1995a,b; Duval and Giger, 1996). Although some of these problems have been traced to contaminated or poorly attenuated batches of vaccine that revert to virulence, others apparently reflect the host's genetic predisposition to react adversely on receiving the monovalent or polyvalent products given routinely to animals (Dodds,

1983, 1993, 1995a,b,c, 1997; Oehen et al., 1991; Lumsden et al., 1993; Wilbur et al., 1994; Gloyd, 1995; Smith, 1995; Wynn and Dodds, 1995).

Determining causality for adverse effects of vaccines can be asked as three questions: Can it? (potential causality); Did it? (retrodictive causality); and Will it? (predictive causality) (Stratton et al., 1994). Other factors to be weighed in considering the implications of causality include prevalence and clinical severity of the naturally occurring infectious disease, implementing more effective strategies to control infectious diseases, vaccine-related issues such as dosage in relation to body mass and age, advantages and disadvantages of modified live (attenuated) and killed (inactivated) vaccines, hormonal state during vaccination (Smith et al., 1990), and periodicity of booster vaccinations in relation to duration of immunity (Dodds, 1997). Alternatives to current vaccine practices include measuring serum antibody titers; avoidance of unnecessary vaccines or overvaccinating; caution in vaccinating sick, very old, debilitated, or febrile individuals, and families known to be at high risk for immunologic reactions; and use of homeopathic nosodes either as preventive or therapeutic adjuncts. (This last option is considered an unconventional treatment that has not been proven scientifically to be efficacious. If veterinarians choose to use homeopathic nosodes, their clients should be provided with an appropriate disclaimer and written informed consent should be obtained.) A multifaceted approach is needed to further recognition of this situation, along with implementing alternative strategies to contain infectious diseases and reduce the environmental impact of conventional vaccines.

II. Overview of Adverse Effects of Vaccines

The onset of adverse effects of vaccination can be expressed as an immediate hypersensitivity or anaphylactic reaction; an acute event occurring 24–72 hours afterwards, or 10–28 days later in a delayed-type immunologic response (Dodds, 1983, 1995a, 1997; Tizard, 1990; Phillips and Schultz, 1992; Duval and Giger, 1996), or even later as seen with mortality from high-titered measles vaccine in infants (Garenne et al., 1991), canine distemper antibodies in joint diseases of dogs (May et al., 1994), and feline injection-site fibrosarcomas (Kahler, 1993; Kass et al., 1993). The increasing antigenic load presented to the host individual by modified live virus (MLV) vaccines during the period of viremia is presumed to be responsible for the immunologic challenge that can result in a delayed hypersensitivity reaction (Tizard, 1990; Phillips and Schultz, 1992).

These adverse vaccine reactions typically include fever, stiffness, sore joints and abdominal tenderness, susceptibility to infections, neurologic disorders and encephalitis, collapse with autoagglutinated red blood cells and icterus (autoimmune hemolytic anemia, AIHA), or generalized petechiae and ecchymotic hemorrhages (immune-mediated thrombocytopenia, ITP) (Dodds, 1983, 1993, 1995b, 1997; Jones, 1984; Phillips and Schultz, 1992; Littledike, 1993; May et al., 1994; Gloyd, 1995; Duval and Giger, 1996). Liver enzymes may be markedly elevated, and liver or kidney failure may occur by itself or accompany bone marrow suppression. Furthermore, MLV vaccination has been associated with the development of transient seizures in puppies and adult dogs of breeds or cross-breeds susceptible to immune-mediated diseases especially those of hematologic or endocrine tissues (e.g., AIHA, ITP, autoimmune thyroiditis) (Dodds, 1983, 1993, 1995b). Postvaccinal polyneuropathy is a recognized entity associated occasionally with the use of distemper, parvovirus, rabies, and presumably other vaccines (Tizard, 1990; Phillips and Schultz, 1992; Dodds, 1993; Collins, 1994; Gloyd, 1995). This can result in various clinical signs including muscular atrophy, inhibition or interruption of neuronal control of tissue and organ function, muscular excitation, incoordination and weakness, as well as seizures (Dodds, 1993, 1997). Adverse reactions to vaccination also have been reported recently with increasing frequency in cats (Rosenthal and Dworkis, 1990; Kahler, 1993; Kass et al., 1993). Accordingly, companion animal breeders should be advised of the potential for genetically susceptible littermates and relatives to be at increased risk for similar adverse vaccine reactions (Dodds, 1983, 1993, 1995a,b; Schultz, 1995b).

Among the most alarming adverse reactions to vaccinations are the tragic mortalities from other infections following high-titered measles vaccinations of human infants (Garenne et al., 1991), development of subacute sclerosing panencephalitis in Canadian infants receiving measles vaccine at less than 12 months of age (Stratton et al., 1994), and experiences with refractory injection-site fibrosarcomas in cats (Kahler, 1993; Kass et al., 1993). Commercial vaccines can also be contaminated with other adventitious viral agents, presumably as a result of inadequate quality control during vaccine production (DVM Vaccine Roundtable, 1989; Tizard, 1990; Wilbur et al., 1994; Ellis et al., 1995a). This has been of particular concern in cattle where contamination of bovine respiratory disease and rotacoronavirus vaccines with bovine viral diarrhea virus has occurred with unacceptable frequency (DVM Vaccine Roundtable, 1988; Ellis et al., 1995a; Cortese et al., 1997). Another serious problem arose from a commercial canine parvovirus vaccine that was contaminated by blue tongue virus, as it produced abortion and death when given to pregnant dogs (Wilbur et *al.*, 1994). The authors linked the causality here to the ill-advised but all too common practice of vaccinating pregnant animals. The potential for side effects such as promotion of chronic disease states in male and nonpregnant female dogs receiving this lot of vaccine remains in question, although there have been anecdotal reports of reduced stamina and renal dysfunction in performance sled dogs (J. L. Olson, unpublished observations, 1995). Recently, a commercial manufacturer of distemper vaccines had to recall all of its biologic products containing a distemper component, because the vaccines were associated with a higher than normally observed rate of central nervous system postvaccinal reactions 1–2 weeks following administration (Gloyd, 1995).

Overvaccination raises other issues; the increased cost in time and dollars spent needs to be considered, despite the well-intentioned solicitation of clients to encourage annual booster vaccinations so that pets also can receive a wellness examination (Smith, 1995). Giving annual boosters when they are not necessary has the client paying for a service that is likely to be of little benefit to the pet's existing level of protection against these infectious diseases. It also increases the risk of adverse reactions from the repeated exposure to foreign substances (Smith, 1995; Alderink *et al.*, 1995). Vaccination leads to false-positive serologic test results in viral or bacterial screening assays [e.g., feline leukemia virus (FeLV), feline coronavirus, canine borreliosis]. The experts agree that certain vaccines such as canine coronavirus, Lyme disease, and the commercially available Leptospira bacterins have little justification for their widespread use (Greene, 1992), while others so rarely cause disease today (e.g., infectious canine hepatitis) that their need is questionable (Alderink *et al.*, 1995). Furthermore, only cats at high risk of exposure really need to be vaccinated for feline infectious peritonitis (FIP) or FeLV (Scott and Geissinger, 1997). A controversial canine and feline ringworm vaccine has been marketed, and a canine rotavirus vaccine is being introduced, although there is no recognized canine rotavirus disease beyond the newborn stage. An important point, raised by Dennis W. Macy in the editorial of Smith (1995), is the fallacy of assuming that recommending annual vaccination will cause a greater percentage of the pet population to be vaccinated. What actually happens is that conscientious clients come in regularly and their pets get overvaccinated with the attendant higher risk of adverse reaction.

Polyvalent MLV vaccines that multiply in the host elicit a stronger antigenic challenge to the animal and should mount a more effective and sustained immune response (Greene, 1990; Tizard, 1990; Phillips and Schultz, 1992; Schultz, 1995a,b; Hoskins, 1997). However, this can overwhelm the immunocompromised or even a healthy host that has ongoing exposure to other environmental stimuli as well as a genetic predisposition that promotes adverse response to viral challenge (Brenner et al., 1988; Garenne et al., 1991, Phillips and Schultz, 1992; Dodds, 1993, 1997; Allen et al., 1996). The recently weaned young puppy or kitten being placed in a new environment may be at particular risk. Furthermore, while the frequency of vaccinations is usually spaced 2–3 weeks apart, some veterinarians have advocated vaccination once a week in stressful situations (McDonald, 1992; Smith, 1995). This practice makes little sense scientifically or medically, as the relatively immature immune systems of young animals may be temporarily or more permanently harmed (Schultz, 1995a,b). One could even envision the consequences of increased susceptibility to chronic debilitating diseases in later life.

Dogs with preexisting inhalant allergies (atopy) to pollens have an augmented immune response to vaccination, as a natural example of the "allergic breakthrough phenomenon" (Frick and Brooks, 1981). The increasing current problems with allergic and immunologic diseases has been linked to the introduction of MLV vaccines more than 20 years ago (Tizard, 1990). While other environmental factors no doubt have a contributing role, the introduction of these vaccine antigens and their environmental shedding (Tizard, 1990) may provide the final insult that exceeds the immunologic tolerance threshold of some individuals in the pet population.

III. Breed Study Examples

In the early 1980s, this author began studying families of dogs with an apparent increased frequency of immune-mediated hematologic disease (AIHA and/or ITP) (Dodds, 1983, 1995b). Among the more commonly recognized predisposed breeds were the Akita, American cocker spaniel, German shepherd, golden retriever, Irish setter, Kerry blue terrier, miniature and standard dachshund, toy, miniature, and standard poodle, old English sheepdog, Scottish terrier, Shetland sheepdog, shih tzu, vizsla, and Weimaraner (Dodds, 1983, 1995b). Since then, other investigators have noted the relatively high frequency of AIHA in American cocker spaniels (Duval and Giger, 1996) and old English sheepdogs (Day and Penhale, 1992). A significant proportion of these animals had been vaccinated with monovalent or polyvalent vaccines within the 30- to 45-day period prior to the onset of their autoimmune disease (Dodds, 1983, 1995a,b; Duval and Giger, 1996).

As an example, this author's recent survey of 13 cases of vaccine-associated AIHA included the following descriptors: six males (two neutered) and six females (four spayed) with one case of unknown sex; age at onset ranged from 1 to 10 years with a mean age of 4.9 years; time postvaccination ranged from 3 to 42 days with a mean of 19.5 days; all received polyvalent vaccines and two also received Lyme vaccine; and one was in estrus at the time, one had monthly heartworm preventive, and five had ITP concomitantly (Evan's syndrome). Findings from the author's much larger accumulated database of three susceptible breeds are summarized next.

A. Vaccine-Associated Disease in Old English Sheepdogs

The old English sheepdog apparently is predisposed to a variety of autoimmune diseases (Dodds, 1983, 1995b; Day and Penhale, 1992). Of these, the most commonly seen are AIHA, ITP, thyroiditis, and Addison's disease (Dodds, 1995b; Happ, 1995). Between 1980 and 1990, this author studied 162 cases of immune-mediated hematologic diseases in this breed. One hundred twenty-nine of these cases had AIHA and/or ITP as a feature of their disease. Recent vaccination was the only identified triggering event in seven cases, and was an apparent contributing factor in another 115 cases (Dodds, 1995b). Thyroid disease was recognized as either a primary or secondary problem in 71 cases, which is likely an underestimate of the true incidence, because thyroid function tests were not run or were inconclusive in most of the other cases.

The disease experience with a particular old English sheepdog family illustrates the relationship between autoimmune thyroiditis and the concomitant predisposition to AIHA and/or ITP (Tomer and Davies, 1993; Dodds, 1995b; Happ, 1995). Four of five littermates had severe adverse vaccine reactions between 7.5 and 12 months of age. Three of the four had elevated thyroglobulin autoantibodies, and two had thyroid biopsies, which confirmed lymphocytic thyroiditis. Von Willebrand factor antigen levels were also low (< 50%) or borderline normal (50–69%) in this litter. Other immediate family members were also affected. The sire and two litterbrothers of the dam had thyroid disease, and the dam had low von Willebrand factor antigen (31%), abnormal thyroid function tests, and elevated circulating T3 autoantibody and thyroglobulin autoantibody. The maternal grandsire also had elevated thyroglobulin autoantibodies; and the maternal great granddam produced a daughter with thyroid disease that progressed to thyroid adenocarcinoma at age 10 years. This female's paternal grandsire was the foundation sire of many dogs affected with AIHA and/or ITP and his littersister had died of AIHA. These dogs represented a closely related subset of the larger study summarized by Dodds (1995b).

Pedigrees were available from 108 of the 162 old English sheepdog cases of autoimmune disease; a close relationship was found among all but seven of the affected dogs (Dodds, 1995b). Two of three pedigrees available from the studies of Day and Penhale (1992) were also related to this large North American study group.

B. Vaccine-Associated Disease in a Family of Young Akitas

Akitas are subject to a variety of immune-mediated disorders including Vogt-Koyanagi-Harada syndrome (VKH), pemphigus, and juvenile-onset immune-mediated polyarthritis (IMPA) syndrome (Dougherty and Center, 1991; Wynn and Dodds, 1995). Juvenile-onset IMPA occurs in Akitas less than 8 months of age. This author initially studied eight affected Akitas puppies, in collaboration with Susan Wynn (Wynn and Dodds, 1995), and five of them were closely related. Affected dogs exhibit signs of profound joint pain and cyclic febrile illness lasting 24–48 hours. The mean age of onset was 14 weeks, with all dogs showing signs by 16 weeks of age. Three were male, and five were female. The dogs consistently exhibited cyclic febrile illness with signs of severe pain, usually related to the joints. Most of the dogs had elevated hepatic enzymes, creatine kinase, and blood urea nitrogen. Three of the dogs tested had low thyroid hormone levels (T4, free T4, T3). Screening for rickettsial diseases was negative. One dog was ANA positive at 1:40. Hemograms revealed mild nonregenerative anemia, neutrophilic leukocytosis, and occasional thrombocytopenia. Joint aspiration and radiography of three dogs indicated nonseptic, nonerosive arthritis. Juvenile IMPA in Akitas is a syndrome distinct from the nonerosive, noninfectious, non-neoplastic polyarthritis seen in other breeds. Affected Akitas show signs of the disease at a much earlier age, and the syndrome is heritable (Dougherty and Center, 1991). The mechanism of disease development has not been elucidated, but it shares several features with the inherited renal amyloidosis and recurrent fever of unknown origin syndrome of Chinese shar pei dogs (May *et al.*, 1992; Rivas *et al.*, 1993; Zeiss, 1994). This combination of symptoms is reminiscent of familial Mediterranean fever of humans, which has an autosomal recessive inheritance (Rivas *et al.*, 1993).

Pedigree analysis revealed that all eight dogs were linebred on one popular sire, now deceased, and that there were three sets of littermates involved (Wynn and Dodds, 1995). Treatment was unsuccessful over the long term, because all dogs had relapsing signs despite symptomatic therapy for immune-mediated disease and pyrexia. All dogs died or were euthanized by 2 years of age following progressive systemic disease and renal failure. Necropsies were performed on three dogs, two of which had glomerular amyloidosis and multisystemic inflammatory lesions. In all dogs with known vaccination histories (seven of eight), the initial signs appeared 3–29 days following polyvalent MLV and/or killed virus vaccination with a mean reaction time of 14 days. The history, signs, and close association with immunization suggest that juvenile-onset polyarthritis and subsequent amyloidosis in Akitas may be an autoimmune response triggered by the viral antigens or other components of vaccines (Wynn and Dodds, 1995).

A ninth, related dog became affected 4 months after receiving two killed CPV vaccines. Previously the dog had received only homeopathic nosodes. This dog, a male, had a very high parvovirus HA titer (1:6250), and succumbed at 2 years of age to systemic amyloidosis that affected multiple tissues. A tenth, related male Akita became acutely febrile, and appeared paralyzed and in severe pain after receiving a killed CPV vaccine. As with the sixth, eighth, and ninth cases, only homeopathic nosodes had been given previously by the breeder, who kept meticulous records. Recurring episodes of fever continued in a cyclic fashion. The tenth dog died at 11 months after deteriorating rapidly. Necropsy showed suppurative, eosinophilic enteritis. An eleventh related male Akita began showing clinical signs of high fever and joint pain as a 4-month-old puppy. The dog was euthanized in a moribund state at 2.5 years of age, and necropsy determined the cause to be systemic amyloidosis.

The vaccine-related history of 129 puppies produced by this Akita breeder has been collected. Polyvalent MLV vaccine was given to 104 of them with 10 puppies showing adverse reactions and death (9.8%). Another 6 pups received a polyvalent all-killed vaccine product (no longer commercially available) with no reactors, and 19 pups received homeopathic nosodes initially followed by killed CPV vaccine with one reactor that died (5.6%) and one that became ill but survived.

A genetic basis for immune-mediated diseases is well recognized (Dodds, 1983, 1995b; Carson, 1992; Happ, 1995). A group of inherited immunodeficiencies characteristic of certain breeds already has been described (Felsburg and Jezyk, 1982; Felsburg, 1985; Dodds, 1992). Breed-specific disorders with suspected autoimmune etiologies are being reported with increasing frequency (Dodds, 1983, 1995b; Meric *et al.*, 1986; Scott-Moncrieff *et al.*, 1992). The mechanism for induction of immune-mediated disease in these dogs is poorly understood, but predisposing factors have been implicated. Immune-mediated disease may develop in genetically susceptible individuals when triggered by environmental agents that induce nonspecific inflammation and/or molecular mimicry (Dodds, 1983, 1992, 1995b; Barnett and Fujinami,

1992). The combination of these genetic and environmental factors overrides normal self-tolerance, and is most often mediated by T-cell imbalance or dysregulation (Sinha et al., 1990).

Since Akitas are mostly inbred from a relatively small gene pool, genetic derangement of immunologic function is not unexpected. For owners of existing breeding stock, understanding the possible environmental triggers of juvenile-onset IMPA has immediate importance. Numerous agents have been implicated, including drugs, vaccines, viruses, bacteria, chemicals, and other toxins (Dodds, 1983, 1993, 1995a,c, 1997; Barnett and Fujinami, 1992; Cohen and Shoenfeld, 1996; Duval and Giger, 1996). Although littermates from affected families are usually placed in different environments, all of them undergo relatively standardized immunization procedures at a similar age. The fact that signs of the disease appeared initially during a period of concentrated vaccine exposure could provide the key triggering event, as discussed in Section II.

C. Vaccine-Associated Disease in Young Weimaraners

The Weimaraner appears to be especially prone to both immune deficiency and autoimmune disease, which have been recognized with increasing frequency in the breed during the past decade (Couto, 1988; Dodds, 1995c). Dogs of susceptible genotype are known to have transmitted these problems to some of their offspring. Autoimmune thyroiditis leading to clinically expressed hypothyroidism is probably the most common of these disorders, although an immune deficiency syndrome with low levels of circulating immune globulins (especially IgA and IgM deficiency) is being recognized more often, as is the vaccine-associated disease of young Weimaraners described previously by Couto (1988) and Dodds (1995c).

During the period between 1986 and 1988, Couto (1988) evaluated 170 Weimaraners suspected of having immune deficiency or related to suspected or confirmed cases. Fifty of these dogs were ill at the time or had been chronically ill before evaluation. The clinical signs of the affected dogs included high fevers, polyarthritis with pain, and swelling typical of hypertrophic osteodystrophy (HOD), coughing and respiratory distress from pneumonia, enlarged lymph nodes, diarrhea, pyoderma, and ulcers of the mouth. In most of these cases, clinical signs were first detected shortly after vaccination with a second dose of polyvalent MLV vaccine. Most affected puppies therefore were between 2 and 5 months of age. Laboratory assessment of these puppies showed leukocytosis, low plasma protein, neutropenia, and low levels of IgG and IgM.

A subset of dogs also had low IgA levels, but whether the plasma protein and immunoglobulin levels were below expectations for puppies of this age is unclear. A familial or genetic component was postulated because of the clustering of cases in particular kennels or litters.

In this author's series of Weimaraners with vaccine-associated disease, 24 cases were evaluated (Dodds, 1995c). The mean age of onset was 13.5 weeks with a mean reaction time of 10.5 days postvaccination. The disease syndrome predominantly affected males, although the sex was not reported in 6 of the 24 cases. All affected pups showed high spiking fevers, cyclic episodes of pain, and polyarthritis (HOD)— a group of signs identical to those of the affected young Akitas earlier. Most affected puppies also showed leukocytosis (with neutrophilia or neutropenia), diarrhea, lethargy, anorexia, and enlarged lymph nodes. Some pups also had levels of IgA and/or IgM below those expected for their age, and 1 pup had IgG deficiency as well. Other signs included coughing, pneumonia, depression, seizures or spaced out behavior, refusal to stand or move, and hyperesthesia ("walking on eggshells"). The outcome for half of these cases was good (12 of the 24 are healthy adults), although 2 died, 3 were euthanized as puppies, and 3 remained chronically ill as adults. Another 4 cases were lost to follow-up.

Management of this clinical syndrome in the author's case cohort involved use of parenteral corticosteroids followed by systemic antibiotics and a tapering course of corticosteroids over 4–6 weeks. Recurring episodes were treated by increasing the steroid dosage for a few days until the flareup had subsided. The response to initial corticosteroid treatment was always dramatic. Fever and joint pain subsided within a matter of hours. This experience is contrary to that described by Couto (1988), where corticosteroid use was reserved for refractory cases or used only with extreme caution. He also recommended vitamin C supplementation (500–1000 mg daily) and levamisole given twice weekly.

Instead of revaccination, CDV and CPV serologic titers were measured in the affected surviving puppies (19 of 24). Because all had adequate antibody titers, booster vaccinations were not given. On reaching adulthood, serum antibody titers were reevaluated and detectable CDV- and CPV-specific IgG persisted. Several of these dogs have developed hypothyroidism in the interim and are receiving thyroid replacement (Dodds, 1995c).

IV. Periodicity of Booster Vaccination

The landmark review commentary by Smith (1995) focused the attention of the veterinary research, diagnostic, and clinical communities on the advisability of current vaccine practices—that is, are we overvaccinating companion animals, and if so, what is the appropriate periodicity of booster vaccines? The answers to this provocative topic generally concur in the affirmative to the first question, but lead to another question concerning the duration of immunity conferred by the currently licensed vaccine components (Alderink et al., 1995). Examples of the newly recommended protocols for cats and dogs include giving the kitten and puppy vaccine series followed by a booster at 1 year of age; further boosters to be given every 3 years until geriatric age, at which time booster vaccination may be unadvisable, especially for animals with aging or other diseases and except where vaccination is required by law. In the intervening years between adult booster vaccinations, and in the case of geriatric pets, humoral immunity can be evaluated by vaccine antibody serology as an indication of the presence of "adequate immune memory." This latter terminology is generally preferred over the term "protective immunity" because serum antibody titers may not correlate directly with protection against disease (Olson et al., 1988; Sprent and Tough, 1994; Alderink et al., 1995; Schultz, 1995a,b; Smith, 1995).

Relatively little published information exists about the duration of immunity following vaccination (DVM Vaccine Roundtable, 1988; Olson et al., 1988; Phillips et al., 1989; Tizard, 1990; McDonald, 1992; Phillips and Schultz, 1992; Dodds, 1993, 1997; Alderink et al., 1995; Ellis et al., 1995b; Schultz, 1995a,b; Smith, 1995). Most veterinarians recommend that annual booster vaccinations be given after completion of the initial vaccine series and continue them throughout old age. An increasing number of experts, however, advocate lengthening the interval between boosters, especially for geriatric animals (Frick and Brooks, 1981; Tizard, 1990; Alderink et al., 1995; Schultz, 1995a,b), while other publications reason that the waning immune function of older animals should be boosted by giving vaccinations more frequently. It seems obvious that the latter suggestion is unwise and unnecessary, especially in light of the long-term immunologic memory elicited by earlier vaccination or exposure (Etlinger et al., 1990; Sprent and Tough, 1994; Alderink et al., 1995).

An in-depth study from Sweden (Olson et al., 1988) examined the duration of serum antibody response to CPV, canine adenovirus 1, and CDV immunizations. Only killed CPV vaccine was used, whereas the CDV vaccine was MLV and the adenovirus 1 was either killed or MLV in origin. Several interesting conclusions arose from this work, which examined several hundred dogs. For adult dogs vaccinated with killed CPV vaccine, there was no significant difference in antibody titer between vaccinated and nonvaccinated animals. While protective levels of immunity induced by the killed vaccine were of relatively short duration, two vaccinations with optimal spacing (21–35 days apart) adequately protected against parvovirus disease (Olson et al., 1988). As expected, the MLV CDV and adenovirus 1 vaccines induced more long-lasting protective immunity. Equating the effectiveness of vaccination with humoral antibody concentration alone is fraught with problems, however, because cell-mediated immunity can fully protect against disease in the absence of circulating antibody titers (Schultz, 1995a,b). Regardless of the type of vaccine used, persistence of maternal immunity which interferes with active immunization remains the primary cause of vaccine failures (DVM Vaccine Roundtable, 1988; Greene, 1990, Garenne et al., 1991; McDonald, 1992; Schultz, 1995a,b; Smith, 1995; Hoskins, 1997). Protection afforded by most MLV vaccines and by the MLV vaccines used in the Swedish study lasted at least 3 years (Olson et al., 1988). In humans, once the series of childhood vaccinations is completed, protection against these diseases is generally assumed to be long-lived (Stratton et al., 1994). Furthermore, Etlinger and colleagues (1990) emphasized that long after an individual is vaccinated, immunologic memory will be recalled on renewed exposure to the constituents of the vaccine. Thus, prior immunization can be successfully exploited to elicit memory responses as well as to assist in immunizing individuals against new vaccines (Etlinger et al., 1990; Sprent and Tough, 1994).

In that regard, Olson et al. (1988, 1997) stated the protective serum neutralization (SN) titer for canine distemper virus (CDV) to be ≥1:16, and the protective hemagglutination inhibition (HA) titer for canine parvovirus (CPV) to be ≥1:80, basically in agreement with earlier published studies from Carmichael at Cornell (1997). Hoskins (1997) agrees with the Cornell group for CPV (HA ≥1:80), and further stated that a four-fold increase in titer from before or at vaccination as compared to 3 weeks later affords protection. McMillen et al. (1995) studied humoral and cellular immunity in racing greyhounds given a minimal or intensive vaccination protocol and found little difference in the outcome with respect to titers or immune protection. Both protocols afforded good protective immunity. Their titers for successful immunization were the same as those of the Cornell group for CDV and CPV. Carmichael (1997) stated the ideal protective titer for CDV SN to be >1:100 and for CPV HA to be >1:320. However, he also stated that there is no point or need to booster titers unless HA levels fall below 1:10 or 1:20. Schultz (1995a,b, unpublished observations, 1997) considered a CDV SN titer of 1:40 and a CPV HA titer of 1:160 to be "protective." Finally, for cats, the recent paper by Scott and Geissinger (1997) indicated the following protective titers for three common feline viral diseases: feline panleukopenia virus (FPV) ≥1:8, feline herpesvirus (FHV) ≥1:2, although any titer is adequate; and feline calicivirus (FCV) ≥1:4.

V. Alternative Strategies to Conventional Vaccination

This review, which includes examples of the adverse reactions associated with conventional vaccination, illustrates the rationale and justification for seeking alternative approaches to protection against the common infectious diseases of animals. Several such approaches are discussed next.

A. Monitoring Serum Antibody Titers

Except where vaccination is required by law, animals that previously experienced an adverse reaction to vaccination or are at genetic or physiologic risk for such reactions can have serum antibody titers measured annually instead of revaccination. This approach recently has been recommended to assess the adequacy of protection during the interval between routine adult booster vaccinations, in coordination with the policy change of giving them every 3 years (Alderink et al., 1995; Dodds, 1995a, 1997; Schultz, 1995a,b; Scott and Geissinger, 1997). Examples of the currently available methods are discussed in Section IV. If adequate titers are found, the animal should not need revaccination until some future date. Rechecking of antibody titers can be performed annually thereafter, and can be offered as an alternative to pet owners who object to conventional vaccination.

B. Reducing the Number of Vaccine Antigens Used or Given Simultaneously

An argument can be made for vaccinating well-nourished, healthy pet animals only against the clinically important infectious diseases of their species. For the dog, this would include CDV, CPV, and rabies virus; and for the cat, it would include FPV and rabies virus (Alderink et al., 1995; Schultz, 1995a,b; Scott and Geissinger, 1997). Why, then, are we giving animals so many other antigens in polyvalent vaccines, and is this approach really necessary or safe? For example, with respect to Leptospira bacterins, the clinically important serovars are not contained in the currently licensed products, and the antibodies they elicit only last a few months. Similarly, there have been very few clinical cases of infectious canine hepatitis from adenovirus 1 infection, although the standard polyvalent vaccines all contain adenovirus 2 to afford cross-protection. Other vaccine components such as that for Lyme disease need not be used universally, because the disease is limited to certain geographic areas. Use of FeLV vaccines could be reserved for cats that live mostly outdoors or live both indoors and outdoors, and for catteries where new animals are introduced on a regular basis, as their efficacy is only modest and they have been implicated along with rabies vaccine in producing injection-site fibrosarcomas (Kahler, 1993; Kass *et al.*, 1993). Perhaps one way to address these issues would be to offer more individual or dual vaccine components that could be given on alternating years, in between the 3-year booster vaccinations for the clinically important diseases. The overall risk–benefit ratio of using multiple antigen vaccines given simultaneously and repeatedly should be reexamined, although we have the luxury of asking such questions today only because the risk of disease has been effectively reduced by the widespread use of vaccination programs (Alderink *et al.*, 1995; Schultz, 1995a,b; Dodds, 1997).

C. Avoid Vaccinating or Overvaccinating Certain Populations

Common sense dictates that sick, very old, or debilitated animals should not be vaccinated. It also would be unwise to vaccinate immunocompromised and febrile animals until their physiologic state returns to normalcy. Animals of certain susceptible breeds or families such as old English sheepdogs, Akitas, and Weimaraners, and including those with coat color dilutions (e.g., double-dilute Shetland sheepdogs, harlequin Great Danes, albinos) appear to be at increased risk for severe and lingering adverse reaction to vaccines (Dodds, 1995a,b,c, 1997; Wynn and Dodds, 1995) (see Section III).

Another situation where needless overvaccination occurs is a consequence of the varying state regulatory policies for rabies vaccination. Because the federal U.S. Department of Agriculture has licensed rabies vaccines for 3 years now, there is no legitimate reason for some individual states to insist on annual revaccination. This is particularly worrisome because rabies vaccine is associated with significant adverse neurologic and other immune reactions, as well as producing injection-site fibrosarcomas in cats (Kahler, 1993).

D. Alternative Methodologies

In situations where an animal has experienced a severe adverse reaction to vaccination or when the owner refuses conventional vaccination, there is appropriate justification for selecting alternative methodologies, such as homeopathic nosodes (Pitcairn, 1993; Dodds, 1995a, 1997), to protect against the common infectious diseases of animals. A word of caution is in order here, however, because a recently conducted, preliminary trial with a parvovirus nosode failed to protect puppies against challenge with a street virus strain of CPV (S. G. Wynn and R. D. Schultz, personal communication, 1997). Additional trials are obviously needed to assess the efficacy of the various nosode preparations currently in use.

These alternative techniques must be performed under the supervision of a licensed veterinarian with an established doctor/client/patient relationship, and requested by the owner of the pet after receiving appropriate informed consent. Obtaining a signed disclaimer and release form from the client is also advisable. Finally, minimizing the risk for exposure to infectious diseases should always be kept in mind, by avoiding areas where animals of unknown health status congregate or exercise.

VI. Summary and Future Directions

Veterinary clinicians are increasingly faced with patients exhibiting signs of immunologic dysfunction and disease. In a troublesome number of cases, the onset follows a recent vaccination, therapeutic or preventative drug use, infection, toxic exposure, hormonal change/imbalance, or stress event. The evidence implicating vaccines as triggering agents in genetically susceptible individuals is growing. A multifaceted approach to furthering the recognition of this situation, along with alternative strategies for containing infectious diseases and reducing the environmental impact of conventional vaccines is clearly needed. As a beginning we can increase the periodicity between adult booster vaccinations from 1 to 3 or 5 years, except as required by law, and implement monitoring of serum antibody levels for assessing protection against the clinically important infectious agents.

REFERENCES

Alderink, F. J., Zydeck, F. A., Jacobs, F. S., and Schultz, R. D. (1995). Letters to the Editor. *J. Am. Vet. Med. Assoc.* **207,** 1016–1018.

Allen, L. J., Kabbur, M. B., Cullor, J. S., Gardner, I. A., Stott, J. L., and George, L. W. (1996). Alterations in blood lymphocyte subpopulations and hematologic values in neonatal calves after administration of a combination of multiple-antigen vaccines. *J. Am. Vet. Med. Assoc.* **209,** 683–642.

Barnett, L. A., and Fujinami, R. S. (1992). Molecular mimicry: A mechanism for autoimmune injury. *FASEB J.* **6,** 840–844.

Bloom, B. (1994). The United States needs a national vaccine authority. *Science* **265,** 1378–1380.

Brenner, J., Trainin, Z., Orgad, U., Perl, S., Meirom, R., and Yacobson, B. (1988). A thymic depletion syndrome associated with a combined attenuated distemper parvovirus vaccine in dogs. *Isr. J. Vet. Med.* **44,** 151.

Carmichael, L. E. (1997). Canine viral vaccines at a turning point—a personal perspective. *Proc. Int. Vet. Vaccines Diagn. Conf., 1st,* July 27–31, 1997, p. 26.

Carson, D. (1992). Genetic factors in the etiology and pathogenesis of autoimmunity. *FASEB J.* **6,** 2800–2805.

Cohen, A. D., and Shoenfeld, Y. (1996). Vaccine-induced autoimmunity. *J. Autoimmun.* **9,** 699–703.

Cohen, J. (1994). Bumps on the vaccine road. *Science* **265,** 1371–1372.

Collins, J. R. (1994). Seizures and other neurologic manifestations of allergy. *Vet. Clin. North Am.: Small Anim. Pract.* **24,** 735–748.

Cortese, V. S., Ellis, J., Whittaker, R., Grooms, D. L., and Brock, K. V. (1997). BVD virus transmission following attenuated vaccines to BVDV seronegative cattle. *Large Anim. Pract.* **18**(5), 18–24.

Couto, C. G. (1988). Immunodeficiency in young Weimaraners. *Weimaraner Mag.,* September, pp. 9–10.

Day, M. J., and Penhale, W. J. (1992). Immune-mediated disease in the old English sheepdog. *Res. Vet. Sci.* **53,** 87–92.

Dodds, W. J. (1983). Immune-mediated diseases of the blood. *Adv. Vet. Sci. Comp. Med.* **27,** 163–196.

Dodds, W. J. (1992). Genetically based immune disorders: Autoimmune diseases (Parts 1–3) and immune deficiency diseases (Part 4). *Vet. Pract. Staff* **4** (1, 2, 3 and 5), 8–10; 1, 26–31,35–37; 19–21.

Dodds, W. J. (1993). Vaccine safety and efficacy revisited: Autoimmune and allergic diseases on the rise. *Vet. Forum* **10,** 68–71.

Dodds, W. J. (1995a). More bumps on the vaccine road. *Proc. Am. Holistic Vet. Med. Assoc.,* pp. 74–80.

Dodds, W. J. (1995b). Estimating disease prevalence with health surveys and genetic screening. *Adv. Vet. Sci. Comp. Med.* **39,** 29–96.

Dodds, W. J. (1995c). Vaccine-associated disease in young Weimaraners. *Proc. Am. Holistic Vet. Med. Assoc.,* pp. 85–86.

Dodds, W. J. (1997). Vaccine-related issues. *In* "Complementary and Alternative Veterinary Medicine" (A. M. Schoen and S. G. Wynn, eds.). pp. 701–712. Mosby, St. Louis, MO.

Doughtery, S. A., and Center, S. A. (1991). Juvenile onset polyarthritis in Akitas. *J. Am. Vet. Med. Assoc.* **198,** 849–855.

Duval, D., and Giger, U. (1996). Vaccine-associated immune-mediated hemolytic anemia in the dog. *J. Vet. Intern. Med.* **10,** 290–295.

DVM Vaccine Roundtable (1988). Safety, efficacy heart of vaccine use: Experts discuss pros, cons. *DVM Newsmag.* **19**(12), 16–21.

DVM Vaccine Roundtable (1989). Measles vaccination: A different perspective. *DVM Newsmag* **20**(1), 33, 51.

Ellis, J. A., Hassard, L. E., and Morley, P. S. (1995a). Bovine respiratory syncytial virus-specific immune responses in calves after inoculation with commercially available vaccines. *J. Am. Vet. Med. Assoc.* **206,** 354–361.

Ellis, J. A., Bogdan, J. R., Kanara, E. W., Morley, P. A., and Haines, D. M. (1995b). Cellular and antibody responses to equine herpesvirus 1 and 4 following vaccination

of horses with modified-live and inactivated viruses. *J. Am. Vet. Med. Assoc.* **206,** 823–832.
Etlinger, H. M., Gillessen, D., Lahm, H. W., Matile, H., Schonfeld, H. J., and Trzeciak, A. (1990). Use or prior vaccinations for the development of new vaccines. *Science* **249,** 423–425.
Felsburg, P. J. (1985). Selective IgA deficiency in the dog. *Clin. Immunol. Immunopathol.* **36,** 297.
Felsburg, P. J., and Jezyk, P. F. (1982). A canine model for combined immunodeficiency. *Clin. Res.* **30,** 347.
Frick, O. L., and Brooks, D. L. (1981). Immunoglobulin E antibodies to pollens augmented in dogs by virus vaccines. *Am. J. Vet. Res.* **44,** 440–445.
Garenne, M., Leroy, O., Bean, J.-P., and Sene, I. (1991). Child mortality after high-titre measles vaccines: Prospective study in Senegal. *Lancet* **338,** 903.
Gloyd, J. (1995). Distemper vaccines recalled. *J. Am. Vet. Med. Assoc.* **207,** 1397.
Greene, C. E. (1990). Immunoprophylaxis and immunotherapy. *In* "Infectious Diseases of the Dog and Cat" (C. E. Greene, ed.), pp. 43–48. Saunders, Philadelphia.
Greene, R. T. (1992). Questions "push" for vaccination against *Borrelia burgdorferi* infection. *J. Am. Vet. Med. Assoc.* **201,** 1491–1493.
Happ, G. M. (1995). Thyroiditis—a model canine autoimmune disease. *Adv. Vet. Sci. Comp. Med.* **39,** 97–139.
Hoskins, J. D. (1997). Performance of a new generation canine parvovirus vaccine in Rottweiler puppies. *Can. Pract.* **22**(4), 29–31.
Jones, B. E. V. (1984). Platelet aggregation in dogs after live-virus vaccination. *Acta Vet. Scand.* **25,** 504–509.
Kahler, S. (1993). Collective effort needed to unlock factors related to feline injection-site sarcomas. *J. Am. Vet. Med. Assoc.* **202,** 1551–1554.
Kass, P. H., Barnes, W. G., Jr., Spangler, W. L., Chomel, B. B., and Culbertson, M. R. (1993). Epidemiologic evidence for a causal relation between vaccination and fibrosarcoma tumorigenesis in cats. *J. Am. Vet. Med. Assoc.* **203,** 396–405.
Littledike, E. T. (1993). Variation of abscess formation in cattle after vaccination with a modified-live *Pasteurella haemolytica* vaccine. *Am. J. Vet. Res.* **54,** 1244–1248.
Lumsden, J. S., Kennedy, B. W., Mallard, B. A., and Wilkie, B. N. (1993). The influence of the swine major histocompatibility genes on antibody and cell-mediated immune responses to immunization with an aromatic-dependent mutant of *Salmonella typhimurium*. *Can. J. Vet. Res.* **57,** 14–18.
May, C., Hammill, J., and Bennett, D. (1992). Chinese shar pei fever syndrome: A preliminary report. *Vet. Rec.* **131,** 586–587.
May, C., Carter, S. D., Bell, S. C., and Bennett, D. (1994). Immune responses to canine distemper virus in joint disease of dogs. *Br. J. Rheumatol.* **33,** 27–31.
McDonald, L. J. (1992). Factors that can undermine the success of routine vaccination protocols. *Vet. Med.* **87,** 223–230.
McMillen, G. L., Briggs, D. J., McVey, D. S., Phillips, R. M., and Jordan, F. R. (1995). Vaccination of racing greyhounds: effects on humoral and cellular immunity. *Vet. Immunol. Immunopathol.* **49,** 101–113.
Meric, S. M., Child, G., and Higgins, R. J. (1986). Necrotizing vasculitis of the spinal pachyleptomeningeal arteries in three Bernese mountain dog littermates. *J. Am. Anim. Hosp. Assoc.* **22,** 456–459.
Oehen, S., Hengartner, H., and Zinkernagel, R. M. (1991). Vaccination for disease. *Science* **251,** 195–198.
Olson, P., Klingeborn, B., and Hedhammar, A. (1988). Serum antibody response to canine

parvovirus, canine adenovirus-1, and canine distemper virus in dogs with known states of immunization: Study of dogs in Sweden. *Am. J. Vet. Res.* **49**, 1460–1466.

Olson, P., Klingeborn, B., Bonnet, B., and Hedhammar, A. (1997). Distemper titer study in Sweden. *J. Vet. Intern. Med.* **11**, 148.

Phillips, T. R., and Schultz, R. D. (1992). Canine and feline vaccines. *Curr. Vet. Ther.* **11**, 202–206.

Phillips, T. R., Jensen, J. L., Rubino, M. J., Yang, W. C., and Schultz, R. D. (1989). Effects of vaccines on the canine immune system. *Can. J. Vet. Res.* **53**, 154–160.

Pitcairn, R. H. (1993). Homeopathic alternatives to vaccines. *Proc. Am. Holistic Vet. Med. Assoc.*, pp. 39–49.

Rivas, A. L., Tintle, L., Meyers-Wallen, V., Scarlett, J. M., and van Tassell, C. P. (1993). Inheritance of renal amyloidosis in Chinese shar-pei dogs. *J. Hered.* **84**, 438–442.

Rosenthal, R. C., and Dworkis, A. S. (1990). Incidence of and some factors affecting adverse reactions to subcutaneously administered Leukocell. *J. Am. Anim. Hosp. Assoc.* **26**, 283–287.

Schultz, R. D. (1995a). Theory and practice of immunization. *In* "Small Animal Immunology: New Faces of Immune-Mediated Diseases and Current Concepts in Vaccine Immunology," Proc. San Diego Spring Vet. Conf., May 6–7, 1995, pp. 82–99.

Schultz, R. D. (1995b). Canine vaccines and immunity: Important considerations in the success of vaccination programs. *In* "Small Animal Immunology: New Faces of Immune-Mediated Diseases and Current Concepts in Vaccine Immunology," Proc. San Diego Spring Vet. Conf., May 6–7, 1995, pp. 100–113.

Scott, F. W., and Geissinger, C. (1997). Duration of immunity in cats vaccinated with an inactivated feline panleukopenia, herpesvirus, and calicivirus vaccine. *Feline Pract.* **25**(6), 12–19.

Scott-Moncrieff, J. C. R., Snyder, P. W., Glickman, L. T., Davis, E. L., and Felsburg, P. J. (1992). Systemic necrotizing vasculitis in nine young beagles. *J. Am. Vet. Med. Assoc.* **201**, 1553–1558.

Sinha, A. A., Lopez, M. T., and McDevitt, H. O. (1990). Autoimmune diseases: The failure of self-tolerance. *Science* **248**, 1380–1388.

Smith, C. A., ed. (1995). Are we vaccinating too much? *J. Am. Vet. Med. Assoc.* **207**, 421–425.

Smith, P. C., Nusbaum, K. E., Kwapien, R. P., Stringfellow, D. A., and Driggers, K. (1990). Necrotic oophoritis in heifers vaccinated intravenously with infectious bovine rhinotracheitis virus vaccine during estrus. *Am. J. Vet. Res.* **51**, 969–972.

Sprent, J., and Tough, D. F. (1994). Lymphocyte life-span and memory. *Science* **265**, 1395–1400.

Stratton, K. R., Howe, C. J., and Johnston, R. B., Jr., eds. (1994). "Adverse Events Associated with Childhood Vaccines: Evidence Bearing on Causality." National Academy Press, Washington, DC.

Tizard, I. (1990). Risks associated with use of live vaccines. *J. Am. Vet. Med. Assoc.* **196**, 1851–1858.

Tomer, Y., and Davies, T. F. (1993). Infection, thyroid disease, and autoimmunity. *Endocr. Rev.* **14**, 107–125.

Wilbur, L. A., Evermann, J. F., Levings, R. L., Stoll, L. R., Starling, D. E., Spillers, C. A., Gustafson, G. A., and McKeirnan, A. J. (1994). Abortion and death in pregnant bitches associated with a canine vaccine contaminated with blue tongue virus. *J. Am. Vet. Med. Assoc.* **204**, 1762–1765.

Wynn, S. G., and Dodds, W. J. (1995). Vaccine-associated disease in a family of young Akita dogs. *Proc. Am. Holistic Vet. Med. Assoc.*, pp. 81–84.

Zeiss, C. J. (1994). Amyloidosis in the shar pei. *Adv. Small Anim. Med. Surg.* **7**(4), 7–8.

Vaccine-Induced Autoimmunity in the Dog

HARM HOGENESCH, JUAN AZCONA-OLIVERA,
CATHARINE SCOTT-MONCRIEFF, PAUL W. SNYDER,
AND LARRY T. GLICKMAN

*Departments of Veterinary Pathobiology and Veterinary Clinical Sciences,
Purdue University, West Lafayette, Indiana 47907*

I. Introduction
II. Materials and Methods
  A. Animals
  B. Vaccination Schedule
  C. Viral Serology
  D. Hematology
  E. Endocrinology
  F. Immunology
  G. Lymphocyte Blastogenesis Assay
  H. Enzyme-Linked Immunosorbent Assay (ELISA)
  I. Necropsy
  J. Statistical Analysis
III. Results
  A. Viral Serology
  B. Clinical Observations, Hematology, and Endocrinology
  C. Immunology
  D. Necropsy
IV. Discussion
  Acknowledgments
  References

I. Introduction

Vaccines are widely used in human and veterinary medicine as an effective and economic method to control viral and bacterial diseases. Although generally considered safe, vaccination is occasionally accompanied by adverse effects. Many adverse effects related to vaccination are acute and transient, for example, fever, swelling at the site of inoculation, and allergic reactions. In contrast, reports of autoimmune disease following vaccination are relatively rare. In most instances, it is difficult, if not impossible, to ascertain that vaccination caused or precipitated the autoimmune disease. In a recent report, the Advisory Committee on Immunization Practices in people concluded that there is a causal relation between diptheria-tetanus-pertussis (DTP) and measles-mumps-rubella (MMR) vaccination and arthritis, but no evidence of a causal relationship between these vaccinations and other autoimmune diseases such as autoimmune hemolytic anemia and Guillain-Barré syndrome (Centers for Disease Control and Prevention, 1996). Cohen and Shoenfeld (1996) also stated that the relation between vaccination and autoimmunity is obscure. They added that there is a need for experimental studies to address this subject (Cohen and Shoenfeld, 1996).

There has been a growing concern among dog owners and veterinarians that the high frequency with which dogs are being vaccinated may lead to autoimmune and other immune-mediated disorders (Dodds, 1988; Smith, 1995). The evidence for this is largely anecdotal and based on case reports. A recent study observed a statistically significant temporal relationship between vaccination and subsequent development of immune-mediated hemolytic anemia (IMHA) in dogs (Duval and Giger, 1996). Although this does not necessarily indicate a causal relationship, it is the strongest evidence to date for vaccine-induced autoimmune disease in the dog.

We are investigating the effect of vaccination on dogs in a series of experimental studies. The goals of these experiments are (1) to determine if vaccination of dogs affects the function of the immune system and, in particular, if vaccination results in autoimmunity; (2) to delineate the mechanisms by which vaccination results in autoimmunity if this occurs; and (3) to develop alternative vaccination strategies that will not be accompanied by adverse effects. The issue that is the focus of this and ongoing studies in our laboratory is somewhat different from that examined by Duval and Giger (1996). In their study, a statistically significant temporal relationship between the onset of IMHA and prior vaccination suggested that vaccination caused IMHA or accelerated preexisting IMHA in *adult* dogs. Although not documented, it is likely that these middle-aged dogs had received multiple vaccines prior to the last vaccination. Why this last vaccination suddenly triggered the onset of IMHA is unknown. In contrast, our studies examine if vaccination of dogs at a *young* age causes alterations in the immune system, including the production of autoantibodies, that could eventually lead to autoimmune disease in susceptible individuals. In this paper, we report on the findings of the first study in which a group of vaccinated dogs and a group of unvaccinated dogs were followed for 14 weeks after the first vaccination.

II. Materials and Methods

A. Animals

Two pregnant Beagle dogs were purchased from a commercial breeder. The animals whelped in the Animal Facility of the Purdue University School of Veterinary Medicine and the pups were weaned at 6 weeks of age. Five pups were assigned to one of two groups, a vaccinated and an unvaccinated group, based on body weight, gender, and litter of origin. The vaccinated and unvaccinated group of dogs were housed in separate rooms.

The dogs were examined daily. Rectal temperature and body weight were recorded twice a week. Blood samples were collected from the jugular vein prior to each vaccination and 2, 5, 7, and 14 days following vaccination for hematology, endocrinology, and viral serology. Blood samples collected on days 5 and 14 following vaccination were also used for lymphocyte phenotyping and lymphocyte proliferation assays, and blood samples collected at 7 days following vaccination were used for the detection of autoantibodies.

B. Vaccination Schedule

The dogs in the vaccinated group were injected subcutaneously with a commercially available multivalent vaccine, Vanguard-5 CV/L (Pfizer, Groton, CT) at 8, 10, 12, 16, and 20 weeks of age according to the instructions of the manufacturer. They were injected subcutaneously with an inactivated rabies vaccine, Imrab-3 (Rhone-Merieux, GA) at 16 weeks of age. The unvaccinated group of dogs received subcutaneous injections of sterile saline at the same time points.

Both groups of dogs were injected subcutaneously with 1 mg of keyhole limpet hemocyanin (KLH, Calbiochem) in RIBI-adjuvant at week 20.

C. Viral Serology

Serum samples collected at 6 weeks of age and 0, 2, 5, 7, and 14 days after each vaccination were assayed for the presence of antibodies to canine distemper virus by serum neutralization test, and for antibodies against canine parvovirus by hemagglutination inhibition test. Serum samples were analyzed for antibodies against rabies virus at 16 and 20 weeks of age by a rapid fluorescent focus inhibition test.

D. Hematology

Blood samples were collected at 0, 2, 5, 7, and 14 days after each vaccination for hematocrit, corrected white blood cell count and differential, and platelet counts.

E. Endocrinology

Plasma and serum samples collected at 0, 2, 5, 7, and 14 days after each vaccination were assayed for cortisol, triiodothyronine (T3), and thyroxine (T4) by radioimmunoassay.

F. Immunology

Lymphocyte phenotyping was used. Whole blood was stained with a panel of mouse monoclonal antibodies, followed by $F(ab')_2$ goat anti-mouse IgG (Jackson Research Laboratories). The monoclonal antibodies used were CA2.1D6 (anti-CD21), CA15.8G7 (anti-TCR$\alpha\beta$), CA20.8H1 (anti-TCR$\gamma\delta$), 12.125 (anti-CD4), and 1.140 (anti-CD8). The characteristics of these monoclonal antibodies have been described (Gebhard and Carter, 1992; Moore et al., 1996). Following red blood cell lysis and fixation in 2% paraformaldehyde, the cells were analyzed by flow cytometry.

G. Lymphocyte Blastogenesis Assay

Heparinized blood samples were diluted 1:10 in RPMI-1640 and distributed in the wells of a 96-well plate. Triplicate samples were incubated for 96 hours in the presence of medium only, 2.5 and 5 µg/ml PHA, 5 and 10 µg/ml Concanavalin A (Con A) and 1 and 10 µg/ml PWM. During the last 24 hours of incubation the wells were pulsed with 0.5 µCi of ^{3}H-thymidine. The cells were harvested with a 96-well cell harvester, and the incorporation of radioactivity was measured in a TopCount scintillation counter (Packard Instrument Co., Meriden, CT).

H. Enzyme-Linked Immunosorbent Assay (ELISA)

The presence of antibodies reactive with homologous and heterologous antigens in serum samples collected at 22 weeks of age was analyzed by an indirect ELISA. High-binding ELISA plates (Costar, Cambridge, MA) were coated with 10 μg/ml of antigen in 0.1 M bicarbonate buffer. The wells were rinsed and incubated for 1 hour with phosphate-buffered saline (PBS)/0.1% Tween. Serum samples were diluted 1:10 in PBS and added to the wells in triplicate. Following incubation, the wells were rinsed and incubated with alkaline phosphatase labeled goat anti-dog IgG (Kirkegaard and Perry, Gaithersburg, MD). Alkaline phosphatase activity was measured after addition of p-NPP substrate at 405 nm in a microplate reader (Molecular Devices, Menlo Park, CA).

Essentially the same procedure was used to measure the presence of antibodies against KLH. Alkaline phosphatase labeled anti-dog IgM and IgG were used as secondary reagents.

I. Necropsy

At 22 weeks of age, the dogs were killed by intravenous injection of barbiturates, and a complete necropsy performed. Tissue samples were collected in 10% buffered formalin and processed for light microscopic examination. The tissues that were examined included the spleen, lymph nodes, tonsils, thymus, Peyer's patches, adrenal glands, thyroid glands, pituitary gland, pancreas, heart, lung, kidney, liver, and brain.

J. Statistical Analysis

Data were analyzed for significant differences between groups by Student's t test or repeated measures ANOVA and a significant change over time using a repeated measures ANOVA.

III. Results

A. Viral Serology

None of the pups had detectable antibodies against canine distemper virus and canine parvovirus at 6 weeks of age and against rabies virus at 16 weeks of age. The unvaccinated dogs remained seronegative for these three viruses during the course of the study. The dogs that were immunized developed titers against CDV (maximum titers ranged from 1:48 to 1:1024), CPV-2 (1:320 to 1:1280), and rabies (1:25 to 1:1000).

B. Clinical Observations, Hematology, and Endocrinology

No differences between the unvaccinated and vaccinated groups were found for rectal temperature, body weight, and hematologic values.

There were no significant differences between unvaccinated and vaccinated dogs for concentrations of cortisol, T3, and T4. However, a significant ($p < 0.02$) change was observed over time for each of these three hormones. The plasma concentration of cortisol decreased from a mean of 41.1 ng/ml at 8 weeks of age to 17.6 ng/ml at 22 weeks of age. The concentration of T4 also decreased, from 31.1 ng/ml at 8 weeks of age to 22.8 ng/ml at 22 weeks of age. The concentration of T3 increased from 0.63 ng/ml at 8 weeks of age to 1.1 ng/ml at 22 weeks of age.

C. Immunology

No differences were observed between the unvaccinated and vaccinated dogs for lymphocyte subpopulations or for the proliferative response to any of the mitogens tested.

The response of both groups of dogs to KLH was similar. There was no statistically significant difference in the KLH-specific IgM and IgG concentrations in the serum (not shown).

At 8 weeks of age, antibodies against homologous and conserved heterologous antigens were negligible in the serum of the dogs. At 22 weeks of age there was a significant increase of IgG antibodies reactive with 10 of 17 antigens in the vaccinated dogs versus no increase in the unvaccinated dogs (Table I). The increase of optical density was modest for 8 of these 10 antigens, but a large increase was observed for fibronectin and laminin. All vaccinated dogs developed high levels of fibronectin-specific IgG antibodies. Similar levels of IgG anti-fibronectin antibodies were observed when bovine fibronectin was substituted by human or mouse fibronectin (not shown). The concentration of anti-fibronectin antibodies began to increase after the second vaccination in three dogs and after the third vaccination in the other two vaccinated dogs, and reached a maximum level after the fourth vaccination (Fig. 1). To determine if the antibodies had a preferential reactivity with a particular part of the fibronectin molecule, we tested the reactivity of serum samples with two fragments of fibronectin. The 30-kDa fragment contains the heparin-binding domain of fibronectin, whereas the 45-kDa fragment contains the collagen-binding domain. As shown in Fig. 2, little reactivity was observed with the 45-kDa fragment, but significant reactivity was observed with the 30-kDa fragment.

High levels of anti-laminin antibodies were observed in the serum of

TABLE I

Reactivity of Serum IgG Antibodies of Vaccinated and Unvaccinated Dogs with Homologous and Heterologous Antigens at 22 Weeks of Age[a]

Antigen[b]	Unvaccinated dogs	Vaccinated dogs	p value
c. albumin	0.119 ± 0.017	0.223 ± 0.046	0.001
c. cytochrome C	0.168 ± 0.023	0.200 ± 0.034	NS[c]
c. hemoglobin	0.171 ± 0.029	0.200 ± 0.019	NS
c. myocardial myoglobin	0.118 ± 0.018	0.157 ± 0.039	NS
c. skeletal muscle myoglobin	0.675 ± 0.303	0.745 ± 0.510	NS
c. thyroglobulin	0.191 ± 0.089	0.190 ± 0.040	NS
c. transferrin	0.172 ± 0.015	0.230 ± 0.042	0.021
b. cardiolipin	0.112 ± 0.014	0.256 ± 0.057	0.001
b. collagen I	0.106 ± 0.012	0.164 ± 0.049	0.032
b. DNA	0.125 ± 0.031	0.223 ± 0.058	0.010
b. fibronectin	0.159 ± 0.046	2.811 ± 0.514	<0.001
b. sphingomyelin	0.105 ± 0.018	0.159 ± 0.046	0.040
p. insulin	0.094 ± 0.010	0.142 ± 0.041	0.037
p. myocardial myosin	0.179 ± 0.016	0.234 ± 0.061	NS
p. skeletal muscle myosin	0.367 ± 0.149	0.522 ± 0.165	NS
m. collagen IV	0.117 ± 0.008	0.178 ± 0.043	0.014
m. laminin	0.100 ± 0.006	0.761 ± 0.642	0.050

[a]Values represent the mean ± SD of the OD_{405} as measured by ELISA.
[b]c, canine; b, bovine; p, porcine; m, murine.
[c]NS, not significant ($p > 0.05$).

three of the five vaccinated dogs at 22 weeks of age. One dog had high levels at 17 weeks of age, whereas the other two dogs did not develop high levels until the end of the study.

High levels of antibodies reactive with skeletal muscle myosin and myoglobin were observed in both groups of dogs at 22 weeks of age. The antibody levels increased at 11 weeks of age in three dogs, at 13 weeks of age in another three dogs, and at 17 weeks of age in the remaining four dogs.

D. Necropsy

Gross and light microscopic examination of the tissues of the dogs revealed no significant lesions. The thyroid gland of one of the vaccinated dogs had a small lymphoid nodule with obliteration of adjacent thyroid follicles.

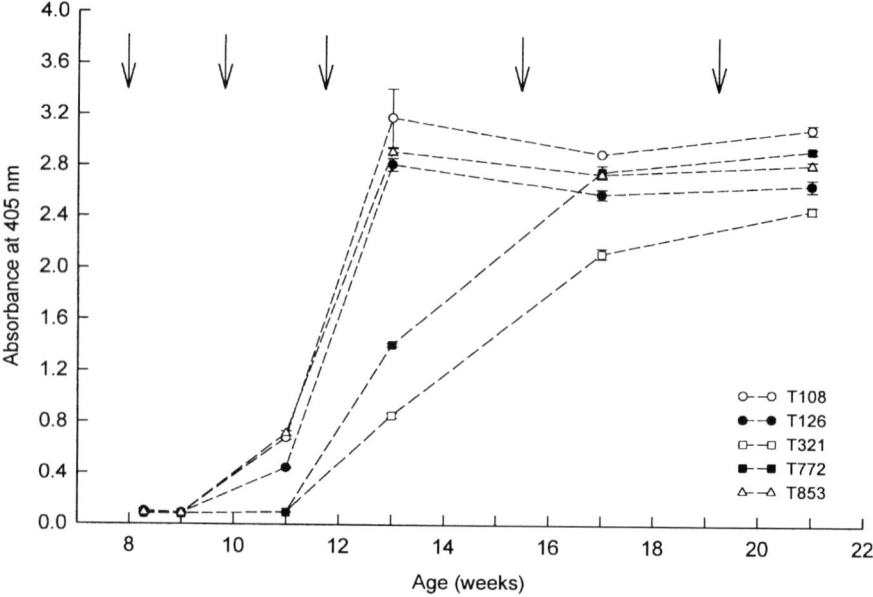

Fig. 1. Fibronectin-reactive IgG antibodies in the serum of vaccinated dogs as measured by ELISA. The vertical arrows indicate the days at which vaccines were administered. The numbers in the legend represent the individual dog numbers.

IV. Discussion

In this study, we exhaustively evaluated the effects of vaccination with a multivalent vaccine and a rabies vaccine on the immune system of young dogs. Vaccination did not cause immunosuppression or alter the response to an unrelated antigen (KLH). In contrast to an earlier study (Mastro et al., 1986), but in agreement with other work (Phillips and Schultz, 1987), we did not observe a transient lymphopenia in the dogs at any time. However, vaccination did induce autoantibodies and antibodies to conserved heterologous antigens. The pathogenic significance of these autoantibodies is presently uncertain. We did not find any evidence of autoimmune disease in the vaccinated dogs, but the study was terminated when the dogs were 22 weeks of age, well before autoimmune diseases usually become clinically apparent. It is likely that genetic and environmental factors will trigger the onset of clinical autoimmune disease in a small percentage of the animals that develop

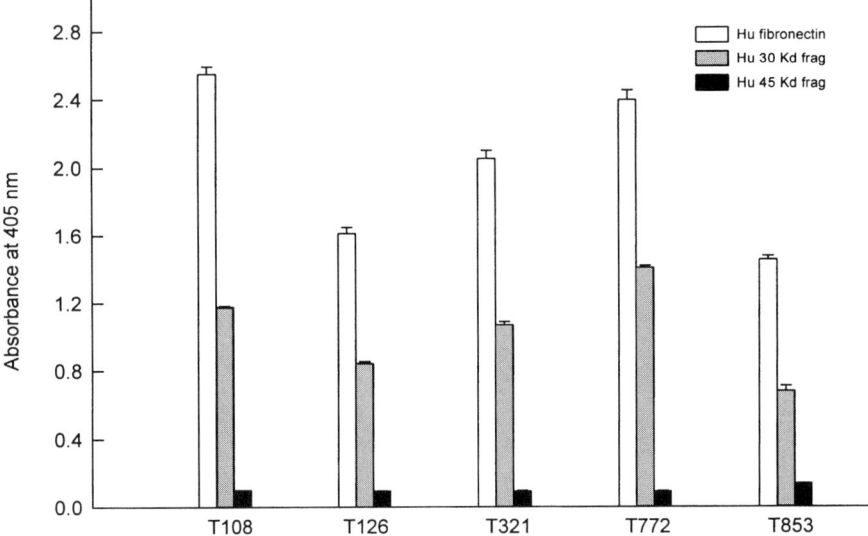

FIG. 2. Specificity of serum IgG antibodies for human fibronectin, and the heparin-binding 30-kDa fragment and the collagen-binding 45-kDa fragment of human fibronectin as measured by ELISA. The numbers below the horizontal axis represent individual dog numbers.

autoantibodies. For practical and economic reasons, only a small number of dogs can be followed in an experimental study, and clinical autoimmune disease may, therefore, never be observed. The principal value of an experimental study is that it enables us to determine the frequency of autoantibody responses and the mechanism(s) that cause vaccines to induce autoantibodies.

We used two vaccines, a multivalent vaccine and an inactivated rabies vaccine of a particular commonly used brand. We consider it unlikely that the observed autoantibodies were specifically induced in response to these brands of vaccine and this phenomenon will likely occur with other commercial vaccines. In a follow-up study, we have observed similar autoimmune phenomena in dogs immunized with the multivalent vaccine only and in dogs immunized with the rabies vaccine only (unpublished observations).

There was a marked increase of autoantibodies to the skeletal muscle proteins, myoglobin and myosin, in both groups of dogs. The reason for the appearance of these antibodies is uncertain, but it may be the result of the frequent blood sampling of the dogs. The dogs were bled five times following each vaccination, and some tissue trauma was unavoidable.

We examined the thyroid and adrenal cortical function in the dogs, and did not find evidence of any abnormality. Autoimmune thyroiditis is one of the most common autoimmune diseases of dogs, and it has been suggested that the apparent increase of this condition in dogs is related to the increased frequency of vaccination with modified live vaccines. There was no increase of anti-thyroglobulin antibodies in the vaccinated animals, or other evidence of thyroid dysfunction. However, the lymphoid nodule found in the thyroid gland of one of the vaccinated dogs may be an early manifestation of thyroiditis, a common lesion in purpose-bred Beagles (Fritz et al., 1970).

The most strikingly increased concentrations of autoantibodies were directed against fibronectin and laminin. Fibronectin is widely distributed in the body as a component of the extracellular matrix and plasma. The anti-fibronectin antibodies were reactive with fibronectin of bovine, murine, and human origin. Although we have not yet demonstrated that they also react with canine fibronectin, this is very likely, since fibronectin is highly conserved between species. Anti-fibronectin antibodies have been found in human patients with systemic lupus erythematosus (SLE) and rheumatoid arthritis, and a patient with a poorly defined connective tissue disease (Henane et al., 1986; Atta et al., 1994, 1995; Girard et al., 1995). The anti-fibronectin antibodies in four human SLE patients were directed against the collagen-binding domain (Atta et al., 1994), in contrast to the anti-fibronectin antibodies in the vaccinated dogs, which showed no affinity for this domain. The anti-fibronectin antibodies in the human patient with connective tissue disease showed reactivity with the cell-binding domain of fibronectin (Girard et al., 1995).

Anti-fibronectin antibodies have been experimentally induced in rabbits by immunization with human fibronectin in complete Freund's adjuvant (Murphy-Ullrich et al., 1984). The antibodies were reactive with both human and rabbit fibronectin. The rabbits subsequently developed a glomerulopathy with granular deposits suggestive of immune complexes in the glomerular basement membrane. Anti-fibronectin antibodies have been induced in mice by multiple injections of homologous fibronectin without adjuvant (Murphy-Ullrich et al., 1986). The titer of anti-fibronectin antibodies was much lower in mice immunized with native fibronectin than in mice immunized with denatured fibronectin. However, in both groups, immune complexes were present in the serum and in the glomeruli (Murphy-Ullrich et al., 1986). Light microscopic examination of the glomeruli of the kidneys of vaccinated dogs did not reveal evidence of glomerulopathy, but we cannot exclude the possibility of sub-light microscopic lesions.

Anti-laminin antibodies were prevalent in the serum of three of the five vaccinated dogs. Anti-laminin antibodies are increased in human patients with SLE, rheumatoid arthritis, and vasculitis. Injection of polyclonal anti-laminin antibodies into rats resulted in glomerulopathy and proteinuria (Abrahamson and Caulfield, 1982). Anti-laminin antibodies have also been implicated in glomerular disease in rats induced by mercuric chloride (Aten *et al.*, 1995).

The mechanisms that may underlie the production of autoantibodies following vaccination are unknown, but at least four mechanisms can be proposed: cross-reactivity with vaccine-components, somatic mutation of immunoglobulin variable genes, "bystander activation" of self-reactive lymphocytes, and polyclonal activation of lymphocytes. Perhaps the simplest and most likely mechanism is that of cross-reactivity of vaccine and self-antigens. Although certain microbial antigens may cross-react with self-antigens (Schattner and Rager-Zisman, 1990), the most likely sources of cross-reactive epitopes are bovine serum and cell culture components. These are present in almost all vaccines as residual components of the cell culture necessary to generate vaccine viruses and may purposely be added to the vaccine as a stabilizer. In the presence of an adjuvant, these bovine products stimulate a strong immune response and induce antibodies that cross-react with conserved canine antigens. Thus, the strong response to fibronectin in the vaccinated dogs is most likely the result of the injection of bovine fibronectin contaminants in the vaccine. Indeed, this is essentially identical to the protocol used to produce anti-fibronectin antibodies in rabbits with human fibronectin in complete Freund's adjuvant (Murphy-Ullrich *et al.*, 1984), as mentioned above. The lower response to other antigens (e.g., cardiolipin and laminin) may be due to a lower concentration of these antigens in the vaccine or lower immunogenicity.

During every immune response, self-reactive B and T lymphocytes are generated and activated. This is the result of somatic mutation and bystander activation. Under normal conditions, this will not lead to significant production of autoantibodies, because of the selection process in the germinal centers of lymph nodes. In the germinal centers, only B cells that successfully compete for interaction with antigen presented on the surface of follicular dendritic cells will be allowed to survive (MacLennan, 1994). These B cells generally have high-affinity receptors for the antigen to which the immune response was induced. B cells with low affinity for the antigen or affinity for other antigens, including self-antigens, will undergo programmed cell death. The B cells with high-affinity receptors express bcl-2, which may rescue them from programmed cell death (MacLennan, 1994). This mechanism was elegantly demonstrated in mice immunized with a nominal antigen, phosphorylcholine (Ray et al., 1996). A single point mutation in the hypervariable region of the expressed immunoglobulin genes was sufficient for the phosphorylcholine-specific B cells to acquire specificity for DNA. However, it was only possible to demonstrate DNA-specific B cells by fusing germinal center B cells with cells that expressed high levels of bcl-2, thereby rescuing them from programmed cell death (Ray et al., 1996). An increased expression of bcl-2 was observed in thymic lymphoid follicles of patients with myasthenia gravis, suggesting that failure to delete self-reactive B cells in these patients may lead to autoimmune disease (Shiono et al., 1997). While this may seem an attractive hypothesis to explain autoimmune phenomena in human beings and dogs, there is currently no evidence that this is a common mechanism.

Finally, polyclonal activation of lymphocytes, including activation of self-reactive lymphocytes, is a possible mechanism of vaccine-induced autoimmunity. Certain viruses and bacteria have superantigen or mitogen activity (Schwartz, 1993). This could also be the case for the microbial products included in the vaccines. The present study does not support this mechanism. Firstly, antibodies were observed against 10 of 17 antigens tested. Secondly, the anti-fibronectin antibodies did not react with any portion of the fibronectin molecule, but, instead, reacted most strongly with the heparin binding domain. These observations indicate that the appearance of autoantibodies in the serum of vaccinated dogs is an antigen-driven process and not caused by polyclonal activation. As argued earlier, the main antigens implicated are cell culture contaminants and bovine serum components.

In the dog, certain autoimmune diseases occur more frequently in particular breeds of dogs, indicating genetically determined susceptibility (Dodds, 1983; Happ, 1995). There is abundant evidence from studies in rodents and human beings that the magnitude of the antibody response and the susceptibility to autoimmune disease are in part genetically determined (Schwartz, 1993). It is likely that genetic factors also determine the susceptibility to vaccine-induced autoimmunity. That this is indeed the case is suggested by the finding that only three of the five vaccinated dogs developed a strong anti-laminin antibody response and that the kinetics of the anti-fibronectin response differed between individual animals. Identification of susceptibility genes will be important, because it may shed light on the pathogenesis of the autoimmunity. In addition, it will provide genetic tests that will enable dog breeders to monitor the susceptibility of their breeding stock to vaccine-induced autoimmunity.

Although the pathogenic significance of the vaccine-induced autoantibodies is still unclear, there are a number of ways to prevent their induction. Not vaccinating dogs is not a viable option, because the benefits of vaccination clearly outweigh the still uncertain risks of immune-mediated disease. However, since bovine serum components in the vaccine may be responsible for the majority of autoantibodies, elimination of these bovine components may avoid this problem. This could be accomplished by substituting homologous serum for bovine serum. However, as mentioned earlier, anti-fibronectin antibodies may still be induced by immunization with homologous fibronectin. New generations of vaccines, especially naked DNA vaccines, are free of serum components, and these should not induce autoantibodies. A recent study in mice indicates that DNA vaccination does not induce or accelerate autoimmune disease (Mor et al., 1997). Finally, mucosal vaccines are less likely to induce autoantibodies than parenterally administered vaccines. Depending on the formulation of the vaccine, soluble serum components are less likely to be absorbed via the mucosal surface, and, in fact, may induce tolerance instead of autoantibodies (Weiner et al., 1994).

In conclusion, we have demonstrated that vaccination of dogs using a routine protocol and commonly used vaccines, induces autoantibodies. The autoantibody response appears to be antigen driven, probably directed against bovine antigens that contaminate vaccines as a result of the cell culture process and/or as stabilizers. The pathogenic significance of these autoantibodies has not yet been determined.

Acknowledgments

The authors thank Cheryl Anderson and Julie Tobolski-Crippen for animal care and technical support, and Nita Glickman for data management. This work is supported by the John and Winifred Hayward Foundation.

References

Abrahamson, D. R., and Caulfield, J. P. (1982). Proteinuria and structural alterations in rat glomerular basement membranes induced by intravenously injected anti-laminin immunoglobulin G. *J. Exp. Med.* **156,** 128–145.

Aten, J., Veninga, A., Coers, W., Sonnenberg, A., Timpl, R., Claessen, N., Van Eendenburg, J. D. H., De Heer, E., and Weening, J. J. (1995). Autoantibodies to the laminin P1 fragment in $HgCl_2$-induced membranous glomerulopathy. *Am. J. Pathol.* **146,** 1467–1480.

Atta, M. S., Powell, R. J., Hopkinson, N. D., and Todd, I. (1994). Human anti-fibronectin antibodies in systemic lupus erythematosus: occurrence and antigenic specificity. *Clin. Exp. Immunol.* **96,** 20–25.

Atta, M. S., Lim, K. L., Ala'aldeen, D. A., Powell, R. J., and Todd, I. (1995). Investigation of the prevalence and clinical associations of antibodies to human fibronectin in systemic lupus erythematosus. *Ann. Rheum. Dis.* **54,** 117–124.

Centers for Disease Control and Prevention (1996). Update: Vaccine side effects, adverse reactions, contraindications, and precautions—recommendations of the Advisory Committee on Immunization Practices (ACIP). *Morbid. Mortal. Wkly. Rep* **45,** 1–35.

Cohen, A. D., and Shoenfeld, Y. (1996). Vaccine-induced autoimmunity. *J. Autoimmun.* **9,** 699–703.

Dodds, J. W. (1983). Immune-mediated disease of the blood. *Adv. Vet. Sci. Comp. Med.* **27,** 163–196.

Dodds, J. W. (1988). Contributions and future directions of hemostasis research. *J. Am. Vet. Med. Assoc.* **193,** 1157–1160.

Duval, D., and Giger, U. (1996). Vaccine-induced immune-mediated hemolytic anemia in the dog. *J. Vet. Intern. Med.* **10,** 290–295.

Fritz, T. E., Zeman, R. C., and Zelle, M. R. (1970). Pathology and familial incidence of thyroiditis in a closed Beagle colony. *Exp. Mol. Pathol.* **12,** 14–30.

Gebhard, D. H., and Carter, P. B. (1992). Identification of canine T-lymphocyte subsets with monoclonal antibodies. *Vet. Immunol. Immunopathol.* **33,** 187–199.

Girard, D., Raymond, Y., Labbe, P., and Senecal, J.-L. (1995). Characterization of a novel human antibody xenoreactive with fibronectin. *Clin. Immunol. Immunopathol.* **77,** 149–161.

Happ, G. M. (1995). Thyroiditis—A model canine autoimmune disease. *Adv. Vet. Sci. Comp. Med.* **39,** 97–139.

Henane, T., Tigal, D., and Monier, J. C. (1986). [Anti-fibronectin autoantibodies in systemic lupus erythematosus, rheumatoid polyarthritis, and various viral or bacterial infectious diseases] in French. *Pathol. Biol.* **34,** 165–171.

MacLennan, I.C. M. (1994). Germinal centers. *Annu. Rev. Immunol.* **12,** 117–139.

Mastro, J. M., Axthelm, M., Mathes, L. E., Krakowka, S., Ladiges, W., and Olsen, R. G. (1986). Repeated suppression of lymphocyte blastogenesis following vaccinations of CPV-immune dogs with modified-live CPV vaccines. *Vet. Immunol. Immunopathol.* **12,** 201–211.

Moore, P. F., Schrenzel, M. D., Affolter, V. K., Olivry, T., and Naydan, D. (1996). Canine cutaneous histiocytoma is an epidermotropic Langerhans cell histiocytosis that expresses CD1 and specific β_2-integrin molecules. *Am. J. Pathol.* **148,** 1699–1708.

Mor, G., Singla, M., Steinberg, A. D., Hoffman, S. L., Okuda, K., and Klinman, D. M. (1997). Do DNA vaccines induce autoimmune disease? *Hum. Gene Ther.* **8,** 293–300.

Murphy-Ullrich, J. E., Oberley, T. D., and Mosher, D. F. (1984). Detection of autoantibodies and glomerular injury in rabbis immunized with denatured human fibronectin monomer. *Am. J. Pathol.* **117,** 1–11.

Murphy-Ullrich, J. E., Oberley, T. D., and Mosher, D. F. (1986). Serologic and pathologic studies of mice immunized with homologous fibronectin. *Am. J. Pathol.* **125,** 182–190.

Phillips, T. R., and Schultz, R. D. (1987). Failure of vaccine or virulent strains of canine parvovirus to induce immunosuppressive effects on the immune system of the dog. *Viral Immunol.* **1,** 135–144.

Ray, S., Putterman, C., and Diamond, B. (1996). Pathogenic autoantibodies are routinely generated during the response to foreign antigen: A paradigm for autoimmune disease. *Proc. Natl. Acad. Sci. USA* **93,** 2019–2024.

Schattner, A., and Rager-Zisman, B. (1990). Virus-induced autoimmunity. *Rev. Infect. Dis.* **12,** 204–222.
Schwartz, R. S. (1993). Autoimmunity and autoimmune diseases. *In* "Fundamental Immunology" (W. E. Paul, ed.), 3rd ed., pp. 1033–1097. Raven Press, New York.
Shiono, H., Fujii, Y., Okumura, M., Takeuchi, Y., Inoue, M., and Matsuda, H. (1997). Failure to down-regulate Bcl-2 protein in thymic germinal center B in myasthenia gravis. *Eur. J. Immunol.* **27,** 805–809.
Smith, C. A. (1995). Are we vaccinating too much? *J. Am. Vet. Med. Assoc.* **207,** 421–425.
Weiner, H. L., Friedman, A., Miller, A., Khoury, S. J., Al-Sabbagh, A., Santos, L., Sayegh, M., Nussenblatt, R. B., Trentham, D. E., and Hafler, D. A. (1994). Oral tolerance: Immunologic mechanisms and treatment of animal and human organ-specific autoimmune diseases by oral administration of autoantigens. *Annu. Rev. Immunol.* **12,** 809–837.

An Introduction to Analytical Methods for the Postmarketing Surveillance of Veterinary Vaccines

DAVID SIEV

Center for Veterinary Biologics, U.S. Department of Agriculture, 510 South 17th Street, Ames, Iowa 50010

I. Introduction
II. Postmarketing Surveillance and Public Policy
III. The Postmarketing Surveillance Datum
IV. Fathoming Spontaneous Adverse Event Report Data
   A. Causality
   B. Bias
   C. Interpretation
V. Quantitative Analysis of Adverse Event Report Data
   A. Adverse Event Report Process
   B. Overdispersion
   C. Some Methods for Adverse Event Report Data
VI. Summary
   References

I. Introduction

Information about adverse events after the vaccination of animals in the United States is available largely from voluntary reports spontaneously submitted by vaccine users. Interpreting spontaneous adverse event report (AER) data is complicated by biases inherent in the observation of vaccine adverse events, the use of surrogate measures, and the lack of a clear probability structure. The quantitative analysis of AER data often encounters a degree of variability exceeding that explained by the standard sampling models which justify many statistical procedures. This paper considers the implications of these phenomena and their impact on both the analysis and interpretation of AER data. It reviews and illustrates several statistical methods, and introduces a modification of a useful rate ratio formula. The final section considers issues in the practice of postmarketing surveillance of veterinary products.

II. Postmarketing Surveillance and Public Policy

Early this year, newspaper headlines screamed the alarming news that human poliomyelitis is nearly always caused by the vaccine intended to prevent it (e.g., USA Today, 1/31/97). In fact, since 1980 the vaccine has been the only source of indigenous polio disease in the United States (Strebel et al., 1992).

Although such headlines had the desired shock effect, the fact that poliomyelitis results only from the vaccine is actually very good news. The good news is that the vaccine has been highly effective. Disease due to wild poliovirus has been eradicated from the United States, having dropped from more than 20,000 cases of paralytic polio in 1952, the year the vaccine was introduced. The flood of newspaper stories about polio vaccine in early 1997 was prompted by the introduction of a revised vaccination regimen aimed at halving the vaccine-induced disease rate of about two cases per million doses (Advisory Committee on Immunization Practices, 1997).

It is evident that whenever the risk of naturally occurring disease diminishes, the hazards of vaccination bear greater scrutiny. Often vaccination coverage in the population will drop, as increased confidence about the reduced disease incidence coincides with increasing concern over the risks of vaccination. A drop in the fraction of children vaccinated did, in fact, contribute to a small outbreak of polio disease in 1979.

This also highlights why it is important that vaccination risk–benefit decisions be made in the arena of public policy rather than on an individual basis. A parent may accurately conclude that the risk of vaccine-induced polio, however small, outweighs the risk of natural infection, which is virtually nil in the United States. A risk analysis on an individual basis dictates the decision not to vaccinate one's child. Yet if all parents were to make such a decision, vaccination would cease, the disease would be reintroduced, and the goal of worldwide eradication would recede into impossibility. Because vaccination exerts its effects in the population, not just in the individual, vaccination policy must be determined at the population level as well.

Postmarketing surveillance ideally seeks to monitor the relationship between vaccine effectiveness, vaccine adverse events, and disease prevalence. Vaccine efficacy, often expressed as the risk ratio complement, may be reformulated as the attributable fraction [which is the way Greenwood and Yule (1915) originally defined it]. If the expected sick fraction of a group of vaccinated animals is designated p_v and the sick fraction of those not vaccinated is p_u, then vaccine efficacy is VE = $(p_u - p_v)/p_u$. Vaccine efficacy is thus the risk difference in the reduced sample space of those who would be expected to be sick had they not been vaccinated. If p_u estimates the prevalence, then vaccine efficacy is the expected risk difference (RD) as a fraction of the expected prevalence (VE = RD/P), as depicted in Fig. 1a.

If VE is presumed to be a constant property of the vaccine, then the risk difference would drop proportionately with a drop in prevalence. Now it is the risk difference which describes the effect of the vaccine in the population. RD is often omitted from reports of vaccine studies, because, besides lacking VE's presumed invariance to prevalence, it is always smaller and less dramatic than VE.

The effect of this relationship in practice is illustrated in Fig. 1b. A field study of a canine borreliosis vaccine (Levy *et al.*, 1993) estimated vaccine efficacy at 79% and disease prevalence at 4.7%. The 3.7% risk difference exceeded the reported 1.9% rate of adverse reactions. With a drop in prevalence of just a few percent, however, adverse reactions would exceed the risk difference, and the costs of vaccination relative to its benefits would need to be reconsidered.

A distinction is often made between the term *vaccine efficacy*, referring to the protection of an individual, and the term *vaccine effectiveness*, referring to the vaccine's effect in the population. In prelicensing trials of veterinary vaccines, the subjects are usually challenged with the virulent pathogen, virtually ensuring equivalent exposure. Although vaccine efficacy may thus be legitimately computed from properly designed veterinary vaccine trials, the U.S. Department of Agriculture (USDA) Center for Veterinary Biologics (CVB) does not specifically require the estimation of vaccine efficacy for licensure. Challenge trials of animal vaccines provide a significant advantage over trials of human vaccines, but further distance the conditions of experimental trials from field conditions of actual use, which observational work aims to study.

III. The Postmarketing Surveillance Datum

The postmarketing surveillance of veterinary biologics involves monitoring a vast market of billions of doses of thousands of products. Active surveillance, in which each individual vaccinated animal would

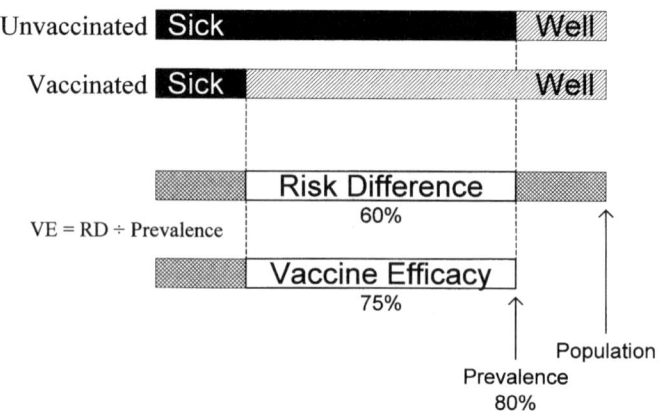

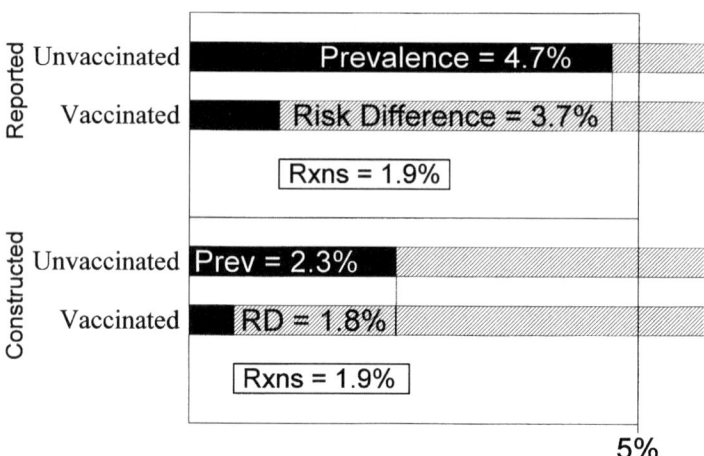

FIG. 1. Vaccine efficacy, risk difference, and adverse reactions. (a) Vaccine efficacy as the attributable (prevented) fraction. In this illustration, the sick fractions are 0.8 of unvaccinated and 0.2 of vaccinated. The vaccine prevents disease in 0.6 of the population (risk difference), or 0.75 of those that would be sick without vaccination (attributable fraction). (b) The effect of changing prevalence on the relation between risk difference and adverse reactions (based on data from Levy et al., 1993).

be actively followed up, is clearly out of the question. Both human and veterinary surveillance systems have had some success with strictly passive surveillance schemes, which depend on voluntarily submitted reports of adverse events after vaccination.

To ameliorate the deficiencies of a strictly passive system, reporting may be encouraged by what might be termed stimulated spontaneous surveillance. Reports to the human Vaccine Adverse Event Reporting System are stimulated by the requirements of the National Childhood Vaccine Injury Act of 1986. Regulations proposed by the USDA would require a milder stimulus, such as a label statement informing vaccine users where to submit a report. In any case, every stimulus is an intervention with the potential for introducing bias. In late 1996, the recommendation was made by the Feline Sarcoma Task Force to report injection-site lumps to the United States Pharmacopeia (USP). Such reports to the USP promptly increased to about half of all reports involving cats. While most were current, some reports were of events occurring years before.

The very first step in any analytical approach is to identify explicitly the random quantity one is able to study. It is at this initial stage that many attempts to analyze data from veterinary vaccine AERs go wrong. All too often even specialists use terms such as *events, cases, reaction rate,* and the like. Yet surveillance systems for AERs only observe the *report* of an adverse event, not the putative event itself (Fig. 2). The role of the vaccine is even more distant. While apparently obvious, this crucial distinction is often overlooked.

Furthermore, to minimize bias, reports should ideally be of "adverse events unqualified by any suspicions" (Finney, 1982). Thus, every adverse event occurring after vaccination would be reported, whether or not the reporter suspected the vaccine. This is rarely the case, and reports are more frequently submitted when the reporter has a high level of suspicion. Those receiving the reports, at least, should be careful to record them objectively without additional filtering.

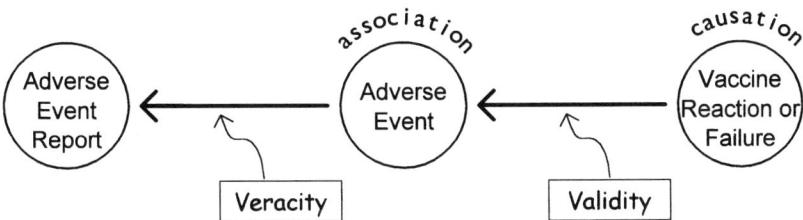

Fig. 2. Schematic diagram of adverse event reporting. From the perspective of the surveillance system, neither the veracity of the report nor the validity of the inference can be ascertained.

IV. Fathoming Spontaneous Adverse Event Report Data

A. Causality

The quality and accuracy of AERs vary widely, particularly when received from lay consumers rather than veterinary health professionals. From the perspective of AER surveillance, when the veracity of a report is undetermined, it would be foolhardy to assess the validity of the causal relationship between vaccination and the reported event. Association is observable (and any adverse event observed after vaccination is temporally associated with it), while causation requires inference. Such inference is rarely straightforward, except in cases where there is close proximity of onset (e.g., peracute anaphylaxis) or location (e.g., injection site abscess). In any case, assessing the possibility of a causal relationship for a specific individual AER is often an inappropriate activity for vaccine surveillance systems (Chen, 1994), which are focused on studying the action of the vaccine in the population.

Appropriate causality assessment is determined by the arena of activity in which one is engaged. A clinician, focused on the patient and observing the event, is often obligated to perform an ad hoc assessment of retrodictive causality. An answer to the question "Did the vaccination cause this event?", however presumptive, may be necessary for the initiation of treatment. The regulatory epidemiologist, by contrast, observes the report, not the event, and is focused on the vaccine, not the patient. Appropriate to this activity is an assessment of predictive causality, and the question becomes "How likely is it that the vaccine will cause this type of event?" Hence, the postmarketing surveillance of veterinary vaccines is often engaged in the estimation of rates.

The distinction between reasoning about causality in the epidemiologic arena and the approach taken in the clinical arena begins with the fundamental meaning given to causality. The causality criterion most often used by the clinician is that of necessity. Necessary causality implies that without the vaccination, the event would not have occurred, as distinct from sufficient causality, which means that all vaccinations result in the event. While the conjunction of necessity and sufficiency together underlie disease causality in the sense of the Henle-Koch postulates, for the clinician the criterion of necessity compels the formation of a belief about something that has not occurred ("had the animal not been vaccinated, it would not have experienced the event"). This proposition, a counterfactual conditional (Glymour, 1986), may be tenuous but is often appropriate in the clinical context.

The Henle-Koch postulates break down in the face of multifactorial causal relationships, such as those often studied by epidemiologists. A vaccine may, for example, potentiate an adverse response in an individual with a hereditary disposition under certain environmental conditions. In regulatory or public health contexts, the recognition of multifactorial causality is indispensable.

Risk factor epidemiology is an interdisciplinary mosaic which relies on judgment based on the weight of evidence to solve problems pragmatically (Charlton, 1996). Although Charlton does a good job of articulating its exigencies, he focuses on its deficiencies, claiming that "risk factor epidemiology cannot be regarded as a scientific discipline because it aims at concrete usefulness rather than abstract truthfulness." In fact, this approach serves the regulatory epidemiologist far better than a useless abstract truth.

B. Bias

Observational data are an essential complement to the experimental data generated in the licensing process. They reflect the actual use of the vaccine under the intended field conditions in all subsets of the target population, and they provide the quantities necessary to detect relatively rare events. Hence postmarketing surveillance is "not . . . an optional activity" (Finney, 1982).

We would like to know something about the incidence rate of adverse events after vaccination. To estimate the incidence rate, however, we need information about the number of affected individuals and the number of vaccinated individuals, neither of which is known in postmarketing surveillance. Instead, we estimate the reporting rate using information available to us, the number of AERs over the number of doses sold.

The use of surrogates introduces bias into both the denominator and numerator. Sales volume is the usual denominator surrogate for the population at risk (Praus *et al.*, 1993). In general, the sales volume of the reporting period is used, although seasonal or other variation may suggest the use of a lagged denominator, which utilizes sales volume during an earlier period. Marketing patterns of veterinary immunobiologics vary widely and will affect how closely vaccine use parallels sales volume. The discrepancy is reduced when considering historical data by production lot rather than time interval.

The interpretation of the reporting rate as a surrogate for incidence rate depends on defining the relationship between the two. This is not easily done, and it appears that this relationship varies by factors such as species, manufacturer, event type and severity, and it may even change over time. The relationship between reporting rate and incidence rate for human vaccines has been found to have a wide range. One study estimated the reporting rate to be about 70% of the incidence rate of vaccine-induced polio disease, a severe and rare event, while the reporting rate for rash after measles-mumps-rubella vaccination was estimated at less than 1% of the incidence rate (Rosenthal and Chen, 1995). It is likely that reporting fractions are even lower for veterinary vaccines, which lack the mandatory reporting requirements of human vaccines.

Whether designed to study the vaccination history of groups of cases and noncases, or to observe vaccinated and nonvaccinated cohorts for adverse events, observational studies generally aim to provide information on all four cells of a typical 2×2 table, which exclusively and exhaustively partitions reality. Spontaneous AER data, on the other hand, yield no information for three of the cells (the two for nonvaccinated and the one for vaccinated noncases), and incomplete information for the fourth (vaccinated cases if observed). In fact, a suitable partition for considering AER data requires at least a $2 \times 2 \times 2$ cube rather than a flat table, since any observation may or may not be reported. Herein lies the most important bias in spontaneously reported data: self-selection.

The serious biases introduced by self-selection are manifestly evident. In addition, self-selection means that there is no clear probabilistic structure to AER data, such as would be imposed through probability sampling or random allocation.

The study of vaccine adverse events is also complicated by other specific biases. Although different vaccination rates between risk groups is often the norm in veterinary medicine, the result may be naively interpreted as a failure of the vaccine's efficacy. It should be understood that vaccinating individuals with a high risk of exposure while not vaccinating low-risk individuals makes it likely that most cases of the disease will be seen in vaccinated individuals. In addition, the decision to vaccinate may be associated with a factor that is also associated with an adverse event. Such a confounder may result in the unwarranted inference that the vaccine causes a particular adverse event or even protects against it (Fine and Chen, 1992).

Neither exposure to the pathogen nor the observation of animals may be equivalent between vaccinated and unvaccinated individuals. Exposure or event ascertainment are often sloppy. In addition to these are the usual biases which dog observational work. The possibility of an undetected confounder, in particular, should never be overlooked, as it may result in an apparent association in the direction opposite the true association, known as Simpson's paradox.

While a major thrust of this paper is on the variability of AER data, remember that bias can have a far greater impact on the validity of one's interpretation of such data. The relative significance of bias and variance appears explicitly in many statistical procedures through the mean squared error loss function, where the contribution of bias to error is seen to be of a higher order than the contribution of variance.

C. INTERPRETATION

Surveillance with spontaneous AER data labors under two major handicaps. The first is the constant omnipresence of bias, and the second is the absence of a clear probabilistic structure to the data. The latter invalidates many popular statistical procedures, while the former subverts the interpretation of any apparent feature of the data. In view of these considerable hurdles, any analysis should maintain "conceptual simplicity and inferential modesty" (Hall, 1989). We cannot even routinely assume that summary descriptions may be extended to a parent population. While these hurdles must never be overlooked, they should not be allowed to deter us from the appropriate study of AER data.

Several mistakes in the use of AER data are widely prevalent: *misrepresenting* an AER as what it is not, an adverse event or a vaccine-related event; *misunderstanding* the random nature of the data, which have not arisen from sampling or allocation; and *misinterpreting* the data by ignoring biases.

Making sense of AER data is clearly challenging. The cautious interpretation of AER rates requires experience with spontaneously reported data, familiarity with the animal vaccine industry, and great discretion. Conclusions are often tentative and, in any case, incomplete without sensitivity analysis and consideration of alternative explanations. In general, the analysis of AER data should be viewed as exploratory, rather than confirmatory, for which other types of information may be needed.

The initial objectives of postmarketing surveillance are first, the description of typical activity, and second, the detection of unusual activity. The most useful analytical methods are often graphical. Estimation and modeling may be helpful descriptive tools, although they should be used in an empirical and heuristic spirit rather than pivoting on a specific stochastic basis. Surveillance ends with detection, since confirmation usually relies on further investigation.

Comparisons for detecting differences between subsets of the data should also have a descriptive tenor. Estimation is preferable to hypothesis testing in that attention is focused on the magnitude of an apparent difference rather than hinging on an unverified probability structure. Even under the most proper conditions, statistical hypothesis tests are too often abused, misused, and misunderstood. With AER data they must never be interpreted as formal tests of significance, but as detection tools providing indices of suspicion (Finney, 1982). Their use should be limited to internal comparisons between subsets of the data, and some advocate permutation methods as most appropriate for this purpose (Hall, 1989).

V. Quantitative Analysis of Adverse Event Report Data

A. Adverse Event Report Process

Having identified the random variable we wish to study, the AER rate, we must investigate its nature more closely. Figure 3 gives a quantitative depiction of the process outlined in Fig. 2. The number of AERs is, in fact, a compound of several other random quantities. Many

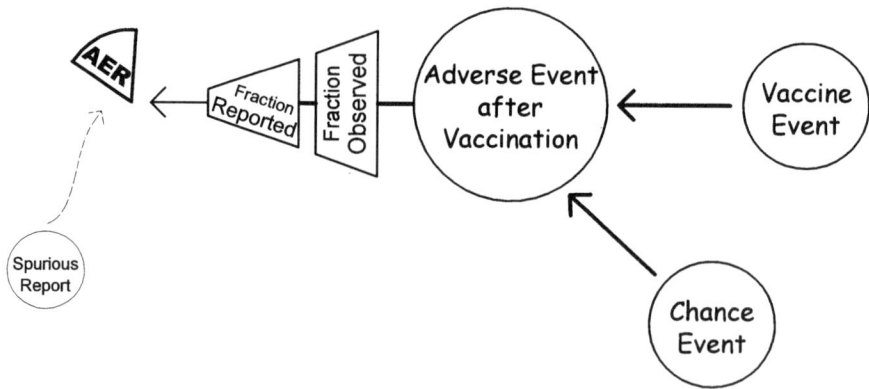

FIG. 3. Adverse event reports as a compound quantity. Both vaccine-related events and coincidental events contribute to adverse events occurring after vaccination. Only a fraction of adverse events are observed, and of those only a fraction are reported.

vaccine adverse events are never observed, and only a small fraction of those observed are reported. An adverse event after vaccination may be either vaccine related or coincidental. For example, while it is probably a chance event when a child dies of sudden infant death syndrome (SIDS) the day after immunization, even under optimal conditions it may not be possible to differentiate a coincidental from a vaccine-related event. (It is noteworthy that concern over postvaccination SIDS arising from postmarketing surveillance led to studies that exonerated the vaccine and identified genuine risk factors for the syndrome. As a result, the incidence of SIDS has diminished.) Vaccine-related events themselves may be the manifestation of a complex interaction between vaccine, vaccinee, and environmental factors.

Surveillance is the continual observation of a sequence of occurrences in order to study the process underlying that sequence. Vaccine AERs may be viewed in two types of sequences, the production sequence and the vaccination sequence. The production sequence, in which the vaccine is manufactured, has a natural unit, the production lot, a single batch of vaccine assembled and filled at one time. (Production lots of USDA regulated biologicals are called *serials.*) By contrast, the doses comprising the sequence of vaccine administration must be aggregated into practical clusters by an appropriate time interval, such as month or quarter. Each of the two types of sequences may highlight different features of the data, as illustrated in example 3 (in a later section).

The process which our surveillance aims to elucidate is one in which AERs result from vaccination. It may be formulated

$$\text{AERs} = \text{exposure} \times \text{intensity}$$

The intensity of the process is the probability that an AER results from a vaccination with a single dose. The number of AERs is then the product of the intensity and the exposure in doses. The intensity multiplied by a standard exposure, such as a million doses, is a reporting rate.

In studying AERs, we are often interested in the rate intensity parameter or some function of it. Parameter estimation is straightforward under certain assumptions. If the outcome of each vaccination is stochastically independent and identically distributed with a common intensity, then the number of AERs is generally assumed to follow the Poisson distribution. That distribution is completely specified by a single parameter, whose maximum likelihood estimator is the empirical cumulative AER rate. The cumulative rate is conveniently computed by taking the total number of AERs summed across all individual rates and dividing by the total number of doses summed across all the rates.

If these assumptions do not hold, lumping the data into a cumulative rate may be unwarranted. The precision of estimates may be inaccurately evaluated, and inferences may be suspect. It is essential to identify the assumptions underlying one's methods and to scrutinize those assumptions closely.

B. OVERDISPERSION

At each stage from vaccine manufacture to adverse event report, the assumptions of independence or a common distribution can be questioned. Vaccines are produced in lots, each of which is a batch of hypothetically homogeneous doses. Each lot may be sold in a different geographic region and, within each region, to many veterinarians (or lay consumers). Each veterinarian vaccinates different herds, and within each herd there may be units such as litters of pigs. Is it reasonable to assume that the outcome in a pig is independent of the outcome in its adjacent littermate? Or that it is identical to a pig in the next litter? The vaccine chain forms clusters within clusters, and clustering is often at the root of the failure of these assumptions. Thus, even before looking at the data we need to be skeptical about them.

Once the data are before us, they should be closely examined for consistency with the assumptions. In example 2 (Fig. 7; see later section), we consider a sequence of anaphylaxis AER rates for one year's production lots of a vaccine. There is an increase toward the end of the period. At what point could one say that a rate is different than the others? It is apparent that the data are somewhat variable, and the challenge is differentiating an observation that is truly different from one simply manifesting the random variation typical of the type of data we are observing.

In what comes, notation is as follows. Let y_i represent the number of AERs for lot or interval i, with h_i the number of doses sold of that lot or in that interval; λ is the rate parameter governing the intensity of the process, and a circumflex ($\hat{}$) over a parameter indicates an estimate of that parameter. The cumulative AER rate intensity can thus be written $\hat{\lambda} = \Sigma y_i / \Sigma h_i$, where summation is over all i. The expected value of y_i will be designated μ_i, and $\mu_i = \lambda h_i$.

In Fig. 4, the data of Figure 7 (shown later) are shown in the upper panel, while the lower panel shows a random sample generated under the Poisson assumptions. The cumulative AER rate was estimated as described, by lumping the data across all of the individual lot rates, giving the estimate $\hat{\lambda} = 3.3 \times 10^{-6}$. The illustrated random sample is

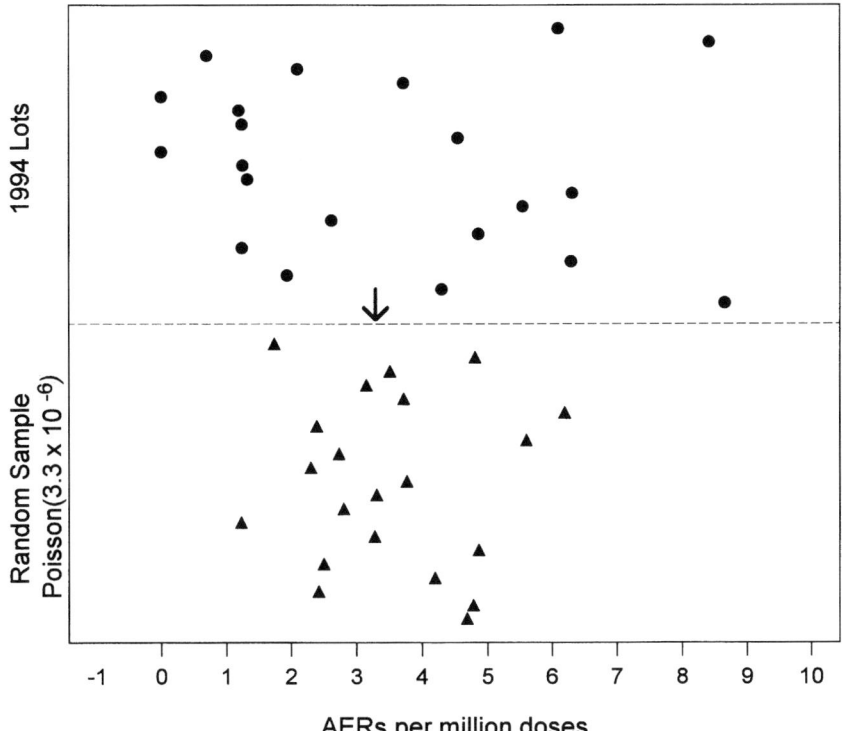

FIG. 4. Reporting rates for one year's production lots of a vaccine compared with a random sample drawn from a Poisson distribution with mean equal to the observed cumulative average rate (arrow).

$\{(x_i/h_i) \times 10^6\}$ where the x_i were randomly drawn as Poisson variates with expectation $\hat{\lambda} h_i$.

It is clear that the data are more spread out than the Poisson random sample; they manifest extra-Poisson variation, also known as overdispersion. If we were to apply standard procedures blindly for count data, say, by clicking the apparently relevant menu item in our favorite statistical software package, our inferences could be suspect.

Those uncomfortable with graphical analysis may prefer a formal hypothesis test for overdispersion. A test statistic appropriate to most simple single-sample problems with an adequate sample size is

$$\left(2\sum \hat{\mu}_i^2\right)^{-1/2} \sum \left[(y_i - \hat{\mu}_i)^2 - y_i + \hat{\mu}_i^2 / \sum \hat{\mu}_i\right]$$

(For details on this and other tests, see Dean and Lawless, 1989, and Dean, 1992.)

The first step in handling apparent overdispersion should be the search for systematic factors that have been overlooked in constructing the model. If such factors can be observed, the model's systematic formulation can be refined. If we believe such factors exist but cannot be observed, they are termed *latent*. It may be possible to take latent variables into account in the analysis, as is done, for example, with finite mixture models. Often, however, we have no choice but to consider the unexplained variation to be random and use methods appropriate for overdispersed data. In postmarketing surveillance, overdispersion appears to be more common with biological products rather than drugs, veterinary rather than human products (possibly due in part to reports by lay consumers as well as health professionals), and in data grouped by lot rather than time interval.

The source of overdispersion may influence our selection of the random component of our model. Among the contributors to overdispersion may be intracluster correlation, intercluster variation, or a latent mixture of populations. A compound process may exist (Fig. 3) where the probability that an event occurs, the probability that it is observed, and the probability that it is reported may be governed by different random processes, any of which may be subject to overdispersion. Where there is no substantive basis for assuming the data have arisen from a particular parametric distribution, it may be difficult to determine which distribution provides the best fit to the data.

Figure 5 illustrates empirical histograms of AER rates compared with the estimated probability mass functions of a number of distributions. Most are plausible, offering greater dispersion than the Poisson, but the data are not sufficient for us to select among them confidently.

The dilemma of selecting a parametric model for the AER data we have examined, as well as computational and other considerations, has led us to rely on quasi-likelihood (QL) methods (Wedderburn, 1974). QL requires the specification of only the mean and variance rather than the entire distributional structure of the data, providing an appealing flexible approach to modeling overdispersion.

C. Some Methods for Adverse Event Report Data

The several methods illustrated here are not intended to survey surveillance methodology, but were selected to illustrate some procedures for overdispersed data. While approaches based only on means are inadequate, and describing the full distributional structure of the

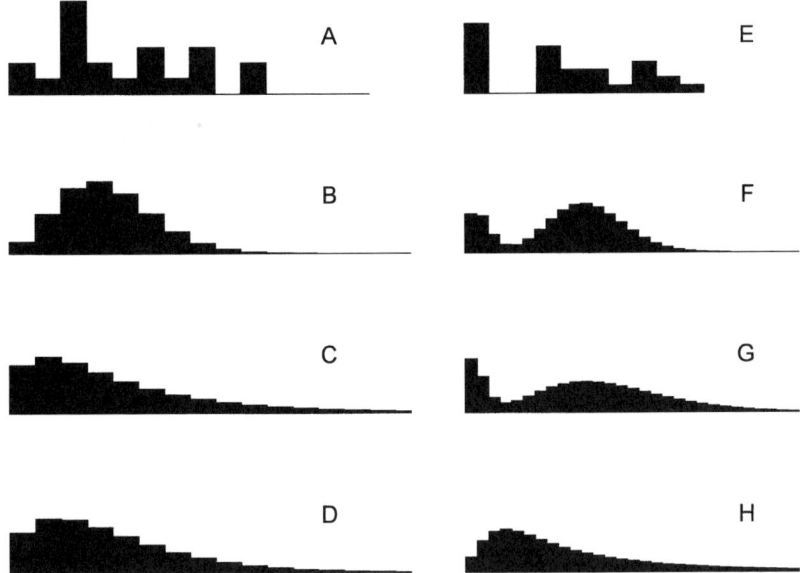

FIG. 5. Selecting the parametric model that best fits a set of reporting rates may be difficult. Probability mass functions of several distributions are compared with empirical histograms. (A) Empirical histogram for the reporting rates of a dog vaccine. (B) Poisson. (C) Generalized poisson. (D) Negative binomial. (E) Empirical histogram for a cattle vaccine. (F) Poisson-Poisson finite mixture. (G) Poisson-negative binomial finite mixture. (H) Poisson-inverse Gaussian hierarchical mixture.

data is not always possible, the variability of the data must always be considered. The data should always be examined directly, before proceeding to summaries and reduction. The utility of graphical methods cannot be overemphasized.

1. Interval Estimation: Modified Formula for Rate Ratio

For the comparison of AER rates, such as a current rate with one during some preceding period, an interval estimate of the rate ratio allows us to judge both the magnitude of a change as well as its plausibility.

An interval estimate of the rate ratio for count data which are not overdispersed may be computed from the following formula:

$$R = \hat{R}\left[1 + \frac{Z_{\alpha/2}^2}{2}\left(\frac{1}{y_1} + \frac{1}{y_2}\right) \pm \frac{Z_{\alpha/2}}{2y_1 y_2}\sqrt{y.\left(y.Z_{\alpha/2}^2 + 4y_1 y_2\right)}\right]$$

where $y_. = y_1 + y_2$, $\hat{R} = (y_1/h_1)/(y_2/h_2)$, and Z_α is the α quantile of the standard normal distribution. This formula is derived from a standard Poisson likelihood function, which is also the basis for the score test below. (The derivation is given in Siev, 1994, Appendix 1, and closely parallels one for the binomial case given by Gart, 1985.)

In the presence of overdispersion, the following QL approximation can be used:

$$R = \hat{R}\left[1 + \frac{\hat{\phi}Z_{\alpha/2}^2}{2}\left(\frac{1}{y_1} + \frac{1}{y_2}\right) \pm \frac{Z_{\alpha/2}}{2y_1y_2}\sqrt{\hat{\phi}y_.\left(y_.\hat{\phi}Z_{\alpha/2}^2 + 4y_1y_2\right)}\right]$$

The dispersion parameter (ϕ) may be estimated by the method of moments as follows (Wedderburn, 1974). For the comparison of a single rate with an historical period, use

$$\hat{\phi} = (n-1)^{-1}\sum_{i \in H}(y_i - \hat{\mu}_i)^2/\hat{\mu}_i$$

where summation is usually over the rates during the historical period only (indicated by H). For other comparisons involving, say, sets of rates for two vaccines, estimate

$$\hat{\phi} = \left(\sum n_j - 2\right)^{-1}\sum_{j=1}^{2}\sum_{i=1}^{n_j}(y_{ij} - \hat{\mu}_{ij})^2/\hat{\mu}_{ij}$$

where $j = 1, 2$ depending on the vaccine, n_j is the number of rates for vaccine j, and $\hat{\mu}_{ij} = \hat{\lambda}_j h_{ij}$ with $\hat{\lambda}_j$ estimated by the cumulative rate for each vaccine separately.

2. Hypothesis Testing: Suitability of Usual Procedures

Although estimation is preferable, certain types of hypothesis test procedures may occasionally be useful to guide one's suspicions. Remember that they are just guides and not formal significance tests. Adjustments for multiple comparisons may be needed.

A test for of the hypothesis that a current AER rate is no greater than the rate during some comparison period has been proposed (Norwood and Sampson, 1988). The basis of the Norwood-Sampson test is that, if the reporting rate has not increased, the proportion of AERs received during the current period, conditioned on the total number of AERs, would not be greater than the proportion of doses sold. The comparison is thus internal, and it could be interpreted as a minimalist test in the spirit advocated by Hall (1989) if it did not specify a one-sided alternative.

The Norwood-Sampson test assumes there is neither any systematic trend during the historical comparison period nor overdispersion during the current and comparison periods. Rejection of the null hypothesis would then imply a systematic difference between the rate intensities of the current and comparison periods. The one-sided hypothesis is tested using the complement of the binomial cumulative distribution function:

$$1 - \sum_{x=0}^{y_c - 1} p^x (1-p)^{n-x} n!/(x!(n-x)!)$$

where $n = y_C + y_H$, $p = h_C/(h_C + h_H)$, and the subscripts are C for current period and H for historical comparison period.

The Norwood-Sampson test is not appropriate when there is overdispersion. Suppose we are comparing lot 21 with lots 1–20 in example 2 (see Fig. 7 later). The hypothesis being tested is really that $\lambda_i = \lambda$ for all i, that is, all the rates are the same. Conditioned on the total, Σy_i, the sequence of AERs by lot $\{y_i: i = 1, \ldots, 21\}$ has a Multinomial(Σy_i, $\{p_i: i = 1, \ldots, 21\}$) distribution. The multinomial probabilities are $p_i = E(y_i)/E(\Sigma y_j) = \lambda_i h_i / \Sigma \lambda_j h_j$ where summation is over all lots. If all $\lambda_i = \lambda$, then $p_i = h_i/\Sigma h_i$, the sales fraction, and partitioning as a binomial into current rate and historical period is legitimate and results in the Norwood-Samson test. If, however, this is not the case, the p_i are not the sale fractions, and the test is invalid.

The null hypothesis that the rate ratio is unity could be tested with the statistic $(y_1 - (h_1 y_.)/h_.)\bigl((1/h_1 + 1/h_2)h_. /(\hat{\phi} y_.)\bigr)^{\frac{1}{2}}$, where $y_. = y_1 + y_2$ and $h_. = h_1 + h_2$. If overdispersion is not present, assume $\hat{\phi} = 1$; otherwise use the moment estimator shown earlier. Values of this statistic may be compared to the standard normal distribution, with extreme values offering evidence that the rate ratio is not one. Results will agree with those from the formula for interval estimation of the rate ratio, from which it is developed, and there is thus little justification for using the test rather than the estimate, other than pedagogical grounds.

a. Example 1. In the sequence of quarterly anaphylaxis AER rates illustrated in Fig. 6, we may wish to compare the final rate to the preceding eight. Most of the points during the comparison period fall within a 95% profile likelihood interval for the cumulative rate, and Dean's test gives no evidence of overdispersion ($p = 0.80$). The rate ratio for the ninth quarterly rate relative to preceding 2 years is 3.14 with a 95% interval estimate of 1.76 to 5.61. By the score test $p < 10^{-4}$, and by Norwood-Sampson test $p < 10^{-3}$.

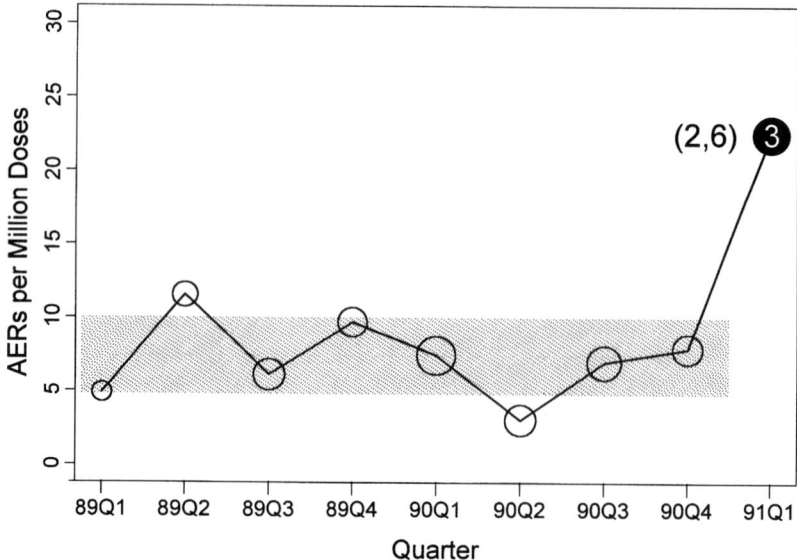

FIG. 6. Adverse event reporting rate sequence with no evidence of overdispersion. Circles are proportional to the number of doses. The shaded band shows a 95% profile likelihood interval for the observed cumulative AER rate over the first eight quarters. The rate ratio for the ninth quarter is 3.14 (95% interval 1.76 to 5.61).

b. Example 2. Here we see a sequence of anaphylaxis AER rates by production lot (Fig. 7). A fair bit of variation is evident. If lots 1–20 are the comparison interval, most of its points fall outside a 95% profile likelihood interval for the cumulative rate. By Dean's test, $p < 10^{-5}$, supporting our impression of overdispersion. Interval estimates of the rate ratios for lots 21–23 all exceed 1.0, with lot 23 substantially so. If the formula neglecting overdispersion were used, intervals for lots 13 and 18 would also exceed 1.0 (Table I,B). The quasi-score test finds lots 21–23 significantly different than the comparison period (Table I,C). Now, suppose we had applied the Norwood-Sampson test to these data. In addition to the final three lots, the rates of lots 13, 14, and 18 (indicated by diamonds in the figure) would also have p values ≤0.05 when compared with the preceding lots.

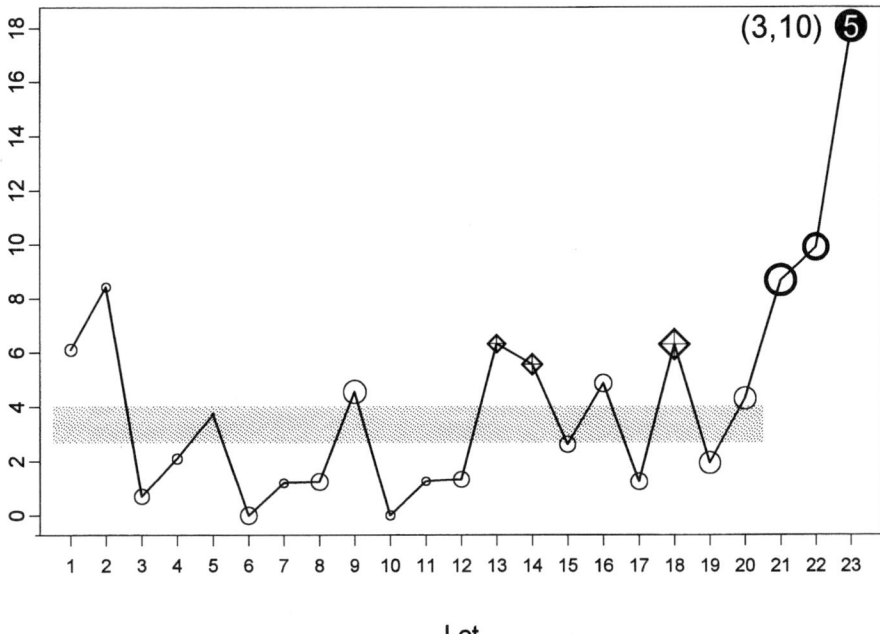

FIG. 7. Adverse event reporting rate sequence with overdispersion. The shaded band shows a 95% profile likelihood interval for the observed cumulative AER rate over lots 1–20. A rate ratio interval for lot 23 supports the position that its AER rate is substantially elevated compared with lots 1–20, and intervals for lots 21 and 22 indicate moderate elevation. Diamonds indicate lots that could be found significantly elevated by the inappropriate use of the Norwood-Sampson test. Circles and diamonds are proportional to the number of doses.

3. Loglinear Modeling: Quasi-Likelihood

A large set of data, with many concomitant factors and covariates, cannot be adequately handled with just univariate procedures for individual estimates or comparisons. What is needed is a unified model for the entire data set. Loglinear models for counts have been extensively developed (e.g., Agresti, 1990). The multiplicative process identified earlier, AERs = Exposure × Intensity, is expressed in our notation as the loglinear model $\log(\mu) = \log(h) + \lambda$, where $\lambda = \Sigma x\beta$ (subscripts

TABLE I

SEQUENCE OF ANAPHYLAXIS AERs USED IN EXAMPLE 2

A. Data[a]

7(114,600), 7(83,000), 1(142,250), 2(95,100), 1(26,875), 0(161,300), 1(82,450), 2(160,375), 10(219,125), 0(86,700), 1(79,575), 2(150,900), 9(142,325), 9(161,825), 4(152,425), 8(164,250), 2(159,700), 15(238,125), 4(206,375), 9(208,875), 24(276,925), 23(232,625), 29(160,675)

B. Rate ratio intervals[b]

Lot	Adjusted formula		Standard formula		$\hat{R}$
13	0.78	8.74	**1.27**	**5.36**	2.61
14	0.61	6.50	0.99	4.04	2.00
18	0.88	4.96	**1.20**	**3.63**	2.08
21	**1.32**	**5.18**	**1.68**	**4.08**	2.62
22	**1.49**	**5.98**	**1.90**	**4.69**	2.98
23	**2.88**	**10.30**	**3.60**	**8.24**	5.45

C. p values[c]

Lot	Quasi-score	Norwood-Sampson
13	.13	**.01**
14	.27	**.05**
18	.10	**.01**
21	**.005**	**10^{-4}**
22	**.002**	**$<10^{-4}$**
23	**$<10^{-8}$**	**$<10^{-10}$**

[a] Number of AERs and number of doses (in parentheses) for each of 23 production lots.

[b] Interval estimates (95%) of rate ratios by the formula adjusted for overdispersion compared with the standard formula. Intervals not covering 1.0 are in bold type. Rate ratios for lots 21–23 are each relative to lots 1–20. Rate ratios for lots 13, 14, and 18 are each relative to all preceding lots.

[c] p Values by the quasi-score and Norwood-Sampson tests. Values ≤ 0.05 are in bold type. Adjustments for multiple comparisons were not made.

suppressed) so that the intensity is a function of explanatory factors. The model's random component is implied by μ, the expectation of y, and y is typically taken to follow the Poisson distribution.

When overdispersion is present, the random component is no longer strictly Poisson. Parametric regression models have been applied to overdispersed count data (e.g., Lawless, 1987; Dean et al., 1989; Consul and Famoye, 1992). We have chosen models based on the QL approach, which requires fewer assumptions. QL models for count data include a variance specification that allows for adjustment by dispersion. The models used here specify the variance as $\phi\mu$ (which is the Poisson variance when $\phi = 1$). Estimation can then be done as usual, using standard loglinear modeling software, and ϕ estimated as above. [Another widely used variance formulation is $\mu(1 + \tau\mu)$, and an iterative procedure is used for estimation (Breslow, 1984)]. It may also be that ϕ is not constant but varies as function of the covariates. Estimation of ϕ

along with the regression coefficients can be done by maximizing the extended quasi-likelihood (EQL) function (Nelder and Pregibon, 1987).

a. Example 3. A set of AERs is illustrated in Fig. 8a plotted by type-specific rate per month, and in Fig. 8b by type-specific rate per lot. Only marginal AER totals were available (AER counts by lot or by month, but not jointly by lot and month), so that separate models for each sequence had to be fit, rather than the more desirable joint model. Each type of sequence may highlight particular features of the data. The vaccination sequence may detect temporal changes and hence follow field conditions more closely. The production sequence may reflect the influence of manufacturing circumstances.

EQL models were fit to each sequence. For the vaccination sequence (Fig. 8a), a polynomial regression curve models the spike in reporting rates well, with little temporal effect evident before the spike. It is interesting that there is not just a change in the magnitude of the AER rate, but in its variability as well. Estimates of ϕ are 1.49 before the spike and 0.43 during the spike. This sharp reduction in dispersion during a period when reporting rates increased supports the view that the process was altered. Overdispersion during the "typical" period makes it difficult to model underlying temporal patterns that may be present, at least without much more data. In the production sequence (Fig. 8b), the effect of a formulation change is modeled using four covariates related to the new formulation. The shaded bars show the percent of each lot by the new formulation. AER rates appear to be lower with greater amounts of the new formulation.

VI. Summary

Any analysis of spontaneous AER data must consider the many biases inherent in the observation and reporting of vaccine adverse events. The absence of a clear probability structure requires statistical procedures to be used in a spirit of exploratory description rather than definitive confirmation. The extent of such descriptions should be temperate, without the implication that they extend to parent populations. It is important to recognize the presence of overdispersion in selecting methods and constructing models. Important stochastic or systematic features of the data may always be unknown.

Our attempts to delineate what constitutes an AER have not eliminated all the fuzziness in its definition. Some count every event in a report as a separate AER. Besides confusing the role of event and report, this introduces a complex correlational structure, since multi-

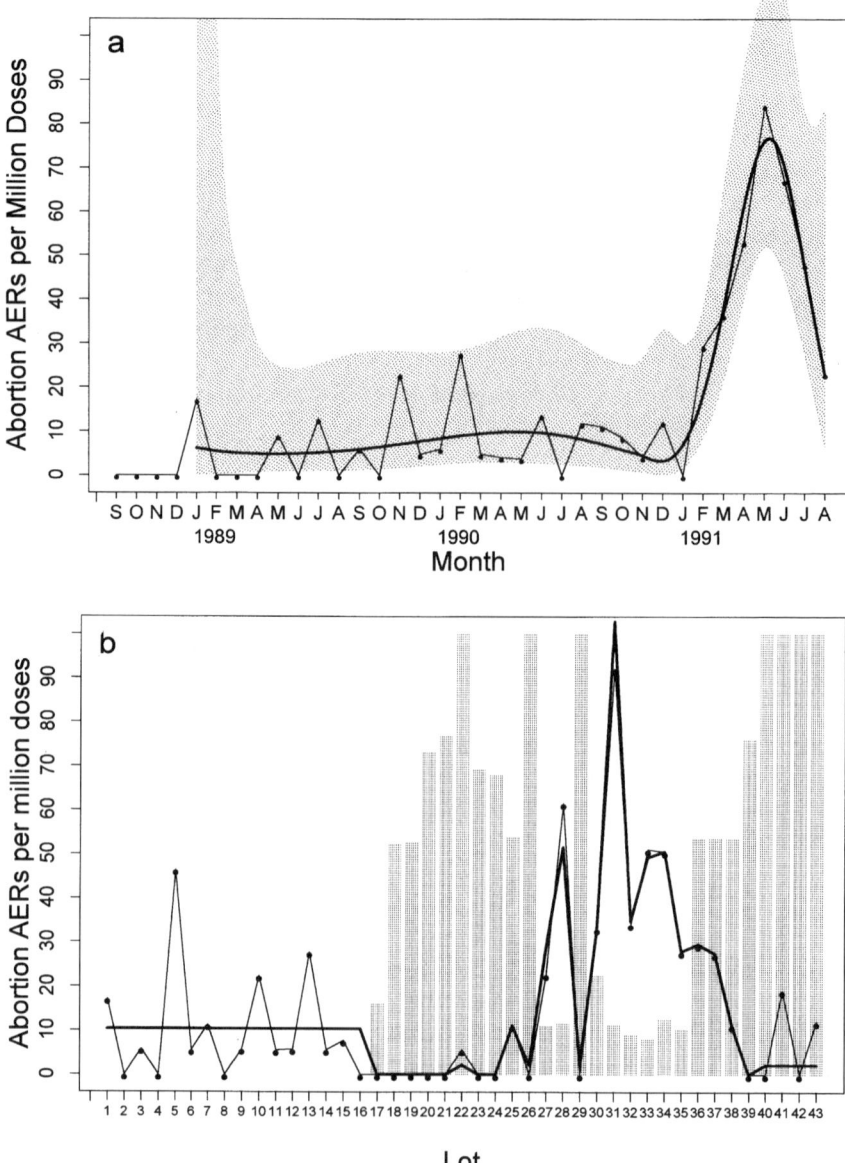

FIG. 8. Observed abortion AER rates (thin line) for a cattle vaccine and the fit of extended quasi-likelihood models to the data (thick line). (a) AER rates plotted by month. The shaded area is a joint 95% confidence band for the regression curve. (b) AER rates plotted by production lot. The shaded bars show the percent of each lot comprised of material manufactured according to a new formulation.

ple event descriptions received in a single report can hardly be considered independent. The many events described by one reporter would then become inordinately weighted. The alternative is to record an AER once, regardless of how many event descriptions it includes. As a practical compromise, many regard the simultaneous submission of several report forms by one reporter as a single AER, and the next submission by that reporter as another AER. This method is reasonable when reporters submit AERs very infrequently. When individual reporters make frequent reports, it becomes difficult to justify the inconsistency of counting multiple events as a single AER when they are submitted together, but as separate AERs when they are reported at different times.

While either choice is imperfect, the latter approach is currently used by the USDA and its licensed manufacturers in developing a mandatory postmarketing surveillance system for veterinary immunobiologicals in the United States. Under the proposed system, summaries of an estimated 10,000 AERs received annually by the manufacturers would be submitted to the USDA. In quantitative summaries, AERs received from lay consumers are usually weighted equally with those received from veterinary health professionals, although arguments have been advanced for separate classifications.

The emphasis on AER rate estimation differentiates the surveillance of veterinary vaccines by the USDA CVB from the surveillance of veterinary drugs as practiced by the Food and Drug Administration (FDA) Center for Veterinary Medicine (CVM). The FDA CVM does, in fact, perform a retrodictive causality assessment for individual AERs (Parkhie et al., 1995). This distinction reflects the differences between vaccines and drugs, as well as the difference in regulatory philosophy between the FDA and the USDA. The modified Kramer algorithm (Kramer et al., 1979) used by the FDA relies on features more appropriate to drug therapy than vaccination, such as an ongoing treatment regimen which allows evaluation of the response to dechallenge and rechallenge. In tracking AERs, the FDA has emphasized the inclusion of clinical manifestations on labels and inserts, while the USDA has been reluctant to have such information appear in product literature or to use postmarketing data for this purpose.

The potential for the misuse of spontaneous AER data is great. Disinformation is likely when the nature of this type of data is misunderstood and inappropriate analytical methods blindly employed. A greater danger lies in the glib transformation of AER data into something else entirely. Since approval before publication is not required, advertisements for veterinary vaccines appear with claims such as "over 3 million doses, 99.9905% satisfaction rating," or "11,500,000 doses, 99.98% reaction free." These claims, presumably based on spontaneous AERs, are almost fraudulent in their deceptiveness. Are we to suppose that 11.5 million vaccinations were observed for reactions?

In comparing the two advertisements, we find the second presumed AER rate is double the first. There is no basis for supposing that a comparison of this type provides any information whatsoever about the relative safety of the two vaccines. Comparisons of AER rates must be done with great caution, and the comparison between manufacturers is particularly tricky. Figure 9 shows the reporting rate ratios for the cat vaccines of three manufacturers relative to a fourth. Two of the manufacturers have about the same AER rate, but manufacturer C's rate is about three times that of manufacturer D. Should we conclude that C's vaccines are three times as reactive? This is certainly not so, since the vaccines of C and D were identical, produced by the same manufacturer who simply labeled one lot for C and another lot for D. There is a threefold difference in AER rates between vaccines that were identical except for their label or some factor confounded with the label.

It is apparent that the human element is the most critical one in the analysis of rates of spontaneous reports of adverse events. The specific

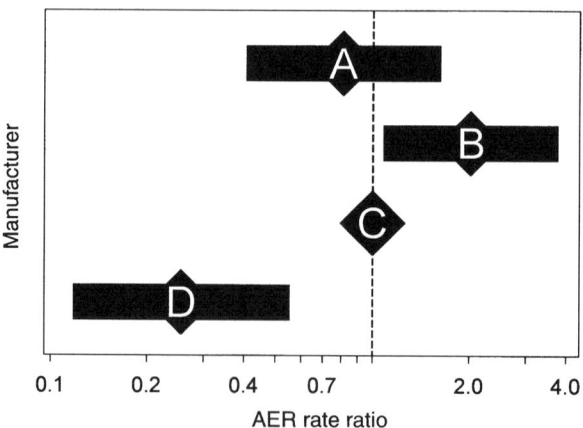

Fig. 9. Adverse event reporting rate ratios for four manufacturers' cat vaccines relative to manufacturer C. Rate ratios and 95% intervals were estimated from a quasi-likelihood model. Manufacturer A's interval overlaps the vertical line (rate ratio = 1), indicating no evident difference between the reporting rates of A and C.

features of spontaneous AER data, combined with the complexity of the U.S. veterinary biologics industry, cloud the interpretation of any unusual phenomenon observed in the surveillance of reporting rates. Automated procedures alone will not give consistently meaningful results. In view of their distinctive characteristics, reporting rates must be interpreted cautiously by experienced analysts using great discretion.

References

Advisory Committee on Immunization Practices (1997). Poliomyelitis prevention in the United States: Introduction of a sequential vaccination schedule of inactivated poliovirus vaccine followed by oral poliovirus vaccine. Recommendations of the Advisory Committee on Immunization Practices. *Morbid. Mortal. Wkly. Rep.* **46,** 1–25.

Agresti, A. (1990). "Categorical Data Analysis." Wiley, New York.

Breslow, N. E. (1984). Extra-Poisson variation in loglinear models. *Appl. Stat.* **33,** 38–44.

Charlton, B. G. (1996). Attribution of causation in epidemiology: Chain or mosaic? *J. Clin. Epidemiol.* **49,** 105–107.

Chen, R. T. (1994). Special methodological issues in pharmacoepidemiology studies of vaccine safety. In "Pharmacoepidemiology" (B. L. Strom, ed.), 2nd ed., pp. 581–594. Wiley, Chichester.

Consul, P. C., and Famoye, F. (1992). Generalized Poisson regression model. *Commun. Stat. Theory Methods* **21,** 89–109.

Dean, C. B. (1992). Testing for overdispersion in Poisson and binomial regression models. *J. Am. Stat. Assoc.* **87,** 451–457.

Dean, C. B., and Lawless, J. F. (1989). Tests for detecting overdispersion in Poisson regression models. *J. Am. Stat. Assoc.* **84,** 467–472.

Dean, C. B., Lawless, J. F., and Willmot, G. E. (1989). A mixed Poisson-inverse Gaussian regression model. *Can. J. Stat.* **17,** 171–181.

Fine, P. E. M., and Chen, R. T. (1992). Confounding in studies of adverse reactions to vaccines. *Am. J. Epidemiol.* **136,** 121–135.

Finney, D. J. (1982). The detection of adverse reactions to therapeutic drugs. *Stat. Med.* **1,** 153–161.

Gart, J. J. (1985). Approximate tests and interval estimation of the common relative risk in the combination of 2 × 2 tables. *Biometrika* **72,** 673–677.

Glymour, C. (1986). Statistics and metaphysics. *J. Am. Stat. Assoc.* **81,** 964–966.

Greenwood, M., and Yule, U. G. (1915). The statistics of anti-typhoid and anti-cholera inoculations, and the interpretations of such statistics in general. *Proc. R. Soc. Med.* **8,** 113–194.

Hall, D. A. (1989). Analysis of surveillance data: A rationale for statistical tests with comments on confidence intervals and statistical models. *Stat. Med.* **8,** 273–278.

Kramer, M. S., Leventhal, J. M., Hutchinson, T. A., and Feinstein, A. R. (1979). An algorithm for the operational assessment of adverse drug reactions. 1. Background, description, and instructions for use. *JAMA, J. Am. Med. Assoc.* **242,** 623–632.

Lawless, J. F. (1987). Negative binomial and mixed Poisson regression. *Can. J. Stat.* **15,** 209–225.

Levy, S. A., Lissman, B. A., and Fick, C. M. (1993). Performance of a *Borrelia burgdorferi* bacterin in borreliosis-endemic areas. *J. Am. Vet. Med. Assoc.* **202,** 1834–1838.

Nelder, J. A., and Pregibon, D. (1987). An extended quasi-likelihood function. *Biometrika* **74,** 221–232.
Norwood, P. K., and Sampson, A. R. (1988). A statistical methodology for postmarketing surveillance of adverse drug reaction reports. *Stat. Med.* **7,** 1023–1030.
Parkhie, M., Sessions, W., Sharar, M., and Keller, W. (1995). Animal adverse drug experience monitoring program. *Annu. Meet., Am. Vet. Med. Assoc.* Pittsburgh, PA, 1995.
Praus, M., Schindel, F., Fescharek, R., and Schwarz, S. (1993). Alert systems for postmarketing surveillance of adverse drug reactions. *Stat. Med.* **12,** 2383–2393.
Rosenthal, S., and Chen, R. (1995). The reporting sensitivities of two passive surveillance systems for vaccine adverse events. *Am. J. Public Health* **85,** 1706–1709.
Siev, D. (1994). Estimating vaccine efficacy in prospective studies. *Prev. Vet. Med.* **20,** 279–296.
Strebel, P. M., Sutter, R. W., and Cochi, S. L. (1992). Epidemiology of poliomyelitis in the United States one decade after the last reported case of indigenous wild virus-associated disease. *Clin. Infect. Dis.* **14,** 568–579.
Wedderburn, R. W. M. (1974). Quasi-likelihood functions, generalized linear models, and the Gauss-Newton Method. *Biometrika* **61,** 439–447.

Index

A

Abortion
  bovine genital tract infections and, 218
  bovine trichomoniasis and, 224
  brucellosis and, 218
  equine abortion virus and, 361–363
  equine arteritis virus and, 363–364
  Haemophilus somnus infection and, 219
$\alpha 1$-Acid glycoprotein, 644, 646, 647, 648
Acquired immunity, *see also* Active immunity; Protective immunity
  cytokine modulation of, 449–450, 458
Actinobacillus pleuropneumoniae, 46, 421, 423, 427
Active immunity, *see also* Acquired immunity; Protective immunity
  inhibition of, immunoglobulin G and, 172
  in poultry, 483
  pseudorabies vaccines and, 621–622
Active substance, new European definition of, 599–600
Acute phase protein assays
  biochemical, 647–648
  immunochemical, 648–649
  international reference preparations and, 651–652
  need for international standardization in, 644–645, 647, 649–650
Acute phase proteins, 643–647
Adenovirus vector, 151
Adhesion molecules, *see also individual molecules*
  effects of glucocorticoids on, 74
  leukocyte trafficking and, 65, 68–69
  lymphocyte trafficking and, 62–63

Adjuvants, 4–5; *see also* Freund's adjuvant; Mucosal adjuvanticity
  antigen presenting cells and, 407
  cholera toxin and, 85–98, 99, 106
    B subunit, 106, 110, 111
  for culture filtrate fraction vaccine, 140
  cytokines as, 448–449, 450, 457–458, 486–487
  in DNA immunization, 168, 169
  in fish immunization, 541
  for foot-and-mouth disease vaccine, 198
  heat labile toxin and, 106, 110
  in helminth vaccines, 246, 247
  immune stimulating complexes and, 405, 406, 408–409
  injection site reactions and, 169, 620
  modulation of immune responses by, 448–449
  for transmissible gastroenteritis virus vaccines, 440
  in USDA type I recombinant vaccines, 573
  virus vectors and, 148–149
ADP ribosylation, cholera toxin and, 84, 85, 106
Adrenocorticotropic hormone, 63, 70
Adverse event report data
  bias and, 755–757, 769
  concepts of causality and, 754–755
  confounding events and, 756–757
  defining single events in, 769, 771
  difficulties in interpreting, 749–750, 757–758, 772–773
  event reporting process and, 753
  hypothesis testing and, 758, 764–766
  interval estimation and, 763–764
  loglinear modeling and, 766–769

Adverse event report data (cont.)
  misused in marketing, 771–772
  overdispersion and, 760–762, 769
  probability and, 757, 769
  quantitative analysis of, 758–769
  reporting rate vs. incidence rate in, 755–756
Adverse reactions
  autoimmunity and, 734
  causes of, 684–691, 696
  clinical overview of, 716–717
  immune-mediated disease, 715
  overvaccination and, 701–702, 718
  polyvalent vaccines and, 718–719
  postmarketing surveillance and, 708–709, 710
  sources of data on, 704–706
  types of, 703–704
  vaccine contamination and, 717–718
Advertisements, misuse of adverse event report data in, 771–772
Aeromonas salmonicida, 525, 547–549; see also Furunculosis
Aerosolization, mink distemper vaccine and, 562
Afferent lymph veiled cells, 284–285
African cape hunting dogs, 552
African horse sickness, 365–366
Agammaglobulinemia, 452
AgI/II adhesin, 106–108, 109
AIDS, see also HIV-1
  Bartonella henselae and, 340
  infectious disease and, 334–335
  Toxoplasma gondii and, 335, 336
Ailurus fulgens, 552
Akitas, vaccine-associated disease in, 721–723
Albumin, acute phase proteins and, 645
Aleutian disease parvovirus, 558, 563
Allergies, adverse vaccine reactions and, 690, 719
Alpha herpesviruses, 361–362, 451
Alphavirus, 364
Aluminum hydroxide, in helminth vaccines, 246, 247
Amblyomma americanum, 662
American Association of Feline Practitioners, on feline infectious peritonitis vaccination, 355–356

American Veterinary Medical Association, 658
Aminopeptidases, 245
Amyloid A protein, 646
Amyloidosis, 646, 722
Anamnestic eosinophilia, 391, 400
Anamnestic recall response, see also Secondary immune response
  HIV-1 infection and, 127
Anaphylaxis, 304, 637, 688–689
Animal and Plant Health Inspection Service, 586, 587, 590, 591, 592, 593, 627, 682
Animal breeding
  genetic health enhancement and, 40, 41–49
  value of acute phase protein assays to, 644
Animal growth
  conjugated linoleic acid and, 56
  immune-induced wasting and, 53–57
Animal health
  effects of vaccination on, 664–665
  genetics and, 40–41
  genetic enhancement of, 39–40, 41–49
  partnership model for, 667
Animal management, occurrence of infectious disease and, 417
Animal owners
  responsibilities of, 6
  vaccine label information and, 634
Animal products, antimicrobial residues and, 664
Animals
  age of, vaccine label statements on, 639–640
  conjugated linoleic acid and, 56
  DNA immunization and, 165, 175
  geriatric, frequency of vaccination and, 725–726, 728
  immune-induced wasting and, 53–57
  nonvaccinated, see Nonvaccinated animals
  young, see Neonates
Animal side tests, 210
Annual vaccinations, see also Revaccination
  adverse reactions and, 701–702, 718
  canine vaccines and, 295, 303

INDEX

for dogs, 318
vaccine label statements on, 639
Anorexia, immune-induced, 55, 56–57
Anterior uveitis, 689
Anthelmitics, 390
Anthrax vaccines, 11–17, 22
Antibiotics
  in aquaculture, 540, 542–543
  effects on animal performance, 54
  residues in animal products, 664
  resistance to, 664
Antibodies, see also Autoantibodies; Maternal antibodies
  anti-DNA, 173
  antigen recognition and, 182–184
  in bovine respiratory syncytial virus immunity, 188
  in deceptive imprinting, 118, 119, 128
  DNA immunization and, 165, 168–169, 173
  passive, see Passive immunity
Antibody-dependent enhancement, 351, 352–353, 354, 355
Antibody detection
  for bovine herpesvirus 1, 205–206
  for bovine virus diarrhea virus, 209
  enzyme-linked immunoabsorbent assays and, 372
  for foot-and-mouth disease, 200
  in poultry, 488–489
Antibody-mediated immune response, see Humoral immune response
Antibody-secreting cells
  in mucosal immune response, 435, 436, 438, 439
  parasite antigens and, 243–244
  porcine group A rotaviruses and, 441–442
$\alpha 1$-Antichymotrypsin, 648
Antidistemper vaccines, see Canine distemper vaccines
Anti-DNA antibodies, 173
Anti-fibronectin antibodies, 738, 742–743, 744, 745
Antigastrointestinal peptides, 56–57
Antigen(s), see also Epitopes; Novel antigens; Protective antigens
  ASC probes and, 243–244
  in B cell evolution, 123

classes of, 122–123
conventional and unconventional, 121
effects on DNA immunization, 167–168, 170
of *Haemophilus somnus*, 220
hypersensitivity reactions to, 688–690
immune stimulating complexes and, 405, 406, 407, 408–409
multideterminant, 115
produced by virus vectors, 148, 467, 473
recognition, 182–184
T cell-recognized, identification of, 186–187
unconventional interactions with antibodies, 128
VH chains and, 121
Antigen-antibody complexes, 510
Antigenic competition, 121–122
Antigenic diversity
  in bovine virus diarrhea virus, 208
  factors effecting vaccination, 660
  vaccine failure and, 129, 695
Antigen presentation
  cholera toxin adjuvanticity and, 92
  DNA immunization and, 165
  in vaccine development, 665
Antigen presenting cells, see also Afferent lymph veiled cells; Dendritic cells
  adjuvants and, 407
  cholera toxin and, 93
  immune stimulating complexes and, 410
  impact on T cell responses, 186
  plasmid-induced immunity and, 167
  T cell responses and, 275–276
Antigen processing, intracellular pathways, 184–185
Anti-laminin antibodies, 738–739, 742, 743, 744
$\alpha 1$-Antiprotease, 648
Antiserum, in acute phase protein assays, 649
Aphtovirus, 198; see also Foot-and-mouth disease virus
Aquaculture, see also Fish; Fish vaccines
  antibiotics and, 540, 542–543
  infectious diseases and, 539–540
  productivity of, 539
  vaccines and, 540, 550
Arboviruses, 29

Arteritis, 390
Arteriviruses, 363, 448
Arthus reaction, 689
ASC probes, 243–244
Aspergillosis, 499
Aspergillus flavus, 499
Aspergillus fumigatus, 499
Aspergillus niger, 499
Assays, *see specific types*
Astroviruses, 430
Atheresclerosis, 56
Atlantic salmon, *Vibrio viscosus* and, 544, 546–547
Atrophic rhinitis vaccines, 420, 421, 422–423, 427
Attenuation
  in early vaccinology, 12–16
  live, virus vectors and, 149–151
Aujeszky's disease, 451, 464; *see also* Pseudorabies virus
Autoantibodies, in vaccine-induced autoimmunity experiments, 738–745
Autogenous vaccines, licensing of, 592–593
Autoimmune disease, *see also* Immune-mediated disease; *specific diseases*
  vaccine-induced, 690, 705–705, 720–721, 723, 734–745
Autoimmune hemolytic anemia, 717, 719–720, 734; *see also* Immune-mediated hemolytic anemia
Autoimmune thyroiditis, 720, 723, 742
Auxotrophic vaccines, tuberculosis and, 138–139
Avian botulism, 554–555
Avian immune system, 487–490
Avian influenza vaccines, 521
Avian leukosis viruses, 482
Avianpox virus vectors
  in mammalian hosts, 476
  in recombinant vaccines, 467–469
Avian Products Standardization Committee, 612
Avipoxvirus, 518

B

Bacille Calmette-Guérin vaccine, 135, 136, 141
Bacillus anthracis, 17; *see also* Anthrax

Bacteremia, in *Bartonella henselae* infection, 340
Bacterial antigens, in B cell evolution, 123
Bacterial polysaccharides, homology with HIV-1 envelope, 127–128
Bacterial vectors, *see also* Salmonella vectors
  for mucosal vaccines, 108–109, 110–111
  for transmissible gastroenteritis vaccines, 440
Bacterins
  for clinical coliform mastitis, 265–267
  hypersensitive reactions to, 689–690
  licensing procedures, United States, 587–593
  in ovo vaccination and, 511–512
  for *Staphylococcus aureus,* 258–260, 261–262
Badgers, 552
Bartonella henselae, 335, 338–340, 342
Batch control/release
  European regulation of, 602
  in pseudorabies vaccines, 624
Bat-eared foxes, 553
Bats, as rabies carriers, 29, 34
B cells, *see also* Antibody-secreting cells
  antigenic competition and, 121–122
  in avian immune system, 487, 488, 490
  cholera toxin and, 85
  clonal dominance and, 120
  evolution in, bacterial antigens and, 123
  gangliosides and, 85
  HIV-1 infection and, 119–120, 123
  immune stimulating complexes and, 408, 409
  immunodominant epitopes and, 128
  impairment of, in periparturient cows, 72
  isotype differentiation in, cholera toxin adjuvanticity and, 92–93
  in mucosal immune response, 434, 435
  self-reactive, in vaccine-induced autoimmunity, 743–744
  trafficking and, 63, 69
  VH genes and, 120–121
Beagles, vaccine-induced autoimmunity and, 735–745

Bias
    in adverse event report data, 755–757, 769
    in postmarketing surveillance, 753
Bicinchoninic acid reaction, 648, 649
Bicistronic plasmids, 168
Biogeography, effects on vaccination, 662
Biologicals (journal), 611
Biotechnology
    recombinant vaccines and, 521–522, 581–582
    vaccine development and, 182, 517, 665
Birds, *see also* Poultry
    type C botulism and, 554–555
Black-footed ferrets, 552, 559
Blastogenesis, suppression, in blood lymphocytes, 687
Blood lymphocytes, suppression of blastogenesis in, 687
Blowflies, 554
Blue catfish, *see* Catfish
"Blue eye," 314, 689
Blue tongue virus, 718
Bone marrow, immune response in, 408, 409
Booster vaccinations, *see* Annual vaccinations; Revaccination
Bordetella bronchiseptica
    canine distemper and, 20–21
    kennel cough and, 315–316
Bordetella bronchiseptica vaccines, 316, 420, 423, 427
Borrelia burgdorferi, 316–317, 368–369; *see also* Lyme disease
Borrelia hermsii, 369
Borrelia theileri, 369
Botulism, Type C, in wild animals, 553–555
Bovine herpesvirus 1
    diagnostics for, 205–206
    distribution, 659
    early detection of, 31
    economics of vaccination for, 661
    eradication of, 197, 206–207
    herd immunity and, 210–211
Bovine herpesvirus 1 vaccines
    administering with other vaccines, 695
    efficacy of, 202–204
    marker vaccines, 201–202
    overview of, 201
    residual virulence and, 686
    safety of, 204–205
    vaccine-induced immunosuppression and, 687
Bovine leukemia, 40
Bovine leukosis virus, 29, 31
Bovine mastitis
    diagnostics for, 267–268, 269
    economics of, 257, 266
    etiologic agents of, 258
Bovine mastitis vaccines
    for clinical coliform mastitis, 264–267
    difficulties in, 257–258, 268
    economics of, 257, 266
    Mycoplasma bovis and, 263
    overview of, 268–269
    polyvalent, 261
    for *Staphylococcus aureus*, 258–262
    for *Streptococcus agalactiae*, 261, 262
    Streptococcus dysgalactiae and, 262–263
    for *Streptococcus uberis*, 261, 263–264
Bovine respiratory disease, 207–208
Bovine respiratory syncytial vaccines, 689
Bovine respiratory syncytial virus
    CD8 T cells and, 277
    characteristics of, 187–188
    immune response to, 188–191, 192
Bovine serum, in vaccines, cross-reactivity and, 743, 745
Bovine trichomoniasis, *see* Tritrichomonas foetus
Bovine tuberculosis, *see* Tuberculosis
Bovine vaccines, *see also* specific vaccines
    contamination of, 717
Bovine venereal campylobacteriosis, *see* Campylobacteriosis
Bovine viral diseases
    ecology of infection and, 29
    eradication of, 197–198
    posteradication period and, 210–211
Bovine virus diarrhea vaccines
    efficacy of, 208
    onset of immunity and, 211
    overview of, 207–208
    safety of, 208–209
    triggering of mucosal disease by, 690–691

Bovine virus diarrhea vaccines (*cont.*)
   vaccine-induced immunosuppression and, 207, 687
Bovine virus diarrhea virus
   antigenic diversity and, 660
   CD4 T cells and, 277
   in contamination of bovine vaccines, 717
   diagnostics for, 209
   distribution, 659
   eradication of, 197, 209–210
Breeders, canine, vaccine use and, 291
Breeds, *see* Dogs
Bronchial lymph nodes
   enteropathogenic viruses and, 433, 436–439
   functional compartmentalization in, 439
"Brooder pneumonia," 499
Brucella abortus, 218–219, 232
Brucella abortus vaccines, 524–525, 532
Bulls
   campylobacteriosis and, 230
   trichomoniasis and, 225
Burdon-Sanderson, J.S., 15, 22
Bursa of Fabricius, 487, 488, 507, 511
Bursaplex, 507–509

C

Calciviruses, 3
Camplylobacter fetus subsp. *venerealis,* 221–224
Campylobacteriosis, *see also* Vibrosis
   diagnosis of, 223–224
   evasion of host defenses by, 223, 231
   genital mucosal inductive sites of, 228–229
   immune response to, 222–223, 228, 230
   pathogenesis of, 221–222
   similarity with trichomoniasis, 224
   vaccine development and, 233
Campylobacter jejuni, 512
Canarypox virus vector, 151, 153
Canine adenovirus type 1, 314, 726
Canine adenovirus type 1 vaccines, 294–295
   "blue eye" and, 314, 689

Canine adenovirus type 2, 314, 315
Canine adenovirus type 2 vaccines
   for canine adenovirus type 1, 314
   canine hepatitis and, 302–303
   overadministration of, 728
   overview of, 315
Canine coronavirus, 349
   antibody-dependent enhancement and, 351
   in canine parvovirus mixed infections, 301, 313
   outbreaks of, 302
   overview of, 300–301
Canine coronavirus vaccines
   efficacy and, 291
   overview of, 301–302, 313–314
   in polyvalent canine distemper vaccines, 294
   safety issues and, 293
Canine distemper encephalitis, 294–295, 296, 301–302, 312
Canine distemper vaccines, *see also* Mink distemper vaccines
   annual vaccination and, 5, 295, 303
   canarypox virus vector and, 153
   development of, 8
   duration of immunity and, 295, 296, 726
   early vaccines, 17–22
   efficacy issues and, 293
   maternal antibodies and, 294, 295
   minimal immunizing dose for, 294, 295, 296
   overview of, 311–312
   in polyvalent vaccines, 294–295, 301–302, 314
   postvaccinal encephalitis and, 294–295, 296, 301–302, 312
   production of, 293–294
   protective titers for, 726–727
   recombinant, 296, 312
   residual virulence and, 686, 687
   safety issues and, 293
   success of, 302
   wild animals and, 552–553
Canine distemper virus, *see also* Mink distemper
   clinical course of, changes in, 568
   distribution, 659
   first isolation of, 19

future research and, 567–568
induced outbreaks and, 566–567
infectious period in, 565–566
production of inclusion bodies by, 561
in wild animals, 551–553
Canine hepatitis, 302–303, 314
Canine herpesvirus, 26
Canine measles, *see* Canine distemper
Canine parainfluenza virus, 315
Canine parvovirus, 411, 659, 661
Canine parvovirus type 2, 298–299, 301, 312–313
Canine parvovirus type 2 vaccines
  administration to pups, 297–298, 299–300
  efficacy and, 291, 298–300
  immunity duration in, 296–297, 300
  immunosuppression and, 298
  maternal antibodies and, 297, 313
  in polyvalent vaccines, 314
  revaccination and, 303
  safety and, 300
  types of, 296–297
Canine parvovirus vaccines
  adverse reactions to, 722
  age at first administration, 661
  contamination of, 718
  duration of immunity and, 726
  efficacy and, 293
  overview of, 312–313
  protective titers for, 726–727
  virus-antibody complexes and, 510
Canine rotavirus, 718
Canine vaccines, *see also specific vaccines*
  disease outbreaks, nonvaccinated populations and, 291, 319
  efficacy of, 291–292, 293
  frequency of administration, 303, 310, 318
  history of, 309
  intranasal administration, 310
  modified live virus strains and, 292–293, 309–310
  overview of, 309–310, 318–319
  public awareness of, 289–291, 304
  quality of, 304
  recombinant, 292
  safety issues and, 291, 293
  in small animal practice, 290
  successes of, 302–303

Caprine arthritis-encephalitis, 29
Capsular polysaccharide, type 5, 261
Capsular proteins, 260, 261
Carbohydrate antigen type 1, 46, 48
Carbohydrate antigen type 5, 46, 48
Carré, H., 21, 22
Carrier organisms, *see* Bacterial vectors; Virus vectors
Caseous necrosis, 140
Catfish, *Edwardsiella ictaluri* and, 524, 525–535
Cats
  adverse vaccine reactions in, 717
  Bartonella henselae and, 338–340
  factors affecting vaccination in, 660–661
  feline infectious peritonitis vaccination and, 353
  Helicobacter pylori and, 341
  overvaccination and, 718
  protective titers for, 727
  Toxoplasma gondii and, 336–338
  vaccine-induced fibrosarcomas and, 5, 704–705, 709
  Yersinia pestis and, 342
  zoonotic pathogens and, 335–342
Cat scratch disease, 338–340; *see also Bartonella henselae*
Cattle
  acute phase proteins in, 646
  botulism and, 554
  class I MHC genes in, identification of, 190
  DNA vaccines and, 175
  immunization development and, 285
  stress-related immunosuppression in, 71–74
  T cells in
    CD45 subpopulations, 278–282
    dendritic cell-induced responses, 284–285
    γδ T cell activation and function, 282–284
    γδ T cell trafficking, 64
    identification of subpopulations, 276
    in vivo roles of, 276–278
  vaccine contamination and, 717
  vaccine costs and, 175
  vaccine-induced mucosal disease in, 690–691

Causality, concepts of, 754–755
CD11a/CD18 integrin, 67–68, 69
CD11b/CD18 integrin, 62, 68
CD31 immunoglobulin, 69
CD45 antigen, in T cell subpopulations, 278–282
CD62E adhesion molecule, 63, 66
CD62L adhesion molecule
  γδ T cells and, 64, 73
  glucocorticoids and, 73–74
  leukocyte trafficking and, 62, 66, 67, 69, 70
CD62P adhesion molecule, 63, 66
Cell death, in B cells, rescue from, 744
Cell-mediated immune response, *see also* Cytotoxic T cell immune response
  in avian immune system, 489–490
  in disease resistance, 42
  DNA immunization and, 165
  feline infectious peritonitis and, 350
  heritability and, 43
  herpesviruses and, 454–455
  immune-response selected pigs and, 45–46, 48–49
  modulation by interleukin 12, 450, 455–456
  in protective immunity, 524
  pseudorabies virus vaccines and, 451–452, 455–456
  suppression of during pregnancy, 71–72
  Toxoplasma gondii and, 337
  tuberculosis and, 137–138
Cell surface receptors, *see* Integrins
Cellular compartments, plasmid-induced immunity and, 168
Center for Veterinary Biologics, 627, 682
  adverse event report data and, 771
  licensing procedures and, 587–588
  units within, 586–587
  vaccine efficacy and, 751
Center for Veterinary Medicine, 771
Cerebellar hypoplasia, 687
Ceruloplasmin, 648
Challenge trials
  bovine herpesvirus 1 vaccines and, 202–203
  for clinical coliform mastitis vaccines, 266
  leading to vaccine failure, 694

  in licensing procedures, 589
  for mink distemper vaccine, 560–561
  for *Streptococcus uberis* vaccines, 263–264
  swine vaccines and, 420
  vaccine efficacy and, 751
Channel catfish
  annual production of, 539
  Edwardsiella ictaluri and, 524, 525–535
beta-Chemokines, 411
Chick cell agglutination test, 603
Chicken embryo propagated vaccines, for mink distemper, 558–564
Chickens, *see also* Poultry
  immune system of, 487–490
Chinese Shar Pei dogs, 721
Chlamydia, 29, 31
Chlamydia psittici, 31; *see also* Psittacosis
Chloride ion concentration test, 267
Cholecystokinin, immune-induced anorexia and, 56–57
Cholera toxin
  adjuvanticity of, 85–98, 99, 106
  dual-signal integration and, 94
  interleukin 1 and, 94–96
  interleukin 12 and, 97–98
  SBR-CTA2/B, 106–109
  structure of, 84
  A subunit, 84–85, 106
Cholera toxin, B subunit
  activity of, 84, 85, 92, 93
  adjuvanticity of, 106, 110, 111
  cholera toxin adjuvanticity and, 86–92, 99
  dual-signal integration and, 94
  immunogenicity of, 85, 86
  interleukin 1 and, 94–96
  interleukin 12 and, 97–98
Chromogen, 648
Chromogenic limulus amebocyte lysate test, 267
Chronic fatigue syndrome, 73
Clarius batrachus, 525
Class I antigen processing, 184–185
Class II antigen processing, 184–185
Clearance
  Edwardsiella ictaluri RE-33 vaccine and, 529, 531, 532, 534
  role of antibodies in, 168–169

INDEX

Climate, effects on vaccination, 662
Clonal dominance, deceptive imprinting and, 118, 120, 122
Clonal expansion, in deceptive imprinting, 119
Clostridiosis vaccine, 421
Clostridium botulinum Type C, 553–555
Clostridium perfringens C vaccine, 420
Coccidiosis, 512
Codified efficacy testing, 635
Cold-water vibrosis, 547
Colibacillosis vaccine, 421, 422, 426
Colic, 395, 399
Coliform mastitis, clinical, 264–267, 269
Color genes, in mink, distemper mortality and, 567–568
Combined vaccines, *see* Polyvalent vaccines
Committee for Veterinary Medicinal Products, 385–386, 596–597
Common mucosal immune system, 434–435
Companion animals
  benefits of DNA immunization and, 175
  risk management for, 709
  use of vaccines for, 290
  vaccine safety concerns and, 701–702
Compartments, cellular, plasmid-induced immunity and, 168
Complete Freund's adjuvant, *see* Freund's adjuvant, complete
Concomitant immunity, 117
Conditional licenses, 591
Conjugated linoleic acid, immune-induced wasting and, 55–56
Contagious equine metritis, 370–371
Conventional antigens, 121
Core antigen vaccines, for clinical coliform mastitis, 265–267, 269
Corneal edema, 689
Cornyebacterium equi, 372
Coronaviruses, 349, 430, 431; *see also* Canine coronavirus; Feline coronavirus; Porcine respiratory coronavirus
Corticosteroids, in treatment of vaccine-associated disease in dogs, 724

Cortisol, vaccine-induced autoimmunity and, 738
Costimulatory molecules, *see* Adjuvants
Cowpox, 8, 9, 10, 11
Coxiella burnetti, 342
Coyotes, controlling rabies in, 580–581
Crohn's disease, 96
Cryptosporidium parvum, 96, 342
Culicoides spp., 365
Culture filtrate fraction vaccine, 140
Cutaneous lymphocyte-associated antigen, 62, 70
Cyclic AMP (cAMP), cholera toxin and, 84–85
Cysteine proteases, 231
  of liver fluke, 244–245, 246
Cytokine assays, acute phase protein assays and, 646
Cytokines
  acute phase proteins and, 645–646
  as adjuvants, 5, 448–449, 457–458, 486–487
  CD45 T cell subpopulations and, 279–280
  cholera toxin induction of, 85, 93
  effects of glucocorticoids on, 73–74
  excessive induction of, 687–688
  in helminth-induced responses, 246
  ICAM-1 expression and, 68
  immune-induced wasting and, 54, 55
  in immune-response selected pigs, 46
  induction by immune stimulating complexes, 410
  in mucosal adjuvanticity, 94–99
  in periparturient immunosuppression, 71–72
  SBR-CTA2/B chimeric cholera toxin protein and, 108
  stimulation of selectin expression by, 66–67
  in *Strongylus vulgaris* infections, 391, 397, 400–402
Cytomegalovirus, 450
Cytotoxic T cell immune response, *see also* Cell-mediated immune response
  in bovine respiratory syncytial virus immunity, 189, 190–191, 192
  DNA immunization and, 166, 175
  epitope recognition in, MHC polymorphism and, 183

Cytotoxic T cell immune response (*cont.*)
  feline immunodeficiency virus vaccines and, 327
  immune stimulating complexes and, 407
  pseudorabies virus vaccines and, 451–452

D

Dairy cattle, *see also* Cattle
  disease resistance traits, heritability of, 41
Deceptive imprinting
  B cell abnormalities in, 119–122
  described, 117, 118
  function of, 129
  HIV-1 modeling of, 117–119
  primary response to HIV-1 infection and, 123–128
  suppression of T cell polyclonal response in, 122
  vaccine failure and, 129–130
Delayed-type hypersensitivity immune response
  cutaneous, in immune-response selected pigs, 43, 44
  immune stimulating complexes and, 410
  tuberculosis and, 138
Deletion mutations
  in live attenuation of virus vectors, 150
  in marker vaccine design, 201
Dendritic cells, *see also* Afferent lymph veiled cells
  induction of T cell responses by, 282–284
  plasmid-induced immunity and, 167
  virus-antibody complexes and, 511
Depression, 395
Dermatophilus congolense, 10
Dexamethasone, 72–73, 73, 74
Diagnosis
  concepts of causality and, 754–755
  distinguished from infection detection, 27–31
Diagnostic medicine
  future of, 34–36
  infection detection and, 25–26, 28–31

  methods in, 31–32
  predictive value of preclinical testing in, 33–34
  traditional approaches to, 26–28
Diagnostic tests, *see also individual tests;* Preclinical testing
  controlling disease transmission and, 360, 372
  future of, 34–36
  indirect, enzyme-linked immunoabsorbent assays and, 372
  in infection detection, 26
  international reference reagents and, 676–677
  levels of sensitivity in, 32
  OIE Standards Commission and, 672–677, 678
Diapedesis, 65, 68
Diarrheal disease
  enterpathogenic viruses and, 430, 433
  in swine, economic costs of, 464–465
DiGeorge's syndrome, 452
Dilutents, 693
Diptheria-tetanus-pertussis vaccine, 734
Disease, *see* Autoimmune disease; Diarrheal disease; Exotic disease; Immune-mediated disease; Infectious disease; Sexually transmitted disease; Zoonotic disease
Disease prevention, *see also* Preventive medicine
  epidemiological approach to, 415–417
  role of vaccines in, 666
Disease resistance, genetic predisposition and, 39–49
Disease transmission, *see also* Mosquito-transmitted viruses; Sexually transmitted disease; Tick-borne disease; Transmissibility; Virus shedding
  controlling with vaccination, 664
  of distemper in ferrets, 564
  eradication of carrier states and, 175–176
  feral swine and, 464
  fleas and, 339–340
  long-distance, horses and, 359–360, 372, 379–380
Distemper, *see* Canine distemper; Mink distemper

Distemper vaccines, *see* Canine distemper vaccines
DNA, immunomodulatory effects of, 170
DNA immunization
  antibody isotypes and, 168–169
  development of, 163–164
  DNA integration and, 164, 173
  DNA persistence and, 173
  DNA shedding and, 173
  economics of, 175
  efficacy of, 164
  exotic diseases and, 175–176
  future of, 174–176
  humoral immune response and, 164, 165
  induction of immunity and, 166–168
  injection site reactions and, 169, 175
  mucosal immune system and, 169–170, 175
  passive antibodies and, 171–172, 174
  polyvalent, 174–175
  regulatory issues and, 172–173
  universality of, 164–166
  vaccine delivery in, 169–171
DNA technology, *see* Biotechnology; DNA immunization
DNA vaccines, 4
  for feline immunodeficiency virus, 329
  for poultry, 520–521
  tuberculosis and, 140–141
DNA virus vectors, 147
Doberman pinschers, 661
Dog owners, vaccine use and, 291
Dogs
  adverse vaccine reactions in, 719–725, 728
  overvaccination and, 718
  susceptibility to disease and, 661
  vaccine-induced autoimmunity and, 705–706, 734–745
Dosages, *see* Minimal immunizing dose
Drenching, 250
Drug residues, 664
Drug resistance, 664
Dual-signal integration immune response, 93–94
Dunking, G.W., 22
Duration of immunity, *see* Immunity, duration of

Dusting, mink distemper vaccine and, 562

E

Ear, DNA vaccine delivery and, 170
Eastern equine encephalitis, 364
Ecchymotic hemorrhages, *see* Immune-mediated thrombocytopenia
Ecology of infection, 29–31
Ecosystems, effects on vaccination, 662
Edwardsiella ictaluri, 523; *see also* Enteric septicemia of catfish
  bacterins, 524, 525
  EILO isolate, 525, 532
Edwardsiella ictaluri RE-33 vaccine
  characteristics of, 527–528
  clearance and, 529, 531, 532, 534
  production of, 525–527
  protective immunity and, 531–531, 534–535
  reversion to virulence and, 529
  safety and, 528–529, 534
Effector cells, in helminth-induced responses, 245
Efficacy, *see* Vaccine efficacy; Vaccine efficacy testing
Egg injection equipment, 497–498; *see also* Inovoject egg injection system
Eggs, *see also* In ovo vaccination
  hatchability, 500–502
  in ovo vaccine delivery and, 485, 490
Ehrlichia equi, 368
Ehrlichia ristcii, 368
Ehrlichiosis, 26, 368, 662
EIA test, 124, 126, 127
Eimeria tenella, 512
ELISAs, *see* Enzyme-linked immunoabsorbent assays
ELISPOT assay, 455
Embryos, *see In ovo* vaccination
Encephalitis
  Bartonella henselae and, 339
  caprine arthritis-encephalitis, 29
  equine, 364
  polyvalent parvodistemper vaccine and, 5
  postvaccinal, canine distemper vaccines and, 294–295, 296, 301–302, 312

Encephalitis (cont.)
  Toxoplasma gondii and, 336
Encephalomyelitis, 464
Endogenous antigen processing, 184–185
Endometrium, bovine trichomoniasis
    and, 229
Endothelium
  CD31 immunoglobulin and, 69
  leukocyte trafficking and, 61–62, 65
  selectins and, 67
Endotoxins, see also Lipopolysaccharides
  excessive cytokine induction and, 688
  Haemophilus somnus infection and, 220
Enteric infections, see also Enteropathogenic viruses
  oral immunization and, 476
  in swine, economic costs of, 464
  viral, vaccination and, 433
Enteric redmouth disease, 540, 543–544
Enteric septicemia of catfish
  causative agent, 523
  Edwardsiella ictaluri RE-33 vaccine
    and, 525–535
  protective immunity and, 525
Enteric vaccines, 433–434, 475
Enterocytes, 410, 429, 430
Enteropathogenic viruses, see also Enteric infections
  characteristics of, 430–434
  diarrhea and, 430
  enteric vaccines and, 433–434
  group A rotavirus vaccines and, 440–442
  mucosal immune response to, 434–442
  oral vaccines and, 429–430
  pathogenesis of, 431–433
  transmissible gastroenteritis virus vaccines and, 439–440
Enterotoxins, see also Cholera toxin; Heat-labile toxin
  mucosal adjuvanticity and, 106
Envelope glycoproteins, see Glycoproteins
Enzyme-linked immunoabsorbent assays
    (ELISAs)
  antibody detection and, 372
  in bovine herpesvirus 1 diagnostics, 205–206
  in bovine mastitis diagnostics, 268, 269
  in bovine virus diarrhea virus diagnostics, 209

  in foot-and-mouth disease diagnostics, 200
  in Haemophilus somnus diagnostics, 231
  in poultry antibody detection, 488, 489
  in Strongylus vulgaris antibody detection, 396–397
  in trichomoniasis diagnostics, 231
Eosinophilia, 246
  interleukin 5 and, 400
  in Strongylus vulgaris infections, 390–391, 396
Epidemiology
  concepts of causality and, 754–755
  impact on vaccination, 660, 666
  input to immunoprophylaxis, 415–417
  in vaccine risk assessment, 704–706, 710
Epinephrine, 637
Epithelium
  genital, bovine trichomoniasis and, 230
  gut-associated, interleukin-1 and, 94
  mucosal, immunoglobulin A and, 111
Epitopes, 115; see also Immunodominant epitopes
  diversity and, 660
  effects on DNA immunization, 170
  masking of, 129
  T cell recognition of, 183
  T cell-recognized, identification of, 186–187
Epizootics
  equine influenza and, 379
  equine salmonellosis and, 371
  rabies and, 576
Epstein-Barr virus, 411
Equine abortion virus, 361–363
Equine arteritis virus, 363–364
Equine coital exanthema, 362
Equine ferritin, 73
Equine herpesviruses, 361–363, 411
Equine infectious anemia, 29, 33, 366
Equine Influenza Surveillance Programme, 380–386, 603
Equine influenza vaccines, 361
  antigenic drift and, 381–382, 384
  efficacy of, 380
  Equine Influenza Surveillance Programme and, 380–386
  licensing of, 382–383, 385–386, 602–604

Equine influenza virus
  causative agent of, 360
  diagnosis of, 361
  emerging strains, 381–382, 384, 603
  epizootics and, 379
  Equine Influenza Surveillance Programme and, 380–386, 603
  long-distance transmission of, 379–380
Equine metritis, contagious, 370–371
Equine monocytic ehrlichiosis, 368
Equine rabies, 367
Equine rhinopneumonitis virus, 361–363
Equine rotaviruses, 367
Equine variola, 11; see also Horsepox
Eradication
  of bovine herpesvirus 1, 197, 206–207
  of bovine virus diarrhea virus, 197, 209–210
  of foot-and-mouth disease virus, 197, 200–201
  as goal of vaccination, 474
  posteradication period and, 210–211
Erysipelas vaccines, 420, 421, 422, 426
Escherichia coli, see also Heat labile toxin
  O157:H7, preclinical detection and, 26
  vector for transmissible gastroenteritis virus vaccine, 440
Escherichia coli bacterins, 420
  for clinical coliform mastitis, 265–266, 269
Europe, wildlife reservoir species of rabies in, 572
European Agency for the Evaluation of Medicinal Products, 596–597
European ferrets, 559
European Pharmacopoeia, 604–606
  equine influenza vaccines and, 382, 385
  pseudorabies vaccine licensing and, 616, 624
European Union, 659
  clinical swine fever and, 661–662
  economics of bovine herpesvirus 1 vaccination and, 661
  licensing procedures, 595–596, 607
    centralized and decentralized registration in, 597–598
    definitions of veterinary biologicals in, 598–600
    for equine influenza vaccines, 602–604
    European Agency for the Evaluation of Medicinal Products and, 596–597
    for manufacturing authorization, 601–602
    pre-existing products and, 600–601
    for pseudorabies vaccines, 615–625
    role of European Pharmacopoeia in, 604–606
Evan's syndrome, see Immune-mediated thrombocytopenia
Evolution
  in B cells, 123
  in viruses, 2–3
Exogenous antigen processing, 184–185
Exotic animals, canine distemper virus and, 552, 553
Exotic disease, eradication of, 175–176, 210–211
Extended quasi-likelihood function, 769

F

Falvivirus, 364
Fasciola hepatica, see Liver flukes
Feedback loops, positive, in host-parasite relations, 249–250
Feline calicivirus, 666, 727
Feline coronavirus, 348, 349–350, 351, 355; see also Feline infectious peritonitis
Feline coronavirus vaccine, 351–356
Feline enteric coronavirus, 301, 349
Feline herpesvirus, 5–6, 727
Feline immunodeficiency vaccines, 326–329
Feline immunodeficiency virus, 325–326
Feline infectious peritonitis vaccine, 348
  benefits of, 353–356
  description of, 351
  hypersensitive reaction to, 689
  overvaccination and, 718
  risks of, 352–353
Feline infectious peritonitis virus, 348–349
  antibody-dependent enhancement and, 351, 352–353, 354, 355

Feline infectious peritonitis virus (*cont.*)
  canine coronavirus and, 301
  causative agent of, 348–349
  current status of, 348
  exposure dose for, 355
  forms of, 350
  historical perspectives on, 347–348
  immune response to, 350
  overvaccination and, 718
  pathogenesis of, 349–350
  preclinical detection and, 26, 33
Feline leukemia vaccines, 573, 636, 728
Feline leukemia virus, 27, 411, 718
Feline panleukopenia, 313, 687, 727
Feline parvovirus, 659
Feline Sarcoma Task Force, 753
Feline upper respiratory disease, 3
Feline vaccines, vaccine label safety statements and, 638
Feline viral rhinotracheitis, 686
Fennec foxes, 552, 553
Feral swine, disease transmission and, 464
Ferrets
  distemper and, 558, 564, 565, 566–567
  maternal antibodies and, 562–563
  mink distemper vaccine and, 559, 560, 561–563, 564
  neonatal and transplacental immunization in, 564
Ferry, N.S., 19–21, 22
Ferry vaccine, 19–21
Fetlocks, disease of, 10
Fibronectin, 742; *see also* Anti-fibronectin antibodies
Fibronectin-binding receptors, 261
Fibrosarcomas, vaccine-induced
  aluminum hydroxide adjuvant and, 4–5
  epidemiological surveillance of, 704–705
  risk management for, 709
  vaccine label safety statements on, 638
Field trials
  bovine herpesvirus 1 vaccines and, 204
  in licensing procedures, 589
  of pseudorabies vaccines, 620, 623
Fish, *see also* Aquaculture; Catfish
  Aeromonas salmonicida and, 547–548
  Edwardsiella ictaluri and, 523
  Vibrio anguillarum and, 548
  Vibrio salmonicida and, 547

Fishes Disease Commission (Office International des Epizooties), 672
Fish vaccines
  advantages of, 550
  in aquaculture, 540
  benefits of, 542–543
  for enteric redmouth disease, 543–544
  for enteric septicemia of catfish, 525–535
  immune response to, 550
  methods of administering, 541–542
  polyvalent, 547–549
  for *Vibrio viscosus,* 544, 546–547
Fleas, transmission of *Bartonella henselae* and, 339–340
Flesh flies, 554
"Floppy flipper disease," 554
Florida, controlling raccoon rabies in, 579–580
Foals, *Rhodococcus equi* and, 372
Food and Agriculture Organization (United Nations), 659, 677
Food and Drug Administration (United States), 771
Food Security Act (United States), 627
Foot and Mouth Disease and other Epizootics Commission (Office International des Epizooties), 672
Foot-and-mouth disease vaccine, 198–200
Foot-and-mouth disease virus
  antigenic diversity and, 660
  diagnostics for, 200
  eradication of, 197, 200–201
  failure to kill vaccine organism in, 685
  herd immunity and, 210
  synthetic peptide-stimulated immunity and, 184
Formaldehyde, 685
Fowl cholera, 13
Fowlpox vaccine, 468
Fowlpox virus vector, 153, 468–469, 486, 487, 518–519
Foxes, 552, 553
  controlling rabies in, 580–581
Francisella tularensis, 342
Frenkel, H.S., 198
Frenkel vaccine, 198

Freund's adjuvant
  complete
    compared to ISCOMs, 408–409
    in helminth vaccines, 246, 247
    Staphylococcus aureus vaccines and, 261, 262
  DNA vaccines and, 173
  incomplete, 227
    transmissible gastroenteritis virus vaccines and, 440
Furunculosis, 525, 543

G

Gammaherpesvirinae, 362
Gangliosides
  B cells and, 85
  cholera toxin and, 84, 92, 95
Gastric ulcers, *Helicobacter pylori* and, 341
Gastroenteritis, 433
Gastrointestinal immunization, 229
Gastrointestinal nematode infections, 250–251; *see also* Trichostrongylid nematodes
Geison, Gerald, 14
Gene delivery, DNA immunization and, 165–166
Gene expression, in virus vectors, 467, 468
Gene gun, DNA vaccines and, 165, 170
Genes, major, disease resistance and, 42, 43
Genetics
  effects on health and vaccinations, 40–41
  health enhancement and, 39–40, 41–49
  susceptibility to vaccine-induced autoimmune disease and, 744–745
Genetic testing, 32
Gene vaccines, *see* DNA vaccines
Genital tract infections, bovine
  Brucella abortus, 218–219
  campylobacteriosis, 221–224
  Haemophilus somnus, 219–221
  overview of, 217–218
  Tritrichomonas foetus, 224–232
Giardia lamblia, 342
Glässer's disease vaccine, 421

Glomerular amyloidosis, 722
Glomerulopathy, 742–743
Glucocorticoids
  bovine $\gamma\delta$ T cells and, 64, 73
  in immunosuppression, 63–64, 70, 73–74
  stress-released, dexamethasone models of, 72–73
Glutathione *S*-transferase, 244, 246, 247
Glycoproteins
  acid soluble, assay for, 648
  clonal dominance and, 122
  in feline immunodeficiency virus vaccines, 326, 328
  in HIV-1 deceptive imprinting, 118, 120
  in Raboral V-RG vaccine, 574
  in swinepox recombinant vaccine, 474
Good Manufacturing Practice, 601
gpt gene, 471
Gray foxes, 552
  controlling rabies in, 580–581
Gray seals, 552
Grease, 9, 10–11
Greenfield, William, 8, 15–17, 22
Green sea turtles, 554
Guaiacol, 648
Guillain-Barré syndrome, 734
Gumboro disease, *see* Infectious bursal disease
Gut-associated lymphoid tissue
  characteristics of, 83–84
  dual-signal integration immune response in, 93–94
  enteropathogenic viruses and, 433, 436–439
  functional compartmentalization in, 439
  interleukin 1 and, 94
  interleukin 12 and, 97
  mucosal adjuvanticity in, 92, 98–99
  in mucosal immune response, 435
  mucosal vaccines and, 108–109, 110–111

H

H11 aminopeptidase, 245
Haemonchus contortus, 242; *see also* Trichostrongylid nematodes

Haemonchus contortus (cont.)
  antigen identification in, 243–244
  H11 vaccine and, 245
  modeling of vaccines for, 251–253
Haemonchus placei, 242
Haemophilus somnus
  diagnosis of, 221, 232–233
  immune response to, 220–221
  pathogenesis of, 219–220
  virulence factors in, 220
Haemophilus somnus vaccines, 232, 685–686
Halichoerus grypus, 552
Haptoglobin, 644, 650
  assays for, 647–648, 649
  in cattle, 646
  in pigs, 646
Harbor seals, 552
Hatchability, *in ovo* vaccination and, 500–502
Hc-sL3 antigen, 243–244, 246
Hc-sL3 vaccine, 252, 253
Health, *see* Animal health; Public health
Heat attenuation
  Pasteur and, 13–14
  Toussaint and, 12
Heat-labile toxin
  adjuvanticity and, 106, 110
  immuogenicity of, 85–86
  structure and activity of, 84, 85
  in vaccines for transmissible gastroenteritis virus, 440
Helicobacter acinomyx, 341
Helicobacter mustelae, 341
Helicobacter pylori, 341
Heligmosomoides polygyrus, 401
Helminthic infections, *see also Strongylus vulgaris*
  eosinophilia and, 391, 400
  interleukin 4 and, 400–401
  overdispersion and, 251
  Th2 immune response and, 391
Helminthic vaccines
  adjuvants in, 246, 247
  development in, 241–242
  dynamics of gastrointestinal nematode infections and, 250–251
  economic importance of, 242
  future trends in, 253–254
  immune responses, 245–247

  modeling host-parasite population dynamics and, 247–250, 254
  modeling of *Haemonchus* vaccines, 251–253
  novel antigens and, 244–245
  protective antigens and, 242–245
Hemagglutination inhibition test
  equine influenza and, 361
  protective titer levels and, 726–727
Hemagluttinin gene, in equine influenza virus, 602
Hematologic disease, *see* Autoimmune hemolytic anemia; Immune-mediated hemolytic anemia; Immune-mediated thrombocytopenia
Hemoglobin, in assays for haptoglobin, 647–648
Hemolytic anemia, immune-mediated, 705–706, 734
Hemolytic assays, *see also* Single radial hemolysis test
  in trichomoniasis diagnostics, 231
Hemorrhagic enteritis virus, 483
Henle-Koch postulates, 754, 755
Hepatitis contagious canis, *see* Canine hepatitis
Hepatocytes, acute phase proteins and, 645
Herd immunity, 210–211, 664, 711
Heritability, disease resistance traits and, 41
Herpes simplex virus, 452, 454
Herpesvirus(es)
  cell-mediated immune response and, 454–455
  equine, 361–363
  feline, 5–6, 727
  immune stimulating complexes and, 411
  porcine, economic costs of, 464
  of turkeys
    in Marek's disease vaccine, 496, 497, 498, 500, 502–503, 519
    in recombinant poultry vaccines, 486, 519
Herpesvirus vaccines, 150
Herpesvirus vectors, 150, 151, 475, 486, 519
Heterotypic immunity, 117
Hidden antigens, *see* Novel antigens

INDEX 791

High endothelial venules, 65
High immune response phenotype, 43–49
Hitra disease, 547; *see also Vibrio salmonicida*
HIV-1, *see also* AIDS; Lentiviruses
  anamnestic recall response and, 127
  auxotrophic vaccines and, 139
  B cell abnormalities and, 119–120, 123
  deceptive imprinting model and, 117–119
  epitope masking and, 130
  feline immunodeficiency virus DNA vaccines and, 329
  immune stimulating complexes and, 411
  modeling with feline immunodeficiency virus, 325
  OAS phenomenon and, 116–117
  tuberculosis mortality and, 135
  viral envelope of, 127–128
HIV-2, 411
Hog cholera vaccine, 423
Homeopathic nosodes, 305, 318–319, 716
Homologous recombination
  in DNA integration, 173
  in recombinant vaccine formation, 471, 518, 574
  reversion to virulence and, 475
Horsepox, 8, 9, 10–11, 22
Horses
  bacterial diseases of, 368–372
  long-distance disease transmission and, 359–360, 372, 379–380
  Uasin Gishu pox and, 11
  vaccination and, 360, 373
  viral diseases of, 360–367, 371
Host-parasite relations
  in gastrointestinal nematode infections, 250–251
  modeling vaccination in, 247–250, 254
Huidekoper, R.S., 18
Human adenovirus type 5, 151
Human alpha-1 antitrypsin, 164
Human growth hormone, 164
Human immunodeficiency virus type 1, *see* HIV-1
Human parainfluenza virus 2, 315
Human reovirus, 510
Humans
  Bartonella henselae and, 338–339, 340

Helicobacter pylori and, 341
Microsporum canis and, 342
rabies and, 335, 571–572
rotaviruses and, 430
Toxoplasma gondii and, 335, 336–337, 338
Yersinia pestis and, 342
Human vaccines, *see also specific vaccines*
  as models for veterinary vaccines, 659
  reporting of adverse events in, 755–756
  success of, 657
Humoral immune response
  in avian immune system, 488–489, 490
  cytopathic viruses and, 448
  in disease resistance, 42
  DNA immunization and, 164, 165
  feline infectious peritonitis and, 350
  heritability and, 43
  HIV-1 infections and, 119–120, 123–128
  immune-response selected pigs and, 45–49
  pregnancy and, 71
  pseudorabies virus vaccines and, 451–452
HyMast test, 267–268
Hypersensitivity reactions, 688–690
  clinical overview of, 716–717
Hypertrophic osteodystrophy, 724
Hypothesis testing, of adverse event report data, 758, 764–766
Hypothyroidism, 725

I

Ictalurus punctatus, 524; *see also* Channel catfish
IgA, *see* Immunoglobulin A
IgE, *see* Immunoglobulin E
IgG, *see* Immunoglobulin G
IgM, *see* Immunoglobulin M
Immersion vaccination, 541, 543–544, 549
Immune complex type response, in hypersensitive reactions, 689–690
Immune deficiency syndrome, in Weimaraners, 723

Immune enhancement, *see* Antibody-dependent enhancement
Immune-induced wasting, 53–57
Immune-mediated disease, *see also* Autoimmune disease; *specific diseases*
  mechanisms of, 723
  vaccine-related, 705–706
Immune-mediated hemolytic anemia, 705–706, 734; *see also* Autoimmune hemolytic anemia
Immune-mediated polyarthritis, 721
Immune-mediated thrombocytopenia, 717, 719, 720–721
Immune response, *see also* Cell-mediated immune response; Cytotoxic T cell immune response; Delayed-type hypersensitivity immune response; Humoral immune response; Secondary immune response
  to carrier organisms, 111
  dual-signal integration in, 93–94
  effects of stress on, 70–74
  gene-related, 40–41
  genetic enhancement of, 39–40, 41–49
  induction, DNA immunization and, 166–168
  in mucosal immune system, 111–112
Immune response phenotype, 43–49
Immune stimulating complexes (ISCOMs), 326
  antigen presentation and targeting by, 407
  antigen transport by, 408–409
  conceptual design of, 406
  formation of, 406–407
  induction of cytokine Th1-Th2 response, 410
  induction of immune response in neonates, 410–411
  induction of mucosal immunity, 409–410
  induction of protective immunity, 411
  overview of, 405–406
  for rabies, 311
Immune systems, avian, 487–490
Immune tolerance, DNA immunization and, 171
Immunity, *see also* Acquired immunity; Active immunity; Herd immunity; Protective immunity
  concomitant or heterotypic, 117
  inhibition of, immunoglobulin G and, 172
  mucosal, duration of, 663
  sterile, DNA vaccines and, 171
Immunity, duration of
  for canine vaccines, 295, 296–297, 300, 310
  inadequate, vaccine failure and, 694–695
  for mink distemper vaccines, 559–560
  revaccination and, 663, 725–727
Immunization, *see also* DNA immunization; Vaccination
  distinguished from injection, 5
  in poultry, 483
  prophylactic, *see* Immunoprophylaxis
  sexually transmitted diseases and, 223
  use of synthetic peptides in, 183–184
Immunoassays, *see* Acute phase protein assays
Immunocompromised systems, *see also* AIDS
  Bartonella henselae and, 339, 340
  infectious disease and, 334–335
Immunodiffusion, 648–649
Immunodominant epitopes
  in HIV-1 infection, B cell abnormalities and, 123
  masking of, 130
  OAS phenomenon and, 116–117
  pre-existing B cell clones and, 128
  in unconventional reactions with antibodies, 128–129
  vaccine failure and, 129
Immuno-electron microscopy, 407
Immunofluorescence, inclusion bodies and, 561–562
Immunoglobulin A (IgA)
  in avian immune system, 488
  bovine trichomoniasis and, 226–228, 229
  campylobacteriosis and, 222–223
  cholera toxin induction of, 87, 93
  enteropathogenic viruses and, 436–439
  heat labile toxin and, 85–86
  mucosal epithelium and, 111
  mucosal immune response and, 112, 434, 435
  porcine group A rotaviruses and, 441–442

respiratory syncytial virus and, 410
SBR-CTA2/B chimeric cholera toxin
  protein and, 107
Immunoglobulin A plasma cells, see also
  Antibody-secreting cells
  in mucosal immune response, 435
Immunoglobulin E (IgE)
  in helminthic infections, 245, 400–401
  in hypersensitivity reactions, 689
Immunoglobulin G (IgG)
  inhibited by active immunity, 172
  in avian immune system, 488
  bovine trichomoniasis and, 226–228, 229–230
  campylobacteriosis and, 222–223
  CD4 T cells and, 185
  DNA immunization and, 169
  Haemophilus somnus infections and, 220–221
  HIV-1 infections and, 124–127
  monoclonal populations in HIV infections, 119
  mucosal immune response and, 112
  plasmid-induced immunity and, 168
  SBR-CTA2/B chimeric cholera toxin protein and, 107
  Staphylococcus aureus infections and, 260–261
  Strongylus vulgaris infections and, 396, 399–400, 402
  suppression of T cell polyclonal response and, 122
  thymus-independent antigens and, 123
Immunoglobulin M (IgM)
  in avian immune system, 488
  campylobacteriosis and, 223
  HIV-1 infections and, 124, 127
  thymus-independent antigens and, 123
Immunoglobulins
  assays for, international standardization and, 652
  in avian immune system, 488
  cholera toxin induction of, 87, 93
  dexamethasone repression and, 73
  heat labile toxin and, 85–86
  HIV-1 infections and, 123–127
  leukocyte trafficking and, 68–69
  mucosal immune response and, 112

SBR-CTA2/B chimeric cholera toxin protein and, 107
in *Strongylus vulgaris* infections, 396–397, 399–400, 402
Immunological veterinary medicinal product, see Veterinary biological products
Immunological Veterinary Medicinal Working Party, 597, 598
Immunologic memory, see also Secondary immune response
  duration of immunity and, 726
  vaccine development and, 666
Immunology
  origins of, 7–8
  vaccine development and, 182–187
  William Greenfield on, 15–16, 17
Immunomodulators
  DNA as, 170
  licensing procedures for, 590–591
  virus vectors and, 148–149
Immunoprophylaxis
  bovine campylobacteriosis and, 223
  bovine trichomoniasis and, 230
  epidemiological approach to, 415–417
Immunosuppression
  avian viruses and, 483
  bovine virus diarrhea virus and, 207–208
  canine parvovirus type 2 vaccines and, 298
  causing vaccine failure, 694
  infectious disease and, 334–335
  residual virulence of modified live vaccines and, 686–687
  stress-related, 70–74
  vaccine-induced, 687
Immunoturbidimetry, 649
Inactivated bacterins
  excessive cytokine induction and, 688
  failure to kill vaccine organism in, 685–686
Inactivated vaccines
  for canine parvovirus type 2, 296, 297, 313
  for canine viruses, 310
  cytokine adjuvants and, 457–458
  for equine influenza, 380
  failure to kill vaccine organism in, 685–686

794   INDEX

Inactivated vaccines (cont.)
  for feline immunodeficiency virus, 326–327, 328
  importance of, 457
  porcine pseudorabies virus and, 447–448, 451–452, 454, 455–456, 457
  for rabies, 310, 311
  risks of reversion and, 466
Inclusion bodies, 561–562
Incomplete Freund's adjuvant, see Freund's adjuvant, incomplete
Indications, on vaccine labels, 637
Infection
  acute phase proteins in, see Acute phase proteins
  distinguished from disease, 29
  ecology of, 29–31
  genetic predisposition to, 32
Infection detection
  in diagnostic medicine, 25–26
  ecology of infection and, 29–31
  genetic predisposition to infection and, 32
  methods in, 31–32
  preclinical, 26, 28–29
  predictive value of, 33–34
  traditional approach to, 27–28
Infectious bovine keratoconjunctivitis, 687
Infectious bovine rhinotracheitis, 197, 201–207, 210–211; see also Bovine herpesvirus 1
Infectious bronchitis vaccines
  DNA vaccines, 521
  maternal antibodies and, 506
  in ovo vaccination and, 503–504
Infectious bronchitis virus, 482
Infectious bursal disease vaccines, 507–509, 511
Infectious bursal disease virus, 483
Infectious canine hepatitis, see Canine hepatitis
Infectious disease, see also specific diseases
  animal management and, 417
  in aquaculture, 539–540
  breed susceptibility and, 661
  in commercial poultry flocks, 482–483
  control of, diagnostics and, 360, 372
  distinguished from infection, 29
  economic costs of, in swine, 463–465
  emerging, 333, 334–335
  gene-related resistance to, 40–41
  "global," 659
  immunosuppression and, 334–335
  reemergence of, 333
Infectious laryngotracheitis virus, 482
Infectious pancreatic necrosis, 543
Inflammation
  acute phase proteins and, see Acute phase proteins
  leukocyte trafficking and, 61–62, 66–67, 68–69, 74
  mucosal immune response and, 111–112
Influenza vaccines, see also Equine influenza vaccines
  avian, 521
  porcine, 421, 427
Influenza virus, see also Equine influenza virus
  immune stimulating complexes and, 410
Injection, distinguished from immunization, 5
Injection site reactions
  adjuvants and, 169, 620
  Arthus reaction and, 689
  causes of, 691
  DNA vaccines and, 169, 175
  with pseudorabies vaccines, testing for, 620
Innocuity, 4–5
Inovoject egg injection system, 495, 497–503
In ovo vaccination
  aspergillosis and, 499
  automated equipment for, 497–498
  with bacterial vaccines, 511–512
  benefits of, 496
  field efficacy trials in, 498–499
  with fowlpox vaccines, 505
  history of, 495, 496
  with infectious bronchitis vaccines, 503–504
  with infectious bursal disease vaccine, 504–505
  with Marek's disease vaccines, 497–503, 512

maternal antibodies and, 505–506
moisture loss and, 502
with Newcastle disease vaccines, 504
overview of, 485, 490
Rispens-type vaccines and, 502–503
time of injection, 500–501
upside down eggs and, 501–502
vaccine preparation for, 499–500
with virus-antibody complex vaccines, 506–511
Integrins
β2 family, 67–68
α4 family, 69–70
in leukocyte trafficking, 62–63, 67–68, 69–70
Inter-American Institute for Cooperation on Agriculture, 677
Intercellular adhesion molecules
effects of dexamethasone on, 74
leukocyte trafficking and, 62, 68–69
Interferon, gamma
as adjuvant, 458
CD45 T cell subpopulations and, 279–280
in cell-mediated immune response, 453–455
inactivated pseudorabies virus vaccine and, 455–456
induction by immune stimulating complexes, 410
inhibition by interleukin 10, 401
interaction with interleukin 12, 450
in *Strongylus vulgaris* infections, 401–402
Th1 immune response and, 185
Interferons
adjuvanticity in DNA immunization, 168
DNA tuberculosis vaccines and, 140
interleukin 12 and, 97
in mucosal adjuvanticity, 98–99
in periparturient immunosuppression, 71–72
recombinant poultry vaccines and, 487
Interleukin 1
immune-induced wasting and, 54, 55
in mucosal adjuvanticity, 94–96, 98–99
Interleukin 2
CD45 T cell subpopulations and, 279

induction by immune stimulating complexes, 410
Th1 immune response and, 185
Interleukin 4
CD45 T cell subpopulations and, 279
helminthic infections and, 400–401
Strongylus vulgaris infections and, 397, 400–401, 402
Interleukin 5
eosinophilia and, 400
Strongylus vulgaris infections and, 397, 400, 402
Interleukin 6
acute phase proteins and, 645
excessive induction of, 687
Interleukin 10
inhibition of gamma interferon by, 401
inhibition of interleukin 12, 450
Strongylus vulgaris infections and, 401
Interleukin 11, acute phase proteins and, 645
Interleukin 12
adjuvanticity of, 448, 449, 450, 455–456, 458
modulation of acquired immune response, 449–450
modulation of immune response to pseudorabies virus vaccine, 455–456
in mucosal adjuvanticity, 96–99
production and regulation of, 450
structure of, 449–450
Interleukins
CD4 T cells and, 185
cholera toxin-induced, 85
dexamethasone repression and, 73
helminthic infections and, 246
inhibition of gamma interferon by, 401
in mucosal adjuvanticity, 94–99
in periparturient immunosuppression, 71–72
Strongylus vulgaris infections and, 397, 400–402
toxin induction of, 93
Toxoplasma gondii infections and, 337
International Animal Health Code, 673
International Animal Health Code Commission (Office International des Epizooties), 672
International Association of Biological Standardization, 609–614

International organizations, *see also* European Union; Office International des Epizooties; World Health Organization
  in veterinary standardization, 659
International trade
  diagnostic tests and, 673
  disease transmission in horses and, 359–360, 372, 379–380
International Union of Microbiological Societies, 610
Interval estimation, of adverse event report data, 763–764
Intestines, pathogenesis of enteric viruses and, 431, 433
Intradermal vaccination, DNA vaccines and, 165–166
Intragastric vaccination
  mucosal immune system and, 106
  SBR-CTA2/8 chimeric cholera toxin protein and, 109, 110
Intramammary infections, *see also* Bovine mastitis
  coliform vaccines and, 265
  Staphylococcus aureus vaccines and, 259, 260, 261
Intramuscular vaccination, DNA vaccines and, 165, 170
Intranasal vaccination
  canine vaccines and, 310
  mucosal immune system and, 106
  SBR-CTA2/8 chimeric cholera toxin protein and, 109, 110
Intraperitoneal vaccination
  for fish, 541–542, 549
  of transmissible gastroenteritis virus vaccine, 440
ISCOMs, *see* Immune stimulating complexes
Isoelectric focusing, 120
Ivermectin, 390
Ixodes dammini, 662
Ixodes spp., 316, 368

J

Japanese encephalitis vaccine, 423
Japanese equine encephalitis, 364

Jenner, Edward, 465
  canine distemper and, 17–18
  cowpox vaccine and, 9
  origin of vaccinology and, 7, 22
  vaccinia and, 8
Jenner's vaccine, 9, 13

K

Kennel cough
  Bordetella branchispetica and, 315–316
  canine adenovirus type 2, 314
  canine distemper and, 20–21
  canine parainfluenza virus and, 315
  polyvalent nasal vaccination for, 315, 316
Keyhole limpet hemocyanin
  cholera toxin adjuvanticity and, 87, 91, 92, 93
  in vaccine-induced autoimmunity experiments, 735, 737, 738, 740
Kidney failure, *see* Renal failure
Kinkajous, 552
Kramer algorithm, 771

L

L5 phage, 139
Labels, *see* Vaccine labeling
Labrador retrievers, 661
lacZ gene, 471
Laidlaw, P.P., 22
Lamina propria, 436, 438, 439
Laminin, *see* Anti-laminin antibodies
Langerhan cells, 167
Latency testing, 636
  in pseudorabies vaccines, 619
Latent variables, 762
Lebailly, C., 21
Leishmania major, 185
Lentiviruses, *see also* HIV-1
  vaccine development for, 366
Leopards, 553
Leptospira bratislava, 371
Leptospira pomona, 371
Leptospirosis, 304, 317–318
  equine, 371–372

Leptospirosis vaccines
  duration of immunity in, 663
  efficacy and, 291, 303
  for swine, 420, 423, 426
"Lesionless pathology," 31
Lesions, at injection sites, see Injection site reactions
Leukocyte adhesion deficiency type 2, 67
Leukocyte trafficking
  effects of stress on, 63–64, 70, 72–73, 74
  endothelium and, 61–62, 65
  importance of, 61
  integrins and, 62–63, 67–68, 69–70
  intercellular adhesion molecules and, 68–69
  molecular basis of, 62–63
  overview of, 65
  in periparturient immunosuppression, 72
  selectins and, 62, 65–67
  stress-released glucocorticoids and, 72–73
Leukosis
  avian, 482, see also Marek's disease virus
  bovine, 29, 31
LFA-1 integrin, 67–68, 69
Licensing procedures (European Union), 595–596, 607
  centralized and decentralized registration, 597–598
  definitions of veterinary biologicals in, 598–600
  for equine influenza vaccines, 382–383, 385–386, 602–604
  European Agency for the Evaluation of Medicinal Products and, 596–597
  for manufacturing authorization, 601–602
  pre-existing products and, 600–601
  for pseudorabies vaccines, 615–625
  role of European Pharmacopoeia in, 604–606
Licensing procedures (United States)
  for autogenous products, 592–593
  conditional licensing and, 591
  for conventional vaccines and bacterins, 587–590
  exemptions to, 592
  for further manufacturing, 591–592
  for immunomodulators, 590–591
  for modified live pseudorabies vaccines, 628–632
  for recombinant products, 590
  sublicensing and, 592
  Virus Serum Toxin Act and, 585–586, 627
Lifestyle, effects on vaccination, 660–661
Limast test, 267
Linoleic acid, conjugated, immune-induced wasting and, 55–56
Lions, 553
Lipophilic muramyl dipeptides, 169
Lipopolysaccharide antigen type 1, 46, 48
Lipopolysaccharides, see also Endotoxins
  excessive cytokine induction and, 688
  immune-induced wasting and, 56
Live attenuated vaccines, see also Modified live vaccines
  described, 466
  for *Edwardsiella ictaluri,* 525–535
  for rabies, 311
  virus vectors and, 149–151
Liver failure, vaccine-induced, 717
Liver flukes, 242, 244–245, 246–247
Loglinear modeling, of adverse event report data, 766–769
Low immune response phenotype, 43, 45–48, 49
Lowry reaction, 648, 649
Loy, John, 10
Lupus erythematosus, systemic, 706, 742, 743
Lycaon pictus, 552
Lyme borreliosis, see Lyme disease
Lyme disease, 662
  etiologic agent, 316
  horses and, 368–369
Lyme disease vaccine, 303–304, 316–317
  overadministration of, 728
  recombinant, 573
Lymph, dendritic cells in, 284; see also Afferent lymph veiled cells
Lymph nodes
  AgI/II adhesin and, 108
  ISCOM antigen transport and, 408
  mesenteric, in mucosal immune response, 435, 436
  plasmid-induced immunity and, 167

Lymphocytes (cont.)
  immune stimulating complexes and, 407
  proliferation, cholera toxin and, 93
  self-reactive, in vaccine-induced autoimmunity, 744
  suppression of blastogenesis in, 687
Lymphocyte trafficking
  adhesion molecules and, 62–63
  integrins and, 69–70
Lymphoid nodules, genital, bovine trichomoniasis and, 229
Lymphoid system, avian, 487–4488
Lymphoid tissue
  gut-associated, see Gut-associated lymphoid tissue
  nasal, immunization and, 110
Lymphopenia, 687
Lymphoproliferative assays, 436, 438, 439
Lymphoproliferative response, to pseudorabies virus vaccines, 451–452
Lyssavirus, 367, 371

M

Mac-1 integrin, leukocyte trafficking and, 62, 68
Macaques, HIV-1/SIV protection in, 411
α2-Macroglobulin, 646, 648
Macrophage cytokines, in mucosal adjuvanticity, 94–99
Macrophage inflammatory protein, acute phase proteins and, 645
Macrophages
  Edwardsiella ictaluri RE-33 vaccine and, 534
  in feline infectious peritonitis, 350, 351
  inhibition of by interleukins, 401
  ISCOM uptake by, 408
Maggots, botulism and, 554
Major acute phase protein, 647
Major histocompatibility complex (MHC)
  cholera toxin adjuvanticity and, 92
  class I
    in antigen processing, 184, 185
    in antigen recognition, 182–183
    bovine respiratory syncytial virus immunity and, 190–191
  class II
    antigenic competition and, 121
    in antigen processing, 184–185
    in antigen recognition, 182–183
    gamma interferon and, 453
    immune response and, 40, 43
    in mucosal adjuvanticity, 99
    polymorphisms in, 183
    T cell-recognized epitopes and, 187
Mammary glands, in mucosal immune system, 435
Mannose receptor, 121
Manual of Standards for Diagnostic Tests and Vaccines (Office International des Epizooties), 672–674, 676, 677, 678
Marek's disease vaccines, 485, 490
  bivalent, 497
  herpesvirus of turkeys and, 496, 497, 498, 500, 502–503, 519
  International Association of Biological Standardization and, 612–613
  maternal antibodies and, 505
  in ovo vaccination and, 496, 497–503, 512
  Rispens type, 502–503
Marek's disease virus, 482
  distribution, 659
  gene-related resistance to, 40, 41
  in ovo vaccination and, 496
  Rispens-type vaccines and, 502–503
Marginal metallophilic macrophages, 408
Marginal zone macrophages, 408
Marine mammals, canine distemper virus and, 552
Marker vaccines
  for bovine herpesvirus 1, 201–202
  defined, 201
  in disease eradication, 210
Massachusetts, controlling raccoon rabies in, 578–579
Master Cell Stock
  in licensing procedures, 588
  in pseudorabies vaccine licensing, 629
Master Seeds
  in licensing procedures, 588, 590
  in pseudorabies vaccine licensing, 628–629

Mastitis, *see* Bovine mastitis; Coliform mastitis
Maternal antibodies, *see also* Passive immunity
  canine distemper vaccines and, 294, 295
  canine parvovirus type 2 vaccines and, 297, 313
  canine vaccines and, 310
  DNA immunization and, 171, 174
  effects on vaccination, 661
  feline immunodeficiency virus vaccines and, 327
  immune stimulating complexes and, 410–411
  infectious bursal disease and, 507
  kennel cough vaccines and, 315, 316
  mink distemper vaccine and, 562–563
  poultry vaccines and, 505–506
  recombinant Newcastle disease vaccine and, 519
  vaccine failure and, 693–694
  virus-antibody complex vaccines and, 506–511
Maximum residue limits, 596
M cells, 110, 435
Measles vaccine, adverse reactions to, 717
Measles virus, 312
Meat quality
  DNA vaccines and, 175
  Toxoplasma gondii and, 338
Memory immune response, *see* Immunologic memory; Secondary immune response
Memory lymphocytes, *see also* T cells, memory
  trafficking and, 63, 70
Memory T cells, *see* T cells, memory
Mesenteric lymph nodes, in mucosal immune response, 435, 436
MHC, *see* Major histocompatibility complex
Microorganisms, environmental, effects on animal performance, 53–54, 57
Microsporum canis, 342
Milk
  bovine herpesvirus 1 antibody detection and, 206

  bovine virus diarrhea virus detection and, 209
Minimal immunizing dose
  for caninea vaccines, 294, 295, 296
  for mink distemper vaccines, 559
Mink
  botulism and, 554
  botulism vaccine and, 555
  distemper vaccinces and, 558–564
  distemper virus and, 552, 557–558, 565–568
  enteritis vaccine and, 553
  enteritis virus and, 313
Mitogen proliferation assays, 489
Modified live vaccines, *see also* Live attenuated vaccines
  adverse reactions to
    clinical overview of, 716–717
    in polyvalent vaccines, 718–719, 722, 724
  canine, 296, 300, 309–310, 312, 314
  described, 292
  licensing of, 615–625, 628–632
  nonimmunizing mutants and, 293
  pseudorabies virus and, 447–448, 451–452, 454, 455
  residual virulence and, 686–687
  reversion and, 466
Monocytes, afferent lymph veiled cells and, 284
Monophosphoryl lipid A, 169
Morbillivirus, 410, 551
Mortality
  canine coronavirus and, 301
  human, rabies and, 571–572
  pseudorabies virus and, 464
Mosquito-transmitted viruses, equine, 364, 365
Moss, Bernard, 147
Mouse hepatitis virus, mode of release in, 431
Mucosal addressin cell adhesion molecule 1, 62, 70
Mucosal adjuvanticity
  cholera toxin and, 84–92, 106
  dual-signal integration in, 93–94
  immune stimulating complexes and, 409–410
  interleukin 1 in, 94–96

Mucosal adjuvanticity (cont.)
  interleukin 12 in, 94–96
  mechanisms of, 92–98
Mucosal disease, 207, 209
  vaccine-induced, 690–691
Mucosal epithelium, immunoglobulin A and, 111
Mucosal immune system
  adjuvanticity in, see Mucosal adjuvanticity
  characteristics of, 83–84
  DNA vaccines and, 169–170, 175
  immune responses in, 111–112
  immune stimulating complexes and, 409–410
  oral vaccines and, 105–106, 430
  overview of, 434–435
  response to enterpathogenic viruses and vaccines, 436–442
  virus vectors and, 149
Mucosal immunity, duration of, 663
Mucosal vaccines
  carriers organisms and, 110–111
  gut-associated lymphoid tissue and, 110–111
  Salmonella expressing SBR-CTA2/B and, 108–109
  vaccine-induced autoimmunity and, 745
Multideterminant antigens, 115
Multivalent vaccines, see Polyvalent vaccines
Muramyl dipeptides, lipophilic, 169
Murine cytomegalovirus, 450
Muscle cells, transfected, plasmid-induced immunity and, 166
Mustela nicripes, 552
Mustela putorius, 559
Mutants, nonimmunizing, in viral vaccines, 293
Mutation
  in B cells, autoimmunity and, 744
  in canine parvovirus type 2, 313
  in live attenuation of virus vectors, 150
  in marker vaccine design, 201
  in viruses, 2–3
Myasthenia gravis, 744
Mycobacteria, as vectors, 138–139
Mycobacterium bovis, see Tuberculosis

Mycoplasma bovis, 263
Mycoplasma hyopneumoniae vaccine, 421, 422
Mycoplasma hyorhinis, 46
Myoblasts, transfected, plasmid-induced immunity and, 166
Myoglobin, autoantibodies and, 739, 741
Myosin, of skeletal muscle, autoantibodies and, 739, 741

N

NAGase test, 267
Naive lymphocytes, trafficking and, 63, 70
Nasal lymphoid tissue, 110
Nasal shedding, see also Virus shedding
  bovine virus diarrhea virus and, 277
National Environmental Policy Act, 590, 628, 631
National Institute for Biological Standards and Controls, 382, 386
Natural immunity, see Passive immunity
Natural killer cells, see also Cytotoxic T cell immune response; T cells
  in avian immune system, 489–490
Negri bodies, 367
Nematodes, see also Helminthic vaccines; Trichostrongylid nematodes
  gastrointestinal infections and, 250–251
  hidden antigens and, 245
  intestinal, interleukin-12 and, 96
Neonates
  administration of canine vaccines and, 297–298, 299–300, 661
  adverse reactions and, 719
  diarrheal disease and, 464–465
  DNA vaccines and, 171
  gastrointestinal nematode infections and, 250–251
  immune stimulating complexes and, 410–411
  immunosuppression and, 71
  maternal antibodies and, 174
  swinepox virus and, 475–476
  timing of vaccination for, 661, 663, 694, 719

Neoplastic disease
  avian, 482
  vaccine-induced, 690
Neutralization tests
  in bovine herpesvirus 1 diagnostics, 205, 206
  in bovine virus diarrhea virus diagnostics, 209
Neutrophilia, 67, 72
Neutrophils
  CD31 immunoglobulin and, 69
  CD62L adhesion molecule and, 66, 74
  glucocorticoids and, 63
  in inflammatory response, 67
  in leukocyte trafficking, 62
  stress-released glucocorticoids and, 72–73
Newcastle disease vaccines
  fowlpox virus vector and, 153
  maternal antibodies and, 506, 519
  in ovo vaccination and, 504
  recombinant, 486, 487, 518–519
  virus-antibody complexes and, 510
Newcastle disease virus, 482, 518
New Jersey, controlling raccoon rabies in, 576–578
New York, controlling raccoon rabies in, 578, 580
Nippostrongylus brasiliensis, 96
Nonamphipathic proteins, 406
Noncodified efficacy testing, 635–636
Nonvaccinated animals
  disease outbreaks and, 291, 319
  herd immunity and, 711
Norwood-Sampson test, 764–766
Nosodes, 305, 318–319, 716
Novel antigens, helminth vaccines and, 244–245, 246–247, 253
Novel antigen vaccines, modeling of, 253
Nucleic acid vaccines, *see* DNA vaccines
Nutrition, immune-induced wasting and, 55–57

O

OAS phenomenon, 116–119
Office International des Epizooties (OIE)
  Equine Influenza Surveillance Programme and, 380–386
  founding of, 670
  guidelines for veterinary laboratories, 676
  international reference reagents and, 676–677
  objectives of, 670
  organizational structure of, 670–672
  other international organizations and, 677
  reference laboratories and collaborating centers of, 675–676
  Sanitary and Phytosanitary Agreement and, 677, 678
  specialist commissions of, 672
  Standards Commission of, 670, 672–677, 678
OIE, *see* Office International des Epizooties
OIE Bulletin, 383–384, 603
Old English sheepdogs, 720–721
Omega-3 fatty acids, immune-induced wasting and, 55
Oncogenes, in disease prevention, 32
Oncorhynchus mykiss, 525; *see also* Rainbow trout
Ondersteepoort canine distemper vaccine strain, 293, 295, 312
Oocyst shedding, *see also* Virus shedding
  in *Toxoplasma gondii* infections, 337–338
Opiods, endogenous, in periparturient cattle, 71
Oral vaccination
  for enteric pathogens, 476
  for fish, 542, 543–544, 549
  immune stimulating complexes and, 409–410
  mucosal immune system and, 105–106
  Raboral V-RG vaccine and, 572, 582
  recombinant viral vaccines and, 475
  for *Strongylus vulgaris,* 390
Oral vaccines
  cholera toxin and, 84, 85, 86, 87–92
  enterpathogenic viruses and, 429–430
  heat labile toxin and, 85–86
  mucosal immune system and, 430
Orbivirus, 365

"Original antigenic sin," 118; *see also* OAS phenomenon
Ormethoprim-sulfamethoxine, 524
Orthopoxvirus, 11
OspA vaccine, 303, 317, 369
Ostertagia circumcincta, 242
Ostertagia ostertagi, 242; *see also* Trichostrongylid nematodes
Ostertagia spp., 245
Ostriches, 481
Ovalbumin, cholera toxin adjuvanticity and, 92, 93
Overdispersion
  of adverse event report data, 760–762, 769
  in helminth infections, 251
Overvaccination, 663, 701–702, 718, 728; *see also* Revaccination
Oxytetracycline, 524, 540

P

p150,95 integrin, 68
Pan-American Health Organization, 659, 677
Parasites, *see also* Helminthic vaccines
  economically important, 242
  overdispersion and, 251
Parenteral vaccination, local immunity and, 475
Parinaud's syndrome, 339
Parturition, administering vaccines near, 74
Parvodistemper vaccine, 5
Parvovirus, Aleutian disease and, 558, 563; *see also* Canine parvovirus; Canine parvovirus type 2
Parvovirus vaccines, 420, 421, 422, 426; *see also* Canine parvovirus vaccines; Canine parvovirus type 2 vaccines
Passive immunity, *see also* Maternal antibodies
  DNA immunization and, 171–172, 174
  immune stimulating complexes and, 410–411
  in poultry, 483
  pseudorabies vaccines and, 621
Pasteur, Louis
  anthrax vaccines and, 11, 12, 13–14
  origin of vaccinology and, 8, 22
  rabies vaccine and, 311
Pasteurella canina, 19; *see also* Canine distemper
Pasteurella haemolytica vaccine, 689–690, 695
Pasteurella multocida, see Atrophic rhinitis
Pasture contamination, in host-parasite relations, 249, 254
Pasture management, in host-parasite relations, 250
Pathogens, resistance to, 42
PauA plasminogen activator, 263–264
PCR, *see* Polymerase chain reaction (PCR)
Peptides
  antigastrointestinal, immune-induced anorexia and, 56–57
  synthetic, immune stimulation by, 183–184
Periarteriolar lymphoid sheats, 408
Periparturient immunosuppression, 71–72
Peripheral blood lymphocytes, 441
Peripheral blood mononuclear cells
  in cattle, CD45 T cell subpopulations, 276, 279, 280, 281
  γδ T cells in, 283
  in *Strongylus vulgaris* infections, 391, 400, 401
Peripheral lymph node addressin complex, 63
Periportal fibrosis, 398
Peroxynitrite, 136
Pertussis toxin, 84
Pestiviruses, 207, 660
Peyer's patches
  AgI/II adhesin and, 108
  cholera toxin, B subunit and, 110
  immune stimulating complexes and, 409
  interleukin-1 and, 94
  leukocyte trafficking and, 69–70
  in mucosal immune response, 435
  oral vaccines and, 105–106
Pharmaceutical companies, *see* Vaccine industry
Pharmaceutical legislation, *see* Virus Serum Toxin Act
PHARMEUROPA, 606
Pheasants, 554

Phoca sibirca, 552
Phoca vitulina, 552
Phocine distemper virus, 552
Physalix vaccine, 19
Pica, 554
Pigs, *see* Swine
Plague, 342
Plasma cells, *see* Antibody-secreting cells
Plasmid immunization, *see* DNA immunization
Plasmids, *see* Virus vectors
Plasminogen activators, 263–264
Platelet/endothelial cell adhesion molecule 1, 62, 69
Platelets, CD62P adhesion molecule in, 66
Pneumonia, equine, in foals, 372
Poliomyelitis vaccine, 429–430, 684–685, 750
Poliovirus, 429
Polyarthritis, immune-mediated, 721
Polyclonal restriction, in HIV infected B cells, 120
Polygenes, disease resistance and, 43–44
Polyimmunoglobulin receptor, 434
Polylactosamine-1 antigen, 128
Polymerase chain reaction (PCR), *see also* Reverse transcriptase-polymerase chain reaction
  in diagnostic assays, 26, 205, 231, 372
Polyneuropathy, postvaccinal, 717
Polysaccharides, bacterial, homology with HIV-1 envelope, 127–128
Polyvalent vaccines, 5–6
  advantages of, 467
  adverse vaccine reactions and, 718–719, 722, 724
  for bovine mastitis, 261
  bovine virus diarrhea virus vaccines in, 208
  for canine coronavirus, 301–302, 314
  for canine distemper, 294–295
  in DNA immunization, 174–175
  for fish, 547–549
  for kennel cough, 315, 316
  limitations of, 174
  overadministration of, 728
  safety and, 291
  in vaccine-induced autoimmunity experiments, 735–745
  virus vectors and, 148

Ponies, *Strongylus vulgaris* infections and, 390–402
Porcine group A rotaviruses, 433
Porcine group A rotavirus vaccines, 440–442
Porcine reproductive and respiratory syndrome, 458
Porcine reproductive and respiratory syndrome vaccine, 421, 427
Porcine respiratory coronavirus
  mucosal immune response to, 436–439
  pathogenesis of, 431–433
Porcine respiratory coronavirus vaccines, 433–434
Positive feedback loop, in host-parasite relations, 249–250
Postcapillary venules, 61–62, 65
Postmarketing surveillance, 706–709, 710
  bias and, 753, 755–757
  concepts of causality and, 754–755
  data collection in, 751–753
  difficulties in interpreting, 757–758
  implications for public policy, 750–751
  objectives of, 750–751, 757
Post-translational modification, virus vectors and, 148
Postvaccinal encephalitis, 294–295, 296, 301–302, 312
Postvaccinal polyneuropathy, 717
Potassium bichromate, 14
Potency testing, in licensing procedures, 589
Potos flavus, 552
Poultry
  effects of conjugated linoleic acid in, 56
  immune-induced wasting and, 54, 56–57
  immune system of, 487–490
  specific pathogen free, 504, 505, 507, 511, 612–613
  type C botulism and, 554
Poultry industry, *see also In ovo* vaccination
  bacterial infections and, 511–512
  coccidiosis and, 512
  common infectious diseases of, 482–483
  commonly used vaccines in, 483, 490
  disease prevention in, 482–483, 490

Poultry industry (cont.)
  intensive rearing system in, 481–482
  output of, 481, 490
  recombinant vaccines in, 486–487
  vaccination programs in, 484–485
  vaccine costs in, 175
  virus vectors and, 148–149
Poultry vaccines
  commonly used, 483, 490
  delivery methods in, 485, 490
  DNA, 520–521
  recombinant, 486–487, 519–521
  subunit, 519
  synthetic peptides and, 519–520
  in vaccination programs, 484–485
  virus vectors and, 518–519
Poxviridae, 518
Poxvirus vectors, 146, 150, 151, 153, 467–469
Preclinical testing
  ecology of infection and, 29–31
  future of, 34–36
  genetic testing and, 32
  methods in, 31–32
  overview of, 26
  predictive value of, 33–34
Pregnancy
  suppression of cell-mediated immune response and, 71–72
  vaccination and, 74, 718
Premunition, 117
Prepuce, bovine trichomoniasis and, 230
Preventive medicine
  genetic testing and, 32
  preclinical testing and, 33–34
Primucell vaccine, 351–355
Probability, adverse event reports and, 757, 769
Procyon lator, 553
Professional organizations, in veterinary standardization, 658–659
Proinflammatory cytokines, excessive induction of, 687–688
Promoters, in recombinant virus vectors, 471
Prostaglandin E2, 55
Protective antigens, helminthic vaccines and, 242–245
Protective immunity, *see also* Acquired immunity; Active immunity
  cell-mediated immune response and, 524
  DNA vaccines and, 171
  Edwardsiella ictaluri RE-33 vaccine and, 531–531, 534–535
  enteric septicemia of catfish and, 525
  immune stimulating complexes and, 411
  pseudorabies vaccines and, 452–455, 457
  time to develop, 692
Protein assays, *see also* Acute phase protein assays
  human, international standardization in, 651–652
Proteins, *see* Acute phase proteins
Proteinuria, 743
P-selectin glycoprotein ligand, 62
Pseudocapsular proteins, 260
Pseudomonas aeruginosa vaccine, 423
Pseudorabies vaccines, 421
  European Union licensing procedures for, 615–625
  immunes responses to, 447–448, 451–452
  interluekin 12 and, 455–456, 458
  protective immunity and, 452–455, 457
  recombinant, 469, 573
  residual virulence and, 687
Pseudorabies virus
  attenuated, 457
  Aujeszky's disease and, 451
  distribution, 659
  economic costs of, 464
  gamma interferon and, 454
  glycoproteins, in swinepox recombinant vaccine, 474
  immune response to, 457
  mortality rates, 464
Psittacosis, 31
Public health
  Bartonella henselae and, 340, 342
  leptospirosis and, 318
  Toxoplasma gondii and, 338, 342
  zoonotic pathogens and, 334, 335, 342
Public policy, vaccination risk-benefit analysis and, 750–751
Puntoni, V., 21, 22
Pups
  canine distemper vaccines and, 294, 295

INDEX   805

canine parvovirus type 2 vaccines and, 297–298, 299–300
timing of vaccination and, 661, 663
Pustular vulvovaginitis, 197
Pyrexia, 395, 398, 399

Q

QTL, *see* Quantitative trait loci
Quantitative trait loci (QTL), disease resistance and, 40, 42, 43–44
Quasi-likelihood methods, adverse event report data and, 762, 768–769
Quil A, in helminthic vaccines, 246, 247
Quillaja saponaria, 405, 406

R

Rabies
  ecology of, 29
  equine, 367
  human mortality and, 571–572
  immune stimulating complexes and, 408, 409
  overview of, 311
  preclinical detection and, 26
  transmission to humans, 335
  viral glycoprotein in Raboral V-RG vaccine, 574
  wild reservoirs of, 29, 34, 572
Rabies vaccines, *see also* Raboral V-RG vaccine
  feline, risk management for, 709
  recombinant, 475
  residual virulence and, 686, 687
  in vaccine-induced autoimmunity experiments, 735–745
  vaccinia virus vector, 153
Raboral V-RG vaccine
  construction of, 574
  controlling coyote and gray fox rabies with, 580–581
  controlling raccoon rabies with, 573, 576–580
  development of, 572–573
  mechanism of action in, 574–575
  oral administration and, 572, 582

  success of, 582
  USDA testing of, 575–576
Raccoonpox virus, 475
Raccoons
  polyvalent vaccines and, 553
  rabies control in, 573, 576–580
Rainbow trout
  furunculosis and, 525
  Vibrio viscosus and, 546–547
Ratites, 481
Reactivation experiments, bovine herpesvirus 1 vaccines and, 203–204
C-Reactive protein, 646, 650
Recombinant DNA technology, *see* Biotechnology; DNA immunization
Recombinant vaccines, *see also* Virus vectors
  antigen production and, 467
  biotechnology and, 521–522, 581–582
  canine, 292
  for canine distemper, 296, 310, 312
  for feline immunodeficiency virus, 328–329
  foreign hosts and, 476
  future of, 522
  immune responses and, 475
  licensing procedures for, 590
  multivalency and, 476
  for poultry, 486–487
  for rabies, 311, 572–573
  for transmissible gastroenteritis virus, 440
  for tuberculosis, 139
  USDA classification of, 573–574
  virus-like particles and, 442
Recombination, homologous
  DNA integration and, 173
  in recombinant vaccine formation, 471, 518, 574
  reversion to virulence and, 475
Red pandas, 552
Reference reagents, international standards, 676–677
Registration, *see* Licensing procedures (European Union); Licensing procedures (United States)
Regulatory agencies, *see also* Animal and Plant Health Inspection Service; Center for Veterinary Biologics; European Pharmacopoeia; Office Inter-

Regulatory agencies (cont.)
  national des Epizooties; U.S. Department of Agriculture
  DNA immunization and, 172–173
  vaccine label information and, 634
Relative percent survival
  defined, 527, 542
  Edwardsiella ictaluri RE-33 vaccine and, 531–532, 534
Renal amyloidosis, inherited, 721
Renal failure
  leptospirosis and, 317
  Lyme disease and, 317
  vaccine-induced, 717
Reoviridae, 365
Reovirus, human, 510
Respiratory disease, *see also* Porcine reproductive and respiratory syndrome; Porcine respiratory coronavirus; Respiratory syncytial virus
  avian, 482
  bovine, 207–208, *see also* Bovine respiratory syncytial virus
  equine rhinopneumonitis, 361–363
  feline, 3
  in swine, economic costs of, 464
Respiratory syncytial virus, *see also* Bovine respiratory syncytial virus
  human, 187
  immune stimulating complexes and, 410
  vaccines, enhanced disease and, 188
Respiratory tract immunization, bovine trichomoniasis and, 229
Retroviruses, 366
  immune stimulating complexes and, 411
  oncogenes and, 32
Revaccination, *see also* Annual vaccinations; Overvaccination
  adverse reactions and, 718
  canine vaccines and, 303, 318
  frequency of, 663, 725–727
  mink distemper vaccine and, 560
  vaccine label statements on, 639
Reverse transcriptase-polymerase chain reaction, in diagnostics, 209, 361, 372

Reversion
  Edwardsiella ictaluri RE-33 vaccine and, 529
  in pseudorabies vaccines, testing for, 618
  risks of, 466
  via homologous recombination, 475
Rhabdoviridae, 367, 371
Rhesus macaques, HIV-1/SIV protection in, 411
Rheumatoid arthritis, 706, 742, 743
Rhinopneumonitis virus, *see* Equine rhinopneumonitis virus
Rhinotracheitis, *see* Infectious bovine rhinotracheitis
Rhinovirus type 2, 127
Rhodococcus equi, 372
Rickettsiosis
  ecology of infection and, 29, 31
  preclinical detection and, 26
Rifampicin, 527, 532
Rinderpest, 669–670
Risk assessment
  postmarketing surveillance and, 706–709, 710
  public *vs.* private, 750–751
  sources of data on, 704–706
  steps in, 702–703
Risk communication, 709–710
Risk difference, 751
Risk factor epidemiology, 755
Risk management, 709
RNA viruses
  mutation in, 2–3
  synthetic peptide-stimulated immunity and, 184
  as virus vectors, 147
Rockborn canine distemper vaccine strain, 293–294, 295, 296, 312
Rodococcus equi, 411
Romet, 540
Rotaviruses, 277–278
  diarrhea and, 430
  equine, 367
  pathogenesis of, 429, 433
Rotavirus vaccines, 421, 427, 433, 440–442
Rottweilers, 661
Roux, Emil, 13, 14
Ruminants, *see also* Cattle
  immunization development and, 285

S

Safety, *see* Vaccine safety; Vaccine safety testing
Saliva-binding region, 106
Salmon, *Vibrio viscosus* and, 544, 546–547
Salmonella cholerasuis vaccine, 421, 427
Salmonella dublin, 96
Salmonella spp.
  ecology of, 29, 31
  equine infections and, 371
  poultry infections and, 512
  zoonotic infections and, 342
Salmonella typhimurium, preclinical detection and, 26
Salmonella vaccines
  in ovo vaccination and, 512
  Re-17, for clinical coliform mastitis, 265, 269
  for swine, 421, 427
Salmonella vectors
  for mucosal vaccines, 106, 108–109, 111
  for SBR-CTA2/B chimeric cholera toxin, 106, 108–109
  for transmissible gastroenteritis virus vaccine, 440
Salmonids, antibiotics and, 543
Sanitary and Phytosanitary Agreement (World Trade Organization), 677, 678
Saponins, 169, 405, 406
SBR-CTA2/B (chimeric cholera toxin protein), 106–109
Seal morbillivirus, 410
Secondary immune response, *see also* Anamnestic recall response; Immunologic memory
  HIV-1 infection and, 127
  in mucosal immune system, 111
  tuberculosis and, 136
  virus-antibody complexes and, 510
Seed stock, *see* Master seeds; Vaccine seed stock
Seizures, vaccine-induced, 717
Selectins, in leukocyte trafficking, 62, 63, 65–67
Self-selection, of adverse event report data, 756
Sendai virus, 411

Septicemia, *see also* Enteric septicemia of catfish
  bovine genital tract infections and, 218
  Haemophilus somnus infection and, 219
Serengeti lions, 553
Serials
  defined, 750
  prelicensing, 589–590
Serum amyloid A, 644, 646
Serum antibody titers, using to determine revaccination, 727
Serum neutralization titer, protective levels of, 726–727
Sexually transmitted disease
  bovine genital tract infections and, 218, 233
  campylobacteriosis and, 222
  contagious equine metritis and, 370
  systemic immunizations and, 223
Shar Pei dogs, 721
Shedding, *see* Nasal shedding; Oocyst shedding; Transmissibility; Virus shedding
Sheep
  botulism and, 554
  Toxoplasma gondii and, 337
Sheepdogs, *see* Old English sheepdogs
Siberian seals, 552
Simian immunodeficiency virus, 411
Simian virus 5, 315
Simian virus 40, 685
Simulation models
  design and use of, 247–248
  of *Haemonchus* vaccines, 251–253
  of vaccination in host-parasite populations, 248–250, 254
Single radial diffusion test, equine influenza and, 381, 385, 603
Single radial hemolysis test, equine influenza and, 361, 381, 386, 603
Skeletal deformities, conjugated linoleic acid and, 56
Small animal practice
  polyvalent preparations and, 5–6
  use of vaccines in, 290
Smallpox, 9, 10
"Snyder Hill" canine distemper vaccine strain, 293, 296
Somatic cell counts, in bovine mastitis diagnosis, 267, 269

Specific pathogen free poultry, *see* Poultry, specific pathogen free
Spleen, virus-antibody complexes and, 511
Sporothrix schenckii, 342
Spotted hyenas, 553
Stability testing, 632
Stallions, contagious equine metritis and, 370
Standards Commission (Office International des Epizooties), 670, 672–677, 678
Staph A antigen, 128
Staphylococcus aureus vaccines, 258–262, 268
Sterile immunity, DNA vaccines and, 171
Strangles, 369–370
Streptococcus agalactiae vaccines, 261, 262
Streptococcus dysgalactiae, 262–263
Streptococcus equi, 369–370
Streptococcus equismilis vaccines, 423, 427
Streptococcus suis vaccines, 421, 423, 427
Streptococcus uberis vaccines, 261, 263–264, 268, 269
Stress
  effects on immunity, 70–74
  effects on leukocyte trafficking, 63–64
Strongylus vulgaris
  control of, 390
  eosinophilia and, 390–391, 396
  equine, immune response to, 390–402
  pathogenesis, 390
"Subclincal infections," 31
Sublicensing, 592
Substance P, 96
Subunit vaccines
  biasing T cell responses and, 186
  cytokine adjuvants and, 457–458
  for feline immunodeficiency virus, 328–329
  importance of, 457
  for poultry, 519
  risks of reversion and, 466
  tuberculosis and, 139–140
  USDA classification of, 573
Sudden infant death syndrome, 759
Suipoxvirus, 469
Surface proteins, in *Staphylococcus aureus* vaccines, 261

Surveillance
  postmarketing, *see* Postmarketing surveillance
  in vaccine risk assessment, 704–706, 710
Swine
  acute phase proteins and, 646–647
  cholera toxin adjuvanticity in, 86–92
  enteropathogenic viruses and
    enteric vaccines, 433–434
    group A rotavirus vaccines, 440–442
    mucosal immune response, 434–442
    pathogenesis of, 430–433
    transmissible gastroenteritis virus vaccines, 439–440
  feral, disease transmission and, 464
  immune-response selected, 43–49
  infectious diseases of, 463–467, 469
  neonates
    diarrheal disease and, 464–465
    swinepox virus and, 475–476
  pseudorabies virus and, 451
  pseudorabies virus vaccines and, 447–448, 451–452, 454, 455–456, 457, 458
  respiratory infections and, economic costs, 464
  Toxoplasma gondii and, 337
Swine fever, 661–662
Swine fever vaccine, 469
Swine industry
  infectious diseases and, 463–465
  protein output of, 463
Swinepox vaccine, 474–476, 477
Swinepox virus, 469–470, 475–476
Swinepox virus vector, 470–473
Swine vaccines, overview of, 419–427
Synthetic peptides
  immune stimulation by, 183–184
  poultry vaccines and, 519–520
Systemic lupus erythematosus, 706, 742, 743

T

T-263 vaccine, 337–338
Taxidea taxus, 552
Taylorella equigenitalis, 370

INDEX   809

T cell clones, DNA tuberculosis vaccines and, 140
T cell immunodeficiency, 452
T cell receptors
  activation of γδ T cells and, 283
  antigen recognition and, 182, 183
  in avian immune system, 489
T cells, see also Cytotoxic T cell immune response
  antigen presenting cells and, 275–276
  antigen recognition and, 182–184
  antigens recognized by, identification of, 186–187
  in avian immune system, 487, 488, 489, 490
  in cattle
    bovine respiratory syncytial virus immunity, 188–191, 192
    dendritic cell-induced responses and, 284–285
    identification of subpopulations in, 276, 278–282
    in vivo roles of, 276–278
  cholera toxin and, 93
  in deceptive imprinting, 119
  dual-signal integration and, 93–94
  immune stimulating complexes and, 408–409
  in mucosal adjuvanticity, 98–99
  in mucosal immune response, 434, 435
  response determining mechanisms and, 186
  SBR-CTA2/B chimeric cholera toxin protein and, 108
  suppression of polyclonal response in, 122
  trafficking and, 63, 64, 69
  in tuberculosis immunity, 136, 137
  workshop cluster numbers and, 276
T cells, CD4, see also Th1 immune response; Th2 immune response
  antigen processing and, 184–185
  in avian immune system, 489
  in cattle
    afferent lymph veiled cells and, 284–285
    bovine respiratory syncytial virus immunity, 188–190, 192
    bovine virus diarrhea virus and, 277
    identification of, 276
    identifying CD45 subpopulations in, 278–282
    in vivo roles of, 277
  cytokine profiles of, 185–186
  interleukin 12 and, 450
  response determining mechanisms and, 186
  in tuberculosis immunity, 136, 138
T cells, CD8
  antigen processing and, 184–185
  antigens recognized by, identification of, 186–187
  in avian immune system, 489
  in cattle
    afferent lymph veiled cells and, 284–285
    bovine respiratory syncytial virus immunity, 188, 189
    identification of, 276
    identifying CD45 subpopulations in, 278–282
    respiratory syncytial virus and, 277
    in vivo roles of, 277–278
  response determining mechanisms and, 186
T cells, γδ
  in cattle
    activation and function of, 282–284
    bovine respiratory syncytial virus immunity, 188
    identification of, 276
    trafficking of, 64
  glucocorticoids and, 73
T cells, memory, see also Memory lymphocytes
  CD45 antigen and, 279, 280, 281–282
  immune stimulating complexes and, 407
Teladorsagia circumcincta, 242, 243
Tetanus toxoid, 73, 93
Tetra methyl benzidine, 648
Texas, controlling wild animal rabies in, 580–581
Th1 immune response
  characteristics of, 185–186
  defined, 42
  in DNA immunization, 167
  in hypersensitive reactions, 689

Th1 immune response (cont.)
  immune stimulating complexes and, 410
  inhibition by interleukin 10, 401
  in tuberculosis immunity, 137
Th2 immune response
  characteristics of, 185–186
  cholera toxin induction of, 93
  defined, 42
  in DNA immunization, 167
  helminthic infections and, 246, 391
  in hypersensitive reactions, 689
  immune stimulating complexes and, 410
  Strongylus vulgaris infections and, 401
Theileria parva, 183
Thimerosol, 685
Thrombocytopenia, *see* Immune-mediated thrombocytopenia
Thromboembolic disease, *Strongylus vulgaris* infections and, 390
Thromboembolic meningoencephalitis, 686
"Thuga," 305
Thymidine kinase gene, 150
  in fowlpox virus vector, 518
  in recombinant virus vectors, 471, 473
  in swinepox virus, 471
Thymus-dependent antigens, 122–123
Thymus gland, in avian immune system, 487, 488
Thymus-independent antigens, 122–123
Thyroglobulin autoantibodies, 720–721
Thyroid disease, vaccine-induced, 720–721
Thyroiditis, autoimmune, 723, 742
Tick-borne diseases, 316, 360, 662
Togaviridae, 364
Toussaint, Henri, 8, 12, 13, 14, 22
Toxoplasma encephalitis, 336
Toxoplasma gondii, 335, 336–338, 342
Trade, *see* International trade
Transcriptional regulation, glucocorticoids and, 73
Transmissibility, *see also* Virus shedding
  of bovine herpesvirus 1 vaccines, 203
  of pseudorabies vaccine strains, 619
Transmissible gastroenteritis vaccines, 420, 421, 423, 427, 433–434, 439–440

Transmissible gastroenteritis virus, 349
  attenuated, 431–434, 439
  mode of release in, 431
  mucosal immune response to, 436–439
  pathogenesis, 429, 431–433, 465
  in swine, economic costs of, 465
Transmission, *see* Disease transmission
Transplacental immunization, 564
Transporter for antigen presentation, 184
Trichomoniasis, *see Tritrichomonas foetus*
Trichostrongylid nematodes, 242; *see also* Helminthic vaccines
  aminopeptidases in, 245
  host immunity and, 250–251
  modeling of vaccines for, 251–253
  overdispersion and, 251
Trichuris muris, 186, 401
Tritrichomonas foetus, 224–232
  diagnosis of, 231–232
  evasion of immune response by, 231
  genital mucosal inductive sites and, 228–230
  immune response to, 226–228, 230
  pathogenesis of, 224
  prevalence of, 224–225
  protective antigens of, 225-226
  vaccine development and, 233
Trout, *see* Rainbow trout
Trovac-NDV, 519
Trypanosoma congolense, 278
Tuberculosis
  acquired cell-mediated immunity and, 137–138
  γδ T cells and, 284
  human mortality and, 135, 141
  memory immunity and, 136
  preclinical detection and, 33–34
  process of infection in, 136–137
Tuberculosis vaccines, 135, 136, 138–141
Tumor necrosis factor
  acute phase proteins and, 645
  adjuvanticity in DNA immunization, 168
  excessive induction of by vaccines, 687
  tuberculosis and, 138
Turkey herpesvirus, in Marek's disease vaccine, 496, 497, 498, 500, 502–503, 519
Turkey herpesvirus vectors, 486, 519

INDEX 811

Turkeys, see also Poultry industry
  recombinant vaccines and, 486
Turpentine injection, 646, 647
Type III response, in hypersensitive reactions, 689–690
Typing systems, in bovine mastitis diagnostics, 268

U

Uasin Gishu pox, 11
Ulcers, *Helicobacter pylori* and, 341
Unconventional antigens, 121
Underdeveloped countries, rabies in, 571–572
United States
  controlling wildlife rabies in, 576–581
  licensing procedures in
    for modified live pseudorabies vaccine, 628–632
    for veterinary biological products, 586–593
    Virus Serum Toxin Act and, 585–586, 627
  regulation of veterinary vaccines in, 682–684
  swine vaccine use in, 422, 423, 425, 426, 427
  wildlife reservoir species of rabies in, 572
United States Pharmacopeia, 753
United States Veterinary Biologics Establishment License, 587–588
Unvaccinated animals, *see* Nonvaccinated animals
Uremia, 317
Urocyn cineieoargenteus, 552
U.S. Department of Agriculture, *see also* Animal and Plant Health Inspection Service; Center for Veterinary Biologics; Licensing procedures (United States)
  adverse event report data and, 771
  annual revaccination regulations, 639
  classification of recombinant vaccines, 573–574
  collection of adverse event reports and, 753
  efficacy testing regulations, 635–637
  equine influenza vaccines and, 386
  licensure exemptions and, 592
  regulation of veterinary vaccines, 682–684
  safety testing regulations, 638
  testing of Raboral V-RG vaccine, 573, 575–576
  vaccine label information and, 634–641
  Veterinary Biologics program and, 586–587
Uterus
  bovine campylobacteriosis and, 222–223
  bovine trichomoniasis and, 228–230

V

Vaccination, *see also* Annual vaccination; DNA immunization; *In ovo* vaccination; Nonvaccinated animals; Revaccination
  adverse reactions and, *see* Adverse reactions
  benefits of, 663–665
  conventional, alternatives to, 716, 727–729
  cross-species, 551–555
  in disease prevention, 465–466
  economics of, 1–2, 661
  emerging infectious agents and, 2–3, 333, 334–335
  epidemiology and, 415–417, 666
  genetic predispositin and, 39–49
  of geriatric animals, 6, 725–726, 728
  goal of eradication in, 474
  holistic perspective and, 290–291
  immune-induced wasting and, 53, 54, 57
  of neonates, 297–298, 299–300, 661, 663, 694, 719
  origins of, 7–8
  overuse of, 663, 701–702, 718, 728
  overview of, 668
  of pregnant animals, 74, 718
  professional knowledge of, 2
  purpose of, 136
  reemerging infectious agents and, 333
  risks and, 662–665, 702–792
  standardization and, 658–662
  success of, 1

Vaccine Adverse Event Reporting System, 753
Vaccine delivery
  DNA immunization and, 165–166, 169–171
  intradermal, 165–166
  intragastric, 106, 109, 110
  intramuscular, 165, 170
  intranasal, 106, 109, 110, 310
  intraperitoneal, 440, 541–542, 549
  in ovo, 485, 490
Vaccine development
  advances in immunology and, 182–187
  antigen processing and, 184–185
  antigen recognition and, 182–184
  biotechnology and, 182, 517, 521–522, 581–582, 665
  for bovine genital tract infections, 233
  for bovine respiratory syncytial virus, 192
  current trends in, 182
  deceptive imprinting and, 129
  epitope masking and, 130
  for equine diseases, 373
  for foot-and-mouth disease virus, 199
  future of, 665–666
  in helminthic vaccines, 242–254
  identification of T cell-recognized antigens and, 186–187
  for lentiviruses, 366
  in poultry vaccines, 486–487
  recombinant vaccines and, 467
  synthetic peptides and, 183–184
  T cell response pathways and, 185–186
  traditional approaches to, 181–182, 466
  virus vectors and, 146, 475
Vaccine effectiveness, distinguished from vaccine efficacy, 751
Vaccine efficacy
  distinguished from vaccine effectiveness, 751
  exposure of adverse effects and, 703
  interpreting, 3–4
  public concern with, 290
  quantitative definition of, 751
  U.S. regulations on, 684
Vaccine efficacy testing
  of bovine herpesvirus 1 vaccines, 202–204
  of bovine virus diarrhea virus vaccines, 208
  of canine coronavirus vaccines, 301, 302
  of canine parvovirus type 2 vaccines, 298–300
  of canine vaccines, 291–292, 293
  codified, 635
  of conventional vaccines, 466
  disease prevention and, 416–417
  of enteric vaccines, 433
  of equine influenza vaccines, 380
  of feline infectious peritonitis vaccines, 354–355
  of fish vaccines, 542
  of foot-and-mouth disease vaccines, 199–200
  in licensing procedures, 589
  noncodified, 635–636
  of *in ovo* vaccination, 498–499
  of pseudorabies vaccines, 620–623, 629
  of Raboral V-RG vaccine, 575–576
  reported on vaccine labels, 635–637
  of swine vaccines, 420, 421, 423, 425
  USDA regulations, 635–637
Vaccine industry, *see also* Licensing
  loyalty of veterinarians to, 3
  postmarketing surveillance and, 706–709
  recommendations for, 6
  research and development in, 3
Vaccine labeling
  animal age information on, 639–640
  annual revaccination information on, 639
  directions for clinical use on, 638–639
  efficacy information on, 635–637, 684
  expectations for, 634
  improving standardization in, 640–641
  indications information on, 637
  information on multiple syndromes, 640
  in pseudorabies vaccine licensing, 632
  safety information on, 637–638, 684
  standardization of, 658
  veterinarian needs and, 633, 634
Vaccine marketing, misuse of adverse event report data in, 771–772
Vaccine production lots, *see* Serials

Vaccines, *see also specific types;* Veterinary biological products
  administering concurrently, dangers of, 691, 693, 695
  administering near parturition, 74, 718
  adverse reactions to, *see* Adverse reactions
  clinical use, directions for, 638–639
  contamination of, 684–685, 717–718
  conventional, overview of, 465–467
  costs of, 175
  cross-reactivity and, 743, 745
  failures, causes of, 129–130, 146, 595, 691–695
  holistic perspective and, 290–291
  human vaccine models and, 659
  hypersensitivity reactions and, 688–690
  immunogenic qualities of, 456–457
  improper handling of, 692
  licensing and registration of, *see* Licensing procedures
  loyalty of veterinarians to, 3
  need for improvement in, 145–146
  partnership model for, 667
  risk assessment for, 702–703
  strategies for use, 666
  success of, 657–658, 682
  target species and, 660
Vaccine safety, 710
  characteristics of, 703
  frequency of administration and, 701–702
  problems in, 662–663
  U.S. regulations on, 683–684
Vaccine safety testing
  of bovine herpesvirus 1 vaccines, 204–205
  of bovine virus diarrhea virus vaccines, 208–209
  of canine coronavirus vaccines, 301–302
  of canine parvovirus type 2 vaccines, 300
  of canine vaccines, 291
  of *Edwardsiella ictaluri* RE-33 vaccine, 528–529, 534
  of foot-and-mouth disease vaccines, 200
  in licensing procedures, 588–589
  of modified live virus vaccines, 466
  of nonimmunizing mutant viral vaccines, 293
  of pseudorabies vaccines, 616–620, 629, 630–632
  public concern with, 290
  reported on vaccine labels, 637–638
Vaccine seed stock, *see also* Master seeds
  canine parvovirus type 2 vaccines and, 298
Vaccinia, 8, 9–11, 22
Vaccinia virus vector, 146, 147, 153, 467–469, 475, 572, 574
Vaccinology
  adjuvants and, 4–5
  DNA injections and, 4
  early anthrax vaccines, 11–17
  early discoveries in, 9–11
  origins of, 7–8
  polyvalent vaccines and, 5–6
  professional knowledge of, 2
  recommendations for, 6
  William Greenfield on, 15–16, 17
Vaccinosis, 290, 715
Vaculitis, 219
Vagina
  bovine campylobacteriosis and, 222–223
  bovine trichomoniasis and, 228–230
Valgus deformity, 56
Vanguard-5 CV/L vaccine, 735
Varicellovirus, 201; *see also* Bovine herpesvirus 1
Varus deformity, 56
Vascular cell adhesion molecule 1, 63
Vectors, *see* Bacterial vectors; Virus vectors
VectorVax FP-N, 519
Venereal campylobacteriosis, *see* Campylobacteriosis
Venezuelan equine encephalitis, 364
Venezuelan equine encephalitis vaccine, 685
Venules, leukocyte trafficking and, 61–62, 65
Verminous arteritis, 390, 398
Vesicular stomatitis, 371
Veterinarians
  animal welfare and, 664–665
  attitudes towards vaccines, 3

Veterinarians (cont.)
  in postmarketing surveillance of vaccines, 707
  risk communication and, 709–710
  risk management and, 709
  swine vaccines and, 420, 425, 427
  vaccine labels and, 633, 634
Veterinary biological establishments, licensing of, 587–588
Veterinary biological products
  defined, 586
  licensing of
    European Union, 595–607
    United States, 587–593, 627
    new European definitions of, 598–600
    Virus Serum Toxin Act and, 585–586
Veterinary Biologics program, 586–593
Veterinary diagnostics, see Diagnostic medicine
Veterinary laboratories, as OIE reference laboratories, 675–676
Veterinary medicinal products, see Veterinary biological products
Veterinary microbiology, 6
Veterinary practices
  corporate, postmarketing surveillance of vaccines and, 708
  standardization and, 658–659
Veterinary Services, 627
VetSmart, 708
VH genes, in deceptive imprinting, 120–121
Vibrio anguillarum, 548–549
Vibrio cholerae, see Cholera toxin
Vibrio salmonicida, 547, 548–549
Vibrio viscosus, 544, 546–547
Vibrio wodanis, 546
Vibrosis, 547, 548; see also Campylobacteriosis
Viral diarrhea, enterpathogenic viruses and, 430
Viral envelope
  deceptive imprinting and, 118
  HIV-1, similarity with other viruses, 127
  immune stimulating complexes and, 406, 410
Viral envelope glycoprotein
  in feline immunodeficiency virus vaccines, 326, 328
  in recombinant swinepox vaccines, 474

Viral neutralizing factor technology, 506–511
Viremia
  bovine virus diarrhea virus and, 277
  in poultry, rispens-type vaccines and, 502–503
Virulence, see also Reversion
  of *Haemophilus somnus*, 220
  modified live vaccines and, 466
  residual, in modified live vaccines, 686–687
Virus-antibody complex vaccines, *in ovo* vaccination and, 506–511
Viruses, see also specific diseases; Virus vectors
  mutation in, 2–3
  "new," 2
  replication limited, 150–151
  strain variability, effects on vaccination, 660
Virus-like particle vaccines, 442
Virus Serum Toxin Act (United States), 585–586, 627, 634, 683
Virus shedding, see also Nasal shedding; Oocyst shedding; Transmissibility
  disease transmission and, 277, 464
  in pseudorabies vaccines, 618–619, 622
Virus vectors, see also Recombinant vaccines; specific vectors
  advantages of, 148–149
  antigen production and, 467
  construction of, 147–148, 518
  disadvantages of, 149
  features of, 148
  formation of, 470–473, 477
  future of, 153, 155
  homologous recombination and, 471, 518, 574
  live attenuation and, 149–151
  overview of, 476–477, 517–518
  for poultry vaccines, 518–519
  poxviruses as, 467–469
  replication limited, 150–151
  risks of DNA integration and, 173
  risks of virulent recombination and, 475
  swinepox virus as, 469–470
  in USDA type II recombinant vaccines, 574

in vaccine development, 146
veterinary use of, 145, 151–153
Vogt-Koyanagi-Harada syndrome, 721

W

Waldeyer's ring, 110
Walking catfish, 525
Waterfowl, type C botulism and, 554–555
WB test, 124, 126
WC1 antigen, 276, 282–284
Weibel-Palade bodies, 66
Weight, in assessing pseudorabies vaccine efficacy, 621–622
Weimaraners, 723–725
Western equine encephalitis, 364
Whole inactivated vaccines, *see also* Inactivated vaccines
  for feline immunodeficiency virus, 326–327, 328
Wild animals
  canine distemper virus and, 551–553
  Clostridium botulinum Type C and, 553–555
  as rabies reservoirs, 29, 34, 572
  rabies vaccination and, *see* Raboral V-RG vaccine
  vaccination and, 664
Winter ulcer, 544, 546–547
Workshop cluster numbers, 276
World Health Organization, 677
  on emerging strains of equine influenza, 381–382
  on equine influenza vaccines, 380–381
  European Pharmacopoeia and, 604
  International Association of Biological Standardization and, 610
  World Trade Organization, 677, 678

X

Xanthine-guanine phosphoribosyltransferase gene, 471

Y

Yersinia pestis, 342
Yersinia ruckeri, 540; *see also* Enteric redmouth disease

Z

Zoonotic disease
  Bartonella henselae and, 338–340
  cats and, 335–342
  controlling with vaccination, 664
  ecology of, 29, 31
  Helicobacter pylori and, 341
  leptospirosis and, 371–372
  Microsporum canis and, 342
  preclinical detection and, 26
  public health and, 334, 335, 342
  Toxoplasma gondii and, 336–338
  Yersinia pestis and, 342
Zoos, type C botulism and, 555

ISBN 0-12-039242-9